C0-ATH-316

# A Global Warming Forum

## Scientific, Economic, and Legal Overview

Edited by

**Richard A. Geyer, Ph.D.**

Professor
Offshore Technology Research Center
Texas A&M University
College Station, Texas

CRC Press
Boca Raton Ann Arbor London Tokyo

363.7387
G5622

**Library of Congress Cataloging-in-Publication Data**

A global warming forum : scientific, economic, and legal overview /
editor, Richard A. Geyer.
p. cm.
Includes bibliographical references and index.
ISBN 0-8493-4419-0
1. Global warming. I. Geyer, Richard A.
QC981.8.G56G5819 1992
363.73'87—dc 20

92-12181
CIP

This book represents information obtained from authentic and highly regarded sources. Reprinted material is quoted with permission, and sources are indicated. A wide variety of references are listed. Every reasonable effort has been made to give reliable data and information, but the author and the publisher cannot assume responsibility for the validity of all materials or for the consequences of their use.

Neither this book nor any part may be reproduced or transmitted in any form or by any means, electronic or mechanical, including photocopying, microfilming, and recording, or by any information storage and retrieval system, without permission in writing from the publisher.

All rights reserved. Authorization to photocopy items for internal or personal use, or the personal or internal use of specific clients, is granted by CRC Press, Inc., provided that $.50 per page photocopied is paid directly to Copyright Clearance Center, 27 Congress Street, Salem, MA, 01970 USA. The fee code for users of the Transactional Reporting Service is ISBN 0-8493-4419-0/93 $0.00 + $.50. The fee is subject to change without notice. For organizations that have been granted a photocopy license by the CCC, a separate system of payment has been arranged.

The copyright owner's consent does not extend to copying for general distribution, for promotion, for creating new works, or for resale. Specific permission must be obtained from CRC Press for such copying.

Direct all inquiries to CRC Press, Inc., 2000 Corporate Blvd., N.W., Boca Raton, Florida, 33431.

© 1993 by CRC Press, Inc.

International Standard Book Number 0-8493-4419-0

Library of Congress Card Number 92-12181

Printed in the United States of America 1 2 3 4 5 6 7 8 9 0

Printed on acid-free paper

# PREFACE

The term ''Global Warming'' consists of only 13 letters, but it encompases a myriad of scientific, technical, economic, and legal endeavors. The difficulties involved in its study and the potential of the problems it creates continue to occupy the minds and activities of an ever-increasing portion of the world population and resources. Efforts to achieve effective and appropriate solutions within a reasonable economic and regulatory framework are both national and international in scope. Unfortunately, they are only in the initial stages.

One measure of the tremendous diversity of the efforts involved is evident in a list of over 200 acronyms listed in Table 1 of the Introduction. These apply to a worldwide network of scientific and technical groups studying this problem in considerable depth. Although this table appears to be quite extensive, it is by no means all-inclusive; however, it can be very useful to facilitate identifying (in this book and elsewhere) the diversified agencies, committees, and other entities actively conducting research toward solving this complex problem.

On the other hand, it should be emphasized that the study of this important process did not begin recently. The literature comprising this subject can be traced back to at least 1827. The renowned physicist and mathematician, Jean Fourier published a paper discussing it. He questioned how the Earth was able to maintain its warm temperature. He believed the Earth's atmosphere acted like the glass in a greenhouse — permitting the sun's energy to enter, but simultaneously serving to slow the escape of the radient heat back into space.

Sixty years later, in 1886, Arrhenius calculated that doubling the amount of $CO_2$ in the atmosphere would raise the averge temperature of the earth by 5.5°C. He subsequently postulated in 1908, that the amount of $CO_2$ and $H_2O$ existing at that time in the Earth's atmosphere absorbed sufficient infrared radiation to increase the temperature by at least 30°C.

Several critical areas must be studied in sufficient detail if a viable solution to the problem of global warming is to be achieved. These include obtaining a better understanding of the need to effectively implement the appropriate crucial economic, legal, and regulatory factors involved. These factors also define why this book differs markedly in its presentation from most concerned with the subject of global warming. The fact is recognized and stressed, that unless some of the effects of these constraints are overcome, this critical problem will never be solved satisfactorily. Thus, it was deemed imperative to include the last two sections of this book. Without them the reader cannot begin to grasp their critical role nor can a well-rounded, integrated, and pragmatic method ever be developed in concert with the successful completion of the associated scientific and technical programs. Some of the basic criteria have been defined and are used to select the individual chapters in five sections comprising this volume. Others include bringing together the ideas, thoughts, and research results from the basic triad of academic, governmental, and industrial laboratories, as presented in the other four sections. This approach helps to provide an integrated assessment of these components in a thematic, as well as a succinct fashion.

Individual components of the scientific and technological subjects germaine to the study of global warming are manifold. These are also so complex that it becomes exceedingly difficult to reach definitive conclusions. Furthermore, many can best be studied using interdisciplinary rather than individual scientific methods to achieve maximum results. A serious question is then raised as to the most effective way to group them for study and evaluation, as well as presentation of the results; however, there is really no unique way in which to accomplish this important task.

It depends to a considerable extent on the specific individual philosophy of the author of a chapter and/or the editor of a volume, as to their respective perspectives and objectives.

University Libraries
Carnegie Mellon University
Pittsburgh PA 15213-3890

To adequately cover the plethora of individual pertinent subjects would require a tome of encyclopedic length. The alternative is to select a series of broader, but basic topics of an interdisciplinary nature that can be presented in an integrated manner. These then should be directed toward solving economic, legal, and regulatory, as well as scientific and geoengineering questions necessary to understand the causes and effects of global warming.

The extreme complexity and diversity characterizing the ramifications of the global warming problem result in a dearth of simple solutions. Nevertheless, some believe that we are reaching the point where the results of scientific research can be integrated successfully with economic, social, and political restraints. This will then provide the objective criteria needed to permit this problem to be solved objectively. At the moment, however, what we do know is much less than what we finally need to know about major changes in climate. Also, based on our present knowledge, there are some lines of evidence which permit the remote chance of an economically or ecologically disasterous episode of global warming.

**Richard A. Geyer, Ph.D.**

# THE EDITOR

**Richard A. Geyer, Ph.D.,** has had a diversified career in geophysics and oceanography in both the industrial and academic sectors. He received his B.S. in geology in 1937 from New York University and his Ph.D. in geophysics from Princeton University in 1951. Beginning at the Woods Hole Oceanographic Institute (1943–1946), he continued in oceanography as head of the Oceanographic section of Humble Oil Co., Technical Director of Oceanography at Texas Instruments, and finally as head of the Department of Oceanography at Texas A&M University (1966–1978). His activities in geophysics include periods as manager of the Gravity and Magnetic Departments of Geophysical Service, Inc., and adjunct professor of Marine Geology and Geophysics at the University of Houston. He is presently a consultant in exploration geophysics and since September 1989 has served as adjunct professor of the Offshore Technology Research Center of Texas A&M University.

He has received national and international recognition for his efforts in these two fields by giving invited papers and was vice-chairman of a presidentially appointed Committee on Marine Science, Engineering, and Resources (1967–1969). He was named an honorary member of the Society of Exploration Geophysicists in 1981 and of the Dallas Geophysical Society in 1980, of which he was president from 1964–1965.

Dr. Geyer also has had extensive experience as an editor of several books on marine pollution and the use of submersibles in oceanography and ocean engineering. He edited the journal *Geophysics* from 1950–1952 and has published 30 papers in geophysics and oceanography. In addition, he was co-director of a NASA-sponsored project (1961–1962) to determine what geological and geophysical experiments should be performed by astronauts upon reaching the moon.

# INTRODUCTION

The philosophical thoughts expressed in the Preface determined the basic criteria for selecting five general subject areas into which this volume has been divided. The fundamental difference is that the last two sections of the five subject areas are directed specifically toward studying and evaluating the role of economic, legal, and political considerations. It is imperative that they be included in order to obtain a more pragmatic, rather than a purely scientific and/or technical, solution to the provocative problems of global warming. These five areas are listed as follows:

1. Role of geophysical and geoengineering methods to solve problems related to global climatic change
2. Role of oceanographic and geochemical methods to provide evidence for global climatic change
3. Global assessment of greenhouse gas production including the need for additional information
4. Natural resource management needed to provide long-term global energy and agricultural uses
5. Legal, policy, and educational considerations required to properly evaluate global warming proposals

It is in order at this point to also present, for comparative purposes, an outline for an alternative systematic format that has been offered and is being conducted in an in-depth and widespread manner. It is the organization of the U.S. Global Change Research Program Priority Framework which is presented in Figure 1.

Before proceeding further into this discussion it would be very helpful to present the list (Table 1) containing almost 200 acronyms used to describe the various major research programs now underway, as well as their objectives. This table provides further evidence of the tremendous diversity of efforts required to study different aspects of global warming. Many international organizations are included, but by far the greatest number refer to U.S. government agencies and academic institutions as well. A few also apply to instruments and instrumentation systems used in this diversified research.

The high degree of complexity inherent in this study is reflected in the tremendous size of the organization's budget of over 1 billion dollars for fiscal year (FY) 1992. Details can be studied in Table 2. The bulk of the major processes and functions, together with the associated problems comprising this research, are grouped into five major scientific and/or technical categories. Each in turn is further subdivided into their major components. These are also developed with respect to their assigned priorities. Some examples of the relative importance of the many interdisciplinary aspects also become evident. However, these are still more-or-less rigidly compartmentalized. It should be noted that no attempt has been made in the organizational chart to provide for any direct interactions between the many aspects, but it is significant to see that some attempt has been made for budgetary purposes.

A somewhat broader and more integrated method has been developed, as seen in Figure 2, in which seven basic science elements are stressed. This compares with the criteria outlined in Figure 3, which focuses on five basic scientific elements. Comparative data for FY 1991 and FY 1992 are also presented in Figure 4; these are characterized with respect to four different scientific objectives.

The chapters in this book are not confined just to scientific aspects of global warming for reasons previously discussed in the Preface. However, it is reasonable to assume that some of the readers will wish to become more acquainted with additional information of a scientific and technical nature. This can of course be obtained in part from the numerous references found in each chapter. Also, the voluminous number of scientific organizations listed in Table 1 are the source of many excellent and diversified publications.

## CONFLICTING THEORIES

One of the difficulties frequently encountered when studying the diversified processes involved in obtaining a better understanding of global warming is the numerous conflicting theories and supporting data presented. In addition, the situation frequently arises where even if it can be demonstrated that a certain pollutant is dangerous to human welfare, if it were to be eradicated the net result would still be a negative one.

One example is that of halogens and chlorofluorocarbons (CFCs). The undesirable effects from these have been demonstrated without a doubt, and some definitive steps have been taken nationally and internationally to mitigate these. However, subsequent research from another perspective has indicated that, even if CFCs are curtailed significantly and the ozone layer would eventually repair itself, the greenhouse effect would be worsened. This conclusion was reached because a thicker ozone layer absorbing more UV will be warmer. This in turn will radiate more heat back toward the earth. Thus, in short, we again have the phenomenon where, by solving one problem, another is caused or an existing one is exacerbated. In spite of this evidence, environmentalists are stridently calling for amending the treaty on ozone completion, sometimes also referred to as the Montreal Protocol, at the conference in Copenhagen in 1992. They propose to further crack down on the use of CFCs and phasing out any of the proposed substitutes. Similarly, in the 1970s some scientists predicted that a fleet of supersonic transport jets would eat away the ozone layer. This helped to doom the U.S. effort to build them, but now NASA is studying a new design because a subsequent study predicted that such a fleet would have little effect on the ozone layer.

Another more recent example involves the reporting by the IPCC on the Scientific Assessment of Climate Change Conference held in 1990 in which 170 scientists from all over the world participated. No scientific consensus was actually reached on greenhouse effects as these relate to the global warming theory. It was only in the much publicized nontechnical summary, that this appeared to be so. It was widely distributed to politicians and the general public. They could easily interpret this summary as an authoritative analysis reached by the ''international community''.

Subsequently, a separate survey was conducted of the conference participants by Dr. Singer, Professor of Environmental Sciences at the University of Virginia and Mr. Winston, former Director of the Climate Analysis Center of the National Weather Service. Their objective was to determine the views of the conference participants independently. The response was sent to 126 scientists, 31 IPCC reviewers, and the 24 members of the Phoenix Group. The results were statistically significant. The response of the Phoenix Group was 60% and 21% for the IPCC contributors.

The Phoenix (or comparison) Group consisted primarily of academicians from the University of Virginia, M.I.T., and other well-known institutions. They concluded that, ''it is not possible to attribute all, or even a large part of the observed global warming since 1890 to an increased greenhouse effect based on currently available data.'' Similarly, on future warming the various theoretical models used to predict climate have not been adequately validated by the existing climatic record. Thus, in theory if this is applied retroactively to 1890, it fails to predict the observed climatic patterns that have actually occurred. Also, global mean-surface air temperatures have increased between 0.3 to 0.6°C over the last 100 years. However, these values are of the same magnitude as those observed for natural climate variability.

Drs. John Moen and Tony Michaels, research oceanographers at the Bermuda Biological Station, have analyzed results of temperature data collected for the last 37 years. Their results indicated that the adjacent oceanic waters have cooled during the past 40 years. This is contrary to the greenhouse effect and the results of climate-change model studies. In addition, the temperatures dropped dramatically between 1950 to 1970. Also, over the past 20 years, the adjoining ocean waters have become steadily warmer, although some depths show even lower temperatures than those recorded in 1954. Additional examples of this difficulty to various degrees of intensity can be found from time to time in some of the chapters in this book, as well as in some of the references.

On the other hand, some models are sufficiently accurate with respect to climatic changes in retrospect that reasonable agreement as to temperature increases can be corroborated by other methods. For example, Overpeck et al., *Science,* October 28, 1991, checked the accuracy of their thermal computer model by running it backwards in time for 18,000 years. The model's prediction agreed with what paleoclimatologists have already deduced from fossil pollen records about past climate that caused vegetation changes.

The computer models of Overpeck and co-workers suggested that spruce populations in the Eastern U.S., as well as populations of northern pines, will decrease significantly in abundance. The southernmost limit of spruce forests would migrate north as much as 1,000 km. This would leave much of New England, New York, Pennsylvania, and the Great Lakes regions without these evergreens. Similarly, the southern pines would also be expected to migrate northward as much as 1,000 km.

This is but another example of the successful use of diversified multidisciplinary methods used to study global warming. Therefore, it is somewhat surprising that more emphasis has not been given to studying the effect of variations in the greatest source of energy for the earth, namely, solar energy. This applies especially to sunspot variations that have been studied in varying degrees of ever-increasing degrees of sophistication for about 200 years. A detailed study of major changes in the intensity of the sun's radiation would appear to be obvious. However, quite recently (Friis-Christensen and Lassen, *Science,* November 11, 1991) demonstrated a well-defined relationship between sunspot activity and global temperature over a 130-year period for which accurate climatic records exist. Actually, this corrrelation does not involve the appearance of sunspots per se, but with the length of the sunspot cycle events which vary between 9 to 11 years. Shorter cycles correlate with higher earth temperatures. Direct satellite measurements of solar radiation itself would be the most definitive method, but, unfortunately, these have only been made for about 10 years, and more positive conclusions could not be expected for at least several decades.

The concept of climate change has been discussed scientifically for over 150 years, as mentioned in the Preface, but, it did not gain much publicity until researchers in the 1960s, working with data from limited and often urbanized stations, identified, from a study of the annual mean temperatures, a warming from about 1880 to 1940, then a slight cooling, and finally another slight warming. This temporal pattern, despite its limited scope and coverage, has been interpreted as evidence of ''global warming''. During the last decade or so it has been realized that, without refuting the physical aspect of potential warming via increased concentration of carbon dioxide in the atmosphere, the observational evidence was tenuous at best.

Recently, the National Climatic Data Center (NCDC), located in North Carolina, has formed two new interlinked units — the Global Climate Perspectives System (GCPS) and the Global Historical Climate Network (GHCN). The purpose of the GCPS is mainly to monitor climate (at different time scales up to near real-time) so as to assist decision-makers. The GHCN unit, which has assistance and input from an international team, is acquiring worldwide data. An essential component of these data will be the identification of Core stations, some a few hundred in number. These will give a representative coverage of the world land area and its climatic zones; additionally, stations will have no, or a minimum of, urbanization pollution, the longest period of record for their area, and little or no missing data. All the data will be subjected to quality checks, both subjective and objective. This undertaking will ensure that eventually the best data set possible will be used in making pronouncements on climatic change (personal communication, Prof. John Griffiths, Department of Meterology, Texas A&M University).

## RECOMMENDATIONS

Tremendous strides have been made, especially during the past several decades, to focus on a multitude of scientific and technical research efforts, both on a national as well as international scale, to study global warming. Much of this effort has been very successful in obtaining a definitive scientific solution to some of the major problems. Nevertheless, some legal and/or regulatory actions will be required in order that the research efforts be implemented. These actions range from national (and frequently to international) efforts to achieve this objective. This in turn requires the concerted and coordinated efforts of these many diversified groups. It also involves significant expenditure of funds, as well as time.

This situation has been recognized by those interested in the problem. It frequently has led to much pessimism regarding the success rate at which these objectives have been achieved. On the other hand, there is ample evidence that significant progress has been achieved for a variety of the objectives. For example, there have been a number of meetings on both national and international levels that have been very successful in furthering the aims of the proponents of studying ameliorating processes and activities designed to check and eventually reduce global warming. These include the Vienna and Montreal Conventions that have been held, as well as those scheduled soon, such as Rio de Janeiro and Copenhagen. An entire section of this book is devoted to this subject and provides much diversified information on the methods used and results achieved. It deals with the broad subject of national and international legal and policy considerations required to control global warming. The authors are all recognized both nationally and internationally in this field. Additional information pertinent to a study of this subject is found in several of the chapters in Section IV: Natural Resource Management Needed to Provide Long-Term Global Energy and Agricultural Uses. The necessary scientific and technical efforts to study global warming as it applies eventually toward implementing objectives of this subject are discussed in detail in other sections.

## TABLE 1
## List of Acronyms

| | |
|---|---|
| **ABLE** | Atmospheric Boundary Layer Experiment |
| **ACRIM** | Active Cavity Radiometer Irradiance Monitor |
| **ADEOS** | Japanese Advanced Earth Observing Satellite |
| **AEDD** | Arctic Environment Data Directory |
| **AID** | Agency for International Development |
| **AIRS-AMSU** | Atmospheric Infrared-Advanced Microwave Sounding Unit |
| **ARISTOTELES** | Applications and Research Involving Space Technologies Observing the Earth's Field from Low Earth Orbiting Satellite |
| **AOL** | Airborne Oceanographic Lidar |
| **ARCSS** | Arctic System Science |
| **ARM** | Atmospheric Radiation Measurements |
| **ARS** | Agricultural Research Service |
| **ASAS** | Advanced Solid State Array Sensor |
| **ASF** | Alaska SAR Facility |
| **AVHRR** | Advanced Very High Resolution Radiometer |
| **AVIRIS** | Airborne Visible-Infrared Imaging Spectrometer |
| **BIOSYNOP** | Biological Synoptic Ocean Prediction |
| **BLM** | Bureau of Land Management |
| **BOREAS** | Boreal Ecosystem-Atmosphere Study |
| **CART** | ARM Cloud and Radiation Testbed |
| **CCN** | Cloud-condensation-nuclei |
| **CDDIS** | Crustal Dynamics Data Information System |
| **CDP** | Crustal Dynamics Project (NASA) |
| **CD-ROM** | Compact Disc-Read Only Memory |
| **CEDAR** | Coupled Energetics and Dynamics of Atmospheric Regions |
| **CEES** | Committee on Earth and Environmental Sciences |
| **CEOS** | Committee on Earth Observation Satellites |
| **CEQ** | Council on Environmental Quality |
| **CERES** | Clouds and Earth's Radiant Energy System |
| **CES** | Committee on Earth Sciences |
| **CFCs** | Chlorofluorocarbons |
| **CGC** | Committee on Global Change |
| **CHAMMP** | Computer Hardware, Advanced Mathematics, and Model Physics |
| **CNES** | Centre National d'Etudes Spatiales |
| **COADS** | Comprehensive Ocean-Atmosphere Data Set |
| **COARE** | TOGA Coupled Ocean-Atmosphere Response Experiment |
| **CRF** | Cloud Radiative Forcing |
| **CZCS** | Coastal Zone Color Scanner |
| **DMS** | Dimethylsulfide |
| **DMSP** | Defense Meteorological Satellite Program |
| **DOC** | Department of Commerce |
| **DOE** | Department of Energy |
| **DOI** | Department of Interior |
| **DOS** | Department of State |
| **DOT** | Department of Transportation |
| **ENSO** | El-Niño/Southern-Oscillation |
| **EO-ICWG** | Earth Observations International Coordination Working Group |
| **EOS** | Earth Observing System |
| **EOSAT** | Earth Observing Satellite |
| **EOSDIS** | EOS Data and Information System |
| **EPA** | Environmental Protection Agency |
| **ERBE** | Earth Radiation Budget Experiment |
| **EROC** | Ecological Rates of Change (NSF) |
| **EROS** | Earth Resources Observation Systems |
| **ERP** | Environmental Research Parks |
| **ERS-1** | European Remote-sensing Satellite |
| **ESA** | European Space Agency |
| **ESTAR** | Electronically Scanned Thinned Array Radiometer |

**TABLE 1 (continued)**
**List of Acronyms**

| | |
|---|---|
| **FCCSET** | Federal Coordinating Council for Science, Engineering and Technology |
| **FIFE** | First Field Experiment |
| **FIRE** | First ISCCP Regional Experiment |
| **FLINN** | Fiducial Laboratories for an International Natural Science Network |
| **FOLD** | Federally Owned Landsat Data |
| **FS** | Forest Service |
| **FWS** | Fish and Wildlife Service |
| **FY** | Fiscal Year |
| **GAGE** | Global Atmospheric Gases Experiment |
| **GCDIS** | Global Change Data and Information System |
| **GCM** | Global Circulation Model |
| **GCRP** | Global Change Research Program |
| **GEM** | Geospace Environment Modeling |
| **GEMS** | Global Environmental Monitoring System |
| **GEOSAT** | Geodetic Satellite Mission |
| **GEWEX** | Global Energy and Water Cycle Experiment |
| **GFO** | GEOSTAT Follow-On |
| **GIS** | Geographic Information System |
| **GISPII** | Greenland Ice Sheet Project |
| **GLOBEC** | Global Ocean Ecosystems Dynamics |
| **GLRS** | Geoscience Laser Ranging System |
| **GMCC** | Geophysical Monitoring for Climate Change |
| **GOES-VISSR** | Geosynchronous Operational Environmental Satellite-Visible Infrared Spin Scan Radiometer |
| **GOFS** | Global Ocean Flux Study |
| **GPS** | Global Positioning System |
| **GRID** | Global Resource Information Database |
| **GSN** | Global Seismic Network |
| **Gt** | Gigaton(s) |
| **GTS** | Global Telecommunications System |
| **GWPs** | Greenhouse Warming Potentials |
| **HAP** | High Altitude Aerial Photography |
| **HCFC** | Hydrochlorofluorocarbon |
| **HIMSS** | High-Resolution Microwave Spectrometer Sounder |
| **HIRIS** | High-Resolution Imaging Spectrometer |
| **HIRS** | High-Resolution Infrared Sounder |
| **ICSU** | International Council of Scientific Unions |
| **IGAC** | International Global Atmospheric Chemistry Program |
| **IGBP** | International Geosphere-Biosphere Programme |
| **IGCP** | International Geological Correlation Programme |
| **IGOSS** | Integrated Global Ocean Science System |
| **IHP** | Intenational Hydrological Programme |
| **IOC** | Intergovernmental Oceanographic Commission |
| **IPCC** | Intergovernmental Panel on Climate Change |
| **IRAP** | Interdisciplinary Retrospective Analysis Program |
| **IRIS** | Incorporated Research Institution for Seismology |
| **ISCCP** | International Satellite Cloud Climatology Project |
| **ISLSCP** | International Satellite Land Surface Climatology Project |
| **JGOFS** | Joint Global Ocean Flux Study |
| **JOI** | Joint Oceanographic Institutions, Inc. |
| **JPL** | Jet Propulsion Laboratory |
| **JSC** | Joint Scientific Committee |
| **LAWS** | Laser Atmospheric Wind Sounder |
| **LMER** | Land Margins Ecosystems Research |
| **LTER** | Long-Term Ecological Research |
| **LTREB** | Long-Term Research in Environmental Biology |
| **LTRRS** | Long-Term Regional Research Site |
| **MAB** | Man and the Biosphere Program |

| | |
|---|---|
| **MAC** | Multisensor Airborne Campaigns |
| **MARS** | Mitigation and Adaption Research Strategies |
| **MESEEC** | Methodologies to Estimate Social, Environmental, and Economic Consequences |
| **MFE** | CNES Magnetic Field Experiment |
| **MISR** | Multi-Angle Imaging Spectro-Radiometer |
| **MODIS** | Moderate Resolution Imaging Spectrometer |
| **MODIS-N** | MODIS-Nadir |
| **MSS** | Landsat Multispectral Scanner System |
| **NABE** | North Atlantic Bloom Experiment |
| **NAS** | National Academy of Sciences |
| **NASA** | National Aeronautics and Space Administration |
| **NCAR** | National Center for Atmospheric Research |
| **NCDS** | NASA Climate Data System |
| **NDSC** | Network for the Detection of Stratospheric Change |
| **NDVI** | Normalized Difference Vegetation Index |
| **NGDC** | National Geophysical Data Center (NOAA) |
| **NIGEC** | National Institute for Global Environmental Change |
| **NMFS** | National Marine Fisheries Service (NOAA) |
| **NOAA** | National Oceanic and Atmospheric Administration |
| **NODC** | National Ocean Data Center (NOAA) |
| **NOZE** | National Ocean Expedition |
| **NPS** | National Park Service |
| **NSCAT** | NASA Scatterometer |
| **NSIDC** | National Snow and Ice Data Center |
| **NSSDC** | National Space Science Data Center |
| **NSF** | National Science Foundation |
| **NWS** | National Weather Service |
| **OAR** | Office of Atmospheric Research (NOAA) |
| **ODPs** | Ozone Depleting Potentials |
| **OFDA** | Office of U.S. Foreign Disaster Assistance |
| **OLR** | Outgoing Longwave Radiation |
| **OMB** | Office of Management and Budget |
| **ONR** | Office of Naval Research |
| **OSTP** | Office of Science and Technology Policy |
| **PCAST** | President's Council of Advisors on Science and Technology |
| **PAGES** | Past Global Changes |
| **PBMR** | Pushbroom Microwave Radiometer |
| **PCMDI** | Program for Climate Model Diagnostics and Intercomparison |
| **PLDS** | Pilot Land Data System |
| **POES** | Polar-Orbiting Environmental Satellites |
| **PSI** | Pacific/Sulfur/Stratus Investigation |
| **RIDGE** | Ridge Inter-Disciplinary Global Experiments |
| **RITS** | Radiatively Important Trace Species |
| **SAC** | Scientific Advisory Committee |
| **SAGE II** | Stratospheric Aerosol and Gas Experiment |
| **SAR** | Synthetic Aperture Radar |
| **SBUV** | Solar Backscatter Ultraviolet |
| **SBUV-2** | Solar Backscatter Ultraviolet-2 |
| **SC-IGBP** | Scientific Committee for the IGBP |
| **SCOR** | Scientific Committee on Oceanic Research |
| **SeaWIFS** | Sea-Viewing, Wide-Field Sensor |
| **SLR** | Satellite Laser Ranging |
| **SMM** | Solar Maximum Mission |
| **SMM/I** | Special Sensor Microwave/Imager |
| **SMRR** | Scanning Multichannel Microwave Radiometer |
| **SNOTEL** | Snow Survey and Remote Telemetry |
| **SOLSTICE** | Solar Stellar Irradiance Comparison Experiment |
| **SPOT** | Satellite Pour l'Observation de la Terre |
| **SSG** | Scientific Steering Groups |
| **SSMR-SSM/I** | Scanning Multichannel Microwave Radiometer-Special Sensor Microwave/Imager |
| **STORM** | Stormscale Operational and Research Meterology |
| **STRESS** | Sediment Transport Events on Shelves and Slopes |

## TABLE 1 (continued)
## List of Acronyms

| | |
|---|---|
| **TIA** | Territorial and International Affairs |
| **TIMS** | Thermal Infrared Spectral Scanner |
| **TIROS** | Television and Infrared Observation Satellite |
| **TM** | Thematic Mapper |
| **TOCSEEG** | Tradeoffs Among Competing Social, Environmental, and Economic Goals |
| **TOGA** | Tropical Ocean and Global Atmosphere |
| **TOGA-COARE** | TOGA Coupled Ocean-Atmosphere Response Experiment |
| **TOMS** | Total Ozone Mapping Spectrometer |
| **TOPEX/Poseidon** | Ocean Topography Experiment |
| **TOVS** | TIROS Operational Vertical Sounder |
| **TRAC** | Science Teacher Research Associates Program |
| **TRMM** | Tropical Rainfall Measurement Mission |
| **U.N.** | United Nations |
| **UARS** | Upper Atmosphere Research Satellite |
| **UNEP** | United Nations Environment Programme |
| **UNESCO** | United Nations Educational, Scientific, and Cultural Organization |
| **USDA** | United States Department of Agriculture |
| **US GCRP** | United States Global Change Research Program |
| **USGS** | United States Geological Survey |
| **USIA** | U.S. Information Agency |
| **UTM** | Universal Transverse Mercator |
| **UVB** | Ultraviolet-B |
| **VAS** | VISSR Atmospheric Sounder |
| **VISSR** | Visible Infra-red Spin Scan Radiometer |
| **VLBI** | Very Long Baseline Interferometry |
| **WPC** | World Climate Programme |
| **WCRP** | World Change Research Programme |
| **WEBB** | Water, Energy, and Biogeochemical Budgets (USGS) |
| **WOCE** | World Ocean Circulation Experiment |

From Policy Implications of Greenhouse Warming, Mitigation Panel, National Academy Press, Washington, D.C., 1991. With permission.

**TABLE 2**

**FY 1991–1992 U.S. Global Change Research Program Focused Budget by Integrating Theme**

| | Total budget | | Climate modeling and prediction | | Global water and energy cycles (inc. sea level change) | | Global carbon cycle | | Ecological systems and population dynamics | | Other research activities | |
|---|---|---|---|---|---|---|---|---|---|---|---|---|
| **Focused program** | **1991** | **1992** | **1991** | **1992** | **1991** | **1992** | **1991** | **1992** | **1991** | **1992** | **1991** | **1992** |
| Agency totals | 953.7 | 1185.5 | 59.8 | 95.4 | 417.3 | 521.2 | 129.9 | 162.8 | 105.8 | 146.9 | 240.9 | 259.2 |
| DOC/NOAA | 47.0 | 78.0 | 11.3 | 20.2 | 28.1 | 44.1 | 5.4 | 7.2 | 0.6 | 2.4 | 1.6 | 4.1 |
| DOD | 0 | 6.3 | 0 | 0 | 0 | 4.8 | 0 | 0.3 | 0 | 1.0 | 0 | 0.2 |
| DOE | 65.6 | 77.0 | 21.9 | 26.5 | 27.2 | 32.0 | 16.5 | 18.5 | 0 | 0 | 0 | 0 |
| DOI | 36.6 | 46.4 | 4.3 | 4.9 | 18.1 | 20.7 | 4.8 | 5.0 | 8.1 | 14.0 | 1.3 | 1.8 |
| EPA | 21.8 | 26.0 | 0 | 0 | 0 | 0 | 12.2 | 14.8 | 6.3 | 7.4 | 3.3 | 3.8 |
| NASA | 651.6 | 772.6 | 14.9 | 27.9 | 312.5 | 377.7 | 57.6 | 70.5 | 51.2 | 72.1 | 215.4 | 224.4 |
| NSF | 87.1 | 118.5 | 5.1 | 10.4 | 29.0 | 39.0 | 21.3 | 30.1 | 13.1 | 19.4 | 18.6 | 19.6 |
| Smithsonian | 5.4 | 7.5 | 0 | 0 | 0.1 | 0.1 | 1.1 | 1.9 | 3.5 | 4.5 | 0.7 | 1.0 |
| USDA | 38.6 | 53.2 | 2.3 | 5.5 | 2.3 | 2.8 | 11.0 | 14.5 | 23.0 | 26.1 | 0 | 4.3 |

*Note:* Dollars are in millions.

From U.S. Global Change Research Program by focused budget source, U.S. Global Change Research Program, 1991–1992 fiscal year budget. With permission.

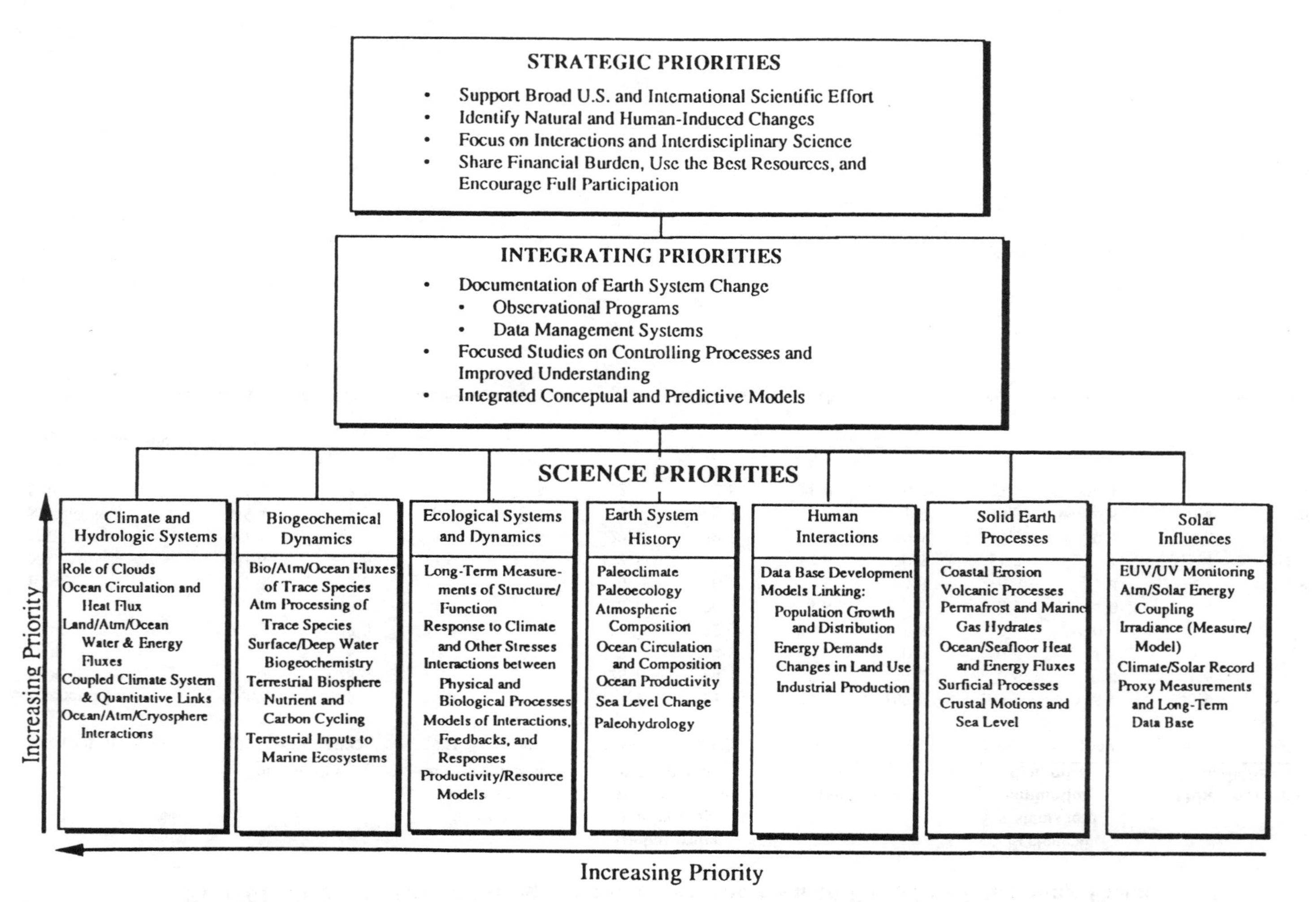


**FIGURE 1.** U.S. Global Change Research Program priority framework. (From "Our Changing Planet", The FY 1992 U.S. Global Change Research Program, Rep. by Committee on Earth & Environmental Sciences; A Suppl. to the U.S. President's FY 1992 Budget. With permission.)

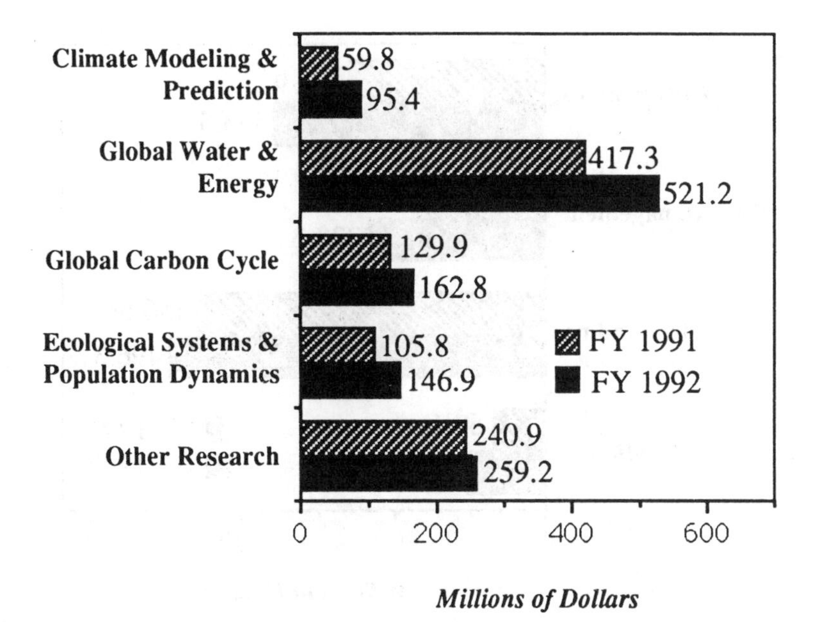


**FIGURE 2.** U.S. Global Change Research Program budget by integrating themes. (From "Our Changing Planet", The FY 1992 U.S. Global Change Research Program, Rep. by Committee on Earth & Environmental Sciences; A Suppl. to the U.S. President's FY 1992 Budget. With permission.)

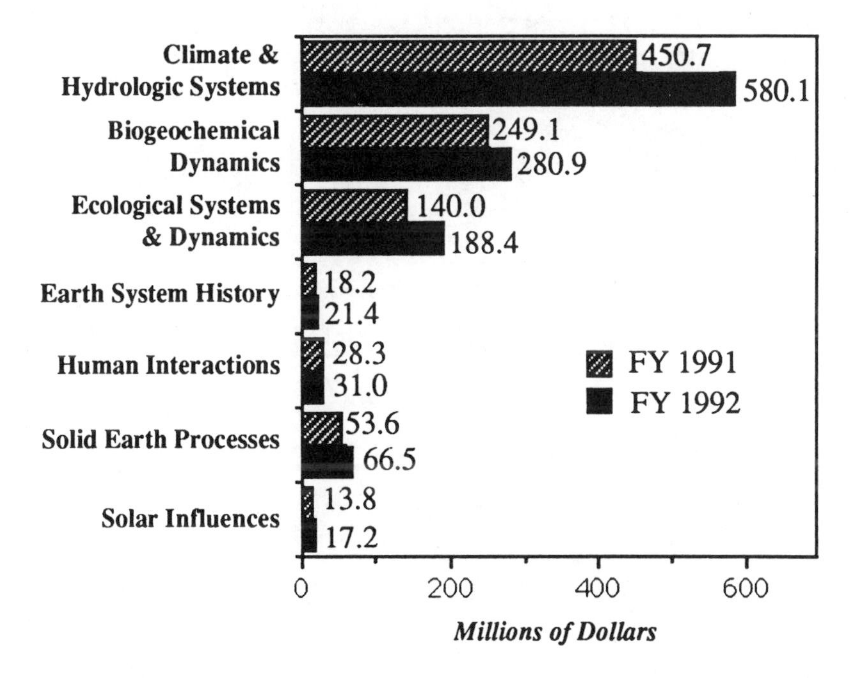


**FIGURE 3.** U.S. Global Change Research Program budget by science element. (From "Our Changing Planet", The FY 1992 U.S. Global Change Research Program, Rep. by Committee on Earth & Environmental Sciences; A Suppl. to the U.S. President's FY 1992 Budget. With permission.)

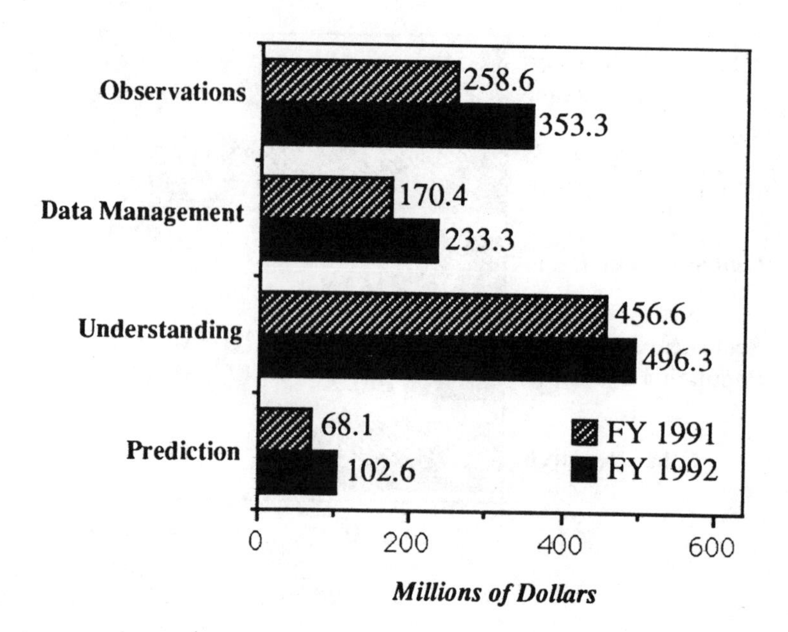


**FIGURE 4.** U.S. Global Change Research Program budget by scientific objective. (From "Our Changing Planet", The FY 1992 U.S. Global Change Research Program, Rep. by Committee on Earth & Environmental Sciences; A Suppl. to the U.S. President's FY 1992 Budget. With permission.)

# CONTRIBUTORS

**M. D. Ackerson, Ph.D.**
Department of Chemical Engineering
University of Arkansas
Fayetteville, Arkansas

**J. G. Baldauf, Ph.D.**
Assistant Professor
Department of Oceanography
Ocean Drilling Program
Texas A&M University
College Station, Texas

**Robert C. Balling, Ph.D.**
Associate Professor
Department of Geography
and
Director
Office of Climatology
Arizona State University
Tempe, Arizona

**Timothy R. Barber, Ph.D.**
Marine Research Associate
Department of Marine Science
University of South Florida
St. Petersburg, Florida

**Daniel J. Basta, Ph.D.**
NOAA
U.S. Department of Commerce
Rockville, Maryland

**Terry Bensel, M.S.**
Complex Systems Research Center
University of New Hampshire
Durham, New Hampshire

**Eugene E. Berkau, Ph.D.**
Research Professor
Department of Chemical Engineering
Vanderbilt University
Nashville, Tennessee

**N. C. Bhattacharya, Ph.D.**
Plant Physiologist
U.S. Department of Agriculture
Agricultural Research Service
Western Cotton Research Laboratory
Phoenix, Arizona

**Denise Blaha, B.S.**
Research Associate
Complex Systems Research Center
University of New Hampshire
Durham, New Hampshire

**Jenifer W. Clark, B.S.**
NOAA/NOS/OPC
U.S. Department of Commerce
Camp Springs, Maryland

**E. C. Clausen**
Department of Chemical Engineering
University of Arkansas
Fayetteville, Arkansas

**David R. Criswell, Ph.D.**
Director
Institute of Space Systems Operations
University of Houston
and
President
Cis-Lunar Industries
Houston, Texas

**Partha R. Dey, Ph.D.**
Senior Environmental Specialist
Resource Consultants, Inc.
Brentwood, Tennessee

**H. E. Dregne, Ph.D.**
Professor
International Center for Arid/Semi-Arid Land Studies
Texas Tech University
Lubbock, Texas

**J. B. Elsner, Ph.D.**
Assistant Professor
Department of Meterology
Florida State University
Tallahassee, Florida

**Penelope Firth, Ph.D.**
Executive Secretary
Committee on Earth and Environmental Sciences
Division of Environmental Biology
National Science Foundation
Washington, D.C.

**Christopher Flavin**
Vice President for Research
Worldwatch Institute
Washington, D.C.

**J. L. Gaddy**
Department of Chemical Engineering
University of Arkansas
Fayetteville, Arkansas

**Thomas J. Goreau, Ph.D.**
President
Global Coral Reef Alliance
Chappaqua, New York

**Robert C. Harriss, Ph.D.**
Professor
Complex Systems Research Center
University of New Hampshire
Durham, New Hampshire

**Raymond L. Hayes, Ph.D.**
Professor and Chairman
Department of Anatomy
Howard University, College of Medicine
Washington, D.C.

**S. Taseer Hussain, Ph.D.**
Professor
Department of Anatomy
Howard University College of Medicine
and
Research Associate
Department of Paleobiology
Smithsonian Institution
Washington, D.C.

**Solomon A. Isiorho, Ph.D.**
Assistant Professor
Department of Geosciences
Indiana University-Purdue University
Fort Wayne, Indiana

**Robert B. Jackson IV, M.S.**
Environmental Engineer
Office of Research and Development
U.S. Environmental Protection Agency
Athens, Georgia

**Lyle M. Jenkins, M.S.**
Aerospace Engineer
New Initiatives Office
NASA, Johnson Space Center
Houston, Texas

**M. A. K. Khalil, Ph.D.**
Professor
Department of Environmental Science and Engineering
Global Change Research Center
Oregon Graduate Institute
Beaverton, Oregon

**Justin Lancaster, Ph.D.**
Research Fellow
Department of Physics
Harvard University
Cambridge, Massachusetts

**Daniel A. Lashof, Ph.D.**
Senior Scientist
Natural Resources Defense Council
Washington, D.C.

**Nicholas Lenssen**
Research Associate
Worldwatch Institute
Washington, D.C.

**Patrick J. Michaels, Ph.D.**
Associate Professor
Department of Environmental Sciences
University of Virginia
Charlottesville, Virginia

**Alan S. Miller**
Center for Global Change
University of Maryland
College Park, Maryland

**Curtis Moore**
Center for Global Change
University of Maryland
College Park, Maryland

**Hayri Önal, Ph.D.**
Assistant Professor
Department of Agricultural Economics
University of Illinois
Urbana, Illinois

**Tsung-Hung Peng, Ph.D.**
Senior Scientist
Environment Sciences Division
Oak Ridge National Laboratory
Oak Ridge, Tennessee

**Ronald C. Pflaum, Ph.D.**
Center for Isotope Geochemistry
Department of Earth and Planetary Sciences
Harvard University
Cambridge, Massachusetts

**Josef Podzimek, Ph.D.**
Professor
Department of Mechanical and Aerospace Engineering
University of Missouri-Rolla
Rolla, Missouri

**P. D. Rabinowitz, Ph.D.**
Professor
Departments of Oceanography and Geophysics
Ocean Drilling Program
Texas A&M University
College Station, Texas

**Roger Revelle, Ph.D.** (deceased)
Professor
Department of Science and Public Policy
University of California
San Diego, California

**Craig N. Robertson, M.S.**
Oceanographer
NOS/ORCA/Sea Division
U.S. Department of Commerce NOAA
Rockville, Maryland

**William M. Sackett, Ph.D.**
Professor
Department of Marine Science
University of South Florida
St. Petersburg, Florida

**Karl B. Schnelle, Jr., Ph.D.**
Professor
Department of Chemical Engineering
Vanderbilt University
Nashville, Tennessee

**S. Fred Singer, Ph.D.**
Professor
Science and Environmental Policy Project
Arlington, Virginia

**Lowell Kent Smith, Ph.D.**
Professor
Department of Biology
University of Redlands
Redlands, California

**Anthony Socci, Ph.D.**
Research Associate
Department of Paleobiology
Museum of Natural History
Smithsonian Institution
Washington, D.C.

**Chauncey Starr, Ph.D.**
President's Office
EPRI (Electric Power Research Institute)
Palo Alto, California

**C. J. Tucker, Ph.D.**
Physical Scientist
NASA/Goddard Space Flight Center
Laboratory for Terrestrial Physics
Greenbelt, Maryland

**A. A. Tsonis, Ph.D.**
Professor
Department of Geosciences
University of Wisconsin-Milwaukee
Milwaukee, Wisconsin

**Robert R. Waldron, Ph.D.**
Staff Engineer
Space Systems Division
Rockwell International
Downey, California

**Dasheng Yang, Ph.D.**
Professor
Department of Geophysics
and
Deputy Chairman
Peking University
Beijing Meterological Society
Beijing, China

**David A. Wirth, J.D.**
Assistant Professor
School of Law
Washington and Lee University
Lexington, Virginia

**Weiyu Yang, Ph.D.**
Associate Professor
Institute of Atmospheric Physics
Chinese Academy of Sciences
Beijing, China

# TABLE OF CONTENTS

# A Global Warming Forum

## Scientific, Economic, and Legal Overview

# *Section I: Role of Geophysical and Geoengineering Methods to Solve Problems Related to Global Climatic Change*

Chapter 1

# RADIATIVE, HYDROLOGICAL, AND DYNAMIC INTERACTION IN THE AIR-SEA SYSTEM

**Dasheng Yang and Weiyu Yang**

## TABLE OF CONTENTS



0-8493-4419-0/93/$0.00+$.50
© 1993 by CRC Press, Inc.

## ABSTRACT

January 1983 was the mature phase of the 1982 to 1983 El Niño event for which the monthly mean net radiation on the sea surface, fluxes of the latent and the sensible heat from ocean to the atmosphere, and net heat gain of the sea surface were calculated over the Indian and the Pacific Oceans in the region 35°N to 35°S and 45°E to 75°W. The results indicated that the upward transfer of the latent and the sensible heat fluxes over the winter hemisphere is larger than that over the summer hemisphere. Slight downward transports of the sensible heat appear in the tropical mid-Pacific of the Southern Hemisphere. The latent heat gained by the air over the eastern Pacific is less than the mean value of the normal year. Moreover, the net radiation on which the cloud amount has considerable impact has a very good correlation with the sea surface temperature (SST), warm SST coinciding with the low net radiation. The axis of the lowest net radiation conforms to the ITCZ very well. The overview of the heat budget is that on the winter hemisphere, the ocean is the energy source of the atmosphere, and on the summer hemisphere the ocean is its energy sink.

The global ocean total cloud amounts in recent 126 years (1854 to 1979) were analyzed and it was found that the cloud cover had a tendency toward increasing. This tendency is in agreement with the global warming and intensification of the $CO_2$ greenhouse effect. By a one-dimensional radiative equilibrium model, we demonstrate that within the last 100 years, the cooling of the air temperature due to the variance of the cloud amount largely exceeds the warming arising from the greenhouse effect by the intensification of the $CO_2$ concentration.

Using the linearized barotropic model, under the action of the surface wind stress, the analytical expressions for the stream function of the ocean current and the perturbed sea surface height (SSH) are deduced. By using the grid data set of the 1000 hPa wind field, January 1980 to 1988, the dynamic adaption of the ocean current to the surface wind stress is investigated. The results indicate that for January 1983, the time scales of the response of the ocean current to the surface wind field are mainly the slow time scale, days 10 to 50, as well as the fast time scale, day 1 to 3. The stream function and the perturbed SSH change rapidly with time, until day 5, when both approximately reach the steady state. Compared to the monthly mean for 1980 to 1988, the monthly mean of 1983 has a large sheet of positive anomaly of the SSH off the west coast of South America in the South Pacific — its maximum value can attain 40 cm.

## I. INTRODUCTION

In recent years, climatologists, atmospheric chemists, and environmental engineers have paid much attention to the problem of global warming, and exchanged ideas on its causes and discussed, in particular, the general opinion that the global warming of the last century is a direct result of industrial pollution and, especially, of the burning of fossil fuels. It is truly a question of whether global warming is a natural tendency, and if so, how to mitigate its severity from man-induced temperature rise. There are also questions regarding the direct consequences of global warming, such as desertification processes and sea-level rise.

It is quite convincing that the main cause of global warming is $CO_2$ emissions, thus, utilization of biomass resources to substitute the fossil resources as the fuel and chemical production can help alleviate the global warming, because all $CO_2$ produced must be balanced by photosynthetic consumption.

The Chinese emissivities of greenhouse gases at present and in the year 2000 have been roughly estimated. In the year 2000, emissions of $CO_2$ and $N_2O$ will reach 1178.35 Tgc and 0.27 to 0.38 TgN, respectively (now the emissions are 809.39 Tgc/yr and 0.25 to 0.32 TgN/yr, respectively).

The continual increase of atmospheric $CO_2$ and other greenhouse gases may be one of the major influential factors of the climatic change of the next century. This viewpoint has been generally accepted, however, whether it will lead to exceptionally impetuous climatic change (e.g., some climatic simulation results indicate that if the present content of the atmospheric $CO_2$ increases onefold, the global mean air temperature near the Earth's surface will rise 1.5 to 4.5°C) is still a problem at issue. One of the reasons is that at present, there exist numerous uncertainties in the climatic simulation of the greenhouse effect, since the treatment of the greenhouse effect presently is not fine enough.

Nevertheless, irrespective of the entire globe, Northern Hemisphere, or China, the climate of the 20th century will evidently get warmer. Its time variation and spatial distribution are not completely consistent with that predicted by the greenhouse effect. Hence, although we are unable to negate the greenhouse effect, at least it should be recognized that in the past 100 years, the greenhouse effect is not the unique factor determining global temperature change, nor can we say it is not the determinant factor. Thus, in the past 100 years, natural variation may still be the principal factor in climate change. Therefore, it is rather early to presume to predict the climatic change simply from the unique factor of the greenhouse effect of $CO_2$.

Vegetation, specifically forests, can be used to store more carbon, and thereby slow (or partially offset) the observed increase in atmospheric $CO_2$. The tropical zones of the world seem particularly attractive for forestation, since mean long-term carbon storage ranges from 8 to 17 tc/ha in arid regions, to as high as 78 tc/ha in humid regions.

Weather is a chaotic system, and due to its sensitivity to initial conditions, it is unpredictable. Regional climate is also highly sensitive to man-made perturbations to the Earth's atmospheric system, thus, making it unpredictable. Man-made perturbations are increasing the climate variability. This is evidenced by the comparison of the previous climate changes with the current greenhouse gases forced changes, and the recently noted changes in ocean circulation as well as the probable thinning of the Arctic ice. Hence, within 20 years there may be many larger-than-expected regional and global climatic changes. This indicates that climatic change is driven partly by human activity. The relationship between population and natural disasters is being recognized, and most scientists agree that population is important in global climatic change. The concept of the human bolide is suggested.

Longer-term solar activity variations could explain some of the increased global temperature observed in the past century. Extrapolating the observed cycles of solar variation into the next century, it should be noted that a solar-forced warming will likely continue until at least the year 2040. However, solar radiation on its path to the Earth's surface incurs many losses through absorption and scattering due to the water vapor ($O_3$) and other greenhouse gases, as well as reflection by the air particles, water vapor, etc., and the Earth's surface. Therefore, the atmospheric part of the hydrosphere (clouds, precipitation, and the ambient moisture distribution) interacts with the radiation field, in which the amount and height of the clouds exert a profound influence on the budgets of the solar and infrared (IR) radiation, and, thus, on the net radiation which is available to the Earth atmosphere system.

It is commonly agreed that clouds have two basic functions with respect to the radiation balance at the Earth's surface. First, the solar contribution to the radiation budget at the Earth's surface of a previously clear column may change when clouds are introduced. This depends on the values of the cloud characteristics and the surface albedo. Over most parts of the globe, clouds cause the solar contribution to decrease by reflecting more radiation back to space. It is frequently said that clouds exhibit an albedo effect. Secondly, clouds generally enhance the IR component of the budget by absorbing radiation that originates from the warmer layers beneath them, and emitting radiation to space at a comparatively colder temperature. This effect, of course, depends on the cloud emissivity and temperature.

It is frequently called the greenhouse effect, because the budget is increased by absorption of the thermal radiation.

Model simulations show that increases in the emissions of greenhouse gases (aerosols, $SO_2$, nitrogen, oxides, etc.) lead to global warming. However, some of the aerosols formed due to the chemical reactions in the atmosphere are incorporated into clouds and modify their microphysics, thus, altering cloud radiative properties. Estimates indicate that an increase of 30% of the cloud droplet concentration will increase cloud albedo, which results in a decrease in temperature of more than 1.0°C at the Earth's surface. This implies that any climate model, in assessing the impact of greenhouse gases emission on the global warming, must include the effect of greenhouse gases on the cloud radiative properties.

In recent years, an international experiment using the treeline as climate change indicator in the extremely climate-sensitive circumpolar sub-Arctic region has been carried out. This study aims to establish hard scientific facts on the impacts of environmental changes on terrestrial ecosystems and on global climate.

Biological long-term evidence of the circumpolar sub-Arctic climate oscillation (BLECSCO) project aims to study the dynamics of ecosystem response to environmental change, and the impact of terrestrial ecosystem on global climate. The near-term objectives of this project consist of the assessment of the potential increase in forest area and forest product resources, as a result of global warming in the most climate-sensitive areas (near the treeline) of the circumpolar zone, and establishing the standards of data needed in utilizing the biological information contained between the forest line and the treeline for the former assessment. A comprehensive approach for assessing the increase in forest resources in the most climate-sensitive areas of the Northern Hemisphere has been carried on.

A one dimensional climate model is designed to estimate global warming arising from increasing concentrations of $CO_2$, methane, nitrous oxide, and the chlorofluorocarbons (CFCs). The model includes a water vapor feedback and an ocean-mixed layer. A preliminary result shows thus far that rapid changes in trace gas compositions cause a delayed response of global warming.

In recent years, numerous ocean-atmosphere coupled characteristic systems were formulated, for which there are abrupt and finite amplitude nonlinear characteristic behaviors, due to the nonlinear interaction between a linear unstable low frequency primary eigen component and its higher order harmonic components separately corresponding to a strong and weak ocean-atmospheric coupling.

Climatic change has consequences on water resources, agricultural production, and ecosystems. Scientists have thus proposed, that research must be undertaken to translate the general circulation model (GCM) scale predictions of climate change to regional, social, and economic impacts before appropriate policy decisions can be taken.

Scientists made use of remote sensing for modeling the impacts of sea-level rise due to the greenhouse effect, which predicts response of wetlands and lowlands to inundation and erosion by sea-level rise. Significant changes in coastal wetland, lowlands, and fisheries were predicted.

In this chapter, we will begin by discussing the heat budget on the air-sea interface over the Pacific and Indian Oceans, taking into account the radiative and cloud effect. We will then proceed to analyze the variance of 126 years (1854 to 1979) of global cloud amounts. It was found that in recent 126 years, the global cloud amounts have a tendency toward increasing, which is in accord with the intensifying $CO_2$ greenhouse effect. Finally, we will given an account of the dynamic adaption of the ocean to the wind stress exerting on the oceanic surface, in which the observed distribution of wind stress for January 1980 to 1988 is used to calculate the ocean circulation by simplified linear barotropic model.

# II. THE ENERGY BUDGET ON THE OCEANIC SURFACE IN THE MATURE PHASE OF 1982 TO 1983 EL NIÑO EVENT

The El Niño phenomenon has been investigated in many aspects, such as the synthetic statistical analysis of the numerous El Niño events, the search for the correlation factors for predicting its genesis and development, and discussing the global or regional climatic change of its duration. In this section we endeavor to discuss the energy budget on the surface of Pacific and Indian Ocean during January 1983, the mature phase of the 1982 to 1983 El Niño phenomenon.

We selected the 1982 to 1983 event as an example not only because the SST and intensity of anomalies of other elements made it the most striking phenomenon of this century, but also because of its notable difference of genesis and development compared with the multitude of typical El Niño events.

## A. DATA SET AND CALCULATION SCHEME

The energy budget is calculated within the scope 45°E to 75°W and 35°N to 35°S, which covers the low latitudinal region of the Indian and Pacific Oceans. The global network data set from ECMWF with grid length 2.5° × 2.5° latitude and longitude, the monthly mean SST grid data from CAC, U.S., and the 5-day mean values of SST at the grid points manipulated by the National Weather Service (NWS) from daily SST data of KWBC, U.S., of January 1983 are used. Since SST varies very little in a short period and also because of the scantiness of the daily SST data, thus the 5-day mean SST are taken as the daily SST data within each 5-day period. Owing to the insufficiency of the atmospheric data near the sea surface, the datum on 1000 hPa level is used.

We first use the formulas by Ramanadham et al. (1981) to calculate the net radiation absorbed by oceanic surface and the sensible heat flux to the atmosphere, then compute the latent heat flux, referring to formulas by Sutton (1953) and Gurmer (1983) then finally calculate the net heat gain of the sea surface. Later we do the same calculation by the scheme of Reed (1983) and Weare et al. (1981). It is found that the overall distributions are essentially consistent, only with difference in numerical value, however, without difference in the order of magnitude. Nonetheless, it is to be noted that the results by the scheme of Reed and Weare et al., especially the fine structure of the radiation flux field and hence the distribution of the net heat gain over the sea surface, are more discernible and revealed some meaningful features. The formulas given by the scheme of Reed and Weare et al. are enumerated as follows:

$$Q_0 = A_0 + A_1 \cos \varphi + B_1 \sin \varphi + A_2 \cos \varphi + B_2 \sin \varphi$$

$$Q_h = 0.94Q_0(1 - 0.062C + 0.0019\alpha)$$

$$Q_b = \epsilon\sigma T_s^4(0.254 - 0.00495e_a)(1 - 0.07C)$$

$$Q_e = L\rho_a C_D U(q_s - q_a)$$

$$Q_s = \rho_a C_D C_p U(T_s - T_a)$$

$$Q_N = Q_h - Q_b$$

$$Q = Q_N - (Q_e + Q_s)$$

where $Q_0$ denotes the downward solar radiation flux under the condition without cloud; $Q_h$ the net downward solar radiation flux, namely the solar radiation flux subtracts the reflected shortwave solar radiation by the atmosphere and Earth's surface under the condition with cloud; $Q_b$ the upward effect longwave radiation; $Q_e$, $Q_s$ are, respectively, the latent and sensible heat flux over the sea surface, positive from ocean to the atmosphere; thus $Q_N$ is the net radiation flux obtained by the ocean; and Q the net heat gain of the ocean. The fluxes are in units of $Wm^{-2}$. The parameters in the formulas are separate as:

$$A_0 = -15.82 + 326.87 \cos(LA \times 3.14159/180)$$

$$A_1 = 9.63 + 192.44 \cos[(LA + 90) \times 3.14159/180]$$

$$B_1 = -3.27 + 108.70 \sin(LA \times 3.14159/180)$$

$$A_2 = -0.04 + 7.80 \sin[2(LA - 45) \times 3.14159/180]$$

$$B_2 = -0.50 + 14.42 \cos[2(LA - 5) \times 3.14159/180]$$

$$\varphi = (t - 21) \times 2 \times 3.14159/365, \text{ unit in radian}$$

where LA represents the geographical latitude in degree; $t = 1, 2, \ldots 365$, and 1 denotes 1 January, then in order of days of the year.

The $1 - 0.07$ C denotes cloud amount in units of 1/10; $\alpha$ the solar altitude angle at noon in unit of degree; $\epsilon$ the specific radiation power over the sea surface, which is set to be 0.97; $\sigma$, the Stefan-Boltzman constant, is equal to $5.67 \times 10^{-8} m^{-2} K^{-4}$; $T_s$, $q_s$ are, respectively, the temperature (unit in K) and saturated specific humidity (unit in Kg/Kg) over the sea surface; $e_a$ the vapor pressure of the air near the oceanic surface, unit in hPa; L the vaporization heat, taken as $2.45 \times 10^{-7} J\ Kg^{-1}$; $\rho_a$ and U are, respectively, the air density, unit in $Kgm^{-3}$; and the wind velocity near the sea surface, unit in $ms^{-1}$; $C_D$ the drag coefficient, a dimensionless quantity, of which the value depends on U and temperature difference near the air-sea interface enumerated by Zhu et al. (1990); $T_a$, $q_a$ are separately the air temperature (in K) and the atmospheric specific humidity (in Kg/Kg) near the sea surface; $C_p$ the specific heat under constant pressure equal to $1004.0\ J\ Kg^{-1}K^{-1}$.

Due to the incomplete observational data of the cloud amount, the numerically simulated cloud amount and its height by the method proposed by Krishnamurti et al. and Yin, (1983) are adopted. Getting the amount of high, middle, and low cloud $C_h \cdot C_m$ and $C_l$, the following expression is used to obtain the total cloud amount C:

$$C = 1.0 - (1.0 - C_h)(1.0 - C_m)(1.0 - C_l)$$

Observational cloud amount data covering the Indian and Pacific Oceans are incomplete; the cloud-amount data, probed by satellite and published by the Japanese Meteorological Satellite Center, only covers the domain 90°E to 170°W and 50°S to 50°N, which for January 1983 is shown in Figure 1. The differences between these observed cloud amounts and the cloud amounts computed by the scheme of Krishnamurti et al. are depicted in Figure 2. As is shown in Figure 1, within the realm 90°E to 170°W, 40°N and 40°S, the maximum cloud amounts are observed over Japan, the coasts of China, the Indo-Chinese peninsula, Indonesia, and from the equator to 10°S, cloud amount can attain to above 8/10. The least cloud amounts are over the South China Sea, the Philippines and to its east, and the east of 160°E, between the latitudes 15° and 20°N, and can be less than 2/10. This least cloud-amount belt coincides with the position of the boreal winter subtropical high pressure. Figure 2 indicates that the value of the difference between the computed and the satellite-observed cloud amount can seldom reach about 4/10 in the region of high value of cloud amount. It is quite small over the subtropical high pressure belt and the region of least cloud amount. Moreover, this

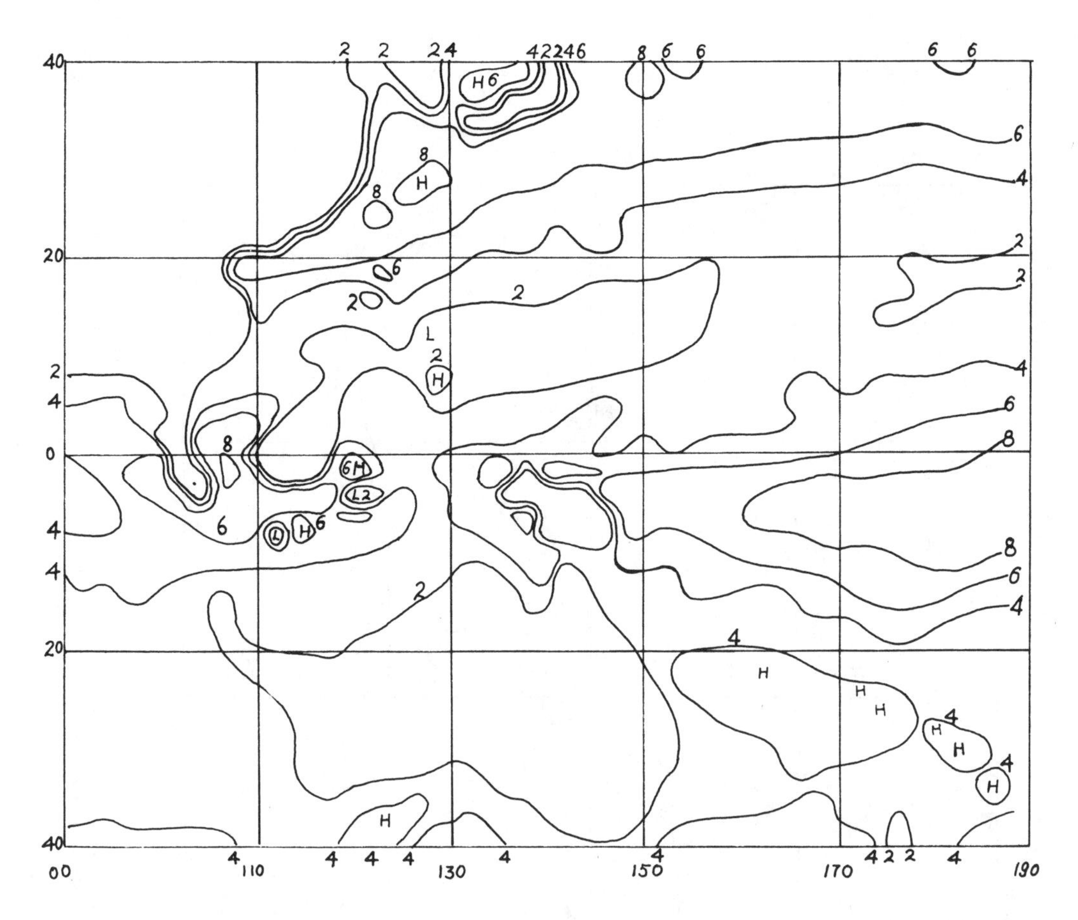

**FIGURE 1.** Satellite observed cloud amount; unit: 1/10. (From Yang Dasheng, Yang Bai, and Pan Zhi, *Acta Oceanologica Sinica* (China Ocean Press), 10:4, 539–554, 1991. With permission).

difference is comparatively systematical. Over the region between 20° and 40°N, as well as between 20° and 40°S, the difference has negative value, whereas that between 20°N and 20°S, the difference is small and mostly positive. Therefore, it is feasible in the numerical forecast model to calculate the cloud amount by means of the relative humidity. In addition, if the parameters in the model are suitable, the computed and observed cloud amount will be in better accord with each other.

## B. NET RADIATION HEAT FLUX ABSORBED BY OCEANIC SURFACE

The distribution of the monthly mean net radiation absorbed by the ocean in January 1983 is exhibited in Figure 3. It is to be noted that the net radiation is generally distributed zonally, which is most evident in the subtropical belt of the Northern Hemisphere. In tropical regions near the equator, there are some discrete centers of high and low value. In the subtropical belt of the Southern Hemisphere, there is wide distribution of high value irrespective of the Indian or Pacific Ocean. In the Pacific Ocean, 130°W is the longitude of demarcation, of which on both sides are regions of high value, and a low value region is distributed near 130°W without centers. Net radiation increases notably from north to south, from 100 $Wm^{-2}$ along about 30°N latitudinal circle of the Northern Hemisphere and increases gradually to the maximum at about 30°S of the Southern Hemisphere. Evidently the magnitude of the gradient of the net radiation over Southern Hemisphere is larger than that over Northern Hemisphere.

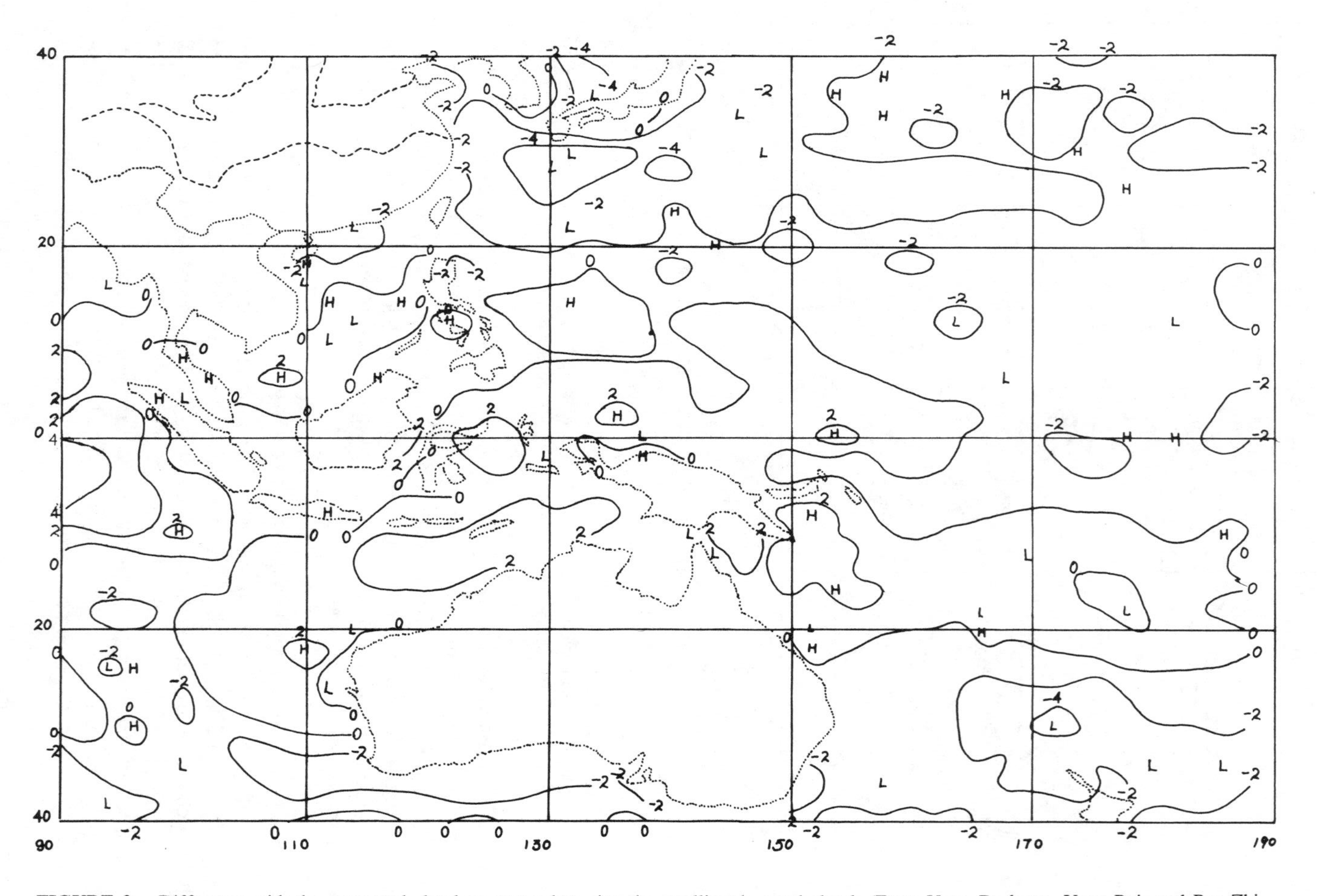


**FIGURE 2.** Difference with the computed cloud amount subtracting the satellite-observed cloud. (From Yang Dasheng, Yang Bai, and Pan Zhi, *Acta Oceanologica Sinica* (China Ocean Press), 10:4, 539–554, 1991. With permission).

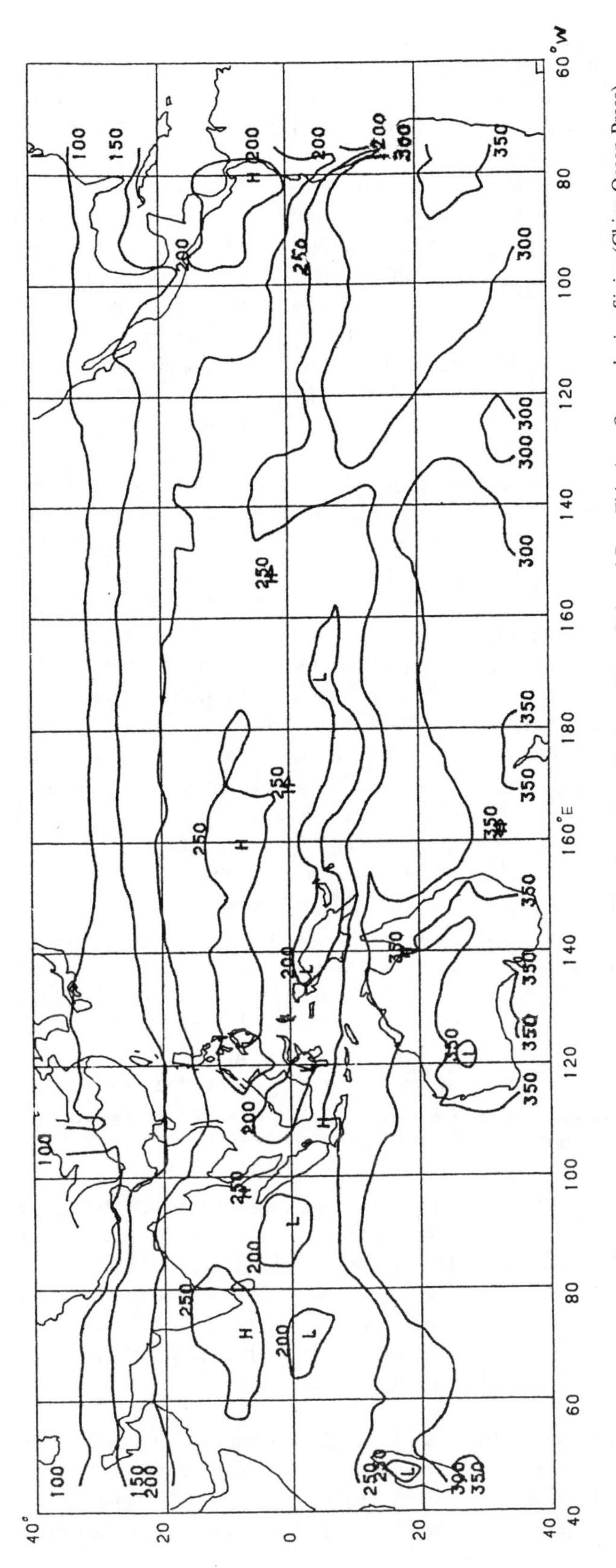


**FIGURE 3.** Monthly mean net radiation of January 1983 ($Wm^{-2}$). (From Yang Dasheng, Yang Bai, and Pan Zhi, *Acta Oceanologica Sinica* (China Ocean Press), 10:4, 539–554, 1991. With permission).

In Figure 3 the distribution of the net radiation flux is very similar to that of the outward longwave radiation (OLR) of January 1983 (Jiang and Zhu, 1990). In connection with the fit for the fine structure of the climatically mean position of ITCZ, the axis of the least value of the net radiation flux coincides with the position of the ITCZ. It seems that the net radiation field has a better accommodation that OLR.

The net radiation depends on the incident and the reflected by air and underlying surface solar shortwave radiation, and the emitting and absorbed longwave radiation. These are chiefly influenced by the solar altitude angle, total cloud amount, the temperature of the underlying surface, and the air humidity in its vicinity, among which solar altitude angle and total cloud amount have a particularly large impact on the radiation. At the same latitude, the solar altitude angle is the same, hence the net radiation is approximately zonally distributed. In January, the sun is scantly incident upon the Earth's surface in the Northern Hemisphere. Due to the abundant cloud amount (Figure 4) over north of 30°N, the net radiation at 35°N assumes a rather small value. The high value centers of net radiation over the Arabian Sea and the west Pacific Ocean near 10°N, are just located in the position of the winter subtropical high, which is also the location of the centers of least cloud amount (Figure 4), so that the sea surface can absorb a large amount of solar shortwave radiation. From the doldrums in the east of the Somali Coast through Indonesia to the central Pacific at about 10°S, there are several centers and a belt of low-net radiation. This belt is in exact accordance with the position of ITCZ, and is in better position to it than the axis of the least value of OLR. The large cloud amount in the convergence zone, where the high value is centered (Figure 4), results in the reduction of the absorbed shortwave radiation. Also owing to the break of the convergence zone in the west of 140°E, the three low centers in the doldrums and Indonesia are not connected in a belt of low value.

In the region of the east and central Pacific, the low centers are biased toward the Southern Hemisphere, because during the El Niño period, the ascending branch of the Walker circulation over the Pacific shifts eastward from Indonesia to near the dateline in the central Pacific and south of the equator. Thus, over this area, the cloud amount is the most abundant (Figure 4) that impedes solar radiation attaining the sea surface. Although the variation of SST has an impact on the strength of longwave radiation, and therefore on net radiation, it is however, not important. The area of warm SST is mainly that of an ascending current of thermal circulation, where the effect of the increase of cloud amount and air humidity near the oceanic surface is much greater than that of the increase of the upward longwave radiation on the warm sea surface. Thus, the upward longwave radiation, or OLR over the sea surface near the center of warm SST, has a low value. At the same time, however, the increased cloud amount also impedes solar shortwave radiation reaching the sea surface, that the net radiation obtained by the sea surface may still be a low value. As shown in Figure 5(a), along 5°N latitude circle, between 60° and 100°E, the relatively warmer SST and plentiful cloud corresponds to the low center.

As described above, during the El Niño event, since the positive anomally propagates along the zonal direction, when the El Niño process develops to its mature phase, the SST of the entire Pacific Ocean is warming up. It is certain that in addition to the direct impact of the SST variation on net radiation, its indirect influence is also very important. For instance, after the SST of the central Pacific gets warm, the original central Pacific condition of drought and scarce clouds becomes cloudy and rainy, hence the net radiation is reduced.

### C. LATENT HEAT FLUX

The formula for computing the latent heat flux by the two schemes by Zhu and Yang (1990) and the results are the same — the latent heat flux in the Northern Hemisphere is generally somewhat higher than that in the Southern Hemisphere at the corresponding latitude (see Figure 1c). There is a large low value area at the equator, which is approximately

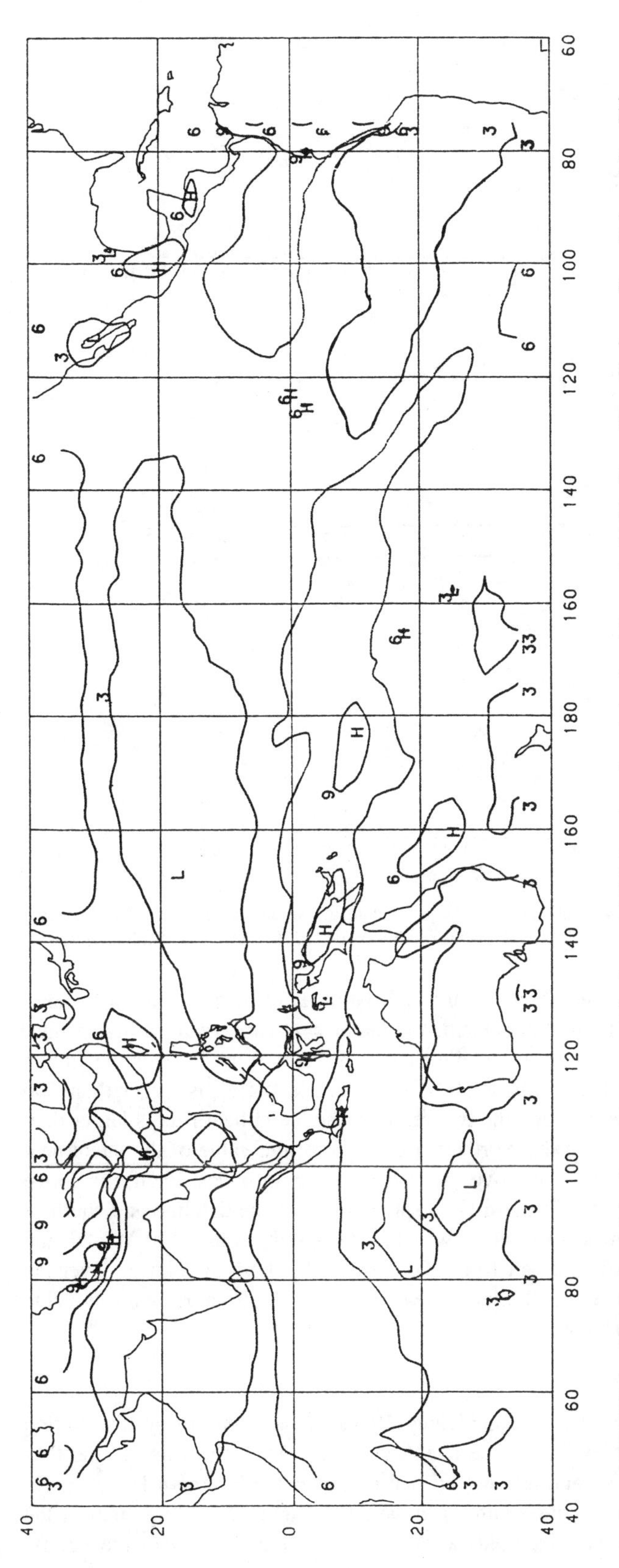


**FIGURE 4.** Distribution of cloud amount computed for January 1983, based on the relative humidity; unit: 1/ 10. (From Yang Dasheng, Yang Bai, and Pan Zhi, *Acta Oceanologica Sinica* (China Ocean Press), 10:4, 539– 554, 1991. With permission).

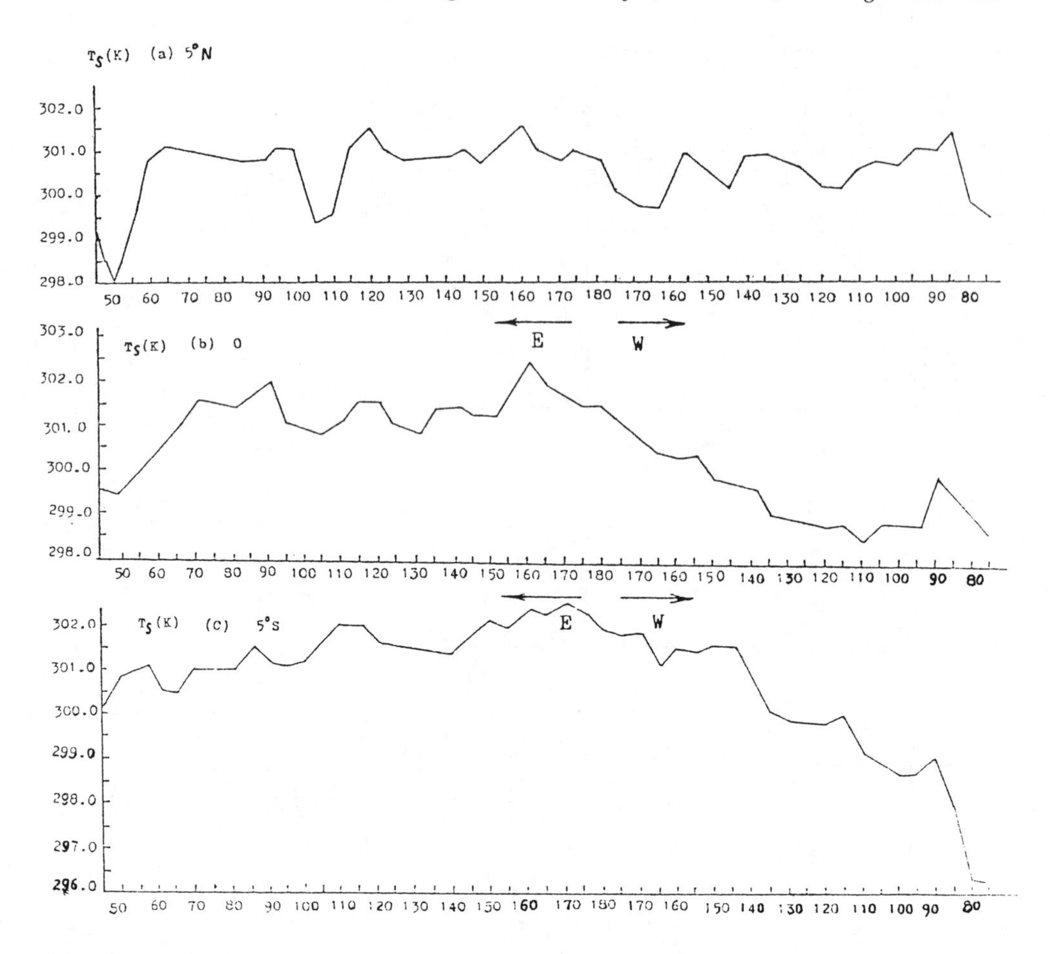


**FIGURE 5.** Distribution of the monthly mean SST near the equator along the latitude circle. (From Yang Dasheng, Yang Bai, and Pan Zhi, *Acta Oceanologica Sinica* (China Ocean Press), 10:4, 539–554, 1991. With permission).

100 $Wm^{-2}$. The latent heat flux increases toward high latitude. Over the Arabian Sea and the south of Bengal in the Northern Hemisphere, the latent heat flux assumes high value, which is larger than 150 $Wm^{-2}$, with a center value above 200 $Wm^{-2}$. Over the Indian Ocean, with 0° to 15°S and in the south of 30°S, the latent heat flux is larger than 100 $Wm^{-2}$. Over the oceanic surface between the two regions, the latent heat flux is larger than 100 $Wm^{-2}$. Over the west Pacific in the Northern Hemisphere, the value of the latent heat flux is above 150 $Wm^{-2}$. The high value over most part of the northwest Pacific is above 200 $Wm^{-2}$. Over the Pacific in the Northern Hemisphere, east of the dateline and south of 35°N, the latent heat flux assumes relatively low value. Generally below 100 $Wm^{-2}$, the lowest value is within the longitudinal belt from 160° to 140°W, about 50 $Wm^{-2}$. Over the sea surface off the west coast of Peru, the latent heat flux has a high value, with its value at the center being larger than 150 $Wm^{-2}$.

## D. SENSIBLE HEAT FLUX

From Figure 1(d) in the paper by Zhu and Yang (1990), the monthly mean distribution of the sensible heat flux resembles that of the latent heat flux. Also, the value of the flux over the Northern Hemisphere is generally larger than that over the Southern Hemisphere. Within the latitudes 5° to 15°S over the Southern Indian Ocean and over a large part of the Southern Pacific, the sensible heat flux holds a value below 15 $Wm^{-2}$. Over the Pacific

Ocean and within the latitudes and longitudes of 0° to 20°S and 170° to 130°W, as well as over the sea surface off the west coast of the North and South American continent, the sensible heat fluxes retain the small negative value.

The factors affecting the sensible heat flux are the wind velocity and the air-sea temperature difference. The sensible heat flux over the Northern Hemisphere is larger than that over the Southern Hemisphere, because over winter boreal hemisphere the SST is warmer than the air temperature near the sea surface, particularly over the subtropical belt of the West Pacific. Owing to the impact of the cold air current over the east Asian continent, the wind is strong, thus, the sensible heat flux can reach 100 $Wm^{-2}$. Over the Southern Hemisphere, the air-sea temperature difference is small, which leads to the small value of the sensible heat flux.

These deductions are consistent with the conclusion by Budyko (1960), namely, over the cold ocean current and the subtropical ocean, the sensible heat flux is rather small and even possesses a small negative value. In tropical region, the sensible heat flux is always small. There are two effects in connection with the negative value appearing in some regions: (1) the cold ocean current or the cold upwelling maintains the sea surface at a lower temperature; and (2) the atmosphere is locally heated and the regional heating is intense, which largely happens near the continent. At 10°N, as shown in Figure 6(a), within the longitudes of 100° to 85°W, the atmospheric warming surpasses that of the sea surface; accordingly the sensible heat flux is transported from the atmosphere to the ocean. This is chiefly because the impact of the cold upwelling off the continent. Figure 6(b) denotes the anomaly curve at 10°S, within the longitudes of 170° to 120°W, where the positive anomaly of the air temperature exceeds that of SST. This downward transfer of sensible heat flux mainly arises from the effect of the local heating of the atmosphere. Within the region of 0° to 20°S and 170° to 120°W, the local heating of the atmosphere is probably because this region is located near the ascending current of the Walker Circulation during El Niño period, thus the precipitation is more abundant than the normal period, and the heat largely released in the tropospheric middle layer also heats the whole troposphere and ocean. However, the heat capacity of the ocean is much greater than that of the atmosphere, hence SST only changes slightly.

## E. NET HEAT GAIN

Net heat gain is obtained by subtracting the upward latent and sensible heat flux over the air-sea interface from the net radiation reaching the underlying surface, thus its distribution characteristics are the synthesis of these quantities.

Figure 7 depicts the distribution of the monthly mean net heat gain. On account of the smallness of net radiation and the largeness of the latent and sensible heat flux in the region north of about 30°N and over the tropical west Pacific, the net heat gain is negative, with the largest value being $-250$ $Wm^{-2}$. This indicates that over the Northern Hemisphere, heat is largely transported from ocean to the atmosphere, i.e., the ocean is the energy source of the atmosphere. Over the Southern Hemisphere, except in the region north of 10°S and 160° to 180°E, owing to the largeness of the net radiation and the smallness of the latent heat flux, the net heat gain is positive, thus there is the energy sink of the atmosphere. Its distribution characteristics are completely similar to those of the net radiation over the Southern Hemisphere. Over the southern Indian and Pacific Oceans, the values of the centers are all above 300 $Wm^{-2}$. The Southern and Northern Hemispheres are in overview, respectively, the energy sink and source of the atmosphere, completing the global energy cycle.

Within the region between the equator and 20°S over the mid- and west Pacific, the net heat gain is less than that in other regions. Thus is just the response to the nonuniform SST

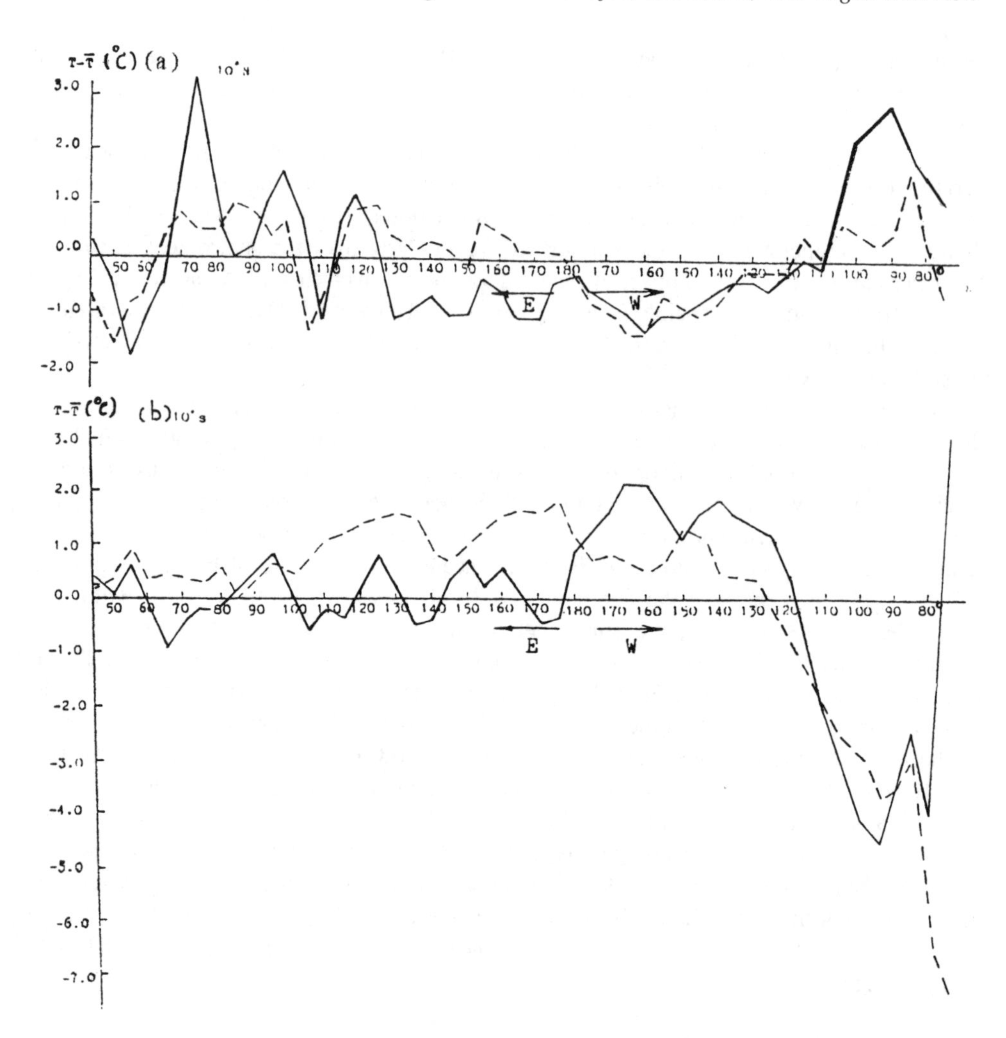

**FIGURE 6.** Anomaly of the sea and air temperature with respect to the zonally mean temperature. (Anomaly of SST; anomaly of the air temperature.) (From Yang Dasheng, Yang Bai, and Pan Zhi, *Acta Oceanologica Sinica* (China Ocean Press), 10:4, 539–554, 1991. With permission).

anomalies. It is worthwhile to point out that the region of negative value of net heat gain over the west Pacific lies exactly in the region of the warm pool which is planned to be investigated in the TOGA-COARE project. In view of the energy budget over the oceanic surface, the warm pool region is very sensitive to the SST and the wind anomaly near the sea surface. During the El Niño period, this region is the source over the tropical west Pacific providing energy from ocean to the atmosphere.

Reed's calculation (1983) of the heat flux over the equatorial east Pacific for the 1972 to 1973 El Niño event shows that in January 1973, within the region of 80° to 85°W and 2°N to 8°S, the heat flux was 145 $Wm^{-2}$, and within the region of 80° to 110°W and 0° to 2°S, it was 140 $Wm^{-2}$. Here the computed net heat gain for this region is roughly 150 $Wm^{-2}$, which is basically consistent with Reed's results. Li and Li (1988), employing the data of Sino-American west Pacific Research Expedition, and by means of the scheme of Ramanadham et al., computed the heat balance over the sea surface for February 8 to 14, 1986, El Niño period. The surface scope in their computation was approximately 140° to 160°E and 18°S to 14°N. They got the 7 day mean and regionally averaged values as follows:

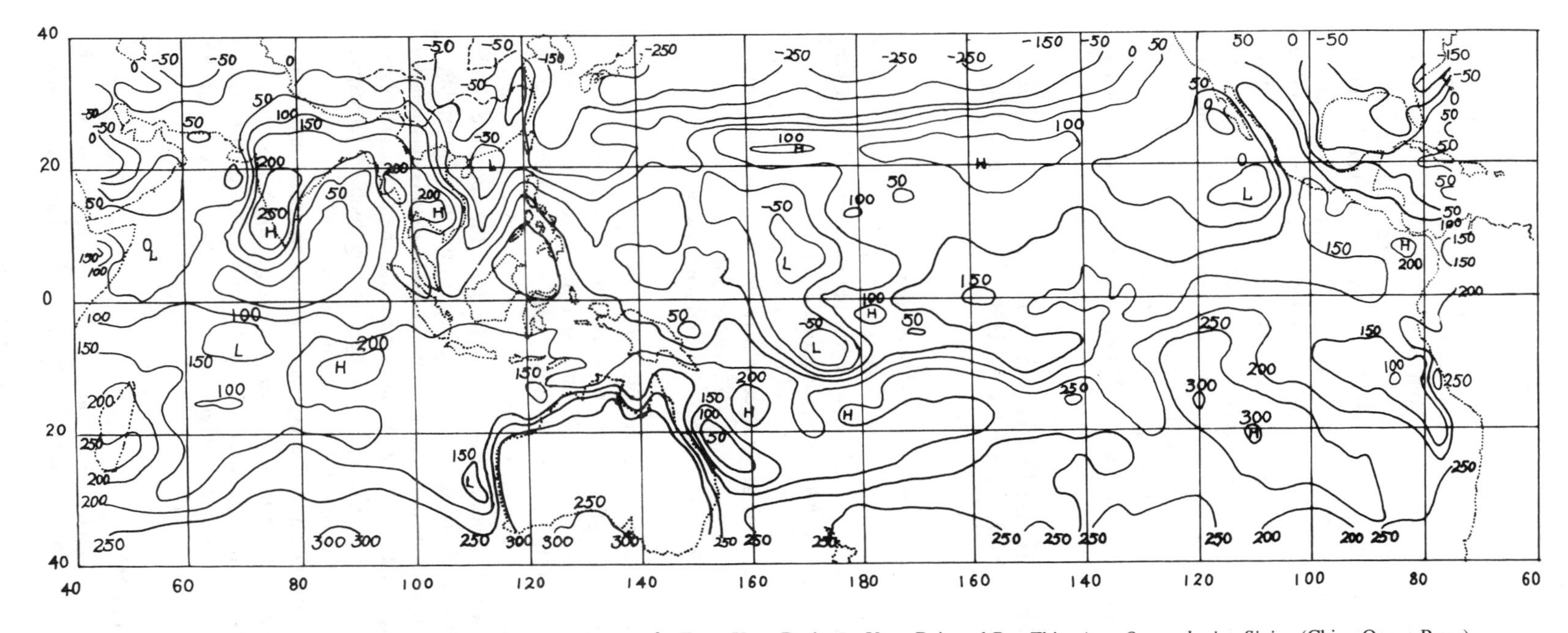

**FIGURE 7.** Monthly mean net heat gain of Jan. 1983, $Wm^{-2}$. (From Yang Dasheng, Yang Bai, and Pan Zhi, *Acta Oceanologica Sinica* (China Ocean Press), 10:4, 539–554, 1991. With permission).

the net radiation flux, 257.9 $Wm^{-2}$; the latent heat flux, 198.5 $Wm^{-2}$; the sensible heat flux, 13.0 $Wm^{-2}$; and the net heat gain of the oceanic surface, 44.6 $Wm^{-2}$. Our computed results for the same region and by the scheme of Reed et al. (1983) are the net radiation flux has values between 200 to 300 $Wm^{-2}$; the latent heat flux about 100 to 200 $Wm^{-2}$, mostly 150 to 200 $Wm^{-2}$; the sensible heat flux about 100 to 200 $Wm^{-2}$, mostly 150 to 200 $Wm^{-2}$; the sensible heat flux roughly 15 to 40 $Wm^{-2}$; and the net heat gain of the sea surface is approximately 40 to 150 $Wm^{-2}$. Our results largely firmed their results. As elucidated before, the calculated results by the scheme of Reed (1983) are larger than that by the scheme of Ramanadham et al. (1981), thus, the difference between Li and Li (1988) and us is rather trivial. In the region covered by the data set of Li and Li, (a considerable part of the oceanic surface) the ocean supplies energy to the atmosphere, in fact, in the region near 10°N and over the west Pacific, this feature is most evident. The computed net heat gain of the sea surface by us for the region 150° to 180°E and 15°S to 18°N is also negative. Therefore it may be recognized that during the El Niño period, for this region, ocean provides energy to the atmosphere, hence contributing notably to the variation of the global atmosphere circulation.

## III. THE TREND OF THE GLOBAL CLOUD AMOUNTS OF RECENT 126 YEARS AND ITS RELATION TO CLIMATE CHANGE

As mentioned in the Introduction, in recent years greenhouse effect has occupied so much of people's attention because it is believed that the greenhouse effect is responsible for global warming. According to the report of Intergovernmental Panel on Climate Change (IPCC), if the present situation keeps on developing, the global temperature will rise 0.3°C, and the sea surface will rise 6 cm every 10 years until the end of the next century. Thus, the temperature will increase 3°C, and the sea surface will rise 65 cm. This can give rise to serious consequences. However, this result may be questionable since in the predictive study of the climate change, the effect of clouds on the climate has yet to be defined accurately.

Cloud has following feedback effect to climate. When the globe gets warm, the evaporation on the underlying surface strengthens, which in turn induces the increase of the atmospheric humidity and then the global cloud amount. As noted before, cloud has both positive and negative feedback effects to climate. The water vapor in clouds is the constituent of the greenhouse gas, which will induce global warming. On the other hand, cloud augments the albedo of the Earth atmosphere system, so it will lower the global temperature. Both of these effects may finally result in a diminution of the tendency of the global warming.

The feedback mechanism depends on the cloud height. The high cloud is biased towards greenhouse effect, whereas the middle and low clouds contribute chiefly to the negative feedback effect. Based on the report of IPCC, the warming effect by doubling atmospheric $CO_2$ concentration is cancelled by increasing 2% of the low cloud cover.

Climate is very sensitive to the cloud feedback effect. In order to understand the inner connection and the mutual restrictive mechanism of the changing climate system, plus to provide basis for improving the climatic model and also to eliminate the uncertainty in climate prediction, it is necessary to have a correct grasp of the variation of the global cloud amount.

### A. DATA AND METHODOLOGY

In this preliminary work, COADS data which was collected and processed by CERES, ERL, NCAR, and NCDC are used. The time series of the number weighted grid points of

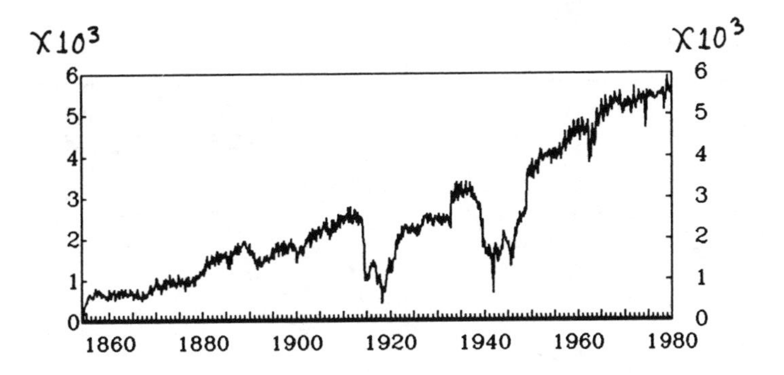


**FIGURE 8.** Time series of the number of weighted grid points of the global marine data without missing observation of the total cloud amounts for the period 1854 to 1979. The network grid area of 2° × 2° latitude longitude at the equator is adopted as unit; unit: × $10^3$.

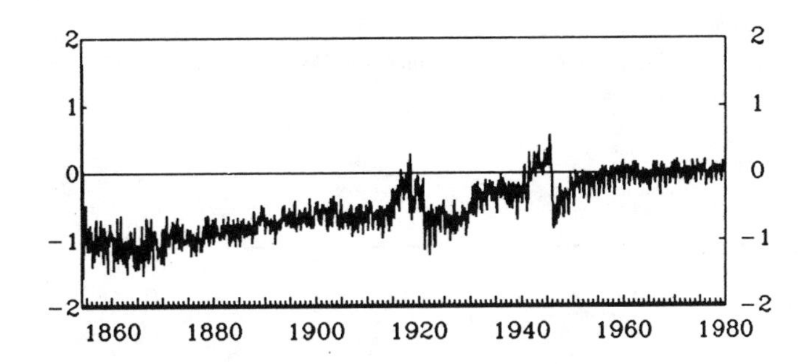


**FIGURE 9.** Time variance of the anomaly of the global total marine cloud amount for the period 1854 to 1979; unit: 10%.

the global marine data (without missing observation of the total cloud amounts) is shown in Figure 8. Using the grid network data, we obtain the zonal weighted mean cloud amounts, then evaluate the 49 months running mean with the results obtained. The time period of the data set is from 1854 to 1979.

## B. RESULTS AND ANALYSIS

Figure 9 depicts the time variance of the global total marine cloud amount in the period of recent 126 years, from which it is noted that there are two jump rises and falls around the year 1915 and 1940, respectively. After 1920 the anomalies recover to the level before 1915; whereas after 1945, the anomalies gradually increase to zero.

By comparison of the two curves in Figure 10, it should be noted that with the increase of the surface air temperature (solid line), the cloud amount has an evident tendency towards increasing (dashed line). However, the increase of the cloud amount has a phase lag of about 0 to 5 years relative to the surface air temperature variance. From the solid line, there occur roughly seven peak values of the surface air temperature before 1970, corresponding to each at which the lag time of the cloud amount attains its relative maximum value is 4.5, 2, 2.5, 3, 5, 3, and 0 years separately. This is probably because the cloud amount curve has been obtained through multiple years running average, the increasing of air temperature gives rise to the increase of cloud amount and may be realized by numerous interacting processes, which has a time scale of several years. Hence, after a certain time of impact, the cloud amounts begin to exhibit increasing effects, i.e., the surface air temperature needs a certain time to exert its influence on the cloud amount.

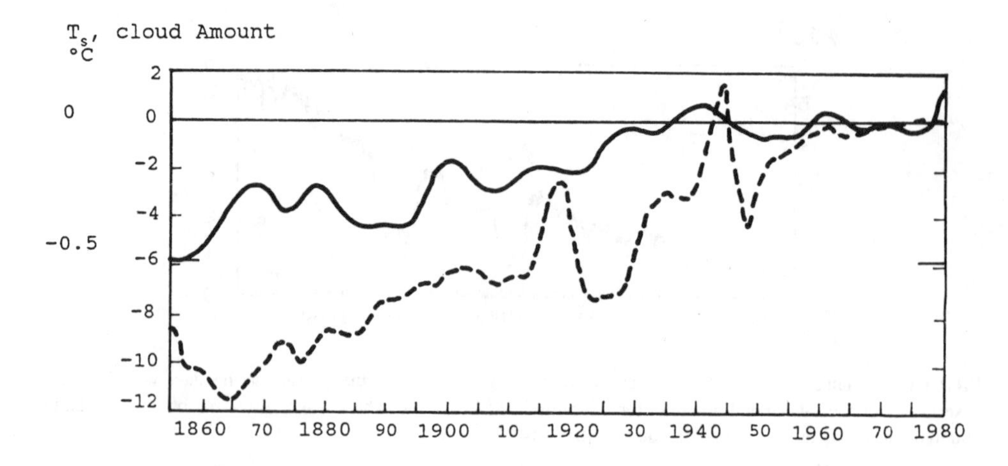


**FIGURE 10.** Time series of the 49 month running mean of the global total marine cloud amount anomaly (dashed line; unit: %), 1854 to 1979. The accompanying solid line is the low-pass filter filtered curve of the global mean surface air temperature by Jones et al.

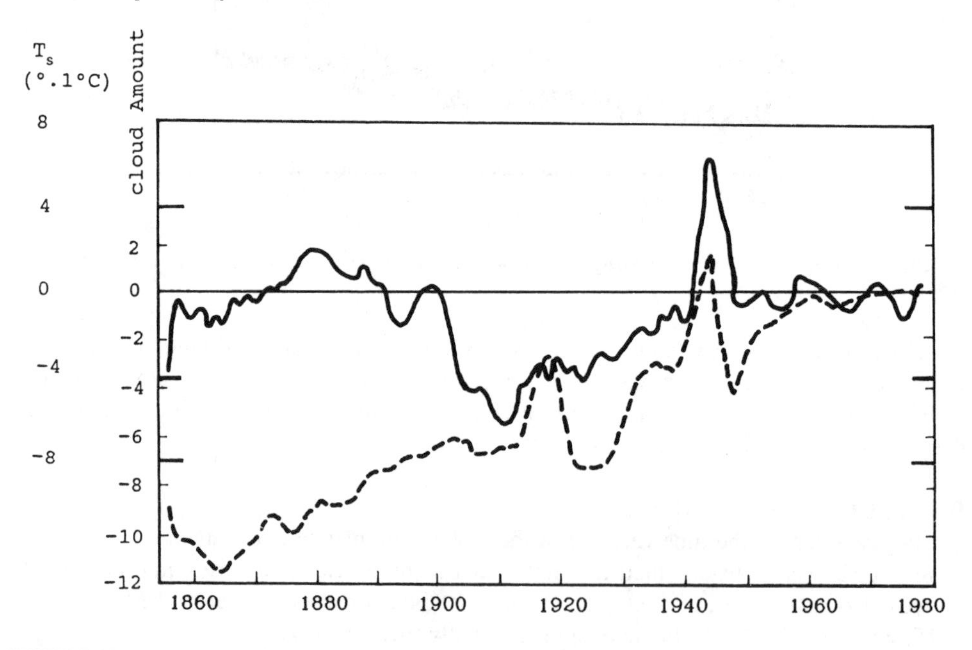


**FIGURE 11.** Time series of the anomaly of the global marine total cloud amount as that in Figure 9 (dashed line), and the air temperature anomaly near the sea surface (solid line).

In Figure 11, the time series of the anomaly of the global mean air temperature near the sea surface is plotted as the thick line, which shows that since 1910, a warming trend is rather evident.

## C. COMPARISON OF THE IMPACT OF THE VARIANCE BETWEEN THE CLOUD AMOUNTS AND THE $CO_2$ GREENHOUSE EFFECT ON CLIMATE CHANGE

We use a one-dimensional radiative equilibrium model (Yin, 1990) to study the difference of the impacts on the climate change of the variance between the cloud amounts and the $CO_2$ greenhouse effect.

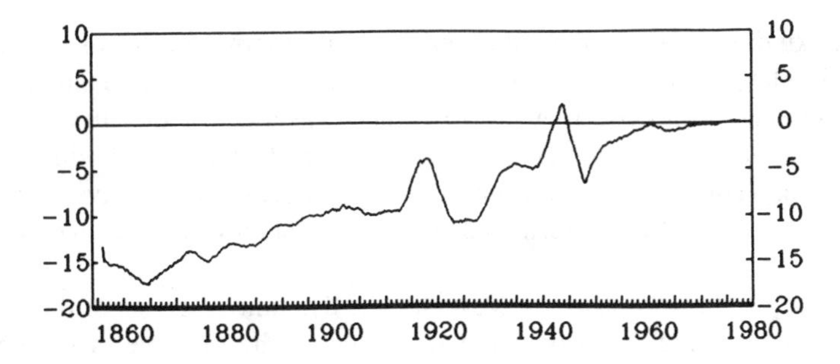


**FIGURE 12.** Time series of the 49 month running mean percentage variation of the global marine total cloud amount, 1854 to 1979. The 10-year mean value for the period 1970 to 1979 is the standard; unit: %.

Physical parameters adapted in the model are as follows: visibility = 20 km, water content in the cloud: high cloud = 0.4, middle cloud = 0.5, low cloud = 0.66, and the top height of the boundary layer = 822 m. There may be some unadapted parameters. Since the variation rate of the global mean air temperature is mainly considered, we do not want to accurately calculate the air temperature itself, moreover, the accuracy of the one-dimensional model is limited. Therefore, even if a rather large error occurs in the calculation, the conclusion still remains true, as the difference (as shown subsequently), is so large.

In the 110 years for the period 1865 to 1975, the total cloud amount varies between 5.54 and 6.74; the $CO_2$ concentration changes within the range 298 to 331 ppm. Adopting different combination ratios of the high-, mid-, and low clouds, the resulted variance of the air temperature by the cloud amount variation is from −3.4°C to −5.8°C, whereas the resulting warming due to the variation of $CO_2$ concentration is only 0.135 to 0.166°C.

IPCC has reached the conclusion that the effect of 2% change of low cloud amount is equivalent to that of double $CO_2$ concentration, which is in accordance with our results. The percentage variation of the global total marine-based cloud amounts is depicted in Figure 12, which indicates that within the recent 100 years, percentage variation of the total cloud amount largely exceeds 2%. Hence, the feedback effect of the cloud is very fundamental, which has an effect much stronger than that of the variation of $CO_2$. In a climate-predictive model, it is certain that the cloud feedback effect must be correctly dealt with, otherwise a slight error in its formulation may lead to large distortion in the results.

## IV. DYNAMIC RESPONSE OF THE PACIFIC OCEAN TO THE WIND STRESS OVER THE SEA SURFACE DURING THE EL NĨNO EVENT OF JANUARY 1983

In the Indian Ocean in summer, when winds north of the equator go in the opposite direction and an extensive region of strong southwest wind sets up, there is an approximate reversal of most current directions in this part of the ocean. This leads oceanographers to believe that wind over the sea surface provides motive force of the ocean current. Moreover, Warren et al. (1966) had interpreted the Somali Current during the Northern Hemisphere summer as part of the ocean's dynamic response to the pattern of wind stress over a large portion of ocean. Therefore, wind-driven ocean current is the fundamental consitutent part of the ocean current.

In recent years, studies of dynamic response between air and sea is drawing great attention, which are helpful in interpreting patterns of mean ocean currents in terms of meteorological elements over the sea surface, and thus understanding the mechanics of current generation. In addition, as to the problem of what variation in meteorological elements induces which kinds of variance about the mean current pattern, analysis of dynamic response is essential. Besides, the interaction between air and sea succeeds in explaining the formation of ocean current and typhoons, as well as the mechanism of the moisture cycle in nature. Another important typical phenomenon of air-sea interaction is the El Niño event. Taking this interaction into consideration, various numerical forecast models have been formulated, and these models will be further refined as the investigation of air-sea interaction is going deeper.

This purpose of this section is to investigate the dynamic response of the Pacific Ocean to the pattern of the wind stress over the sea surface. The multiple years monthly mean wind field over the sea surface and the SST for January 1980 to 1988 and for January 1983 are, respectively, used as the forcing component to compute the dynamic adaption of the oceanic current.

## A. FORMULATION OF THE DYNAMIC MODEL

On the longwave approximation, the linearized dynamic equation system of the simplest model of the response of the ocean to the sea surface wind stress is as follows:

$$\frac{\partial u}{\partial t} - fv = -g\frac{\partial \eta}{\partial x} + \frac{1}{\rho}\frac{\partial \tau_x}{\partial z} \tag{1}$$

$$\frac{\partial v}{\partial t} + fu = -g\frac{\partial n}{\partial y} + \frac{1}{\rho}\frac{\partial \tau_y}{\partial z} \tag{2}$$

$$\frac{\partial \eta}{\partial t} + H\overline{V}_2 \cdot \vec{V}_2 = 0 \tag{3}$$

where $\eta$ denotes the deviation of the disturbed sea surface from its mean (or quiescent state) height, $\tau_x$, $\tau_y$, are respectively the x, y components of the wind shear stress, H represents the depth of the mixed layer, which is about 100 m, and other symbols have the same meaning as usual.

The origin of the z axis is set at the sea surface at rest. Integrating Equations 1, 2 from $-$ H to the height of the perturbed surface $\eta$, if the stress at the depth of the mixed layer is $\tau_{-H}$, and making use of the linearized frictional drag law.

$$\tau_{-H} = -\rho\epsilon\vec{V} \tag{4}$$

in which, $\epsilon$ is a constant, then we get

$$\frac{\partial \bar{u}}{\partial t} - f\bar{v} = -g\frac{\partial \eta}{\partial x} + \frac{\tau_{sx}}{\rho H} - \frac{\epsilon\bar{u}}{H} \tag{5}$$

$$\frac{\partial \bar{v}}{\partial t} + f\bar{u} = -g\frac{\partial \eta}{\partial y} + \frac{\tau_{sy}}{\rho H} - \frac{\epsilon\bar{v}}{H} \tag{6}$$

Where $\bar{u}$, $\bar{v}$ are the mean velocity components averaged with respect to the depth H, and $\tau_{sx}$, $\tau_{sy}$ the wind-stress components over the ocean surface, which are given by the bulk aerodynamic relation

$$\tau_{sy}^{x} = \rho_s C_D \sqrt{u^2 + v^2}\,\frac{u}{v} \tag{7}$$

Where $\rho_s$ is the air density over the ocean surface, $C_D$ is the drag coefficient and u, v are the wind velocity components just above the sea surface.

Integrating Equation 3 in the same interval yields

$$\frac{\partial \eta}{\partial t} + H\left(\frac{\partial \bar{u}}{\partial x} + \frac{\partial \bar{v}}{\partial y}\right) = 0 \tag{8}$$

When the very rapid changes are excluded, then $\partial\eta/\partial t$ in Equation 8 may be ignored, hence,

$$\frac{\partial \bar{u}}{\partial x} + \frac{\partial \bar{v}}{\partial y} = 0 \tag{9}$$

From which we may introduce stream function $\psi$, such that

$$\bar{v} = \frac{\partial \psi}{\partial x}\ \bar{u} = -\frac{\partial \psi}{\partial y} \tag{10}$$

Finally, we get

$$\nabla_2^2\psi_t + \frac{1}{H}\epsilon\nabla_2^2\psi + \beta\psi_x = \frac{1}{\rho H}\left(\frac{\partial \tau_{sy}}{\partial x} - \frac{\partial \tau_{sx}}{\partial y}\right) \tag{11}$$

## B. THE STATIONARY WIND-DRIVEN OCEAN CURRENT

Under steady condition, Equation 11 reduces to

$$\frac{\epsilon}{H}\nabla_2^2\psi + \beta\frac{\partial \psi}{\partial x} = \frac{1}{\rho H}\left(\frac{\partial \tau_{sy}}{\partial x} - \frac{\partial \tau_{sx}}{\partial y}\right) \tag{12}$$

The discrete values of the curl of the monthly mean wind stress are expanded, according to the following two-dimensional Fourier series (Huang et al., 1984).

$$\frac{\partial \tau_{sy}}{\partial x} - \frac{\partial \tau_{sx}}{\partial y} = \sum_{k,l=0} [A_{k,l}\sin(m'kx + n'ly) + B_{k,l}\cos(m'kx + n'ky)$$
$$+ C_{k,l}\sin(m'kx - n'ly) + D_{k,l}\cos(m'kx - n'ky)] \tag{13}$$

Where x, y are, respectively, the zonal and meridional distance in unit of meter (m), k, l is the zonal and meridional wave number, k = 0, 1, 2, . . . , [m/2], l = 0, 1, 2, . . . , [n/2], and m, n are the number of grid points in the zonal and meridional direction, respectively. The parameters m′, n′ are defined as follows:

$$m' = \frac{2\pi}{m \times 5 \times 11 \times 10^4}$$

$$n' = \frac{2\pi}{n \times 5 \times 11 \times 10^4} \tag{14}$$

in unit of $m^{-1}$.

The two-dimensional Fourier coefficients are derived as

$$A_{k,l} = \frac{K}{mn} \sum_{i=0}^{m-1} \sum_{j=0}^{n-1} q_{i,j} \sin(m'kx + n'ly)$$

$$B_{k,l} = \frac{K}{mn} \sum_{i=0}^{m-1} \sum_{j=0}^{n-1} q_{i,j} \cos(m'kx + n'ly)$$

$$C_{k,l} = \frac{K}{mn} \sum_{i=0}^{m-1} \sum_{j=0}^{n-1} q_{i,j} \sin(m'kx - n'ly)$$

$$D_{k,l} = \frac{K}{mn} \sum_{i=0}^{m-1} \sum_{j=0}^{n-1} q_{i,j} \cos(m'kx - n'ly) \quad (15)$$

In which, $q_{i,j}$ is the value of $(\partial\tau_{sy}/\partial x) - (\partial\tau_{sx}/\partial y)$, at the grid point, i, j, and the coefficient K on the right-hand side takes the value

$$K = \begin{cases} 0.5 \text{ (when } k = 0 \text{ and } l = 0) \\ 1.0 \text{ (when } k = 0 \text{ or } l = 0) \\ 2.0 \text{ (when } k > 0 \text{ and } l > 0) \end{cases}$$

Suppose the solution $\psi$ of Equation 12 has the following form:

$$\psi = \sum_{k,l} [A'_{k,l} \sin(m'kx + n'ly) + B'_{k,l} \cos(m'kx + n'ly)$$

$$+ C'_{k,l} \sin(m'kx - n'ly) + D'_{k,l} \cos(m'kx - n'ly)] \frac{1}{\rho H} \quad (16)$$

Substituting Equations 13 and 16 into Equation 12, and comparing the coefficients of the same simple harmonic function with the same zonal and meridional wave numbers on the both sides, we thus have solutions

$$A'_{k,l} = \left(\beta k m' B_{k,l} - \frac{\epsilon}{H} K_{k,l}^2 A_{k,l}\right) \Big/ \left[\beta^2 k^2 m'^2 + \left(\frac{\epsilon}{H} K_{k,l}^2\right)^2\right] \quad (17)$$

$$B'_{k,l} = -\left(\beta k m' A_{k,l} + \frac{\epsilon}{H} K_{k,l}^2 B_{k,l}\right) \Big/ \left[\beta^2 k^2 m'^2 + \left(\frac{\epsilon}{H} K_{k,l}^2\right)^2\right] \quad (18)$$

$$C'_{k,l} = \left(\beta k m' D_{k,l} - \frac{\epsilon}{H} K_{k,l}^2 C_{k,l}\right) \Big/ \left[\beta^2 k^2 m'^2 + \left(\frac{\epsilon}{H} K_{k,l}^2\right)^2\right] \quad (19)$$

$$D'_{k,l} = -\left(\beta k m' C_{k,l} + \frac{\epsilon}{H} K_{k,l}^2 D_{k,l}\right) \Big/ \left[\beta^2 k^2 m'^2 + \left(\frac{\epsilon}{H} K_{k,l}^2\right)^2\right] \quad (20)$$

In which $K_{k,l}^2 = m'^2 k^2 + n'^2 l^2$.

### C. DYNAMIC RESPONSE OF THE OCEAN CURRENT UNDER NONSTATIONARY CONDITIONS

In Equation 11, let the Laplace Transform of $\psi$ be $\tilde{\psi}$, and

$$\tilde{\psi} = \int_0^{\infty} \psi(t)e^{-pt}\,dt \tag{21}$$

Inserting the Fourier expansion of the curl of the wind stress by (13) into (11), then employing the rules in operational calculus we obtain

$$p\left(\frac{\partial^2\tilde{\psi}}{\partial x^2} + \frac{\partial^2\tilde{\psi}}{\partial y^2}\right) + \beta\frac{\partial\tilde{\psi}}{\partial x} = \frac{1}{\rho H p}\sum_{k,l}[A_{k,l}\sin(m'kx + n'ly)$$
$$+ B_{k,l}\cos(m'kx + n'ly) + C_{k,l}\sin(m'kx - n'ly)$$
$$+ D_{k,l}\cos(m'kx - n'ly)] + \nabla_2^2\psi(0) - \frac{\epsilon}{H}\nabla_2^2\tilde{\psi} \tag{22}$$

Where $\psi(0)$ is the stream function at certain initial instant, and similarly expanded into two-dimensional Fourier series, such as

$$\psi(0) = \sum_{k,l}[\bar{A}_{k,l}\sin(m'kx + n'ly) + \bar{B}_{k,l}\cos(m'kx + n'ly)$$
$$+ \bar{C}_{k,l}\sin(m'kx + n'ly) + \bar{D}_{k,l}\cos(m'kx - n'ly)] \tag{23}$$

Inserting Equation 23 into Equation 22,

$$\left(p + \frac{\epsilon}{H}\right)\nabla_2^2\tilde{\psi} + \beta\frac{\partial\tilde{\psi}}{\partial x} = \sum_{k,l}\left[\left(\frac{A_{k,l}}{\rho H p} - K'_{k,l}\bar{A}_{k,l}\right)\sin(m'kx + n'ly)\right.$$
$$+ \left(\frac{B_{k,l}}{\rho H p} - K_{k,l}^2\bar{B}_{k,l}\right)\cos(m'kx + n'ly)$$
$$+ \left(\frac{C_{k,l}}{\rho H p} - K_{k,l}^2\bar{C}_{k,l}\right)\sin(m'kx - n'ly)$$
$$+ \left(\frac{D_{k,l}}{\rho H p} - K_{k,l}^2\bar{D}_{k,l}\right)\cos(m'kx - n'ly) \tag{24}$$

Put

$$\tilde{\psi} = \sum_{k,l}\frac{1}{\rho H}[\tilde{A}'_{k,l}\sin(m'kx + n'ly) + \tilde{B}'_{k,l}\cos(m'kx + n'ly)$$
$$+ \tilde{C}'_{k,l}\sin(m'kx - n'ly) + \tilde{D}'_{k,l}\cos(m'kx - n'ly)] \tag{25}$$

Substitute Equation 25 with Equation 24, and compare the coefficients of the same simple harmonic terms on both sides of the equation. Then by the displacement theorem of the operational calculus, and for convenience putting $K_{k,l} = K$, we obtain the primary function of $\tilde{A}_{k,l}$ as

$$\tilde{A}'_{k,l} \doteqdot \frac{1}{(\beta m'k)^2 + \left(\frac{\epsilon}{H}K^2\right)^2}\left[\left(1 - e^{-\frac{\epsilon}{H}t}\cos\frac{\beta m'k}{K^2}t\right)\left(\beta m'kB_{k,l} - \frac{\epsilon}{H}K^2A_{k,l}\right)\right.$$
$$\left. - e^{-\frac{\epsilon}{H}t}\sin\frac{\beta m'k}{K^2}t\left(\beta m'kA_{k,l} + \frac{\epsilon}{H}K^2B_{k,l}\right)\right]$$
$$- \rho He^{-\frac{\epsilon}{H}t}\left(\overline{B}_{k,l}\sin\frac{\beta m'k}{K^2}t - \overline{A}_{k,l}\cos\frac{\beta m'k}{K^2}t\right) \tag{26}$$

Similarly,

$$\tilde{B}'_{k,l} \doteqdot -\frac{1}{(\beta m'k)^2 + \left(\frac{\epsilon}{H}K^2\right)^2}\left[\left(1 - e^{-\frac{\epsilon}{H}t}\cos\frac{\beta m'k}{K^2}t\right)\left(\beta m'kA_{k,l} + \frac{\epsilon}{H}K^2B_{k,l}\right)\right.$$
$$\left. - e^{-\frac{\epsilon}{H}t}\sin\frac{\beta m'k}{K^2}t\left(\frac{\epsilon}{H}K^2A_{k,l} - \beta m'kB_{k,l}\right)\right]$$
$$+ \rho He^{-\frac{\epsilon}{H}t}\left(\overline{A}_{k,l}\sin\frac{\beta m'k}{K^2}t + \overline{B}_{k,l}\cos\frac{\beta m'k}{K^2}t\right) \tag{27}$$

$$\tilde{C}'_{k,l} \doteqdot \frac{1}{(\beta m'k)^2 + (\epsilon K^2)^2}\left[\left(1 - e^{-\frac{\epsilon}{H}t}\cos\frac{\beta m'k}{K^2}t\right)\left(\beta m'kD_{k,l} - \frac{\epsilon}{H}K^2C_{k,l}\right)\right.$$
$$\left. - e^{-\frac{\epsilon}{H}t}\sin\frac{\beta m'k}{K^2}t\left(\beta m'kC_{k,l} + \frac{\epsilon}{H}K^2D_{k,l}\right)\right]$$
$$- \rho He^{-\frac{\epsilon}{H}t}\left(\overline{D}_{k,l}\sin\frac{\beta m'k}{K^2}t - \overline{C}_{k,l}\cos\frac{\beta m'k}{K^2}t\right) \tag{28}$$

$$\tilde{D}'_{k,l} \doteqdot -\frac{1}{(\beta m'k)^2 + (\epsilon K^2)^2}\left[\left(1 - e^{-\frac{\epsilon}{H}t}\cos\frac{\beta m'k}{K^2}t\right)\left(\beta m'kC_{k,l} + \frac{\epsilon}{H}K^2D_{k,l}\right)\right.$$
$$\left. - e^{-\frac{\epsilon}{H}t}\sin\frac{\beta m'k}{K^2}t\left(-\beta m'kD_{k,l} + \frac{\epsilon}{H}K^2C_{k,l}\right)\right]$$
$$+ \rho He^{-\frac{\epsilon}{H}t}\left(\overline{C}_{k,l}\sin\frac{\beta m'k}{K^2}t + \overline{D}_{k,l}\cos\frac{\beta m'k}{K^2}t\right) \tag{29}$$

Finally, the stream function $\psi$ is given by replacing the coefficients $\tilde{A}'_{k,l}$, $\tilde{B}'_{k,l}$, $\tilde{C}'_{k,l}$, and $\tilde{D}_{k,l}$ in Equation 25 with their primary counterparts Equation 26 to 29, respectively.

### D. VARIATION OF SEA SURFACE WEIGHT (SSW) AND A SOLUTION ADAPTABLE TO LATERAL BOUNDARY CONDITIONS

In early oceanic dynamic models, the upper boundary condition is usually so chosen that a rigid lid is set on the sea surface, so that the height of the oceanic surface is uniform. In actuality, this is evidently not the case; the height of the sea surface changes temporarily and spatially. Its variation along the horizontal distance can give rise to a pressure gradient force in the water to drive the ocean current. The sea is laterally bounded by the continent, so the motion in the ocean basin must obey the lateral boundary condition. The simplest lateral boundary condition is one for which a solid wall is executed on the east or west end of the ocean. Ignoring the variation of the lateral boundary along the meridional direction, the kinematic boundary condition is satisfied (i.e., the zonal component of the velocity of the ocean current at a certain horizontal distance where the solid wall is located equals zero).

#### 1. Variation of the Sea Surface Height (SSH)

Dynamic Equations 5, 6, and 8 are used to acquire the perturbed part $\eta$ of the SSH, differentiating Equation 5 with respect to x and Equation 6 with respect to y, and adding then

$$\nabla_2^2\eta = \frac{f}{g}\nabla_2^2\psi + \frac{\beta}{g}\frac{\partial\psi}{\partial y} + \frac{1}{\rho g H}\left(\frac{\partial t_{sx}}{\partial x} + \frac{\partial\tau_{sy}}{\partial y}\right) \tag{30}$$

In deriving the above equation, the nondivergent approximation has been utilized. Equation 25 gives the Laplace Transform of the stream function $\psi$, by which $\psi$ can be obtained, then it yields

$$\nabla_2^2\psi = -[(m'k)^2 + (n'l)^2\psi = -K^2\psi \tag{31}$$

The divergence of the surface wind stress can be written as

$$\frac{\partial\tau_{sx}}{\partial x} + \frac{\partial\tau_{sy}}{\partial y} = \sum_{k,l}[A^D_{k,l}\sin(m'kx + n'ly) + B^D_{k,l}\cos(m'kx + n'ly) + C^D_{k,l}\sin(m'kx + n'ly) + D^D_{k,l}\cos(m'kx + n'ly)] \tag{32}$$

By use of Equations 25 and 31, we obtain a Poisson equation of $\eta$, in accordance with the nonhomogeneous part on the right side, and the following formal solution for $\eta$ is suggested:

$$\eta = \sum_{k,l}[\eta_1\sin(m'kx + n'ly) + \eta_2\cos(m'kx + n'ly) \quad \eta_3\sin(m'kx - n'ly) + \eta_4\cos(m'kx - n'ly)] \tag{33}$$

From which it follows that

$$\nabla_2^2\eta = -K^2\eta \tag{34}$$

Replacing Equation 34 into the Poisson equation for $\eta$, and comparing the coefficients of the same sinusoidal terms on the two sides of the equation, then

$$\eta_1 = \frac{f}{g}\overline{\overline{A'_{k,l}}} + \frac{\beta n'l}{K^2_{k,l}g}\overline{\overline{B'_{k,l}}} - \frac{1}{K^2_{k,l}\rho Hg}A^D_{k,l} \tag{35}$$

$$\eta_2 = \frac{f}{g}\overline{\overline{B'_{k,l}}} - \frac{\beta n'l}{K^2_{k,l}g}\overline{\overline{A'_{k,l}}} - \frac{1}{K^2_{k,l}\rho Hg}B^D_{k,l} \tag{36}$$

$$\eta_3 = \frac{f}{g}\overline{\overline{C'_{k,l}}} + \frac{\beta n'l}{K^2_{k,l}g}\overline{\overline{D'_{k,l}}} - \frac{1}{K^2_{k,l}\rho Hg}C^D_{k,l} \tag{37}$$

$$\eta_4 = \frac{f}{g}\overline{\overline{D'_{k,l}}} - \frac{\beta n'l}{K^2_{k,l}g}\overline{\overline{C'_{k,l}}} - \frac{1}{K^2_{k,l}\rho Hg}D^D_{k,l} \tag{38}$$

In which $\overline{\overline{A'_{k,l}}}$, $\overline{\overline{B'_{k,l}}}$, $\overline{\overline{C'_{k,l}}}$, and $\overline{\overline{D'_{k,l}}}$ are respectively the primary functions to the Laplace transform $\tilde{A}'_{k,l}$, $\tilde{B}'_{k,l}$, $\tilde{C}'_{k,l}$, and $\tilde{D}'_{k,l}$. From Equations 35 to 38, it follows that except for the factor of gravity force, the variance of the oceanic surface height arises out of three factors. One is due to the Coriolis force, another is due to the $\beta$ effect, and the third is mainly due to the divergence of the wind stress.

**2. Solution Adaptable to the Lateral Boundary Conditions**

Let us set a solid wall at the western end of the ocean, and take this position as the origin of the x axis, which is perpendicular to the wall, then the kinematic boundary condition to be satisfied at the wall is

$$x = 0$$

$$\bar{u} = 0 \tag{39}$$

The solution satisfying the above condition can be found in the following way: decompose the solution into two parts, one part, disregarding the impact of the sea surface wind stress, is the boundary solution, the other is that part of solution driven by the surface wind stress. From Equation 11, the boundary solution $\psi_B$ satisfies the following homogeneous equation

$$\frac{\partial \nabla^2 \psi_B}{\partial t} + \beta\frac{\partial \psi_B}{\partial x} = -\frac{\epsilon}{H}\nabla^2\psi_B \tag{40}$$

Let $\tilde{\psi}_B$ be the Laplace Transform of $\psi_B$, then Equation 40 yields

$$\left(p + \frac{\epsilon}{H}\right)\left(\frac{\partial^2\tilde{\psi}_B}{\partial x^2} + \frac{\partial^2\tilde{\psi}_B}{\partial y^2}\right) + \beta\frac{\partial\tilde{\psi}_B}{\partial x} = 0 \tag{41}$$

In deriving the above equation, the initial value of $\psi_B$ is taken as a constant, usually zero. $\tilde{\psi}_B$ possesses the following formal solution:

$$\tilde{\psi}_B = \sum_l [\tilde{\varphi}_l(x,p)\sin(n'ly) + \tilde{\varphi}'_l(x,p)\cos(n'ly)] \tag{42}$$

Inserting Equation 42 into Equation 41, then we get

$$\left(p + \frac{\epsilon}{H}\right)\frac{\partial^2\tilde{\varphi}_l}{\partial x^2} - (n'l)^2\tilde{\varphi}_l + \beta\frac{\partial\tilde{\varphi}_l}{\partial x} = 0 \tag{43}$$

and the same equation for $\tilde{\varphi}'_1$. Put

$$\tilde{\varphi}_1 \sim e^{\alpha_1 x} \tag{44}$$

Replacing Equation 44 into Equation 43, it gives

$$\alpha_1 = -\frac{1}{2}\frac{\beta}{p + \frac{\epsilon}{H}} \pm \sqrt{\frac{1}{4}\left(\frac{\beta}{p + \frac{\epsilon}{H}}\right)^2 + (n'l)^2} \tag{45}$$

Thus,

$$\tilde{\varphi}_1 = \tilde{A}_1(p)\exp\left[\left(-\frac{1}{2}\frac{\beta}{p + \frac{\epsilon}{H}} + \sqrt{\frac{1}{4}\left(\frac{\beta}{p + \frac{\epsilon}{H}}\right)^2 + (n'l)^2}\right)x\right]$$

$$+ \tilde{B}_1(p)\exp\left[\left(-\frac{1}{2}\frac{\beta}{p + \frac{\epsilon}{H}} - \sqrt{\frac{1}{4}\left(\frac{\beta}{p + \frac{\epsilon}{H}}\right)^2 + (n'l)^2}\right)x\right] \tag{46}$$

If $\tilde{\varphi}_1$ has meaning under present boundary condition, $\tilde{\varphi}_1$ cannot grow exponentially with increase of x. Since on the complex plane of p, then $p = \sigma + is$, where $\sigma$ and s are real numbers, then

$$\frac{\beta}{p + \epsilon/H} = \frac{\beta}{\sqrt{(\sigma + \epsilon/H)^2 + s^2}} e^{i\gamma} \tag{47}$$

In which $tg\gamma = -\frac{s}{\sigma + \epsilon/H}$, then

$$\alpha_1 = -\frac{\beta}{2}\frac{\cos\gamma}{\sqrt{(\sigma + \epsilon/H)^2 + s^2}}$$

$$\pm a + i\left(-\frac{\beta}{2}\frac{\sin\gamma}{\sqrt{(\sigma + \epsilon/H)^2 + s^2}} \pm b\right) \tag{48}$$

It is evident that the growth or diminution of $\tilde{\varphi}_1$ with the increase of x depends on the real part of $\alpha_1$, namely $\pm a - \frac{\beta}{2}\frac{\cos\gamma}{\sqrt{(\sigma + \epsilon/H)^2 + s^2}}$. Since in which

$$a^2 = \frac{1}{2}\left\{\frac{\beta^2 \cos 2\gamma}{4[(\sigma + \epsilon/H)^2 + s^2]} + (n'l)^2\right\}$$

$$\pm \frac{1}{2}\sqrt{\left[\frac{\beta^2 \cos 2\gamma}{4((\sigma + \epsilon/H)^2 + s^2)} + (n'l)^2\right]^2 + \frac{\beta^4 \sin^2 2\gamma}{16[(\sigma + \epsilon/H)^2 + s^2]^2}} \tag{49}$$

Therefore, in order to have real value of a, the plus sign before the square root in the above expression must be taken. Moreover, it can be shown

$$a > \frac{\beta}{2} \frac{\cos \gamma}{\sqrt{(\sigma + \epsilon/H)^2 + s^2}}$$

Hence, in order to have negative value of the real part of $\alpha_1$, i.e., $\pm a - \frac{\beta}{2} \frac{\cos\gamma}{\sqrt{\sigma + \epsilon/H)^2 + s^2}}$, so that $\tilde{\varphi}_1$ decays exponentially with the increase of positive x, the minus sign before the a must be adopted. Therefore,

$$\tilde{\varphi}_1 = \tilde{B}_1(p)\exp\left(-\left[\frac{\beta}{2(p + \epsilon/H)} - \sqrt{\frac{\beta^2}{4(p + \epsilon/H)^2} + (n'l)^2}\right]x\right)$$

Since

$$\sqrt{\frac{\beta^2}{4(p + \epsilon/H)^2} + (n'l)^2} = \frac{\beta}{2(p + \epsilon/H)} \sqrt{1 + \frac{4(n'l)^2}{\beta^2} (p + \epsilon/H)^2}$$

if $\frac{4(n'l)^2}{\beta^2} |(p + \epsilon/H)^2| < 1$, because $p = \sigma + is$, then

$$|(p + \epsilon/H)^2| = (\sigma + \epsilon/H)^2 + s^2 < \frac{\beta^2}{4(n'l)^2}$$

That is

$$|p|^2 + \frac{2\sigma\epsilon}{H} < \frac{\beta^2}{4(n'l)^2}$$

Therefore,

$$|p|^2 < \frac{\beta^2}{4(n'l)^2}$$

Thus,

$$\frac{1}{p^2} > \frac{4(n'l)^2}{\beta^2}$$

Under the condition, we have

$$\sqrt{\frac{\beta^2}{4(p + \epsilon/H)^2} + (n'l)^2} \approx \frac{\beta}{2(p + \epsilon/H)} \left[1 + \frac{2(n'l)^2}{\beta} (p + \epsilon/H)^2\right]$$

In the same way we can get the approximate expression for $\tilde{\varphi}_B$, hence

$$\tilde{\psi}_B = \sum_l [\tilde{B}_l(p)e^{-\left[\frac{\beta}{(p+\epsilon/H)}+\frac{(n'l)^2}{\beta}(p+\epsilon/H)\right]x} \sin n'ly$$

$$+ \tilde{B}_l'(p)e^{-\left[\frac{\beta}{(p+\epsilon/H)}+\frac{(n'l)^2}{\beta}(p+\epsilon/H)\right]x} \cos n'ly] \quad (51)$$

From the boundary condition of Equation 39, it follows that

$$(\tilde{\psi}_B + \tilde{\psi})_{x=0} = 0 \quad (52)$$

Substituting Equation 25 for $\psi$ and Equation 51 into the above relation, and comparing coefficients of sinn'ly and cosn'ly, we get $\tilde{B}_l(p)$ and $\tilde{B}'_l(p)$. Again replacing $\tilde{B}_l(p)$ and $\tilde{B}'_l(p)$ thus obtained into the expression for $\psi_B$. With the acid of relations in operational calculus (Ditkqin and Kyzenkov, 1958); the primary functions of the coefficients of sinn'ly and cosn'ly in the above expression are obtained. We finally then get the boundary solution, from which it is worthy to note that when $t < (n'l)^2x/\beta$ and the vorticity field in the ocean at the initial instant is zero, then $\psi_B(x,y,t) = 0$. This signifies that under this circumstance, the boundary cannot have impact on the stream function.

## E. NUMERICAL EXPERIMENTS AND ANALYSIS OF THE RESULTS

The global grid data set of ECMWF and that of the monthly mean SST of CAC, U.S., for the period 1980 to 1988 with grid length 2.5° × 2.5° latitude and longitude are used to calculate the stream function and the perturbed oceanic surface height. The domain covers most parts of the Pacific and Indian Oceans, i.e., the region 40°E to 60°W and 50°N to 50°S.

The 9 years monthly mean of u, v, and SST for January 1980 to 1988, are considered as the multiple years mean, whereas the monthly mean for January 1983 represents the mean meteorological condition of the mature phase of the 1982 to 1983 El Niño event.

The drag coefficient $C_D$ is the same as that used by Zhu and Yang (1990). The other parameters used in this calculation are as follows: $\epsilon = 0.6 \times 10^{-3}$ ms$^{-1}$, H = 100 m, m = 101, n = 37, $m' = 2.26 \times 10^{-7}$ m$^{-1}$, $n' = 6.18 \times 10^{-7}$ m$^{-1}$, $\beta = 2.289 \times 10^{-11}$ m$^{-1}$ S$^{-1}$, and the grid length $d = 2.75 \times 10^5$ m.

Figures 13, 14, 15, 16 denote, respectively, the monthly mean vorticity and divergence field of the wind stress on 1000 h hPa for January 1980 to 1988, and January 1983. The common traits are that the magnitude of the vorticity of the surface wind stress is larger than that of the divergence, hence, the largest difference can be of an order of magnitude. In addition, irrespective of the vorticity or the divergence field, the general configuration for the multiple years and the 1983 monthly mean are nearly the same. The monthly mean magnitude of the vorticity and divergence for 1983 is rather larger. In the region south of the equator and between 140°E and 140°W, there clearly appears negative value of the vorticity. This denotes that the east wind is weakening, whereas, the west wind is strengthening and the cross equator air current is occurring.

The above four fields are expanded into Fourier series. Fortunately, only a few harmonic terms are needed to simulate the actual fields. Since we pay attention only to the longwave components, thus keeping harmonic components with zonal wave number <15 and meridional wave number <5. Moreover, we set a preassigned number S, and preserve those groups of Fourier coefficients for which the same k and l, at least one of the four Fourier coefficients whose magnitude is larger than S. For the multiple years monthly mean, for

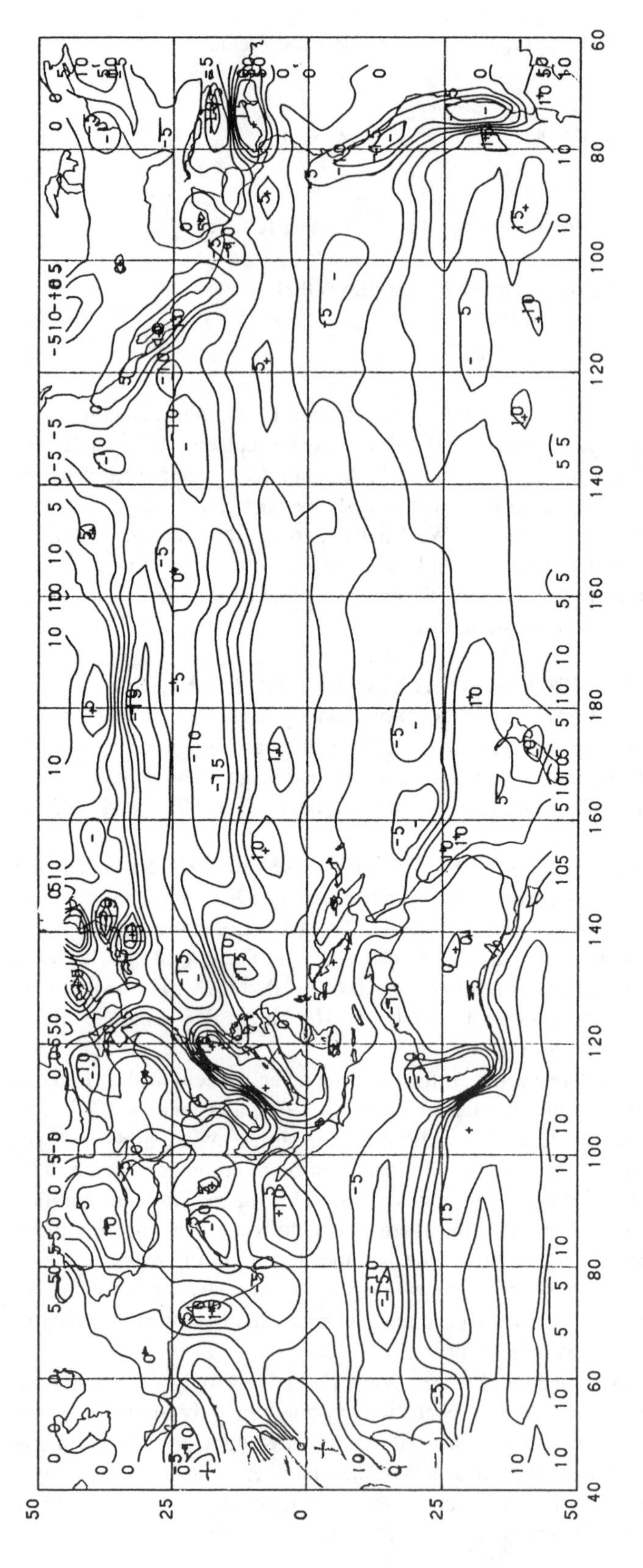

**FIGURE 13.** Distribution of monthly mean $\nabla \times \vec{\tau}$ for January 1980 to 1988; unit: $10^{-8}$ kg$(ms)^{-2}$.

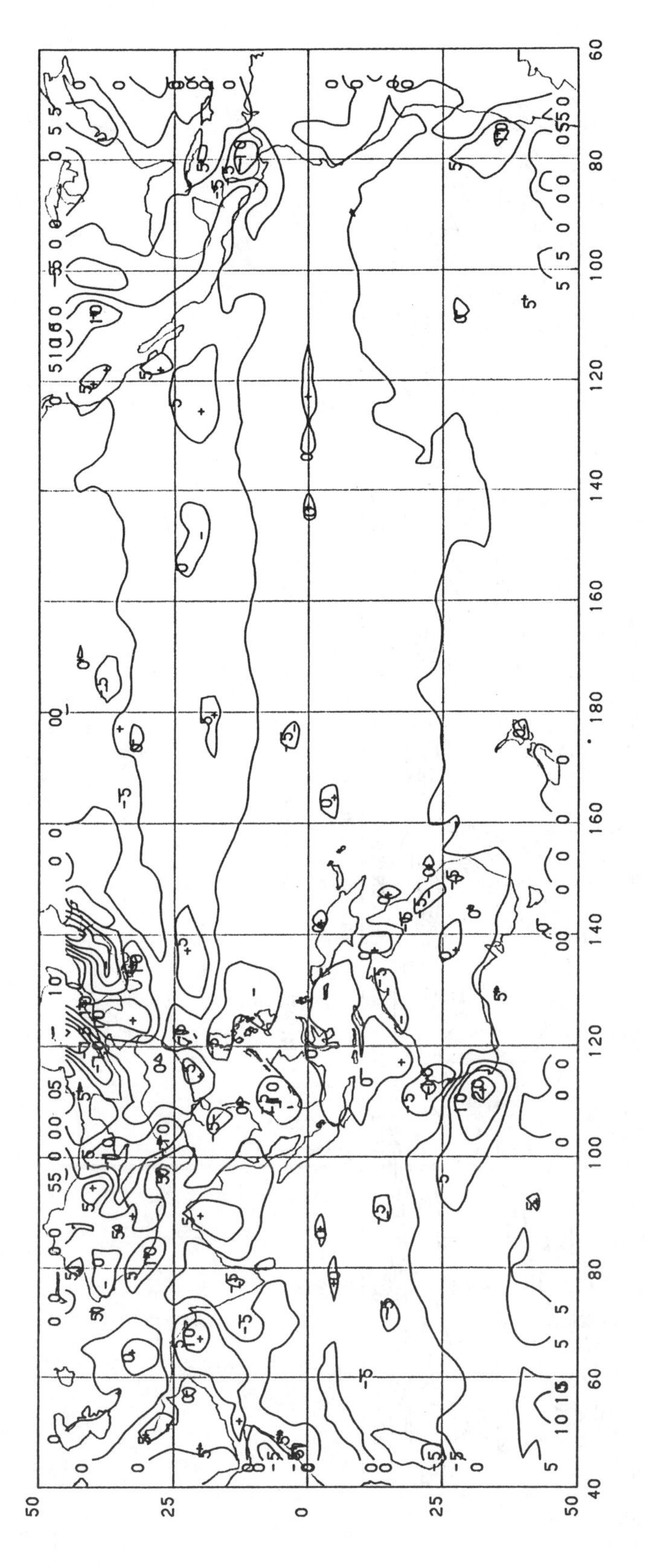


**FIGURE 14.** Distribution of monthly mean $\nabla \cdot \vec{\tau}$ for January 1980 to 1988; unit: $10^{-8}$ kg(ms)$^{-2}$.

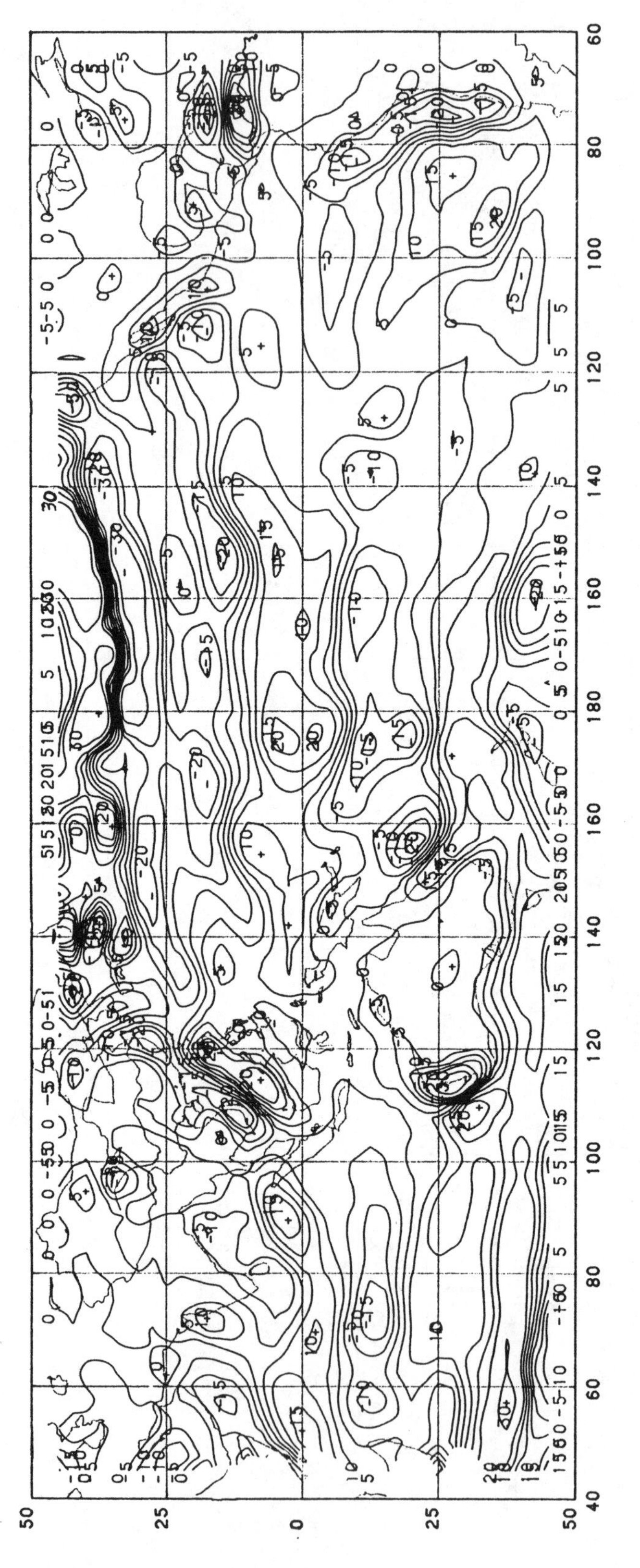

**FIGURE 15.** Same as Figure 13, except it is for January 1983.

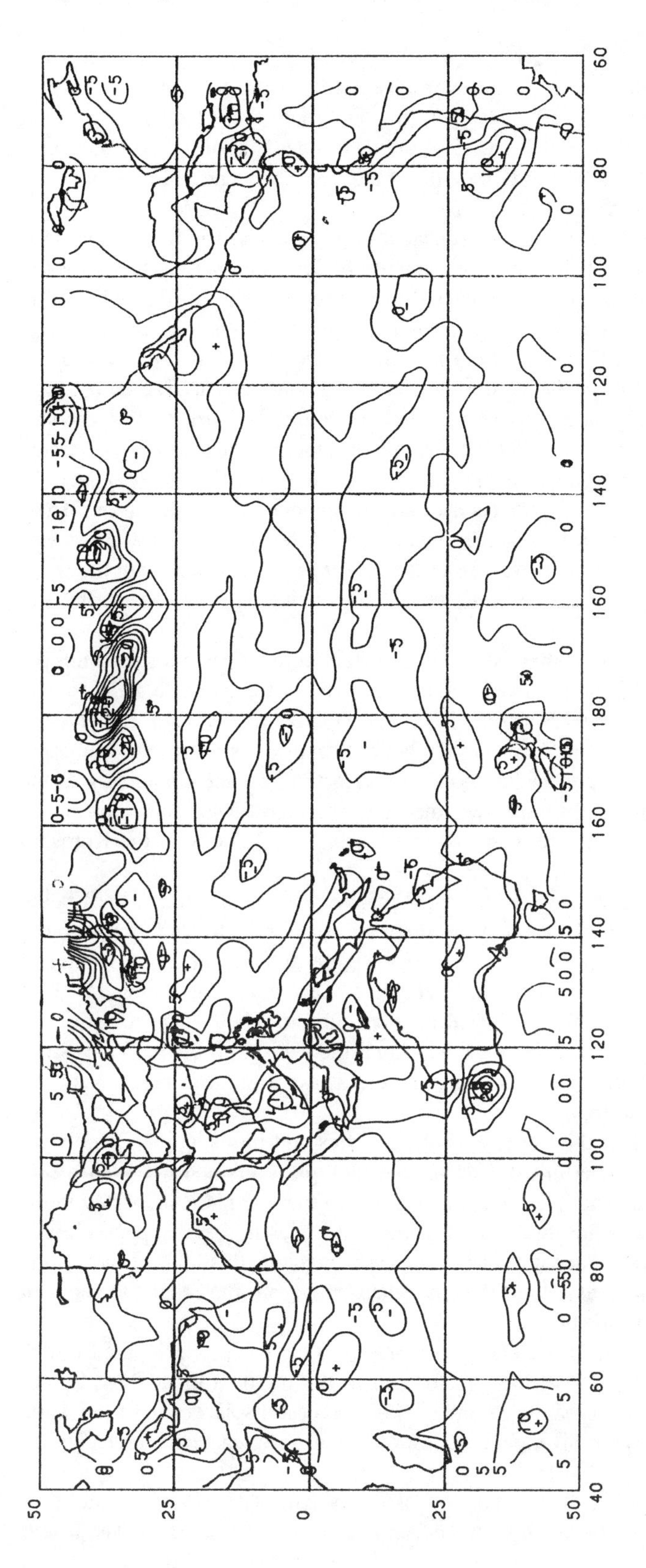


**FIGURE 16.** Same as Figure 14, except it is for January 1983.

the vorticity field, $S = 0.25 \times 10^{-8}$ kg(ms)$^{-2}$; for the divergence field, $S = 0.17 \times 10^{-8}$ kg(ms)$^{-2}$; for the 1983 monthly mean, for the vorticity field, $S = 0.5 \times 10^{-8}$ kg(ms)$^{-2}$; and for the divergence field, $S = 0.17 \times 10^{-8}$ kg(ms)$^{-2}$. Under these circumstances, the simulated fields are shown in Figures 17, 18, 19, and 20. Comparison of these figures with the corresponding Figures 13, 14, 15, and 16 demonstrates that the truncated Fourier series simulate the actual fields quite well.

With the values of the coefficients of the truncated Fourier series, and by Equations 16 to 20, we obtain the $\psi$ field under the stationary condition. The stationary $\psi$ fields for the multiple years and 1983 monthly mean are depicted in Figures 21 and 22, from which it is shown that the coincidence of the oceanic current and the actual sea surface current is generally good. The west drift current in the north Pacific moves eastward to the west coast of North America, turns southward, becoming the California cold current. The latter again turns westward on its way south, and becomes the northern equatorial current upon reaching north of the equator. The current system in the Indian Ocean, the Bengal monsoonal current, and the equatorial counter current south of the equator of the Indian Ocean all appear. In addition, the current system in the South Pacific is also obvious. The distribution of the stream lines possesses the characteristics that regions of large value are biased towards west, and the stream lines are denser in the west than in the east; this is consistent with the well-known fact of west intensification. However, we get this result without considering the impact of the west boundary.

Most of the magnitude of the computed oceanic current component u for the multiple years, as well as for the 1983 monthly mean, conforms to the actuality quite well. Nevertheless, the computed velocity for the Kuroshio is about 30 to 40 cm/s, which is rather small compared to actual velocity of 50 to 75 cm/s. We believe that this is due to the limitation of the model, since we do not consider the boundary effect, if we set a boundary at the west end, then the zonal momentum must be turned into the meridional momentum when the north equatorial current attains the west boundary, thus, the velocity in the Kuroshio region is augmented.

Comparing Figures 21 and 22, it is to be noted that except for the stream lines being denser for 1983 than for the multiple years mean, there are some traits for 1983 to be noted, first, in the equator region between 150°E and 150°W, there appears the closed center of low value, around which the stream function is negative. Second, in the region 150°E to 150°W, 0° to 10° S, there is notable equatorial counter current. These characteristics are connected with the intensification of the west wind and the relaxation of the trade wind. Another important trait is that the stream field off the west coast of South America contracts eastward and extends to the north.

With the few coefficients of the truncated Fourier series and by using the formulas of Equations 33 to 38, we get the perturbed SSH for the multiple years and the 1983 mean, which are shown in Figures 23 and 24, from which it it seen that the ITCZ is located in the region where $\eta$ maintains low value. In the north Pacific, the SSH in the west part is higher than that in the east. If the boundary impact is taken into account, this kind of inclination will be more evident. With an overview of the entire Pacific Ocean, the SSH at the equator is lower than that at the higher latitude.

For 1983, the low region along the equator displaces to the east, as compared to that for the multiple years mean. South of the equator and in the region 140° to 180° E, this displacement is clearer. This is probably associated with the east displacement of the warm SST, plus because the high region off the west coast of South America also contracts to the east and extends northward.

For the sake of investigation of the adaption of the ocean to the surface wind stress under the nonstationary condition, we chose the ocean at rest as the initial state. The results

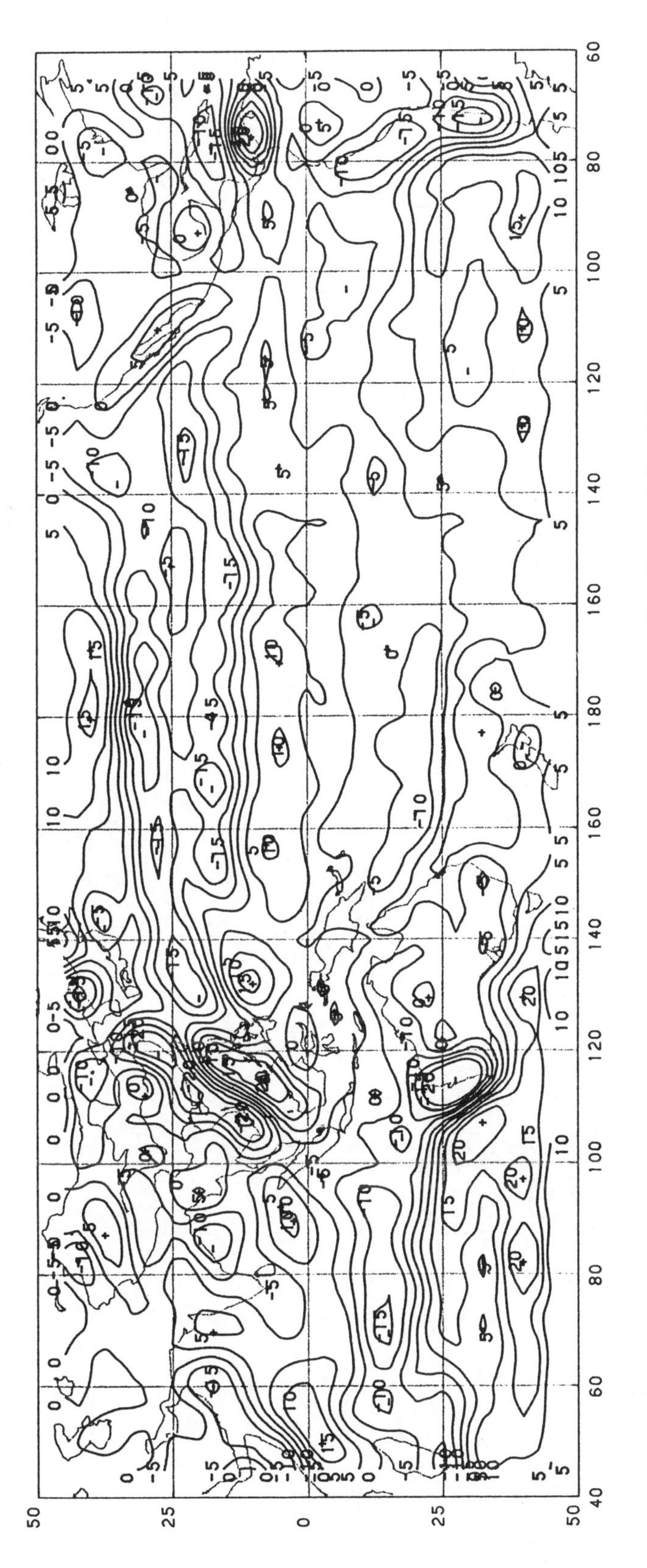


**FIGURE 17.** The simulation of the multiple years monthly mean $\nabla \times \vec{\tau}$; unit: $10^{-8}$ $(ms)^{-2}$.

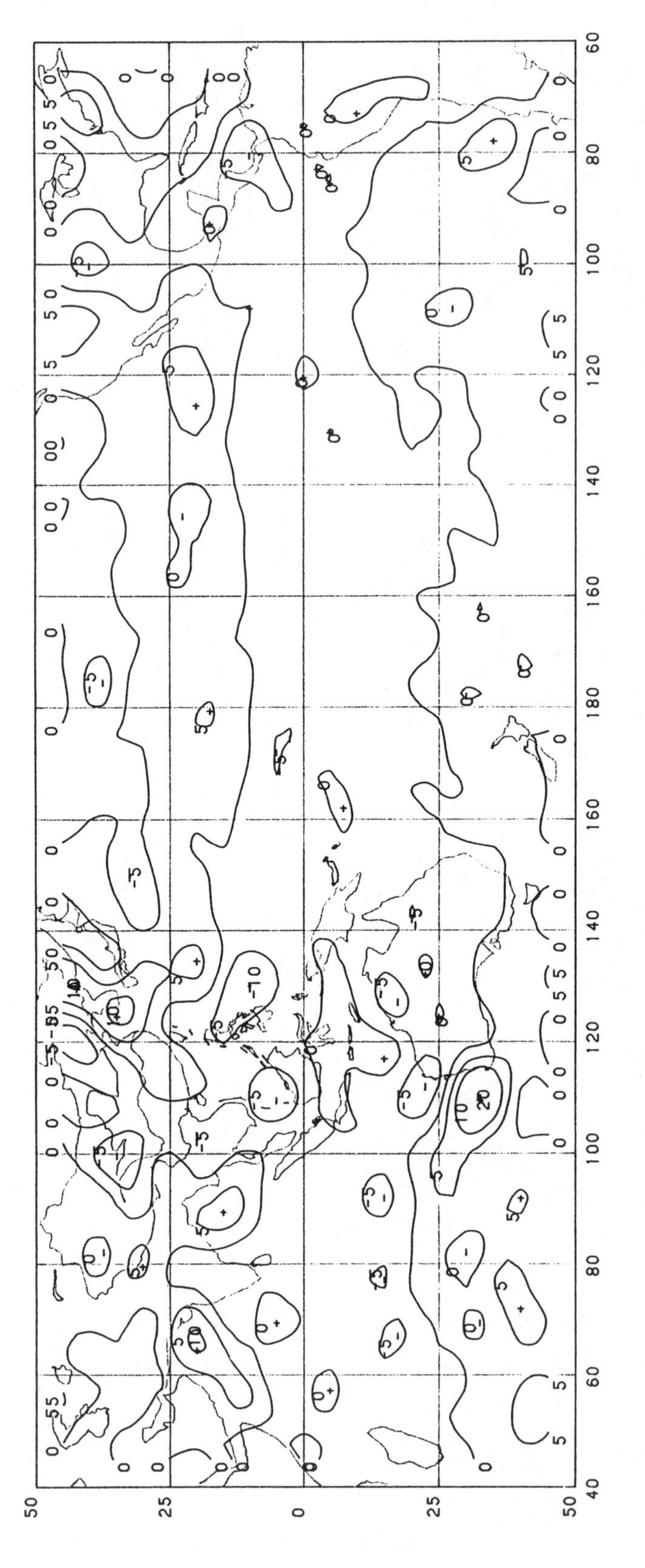


**FIGURE 18.** The simulation of the multiple years monthly mean $\nabla \cdot \vec{\tau}$; unit: $10^{-8}$ $(ms)^{-2}$.

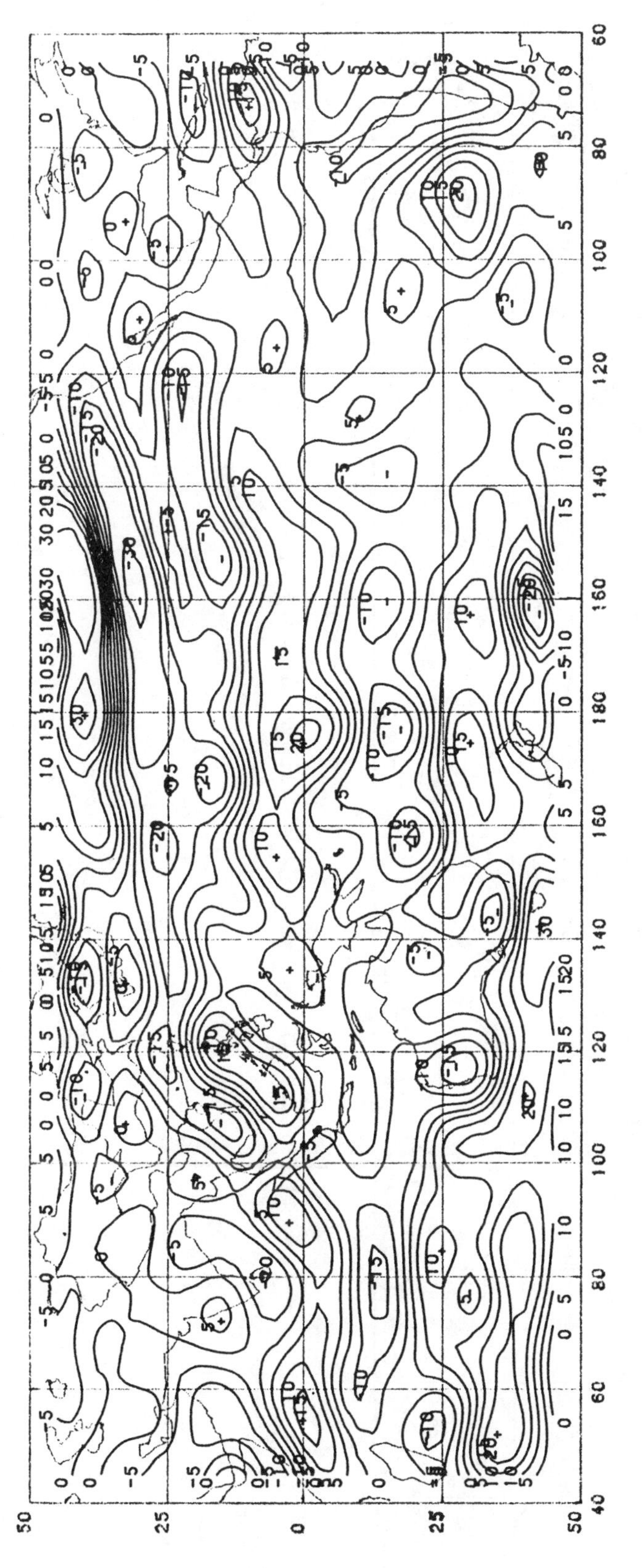


**FIGURE 19.** The simulation of the 1983 monthly mean $\nabla \times \vec{\tau}$; unit: $10^{-8}$ (ms)$^{-2}$.

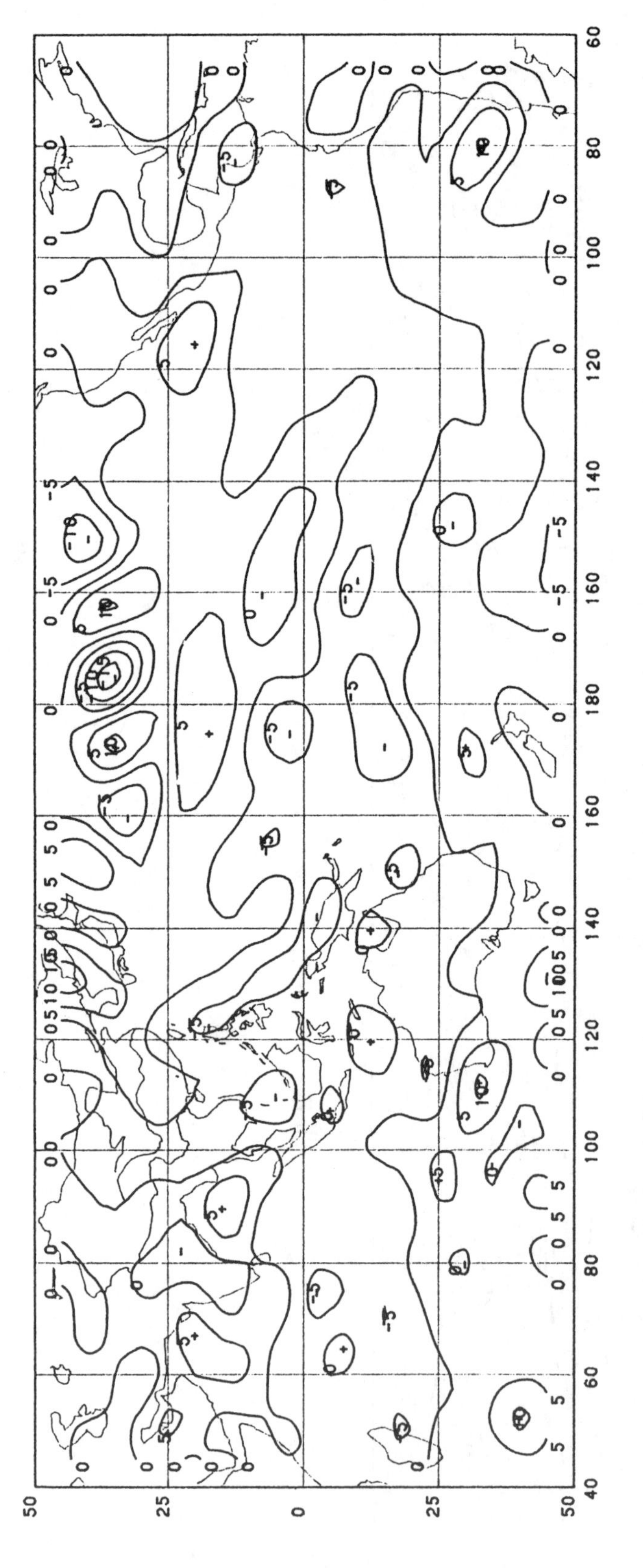


**FIGURE 20.** The simulation of the 1983 monthly mean $V \cdot \vec{\tau}$; unit: $10^{-8}$ $(ms)^{-2}$.

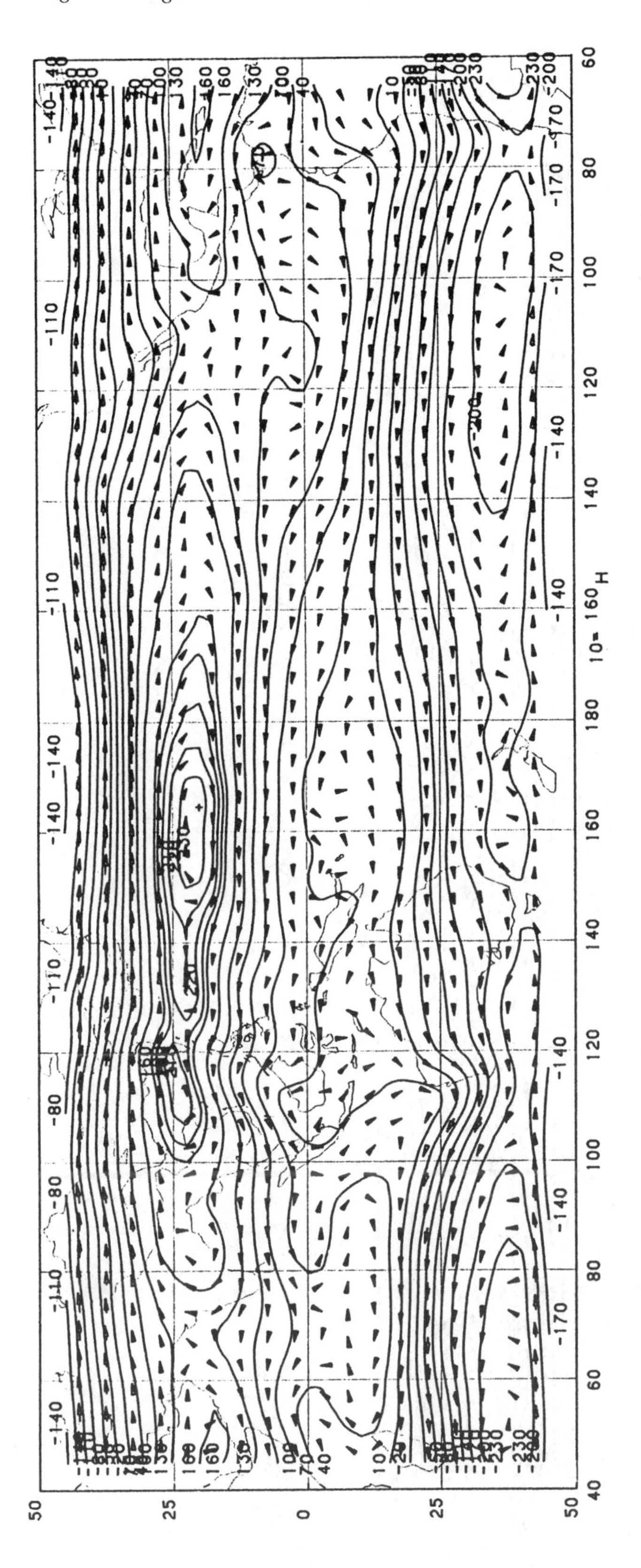

**FIGURE 21.** Stream function for the multiple years monthly mean and under the stationary condition; unit: $10^3$ $m^2$ $s^{-1}$.

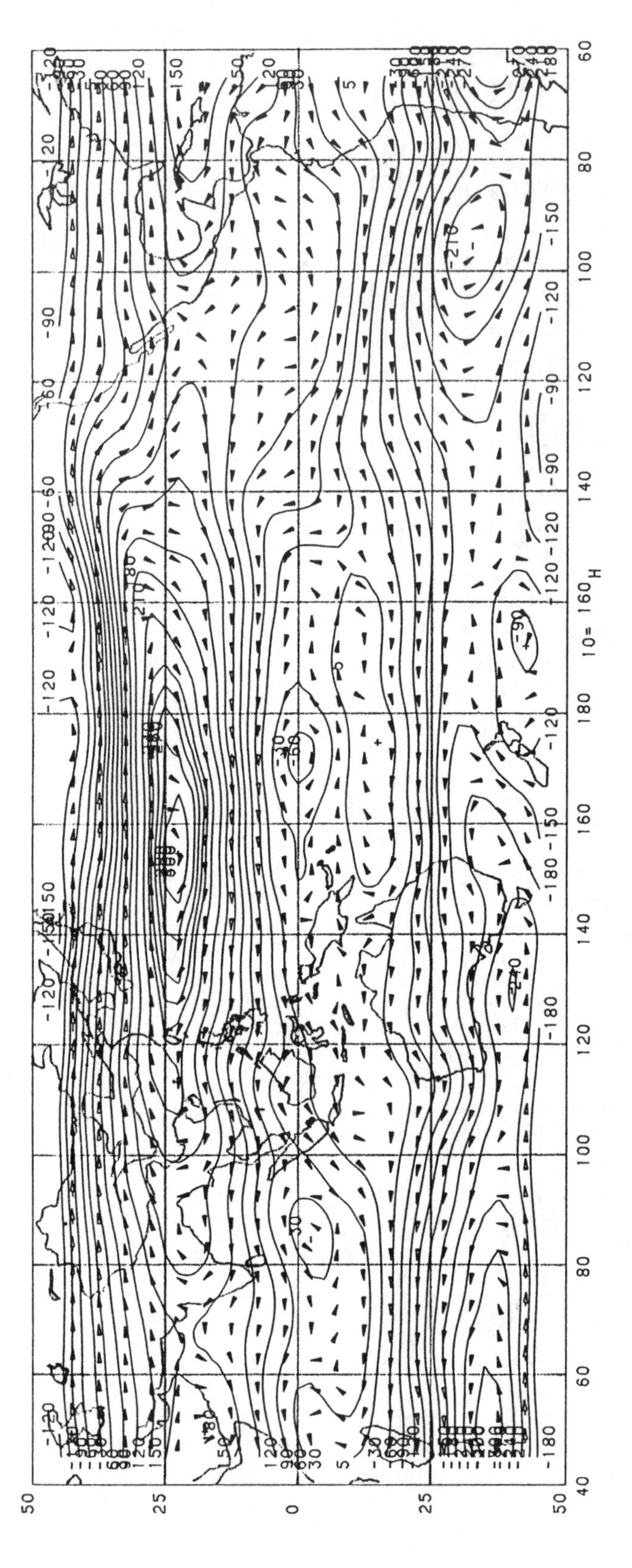


**FIGURE 22.** Same as Figure 21, except it is for 1983; unit: $10^3$ $m^2$ $s^{-1}$.

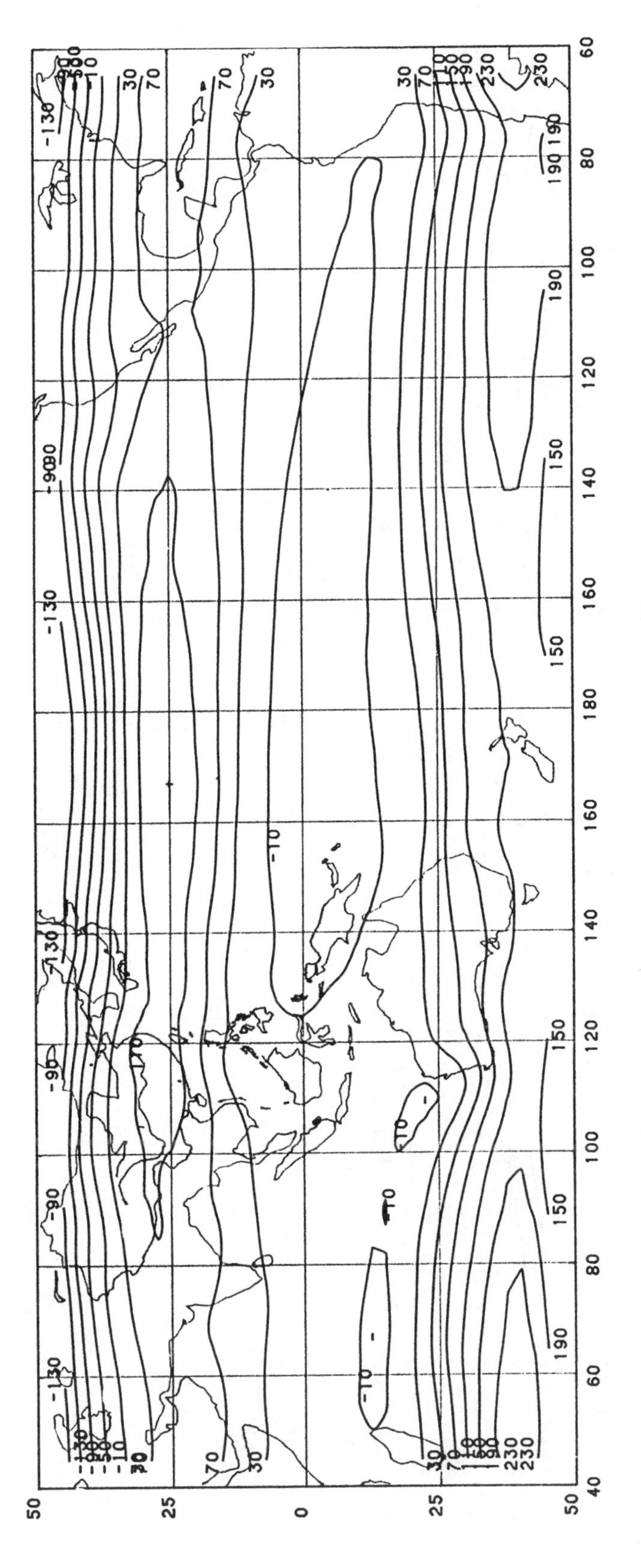


**FIGURE 23.** The stationary perturbed SSH for the multiple years monthly mean; unit: cm.

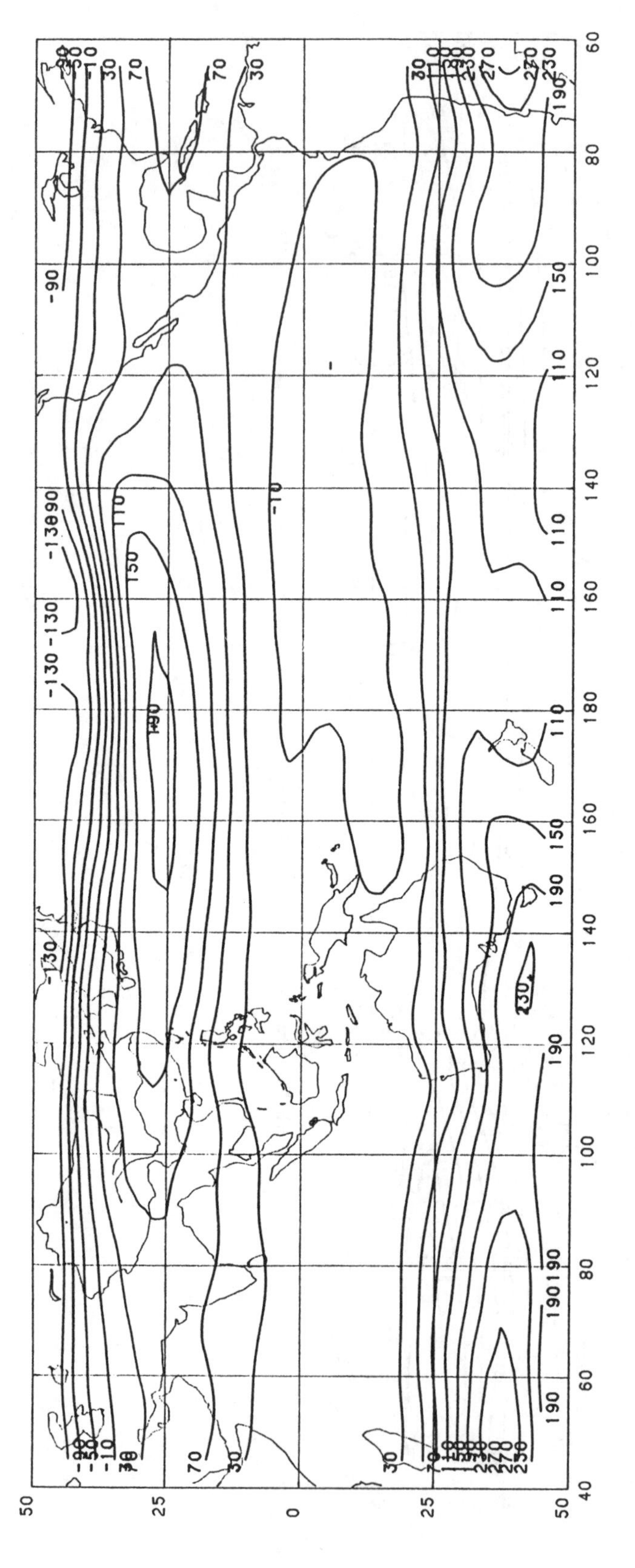


**FIGURE 24.** Same as Figure 23, except it is for the 1983 monthly mean; unit: cm.

indicated that the stream function of the ocean current varies rapidly with time. At 1 h the ocean current already possesses the general pattern, over time the stream function increases its value, until day 5, the stream field tends to be invariable and approximates to the steady state. The variation of the perturbed surface height with time has the same general characteristics as that of the stream function.

# V. CONCLUSIONS

1. The net radiation is fundamentally distributed zonally; its magnitude depends very largely on the cloud amount. Over the equatorial Indian and the west Pacific, as well as within the region 0° to 10°S over the central Pacific, owing to the abundant clouds, the net radiation diminishes. Over the ocean, the net radiation and SST have very good correlation; the warm SST corresponds to the low value of net radiation. The axis of the least net radiation flux coincides with the position of ITCZ, even better than OLR for fitting the fine structure of the latter.
2. For the subtropical and tropical west Pacific, the ocean over the winter hemisphere is the energy source of the atmosphere, whereas that over the summer hemisphere is the energy sink. The region over the tropical west Pacific, where the energy flux is sensitive to the SST and the wind anomaly near the sea surface, just lies in the region of warm pool, where the ocean supplies large amounts of energy to the atmosphere.
3. With global warming, global marine total cloud amounts tend to increase, however, there exists a hysteresis of about 2.9 years mean.
4. In the last 126 years, the offsetting impact due to the increase of the total cloud amounts to the climate warming largely exceeds the greenhouse effect arising from the increase of $CO_2$ for the same period. Hence, the feedback effect of clouds to the climate is very important.
5. Generally the meridional wind stress $\tau_{sy}$ is very small, approximately with zero value in the tropical region, whereas the zonal wind stress $\tau_{sx}$ has considerable magnitude, and both its meridional gradient and the accompanying wave pattern are remarkable.
6. For January 1983, the mature phase of the 1982 to 1983 El Niño event, the time scale of the response of the ocean current to the surface wind stress was 1 to 3 day, 6, 9, 10, 12, 15, 22, 48, and 86 day; the shortest is 0.72 day; the longest can reach 135 day and even 194 day.
7. The stream function $\psi$ changes rapidly with time, starting with an ocean at rest, it approximates to the steady state until day 5.
8. In the Northern Hemisphere, the direction of the oceanic current deviates to the right side from that of the surface wind; in the Southern Hemisphere, it deviates to the left side. With an overview, the ocean current field accords with the air current field near the sea surface. The magnitude of the current velocity, irrespective of the 9 years or 1983 monthly mean, is greater at latitudes 40°N or 40°S than at low latitude. Current velocity along the west coast is greater than that along the east coast.
9. The elevation of the oceanic surface off the west coast of South America for 1983, compared to the 9-years mean, indicates that more sea water accumulated there, which may be associated with the anomalous warm SST.

# REFERENCES


Budyko, M. E. (1960), Heat Balance on the Earth's Surface, Science Press, Beijing (translated into Chinese).

Bunker, A. F. (1976), Computations of surface energy flux and annual air-sea interaction cycles of the North Atlantic ocean, *Mon. Weather Rev.*, 104, 1122–1140.

Ditkqin, V. A. and Kyznekov, L. I. (1958), *Handbook of Operational Calculus*, Science Press, Beijing (translated into Chinese).

Gurmer, T. H. (1983), Transfer processes at the air-sea surface, *R. Phys. Sci.*, 308(1503), 253–273.

Huang, J. and Li, H. (1984), *Spectral Analysis in Meteorology*, China Meteorological Press, Beijing chap. 3 (in Chinese).

Jiang, S. and Zhu, Y. (1990), *The Application of Outgoing Longwave Radiation and its Atlas*, Peking University Press, Beijing, (in Chinese).

Li, Y. and Li, W. (1988), A preliminary analysis of the regional atmospheric energy budget over the tropical, *J. Acad. Meteorol. Sci.*, S. M. A., China, 3, 183–189, 1988 (in Chinese).

Lighthill, M. J. (1969), *Philos. Trans. R. Soc. London Ser. A*, 265, 45–92.

Perry, A. H. and Walker, M. J. (1983), Ocean-Atmosphere System, Science Press, Beijing (translated into Chinese).

Ramanadham, R., Somanadham, S. V. S., and Rao, R. R. (1981), Heat budget of the North Indian Ocean surface during monsoon — 77, in *Monsoon Dynamics*, Lighthill, J. and Pearce, R. P., Eds., Cambridge University Press, 491–508.

Reed, R. K. (1983), Heat fluxes over the eastern tropical Pacific and aspects of the 1972, El Niño, *J. Geophys. Res.*, 88, 9627–9638.

Sutton, O. G. (1953), *Micrometeorology*, McGraw-Hill, New York.

Veronis, G. and Stommel, H. (1956), *J. Mar. Res.*, 15, 43.

Warren, B. A., Stommel, H., and Swallow, J. C. (1966), *Deep-Sea Res.*, 13, 825.

Weare, B. C., Strub, P. T., and Samuel, M. D. (1981), Annual mean surface heat fluxes in the tropical Pacific Ocean, *J. Phys. Oceanogr.*, 11, 705–717.

Yin, H. (1983), Lectures on Atmospheric Radiation, chap. 3, 7 (in Chinese).

Yin, H. (1990), One-dimensional radiative convective climatic model considering the multiple scattering of the high, mid, low cloud and the aerosol, *Tech. Rep. Climatic Res. Inst.*, (in Chinese).

Zhu, Y. and Yang, D. (1990), Analysis of the air-sea heat exchange during the El Niño event in 1983, *Acta Oceanol. Sin.*, 9, 513–526.

Chapter 2

# THE MEASUREMENTS OF ATMOSPHERIC AEROSOLS AND CONDENSATION NUCLEI AND THE POTENTIAL ROLE OF THESE PARTICLES IN CLOUD FORMATION AND TRANSFER OF RADIATIVE ENERGY

**Josef Podzimek**

## TABLE OF CONTENTS



0-8493-4419-0/93/$0.00 + $.50
© 1993 by CRC Press, Inc.

# I. INTRODUCTION

In 1880, John Aitken published a treatise "On Dust, Fogs, and Clouds" in which — based on his laboratory experiments and observations in nature — he clearly stated that "When water vapor condenses in the atmosphere it always does so on some nuclei." He continued to explain that some kinds of dust seem to form better nuclei than others. As an example, he mentioned "sodic chloride dust" which would condense water even in a subsaturated atmosphere and "sulphure in its different forms" produced during the burning process. He rightly pointed to the pollution caused by sulfur compounds (e.g., ammonium sulfate) denoted by him as "fog producers" and at "A possibility of there being some relation between dust and certain questions of climate, rainfall, etc."

For several decades, John Aitken measured, with his expansion counter, the concentration of condensation nuclei on the seashore, in urban and rural areas, and on a mountain observatory in Switzerland. He tried to relate the nuclei concentration to the state of the atmosphere and to the influence of local nuclei sources (see Knott, 1923 or Podzimek, 1989). His investigation of sea-salt nuclei production and the discovery of the gas-to-particle conversion process in the atmosphere was paralleled in other countries by the studies of the nuclei nature and electrical charge and of their potential relationship to the cloud (fog) droplet formation and size distribution (Wegener, 1910a, 1910b). The ground and mountain station nuclei concentration measurements were followed by the balloon measurements in the free atmosphere, which showed very steep nuclei concentration decrease with altitude (e.g., up to 9425 m by Wigand and Lutze, 1914). A unique survey of the nature of atmospheric nuclei, their role in cloud formation, their electrical and optical properties, and their potential role in the climatic changes as it appeared before World War II can be found in the books by Schmauss and Wigand (1929), Landsberg (1938), and Burckhardt and Flohn (1939). Landsberg summarized 13,400 measurements of the concentration of condensation (Aitken) nuclei made on the ground in different countries and regions with expansion type counters (Aitken counters) and found, in mean, the following concentrations in different environmental conditions: metropolitan areas — 147,000 $cm^{-3}$, cities — 34,300 $cm^{-3}$, inland (rural areas) — 9500 $cm^{-3}$, seashore — 9500 $cm^{-3}$, islands — 9200 $cm^{-3}$, oceans — 9400 $cm^{-3}$, and on the mountains — 6000 $cm^{-3}$ at altitudes below 1000 m, 2130 $cm^{-3}$ below 2000 m, and 950 $cm^{-3}$ at altitudes over 2000 m. This steep decrease in nuclei concentration with altitude supports the mentioned pioneering balloon measurements by Wigand (1919) who, after summarizing the results of 14 ascents, deduced an exponential decrease of nuclei concentration with altitude. However, Wigand and other investigators were already aware of the strong deviations from this ideal Aitken nuclei concentration distribution with the altitude due to the temperature inversions, proximity of clouds, etc. Temporary condensation nuclei measurements (performed during supersaturation of several hundred percent with respect to water) with their climatological interpretations, lost their importance due to the fast development of cloud physics in the 1930s.

Cloud physicists, represented by the investigations of A. Wegener, H. Köhler, G. C. Simpson, W. Peppler, T. Bergeron, and W. Findeisen, were primarily interested in condensation or ice nuclei giving origin to cloud elements at low supersaturation prevailing during the natural cloud formation (for a survey on these subjects see the books by Mason, 1957 or Pruppacher and Klett, 1978). One started to pay attention to the largest and most active nuclei. Their size distribution and their physicochemical nature were determined by laboratory spot tests or filter techniques, and by light scattering instruments. Nuclei critical supersaturation was investigated in special cloud chambers. These studies, often supported by the optimistic vision of the weather modification at our wish, brought, shortly after World War II, an enormous information about the most active cloud condensation nuclei of natural or man-made origin. At the same time, leading investigators in this field (like C. E. Junge, H. Dessens, A. H. Woodcock, J. P. Lodge, S. Twomey, and B. J. Mason) learned about the great difficulties of trying to deduce from the many measurements, a coherent picture of the time variability, transformation, and transport of these nuclei before they started to initiate clouds in a large scale (see Junge, 1963 or Pruppacher and Klett, 1978). The investigators found that in most cases, the sea-salt nuclei are not the prevailing cloud condensation nuclei, in spite of their great importance for colloidal instability of the oceanic air masses and the cloud and drizzle element formation over the sea. Most of the large nuclei over the continents have a mixed nature, as was shown by Grabovskii (1951), who used a microcrystalographic method and by Kumai (1951) and Kuroiwa (1951), who employed an electron microscopical observation for the identification of particles found in ice crystals and droplets.

The newer trends in atmospheric aerosol studies are not solely dominated by the wish of cloud or weather modification or by the better understanding of aerosol transport and air pollution propagation. In the foreground of our interest now stands the role of atmospheric natural or man-made aerosols in radiative energy transfer and in the cloud (fog/haze) formation on a large or global scale. In addition, the cloud lifetime and precipitation studies are conducted for modeling more effectively the air cleaning, aerosol, and trace component cycling.

Several notes on the history of the condensation nuclei and cloud condensation nuclei measurements, should introduce a short survey of the main results of condensation nuclei measurements during the past 40 years, and their application to the explanation of cloud formation and transfer of radiative energy. We will first discuss the properties, global distribution and transport of atmospheric particles, their role in cloud and fog element nucleation, their interaction with cloud and precipitation elements, and finally, some of the new findings indicating the future trends in the investigation of the atmospheric aerosol and its role in cloud formation.

## II. GENERATION AND PROPERTIES OF ATMOSPHERIC AEROSOL PARTICLES IMPORTANT FOR CLOUD FORMATION

Atmospheric aerosol consists of particles covering a size range of approximately 0.005 $\mu m$ to 500 $\mu m$ and concentrations in the troposphere of 10 $cm^{-3}$ to almost $10^6$ $cm^{-3}$ (in a highly polluted environment). The chemical composition of atmospheric particles depends on their origin, transport, and interaction with other particles, cloud elements, and gaseous substances in the atmosphere. Table 1 represents the mean aerosol production on our Earth in millions of tons per year. It is a simplified table published by Jaenicke (1984), based on the estimates of the production rate of particles which can be transported intercontinentally ($r < 3$ $\mu m$).

The data from Table 1 lead to the conclusion that the ratio between natural aerosol sources and anthropogenic sources can be 26.5% if the minimal production rates are considered, or 19.3% for maximal production rates. For cloud-forming processes, it is very important to know what percentage of the above-mentioned production rate takes the active cloud condensation nuclei, what is their most effective size range (if we take into consideration different particle removal mechanisms), and what is the most typical activation spectrum (critical supersaturation or undersaturation) of produced nuclei.

The many sources of Aitken nuclei (particles with $r < 0.1$ $\mu m$ active at supersaturations of several hundred percent) which interact show enormous variability of concentrations during the day and year. This depends largely on environmental and meteorological conditions.

**TABLE 1**
**Mean Aerosol Production on Earth**

| Natural aerosol sources | Million tons/yr |
|---|---|
| Sea salt | 180 |
| Mineral dust | 60–360 |
| Volcanoes | 15–90 |
| Forest and brush fires | 3–150 |
| Biogenic material | 80 |
| Converted sulfates | 130–200 |
| Converted nitrates | 140–700 |
| Converted hydrocarbons | 75–200 |

| Anthropogenic aerosol sources | Million tons/yr |
|---|---|
| Particles (production, man's activities) | 6–54 |
| Converted sulfates | 130–200 |
| Converted nitrates | 30–35 |
| Converted hydrocarbons | 15–90 |

Adapted from Jaenicke, R., *Aerosols and their Climatic Effects* (1984).

In Figures 1, 2, and 3, examples of the strong dependence of Aitken nuclei counts (AN $cm^{-3}$) on the location of sampling stations around the Bay of Naples are presented. Close to the harbor downtown, nuclei concentrations surpassed 150,000 AN $cm^{-3}$ (Figure 1) several times, unlike those 20 km westward on the seashore at the small town of Mte Di Procida (Figure 2). Usually during the morning and late afternoon hours, the concentrations were higher due to the air stability and local traffic. Figure 3 shows a comparison of Aitken nuclei concentrations at different wind directions measured at the cliffs 56 m above the small fishing harbor, fully exposed to the polluted air masses coming from the town and from the steel mills and industrial areas located 7 km to the northeast (curve 1). Close to the sea surfaces at the harbor's mole, screened with respect to the polluted air, the measured concentrations were much smaller in mean (curve 2). These samples document well the principal difficulty of attempting to obtain a coherent picture of Aitken nuclei concentrations and their impact on fog and haze element formation on a local scale (Podzimek and DeMaio, 1992; Podzimek and Ianniruberto, 1992).

In spite of the fact that Aitken nuclei are not directly initiating the cloud (fog) element formation, they modify through deposition (together with the diffusion of gaseous substances) the critical supersaturation of larger particles, as will be discussed later. Their high concentrations along the industrialized coastal regions affect the properties of the more active particles of natural or man-made origin (Podzimek, 1980). The dramatic changes in the ultrafine particle concentrations during a brushfire event, however, do not appear to strongly influence the number of activated nuclei at 0.25 and 0.50% supersaturation (Desalmand et al., 1985).

The potential impact of Aitken nuclei produced by aircraft transport and measured in the upper troposphere and low stratosphere seems to be very important. Considerable nuclei concentrations were measured at specific altitudes above the large airports and close to main air-traffic corridors. Figure 4 represents the vertical profiles of Aitken nuclei concentrations (AN $cm^{-3}$), ozone concentrations (relative unit = 3.5 nbar $O_3$), and temperature (T°C) measured above the Houston area in 1974. Several layers of high nuclei concentrations of

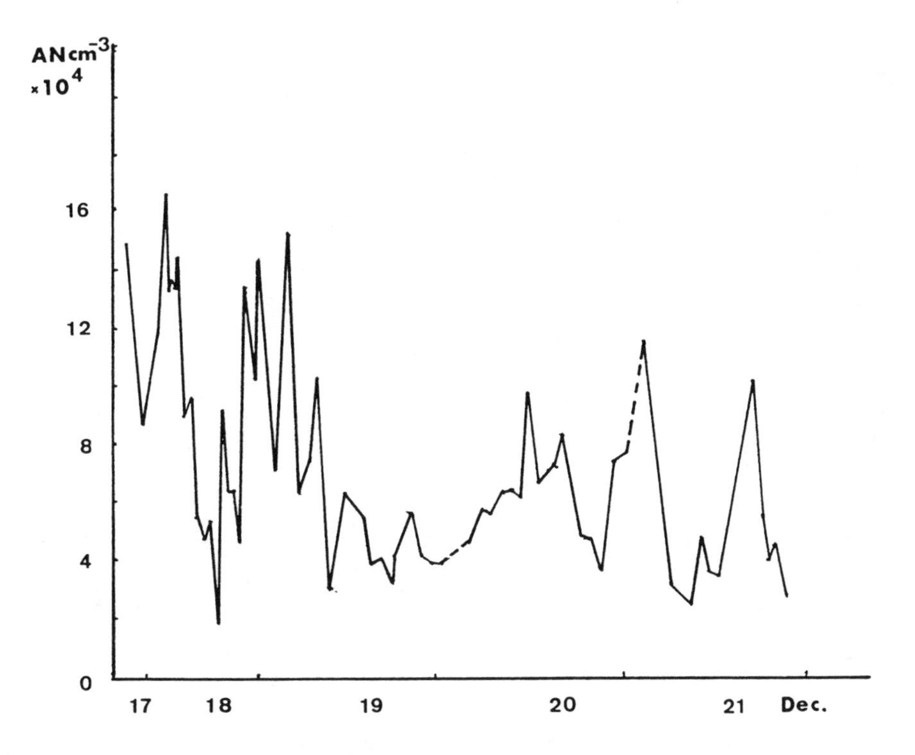


**FIGURE 1.** Aitken nuclei concentration measured at Naples (Italy) harbor during 5 days in December 1987. (From Podzimek, J., *World Res. Rev.*, 3, 221–258, 1991. By permission of the World Resource Review, Chicago, U.S.A.)

around 8 km, 9 km, and 11 km are apparent. A spectacular nuclei concentration peak at around 15 km was explained by Podzimek et al. (1977) as the possible effect of air traffic polluted air exchange around the tropopause probably combined with air chemistry. The negative correlation of nuclei counts with ozone concentration during the aircraft ascents above 15 km has also been found over Panama (Podzimek, 1976). There are several other interesting observations related to particle formation around and above the tropopause due to aircraft pollution or during volcanic activity. Podzimek (1984a) described a clear negative correlation between Aitken nuclei counts and ion flow, and the effect of aircraft pollution, particle formation, and transformation around the tropopause. This study also showed an interesting evolution of bimodal aerosol-size distribution during the interception with the plume of, at that time active volcano, De Fuego in 1974 (14.5°N, 90.9°W). Several other well-documented cases of particle emission and transformation during and after the eruption of Mount St. Helens (26.2°N, 122.2°W) in May 1980 were presented by Hofmann and Rosen (1982) with balloon-borne instruments, and by Hobbs et al. (1982) based on research aircraft measurements.

Several cases of ultrafine particle generation and transformation at the tropopause level which might well have an impact on cirrus cloud formation can be completed by the more numerous measurements in the upper troposphere. The results of many of them are summarized in the review articles by Beard (1987), Cooper (1991), and Vali (1991).

Particles larger than 0.05 μm were investigated intensely in several environments for their great impact on cloud-forming processes. Much attention has been paid to marine, urban, industrial, and rural aerosols. Recently, research was focused on biogenic aerosols

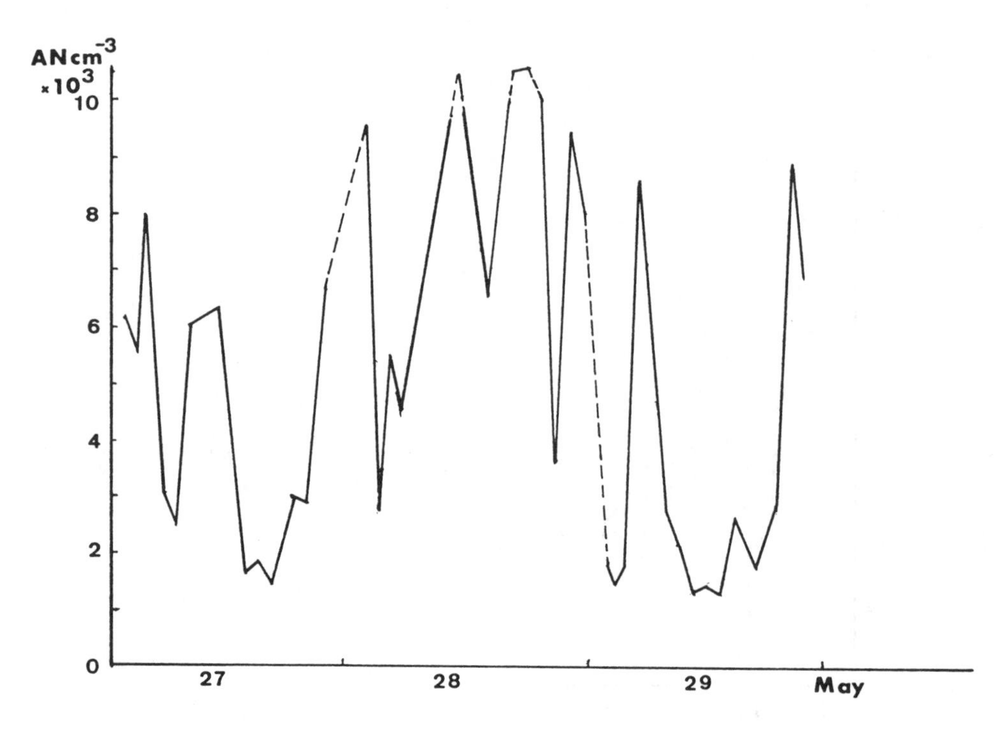


**FIGURE 2.** Aitken nuclei concentration measured at Mte Di Procida (Italy) seashore during 3 days in May 1987. (From Podzimek, J., *World Res. Rev.*, 3, 221–258, 1991. By permission of the World Resource Review, Chicago, U.S.A.)

as a potential source of condensation and ice nuclei, on the role of mixed nuclei in a steady or unsteady condensation process, and on aerosol and nuclei scavenging in the atmosphere.

Marine aerosol usually featured a distinct particle size distribution and nuclei activation described by nuclei supersaturation spectra. In comparison to the continental aerosol, sea-salt particles enabled better predictability in cloud formation and aerosol transformation during their passage over the continent or after cloud droplet evaporation. For the application of marine aerosol studies in modeling the climatic changes speaks the well-defined nuclei activation and formation of clouds which have a large contrast compared to the very low surface albedo of oceans. Surveys on these subjects can be found in the books by Junge (1963), Twomey (1977), Petrenchuk (1979), Ruhnke and Deepak (1984), Monahan and MacNiocaill (1986), and in the survey article of Podzimek (1980).

In this contribution, attention will be paid to some older measurements of the chloride ions containing particles over central Europe (Podzimek, 1959; Podzimek and Cernoch, 1961), over the Caribbean Sea (Podzimek, 1967), over the seashore and mainland in south Texas (Podzimek, 1973, 1980, 1984b), and in the marine-urban atmosphere at the Bay of Naples in Italy. The main conclusions of these studies, which were originally done by the "classical" Liesegang circle technique (spots left after chloride particle impaction on sensitized gelatin layer) and later completed by light-scattering instruments and electron microscopy, are described below.

The flights in northern Bohemia (in total 49 flights) and over the high Tatras mountain region in Slovakia (in total 16 flights) showed that, in general, the effect of sea-derived salt nuclei is small. Most of the large and giant chloride particles ($r > 0.5$ μm) are generated close to the cities and industrial areas, which is demonstrated in Figure 5, obtained as an average salt nuclei concentration vertical profiles over northern Bohemia and the high Tatras mountains. The effect of the exchange in the planetary boundary layer is apparent in both

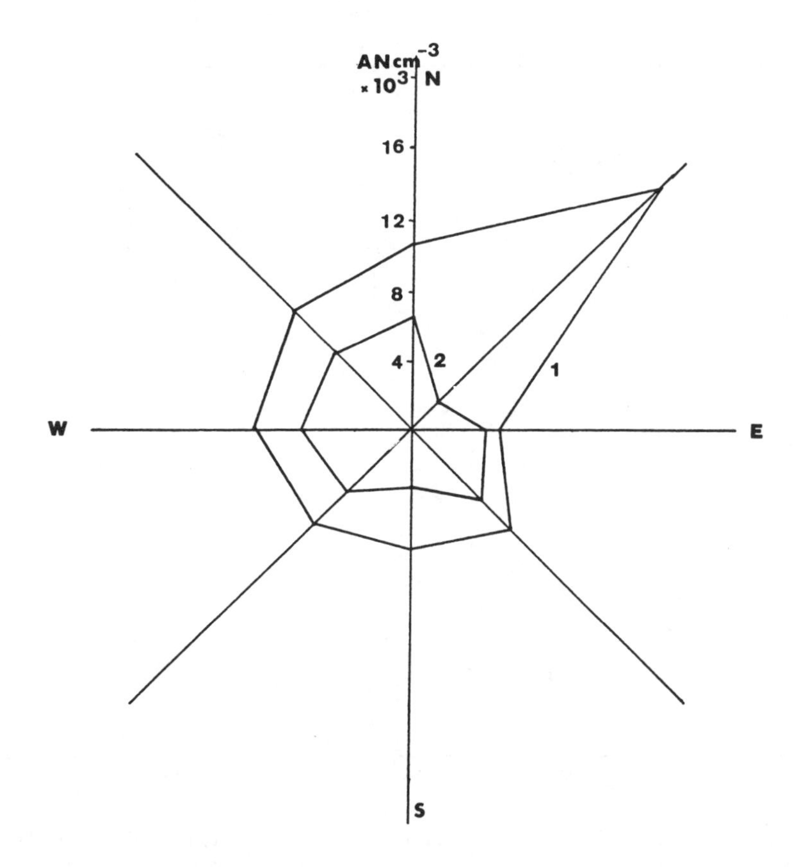


**FIGURE 3.** Aitken nuclei concentrations measured at Mte Di Procida in dependence of wind direction. Measurements were made several times a day in May and June 1987, at the cliffs above the harbor (curve 1) and at the mole (curve 2).

locations, where the maximum chloride nuclei concentrations reached the values of 27.5 or 12.5 nuclei per 1 l.) The nuclei size distribution usually followed Nukiyama-Tanassava formula

$$n = Ar^2e^{-Br^s} \tag{1}$$

where the parameters A, B, s took different values depending on the meteorological parameters mainly on the air humidity (Podzimek, 1973). For most cases, $s = 0.5$ or $s = 0.33$ gave the best results. Also in Figure 5, dashed curves are plotted signifying the nuclei concentration decrease with altitude if simple particle (of density ρ in $g/cm^3$) settling and constant turbulent exchange coefficient (A in $g\ cm^{-1}s^{-1}$) are assumed.

The effect of continental sources on the total chloride particle concentration is shown in Figure 6. The concentrations of salt particles (N/l) are plotted at specific altitudes under the lines showing the aircraft path over Cuba and the Caribbean Sea on January 13, 1966. The aircraft approximately followed the wind direction at the top of the atmospheric boundary layer. The air temperatures (in °C) enabled air stability to be judged on this specific day, which was marked by scattered Cu and Cu con clouds. Convection currents apparently transported the sea or continental aerosol into higher altitudes. In the air passing across the island, the sodium chloride particle concentration at all levels at its lee side was higher. This interaction of salt aerosols of sea and continental origin ought to be considered if a

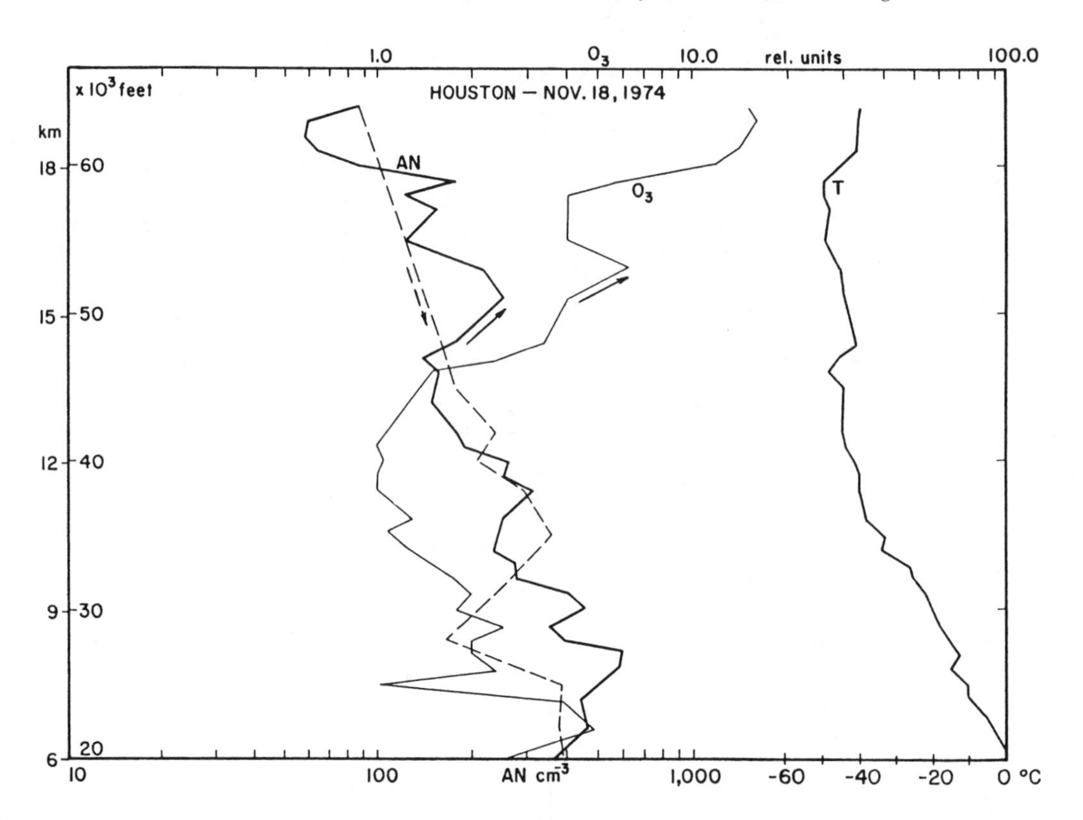


**FIGURE 4.** Vertical profiles of Aitken nuclei concentrations (AN $cm^{-3}$), ozone concentrations (plotted in relative units — 1 relative unit = 3.5 nbar $O_3$) and of temperature (T in °C) measured above the Houston area on November 18, 1974, at altitudes higher than 6 km. (From Podzimek, J., *World Res. Rev.*, 3, 221–258, 1991. By permission of the World Resource Review, Chicago, U.S.A.)

simple mechanism of their production rate at the sea surface is considered (see Toba, 1966; Blanchard, 1969).

The investigation of the generation and transport of sulfate ions containing aerosols prevailing in urban and industrial areas — and also over the sea — was often hampered by the fact that the vast majority of the particles have sizes below the detectability range using the simple spot-test technique. Furthermore, particles with diameters larger than 0.1 μm usually represent mixed particles with an insoluble core on which the active (soluble) nucleus is deposited. Bearing these facts in mind, we understand the low concentrations of sulfate ions containing particles (N/l) in Figure 7. The samples were taken during the same flights as those in Figure 5 (curve 2 over the High Tatras), and show the strong impact of the industrial and urban environment on sulfate aerosol production (Podzimek, 1969; Podzimek and Cernoch, 1961). All our newer measurements over the continents confirmed the predominating mixed particle nature of active larger condensation nuclei in the ground layer and above the atmospheric boundary layer as well.

The coarse particle measurements made on two sites at the Bay of Naples during the summer months at Mte Di Procida (curve MdP) and in December 1987 at Naples, (curve NA) are summarized in Figure 8. The total number size distribution of coarse particles from Naples (curve NA) represents the addition of particles found in the center of a Liesegang circle (curve LCI), of particles found between the cores of a Liesegang circle and its outer edge — "satellites" (curve SAT), and particles found on the impaction slide between individual Liesegang circles (haze or fog elements) — interstitial particles (curve IP). Com-

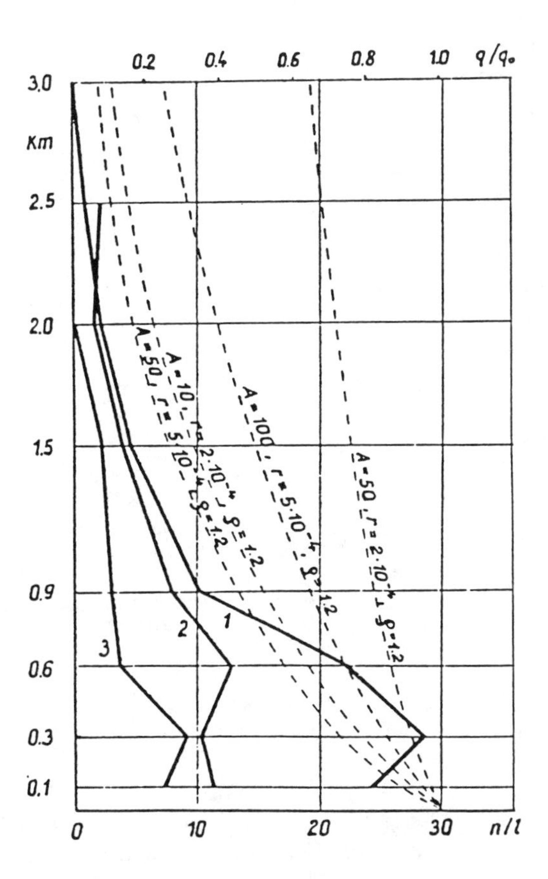


**FIGURE 5.** Total concentration of giant chloride nuclei in 1 l of air measured in northern Bohemia over the city and industrial districts (curve 1), over High Tatras Mountains (curve 2), and over rural areas of northern Bohemia (curve 3). Dashed curves represent the ideal effect of particle settling and of a constant turbulent exchange in the atmosphere (r = particle radius in cm, ρ = particle density in g $cm^{-3}$, A = turbulent exchange coefficient in g $cm^{-1}$ $s^{-1}$).

parison of these curves leads to the conclusion that in the polluted marine-urban atmosphere at Naples, the number of coarse particles deposited in the core and inside of the haze and fog elements is not negligible in comparison to the total number of interstitial particles, and that the prevailing location of the largest particles (r > 10 μm) is in the center of haze elements. Otherwise, the polluted atmosphere at Naples causes the core of many particles to activate slightly at low relative humidities (R.H. ~ 70%) and convert into fog elements at R.H. > 90%). These two forms of Liesegang circles in a gelatin layer are seen in Figure 9a and b.

Some typical coarse aerosol compositions at Naples are demonstrated in Figure 10 (combustion derived aerosol), Figure 11 (soil minerals), Figure 12 (sea-salt particle), and in Figure 13 (biogenic aerosol). Several of these particles can be found in higher levels of the troposphere at the levels of cirrus cloud formation (Figure 14). Plotted X-ray energy spectra indicate (after subtracting the effect of the sampling substrate composition — e.g., content of Si in glass slides) that most of the particles represent a mixture of large carrier particles with a small amount of active or insoluble substance deposited on them. One is able to perform an X-ray energy mapping (location) of small particles deposited on a carrier particle with a resolution power of 0.08 μm of a specific spot diameter.

Recently much attention has been paid to the biogenic aerosol produced in rural areas, over the ocean, in a savannah, or in tropical forests. These particles might serve as active

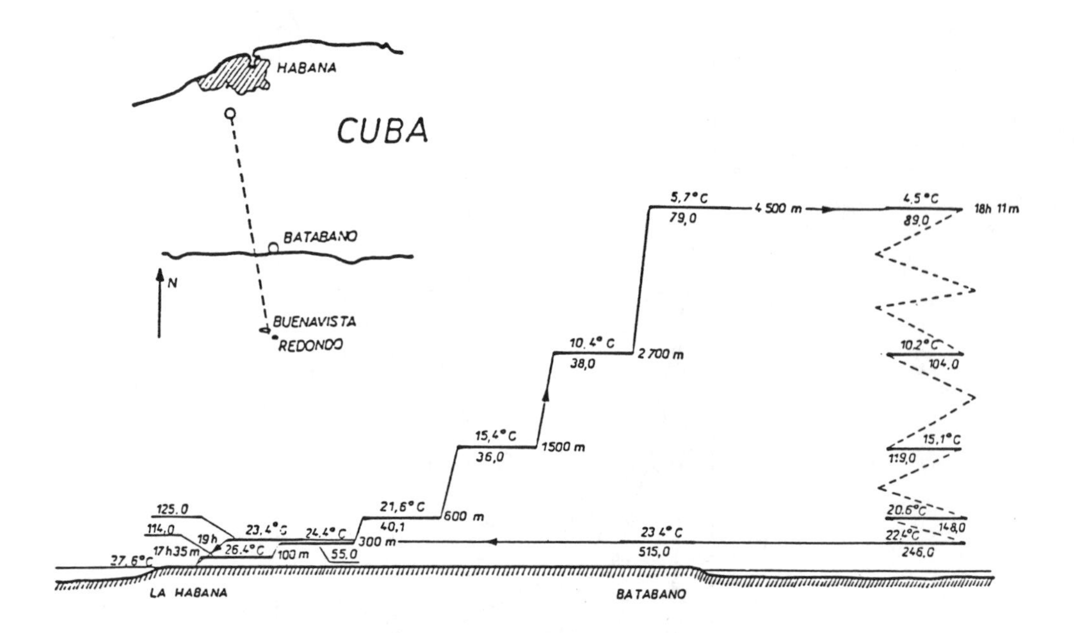


**FIGURE 6.** Vertical profile of measured giant chloride nuclei concentrations (per liter of air) sampled over Cuba on January 13, 1966, along the flight path marked in the left upper corner of the figure. At each altitude, the temperatures (in °C) are also attached.

cloud condensation or ice nuclei and have a specific relation to the ultrafine particles. The measurements at remote sites at the Ivory Coast and close to the metropolitan area of Abidjan, adjacent to lush tropical vegetation and the ocean, revealed several interesting relationships between different particle groups (Desalmand et al., 1982; Desalmand et al., 1985a, 1985b). In the dry season, the Aitken nuclei concentrations were very low in forested savannahs (usually several thousand particles per cm$^3$) and in dense tropical forests. One cannot, however, completely exclude high-condensation nuclei counts measured by other investigators during research flights above the tropical forest (Cros, 1977; Cros et al., 1981). The cloud condensation nuclei (measured at supersaturations of 0.25%, 0.50%, and 1.00%) showed a trend very similar to Aitken nuclei during the dry season, and the counts of $N_{1.00}$ were quite comparable to the concentration of Aitken nuclei (Figure 15). This reveals that during the dry season the aerosol (far from urban and oceanic areas) is well-aged and does not have a well-developed nucleation mode. The nuclei concentration during the day is apparently dominated by the air exchange in the atmospheric ground layer. That explains the low counts during night hours.

The described nuclei and aerosol dry season behavior change dramatically during the rainy season — when freshly generated submicron and large particles appear — and also after the brush fires started. It is assumed that 3 to 4 h were necessary for developing the whole process from the release of gases and aerosol particles during a large scale brush fire to the maximal condensation nuclei concentration found on the ground. Aitken nuclei concentrations larger than 200,000 per cm$^{-3}$ were measured at a distance of several kilometers from the fire zone. This change is also perceived in the cloud condensation nuclei counts (Desalmand, 1985b). The chemical composition of coarse particles collected during the dry season reveals that Al, Fe, Si, and possibly Ca are the most abundant elements which represent the terrestrial origin of particles unlike P, S, K, and Cl which can be derived from terrestrial vegetal and maritime origin.

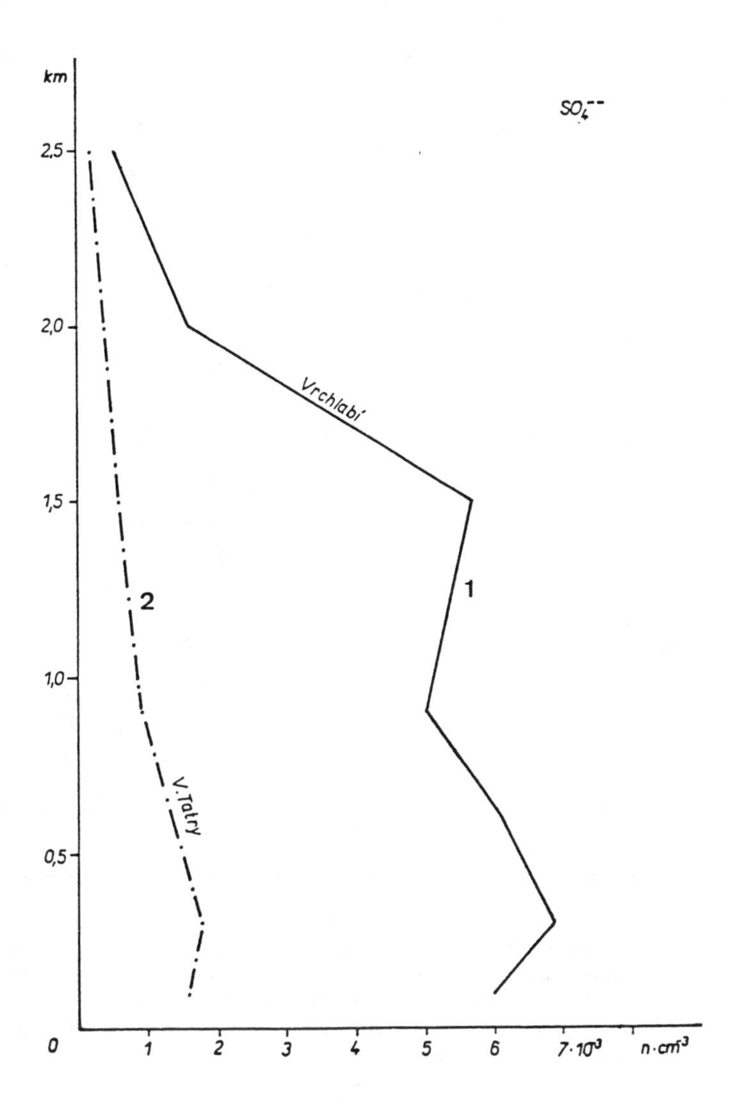


**FIGURE 7.** Vertical profiles of measured concentrations (per cm³) of large and giant particles containing $SO_4$ ions over northern Bohemia (curve 1) and over the High Tatras mountains (curve 2).

In April 1987, systematic measurements of the size distribution of Aitken nuclei were performed in Rolla, a small town of 13,300 population in the center of a rural area close to the Lake of the Ozarks, Missouri (Podzimek, 1988). The main purpose of the investigation was to determine, with the aid of a GE condensation nuclei counter and Zalabsky electrostatic classifier, the size distribution of fine particles and their dependence on the weather situation and time of day. The normalized values (referred to the concentration of AN with $r = 0.005$ μm) of the aerosol concentrations in different particle size intervals are presented in Figure 16. The effect of the air stability (probably combined with the particle irradiation and heterogeneous chemical reactions) is apparent in the shape of curve B drawn from data taken on sunny days in the afternoon. This is also partly observed in the morning hours (around 8:30 a.m.) on sunny days (curve A) and differs from the measurements taken on the days with cloud covered skies (curve C — in the morning; curve D — in the afternoon). The possible fine particle generation by the oxidation of some organic vapors produced by a forest with a subsequent gas-to-particle conversion was already mentioned by Went (1966) and Lopez et al. (1973).

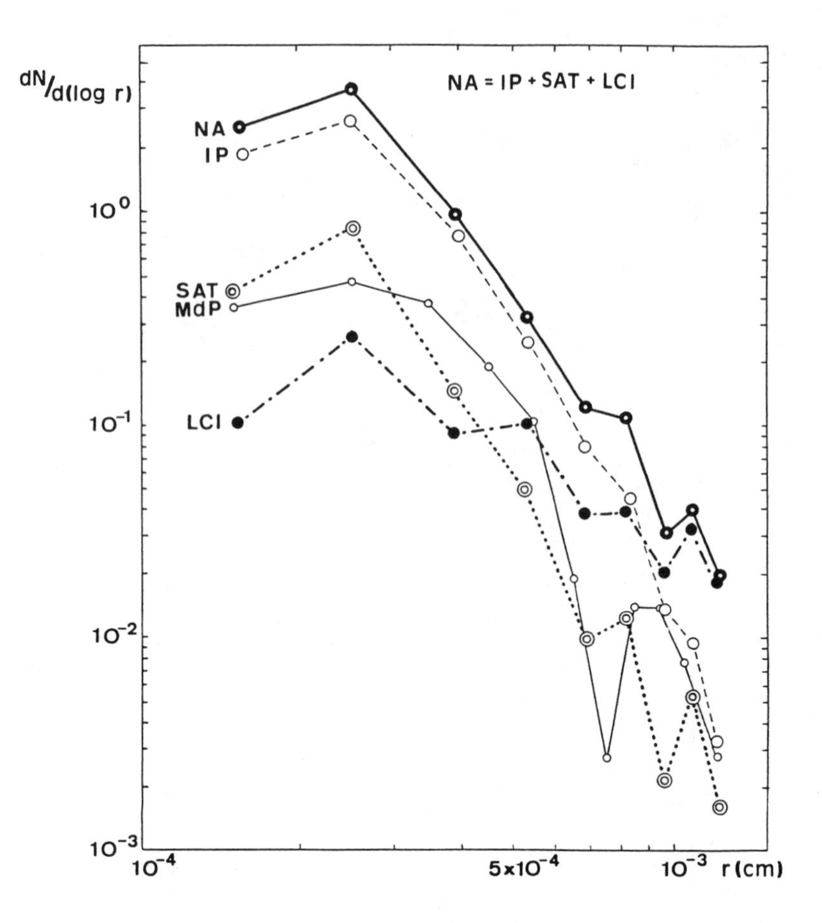


**FIGURE 8.** Number size distribution of coarse insoluble particles found in haze and fog elements sampled during the summer months at Mte Di Procida (curve MdP) and in December 1987, at Naples (curve NA). The distribution curve for NA represents the addition of interstitial particles (IP) found between the fog and haze elements, of particles found in the center of Liesegang circles formed in the sensitized gelatin layer (LCI), and of particles situated inside of a circle between its core and outer edge (SAT).

The other large scale sources of cloud condensation nuclei might be related to the volcanic activity, desert dust, forest fires, and to the combustion aerosols injected by aircraft into the higher troposphere. Systematic research in all these fields is documented in several books and review articles (see Ruhnke and Deepak, 1984; Gerber and Hindman, 1982; Gerber and Deepak, 1984; Hobbs and McCormick, 1988; Beard, 1987; Pruppacher, 1986; and Cooper, 1991).

## III. REMOVAL PROCESS AND TRANSFORMATION OF ATMOSPHERIC AEROSOL PARTICLES

Particle removal in the atmosphere is caused by the following processes: Brownian diffusion of ultrafine particles towards larger soluble or insoluble particles, cloud or precipitation elements, particle deposition under phoretic (diffusiophoretic, thermophoretic, photophoretic) and electrostatic forces, and sedimentation and large particle coagulation with cloud and precipitation elements (under inertial or electrostatic forces or in a turbulent flow). The final result of all these interaction processes is the longest residence time of particles with radii between 0.1 and 1.0 μm in the atmosphere, which is related to the existence of the Greenfield gap or to the Whitby's accumulation mode.

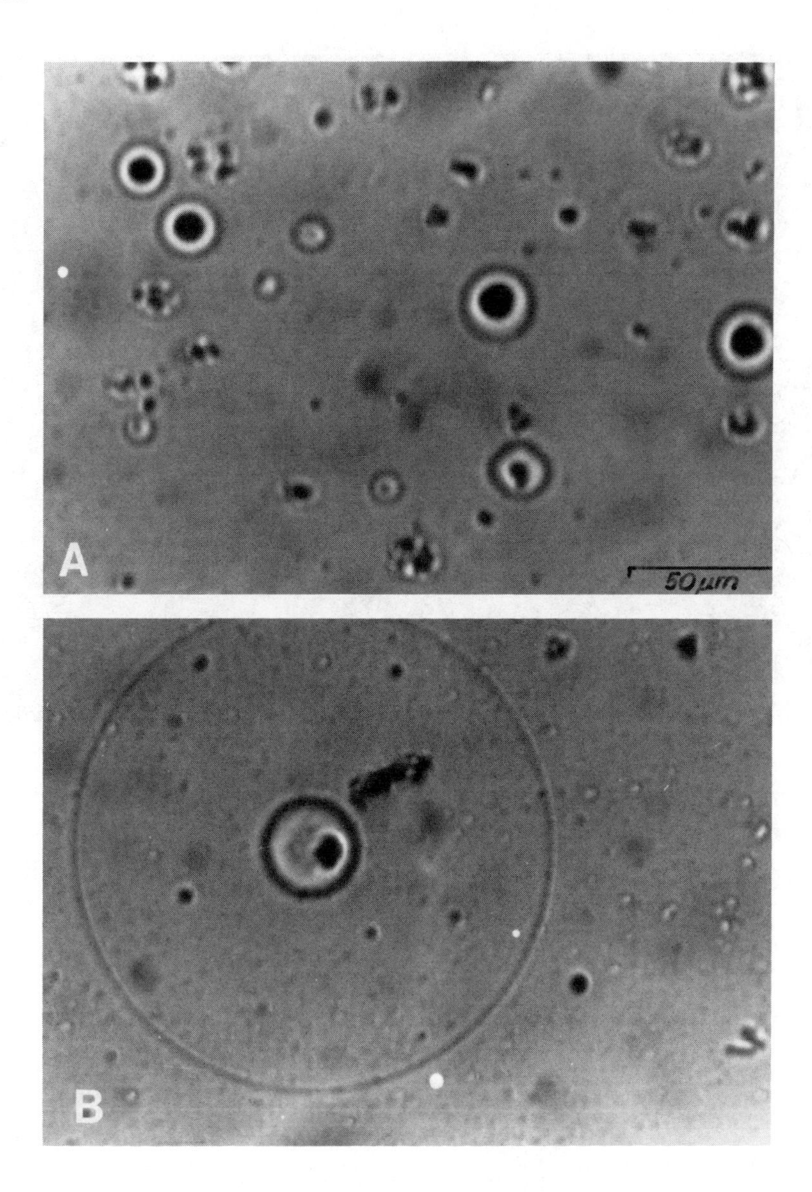


**FIGURE 9.** (A) Haze elements represented by coated insoluble particles sampled at Mte Di Procida at a relative humidity, R.H. = 73%; (B) fog droplets forming a Liesegang circle with insoluble particles in the core at R.H. = 92%. (From Podzimek, J., *World Res. Rev.*, 3, 221–258, 1991. By permission of the World Resource Review, Chicago, U.S.A.)

For cloud formation, the presence and activation of cloud condensation nuclei are the most important factors related to the particle removal process named nucleation scavenging. It might be responsible for the binding and transformation of the majority of sulfur containing particles (Georgii, 1970; Runca-Koeberich, 1979; Dlugi and Jordan, 1982; Hegg and Hobbs, 1982). Some of these studies also stressed the important role of insoluble particles which might catalyze heterogeneous reactions of gaseous and particle substances on their surface or enhance the activation of cloud condensation nuclei. These new trends of research are reflected in the papers by Flossmann et al. (1985), Leaitch et al. 1986, Okada et al. (1990), Podzimek et al. (1991), and Pueschel et al. (1992). These conclusions have also found some

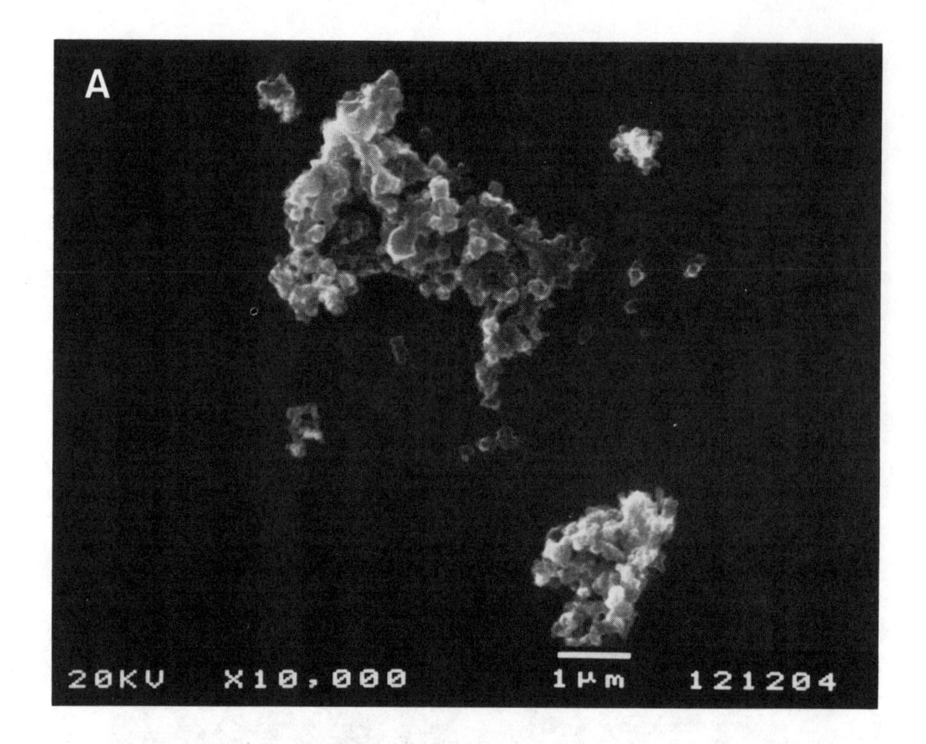


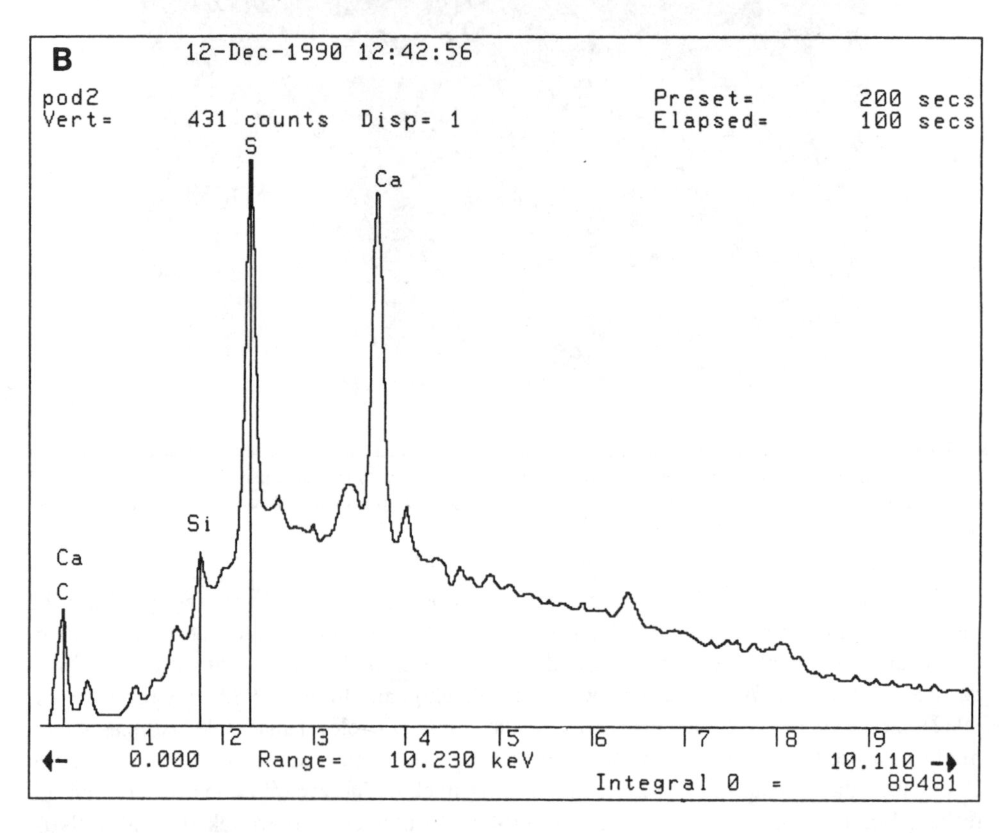


**FIGURE 10.** (A) Typical combustion-derived particles sampled in downtown Naples on December 12, 1987; (B) the X-ray energy spectrum of the largest particle's center.

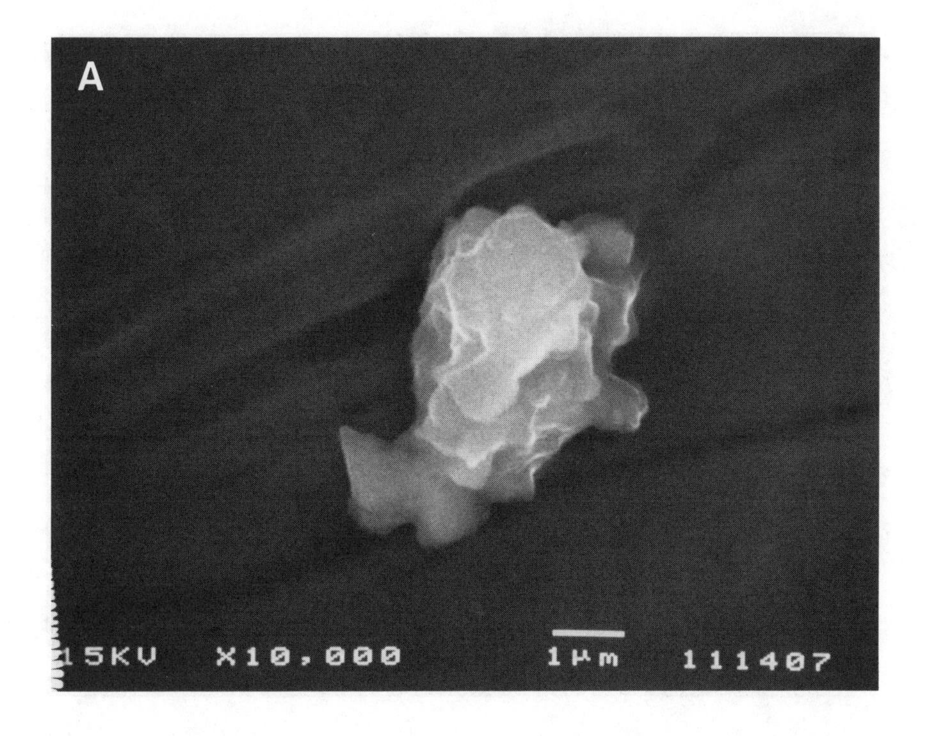


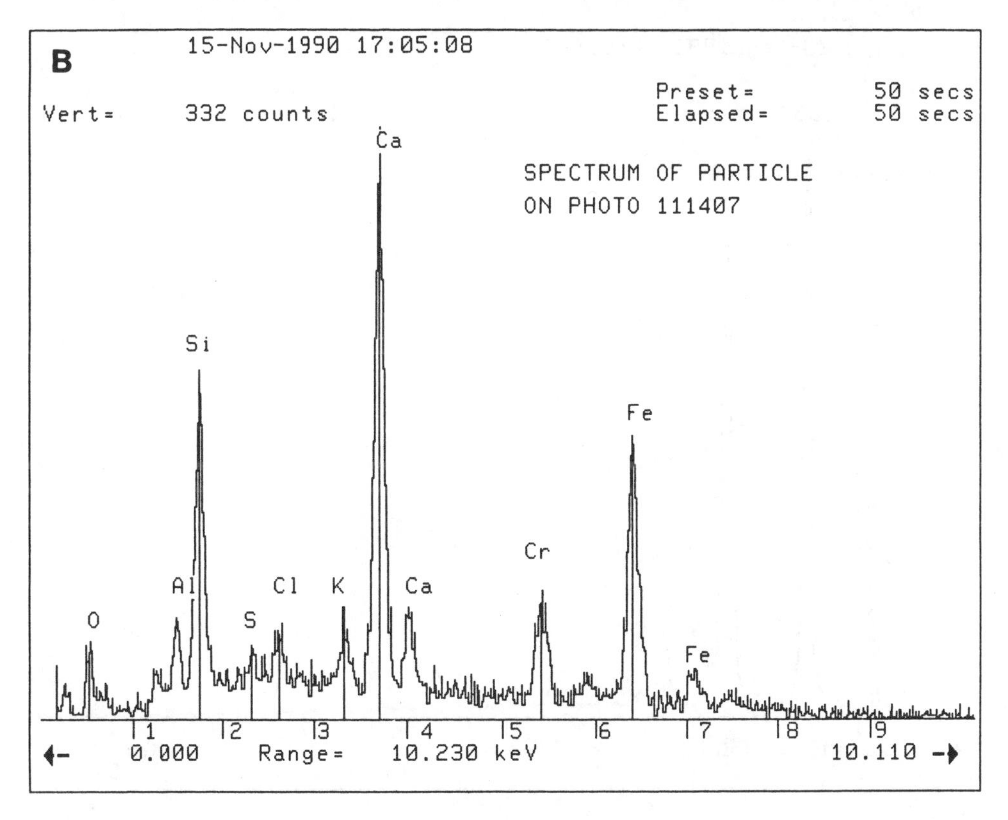


**FIGURE 11.** (A) Composite particle containing the soil minerals shown in (B) the X-ray energy spectrum.

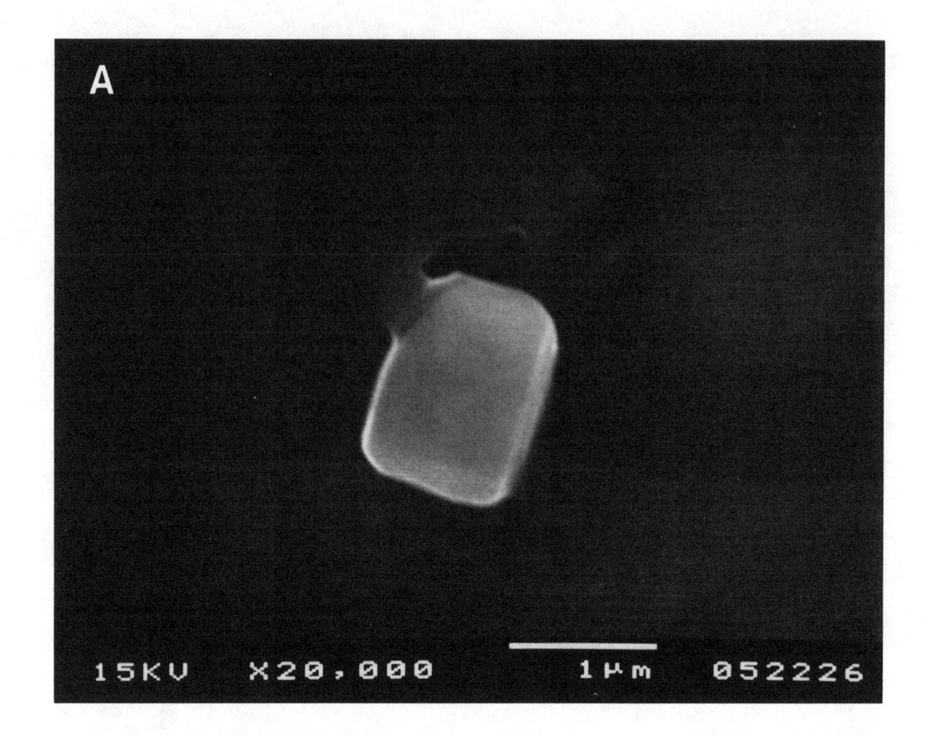


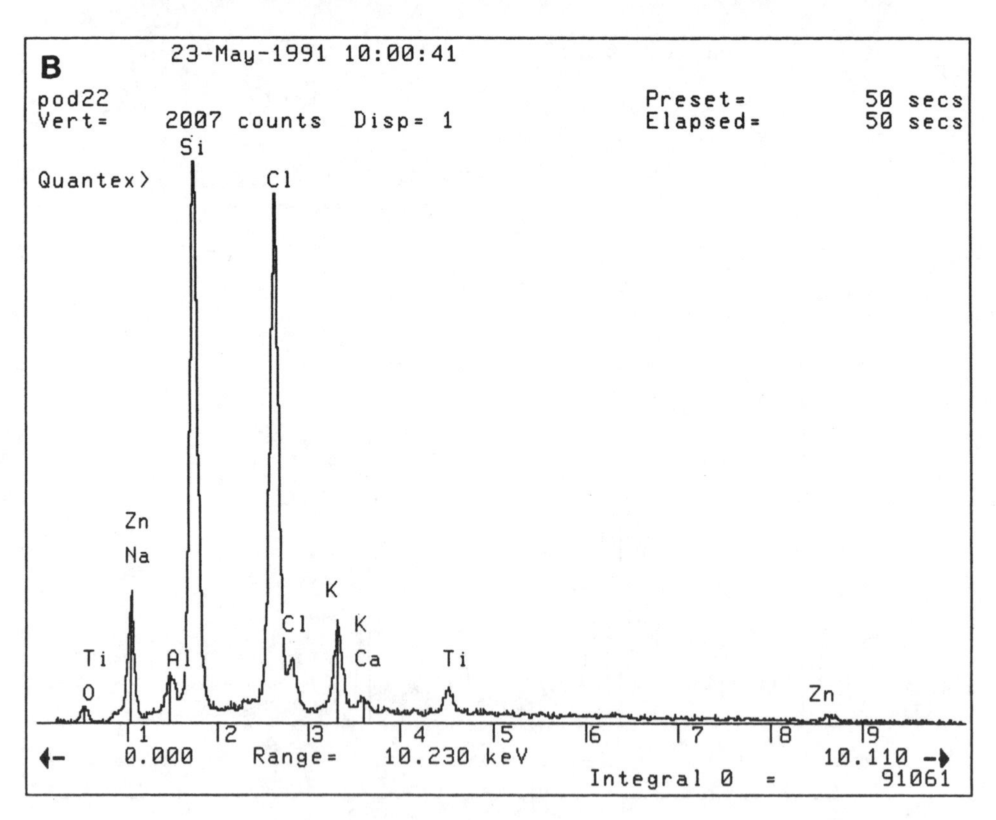


**FIGURE 12.** (A) Sea-salt particle depicted in the scanning electron micrograph with (B) the X-ray energy spectrum.

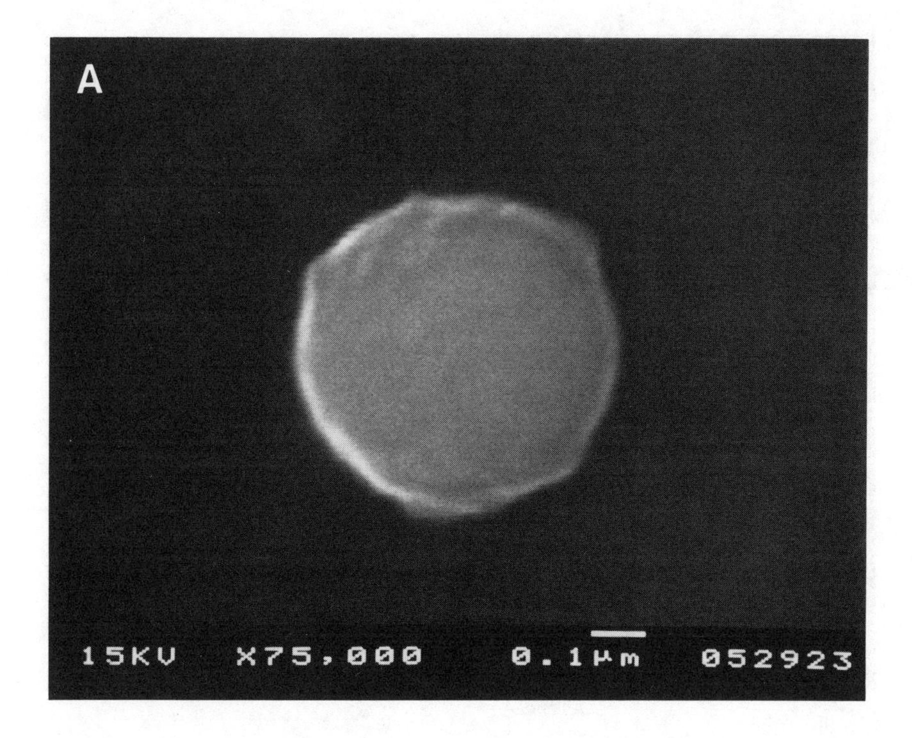


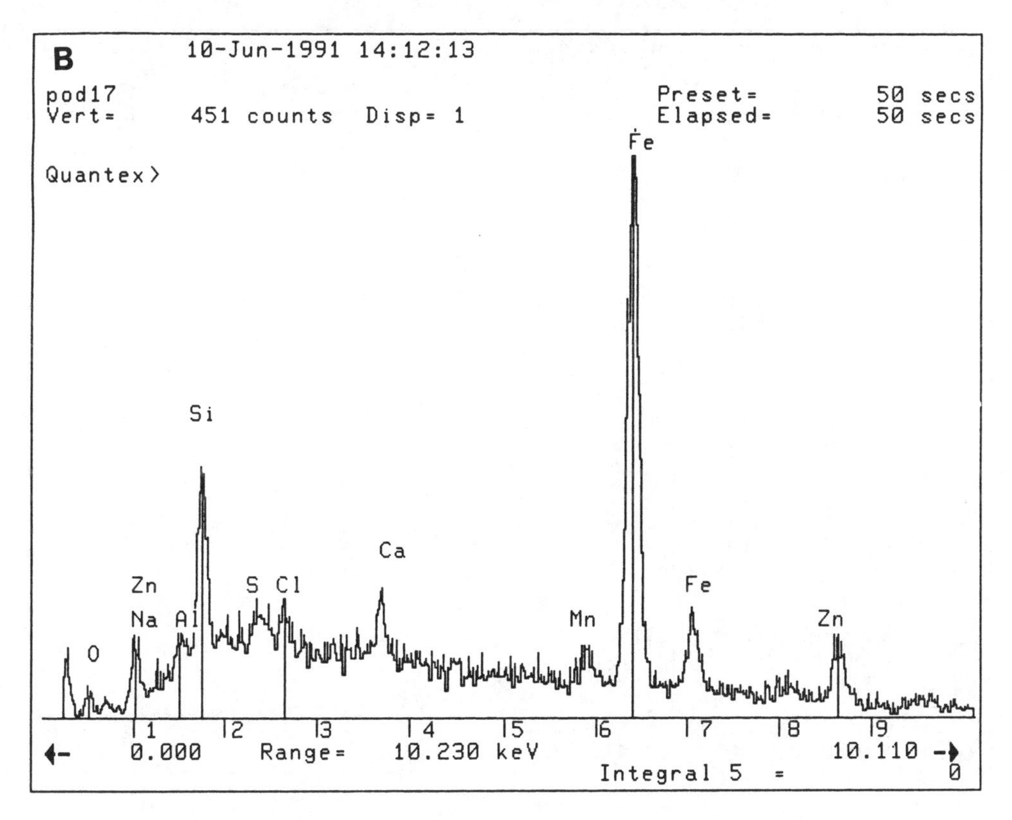


**FIGURE 13.** (A) Electron micrograph of a biogenic spherical particle frequently found in the Bay of Naples. Its composition can be deduced from (B) the X-ray energy spectrum.

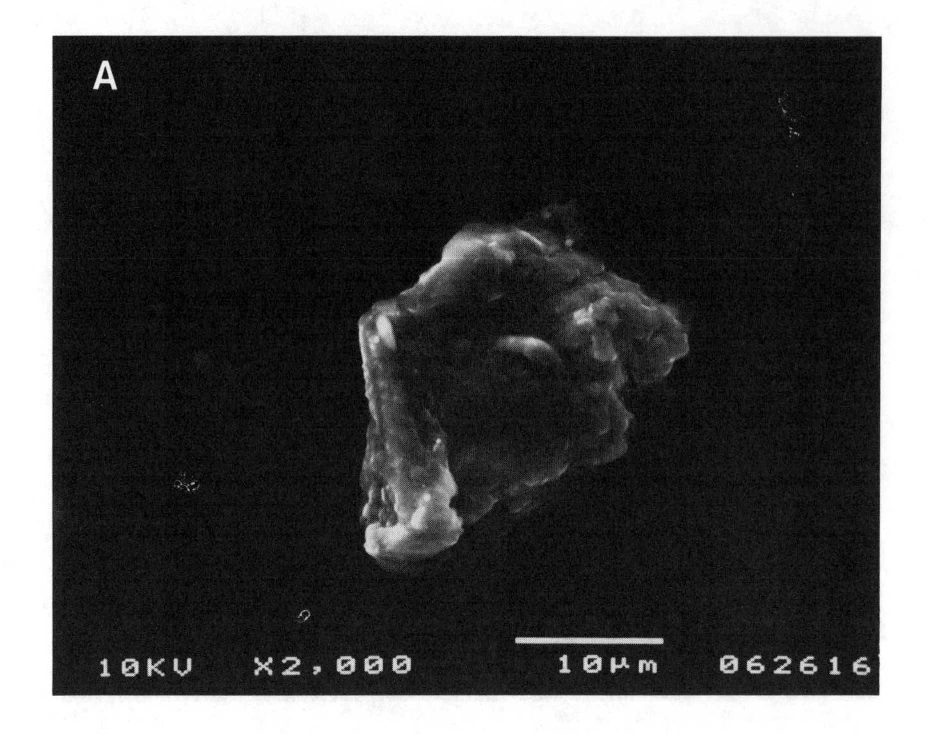


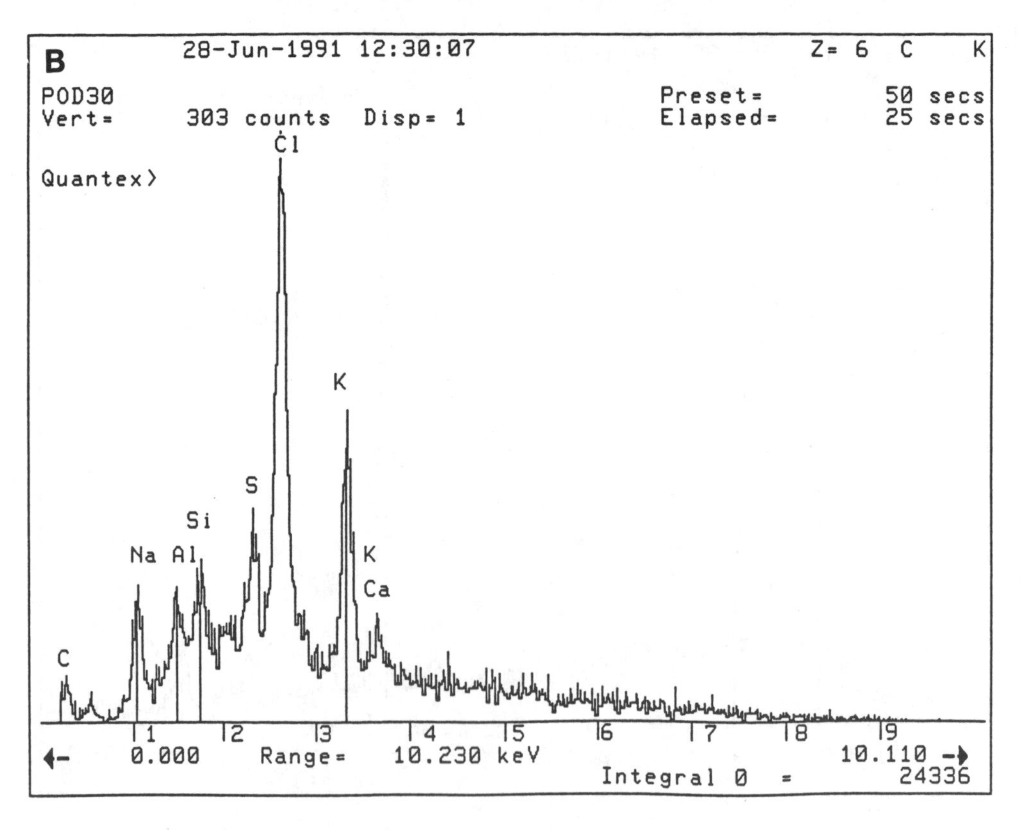


**FIGURE 14.** (A) Mixed coarse particle sampled during a flight over the Midwestern U.S. on November 15, 1990, at an altitude of about 9.5 km. Different components are seen from (B) the X-ray energy spectrum.

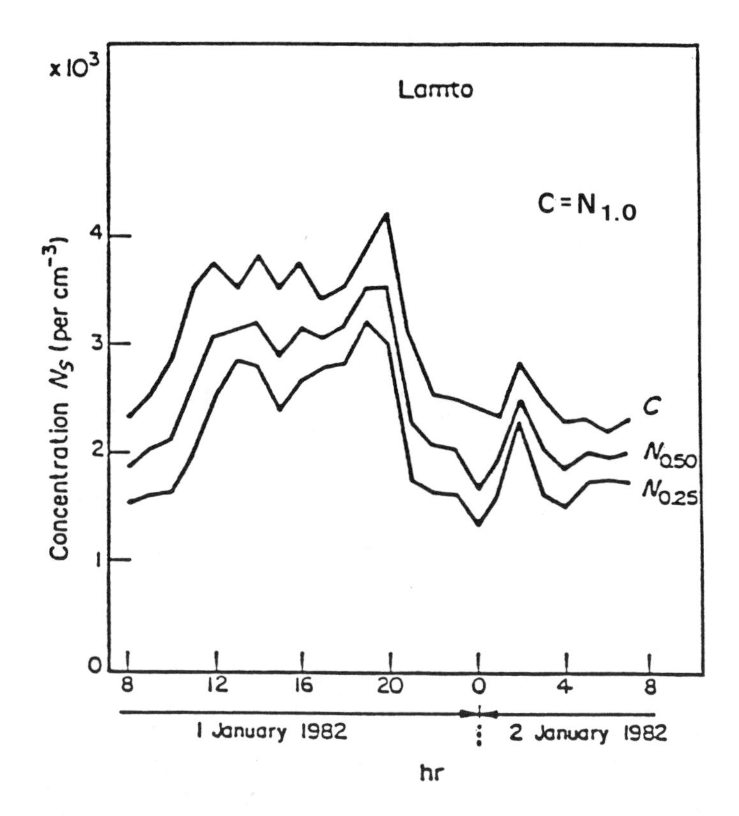


**FIGURE 15.** Cloud condensation nuclei concentrations measured at three different supersaturations (0.25%, 0.50%, and 1.00%) during the dry season in a Guinean savannah (at Lamto) in January 1982.

support by the measurement of cloud condensation nuclei activation at slight supersaturation in coastal and continental regions of the western U.S. (Hudson, 1991).

Clouds are contributing considerably to the change in aerosol size and activation spectra. Our old measurements of the size distribution of large and giant sodium chloride and sulfate particles around cumulus and stratocumulus clouds (Sc) over central Europe clearly show the significant effect of clouds and the relative humidity. Plotted in Figure 17 are the salt nuclei concentrations (N in 1 l air) and size distributions below and above an Sc cloud. The following conclusions can possibly be made: the number of nuclei entering the updrafts ($1.35\ l^{-1}$) is higher than the concentration found above the Sc cloud ($0.82\ l^{-1}$). Most of the nuclei of urban or industrial origin are activated in a subsaturated atmosphere. Above the clouds there is a slight shifting into the larger particle size groups compared to the ground layer.

Recently, the effective scavenging of fine particles in the atmospheric boundary through aerosol nucleation or deposition on haze elements or fog droplets and subsequent scavenging by rain droplets (Podzimek and DeMaio, 1992) was stressed. This three-stage process is documented in Figure 18, in which the measurements of aerosol removal through nucleation and through haze and light rain droplets in the marine-urban atmosphere at the Bay of Naples are summarized. Individual measurements in different size groups are normalized with respect to the particle counts at the maximum rain intensity. The episodic rains did not last more than 2 h. In total, two measurements from Naples (with ten individual scavenging experiments) and seven measurements from Mte Di Procida were summarized. The intensities of rain events accompanied by raindrop size distribution measurements ranged from 0.044 mm $h^{-1}$ to 3.03 mm $h^{-1}$ (Podzimek and DeMaio, 1992). In the mean, the removal of very small

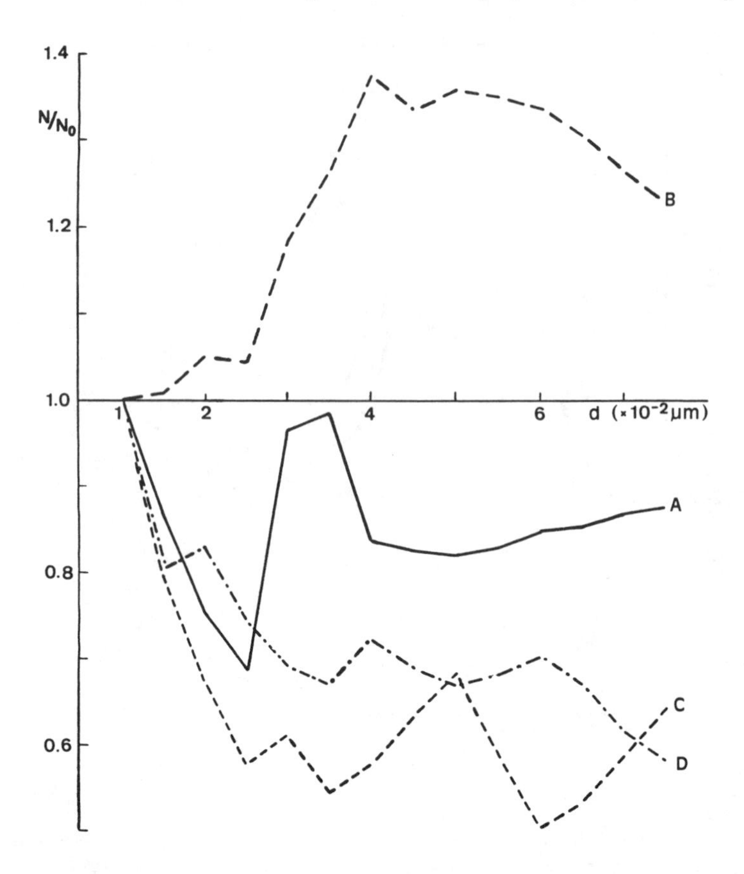


**FIGURE 16.** Aitken nuclei size distribution measured in April 1987, at Rolla, MO, on a sunny day (in the morning — curve A; in the afternoon — curve B) and under a cloudy sky (in the morning — curve C; in the afternoon — curve D). Values are normalized with respect to the concentration of the smallest size of measured nuclei (r = 0.005 μm): $N_{AO}$ = 6100 $cm^{-3}$; $N_{BO}$ = 7000 $cm^{-3}$; $N_{CO}$ = 4750 $cm^{-3}$; $N_{DO}$ = 4700 $cm^{-3}$.

particles [AN ($r < 0.1$ μm; R-4 ($0.08 < r < 0.17$ μm)] and of coarse particles (ranges S-1 and S-2 covering particle radii from 0.5 to 16.0 μm) is very effective, unlike the collection of particles from size ranges R-1, R-2, R-3 ($0.14 < r < 1.5$ μm). An important investigation was made while determining the aerosol concentration recovery time after the rain terminated. In the coarse particle size range, it lasted from half an hour to several hours, depending primarily on the meteorological situation and traffic circulating in downtown Naples. The fast restoring of the aerosol prerain situation — typical for small particle sizes — reveals the interaction with gaseous substances and the fast generation and deposition of ultrafine particles.

The other large particle removal mechanisms, such as particle deposition on the ground and vegetation, are effectively shaping the aerosol size distribution curves mainly in the size range of large and coarse particles. This is well documented in the case of the impact of Sahara aerosols and described in the treatises by Jaenicke (1979, 1984), d'Almeida (1987), or Legrand (1990).

## IV. ACTIVATION OF CLOUD CONDENSATION NUCLEI AND CLOUD FORMATION

In the previous parts of this treatise we described some of our (and other investigators) experiences with the difficult task of determining the mean state of the atmospheric aerosol

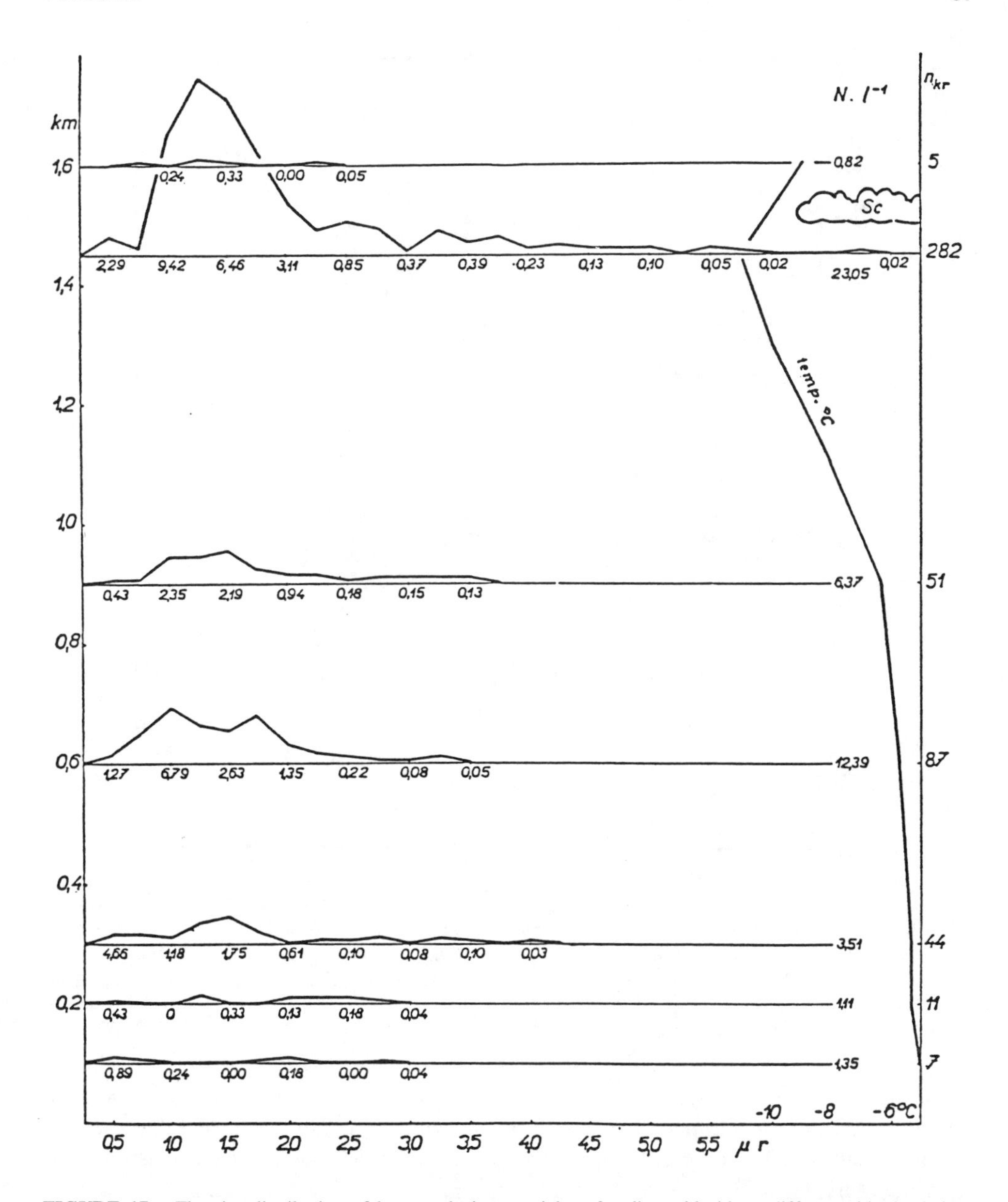


**FIGURE 17.** The size distribution of large and giant particles of sodium chloride at different altitudes below and above the stratocumulus cloud in northern Bohemia in January 1957. Particle concentrations are plotted per liter of air and marked at each measurement level; temperatures are in °C and $n_{kr}$ signifies the total number of particles found under the impactor's slot on the sampling slide.

which would enable us to model the origin and evolution of clouds and fogs on a larger scale. We showed, however, that even in remote areas over the ocean with a well-defined atmospheric boundary layer, many factors might affect the nuclei activation. Most of the nuclei might have a mixed nucleus character and be activated at fluctuating updrafts and supersaturations in a different way from the measurements with thermal gradient diffusion chambers. The situation in continental air masses is much more complicated. Nevertheless, the measurement of the activation spectrum of cloud condensation nuclei is still considered as one of the few means to obtain useful information for judging the conditions of cloud

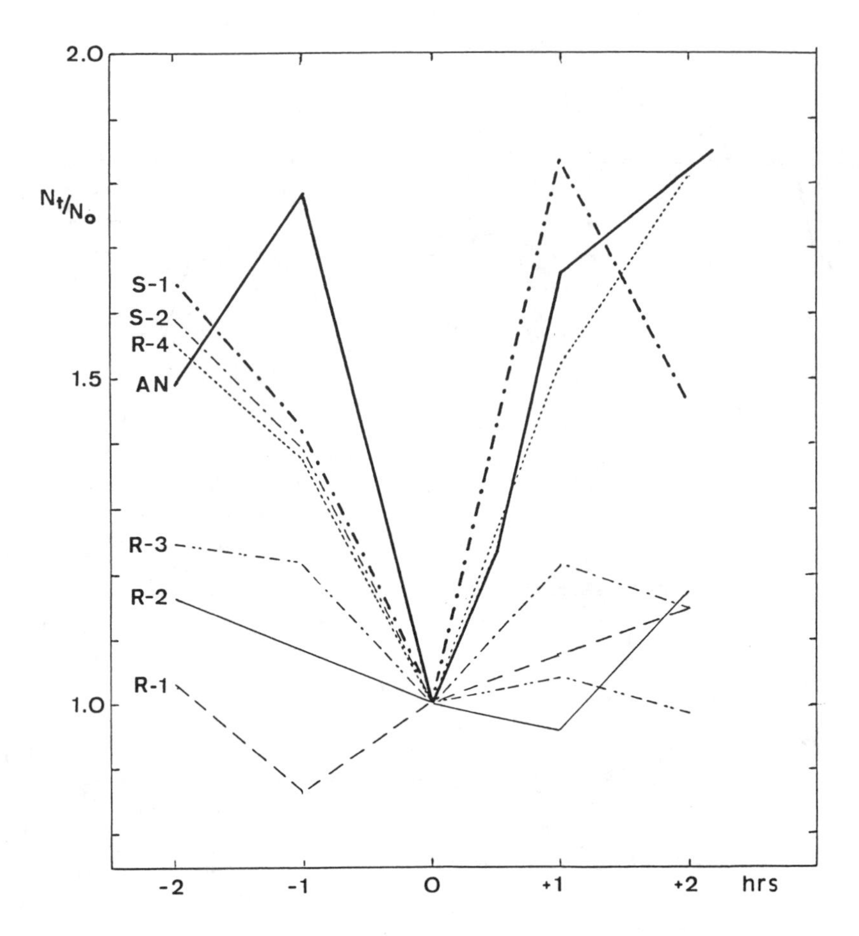


**FIGURE 18.** Particle concentration ratios (referred to the particle concentration at the time of the maximum rain intensity), $N_t/N_o$, before and after the occurrence of light rain. Evaluated were seven summer and two winter measurements at the Bay of Naples, 1987. The meaning of different aerosol size groups is AN ($r < 0.1$ μm); R–4 ($0.08 < r < 0.17$ μm); R–3 ($0.14 < r < 0.38$ μm); R–2 ($0.16 < r < 0.58$ μm); R–1 ($0.32 < r < 1.50$ μm); S–1 ($0.5 < r < 8.0$ μm); and S–2 ($1.0 < r < 16.0$ μm).

formation. We will abstract, for the time being, from the fact that many cloud elements might originate on mixed nuclei at subsaturation, with respect to water, and that we need to know their detailed composition and geometry for the description of their transformation into cloud droplets.

The main premise of the use of a thermal diffusion chamber for nuclei characterization is the definition of a precise supersaturation at which a specific nucleus is activated (Twomey, 1963). Then the concentration of nuclei activated at the supersaturation S (in % with respect to the plane water surface) is given, approximately, by the well-known relation

$$N_s = CS^k \tag{2}$$

where C and k are parameters depending upon the concentration and nature of the nuclei. Equation 1 can be related to the saturation ratio

$$S_r = \exp\left(\frac{A}{r}\right)\exp\left[-\frac{B}{\left(\frac{r^3}{r_0^3}\right) - 1}\right] \cong 1 + \frac{A}{r} - \frac{Br_0^3}{r^3} \tag{3}$$

It is deduced for dry nucleus radius ($r_0$), the assumed very diluted solution in the droplet of size, r, [i.e., $(r^3/r_0^3) - 1 \cong r^3/r_0^3$] and the approximate relation for surface tension of water (of density ρw), $\sigma_w$, and solution, $\sigma_s$, $\sigma_s \cong \sigma_w$. $A = 2\,\sigma_w/(\rho_w R_{vT}) \cong 4.33 \times 10^{-7}\,\sigma_w/T$ (Fitzgerald and Hoppel, 1984). The particle activation (hygroscopicity) parameter

$$B = \frac{\rho_0 \bar{\nu} \bar{\phi} \epsilon \gamma M_w}{\bar{M}_s \rho_w}$$

where $\rho_0$ is the density of dry particle of radius $r_0$, $\bar{\nu}$ is the number of moles of ions formed per mole of salt mixture, $\bar{\phi}$ is the practical osmotic coefficient of the solution, $\epsilon$ is the mass fraction of soluble material, $\gamma$ represents the adsorption effect at the solution-solid interface and $M_w$, $\bar{M}_s$ are the molecular weights of water and soluble material. Usually the assumption is made $B \cong B^\circ = \rho_0 \eta^\circ/\rho_w$ or $B \cong 0.8\ B^\circ$, where $B^\circ$ is the value of B for infinitely diluted solution. This corresponds to the condition that

$$\eta^\circ = \frac{\bar{\nu} \epsilon \gamma M_w}{\bar{M}_s}$$

(for $\bar{\phi} = 1$) and also that

$$a_w = \exp\left[ - \frac{B}{\left(\frac{r^3}{r_0^3}\right) - 1} \right] = 1.0 \qquad (4)$$

The above-mentioned simplifications enable us to determine from Equation 3 and from the relationship $\partial S_r/\partial r = 0$, the critical supersaturation ratio, $S_{rc}$, at which particles of known composition are activated and serve as embryos of the cloud droplets. Our main task is to determine from the measured aerosol parameters (e.g., size distribution, physicochemical properties) how many cloud droplets will originate if the maximal supersaturation, $S_{rc}$, is known. This relationship between $S_{rc}$ and the dry nucleus radius, $r_0$ can be expressed (see Fitzgerald and Hoppel, 1984) in the form

$$S_{rc} = \left( \frac{4A^3}{27B^\circ} \right)^{1/2} r_o^{3/2} \qquad (5)$$

Equation 5 is, to some authors, the basis by which to calculate the maximum concentration of activated cloud condensation nuclei, $N_{max}$, which give origin to cloud droplets. Twomey, in 1959, already suggested the relationship

$$N_{max} = CS^k_{rc\,max} = \frac{2}{k + 2} \left[ \frac{1.63 \times 10^{-3} \cdot V^{3/2}}{kBe\left(\frac{3}{2}, \frac{k}{2}\right)} \right]^{k/k+2} \qquad (6)$$

where $N_{max}$ is the maximum droplet concentration ($cm^{-3}$), V is the updraft velocity ($cm\ s^{-1}$) and Be is the complete beta function. Another formula was introduced by Hänel (1981) in the form

$$N_{max} = N_0\left(\frac{3r_{01}R_w\rho_wT}{2^{5/3}\sigma}\right)^{c_1} (\ln f_{max})^{\frac{2C_1}{3}} (B^\circ)^{\frac{C_1}{3}} \quad (7)$$

$N_0$ is the concentration of nuclei before fog or cloud formation, $r_{01}$ is the smallest particle radius in dry state, $c_1$ is the exponent in simple power law of particle size distribution (e.g., Junge's distribution), and $f_{max}$ is the maximum relative humidity found or expected in the cloud. Equations 2 to 7 enable us to explain some of the peculiarities and problems of the cloud condensation nuclei measurements which last for more than 25 years.

Most of the measurements performed before 1970 have been done at supersaturations higher than 0.10%, or the calculation of the potential cloud droplet formation was based on the knowledge of size distribution of nuclei of known chemical composition. However, the measurements of the nuclei size distribution, e.g., using the impaction technique, suffered several drawbacks, especially when the sampling was done over the continents. Severe minimum size limitation of sampling techniques, uncertainty of the impact of the atmospheric humidity on the particle and sampling substrate, and a very difficult description of the activation of a nucleus composed of an insoluble and soluble part were among the problems encountered over the continents, if one does not consider the laborious sample evaluation. Over the oceans, the situation was better due to the simpler nuclei composition and better defined air dynamics in the planetary boundary layer.

In 1975, Mészáros et al. published an interesting comparison of cloud condensation nuclei sampled over remote areas of the Atlantic and Indian Oceans and the measurements of cloud condensation nuclei by a thermal gradient diffusion chamber in Hawaii (Jiusto, 1966). The comparison of calculated condensation nuclei critical supersaturation spectra shows a similar trend especially close to the overlapping critical supersaturation domains. This conclusion also found some support in the cloud condensation nuclei measurements made by Twomey and Wojciechowski (1969) in higher levels over the Atlantic. The final conclusion by Mészáros et al. was that by considering only two main components of the maritime aerosol (chlorides and ammonium sulfates), the particles containing chlorides prevail only at supersaturations smaller than 0.03%, whereas at higher supersaturation (e.g., 0.3%), ammonium sulfate and other active nuclei might represent more than 80% of the total concentration (e.g., 25 $cm^{-3}$). Similar conclusions actually supporting the use of simple techniques combined with numerical models of nuclei activation were reached by other investigators as discussed in Pruppacher and Klett (1978).

An important improvement in the techniques for measuring the nuclei critical supersaturation spectrum brought the introduction of haze chambers (Laktionov, 1972), enabling us to measure supersaturations down to 0.02%. However, at the same time, great problems related to the interpretation of the measured nuclei critical supersaturation spectra, especially over the continents surfaced. Laktionov (1975) summarized many aircraft measurements made over the European territory of the U.S.S.R., as well as on board of a ship in tropical Atlantic, and came to the conclusion that the simple power relationship (Equation 2) ought to be replaced by two coefficients of k. One ($k_1$) should be used for $0.025 \leq S_c \leq 0.16\%$ and the other ($k_2$) for $0.16 \leq S_c \leq 1.00\%$. In general, the investigators found at altitudes above the atmospheric boundary layer and during air mass exchange (maritime-continental) large deviations from the anticipated slopes of the distribution curves. There is also considerable change in the slope of the mean curves for the surface layer measurement (curve 2, Figure 19) if compared to an altitude of 200 m. The investigations also demonstrated a large statistical error in the mean curve (according to the Equation 2) exponents: $k_1 = 2.3 \pm 0.85$; $k_2 = 0.96 \pm 0.54$. Earlier measurements in the U.S. (Twomey and Wojciechowski, 1969) are represented in Figure 19 by curve 5. The supersaturation curve slope values deduced by Laktionov differ strongly from the values measured by Murty et al. (1978) at

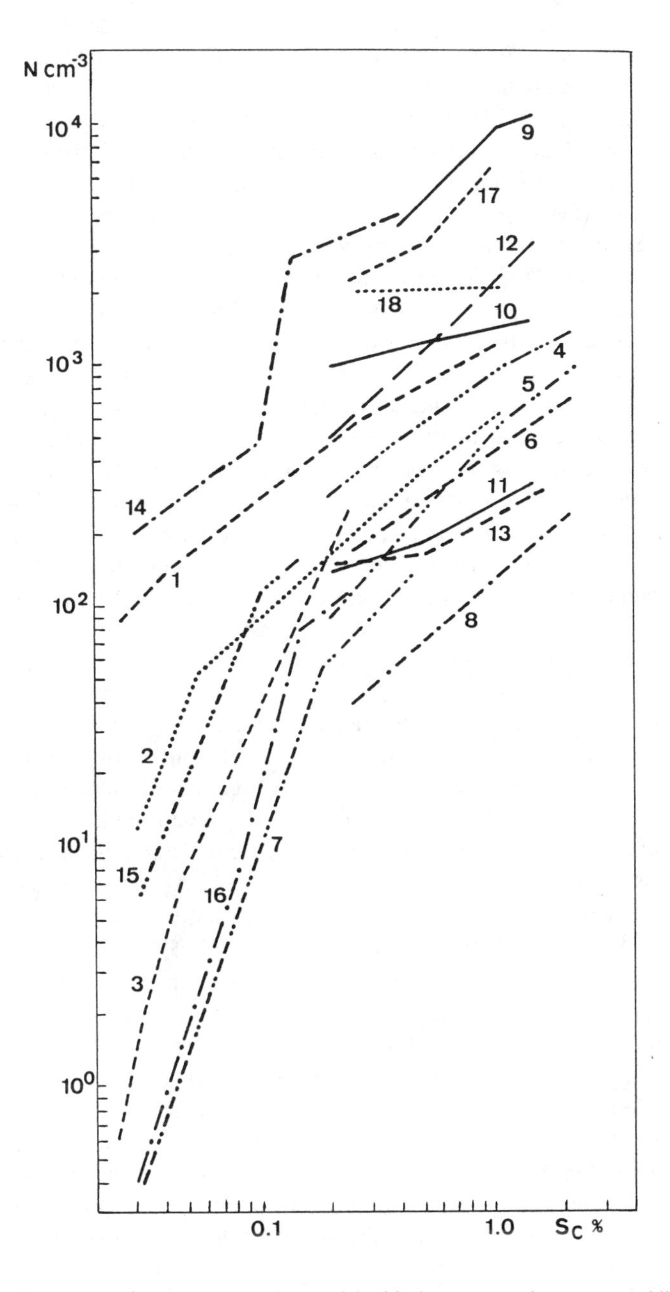


**FIGURE 19.** Measurements of cloud condensation nuclei critical supersaturation spectra published by Laktinov (1975): european territory of the U.S.S.R. (200 m above the ground — curve 1, in the atmospheric boundary layer — curve 2); over southern U.S.S.R. (curve 3); over Australia (curve 4); over Africa (curve 5); over the U.S. (curve 6); over the tropical Atlantic (curve 7); over the northern Arctic region (curve 8). The measurements by Hoppel et al. (1973): over Washington, D.C. (curve 9); 300 m over Arizona (curve 10); over the central Pacific (curve 11). The measurements by Hobbs et al. (1978): over the high plains in the continental air (curve 12) and in the maritime air (curve 13). The measurements by Hudson (1980): at San Diego (curve 14); at the Oregon coast (curve 15); at sea (curve 16). The measurements by Desalmand et al. (1985) on the Ivory Coast: at Abidjan (curve 17); at forested savannah, Lamto, (curve 18).

a supersaturation of 0.1 and 0.3% over India. Close to Bombay k = 0.16 ± 0.15 and $N_{max}$ = 487 ± 161 $cm^{-3}$, at Poona k = 0.37 ± 0.21 ($N_{max}$ = 271 ± 81 $cm^{-3}$), and at Rihand k = 0.62 ± 0.42 ($N_{max}$ = 398 ± 160 $cm^{-3}$) featuring a continental environment. There was considerable nuclei concentration decrease with the altitude above ground up to 1600 m. It is very interesting to compare these measurements to some data obtained over the U.S. and other countries.

A series of research flights have been performed by investigators from the Naval Research Laboratory, Washington, D.C. (Hoppel et al., 1973), at different altitudes (between 0.2 and 5.0 km) over Arizona, Central Pacific, Alaska, and Florida. In addition, five flights were made up to an altitude of 10.5 km off the east and west coast of the U.S. The critical supersaturations ranged from between 0.2 and 1.2%. Also plotted in Figure 19 are three curves representing the measurements at Washington, D.C. (curve 9); over Arizona at an altitude of 300 m above the ground (curve 10); and over the Central Pacific (curve 11). They have several features common to the previous measurements, e.g., the steeper slope for the measurements at low supersaturation over the ocean; however, they differ considerably in measured nuclei concentrations at a specific supersaturation (e.g., the data from Washington, D.C. are certainly influenced by the local sources of pollution). Other characteristics are the steep gradient of cloud condensation nuclei concentration between the surface layer and higher altitudes over the continent. Such a steep gradient was not observed over the ocean and snow-covered territories where the nuclei concentrations often increase with altitude above the planetary boundary layer.

The comparison of the older measurements of cloud condensation nuclei with the data sampled for several years over the U.S. High Plains is very interesting (Hobbs et al., 1978, 1985a, 1985b). Marked in Figure 19 were the mean curves of cloud condensation nuclei concentrations measured in continental air (curve 12) and in maritime air over the High Plains. These curves were obtained from 270 nuclei spectra measurements in the atmospheric mixing layer at supersaturations 0.2, 0.5, 1.0, and 1.5%. All measurements were divided into three groups characterized by nuclei mean concentration at supersaturation 1.0% ($N_0$ $cm^{-3}$), exponent k in the Equation 2 and percentage of a specific group out of the total number of measurements (F in %): maritime nuclei spectrum ($N_0$ = 290, k = 0.7, F = 25%); transitional nuclei spectrum ($N_0$ = 1500, k = 2.8, F = 35%); and aged continental nuclei spectrum ($N_0$ = 2200, k = 0.9, F = 40%). The authors came to several interesting conclusions, such as the relative insensitiveness of the specific type of nuclei spectrum related to the regional air pollutants and contaminants, or the importance of nuclei scavenging for the supersaturation spectra formation. The latter effect might be related to the complex dynamics and evolution of frontal systems affecting mainly the nuclei mode related to the natural or anthropogenic sources superimposed over the background nuclei sources. This is supported by the bimodal frequency distribution of nuclei concentration at different supersaturations.

At very low supersaturation ($S_c < 0.1$), the measured spectra show an even larger dispersion of points than at supersaturation are >1.0%. The nuclei, named by many authors as fog condensation nuclei (FCN), are featured at low supersaturation by very steep slopes of concentrations at increasing humidity in maritime air (see Hudson, 1980; curve 16 in Figure 19 or Laktinov, 1975, curves 3 and 7 in Figure 19). Analysis of these spectra can explain the frequently observed colloidal instability of clouds in marine atmosphere (formation of several large drops — which can eventually initiate the drizzle process — in the presence of low concentration of small activated nuclei).

Peculiar structures have cloud condensation nuclei in forested savannahs and tropical forests, which is a considerable part of the Earth's surface. Their concentration and a specific supersaturation strongly depends on the season and the interaction with other kinds of aerosols (e.g., urban or brush-fire pollution, marine aerosol, desert dust). In the Ivory Coast's forested savannah, close to the tropical forest, the measurements during the dry season revealed very low values of the exponent k in Equation 2 (in mean 0.26), which increased considerably in the suburb of Abidjan (0.67). At the same time, the concentration of nuclei activated at the supersaturation 0.25% was 2060 $cm^{-3}$, and at 0.50% about 2090 (Desalmand et al. 1985a). During these measurements in the forested savannahs, the concentration of Aitken nuclei was very low and close to the number of cloud condensation nuclei activated at a supersaturation of 0.25%. This situation changed considerably during the rainy season and with proximity to a large city, as shown on curves 17 and 18 in Figure 19 (Desalmand 1985a, 1985b; Podzimek and Desalmand, 1985). Similar measurements ought to be done above the savannah and the tropical forest at higher altitudes where several investigators found a high concentration of Aitken nuclei (Cros, 1977; Cros et al. 1981). Also, the measurement at supersaturation $S_c < 0.1$ could reveal the activation threshold and nature of these biogenic nuclei.

Several new investigations of cloud condensation nuclei reflect our interest in their potential impact on cloud formation in a large scale and on the radiative transfer. One of the most important questions is related to the ratio of active cloud condensation nuclei to the nonactive (hydrophobic) particles. Harrison and Harrison (1985) found that more than 50% of submicron particles belong to the hydrophobic fraction. Mazin (1982) called attention to the interaction of clouds and aerosols and, in general, to the transformation of the atmospheric particles during their passage through a cloud. Some of his conclusions were supported by Hudson (1984), who measured the cloud condensation nuclei spectra within and near the clouds. These investigations are related to the nuclei activation, or droplet transformation, in inhomogeneous clouds where relative humidity is, on some sites, lower than 100% (Pruppacher, 1986), and to the interesting nuclei transformation during several passages through the maritime clouds (Hoppel et al. 1986). Also, the old questions of the potential impact of organic surface-active substances on the condensation nuclei activation and droplet growth (e.g., see discussion in Podzimek and Saad, 1975, or Deryaguin et al., 1985) have been revived by the investigation of Saxena et al. (1985) of the aerosol composition over the Ross Sea in Antarctica and by the Bigg's (1986) investigations. These studies represent an important contribution to the still widely disputed problem of nuclei droplet growth in fluctuating flow or supersaturation fields (Merkulovich and Stepanov, 1977; Cooper, 1989; Podzimek and Hopkins, 1990).

Other subjects widely discussed today are the impact of anthropogenic sources of pollution and phytoplankton on the production of cloud condensation nuclei, with their potential impact on cloud formation on a large scale and on the radiative character of clouds. Coakley et al. (1987), Twomey et al. (1987), Scorer (1987), and others called attention to the "ship-trails" effect on the cloud transformation, and many other studies focused on the cloud condensation nuclei production associated with pollutant emission (Pueschel et al., 1986; Herrera and Castro, 1988; Heintzenberg et al., 1989; Ogren et al. 1989). The study of mixed nuclei and their transformation in haze, fog, and cloud droplets (see Podzimek and Ianniruberto, 1992) and the interaction of condensation nuclei with sulfur containing gases, $SO_2$ and DMS (Hoppel, 1987; Kreidenweis and Seinfeld, 1988; Shaw, 1989) seems to be very important. The problem of cloud condensation nuclei formation from gaseous precursors was addressed by Hegg et al. (1990) in the form of repeated cycles of condensation and evaporation of droplets containing nuclei.

The effect of phytoplankton on the condensation nuclei was already mentioned before (Andreae et al., 1983; Bates et al., 1987, Prospero et al., 1991). Recently, the importance of the production of the dimethylsulfide (DMS) with the subsequent cloud condensation nuclei formation and increased reflection of the solar radiation from the increased cloud cover as one of the mechanisms maintaining the stability of the biosphere (Charlson et al., 1987; Mészáros, 1988) was stressed. These ideas are supported by the finding that over the ocean the concentration of DMS is correlated to the sulfate concentration in atmospheric particles, and that there is a diurnal cycle in the small particle concentration which might be explained by the photochemical reaction leading to the particle formation (Parungo et al., 1987). To this hypothesis could be added the fact that Aitken nuclei measurements over the Intertropical Convergence Zone in the Pacific showed a markable nuclei increase during horizonal flights in the upper troposphere (Podzimek et al., 1977) which cannot be explained by the production of larger sea-salt particles or other particles. Another support to the DMS-derived production of condensation nuclei hypothesis comes from the results of the measurements of the concentration of Aitken nuclei and cloud condensation nuclei over the Pacific Ocean (Hegg et al., 1991). A positive correlation between the concentration of cloud droplets, cloud condensation nuclei, and nonsea-salt mass of sulfates was found.

## V. PARTICLES IN THE ATMOSPHERE AND TRANSFER OF THE RADIATION ENERGY

Recently, many articles have been published on the atmospheric aerosol and its impact on the transfer of the radiative energy in which particle shape and nature, size distribution, total concentration, and complex refractive index play an important role. These parameters applied to the Mie theory give the main integral optical characteristics of the polluted or contaminated atmosphere: volume extinction coefficient, $\sigma_e$, the albedo for single scattering, $\tilde{\omega}$ (which characterizes the relative importance of absorption in the aerosol extinction process); and the phase function, $p(\vartheta)$-probability of scattering in a direction at the angle $\vartheta$ from the incident beam. The latter parameter is usually substituted by an asymmetry factor, g, which depends on the mean properties of aerosol particles. For the application in the global aerosol impact on the radiation balance, the aerosol extinction optical depth, $\delta_e$, is introduced in the form

$$\delta_e = \int_0^{\infty} \sigma_e(z)dz \tag{8}$$

In a survey paper, Grassl (1988) analyzed the possible effect of solar (shortwave) and terrestrial (longwave) radiation on a cloud-free and cloudy atmosphere. He emphasized the effect of aerosol particles on the optical properties of cloud (fog) elements and, in general, to better know the optical properties of aerosols and their dependence on relative humidity.

Optical properties of some special types of aerosols, such as marine aerosol, Saharan aerosol, combustion- and fire-derived aerosol, and arctic haze elements, have been studied in more detail in the past. Many interesting results in the domains of these investigations are summarized in several books representing a survey on the potential role aerosols might play in climatic changes (see Gerber and Deepak, 1984). With reference to the subject of this treatise, the most interesting seems to be the study of the optical properties of aerosols, haze, fog, and cloud elements at different relative humidities.

The optical properties of the atmospheric aerosol are usually described by the complex index of refraction. The real part is considered to be in mean $n \cong 1.53$ (Eiden and Eschelbach, 1975) and its imaginary part represents the absorption coefficient varying in the range of 0.004 to 0.07 (with the mean value close to 0.03). The absorption (imaginary) part of the index of refraction is responsible for the planetary albedo in a cloudless sky (Grassl, 1978). Optical properties of marine aerosol were discussed by Gerber and Deepak (1984), those of Saharan aerosol were summarized by Fouquart et al. (1984), and those of carbon particles

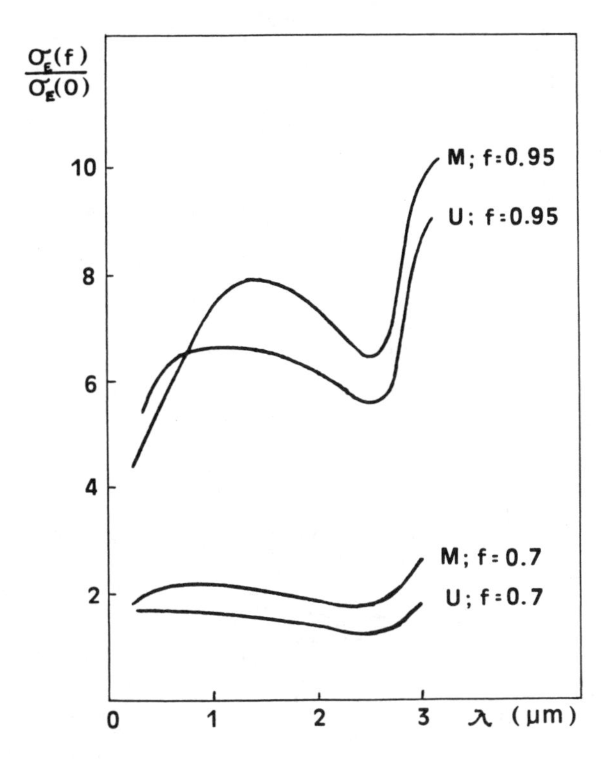


**FIGURE 20.** Ratio of extinction coefficient $\sigma_E(f)/\sigma_E(0)$ for a relative humidity 70% (f = 0.7) and 95% (f = 0.95) and for a typical marine (M) and urban (U) aerosol as a function of a wavelength, λ (in μm). Calculated according to the data published by Hänel (1984).

derived from combustion processes by Rosen and Novakov (1984). The major obstacle still seems to be the effect of large nonspherical particles on the backscatter angle.

Hänel (1984) summarized his fellow workers' investigations of the properties of aerosol particles as a function of relative humidity. These studies related directly to the fog and cloud element formation and are based on the fact that during the uptake of water at increasing relative humidity, the particle size increases; however, its complex refractive index successively approaches that of water. Another assumption was that at relative humidity (R.H.) below 100%, the equilibrium thermodynamics can be used, unlike at R.H. > 100% when nonequilibrium thermodynamics (with the knowledge of the history of the previous course of R.H.) have to be applied. In general, it was found that the R.H. effects on optical properties of aerosol particles are the largest for extinction and absorption coefficients in the IR domain of radiation, which, however, has the smallest effect on the asymmetry factor. In the shortwave region, the effect of humidity is markable for the ratio, r, of the hemispherical forward and backward scattering. For marine, urban, and background model aerosol, the values of the extinction, $\sigma_E$, and absorption, $\sigma_A$, coefficients are expressed as a power function of the water vapor pressure ratio (or relative humidity) for two or three major domains of water vapor pressure ratios (Figures 20 and 21). The same typical aerosol groups were used for the determination of the hemispherical forward and backward scattering ratios, r, and for the asymmetry factor, g, at different water vapor pressure ratios and $\lambda = 0.7$ μm. More detailed tables and analyses of the aerosol effects were later published by Shettle and Fenn (1979), Nilsson (1979), Gathman (1984), and Charlson et al. (1984).

Other investigations focused on the radiative transfer of energy through clouds composed of cloud droplets and aerosol particles, or embedded in droplets or floating between them

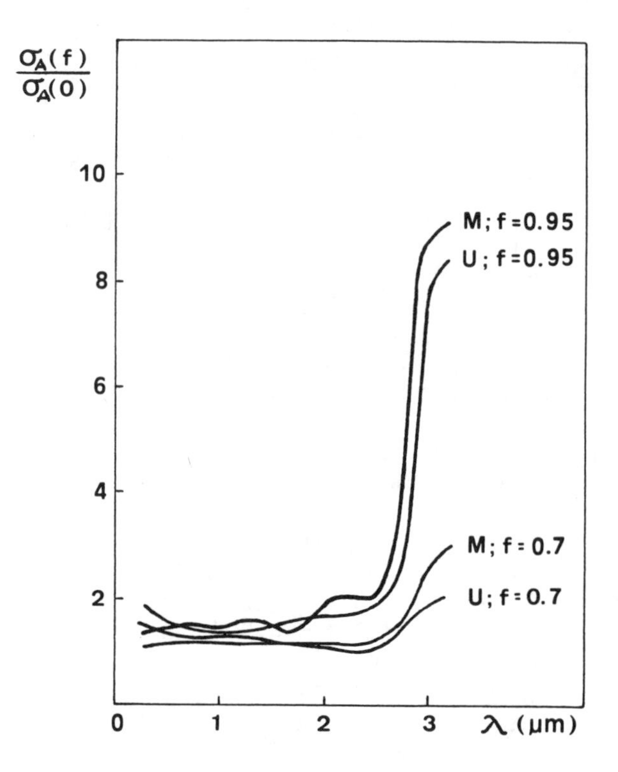


**FIGURE 21.** Ratio of absorption coefficient $\sigma_A(f)/\sigma_A(0)$ for a relative humidity of 70% (f = 0.7) and 95% (f = 0.95) and for a typical marine (M) and urban (U) aerosol as a function of a wavelength, λ (in μm). Calculated according to the data published by Hänel (1984).

as interstitial particles. Kondratyev (1978), Kondratyev et al. (1981), and Feigelson (1984) summarized the main findings made in the U.S.S.R. and other countries. Many studies of optical properties of clouds and interaction with aerosol particles justified the statement of Goisa (in Feigelson, 1984, p. 172) that "It is an undoubted fact that the strong absorption of radiation in clouds exceeds the absorption by pure water or ice." The results of the theoretical studies of the effect of aerosol pollution of clouds on albedo (Twomey, 1977; Ackerman and Baker, 1977) found considerable support by the results of the aircraft measurements of the radiation parameters over the highly industrialized regions of the U.S.S.R. (Zaporozhe and Donetsk). The data clearly shows that the absorption and radiation of clouds contaminated by urban aerosols is approximately twice as high as outside the city's region. The imaginary part of the index of refraction over Zaporozhe was 0.03 for a wavelength of 0.5 μm (Kosarev, 1976). The effects of ship-stack effluents on the radiative properties of marine stratocumulus clouds were already mentioned (Coakley et al., 1987).

For many years, special attention has been paid to the effect of aerosols on haze and fog formation in lower geographical latitudes and also in polar regions. Fogs and haze cover a considerable part of the Earth's surface. According to the climatological data, the number of days with fog occurrence in a year at several meteorological stations in the U.S. might reach between 60 and 80 days; in Rio de Janeiro 164 days; in Bogota 128 days; and in Vienna 43 days (see Mazin and Khrgian, 1989). Yakutsk, in Siberia, has fogs lasting for 499 h in a year. There are regions with long-lasting fogs over the Labrador Sea, the West-African coast (Casablanca), in South America (south of Rio de la Plata), in the Arctics, etc. The occurrence of haze in the U.S. and the physical properties of its elements have been analyzed by Husar and Holloway (1984).

The physical and optical properties of fog and haze elements are largely affected by atmospheric aerosols and gaseous pollutants. However, one can expect stronger deviations from the formula expressing the dependence of the extinction coefficient, $\sigma$, on the droplet concentration, N, and cloud liquid water content, W, (Coakley et al. 1988)

$$\sigma = KN^{1/3}W \tag{9}$$

where K is a constant parameter.

Several measurements of particle hygroscopicity and optical properties (e.g., light scattering extinction) have been done in rural and urban areas as well. Waggoner et al. (1984) found during the measurements on the ground that the aerosol is more acid during daylight hours than at night, and that the always hydrated aerosol in rural Virginia contained the liquid water which was in equilibrium with the ambient humidity. In Houston, however, one third of the sampling time of particles was featured by an excess of liquid water above that expected from the equilibrium state. The light extinction at different relative humidities was also, in mean, higher at Houston as compared to Virginia. This important effect of interacting pollutants on the nuclei activation, dependent on time of day, can be combined with the hygroscopic nuclei activity increase due to the deposition of submicron combustion-derived aerosol (Podzimek et al., 1991). This leads, finally, to the model of a radiative transfer through haze or fog elements composed of soluble and insoluble substance.

Several articles dealt with the determination of the optical parameters such as the volume extinction coefficient, the single scattering albedo, and the asymmetry factor of spherical particles coated with spherical shell (Kerker, 1969; Mita, 1982). These interesting results have especially been applied to solving the optical properties of Arctic haze, which is supposed to contain considerable amounts of anthropogenic compounds like soot, sulfuric acid, and fine ash particles. Blanchet and List (1984) applied the model of mixed haze elements to the prediction of optical properties of Arctic haze. For this study, they used the results of the many measurements of the composition and optical parameters of the haze elements in polar regions by Heintzenberg (1980, 1981, 1982) and others. In conclusion, they found a substantial effect of the pollutants embedded in haze elements on their properties; e.g., the soot concentration of about 0.4 $\mu g\ m^{-3}$ and soot particle size distribution similar to that of Heintzenberg's could increase the absorption coefficient by a factor of between three and ten for the wavelengths in the mid-visible range. These studies ought to be extended into other geographical latitudes where, in a highly polluted marine-urban environment, the prevailing mixed nature of haze and fog elements were found (Podzimek, 1990).

## VI. CONCLUSIONS

This contribution to the investigations of the role atmospheric particles play in cloud, fog, and haze formation and in the related transfer of radiative energy is not a survey article. Rather, it is a collection of personal observations and confrontations of ideas and measuring techniques which evolved during the past 40 years. It is also apparent that the author was more concerned in the past about the immediate impact of atmospheric aerosols on the cloud- and fog-forming process and on our everyday environment than about its climatic impact in a global scale.

The discussion concentrates on the formulation of the theoretical basis of condensation nuclei and cloud condensation nuclei activation and growth into cloud elements. At present, this seems to be adequate for explaining the results of ground and airborne measurements. The article also underscores how difficult it is to interpret many of the ground measurements affected by the interaction of gaseous and particulate substances and subjected to source emission and meteorological factor variability. However, several examples show that the particulate and gaseous substances in marine atmospheres enable a reasonable and reproducible interpretation and calculation of particle effects in a large scale. In the author's opinion, the aerosol transport, activation, and impact on cloud-forming processes in coastal regions and over well-selected islands has not yet been fully investigated. Another more complicated subject seems to be the complex and large-scale study of the cloud condensation nuclei activation over tropical forests which could complete the ongoing research in Brazil and several African countries (see Artaxo et al., 1988).

Cloud, fog, and haze formation over cold regions, often covered by snow, with its impact on radiative transfer, still remains in the forefront of our interest and offers a greater success in coupling the microstructure of cloud, haze, and fog formation with radiative energy transfer and dynamics of the atmosphere. This treatise does not contain a discussion of the impact of ice nuclei on the water-phase transition and ice-crystal formation in the atmosphere. Arctic haze, and cirrus cloud formation, with its impact on the radiation scattering and absorption and on the precipitation process, still represents one of the major spheres of interest for a cloud physicist. In spite of the continuing search for the most suitable technique for counting ice nuclei of natural, anthropogenic, and biogenic origin, there have been many interesting articles published on these subjects since the time when Bergeron-Findeisen precipitation hypothesis started to dominate the minds of cloud physicists after World War II. A survey on these subjects, which might be left for a later discussion, was published by Vali (1985).

Today, airborne aerosol measurements, satellites, and remote sensing remain the major source of data on aerosols in the atmosphere in regional and global scale. However, certain limits are still imposed on the measurements made by satellites or from the ground (e.g., the integrated measurements of the optical thickness at present cannot be sufficiently differentiated for obtaining the detailed vertical distribution of optical depth). For calibrating purposes and measurements of condensation nuclei and their properties, the aircraft and balloon measurements will remain the main source of information. This article also tried to stress some of the more useful results of aerosol and condensation nuclei investigations in the past because the author is aware of the expensive airborne and remote sensing measurements and the necessity to obtain useful information on atmospheric aerosol from remote areas of the Earth. There the measurements on well-selected stations (e.g., ships, island, or mountain observatories) might be extremely worthwhile.

## ACKNOWLEDGMENTS

The author is indebted to Mrs. Vicki Hudgins for her assistance while preparing this article, and to all fellow workers who cooperated, collected, and evaluated the data during many research projects in the past. Thanks to Dr. R. A. Geyer and the personnel of Geophysics Associates, Inc., Bryan, Texas, for their interest in this article.

## REFERENCES

Agee, E. M., Brown, D. E., Chen, T. S., and Dowell, K. E., *J. Appl. Meteorol.*, 12, 409–412, 1973.
Aitken, J., *Proc. R. Soc. Edinburgh, Sect. B*, XI, 14–18, 1880.
d'Almeida, G. A., *J. Geophys. Res.*, 92, 3017–3026, 1987.
Ackerman, T. P. and Baker, M. B., *J. Appl. Meteorol.*, 16, 63–69, 1977.
Andreae, M. O., Barnard, W. R., and Ammons, J. M., *Ecol. Bull.*, 35, 167–177, 1983.

Artaxo, P., Storms, H., Bruynseels, F., and Van Grieken, R., *J. Geophys. Res.*, 93, 1605–1615, 1988.
Bates, T. S., Cline, J. D., Gammon, R. H., and Kelly-Hansen, S. R., *J. Geophys. Res.*, 92, 13,245–13,262, 1987.
Beard, K. V. *Cloud and Precipitation Physics Research 1983–1986,* U.S. Natl. Rep. to IUGG, *Rev. Geophys.*, 25, 357–370, 1987.
Bigg, E. K. *Atmos. Res.*, 20, 82–86, 1986.
Blanchard, D. C. *J. Rech. Atmos.*, 4, 1–6, 1969.
Blanchet, J.-P. and List, R., *Aerosols and their Climatic Effects,* Gerber, H. E. and Deepak, A., Eds., A. Deepak Publ., Hampton, VA, 1984, 179–196.
Burckhardt, H. and Flohn, H., *Die Atmosphärischen Kondensationskerne in ihrer Physikalischen and Bioklimatologischen Deutung*, Springer-Verlag, Berlin, 1939.
Charlson, R. J., Covert, D. S., and Larson, T. V., *Hygroscopic Aerosols*, Ruhnke, L. H. and Deepak, A., Eds., A. Deepak Publ., Hampton, 1984, 35–44.
Charlson, R. J., Lovelock, J. E., Andreae, M. O., and Warren, S. G., *Nature (London),* 326, 655–661, 1987.
Chylek, P., Srivastava, V., Pinnick, R. G., and Wang, R. T., *Appl. Opt.*, 27, 2396–2404, 1988.
Coakley, J. A., Jr., Bernstein, R. L., and Durkee, P. A., *Aerosols and Climate,* Hobbs, P. V. and McCormick, M. P., Eds., A. Deepak Publ., Hampton, 1988, 253–260.
Coakley, J. A., Jr., Bernstein, R. L., and Durkee, P. A., *Science,* 237, 1020–1022, 1987.
Cooper, W. A., *J. Atmos. Sci.*, 46, 1301–1311, 1989.
Cooper, W. A., *Research in Cloud and Precipitation Physics: Review of U.S. Theoretical and Observational Studies, 1987–1990,* U.S. Natl. Rep. to IUGG, *Rev. Geophys.*, 29, 69–79, 1991.
Cros, B., Contribution à l'Etude des Noyaux d'Aitken en Afrique Equatoriale, thèse de Doctorat des Sciences, Universite de Toulouse, 1977, No. 789.
Cros, B., Lopez, A., and Fontan, J., *Atmos. Environ.*, 15, 83–90, 1981.
Deryaguin, B. V., Kurghin, Y. S., Bakanov, S. P., and Merzhanov, K. M., *Langmuir,* 1, 278–281, 1985.
Desalmand, F., Baudet, J., and Serpolay, R., *J. Atmos. Sci.*, 38, 2076–2082, 1982.
Desalmand, F., Podzimek, J., and Serpolay, R., *J. Aerosol Sci.*, 16, 19–28, 1985a.
Desalmand, F., Serpolay, R., and Podzimek, J., *Atmos. Environ.*, 9, 1535–1543, 1985b.
Dlugi, R. J. and Jordan, S., *Idojaras,* 86, 82–88, 1982.
Eiden, R. and Eschelbach, G., *Z. Geophys.*, 39, 189–228, 1975.
Feigelson, E. M., *Radiation in a Cloudy Atmosphere*, D. Reidel Publ., Dordrecht, 1984, 293.
Fitzgerald, J. W. and Hoppel, W. A., *Hygroscopic Aerosols,* Ruhnke, L. H. and Deepak, A., Eds., A. Deepak Publ., Hampton, 1984, 21–34.
Flossmann, A. I., Hall, W. D., Pruppacher, H. R., *J. Atmos. Sci.*, 42, 582–606, 1985.
Fouquart, Y., Bonnel, B., Brogniez, G., Cerf, A., Chaoui, M., Smith, L., and Vanhoutte, *Aerosols and their Climatic Effects,* Gerber, H. E. and Deepak, A., Eds., A. Deepak Publ., Hampton, VA, 1984, 35–62.
Gathman, S. G., *Hygroscopic Aerosols*, Ruhnke, L. H. and Deepak, A., Eds., A. Deepak Publ., Hampton, 1984, 93–114.
Georgii, H. W., *J. Geophys. Res.*, 75, 2365–2371, 1970.
Gerber, H. E. and Deepak, A., *Aerosols and their Climatic Effects*, A. Deepak Publ., Hampton, 1984.
Gerber, H. E. and Hindman, E. E., *Light Absorption by Aerosol Particles,* Spectrum Press, New York, 1982.
Grabovskii, R. I. *Meteorol. Gidrol.*, No. 4, 1951.
Grassl, H. *Man's Impact on Climate,* Bach, W., Pankrath, J., and Kellog, W., Eds., Elsevier, Amsterdam, 1978, 229–241.
Grassl, H. *Aerosols and Climate,* Hobbs, P. V. and McCormick, M. P., Eds., A. Deepak Publ., Hampton, 1988, 241–252.
Hänel, *Hygroscopic Aerosols,* Ruhnke, L. H. and Deepak, A., Eds., A. Deepak Publ., Hampton, 1984, 1–20.
Harrison, L. and Harrison, H., *J. Clim. Appl. Meteorol.*, 24, 302–310, 1985.
Hegg, D. A. and Hobbs, P. V., *Atmos. Environ.*, 15, 1597–1604, 1982.
Hegg, D. A., Radke, L. F., and Hobbs, P. V., *J. Geophys. Res.*, 95, 13,917–13,926, 1990.
Hegg, D. A., Radke, L. F., and Hobbs, P. V., *J. Geophys. Res.*, 96, 18,727–18,733, 1991.
Heintzenberg, J., *Tellus*, 32, 251–260, 1980.
Heintzenberg, J., *Tellus*, 33, 162–171, 1981.
Heintzenberg, J., *Atmos. Environ.*, 16, 2401–2469, 1982.
Heintzenberg, J., Ogren, J. A., Noone, K. J., and Gardneus, L., *Atmos. Res.*, 24, 89–102, 1989.
Herrera, J. R. and Castro, J. J., *J. Clim. Appl. Meteorol.*, 27, 1189–1192, 1988.
Hobbs, P. V., Politovich, M. K., Bowdle, D. A., and Radke, L. F., Airborne Studies of Atmospheric Aerosol in the High Plains and the Structures of Natural and Artificially Seeded Clouds in Eastern Montana, Contrib. Cloud Physics Group, Res. Rep. XIII, University of Washington, Seattle, 1978, 417 pp.
Hobbs, P. V., Bowdle, D. A., and Radke, L. F., *J. Clim. Appl. Meteorol.*, 24, 1344–1356, 1985a.
Hobbs, P. V., Bowdle, D. A., and Radke, L. F., *J. Clim. Appl. Meteorol.*, 24, 1358–1369, 1985b.

Hobbs, P. V. and McCormick, M. P., *Aerosols and Climate*, A. Deepak Publ., Hampton, 1988.
Hobbs, P. V., Tuell, J. P., Hegg, D. A., Radke, L. F., and Eltgroth, M. W., *J. Geophys. Res.*, 87, 11,062–11,086, 1982.
Hofmann, D. J. and Rosen, J. M., *J. Geophys. Res.*, 87, 11,039–11,061, 1982.
Hoppel, W. A., *Atmos. Environ.*, 21, 2703–2709, 1987.
Hoppel, W. A., Dinger, J. E., and Ruskin, R. E., *J. Atmos. Sci.*, 30, 1410–1420, 1973.
Hoppel, W. A., Frick, G. M., and Larson, R. E., *Geophys. Res. Lett.*, 13, 125–128, 1986.
Hudson, J. G., *J. Atmos. Sci.*, 37, 1854–1867, 1980.
Hudson, J. G., *Hygroscopic Aerosols*, Ruhnke, L. H. and Deepak, A., Eds., A. Deepak Publ., Hampton, 1984, 359–372.
Hudson, J. G., *Atmos. Environ.*, 25A, 2449–2455, 1991.
Husar, R. B. and Holloway, J. M., *Hygroscopic Aerosols*, Ruhnke, L. H. and Deepak, A., Eds., A. Deepak Publ., Hampton, VA, 1984, 129–170.
Jaenicke, R. *Saharan Dust: Mobilization, Transport, Deposition*, Morales, C., Ed. SCOPE, Rep. 14, John Wiley & Sons, New York, 1979, 233–242.
Jaenicke, R., *Aerosols and their Climatic Effects*, Gerber, H. E. and Deepak, A., Eds., A. Deepak Publ., Hampton, 1984, 7–34.
Jiusto, J. E., *J. Rech. Atmos.*, 2, 245–250, 1966.
Junge, C., *Air Chemistry and Radioactivity*, Academic Press, New York, 1963.
Kerker, M., *The Scattering of Light and Other Electromagnetic Radiation*, Academic Press, New York, 1969.
Knott, C. G., *Collected Scientific Papers of John Aitken, LL.D., F.R.S.*, Cambridge University Press, London, 1923.
Kondratyev, K. Ya., *The Atmospheric Aerosol and its Effect on Radiation Transfer*, Gidrometeoizdat, Leningrad, 1978.
Kondratyev, K., Ya., Binenko, V. I., and Petrenchuk, O. P., *Izv. Acad. Sci. USSR, Atmos. Oceanic Phys.*, 17, 122–127, 1981.
Kosarev, A. L., et al., Trudy CAO, No. 124, 1976.
Kreidenweis, S. M. and Seinfeld, J. H., *Atmos. Environ.*, 22, 283–296, 1988.
Kumai, M., *J. Meteorol.*, 8, 151–156, 1951.
Kuroiwa, D., *J. Meteorol.*, 8, 157, 1951.
Laktionov, A. G., *Izv. Akad. Nauk SSSR, Fiz. Atmos. Okeana*, 8, 672–677, 1972.
Laktionov, A. G., *Proc. 8th Int. Conf. Nucleation*, Gidrometeoizdat, Moscow, 1975, 437–444.
Landsberg, H., *Gerlands Beitr. Geophys., Suppl. B, (3); Ergeb. Kosm. Phys.*, 155–252, 1938.
Leaitch, W. R., Strapp, J. W., Isaac, G. A., and Hudson, J. G., *Tellus*, 38B, 328–344, 1986.
Legrand, M., *Etude des Aerosols Sahariens au-dessus de l'Afrique à l'aide du Canal à 10 Microns de Meteosat: Visualisation, Interpretation et Modelisation*, thèse de Doctorat d'Etat, Universite des Science et Tech. de Lille, 1990.
Lopez, A., Servant, J., and Fontan, J., *Atmos. Environ.*, 7, 945–965, 1973.
Mason, B. J., *The Physics of Clouds*, Oxford at Clarendon Press, London, 1957.
Mazin, I. P., *Meteorol. Gidrol.* No. 1, 54–61, 1982.
Mazin, I. P. and Khrgian, A. Kh., *Oblaka i Oblachnaia Atmosfera*, Gidrometeoizdat, Leningrad, 1989, 647.
Merkulovich, V. M. and Stepanov, A. S., *Izv. Akad. Nauk SSSR, Fiz. Atmos. Okeana*, 13, 163–171, 1977.
Mészáros, E., *Atmos. Environ.*, 22, 423–424, 1988.
Mészáros, E., Mészáros, A., and Vissy, K., *Proc. 8th Int. Conf. Nucleation*, Gidrometeoizdat, Moscow, 1975, 431–436.
Mita, A., *J. Meteorol. Soc. Japan*, 60, 765–776, 1983.
Monahan, E. C. and MacNiocaill, G., *Oceanic Whitecaps and their Role in Air-Sea Exchange Processes*, D. Reidel Publ., Dordrecht, 1986.
Murty, R. A. S., Kumar, R. V., Selvam, A. M., and Murty, Bh. V. R., *Some Problems of Cloud Physics (Voprosy Fiziki Oblakov)*, Gidrometeoizdat, Leningrad, 1978, 170–179.
Nilsson, B., *Appl. Opt.*, 18, 3457–3473, 1979.
Ogren, J. A., Heintzenberg, J., Zuber, A., Noone, K. J., and Charlson, R. J., *Tellus*, 41, 24–31, 1989.
Okada, K., Tanaka, T., Naruse, H., and Yoshikawa, T., *Tellus*, 42, 463–480, 1990.
Parungo, F. P., Nagamoto, C. T., Rosinski, J., and Haagenson, P. L., *J. Aerosol Sci.*, 18, 277–290, 1987.
Petrenchuk, O. P., *Eksperimentalnye Issledovania Atmosfernogo Aerozolia*, Gidrometeoizdat, Leningrad, 1979.
Podzimek, J., *Stud. Geophys. Geod.*, 3, 393–402, 1959.
Podzimek, J., *Stud. Geophys. Geod.*, 11, 470–476, 1967.
Podzimek, J., *Ann. IUNN*, 38, 3–14, 1969.
Podzimek, J., *J. Rech. Atmos.*, 7, 137–152, 1973.
Podzimek, J., *PAGEOPH*, 114, 925–932, 1976.
Podzimek, J., *J. Rech. Atmos.*, 14, 35–61, 1980a.

Podzimek, J., *J. Rech. Atmos.*, 14, 241–253, 1980b.
Podzimek, J., *PAGEOPH,* 121, 611–632, 1984a.
Podzimek, J., *Tellus,* 36B, 192–202, 1984b.
Podzimek, J., *Aerosols and Climate,* Hobbs, P. V. and McCormick, M. P., Eds., A. Deepak Publ., Hampton, 1988, 153–163.
Podzimek, J., *Bull. Am. Meteorol. Soc.*, 70, 1538–1545, 1989.
Podzimek, J., *J. Aerosol Sci.*, 21, 299–308, 1990.
Podzimek, J. and Cernoch, I., *Geofis. Pura Appl.*, 50, 96–101, 1961.
Podzimek, J. and Desalmand, F., *J. Rech. Atmos.*, 19, 203–211, 1985.
Podzimek, J. and DeMaio, A., *Ann. IUNN,* in print, 1992.
Podzimek, J. and Hopkins, R., *Ann. IUNN,* 59, 145–161, 1990.
Podzimek, J. and Ianniruberto, M., *Ann. IUNN,* in print, 1992.
Podzimek, J. and Saad, A. N., *J. Geophys. Res.*, 80, 3386–3392, 1975.
Podzimek, J., Sedlacek, W. A., and Haberl, J. B., *Tellus,* 29, 116–127, 1977.
Podzimek, J., Trueblood, M. B., and Hagen, D. E., *Atmos. Environ.*, 25A, 2587–2591, 1991.
Prospero, J. M., Savoie, D. L., Saltzman, E. S., and Larsen, R., *Nature (London)*, 350, 221–223, 1991.
Pruppacher, H. R., *NATO ASI Series, Vol. G6, Chemistry of Multiphase Atmospheric Systems,* Jaeschke, W., Ed., Springer-Verlag, Berlin, 1986, 134–190.
Pruppacher, H. R. and Klett, J. D., *Microphysics of Clouds and Precipitation,* D. Reidel Publ., Dordrecht, 1978.
Pueschel, R. F., Van Valin, C. C., Castillo, R. C., Kadlecek, J. A., and Ganor, E., *J. Clim. Appl. Meteorol.*, 25, 1908–1917, 1986.
Pueschel, R. F., Snetsinger, K. G., Goodman, J., Dye, J. E., Baumgardner, D., and Gandrud, B. W., *J. Geophys. Res. Atmos.*, 97, 8105–8114, 1992.
Rosen, H. and Novakov, T., *Aerosols and their Climatic Effects,* Gerber, H. E. and Deepak, A., Eds., A. Deepak Publ., Hampton, VA, 1984, 83–94.
Ruhnke, L. H. and Deepak, A., *Hygroscopic Aerosols,* A. Deepak Publ., Hampton, 1984.
Runca-Koeberich, D. R., Ein Beitrag zur Konstitution atmosphaerischer eisbildender Kerne, *Ber. Inst. Meteorol. Geophys.*, No. 37, Univ. Frankfurt, 1979.
Saxena, V. K., Curtin, T. B., and Parungo, F. P., *J. Rech. Atmos.*, 19, 213–224, 1985.
Schmauss, A. and Wigand, A., *Die Atmosphäre als Kolloid,* Verlag F. Vieweg & Sohn, Braunschweig, 1929.
Scorer, R. S., *Atmos. Environ.*, 21, 1417–1425, 1987.
Shaw, G. E., *Atmos. Environ.*, 23, 2841–2846, 1989.
Shettle, E. P. and Fenn, R. W., *Models for the Aerosols of the Lower Atmosphere and the Effects of Humidity Variations on their Optical Properties, Environ. Res.*, No. 676, Optical Physics Division, Project 7670, Air Force Geophys. Lab., Hanscom AFB, MA, 1979.
Toba, Y., *Tellus,* 18, 132–145, 1966.
Twomey, S., *J. Rech. Atmos.*, 1, 113–119, 1963.
Twomey, S., *J. Atmos. Sci.*, 34, 1149–1152, 1977.
Twomey, S., *Atmospheric Aerosols,* Elsevier, Amsterdam, 1977.
Twomey, S., Gall, R., and Leuthold, M., *Boundary Layer Meteorol.*, 41, 335–348, 1987.
Twomey, S. and Squires, P., *Tellus,* 11, 408–411, 1959.
Twomey, S. and Wojciechowski, T. A., *J. Atmos. Sci.*, 26, 684–688, 1969.
Vali, G., *J. Rech. Atmos.*, 19, 105–115, 1985.
Vali, G., *Report on the Exp. Meet. on Interaction Between Aerosols and Clouds,* February 5–7, 1991, WCRP-59, WMO/TD-No. 423, Hampton, VA, 1991.
Waggoner, A. P., Larson, T. V., and Weiss, R. E., *Hygroscopic Aerosols,* Ruhnke, L. H. and Deepak, A., Eds., A. Deepak Publ., Hampton, 1984, 181–198.
Wegener, A., *Meteorol. Z.*, 27, 354–361, 1910a.
Wegener, A., *Meteorol. Z.*, 27, 451–459, 1910b.
Went, F. W., *Tellus,* 18, 549–556, 1966.
Wigand, A., *Ann. Physik,* 59, 689–742, 1919.
Wigand, A. and Lutze, G., *Beitr. Physik fr. Atmos.*, 6, 173–186, 1914.

Chapter 3

# WHAT DO DATA ANALYSES TELL US ABOUT THE VARIABILITY OF GLOBAL TEMPERATURE?

**A. A. Tsonis and J. B. Elsner**

## TABLE OF CONTENTS



0-8493-4419-0/93/$0.00 + $.50
© 1993 by CRC Press, Inc.

# I. INTRODUCTION

The issue of global warming and its possible connection to the increased concentration of greenhouse gases in the atmosphere has captured the attention of almost everybody in the last decade. The issue arose from recently constructed global annual mean surface air temperature records which show that the temperature of the planet has been on the rise in the last 120 years or so. Since then, a significant amount of research has been performed on these and other relevant data sets, and various results have been reported. Table 1 gives a list of temperature data sets usually employed in studies dealing with the issue of global warming. The data are divided into sets that include only land measurements, only marine sea temperatures, and those that include both. The data sets have been constructed independently, with each group using different averaging techniques, different correction procedures, and in many cases different observational data bases.

The U.K. group (Jones et al., 1986a,b and Jones, 1988) used land surface air temperature and observations aboard fixed-point ships. The U.S. group (Hansen and Lebedeff, 1987, 1988) used only land stations. The U.S.S.R. group (Vinnikov et al., 1987; Gruza et al., 1988) used only Northern Hemispheric land measurements.

Sea surface temperatures (SSTs) and marine air temperatures (MATs), available from observations taken by commercial ships and coastal regions, have been assembled by various groups in the Comprehensive Ocean-Atmosphere Data Set (COADS) (Woodruff et al., 1987), in the U.K. data set (Folland et al., 1984), in the Consolidated Data Set (CDS) (Hsiung and Newell, 1983), and in the Global Ocean Surface Temperature Atlas (GOSTA) (Bottomley et al., 1989). The COADS data set includes 80 million observations and was assembled in the U.S. The U.K. data set includes about 60 million observations assembled by the U.K. Meteorological Office Main Marine Data Bank. Most, but not all, of these observations are contained in the COADS data set. CDS contains data derived at MIT from the U.S. Navy Fleet Numerical Center records. GOSTA includes the U.K. data and the MIT data with several corrections.

Records that combined land air and sea surface temperatures have been compiled by Jones et al. (1986c) and by the International Panel on Climate Change (IPCC). The Jones et al. (1986c) record is based on the Jones land measurements and on average SSTs derived using both the COADS and the U.K. data sets. Figure 1a–e shows some of the above data sets. All records are time series of global annual mean values. All of them exhibit a positive trend. In this paper, we will not deal with the methods and/or corrections applied to the observations in order to produce the aforementioned records. An excellent discussion on this topic is given in Folland et al. (1990). The purpose of this work is to review and assess the significance of results reported from the analysis of those records. In particular, we will discuss the significance of the observed trends and their connection to the increased amounts of greenhouse gases in the atmosphere, the statistical similarity of the various records, and the significance of certain reported periodicities.

# II. SIGNIFICANCE OF THE POSITIVE TRENDS

As was mentioned above, according to the available data, global mean temperature has increased over the past 100 to 120 years. The magnitude of the warming varies with records ranging between 0.3 to 0.6°C/100 years. The fact that every record shows a positive trend makes the existence of a trend unquestionable. Several studies have approached the statistical significance of the observed trends in different ways. These studies can be divided into studies that address the problem directly (using the observed data only) and studies that deal with the problem indirectly (using models).

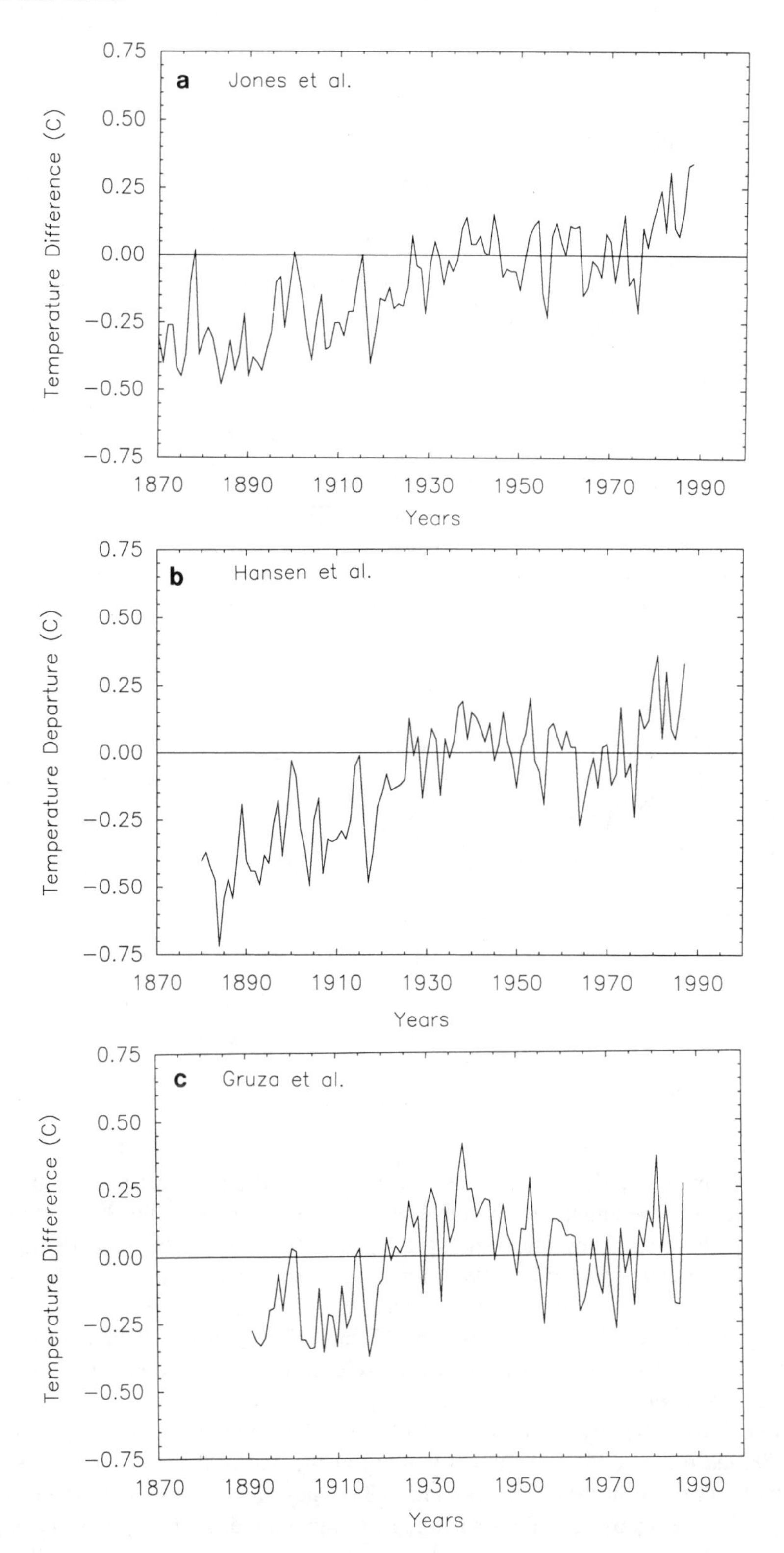


**FIGURE 1.** Global or hemispheric temperature records. (a) Global data set J1; (b) Global data set H; (c) Northern Hemispheric (NH) data set G; (d) The IPCC global data set; (e) Data set GOSTA.

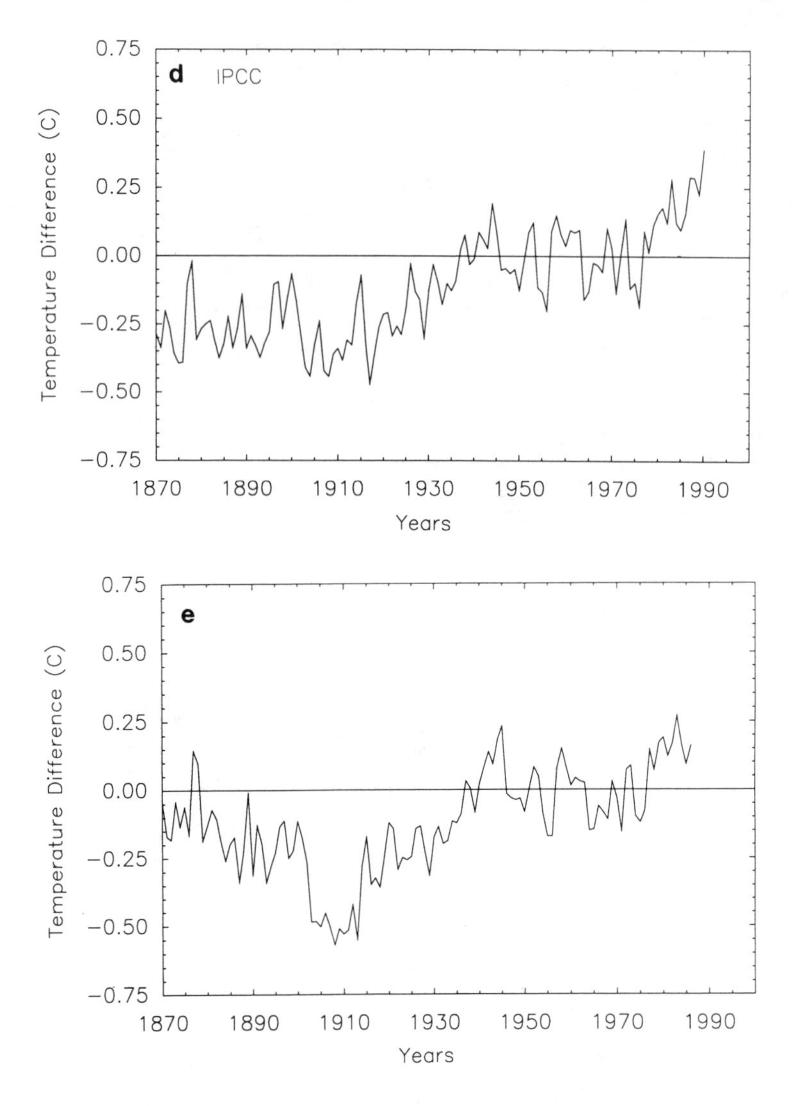


**FIGURE 1 (continued).**

In a direct approach, Hansen and Lebedeff (1988) used data set H and calculated the standard deviation of the annual-mean global-mean temperature about the 30-year mean for the period 1951 to 1980. They found that it is equal to 0.13. The 1987 global temperature deviation of 0.33 relative to 1951 to 1980 climatology presents a warming of between $2\sigma$ and $3\sigma$, and provides a trend significant at about 95% confidence level.

Similarly, Wigley et al., (1989) using the J2 data set, report that both 1987 and 1988 are near the $3\sigma$ level relative to the 1951 to 1980 reference period, and near the $6\sigma$ (!) level relative to the 19th century reference level.

In a more involved analysis, Tsonis and Elsner (1989) considered the J1 record from 1881 to 1988 and estimated the likelihood that six of the warmest years (1980, 1981, and 1985 to 1988) will occur in the period 1980 to 1988. They first described the 1881 to 1988 global temperature series by an optimum autoregressive model using the approach described in Katz (1982). This approach is a parametric modeling of a given time series. The time series under consideration is viewed as a realization of a stochastic process which is taken

**TABLE 1**
**A List of the Available Global Data Sets**

| Data set | Measurements | Ref. |
|---|---|---|
| J1 | Land | Jones et al. (1986a,b); Jones (1988) |
| H | Land | Hansen and Lebedeff (1988); Hansen et al. (1987) |
| G | Land | Vinnikov et al. (1987); Gruza et al. (1988) |
| COADS | Sea | Woodruff et al. (1987) |
| UK | Sea | Folland et al. (1984) |
| CDS | Sea | Hsiung and Newell (1983) |
| GOSTA | Sea | Bottomley et al. (1989) |
| J2 | Land and sea | Jones et al. (1986c) |
| IPCC | Land and sea | Folland et al. (1990) |

to be stationary and having a Gaussian distribition. This assumption has been widely used with the global temperature data. The optimum autoregressive model was found to be of fourth-order. The model seems to reproduce the qualitative properties of the global temperature record very well. Compare, for example, Figure 2 (a simulated record of 108 years in length) and Figure 1a (the actual global temperature record). Subsequently, they performed Monte-Carlo simulations using the derived autoregressive model, thus producing many series of 108 values corresponding to the 1881 to 1988 record. Each simulated record was subsequently searched in order to see if the highest, two highest, or three highest, etc., values fall in the 100 to 108 interval (which will correspond to the period of years 1980 to 1988). This way, from all the simulated records, the probability, $P_x^{n,m}$, that in the record of $\times$ values the n highest values are found in the last m values, was found.

The results from 100,000 Monte-Carlo simulations for $\times = 108$ and $m = 9$ are shown in Table 2. As can be seen, the probability that six of the highest values will be found in the last nine years is 0.012. Note that this (as well as the other values in Table 2) would indicate the minimum chance that such an arrangement will happen. The reason for that is the following: evidence for global warming is based on a temperature record which shows a marked positive trend. Observing n highest values in the last m values will be more likely when a positive trend is present than when a negative trend is present. The Monte Carlo simulations will produce with about 50–50 chance a positive or negative trend. If we only deal with positive trends, this probability may then be as high as 0.024. Thus, $0.012 \leq P_{108}^{6,9} \leq 0.024$. This result was then modified by considering shorter lengths of the record and a more strict definition defining "warmest" years. The final result is that the warming of the 1980s is an event which has only 1.0 to 3.2% chance of happening as a result of the variability of the system. This places the confidence level of the trend higher than 95%.

In an indirect approach, Wigley et al. (1989) and Wigley and Raper (1990) begin by modeling global mean temperature changes ($\Delta T$) using an energy balance model of the form

$$\phi\rho Ch \frac{d\Delta T}{dt} + \lambda\Delta T = \Delta Q - \Delta F$$

where $\phi$ ($\sim$0.7) is a factor accounting for land/sea differences in heat capacity, $\rho$ is the density, C is the specific heat capacity of ocean water, h is the mean depth of the ocean mixed layer, $\lambda$ is the feedback parameter, $\Delta F$ is the flux of heat out of the mixed layer into the deeper ocean, and $\Delta Q$ is "white noise" forcing. Having such a model, one can then generate records of any length and calculate 100-year trends and their probability. Their

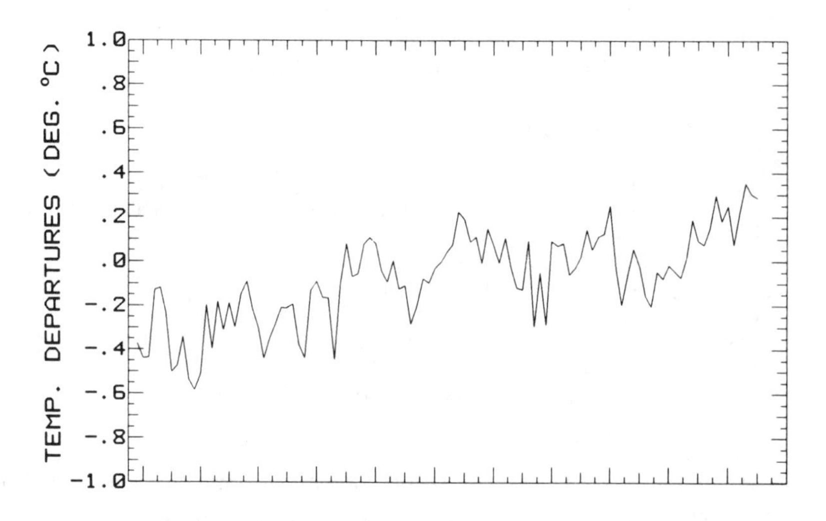


**FIGURE 2.** Simulated global mean temperature record of length 108 years using a fourth-order autoregressive model. This model is an optimum model derived from record J1. Note the similarities between this and Figure 1a.

**TABLE 2**
**Results from 100,000 Monte Carlo Simulations**

| n | 0 | 1 | 2 | 3 | 4 | 5 | 6 | 7 | 8 | 9 |
|---|---|---|---|---|---|---|---|---|---|---|
| x = 108<br>m = 9 | 0.616 | 0.158 | 0.091 | 0.058 | 0.037 | 0.021 | 0.012 | 0.005 | 0.002 | <0.001 |

*Note:* This table shows the probability that in a record x times steps long n highest magnitude events will occur in the last m events. The entries represent the lower bound of such probability, with the higher bound being at most twice the lower bound.

simulations revealed that for 100-year trends, the 90% confidence limits are much less than 0.5, which is the accepted trend over the last 100 years. This result indicates that the observed trend is statistically significant.

## III. THE EXISTENCE OF OSCILLATORY MODES IN THE GLOBAL DATA

The most common procedure to delineate oscillations (periodicities) in a time series is Fourier analysis. Figure 3 shows the spectra of the global IPCC record calculated from the Fourier transform of the autocorrelation function. Several peaks are observed, for example, at around 4.8, 5, 6, 9.2, 15, and 22 years. Just by looking at such spectra many will argue that we simply look at the ''floor'' or noise level. The question, therefore, is how significant are those peaks and how is the significance addressed? A straightforward procedure (Newell et al., 1989) is to compare those peaks to ''white noise'' null spectrum whose 95% confidence level is about 0.03 at all frequencies. In Figure 3, several peaks are above this value and, therefore, could be considered as significant, that is, if you really want to. All peaks are hardly higher than 0.03.

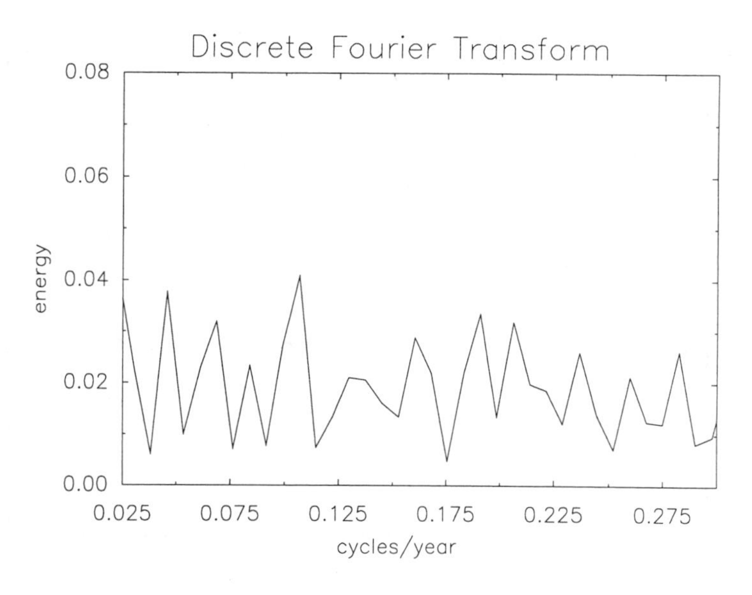


**FIGURE 3.** Frequency spectra of the detrended IPCC global data set.

A fundamental problem in addressing the significance of peaks in frequency spectra is the trends embedded in the data. Fourier analysis is effective only when we are dealing with stationary data. When the data (like the temperature data) exhibit trend(s), detrending the data is required before considering Fourier analysis. Since the data may exhibit quite complicated trends, it is quite difficult to address the significance of the peaks whose amplitude depends heavily on the detrending. Because of this problem and because the peaks are not extremely pronounced, the literature lacks rigorous statistical testing, and many peaks are either taken for granted or are not seriously considered. For this reason, other techniques that "bring the most out" of the data have been employed lately in the search of periodicities in the global temperature data sets.

One very promising technique is the fully nonparametric singular spectrum analysis (SSA) (Broomhead and King, 1986; Fraedrich, 1986). This approach considers M lagged copies of a centered time series X(t) sampled equal intervals $\tau$, $X_i = X(t_o + i\tau)$, $i = 1$, N, and estimates the eigenvalues $\lambda_k$ and eigenvectors $\rho_k$ of their covariance matrix C (here $1 \leq K \leq M$). Often we call the eigenvectors empirical orthogonal functions (EOFs) and the coefficients $\alpha_k$ involved in the expansion of each lagged copy, principal components (PCs). EOFs are of interest when among k eigenvalues there exist a number of distinct ones whose magnitude is appreciable, whereas the rest are close to zero. In such cases this would be a strong indication of "deterministic" parts in the subspace of eigenmodes, with the rest of the modes acting as noise. Thus, this method can be used to separate signal from noise. In the study of Vautard and Ghil (1989), it was observed that pairs of high-variance eigenvalues $\lambda_k = \lambda_{k+1}$ are associated with oscillatory phenomena (both the corresponding EOFs and PCs are in quadrature with each other).

The first application of this approach to a global temperature record is that of Ghil and Vautard (1991). Ghil and Vautard used record J2 and extracted the following oscillatory modes: an oscillation having a period of nearly 20 years (EOFs 3 and 4), an oscillation of 6 years (EOFs 11 and 12), and an oscillation of 5 years (EOFs 6 and 7). The five- and six-year periodicities are attributed to the well-documented El Nino/Southern Oscillation (ENSO). Speculation concerning the 20-year oscillation centers on possible changes in ocean circulation. From these results, Ghil and Vautard proceeded to reconstruct the global climate

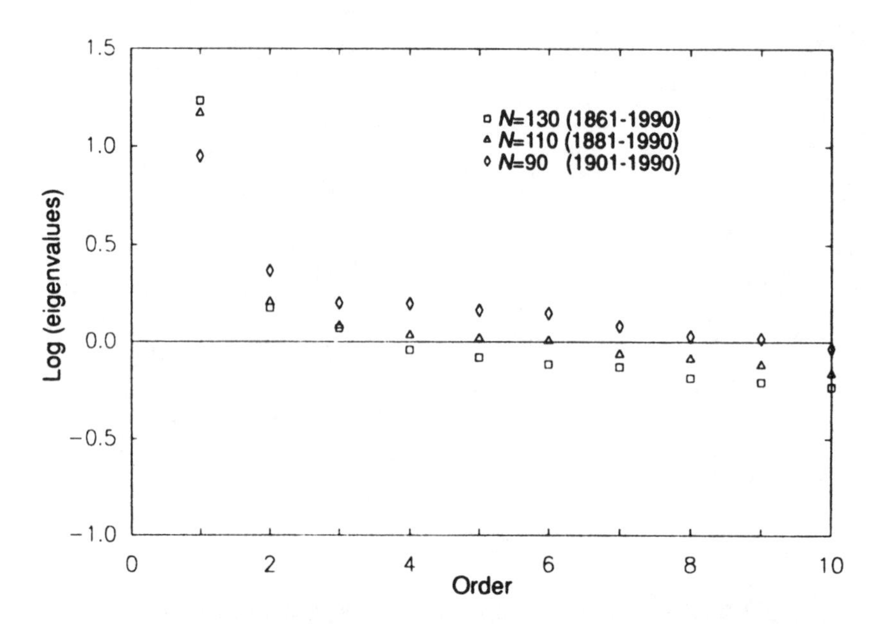


**FIGURE 4.** The estimated eigenvalues, $\lambda_k$, of the lag-covariance matrix C (M = 30) calculated for various lengths of the global surface temperature record J2. Only the first two eigenvalues extend above the noise floor.

record and to investigate the issue of detecting a greenhouse warming signal in the record. They concluded that, as a consequence of the bidecadal oscillation, the warming signal will not be detectable for at least one or two more decades. What was not addressed in that study, however, was the effect to the length and quality of the data on the results. This question was considered in Elsner and Tsonis (1991), where a detailed analysis was performed. Shown in Figure 4 are the eigenvalues of the lag-covariance matrix C against order for the entire record, 1861 to 1990, (N = 130), for the record from 1881 to 1990 (N = 110), and for the record from 1901 to 1990 (N = 90). For any length we observe that the first two eigenvalues ($\lambda_1$ and $\lambda_2$) considerably exceed the rest, explaining 35 to 65% of the variance. The second two eigenvalues ($\lambda_3$ and $\lambda_4$) are quite close to the noise floor, and account for only 7 to 11% of the variance.

The corresponding leading eigenvectors (EOFs 1 to 4) for the different length record are shown in Figure 5. We observe that the estimation of EOFs 1 and 2 is robust; it varies only slightly with record length. EOFs 1 and 2 correspond to the trend in the temperature record. In contrast, however, the estimation of EOFs 3 and 4 is not robust. If the entire record (N = 130) is considered, EOFs 3 and 4 are an oscillatory pair in quadrature with each other and have a period of ~20 years, but for N = 110 and N = 90 this is not the case. Instead, we begin to see an oscillatory pair having a period between five and six years. We note that for N = 110 and N = 90, no higher EOFs (not shown) indicate the 20-year oscillation. The result is not surprising, as the eigenvalues corresponding to EOFs 3 and 4 are clearly very near the noise floor.

From these results, we raise the question of whether a minimum N is required for delineation of the bidecadal oscillations. To answer this question, we repeat the analysis using records that are shortened from the end (most recent years), rather than from the beginning (earlier years) as was done above. Results are shown in Figure 6. We now find that, along with the estimation of EOFs 1 and 2 (not shown), the estimation of EOFs 3 and 4 is robust with bidecadal oscillations present for all data lengths considered. Thus, the comparison of Figures 5 and 6 reveals that it is not the length of the record, but the inclusion of data points roughly before 1881 that makes the difference. Include the earliest years in the analysis and you will detect a bidecadal oscillation, but exclude them and it disappears.

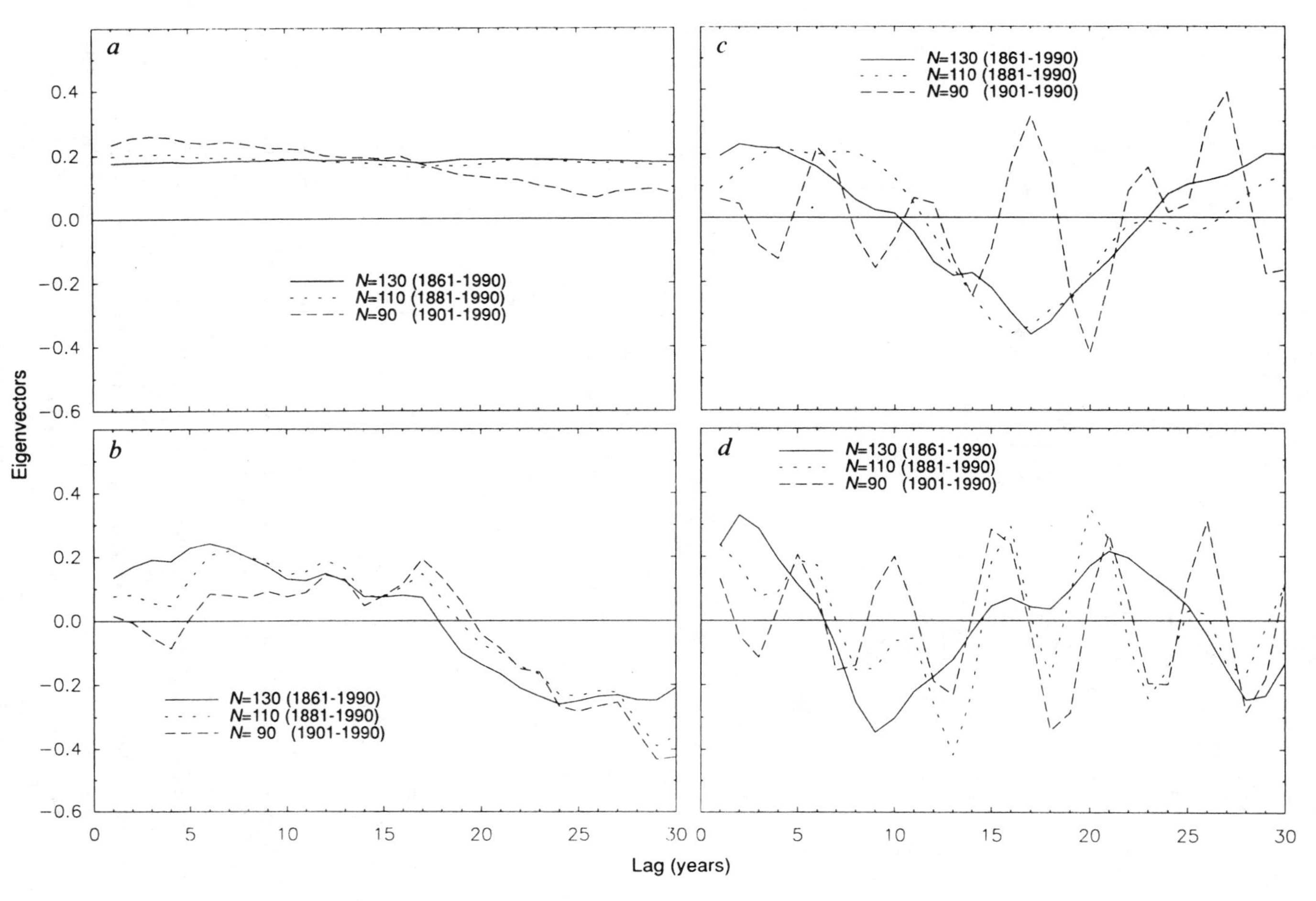


**FIGURE 5.** (a) The first eigenvector (EOF1) of the lag-covariance matrix C (M = 30) computed for various lengths of record J2. (All records end with the year 1990); (b) As in (a), but for EOF 2; (c) as in (a) but for EOF 3; (d) as in (a) but for EOF 4. EOFs 3 and 4 are a pair in quadrature and having a period of ~20 years only for N = 130 (1861 to 1990). EOFs 1 and 2 correspond to the trend and are virtually unchanged for different lengths.

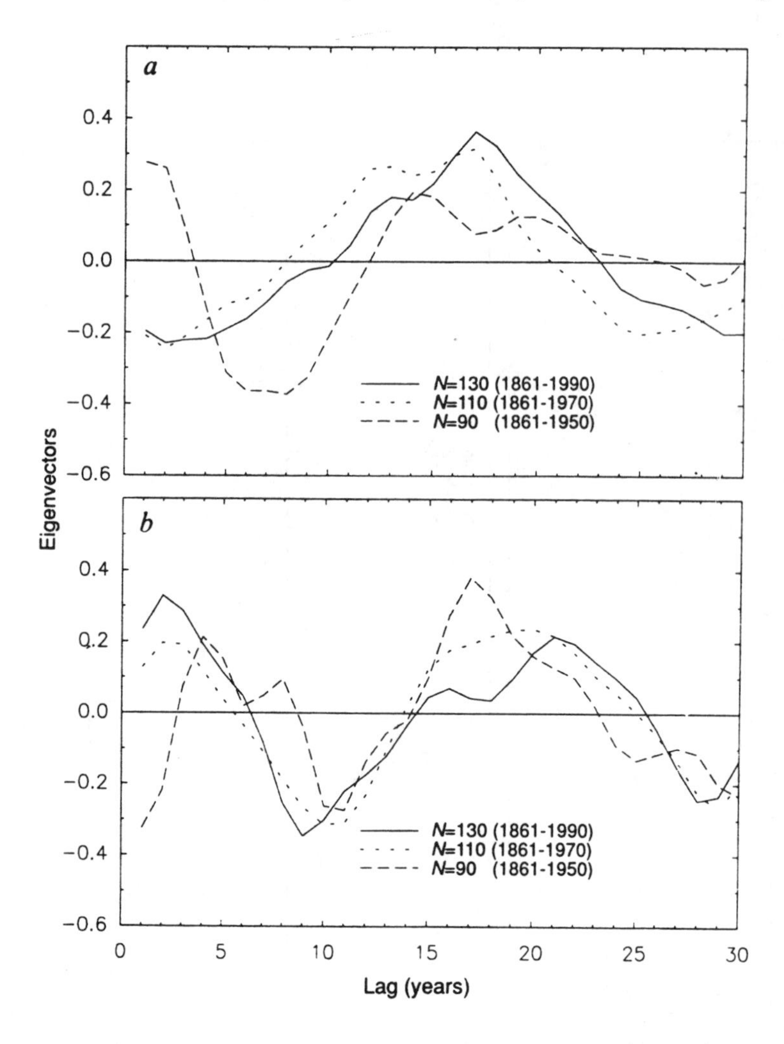


**FIGURE 6.** (a) EOF 3 estimated from the lag-covariance matrix C (M = 30) using various lengths of the record J2. Now all records begin with the year 1861; (b) As in (a) but for EOF 4. EOFs 3 and 4 are an oscillatory pair in quadrature with a period of ~20 years for all record lengths.

Other global surface air temperature records tell a similar story. Estimates of EOFs 3 and 4 from SSA on the U.S. global data set (data set H), on the IPCC global time series, and on the U.S.S.R. data set (data set G) are shown in Figure 7. The U.S. record extends from 1880 to 1987 (N = 108). The IPCC record extends from 1961 to 1990 (N = 130; the 1990 value of the IPCC data was supplied by D. E. Parker). The U.S.S.R. data set represents annual mean surface air temperatures in the Northern Hemisphere, and it covers the period 1891 to 1987 (N = 97). Estimates of EOFs 1 and 2 from all data sets (not shown) are close to those estimated from the U.K. data set. In both the U.S. and U.S.S.R. sets which do not include any values before the 1880s, the 20-year oscillatory pair is not apparent (if anything, EOFs 3 and 4 indicate a five- to six-year oscillatory pair). The results from the IPCC record are similar to the results obtained from the U.K. set: the 20-year oscillation is not robust. It exists when the whole record is considered, but it is absent when somewhat smaller period excluding early values, such as the period 1891 to 1990 (N = 100) shown in Figure 7, are considered. It is worth mentioning here that Schonwiese (1987) showed

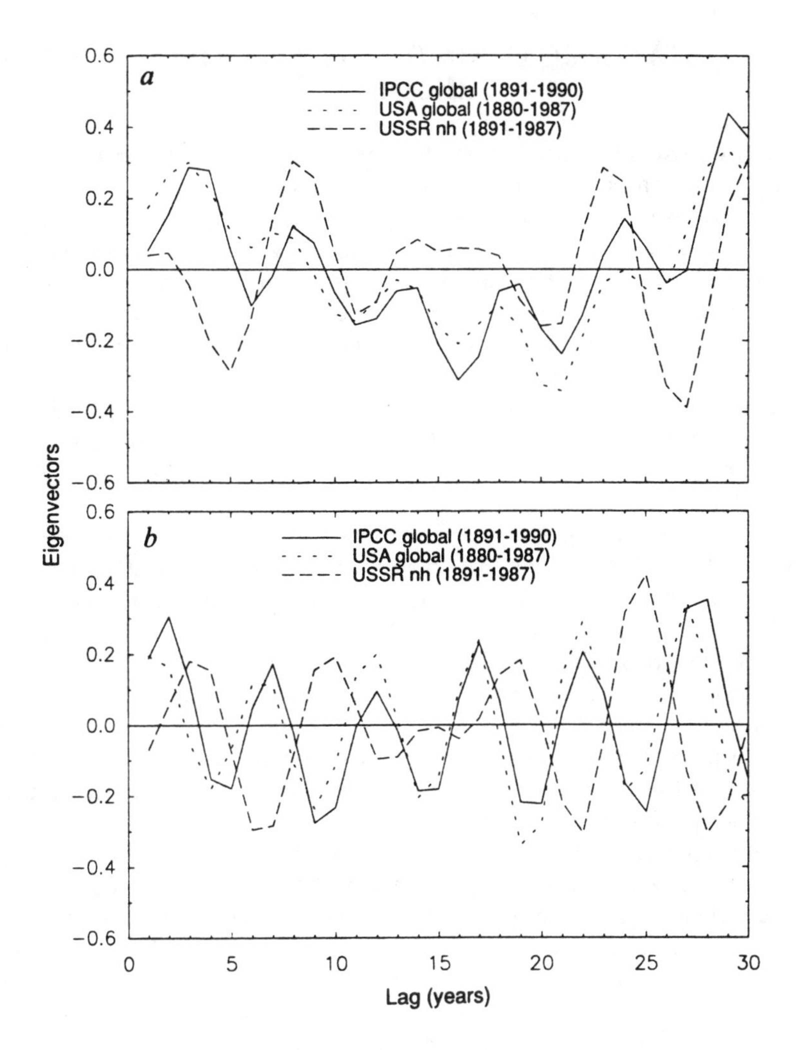


**FIGURE 7.** (a) EOF 3; (b) EOF 4 estimated from the lag-covariance matrix C (M = 30) for the global temperature record H (1880 to 1987), the IPCC record (1891 to 1990), and for data set G (1891 to 1987; nh, Northern Hemisphere). Note the absence of bidecadal oscillations.

that a considerable doubt exists in the reconstructed temperature series prior to 1881. This, together with the results of Elsner and Tsonis (1991), makes the existence of bidecadal oscillations questionable.

In the past, after using Fourier analysis, investigators have come across a peak of 21 to 23 years. For example, there is a pronounced peak at 23 years in the maximum entropy spectrum of a 300-year temperature record (Folland, 1983), and a 21.8-year peak is contained in the Fourier transform of global marine air temperature data (Newell et al., 1989) covering the period 1856 to 1986. The 23-year peak refers to just a small area in central England, and the 21.8-year peak refers to marine data only. Interestingly, estimates of EOFs 3 and 4 from SSA on the marine data set only indicate that a 20-year oscillation is robust. However, although both important, their statistical significance with respect to their connection to global surface air temperature records is not straightforward and was never established. Research in this area is under way.

Therefore, at this point in time we can conclude that only one oscillatory mode exists in the global temperature data: that of a period of about five-to-six years attributed to ENSO.

## IV. THE STATISTICAL SIMILARITY OF THE DIFFERENT RECORDS

As was discussed in the Introduction, there exist many different global records. Several of them have actually been produced from similar sources. Take, for example, the Northern Hemispheric records of J1, H, and G, shown in Figure 8. All of them refer to land data. Do they provide the same information vis-à-vis climate variability? For example, all three show a marked trend, but are the trends similar? A straightforward cross-correlation analysis shows that for any two of those three records, the cross-correlation is very high (around 0.85). Such a result has been used often in the past as an indication that the records are compatible or in agreement with each other.

We have to keep in mind, however, that the cross-correlation measures the linear dependence between two time series. Climate variability is obviously strongly nonlinear (as any spectral analysis will show). How does this affect the comparison between two temperature records?

This problem has been investigated in Elsner and Tsonis (1991) via the boot-strap procedure. The boot-strap procedure is a nonparametric approach used in determining the significance of the statistical properties of small samples. Let us assume that we are given a record of size ten whose mean is $\overline{x}$. According to the procedure, given a record of sample size ten, one generates a boot-strap sample by placing all the values in a ''box'' and then randomly drawing with replacement ten values. This can be repeated many times, thus generating a large number of boot-strap samples each of size ten. For each one of those samples the mean is calculated and the frequency distribution of the mean is subsequently obtained. From that distribution the probability that we will observe a mean equal to $\overline{x}$ in the observed sample can then be inferred.

The boot-strap procedure was applied to the three records as follows. First, difference records were constructed by subtracting two records. This was done in order to remove the heavy autocorrelations present in the individual records. Thus, the records H − J1, J1 − G, and H − G were constructed. The results are shown in Figure 9. If the records were very similar, then one would expect that the difference records would produce a distribution for the mean which will be centered at zero. As Figure 9 indicates, this is not the case for any of the difference records. It was, thus, concluded that all three hemispheric surface records exhibit a significantly distinct mean temperature.

In addition to the mean, Elsner and Tsonis (1991) addressed the statistical similarity between the trends in those three records. In this case, the linear trend in the difference records was subtracted from the actual values of the difference records. The residuals were then boot-straped and as they were ''drawn'' from the box they were added to the value given from the linear regression equation for that year, thus producing a boot-strap sample. The trend of the boot-strap sample was then calculated. The procedure was repeated 10,000 times and the frequency distribution of the trend was obtained. The idea behind this procedure was to see how likely it is for large residuals to produce a trend in the difference record close to zero. Note that zero trend will mean that the two records are similar. If those distributions overlap with the zero value, the probability that a zero trend arising from fluctuations can then be estimated. The results are shown in Figure 10. As can be seen from the results, the distributions do not overlap with zero. Thus, the probability that the two records are similar (in this case, exhibiting statistically similar trends) is close to zero.

Based on these results it is, therefore, concluded that the differences in the observed surface temperature records are significant and that at least two of those three data sets do not represent the true population. The above study raises some interesting questions on the confidence the data sets provide in arguing precise temperature trends, and in general on

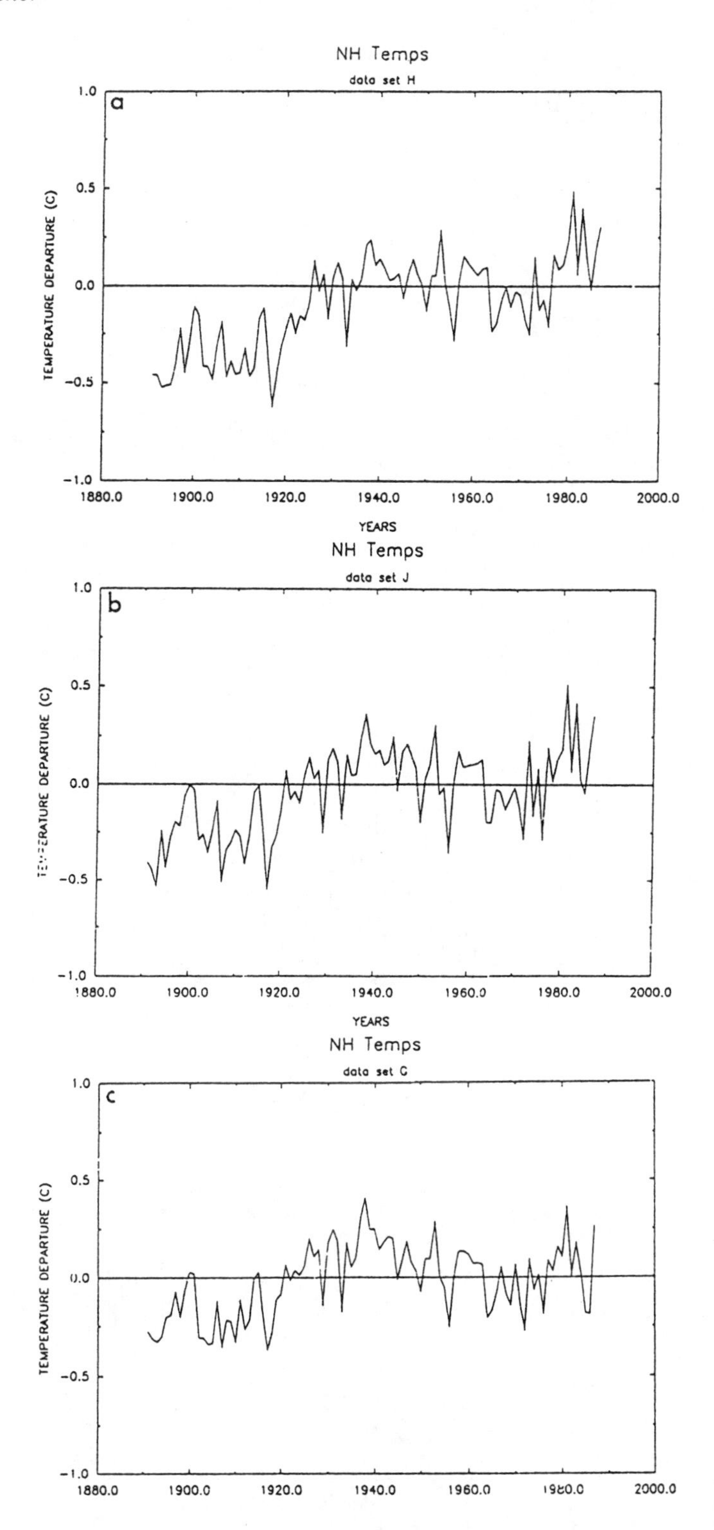


**FIGURE 8.** Northern Hemisphere annual surface air temperature departures in °C from 1891 to 1987. (a) Hansen and Lebedeff, 1987 (data set H); (b) Jones et al., 1986a,b (data set J1); (c) Gruza et al., 1988 (data set G). Each data set represents land data base and different averaging techniques. An overall trend in evident in all three records.

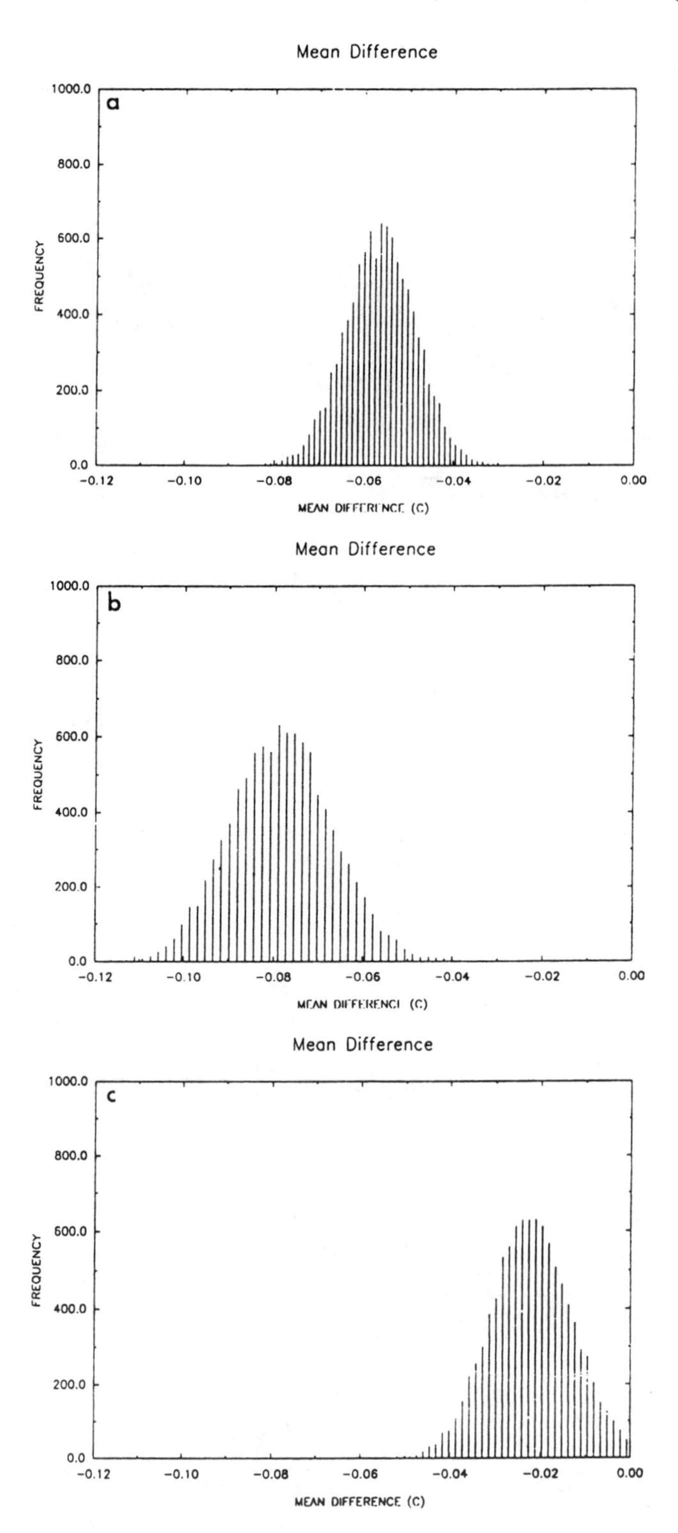


**FIGURE 9.** Frequency distributions of the mean plotted for $10^4$ bootstrap samples for (a) Difference record H − J1; (b) Difference record H − G; (c) Difference record J1 − G. The ordinate scale is the number of times the bootstrap mean fell into a given interval. Note that all three distributions are located to the left of a zero mean difference.

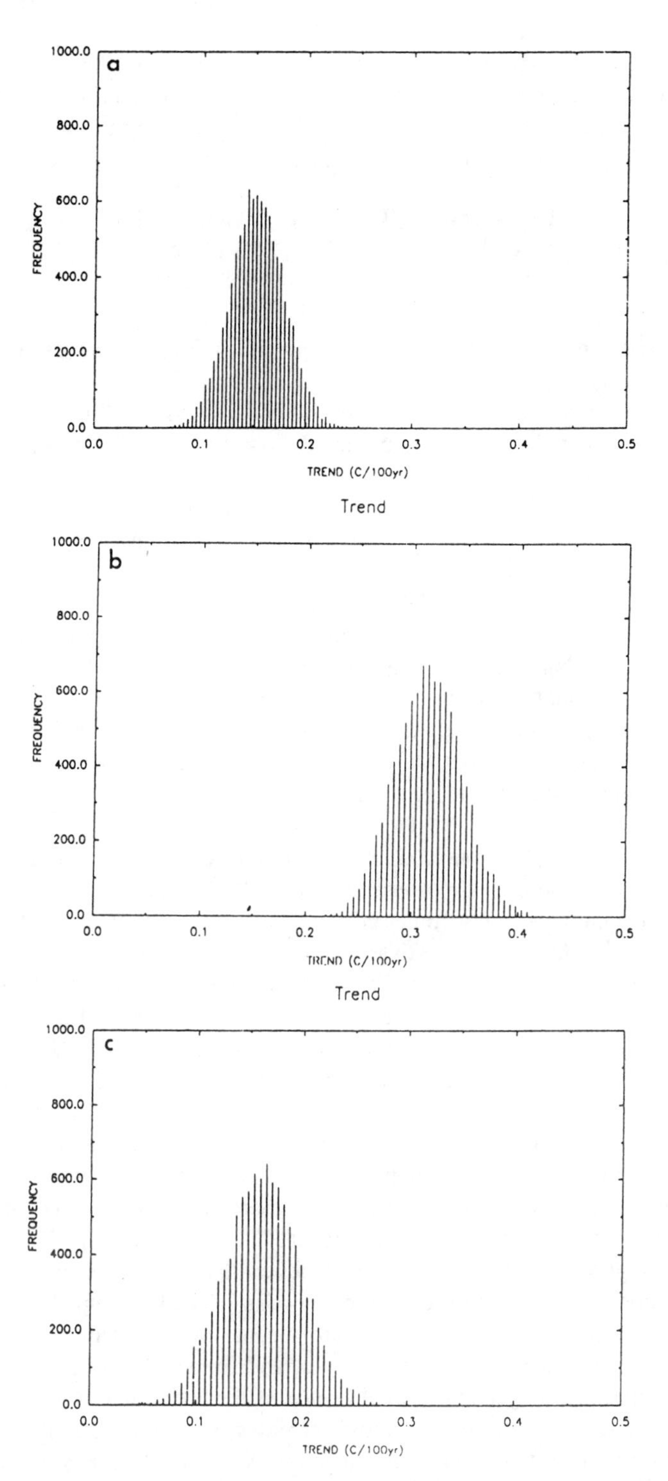


**FIGURE 10.** Frequency distributions of the trend plotted for $10^4$ bootstrap samples for (a) Difference record H − J1; (b) Difference record H − G; (c) Difference record J1 − G. The ordinate scale is the number of times the bootstrap trend fell into a given interval. Note that all three distributions are separated from zero, indicating significant differences between long-term surface temperature trends given by each of the three data sets.

the ability of the records to truly represent the natural variability of the climate system. This does not imply that the data base is useless. It does, however, caution that the production of a global annual-mean temperature record might be affected by the way the data base is manipulated (i.e., correction procedures, averaging techniques, etc.).

## V. IS THERE A RELATION BETWEEN INCREASED GREENHOUSE GASES CONCENTRATION AND THE RISE IN GLOBAL TEMPERATURE?

The first study that attempted to identify possible climate changes in the observed temperature record is that of Solow (1987).

In a parametric analysis, Solow (1987) considered a two-phase linear regression model to test for climate change in a temperature record $T_i = 1 \ldots n$. The model was written as

$$T_i = \begin{cases} a_1 + b_1 i + e_i, & i = 1 \ldots r \\ a_2 + b_2 i + e_i, & i = r + 1 \ldots n \end{cases}$$

where $e_i$ are an independent sequence of normal noise with mean zero and unknown variance $\sigma^2$. The abscissa of the intersection of the two regression lines for some optimum r is $c = (a_1 - a_2)/(b_2 - b_1)$. If there is no change in the record, then $b_2 - b_1 = 0$. Thus, testing for climate change amounts in testing the null hypothesis $H_o$: $b_2 - b_1 = 0$ against the two-sided alternative $H_1$: $b_2 - b_1 \neq 0$. By applying this approach to the J1 Southern Hemisphere data, Solow (1987) concluded that the hypothesis of no change cannot be rejected. Consequently, a climate change due to increased concentration of greenhouse gases cannot be substantiated.

Recently, Kuo et al. (1990) used a rather sophisticated method to infer the coherence between the $CO_2$ concentration time series (1958 to 1988) and the H temperature record. The method is the multiple-window time series method to estimate trends and power spectra. Kuo et al. found that the probability that the slope of the temperature is positive exceeds 99.99%. Furthermore, they found that $CO_2$ time series and the global temperature series from 1958 to 1988 are coherent over much of the Nyquist frequency. They stress, however, that caution must be exercised in interpreting this result as suggesting that $CO_2$ variations cause temperature changes. In short, a definite link between the two time series cannot be justified.

In a semi-direct appraoch, Barnett and Schlesinger (1987) searched for a theoretically predicted $CO_2$ signal in surface air temperature. By comparing model outputs of the $2 \times CO_2 - 1 \times CO_2$ equilibrium climate change to observations like surface air temperature, sea level pressure, sea surface temperature, etc., they concluded that no such signal can be identified with high confidence. They also suggested that the global air temperature data may not be as good a quantity to use for the detection of the $CO_2$ signal.

What we are dealing with are temperature records of approximately 100-years long which shows marked variability from year-to-year and an overall positive trend. If a record were 20-years long, probably nobody would worry if the 20th value was the highest in the record. However, as the length of the record increases, a record-breaking temperature becomes more important, instantly receiving attention. At the same time, we observe a steady increase in amounts of greenhouse gases which theoretically should cause the temperature of the planet to increase. This, together with several recent record breakings, has prompted scientists to argue that there should be a relation between the two observable records. In order to prove that one is causing the other, one should have some information about the behavior of one variable in the absence of the other. Therefore, in order to confirm that increased concentration of greenhouse gases causes the temperature of the planet to increase, the natural variability of the global temperature record should first be established. As with any variable, this can be achieved only if adequate information is available which translates into having a record of adequate length and quality. This length is a function of the factors involved in shaping up the variability of the observable. The question thus arises: is the temperature record long and accurate enough to make the connection? Do we have enough information (data) that will tell us how the mean global temperature varies in the absence of the greenhouse gases forcing? In the absence of that information, can the numerical models of the atmosphere provide the natural variability of the system? The answer to these questions is probably no. The available data do not extend far in the past, and present numerical models are not complete, making comparisons between real data and predictions of atmospheric model difficult to interpret. That is why at this point in time, none of the studies that report on the significance of the positive trend cannot possibly connect the rise in the temperature with the rise in the concentration levels of greenhouse gases.

## VI. CONCLUSIONS

In summary, the following presents the findings of this review:

1. The positive trend in the observed global sets is statistically significant at very high confidence levels.
2. A direct link between temperature changes and variations in greenhouse gases concentrations has not been established.
3. With questionable data prior to 1881, the only significant oscillatory mode is a five-to-six year mode attributed to ENSO.
4. Temperature records constructed differently from the same data base may not be statistically similar.

Even though the above points may not be as definite as some would expect or wish, they do not dismiss the problem of global warming and its ties to a possible enhanced greenhouse effect. It is only logical that by not taking proper care of our atmosphere we will damage it. In this regard we should not wait until the data clearly tell us that the effect is here because by that time it will probably be too late. We should try our best to adjust accordingly in order to avoid future unrepealable damages.

## REFERENCES

Barnett, T. P. and Schlesinger, M. E., Detecting changes in global climate induced by greenhouse gases, *J. Geophys. Res.*, 92, 14,772–14,780, 1987.

Bottomley, M., Folland, C. K., Hsiung, J., Newell, R. E., and Parker, D. E., *Global Ocean Surface Temperature Atlas* "GOSTA", MIT Press, Cambridge, 1989.

Broomhead, D. S. and King, G. P., Extracting qualitative dynamics from experimental data, *Physica D*, 20, 217–236, 1986.

Diaconis, P. and Efron, B., Computer-intensive methods in statistics, *Sci. Am.* 248(5), 115–130, 1983.

Efron, B., The jacknife, the bootstrap and other resampling plans *SIAM*, 1982, 92 pp.

Elsner, J. B. and Tsonis, A. A., Do bidecadal oscillations exist in the global temperature record?, *Nature (London)*, 353, 551–553, 1991.

Elsner, J. B. and Tsonis, A. A., Comparison of observed Northern hemisphere surface air temperature records, *Geophys. Res. Lett.*, 18, 1229–1232, 1991.

Folland, C. K., Regional-scale interannual variability of climate — a north-west European perspective, *Meteorol. Mag.*, 112, 163-183, 1983.

Folland, C. K., Karl, T. R., and Vinnikov, K. Va., Climate change, in *The IPCC Scientific Assessment*, Houghton, J. T., Jenkins, G. J., and Euphraums, J. J., Eds., Cambridge University Press, 1990, chap. 7.

Fraedrich, K., Estimating the dimensions of weather and climate attractors, *J. Atmos. Sci.*, 432, 419–432, 1986.

Ghil, M. and Vautard, R., Interdecadal oscillations and the warming trend in global temperature time series, *Nature (London)*, 350, 324–327, 1991.

Gruza, G. V., Rankova, E. Ya., and Rocheva, E. V., Analysis of global data variations in surface air temperature during instrument observation period, *Meteorol. Gidrol.*, 16–24, 1988.

Hansen, J. E. and Lebedeff, S., Global trends of measured surface air temperature, *J. Geophys. Res.*, 92, 13, 345–13,372, 1987.

Hansen, J. E. and Lebedeff, S., Global surface air temperatures: updated through 1987, *Geophys. Res. Lett.*, 15, 323–326, 1988.

Hsiung, J. and Newell, R. E., The principal non-seasonal modes of variation of global sea surface temperature, *J. Phys. Oceanogr.*, 13, 1957–1967, 1983.

Jones, P. D., Raper, S. C. B., Bradley, R. S., Diaz, H. F., Kelly, P. M., and Wigley, T. M. L., Northern hemisphere surface air temperature variations: 1851–1984, *J. Climate Appl. Meteorol.*, 25, 161–179, 1986a.

Jones, P. D., Raper, S. C. B., and Wigley, T. M. L., Southern hemisphere surface air temperature variations: 1851–1984, *J. Climate Appl. Meteorol.*, 25, 1213–1230, 1986b.

Jones, P. D., Wigley, T. M. L., and Wright, P. B., Global temperature variations between 1861 and 1984, *Nature (London)*, 322, 430–434, 1986c.

Jones, P. D., Hemispheric surface air temperature variations: recent trends and an update to 1987, *J. Climate*, 1, 654–660, 1988.

Katz, R. W., Statistical evaluation of climate experiments with general circulation models: a parametric time series modeling approach, *J. Atmos. Sci.*, 39, 1445–1455, 1982.

Kuo, C., Lindberg, and Thomson, D. J., Coherence established between atmospheric carbon dioxide and global temperature, *Nature (London)*, 343, 709–714, 1990.

Newell, N. E., Newell, R. E., Hsiung, J., and Wu, Z., Global marine temperature variation and the solar magnetic cycle, *Geophys. Res. Lett.*, 16, 311–314, 1989.

Schonwiese, C. D., Moving spectral variance and coherence analysis and some applications on long air temperature series, *J. Climate Appl. Meteorol.*, 26, 1723–1730, 1987.

Solow, A. R., Testing for climate change: an application of two-phase regression model, *J. Climate Appl. Meteorol.*, 26, 1401–1405, 1987.

Tsonis, A. A. and Elsner, J. B., Testing the global warming hypothesis, *Geophys. Res. Lett.*, 16, 795–797, 1989.

Vautard, R. and Ghil, M., Singular spectrum analysis in nonlinear dynamics, with applications to paleoclimatic time series, *Physica D*, 35, 395–424, 1989.

Vinnikov, K. Va., Groisman, P. Ya., Lugina, K. M., and Golubev, A. A., Variations in Northern hemisphere mean air surface temperature over 1881–1985, *Meteorol. Hydrol.*, 1, 45–53, 1987 (in Russian).

Woodruff, S. D., Slutz, R. J., Jenne, R. J., and Steurer, P. M., A comprehensive ocean-atmosphere data set, *Bull. Am. Meteorol. Soc.*, 68, 1239–1250, 1987.

Wigley, T. M. L., Jones, P. D., Kelly, P. M., and Raper, S. C. B., Statistical Significance of Global Warming, Proc. 13th Annu. Climate Diagnostics Workshop, Cambridge, MA, October 31 to November 4 (1988), 1989.

Wigley, T. M. L. and Raper, S. C. B., Natural variability of the climate system and the detection of the greenhouse effect, *Nature (London)*, 344, 324–327, 1990.

Chapter 4

# SPACE-BASED GEOENGINEERING OPTIONS FOR DEALING WITH GLOBAL CHANGE

**Lyle M. Jenkins**

## TABLE OF CONTENTS



This chapter was completed under the auspices of the U.S. Government and is therefore in the public domain.

# I. INTRODUCTION

The exploration of space has provided a new perspective; an overall view of the Earth derived from photographs and the first-hand impressions of astronauts and cosmonauts. As a result, the reality of the limited extent of our environment and its apparent fragility has been communicated to the public, beyond the visions of a few farsighted individuals. The "overview effect," looking at the entire planet Earth, has contributed significantly to the perception of the potential hazards in global environmental change induced by anthropogenic inputs. This vantage point is used effectively in the Earth Observing System (EOS) portion of the U.S. Global Change Research Program.[1] The EOS program is collecting data from satellite-carried instruments to support research on the changing environment. In addition to monitoring environmental parameters, operations in space may offer benefits in mitigating adverse environmental impacts. The complexity of the Earth's ecosystems and the difficulty of developing credible predictions of the rate and magnitude of global change couple with concern for the stability of the climate system. This concern points to the need for developing geoengineering options for interacting with the Earth's environmental system. Environmental groups and many scientists are wary of using technological intervention for dealing with global change. Too often, technical solutions have been implemented without the complete knowledge and consideration of all of the parameters that could affect the outcome. In dealing with the mitigation of the effects of environmental change, there are no simple, fully effective solutions. The potential solutions or modes for dealing with environmental change can be categorized as prohibition, adaptation, and countermeasures or intervention.[2] Geoengineering options for mitigating the effects of change were discussed by the NAS (National Academy of Science) Panel on the Policy Implications of Greenhouse Warming, and published in the report, Policy Implications of Greenhouse Warming.[3] Several of the ideas conveyed in this report are space-based or depend on space systems for implementation. Response actions must be based on an assessment of the direct and indirect consequences of action and trading those consequences against the hazards of inaction.

# II. RESPONSE OPTIONS

Prohibition is undoubtedly the easiest response to identify. It involves the reduction of those insults to environment that are considered to be factors in atmospheric problems. The elimination of an environmentally offensive product is often inhibited by cost or special interests. Economic factors often constrain the implementation of conservation and energy efficiency technology, which are versions of preventive measures. Conservation of resources and energy efficiency should be a basic strategy for addressing anthropogenic factors in global change. Some environmental groups also impede the application of technology. Still, technology in some form is needed to support the world's population. It is unrealistic to advocate a return to a simpler phase in civilization's development. The application of a total systems analysis is an appropriate approach to the issues of energy needs. By weighing the benefits to the environment against the total penalties (both delayed penalties as well as near-term penalties), it may be possible to put economic factors in a more favorable perspective. Nevertheless, a reduction in those activities producing greenhouse gases or depleting ozone in the stratosphere is likely to fall short of an effective goal for inhibiting global changes.

Mankind and the Earth's ecosystems have demonstrated an ability to adapt to change. The future rates of climatic change and of stratospheric ozone depletion are projected to be too rapid for evolutionary adaptation in the usual context. For example, the effect of increased levels of UV radiation may set off a positive feedback loop with photosynthesis that exacerbates climate change. Adaptation implies that there will be severe penalties to ecological systems and to society, despite favorable climate trends in some regions. Improvements in climate are likely to be disruptive just because of the rapidity of change. The severity of these impacts may define the need for active countermeasures or interactive alteration of Earth system processes.

The space community has contemplated the concept of terraforming applied to other planets. In terraforming, technological action is taken to change the environment to produce conditions favorable to life processes. Dr. Carl Sagan proposed a concept of modifying the atmosphere of Venus to counteract the ''runaway greenhouse effect'', an effect that keeps the temperatures on Venus too high for life as it exists here on Earth. McKay suggests that the production of an atmosphere on Mars could make habitation possible. Whether modification of these other planets is feasible or not, knowledge of the current and past conditions on those planets obtained through the Space Exploration Initiative may contribute to our understanding of the Earth's environmental system.

It may be viewed that we are negatively terraforming Earth through the uncontrolled addition of greenhouse gases to the atmosphere. At this time, the concerns are expressed more in terms of alterations in life style and economic impact than in terms of the total disappearance of life, though in fact many species may become extinct. At any level of possibility, preparation for positive action to direct the path of change should be taken.

The NAS Panel on Policy Implications of Greenhouse Warming[3] used the term ''geoengineering'' to descibe concepts for active means of mitigating the rate or the effects of climate change. The examples cited in their report are reforestation, ocean biomass stimulation, removal of chlorofluorocarbons (CFCs),[4] sunlight screening with stratospheric particles, space mirrors, and cloud stimulation. The range of options presented in this report was not very comprehensive because viable ideas have not been well developed at this phase of research. The NAS Policy Implication report recommends undertaking ''research and development projects to improve our understanding of both the potential of geoengineering options to offset warming and their possible side-effects.''[3] An initial phase should include an expansion of the number of options and their evaluation at a conceptual level. Directed by an understanding from the change research, experiments bearing on promising geoengineering concepts would provide a basis for policy decisions.

## III. SPACE-BASED GEOENGINEERING OPTIONS

The evaluation of stratospheric particle or aerosol screens and cloud stimulation to reflect solar radiation was recommended for assessment by the NAS Policy Implication report. The oil fires in Kuwait and the eruption of Mount Pinatubo have provided a full scale laboratory for some aspects of an assessment. Further, concepts involving orbiting sunscreens and laser removal of CFCs were not endorsed for study based on poor benefit-to-cost ratio. The study of solutions with lower ratios of benefit-to-cost may be desirable as the impacts of global change worsen.

Moreover, other space-based ideas not identified by the panel have the potential to affect the course of global change. Examples are space power systems which deliver energy to Earth, orbiting electromagnetic radiation generators to interact with greenhouse gases, space-based sunscreens to shade the earth, and the use of lunar material to aid in the development of fusion energy.

## IV. SPACE POWER SYSTEMS

The concept of collecting power in space for use on Earth was first described in 1967 by Glaser.[5] His idea provides an alternative to the use of fossil fuels for generating electricity.

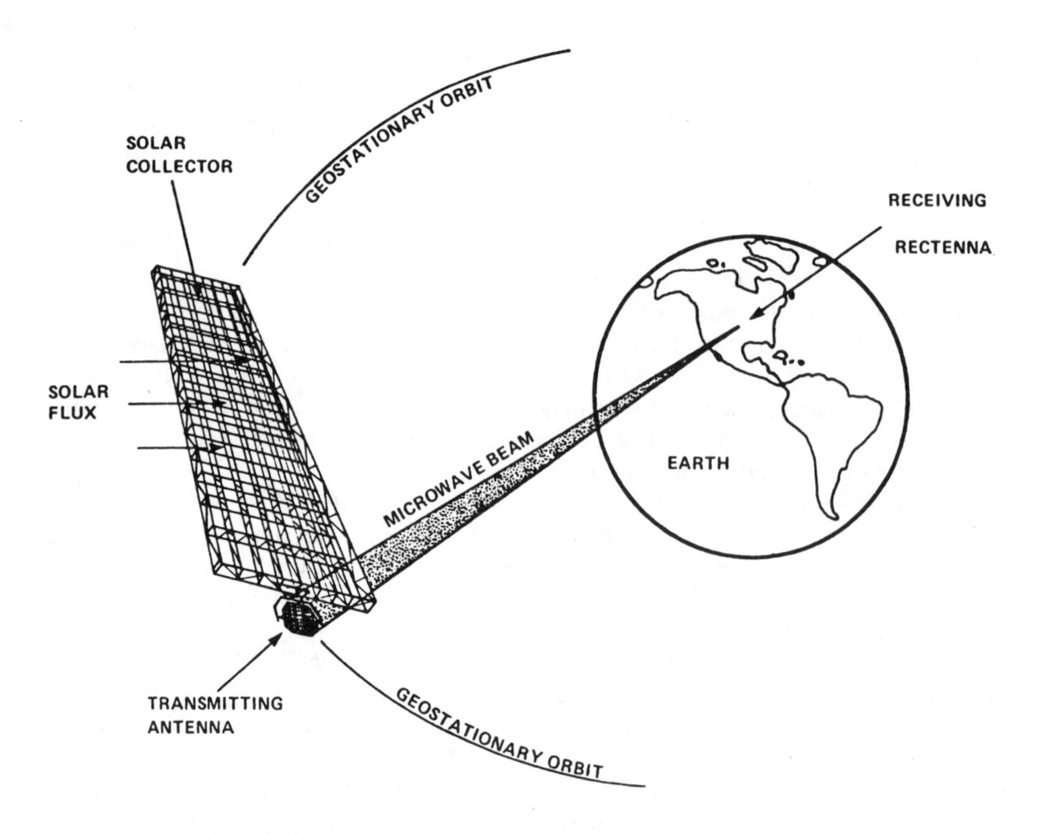


**FIGURE 1.** Major elements of a Satellite Power System (SPS).

Studies of the Satellite Power System (SPS), which evolved from Glaser's theory in the late 1970s, established its technical feasibility and assessed the environmental impact of the project.[6] These system studies were made in the context of supplying growing world energy needs in the face of diminishing fossil fuel reserves. By reducing fossil fuel use and the input of $CO_2$ to the atmosphere, the SPS may provide a benefit by countering a source of greenhouse warming.[7] The SPS approach deserves a fresh look from the perspective of this aid to the global environment.

The fundamental concept of the SPS is the transformation of solar energy into electrical power by huge spacecraft located in geosynchronous orbit (Figure 1). This electricity is converted to beam energy for transmission to the surface of the Earth. There it is intercepted by a rectenna that converts the microwave beam back into electrical energy for distribution in the commercial power grid. Major technological breakthroughs are not required to realize the reference SPS design. It does not need systems for energy storage and does not produce solid or gaseous wastes. The early studies focused on the economic viability of the SPS and on the local environmental risk from microwave power beaming. These issues still exist but need to be assessed in consideration of technological advances and of the potential benefit to the atmosphere.

In fact, hazards from the microwave power beams have been alleviated in the SPS conceptual designs by reducing microwave intensity to levels consistent with safety regulations for leakage from microwave ovens. Trading this local environment concern against the gain to the environment by replacing fossil fuels with a clean, renewable energy source provides a clear choice. The SPS should furnish an overwhelming net gain to the global environment.

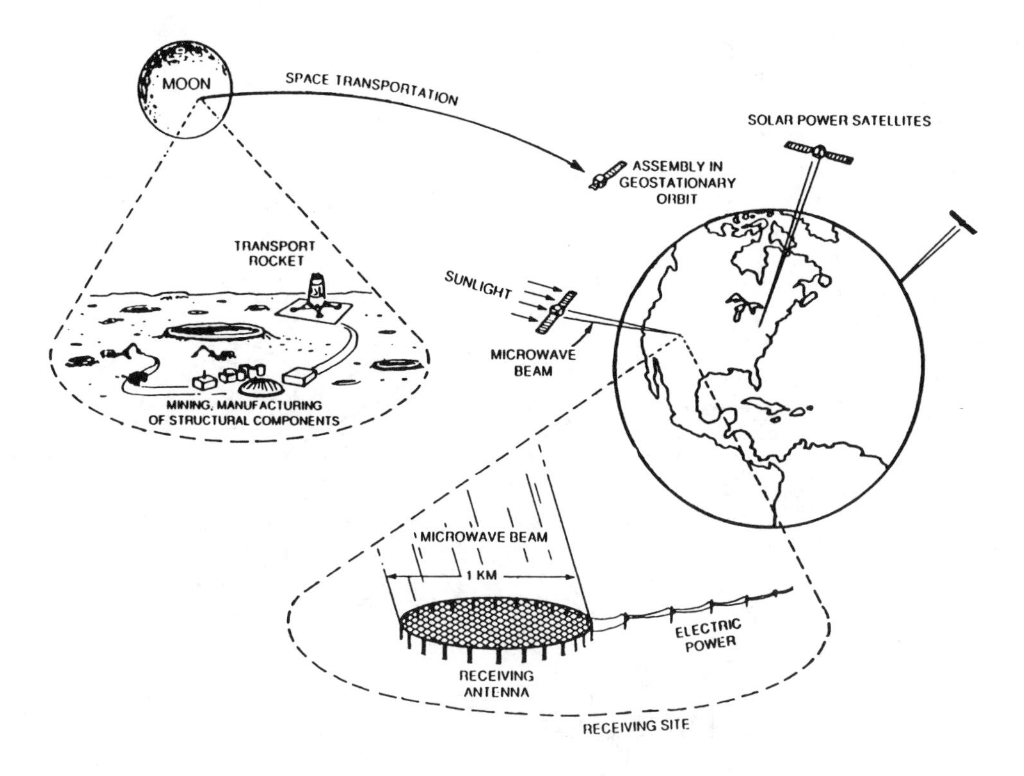


**FIGURE 2.** Satellite Power System (SPS) construction using lunar materials.

Technical options, which were not available during the study or were too uncertain for incorporation in the reference SPS, potentially improve the benefit-to-cost ratio when they are utilized in new SPS designs. For example the efficiency of solar cells has been demonstrated at nearly 40%, vs. the 10% used in the reference system trades. This gain in efficiency would drastically reduce the mass of material in orbit and, consequently, the transportation costs. Advances have also been made in energy beaming efficiency. The combination of technology advances and tests to demonstrate feasibility and refine analysis can be expected to reduce the size of the SPS and the space infrastructure needed to support the endeavor. However, not all aspects of the new SPS cost estimates are positive. Launch costs can be expected to be greater than were forecast in the reference studies. Operational and safety experience now provides a base for realistic assessment of launch and space infrastructure requirements and their price.

Serious consideration should be given to use of lunar materials to fabricate operational space power systems (Figure 2).[8] This option will change the launch-energy requirements for building the orbiting facility. This approach is attractive because less impulse is needed to move material from the lunar surface to geosynchronous orbit than is needed to move the same mass of material from the Earth's surface to orbit. Understanding the costs in this trade-off is an aspect of the Space Exploration Initiative. Obviously, more is involved than just delivery costs to put mass into the operational orbit. Technology development and operational complexity are major factors to be understood. With these considerations, it may be better to place the extensive power generation facilities on the surface of the moon.

## V. LUNAR POWER SYSTEM

The concept of a Lunar Power System (LPS) as conceived by Criswell and Waldron[9] would consist of installations on limbs of the moon (Figure 3). Operating from a moon base,

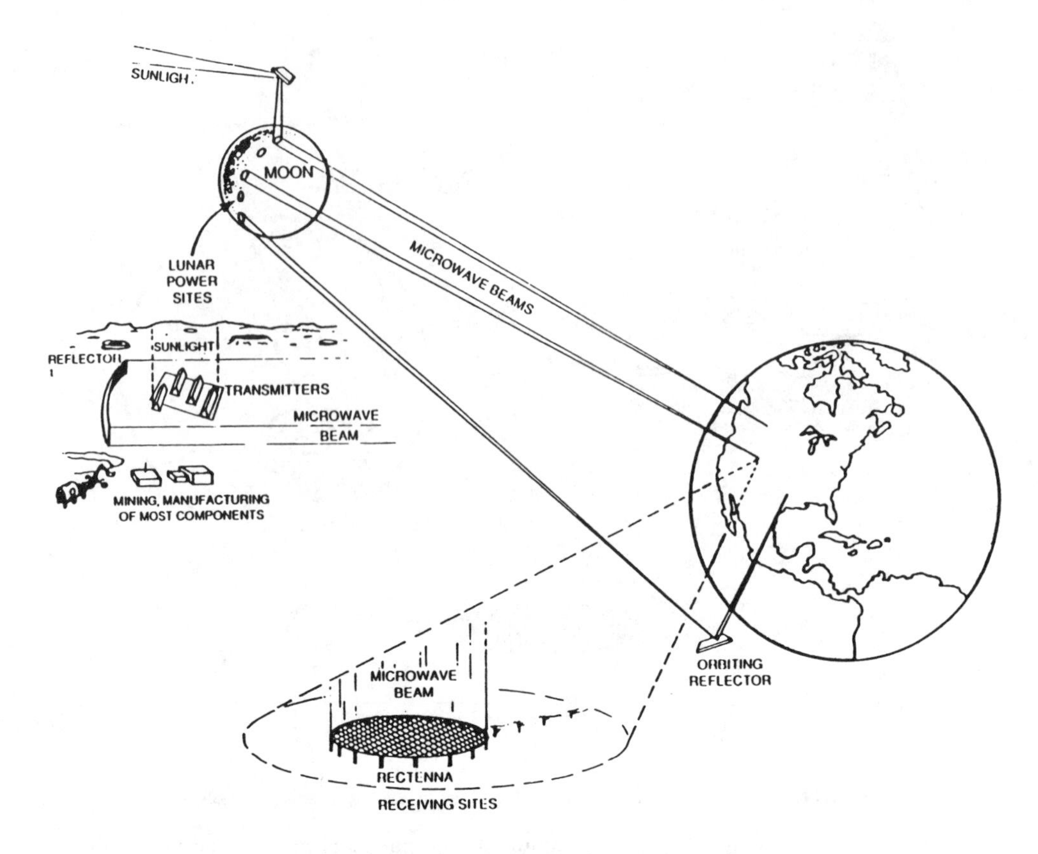


**FIGURE 3.** Lunar Power System (LPS) concept.

lunar materials would be processed into construction elements and functional components to convert solar energy into electrical power and to beam the power to the Earth. Comprehensive coverage of the Earth will require energy storage at the bases and the use of reflectors in lunar and Earth orbit. Such a macroengineering project represents a sizable payoff if implemented.

## VI. HELIUM-3 FOR FUSION FUEL

Resources from the moon are a factor in an alternative global energy source. The fusion of deuterium with a rare isotope of helium, helium-3, has the potential to convert nuclear energy directly into electrical energy with little radioactivity. This fusion reaction has been ignored for research because of the perceived limitations on terrestrial availability of helium-3. The analysis of Apollo lunar samples shows a relative abundance of the isotope in soil on the Moon.[8] With economically significant reserves demonstrated as a product of the space exploration program, fusion research can be focused on the deuterium/helium-3 reaction. Development of fusion reactors can then supply a clean safe alternative to fossil fuels and conventional nuclear reactors.

## VII. OTHER OPTIONS FOR MITIGATING THE EFFECTS OF CHANGE

The SPS represents an enormous power-generating capability in space. The availability of immense quantities of controlled energy in orbit enhances the viability of other concepts for reducing the effects of global change. With increased knowledge of the functioning of the Earth system processes, focusing energy on marginally stable areas may leverage control of some phenomena. Beneficial ideas range from modulating the intensity of hurricanes to nullifying the depletion of the ozone layer. Even with a source of power, these concepts may be limited by physical laws, economics, scale of operations, and risks. The development of a better understanding of these factors is needed to allow policy makers to make informed decisions on future actions.

An idea for modulating the fury of hurricanes was explored by a Texas A&M space systems design class. This concept was stimulated by the predictions that storm intensities will increase if global warming occurs. The approach postulates that heating the upper atmosphere along the leading edge of the storm would create a thermal inversion that would inhibit heated water vapor from rising from the ocean surface. The atmospheric heating might be done by irradiation of critical areas with a battery of lasers tuned to a frequency that causes water vapor to absorb the energy being transmitted. This is intended to disrupt the ring of intense convection that drives the storm. The resultant modifications in flow patterns are expected to reduce wind velocity. Combining radiation heating with cloud seeding to induce rainfall in key areas of a storm may optimize the effect. When a level of insight into storm mechanics, improved observational ability and the capability to focus sufficient energy is attained, control of the Earth's greatest storms may become a reality.

The capability to deliver concentrated energy might reduce the depletion of the ozone layer in the stratosphere using a technique proposed by Wong et al.[10] In the current thinking, chlorine molecules from the breakdown of CFCs are the catalyst in the reaction affecting the balance of ozone in the upper atmosphere. In theory, Wong et al. have shown that chlorine ions are less reactive in the ozone environment than are chlorine molecules. By generating radio waves at electron-cyclotron frequencies, the ionization of chlorine might inhibit the destruction of ozone. The next step might be the attraction of the ions to the lower atmosphere to clear them from the stratosphere. The feasibility of such an ambitious clean-up must depend on the scientific basis for the process, on engineering analysis and demonstration experiments. The need for action will be derived from an assessment of the severity of the effects on photosynthetic processes of reduced shielding from UV sunlight.

Another concept for dealing with ozone layer depletion postulates an interaction with the polar stratospheric clouds. These water crystal clouds are associated with the formation of an area of ozone level reduction near the South Pole often referred to as the "ozone hole." The ideas by another team from the Texas A&M design class proposes to disperse the clouds by heating them with microwave energy from orbiting spacecraft. However, this approach may only shift the period of ozone reactions with chlorine without achieving a net benefit to ozone levels. Even so, the effect of UV light on organisms in the Antarctic regions may be moderated by reducing the depth of the "ozone hole."

## VIII. RISKS AND BENEFITS

In a holistic analysis of geoengineering solutions, rigorous tradeoffs of risks to the benefits will be a key to developing an effective program. As promising trades develop, experiments and demonstrations can be used to refine the analysis and reduce uncertainty. Typical of trade analysis is the issue of transmission of power from orbit to the Earth. The concentration of energy in orbit requires transmission to the Earth's surface by some sort of energy beam. Adverse environmental effects in the vicinity of this beam represent a clear concern that can be addressed with an analysis, test, and demonstration program. Such a program could support development requirements and cost estimates. It also must address social and political issues. For instance, the potential of large space power systems to be used as a weapon has been cited as an objection to such a system. It must be emphasized that the hazard from SPS misuse is not the same as the perceived risk from nuclear sources. If the system is developed under international auspices, potential peril from misuse should be negligible. A well-structured demonstration and development program should allay concerns. In the final analysis, the application and control of the geoengineering level of systems will be complex. These considerations must be factored in to development, testing, and the demonstration of ways to interact with the environment in order to give policy makers the basis needed for implementation decisions.

## IX. POLICY MAKING PROCESS

The technical capability to act depends on a thorough understanding of the Earth's ecosystem processes, on technology and infrastructure, and on the assessment of benefits and risks inherent in any solution. This capability cannot be utilized unless implementation decisions are made by the policy makers. Schneider[2] describes the steps of the policy making process as technical analysis, policy analysis, and policy choice. Any consideration of geoengineering concepts must include the entire staircase to establish a viable program in the event that action must be taken to counter severe shifts in the environment. The products of the technical analysis must define the risks and benefits of a concept as well as its scientific and engineering feasibility. Economic feasibility analyses must take into account the cost of environmental impacts of alternative approaches. These cost projections should include quantification of public health risks as well as ecological damage. Technology development, tests, and demonstrations should address policy implementation issues in addition to basic system requirements. As O'Riordan and Rayner[11] note, "The fusion of science and politics is inescapable if major global change is to be averted before its discovery proves that we have acted too late." Policy analysis should include means for developing international cooperation around global concerns. To be effective, this process should parallel the monitoring and analysis research that is currently under way to synthesize the direction, rate, magnitude, and effect of environmental change.

## X. SUMMARY

Anthropogenic inputs to the atmosphere have been described as an uncontrolled experiment on Earth system processes. Mankind's activities are inducing modifications to the conditions that have evolved to support current planetary life systems. Incontrovertible evidence of the effects of change may not be established until those changes are far advanced. Fortunately, the increased concern over the effects of global climate change and depletion of the ozone layer has resulted in support for the Global Change Research Program. Research to understand Earth system processes is critical, but it falls short of providing ways of reducing the serious effects of change. If global environmental changes are severe in rate and magnitude, the international community will need all of the tools that can be supplied to alleviate the effects on society. To reduce the risk in policy implementation, geoengineering strategies need to be supported by technology development and demonstrations to discern the potential benefits and to assess the risks of implementation. This chapter is not intended as a comprehensive catalog of concepts. Rather, it is to alert decision makers to the potential of space as more than a tool to monitor the course of global change. Space-based concepts for environmental countermeasures are a potential supplement to earth-based actions. Consideration of space-based concepts for environmental geoengineering may provide a necessary adjunct to earth-based actions that enables society to deal with global changes.

# REFERENCES


1. ''Our Changing Planet,'' The U.S. Global Change Research Program, Rep. by Committee on Earth and Environmental Sciences, 1991.
2. Schneider, S. H., *Global Warming,* Sierra Club Books, 1989.
3. ''Policy Implications of Greenhouse Warming—Synthesis Panel,'' U.S. Committee on Science, Engineering, and Public Policy, National Academy Press, Washington, D.C., 1991.
4. Stix, T. H., Removal of chlorofluorocarbons from the Earth's atmosphere, *J. Appl. Phys.,* Dec., 1989.
5. Glaser, P. E., Power from the Sun: its Future, *Science,* 162, 3856.
6. Satellite Power System: Concept Development and Evaluation Program, NASA Reference Publ. 1076.
7. Eisenstadt, M. and Sorenson, J., World-Wide Solar Power Satellite System to Reduce $CO_2$: Do You Have a Better Way?, Center for National Security Studies Papers, No. 22, Nov., 1989.
8. Rep. of NASA Lunar Energy Enterprise Case Study Task Force, NASA Technical Memo. 101652, July, 1989.
9. Criswell, D. R. and Waldron, R. D., Lunar System to Supply Solar Electric Power to Earth, Vol. 1, Proc. 25th Intersociety Energy Conversion Engineering Conf., Reno, NV, 1990, 61–70.
10. Wong, A. Y., Wuerker, R. F., Sabutis, J., Suchannek, P., Hendricks, C. D., and Gottlieb, P., Ion Kinetics and Ozone, Proc. Int. CAGE Workshop, Nuovo Cimento, Jan., 1991.
11. O'Riordan, T. and Rayner, S., Risk management for global environmental change, in *Global Environmental Change, Human and Policy Dimensions,* Vol. 1, No. 2, March, 1991.

Chapter 5

# RESULTS OF ANALYSES OF A LUNAR-BASED POWER SYSTEM TO SUPPLY EARTH WITH 20,000 GW OF ELECTRIC POWER*

**David R. Criswell and Robert D. Waldron**

## TABLE OF CONTENTS



* From Criswell, D. R. and Waldron, R. D., in SPS 91, 2nd Int. Symp., Paris, 1991, a.3.6. With permission.

# I. 21ST-CENTURY POWER NEEDS

The 1 billion people of the developed nations of the world use approximately 6.7 kW/person to enable their high standards of living (kW is total power in kilowatts). On average, the other 4.3 billion people use less than 0.8 kW/person. These 4.3 billion others are strongly motivated, primarily through the examples provided by the developed nations, to raise their standard of living. Meeting their needs will require a vast increase in the production of power. Total world power production, mostly thermal, is approaching 11,000 GW (G = $10^9$). Approximately 30% of this thermal power is used to produce electrical power. World electric systems now generate 1,800 GWe (''We'' is watts of electric power).

It is likely that $10^{10}$ people will inhabit our planet in the year 2050. If today's technologies and energy sources are employed, then the world will have to provide over 60,000 GW to support this population at the level the developed nations currently enjoy. Table 1 shows that this is not possible. Either the primary energy source is inadequate (1 to 7), the system will be interruptible and expensive (8 and 9), or long-lived contaminants become a dominant concern (10).

As technology advances, 2 kWe/person could sustain a higher level of affluence world-wide than now exists in the developed nations. (Criswell and Waldron, 1990). Figure 1 is a simple model of the projected growth and revenue of an electric power system that could supply 20,000 GWe to the world in 2050. The top curve is the growth path the new system must follow. The bottom curve shows the annual cash flow if the electric power is sold to end users at 0.1 $/kW-h. Mature cash flow exceeds $15,000 billion/yr or approximately 10% of the anticipated gross world product (GWP) in 2050. The accumulated gross return would exceed $840,000 billion. The new power system must accommodate growth beyond 2050.

A good world environment can be provided if the source of the 20,000 GWe is clean, reasonably priced, and independent of the biosphere. Even if new technologies such as fusion and large space solar power satellites (Table 1, 11 to 14) are demonstrated early in the next century, adequate systems to produce and maintain them will be much larger than required for the Lunar Power System (LPS) (Criswell, 1991). We maintain that the global needs can be met by providing solar power bases on the moon (Table 1, 15).

# II. LUNAR POWER SYSTEM

P. Glaser (1977) introduced the concept of establishing huge solar power satellites (SPS) in space; the basic challenge is how to produce them economically. In addition, the magnitude of launch operations would lead to environmental contamination (NASA, 1989, p. 73).

NASA funded studies on the production of SPS from lunar materials delivered to factories in space. MIT examined the production and design of the factories (Miller, 1979). General Dynamics developed systems-level engineering and cost models for the production of 1 10-GWe lunar-derived SPS (LSPS) per year over a period of 30 years, and concluded that LSPS would likely be less expensive than SPS after the production of 30 10-GWe satellites (Bock, 1979).

The SPS and LSPS studies raised basic questions. Why transport materials from Earth or the moon to build large platforms in space? Why not build the solar power collectors and transmitters on the moon from lunar materials?

The moon is a far larger platform than any that can be built in space. The same surface of the moon always faces the Earth. Can we build the solar collectors on opposing limbs of the moon, from the local materials, and beam the power from the moon directly to Earth? Those questions lead to the conceptualization of the LPS (Waldron and Criswell, 1991; Mueller, 1984).

**TABLE 1**
**Characteristics of 20,000 GWe Global Power Systems**

| Power system | Maximum energy inventory (GW*yr) | Annual renewal rate (GW) | Maximum useful power (GW) | Limiting factors (@ $2*10^{4}$GWe) | Deplete or exhaust* (yr) | Pollution products | Long-term cost | Feasibility by 2050 |
|---|---|---|---|---|---|---|---|---|
| (1). Bioresources | $<2*10^{5}$ | $<1.5*10^{4}$ | <1000 | Supply<br>Processing<br>Nutrients<br>Water & land | <0.05 | $CO_2$<br>Biohazards–meth-ane-disease<br>Erosion | Up | Not possible |
| (2). Coal | $<1.5*10^{6}$ | 0 | $<2*10^{4}$ | Supply<br>Pollution | <100* | $CO_2$<br>Ash acids | Up | Not possible |
| (3). Oil/gas | $<10^{5}$ | 0 | $<5*10^{3}$ | Supply<br>Lost value | <30* | $CO_2$<br>Acids | Up | Not possible |
| **Renewable terrestrial, noncombustian power systems** | | | | | | | | |
| (4). Hydroelectric | $<2*10^{4}$ | $<2*10^{3}$ | $<2*10^{3}$ | Sites<br>Rainfall | <0.1 | Sediment<br>Flue water<br>Floods | Up | Not possible |
| (5). Tides | 0 | <50 | <50 | Sites<br>Input | <0.003 | Circulation | Up | Not possible |
| (6). Geothermal | $>5*10^{17}$ | $>5*10^{8}$ | <100 | Good sites<br>Local depletion | <0.005 | Minerals<br>Ground cracking | Up | Not possible |
| (7). Ocean thermal | $<1*10^{6}$ | $<1*10^{4}$ | $<1*10^{4}$ | Cold & warm water<br>Build OTECs | <0.5 and <100* | Waste heat<br>$CO_2$ store<br>Corrosion | Up | Not possible |
| (8). Wind | 0 | $<1*10^{6}$ | $<2*10^{4}$ | Diffuseness<br>Storage | <1 | Land use<br>Intrusive | Up | Not possible |
| (9). Terrestrial solar power | 0 | $<1.8*10^{8}$ | $<1*10^{5}$ 5% earth surface (@ 1% effic. overall) | Construction<br>Maintenance<br>Clouds<br>Power–Storage–Distribution | $>10^{8}$ | Waste heat<br>Production wastes<br>Induced climates<br>Land use | Up | Uncertain |

**TABLE 1 (continued)**
**Characteristics of 20,000 GWe Global Power Systems**

| Power system | Maximum energy inventory (GW*yr) | Annual renewal rate (GW) | Maximum useful power (GW) | Limiting factors (@ $2*10^4$GWe) | Deplete or exhaust* (yr) | Pollution products | Long-term cost | Feasibility by 2050 |
|---|---|---|---|---|---|---|---|---|
| **Nuclear power systems** | | | | | | | | |
| (10). Fission | $<1*10^6$ | 0 | <500 (present level) | Acceptance<br>Life cycle & fuel costs | <0.03 and <100* | Radioactive–fuels–parts | Up | Not possible |
| (11). Fusion | $>1*10^9$ (D–T) | 0 | $>2*10^4$ baseload | Feasibility<br>1st wall life<br>Build-up | >1000 | Radioactive parts<br>Waste heat | Up | Not possible |
| (12). Fusion | $0.001–1*10^8$ (D–$He_3$) | 0 (moon) | $>2*10^4$ baseload | Mining $He_3$ | 0 to <100* | Lunar atmosphere | Up | Not possible |
| **Space solar power systems (SPS)** | | | | | | | | |
| (13). Geo-SPS | 0 | $<X*10^4$ X–TBD | $<X*10^4$ | Geo-arc<br>Managing–Satellites–Shadowing | $>10^9$ | Shadowing Earth<br>Visible SPS<br>Debris | Up | Uncertain |
| (14). Nongeo-SPS | 0 | $\geqslant 1*10^6$ | $>1*10^5$ | Production & repair<br>Use on Earth | $>10^9$ | New objects<br>Stray microwave | Down | Uncertain |
| (15). LPS | 0 | $>1*10^6$ (@ 10% effic.) | $>1*10^5$ | Moon's area<br>Power use on Earth | $>10^9$ | Stray microwave | Down | Promising |

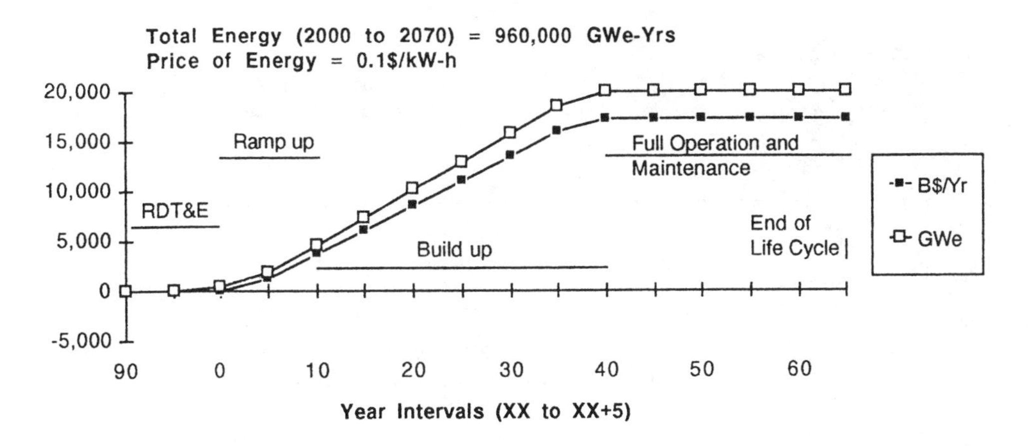


**FIGURE 1.** Capacity and revenue of a New Planetary Power Source. Criswell, D. R. and Waldron, R. D., *Proc. SPS 91 Power from Space: 2nd Intl. Symp.*, Paris, 1991.

The major elements of LPS are shown in Figure 2. The basic system consists of pairs of power basis (1 and 2) on opposite limbs of the moon as seen from Earth, and rectennas on Earth (3, 4, 5, and others). The rectennas convert the microwaves to electric power. Hundreds to thousands of low-intensity microwave beams would be directed from each lunar base to the rectennas on Earth and also to facilities and vehicles in space (6). A microwave power beam is only created, on request, to feed power to a designated rectenna. The beam is shut off when the need for power at that rectenna ceases. Beams would not be swept across the Earth from one rectenna to another.

A given lunar power base would project energy for 12.5 days out of each lunar month. A given terrestrial rectenna receives power approximately 30% of the time when the moon is above its local horizon. This basic LPS requires power storage units on the moon or Earth in order to provide continuous power at Earth. The basic LPS can be extended to provide continuous, load-following power to rectennas. Relatively low mass and simple microwave reflectors (7), in mid-altitude, high-inclination orbits about Earth, can redirect microwave beams to rectennas that can not directly view the moon.

The microwave reflectors can eliminate the need for long-distance power transmission lines on Earth. Also, additional sunlight can be reflected by mirrors (8) in orbit about the moon to bases 1 and 2 around lunar sunrise and sunset and during eclipses. Each limb base can be augmented by a set of photovoltaics just across the lunar limb. Power lines would connect these cells to the transmitters on the Earthward side of the moon.

Figure 3 provides a sequence of five progressively closer-in views of a power base, such as "1" in Figure 2, and that thousands of power plots that constitute one power base (pb). View 1, Figure 3, shows an area of the limb of the moon, as seen from the Earth, that encompasses three power bases. View 2 is from high above that lunar limb. This view reveals that each base is elliptically shaped and approximately six times longer along a moon-Earth line than its apparent diameter as viewed from Earth.

View 3 is a side view of a vertical cut through base pb. It shows a few of the thousands of power plots. View 4 is in the plane of the vertical cut, looking toward the moon from the Earth. It shows that the power plots are arranged within pb so that the microwave reflector grids associated with each power plot appear to fill a circle 10 to 100 km in diameter. That circle of reflector grids forms the huge, segmented phased-array antenna that beams energy back to Earth.

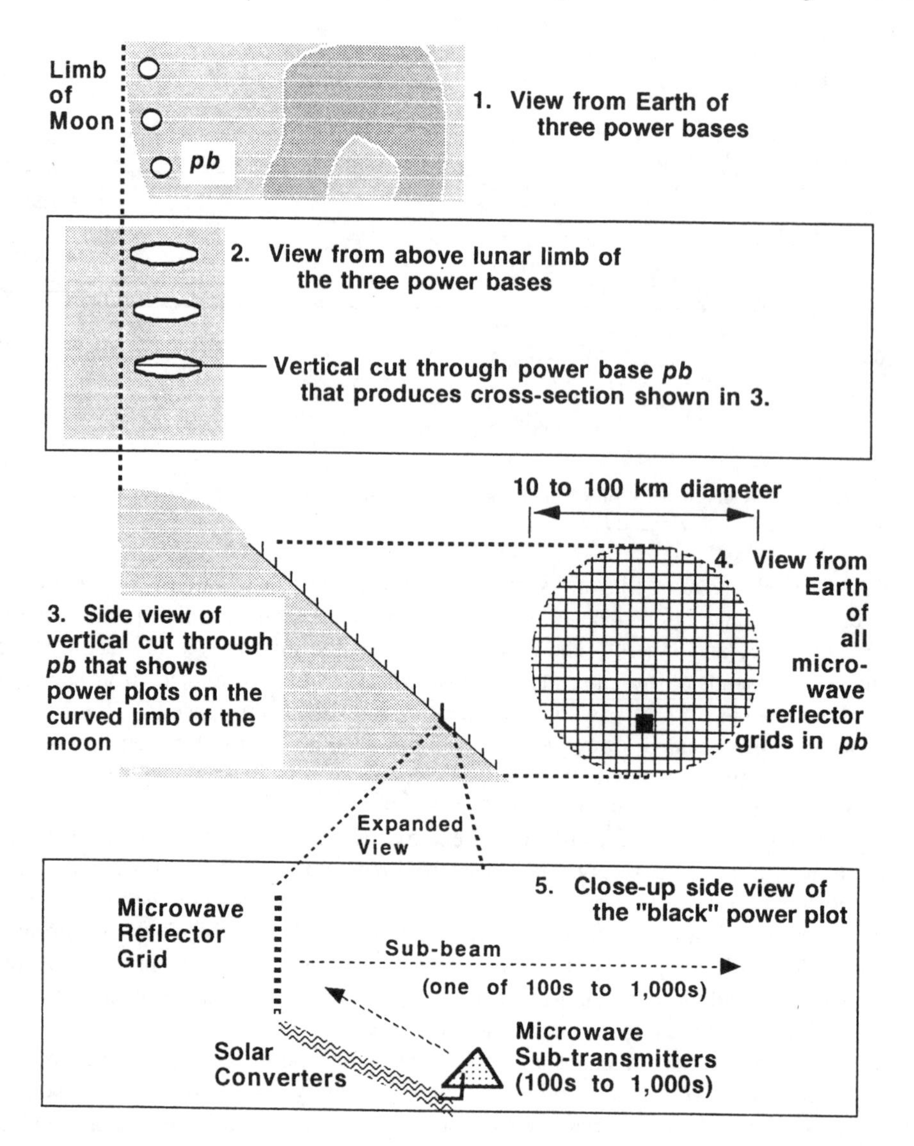


**FIGURE 3.** Progressively closer views of a lunar power base. Criswell, D. R. and Waldron, R. D., *Proc. SPS 91 Power from Space: 2nd Intl. Symp.*, Paris, 1991.

View 5 is a close-up side view of one power plot (the "black" one in View 3, and also near the bottom-center of View 4). The power plot is composed of three major elements: solar converters, microwave subtransmitters, and a microwave reflector grid. The solar converters provide electric power through a grid of buried wires to many solid state microwave subtransmitters. The transmitters illuminate the microwave reflector grid with many sub-beams, and the grid directs each sub-beam toward Earth.

The characteristics of the three elements are considered next. The solar converters will be thin-film photovaltaics formed on thin sheets of lunar-derived glass. Only moderate conversion efficiency is needed (>5 to 10%). The moon is a far better location for intrusive, large-area solar converters toward the microwave reflector grid at the opposite end of the power plot. Power regulation and limited power storage capability can be provided. Some components of the subtransmitters might be imported from Earth.

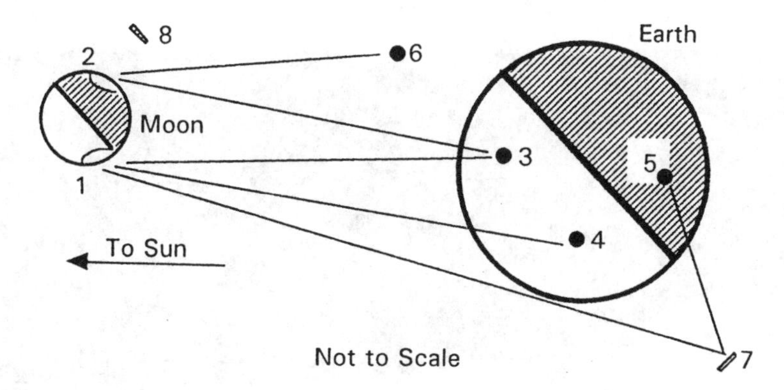


**FIGURE 2.** Major elements of the Lunar Power System (LPS). See text for details. Criswell, D. R. and Waldron, R. D., *Proc. SPS 91 Power from Space: 2nd Intl. Symp.*, Paris, 1991.

Each microwave reflector grid is constructed of foamed or tubular lunar glass beams that support a microwave reflective surface consisting of a cross-grid of glass fibers coated with a metal such as aluminum or iron. The segmented aperture (View 4, Figure 3) could be a few kilometers across for high-frequency beams ($<1$ cm) transmitted to space facilities, and the order of 100 km across for longer wavelength beams (10 cm) transmitted to Earth.

The enormous segmented antenna is possible because the moon is extremely rigid and nonseismic, the antenna always faces Earth, the antenna segments can be constructed of local materials, and the cost of each antenna is shared by thousands of beams of high total power. Thousands of coordinated sub-beams from all the power plots produce one of the power beams depicted in Figure 2. The beams are convergent (near field), but slightly defocused, like a spotlight, to distances ($= D*D/w$) many times that of the Earth-moon distance. $D = 30$ to 100 km; wavelength $(w) = 10$ cm.

Each LPS beam can be fully controlled in intensity across its cross-sectional area to a scale of a few 100 m at Earth. This allows the LPS beams to uniformly illuminate rectennas on Earth that are larger than 200 to 300 m across. The microwave beams projected by the LPS should have very low sidelobe intensity and no grating lobes. The stray power level should be very low and incoherent. LPS could probably operate economically at a lower power density ($\sim$1 mW/cm$^2$) than the leakage allowed under federal guidelines (5 mW/cm$^2$) from microwave ovens used in homes. A beam intensity of 23 mW/cm$^2$ produces little sensible heating in animals. The stray, incoherent power levels of the microwaves on Earth of a 20,000 GWe LPS may be less than the power per unit area radiated thermally by a human or the Earth itself. If so, the power-beaming system can be completely safe. LPS beams can efficiently service rectennas on Earth after they are more than 200 m in diameter and several 10s of MWe output. Thus, as rectennas are enlarged beyond a diameter of 200 m, the additional growth can be paid for by our present cash flow derived from power sales. Smooth growth in capacity from a small starting level provides fundamental financial advantages over all other major power systems.

Figure 4 depicts the operations needed to construct a lunar power plot (De Generes and Criswell 1983). Several tractors are shown smoothing the lunar surface, extracting fine-grained iron, burying wire for power collection, and laying down glass sheets under which are layered solar-converting thin films of moderate efficiency (5 to 10% are adequate).

In the foreground is a mobile glass processor that melts lunar soil to produce foamed glass supports, fiberglass, and glass sheets. The supports and fiberglass are used to make the microwave reflectors. One reflector is shown being erected. Solar electric power is provided to sets of microwave subtransmitters that are buried under the mound at the earthward end of each power plot. Note that the Earth remains in the same general position

**FIGURE 4.** Artists' rendition of the construction of a demonstration lunar power plot. Criswell, D. R. and Waldron, R. D., *Proc. SPS 91 Power from Space: 2nd Intl. Symp.*, Paris, 1991.

in the sky at a given base. Most of the emplacement operations are conducted by a fleet of relatively small and independent machines that move from one construction area to another. The rate of installation of new power is proportional to the number of machines and their productivity. Section III describes costs and profits scaled to deploying sufficient machines to emplace 50 GWe/yr of power; Section IV presents lower emplacement rates.

## III. RESULTS OF MODELS

Engineering and life-cycle cost models of the LPS have been developed for an LPS that meets the power profile shown in Figure 1 and that delivers approximately 960,000 GWe-yr of energy to Earth between the years 2000 and 2070 (Criswell and Waldron, 1990; NASA, 1989). Approximately 400,000 tons of production equipment (miners, materials processors, construction units, habitats, etc.) are taken to the moon between the years 2000 and 2040. Half of the materials arrive the first 10 years. Components, supplies, personnel, and process expendables constitute the additional 600,000 tons transported to the moon on a steady basis between 2000 and 2070. Approximately 18,000 tons of facilities are required on the moon for an average working population of 4,000 people. In low lunar orbit, 12,000 tons of facilities support approximately 340 people in logistics and production of lunar orbital reflectors (No. 8, Figure 2). An additional 170,000 tons of facilities in low Earth orbit support logistics. They also manufacture and maintain the Earth orbital reflectors (No. 6, Figure 2).

Averaged over the entire program, the delivery of 1 GWe-yr of energy to Earth requires the export of 4 tons of mass from the Earth to the moon. The LPS pays back its electric energy of production in 5 days. Over the 70 years of power production, the LPS will return approximately $840,000 billion (B) (B $= 1*10^9$) of net profit if the power is sold for 0.1 $/kWe-h. Total LPS costs, nondiscounted over the 80-year period from 1990 to 2070, are 18,000 B$. This is an enormous total; however, LPS provides such a large return that the energy cost is only 0.002 $/kWe-h.

Figure 5 presents the annual revenue and expenditures for all LPS operations from 1990 to 2015. The model of LPS assumes that all costs are included. The costs are divided into five-year intervals, and the annual levels (B$/yr) are presented for four categories of expenditures: transportation (Trans), Lunar Base (LunB), Earth facilities and rectennas (EF & Rec), and Lunar Power System elements (LPS). The five-year period of research and ground-based demonstrations costs 100 B$. The next five years of development, production, and start of operations of all flight elements cost approximately 500 B$. The elements in this interval are so small that they are barely resolved in Figure 5. It is extremely important to note that after the year 2000, the construction of rectennas dominates all costs of the LPS program. Total cost of rectenna construction and operation between 2000 and 2070 is approximately 18,000 B$. Space operations total only 1,800 B$.

The system will grow to provide Earth with 20,000 GWe by the year 2040. This system would begin at the demonstration stage (50 GWe/yr) and build to the maximum emplacement rate (560 GWe/yr) over the next 10 years. The LPS would be in full operation 40 years after start of construction. This model assumes that maintenance of the LPS requires 50% of the expenditures to build it. The electric energy payback time of 5 days for the lunar installation is much shorter than any other electric power system.

In this model, the revenue from the sale of power at 0.1 $/kWe-h on Earth would begin by the year 2002. Revenue would exceed expenditures within the third year of power production. Within the next three years, the enterprise pays off all previous investments. Annual revenues increase steadily until the maximum power production of 20,000 GWe is achieved in 2040, and the net annual profit exceeds 15,000 B$/yr. An LPS base can be maintained indefinitely. As technology advances, an LPS base can be upgraded to a much higher power level while it is operating. The year 2000 as a date for the start of lunar operations is arbitrary. Delays simply lose profits, defer creation of new markets, push back global environmental benefits, and postpone permanent, self-substained occupancy of the moon.

Figure 6 clearly shows that the LPS can provide a very attractive return on investment under the baseline model. It also shows the effects on LPS profitability of increasing and

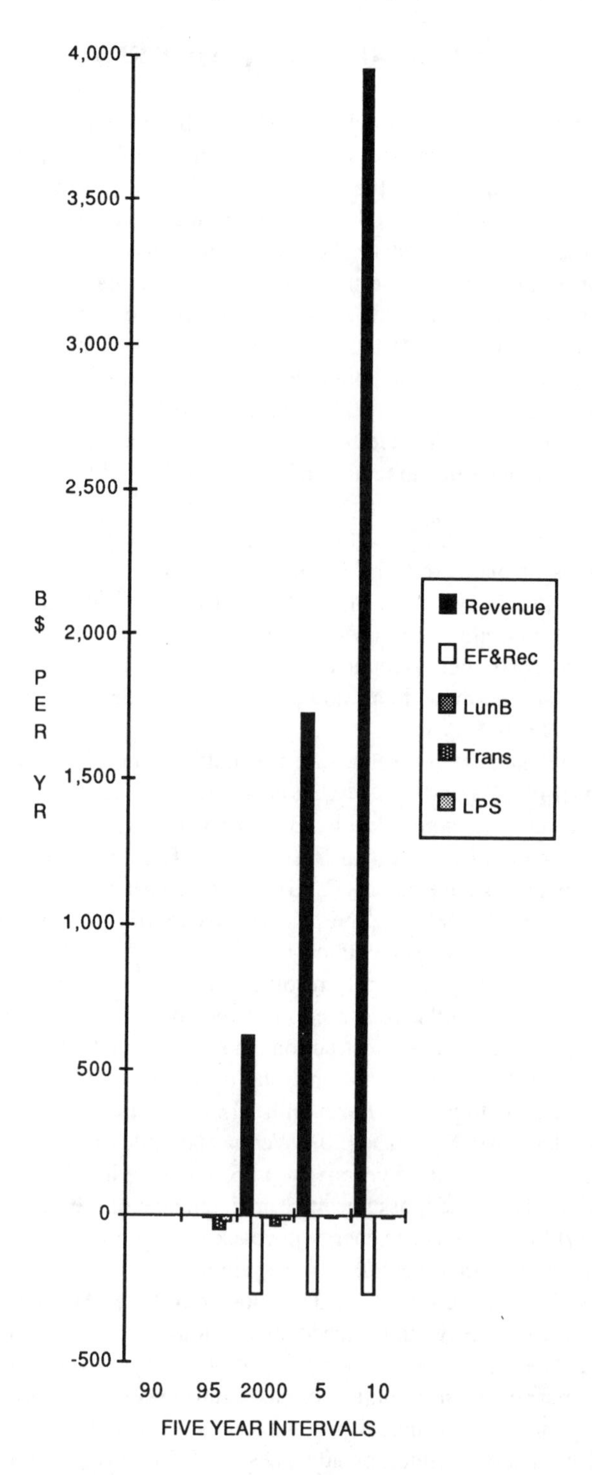


**FIGURE 5.** Annual revenue and expenditures for all LPS operations from 1990 to 2015. Criswell, D. R. and Waldron, R. D., *Proc. SPS 91 Power from Space: 2nd Intl. Symp.*, Paris, 1991.

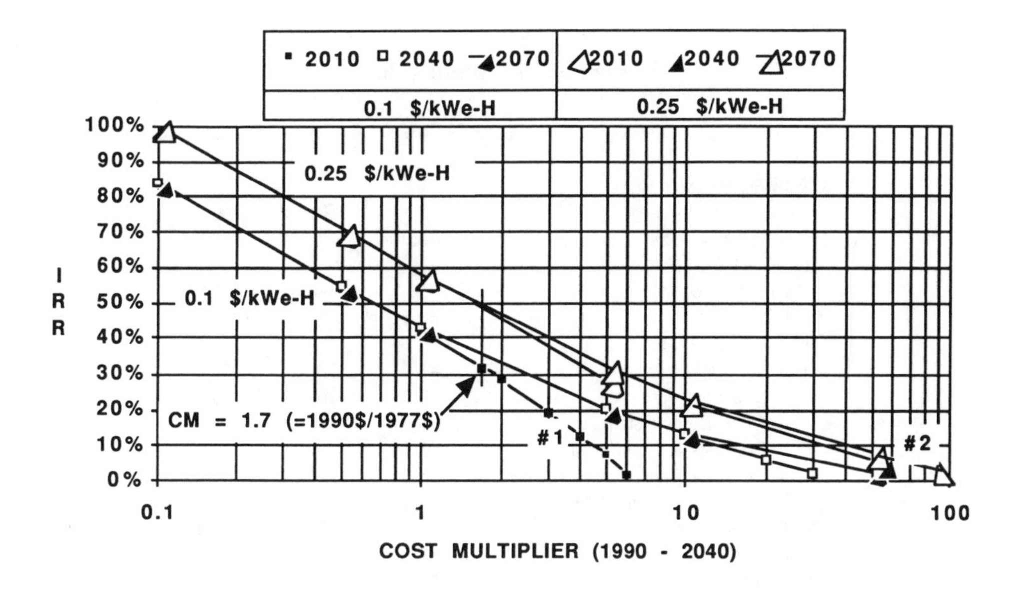


**FIGURE 6.** Internal rate of return (IRR) vs. multiplication of all costs (1990 to 2040). Criswell, D. R. and Waldron, R. D., *Proc. SPS 91 Power from Space: 2nd Intl. Symp.*, Paris, 1991.

decreasing the cost of installing LPS (cost multiplier [CM] applied to all costs between 1990 and 2040) and of a high selling price for power. Internal rate of return (IRR) is used to measure the profitability of LPS. IRR is the annual interest rate a bank would have to pay an investor between the start of profitability and the year 20XX (dates in box), given the profile of investments (expenditures) shown in Figure 5, to match the revenue in Figures 1 and 5.

The studies by General Dynamics were based on 1977 dollars. Prices increased by a factor of approximately 1.7 between 1977 and 1990. The reference case has CM = 1.7, and the selling price of power = 0.1 $/kWe-h. For the reference case, the average IRR = 31%/yr between the start of profitability in 2005 and the year 2010. IRR increases to 35%/yr when the period for return on investment is extended through the two intervals 1990 to 2040 and 1990 to 2070.

Curve 1, Figure 6 (black squares) applies to the reference case for the interval starting with the year of profitability and extending to the year 2010. Note that this IRR decreases as CM increases and crosses IRR = 0 near CM = 6.2. This means that if costs of the reference case rise by a factor of 3.6 (= 6.2/1.7), then the short-term IRR goes to 0. However, revenues are so large that over the longer time periods of 1990 to 2040 (open squares) and 1990 to 2070 (dark triangles) the IRR is 18%/yr at CM = 6.2.

If the world relies on conventional power sources, then the price of electricity will likely rise to 0.25 $/kWe-h or even higher (NASA, 1989). Note the upper curve at CM = 1.7. If power from the reference LPS is sold at 0.25 $/kWe-h, then IRR = 48%/yr when the return period is extended to 2070. Higher selling prices greatly increase the financial robustness of LPS. Note the end point of curve 2 at CM = 100. IRR is still positive at both 2040 (IRR = 0.7%/yr) and 2070 (IRR = 4%/yr) even though costs have increased by 5,900%, or a factor of 59 (= 100/1.7) compared to the reference case.

LPS is very likely to be more robust financially than discussed above. Figure 5 shows that the major cost element of LPS is construction of the rectennas on Earth. The reference case assumes rectennas cost 200 M$/km$^2$ to construct and that they operate at 23 mW/cm$^2$. Rectenna costs should go down as more are produced, and some might be operated at a

**TABLE 2**
**Parameters of Smaller Bases**

| Cases | #1 | #2 | #3 |
|---|---|---|---|
| GWe installed over 10 yr | 1 | 10 | 100 |
| GWe-yr of energy | 5 | 50 | 500 |
| Gross Revenue (B$) (@ 0.1$/kWe-h) | 4.383 | 43.83 | 438.3 |
| Net Revenue (B$) | −55.7 | −46.8 | 194.9 |
| Total Costs (B$) (sum 1 + 2 + 3) | 60.1 | 90.6 | 243.4 |
| (1). R & D (B$) (sum a + b + c + d) | 42.4 | 50.9 | 85.5 |
| a. LPS Hardware | 10.7 | 10.7 | 10.7 |
| b. Consortium system | 1.1 | 2.9 | 10.1 |
| c. Facilities & Equipment | 5.1 | 10.0 | 29.9 |
| d. Transportation | 25.5 | 27.4 | 34.7 |
| (2). Space & Operations (B$) | 17.2 | 34.2 | 102.5 |
| (3). Rectenna (B$) | 0.6 | 5.5 | 55.4 |
| $/kWe-h | 1.37 | 0.21 | 0.06 |
| Moon (ton) | 2,284 | 6,194 | 21,552 |
| Space (ton) | 974 | 2,680 | 9,677 |
| People (moon & space) | 30 | 85 | 300 |

high power density. CM could be substantially <1.7. It is reasonable to anticipate IRR >40%/yr for 0.2 $/kWe-h and IRR >60%/yr for 0.25 $/kWe-h. If rectenna construction costs decrease, then research and development (R & D) and space expenditures should increase by a factor of 60 (0.1 $/kWe-h) to 300 (0.25 $/kWe-h) without significantly degrading the financial robustness of the LPS.

As soon as large transmitting apertures are established on the moon, they can begin to provide many different beams of low power, and each beam can be converged to a few 100 m in diameter. Thus, rectennas on Earth can initially be small, the order of 10s MWe, and then grow smoothly in power output and diameter. Their growth can be paid for out of our current cash flow. All other large power systems require a decade or more of up-front investment before income is generated.

No other global investments on the horizon could provide such enormous returns over many decades. In addition, LPS should not have the hidden environmental costs of conventional power systems.

## IV. STARTING SMALL

The LPS model was used to estimate the costs of establishing much smaller power systems. The results are presented in Table 2 (Case #1, 1 GWe; Case #2, 10 GWe; and Case #3, 100 GWe). The model assumed a 10-year period of R & D, a 3-year deployment of all equipment, and then 10 years of steady-state emplacement of power beaming capacity from the moon to the Earth. These models draw on the analyses by General Dynamics that were scaled to emplace 10 GWe/yr of SPS. Several costs, such as transportation, do not scale linearly to lower rates of power emplacement. Consider the results as encouragement for further work. Case #3 is closest to the situation modeled by Bock.[1]

R & D is the dominant expense for final power levels <100 GWe. R & D that is unique to the equipment to install LPS on the moon is small, 15 to 20 B$, relative to the overall expenditures required to establish a lunar base and the associated transportation system. Costs of space equipment and operations increases sharply between 10 and 100 GWe of final installed capacity. The cost of power drops sharply as the installed power increases.

Between 10 and 100 GWe of final installed power, the net cash flow goes positive (price of power 0.1 $/kWe-h). Net expenditures might be <100 B$ by the time positive cash flow begins. Power could sell for much greater prices to customers on Earth.

There are many possibilities for reducing costs as cis-lunar space becomes energy rich. LPS can support experimenters, supply power for space logistics, and power isolated Earth receivers. Lunar materials can be used to reduce the costs of transportation and logistics and to build portions of the emplacement systems.

## V. ORGANIZING THE LPS PROGRAM

To develop LPS, three types of investors are anticipated: governments, consortia, and local organizations. Between 1990 and 2000, government programs will pay for the development and initiation of the transportation elements and the lunar base. Between 1900 and 2000, expenditures can be comparable to present U.S. government expenditures in aerospace products for the U.S. Department of Defense and NASA. LPS can provide an excellent, peaceful focus for the present defense- and technology-related organizations of the space-faring nations.

A national or international consortium can be formed to develop, procure, and implement the elements for LPS production and do the R D T & E for rectennas. After the year 2000, this consortium can conduct all off-Earth operations. Between 2000 and 2005, it would begin receiving a net positive revenue from the sale of power on Earth. More than one consortium can be formed; many lunar bases are needed.

Rectenna R & D, both for rectennas and their means of production, can involve all the nations of Earth. Rectennas can, as appropriate, be constructed, operated, and paid for by private groups, cooperatives, and countries. Virtually all the costs of rectenna production will be covered by current cash flow. The major challenges are start-up costs and public confidence in LPS.

## VI. LUNAR POWER SYSTEM, SUMMARY

LPS can be developed expeditiously. We have sampled the moon and possess sufficient understanding of the moon to make many uses of its resources. Costs of the initial stages of LPS can be significantly reduced if done with the emplacement of a manned lunar base. Small demonstration units can quickly provide confidence in our abilities to emplace and operate LPS.

The equipment to implement LPS has a very low mass per unit of delivered power compared to all other systems, and can emplace the power system rapidly (Criswell and Waldron, 1990; NASA, 1989). Advances in technology can sharply decrease expenditures and increase the emplacement rate of power for all phases of the LPS program. Many of the key technologies for LPS are developing rapidly because of their value in the terrestrial marketplace. Thin-film solar arrays, solid state microwave electrons, and microcomputers are key examples.

To supply the Earth with 20,000 GWe by the year 2050, we must look beyond the confines of Earth. We must look to the energy production of the sun and the already known natural resources of the moon. LPS appears to be the only option for providing large-scale power systems in the next century without severely distorting the world economy and ecology.

# REFERENCES


Bock, E. (Program Manager), Lunar Resources Utilization for Space Construction, Contract NAS9-15560, DRL No. T-1451, General Dynamics — Convair Division, San Diego, CA, 1979.

Criswell, D. R., Solar power system based on the moon, in *Space Power: A 21st Century Option for Energy Supply to Earth,* Glaser, P. and Davidson, F., Eds., E. Horwood, 1991, in press.

Criswell, D. R. and Waldron, R. D., Lunar system to supply solar electric power to earth, *Proc. 25th Intersociety Energy Conversion Eng. Conf.,* Vol. 1, 61–71, Paper 900279, Reno, NV, 1990, in Nelson, P. A., Schertz, W. W., and Till, R. H., Eds., *Aerospace Power Systems,* American Institute of Chemical Engineers, New York, 72–76.

Criswell, D. R. and Waldron, R. D., Results of Analyses of a Lunar-based Power System to Supply Earth with 20,000 GW of Electric Power, *Proc. SPS 91 Power from Space:* the 2nd Int. Symp., Paris, France, 1991, a36, 186–193.

Criswell, D. R. and Waldron, R. D., International Lunar Base and Lunar-based Power System to Supply Earth with Electric Power, *Proc. 42nd Congr. Int. Astronautical Fed.,* IAF Paper IAA–91–699, Montreal, Canada, Oct. 1991.

De Generes, W. and Criswell, D. R., Unisys Corporation, 1983. With permission.

Glaser, P., Solar power from satellites, *Phys. Today,* 30–38, February, 1977.

Miller, R. Pl., Extraterrestrial Materials Processing and Construction of Large Space Structures, NASA CR-161293, Vol. 1, 2, 3, Space Systems Lab., MIT, Cambridge, 1979.

Mueller, G. The 21st century in space, *Aerosp. Am.,* 84–88, January, 1984.

NASA, *Report of NASA Lunar Energy Enterprise Case Study Task Force,* NASA Tech. Memo. 101652, July, 1989, 179 pp.

Waldron, R. D. and Criswell, D. R., A Power Collection and Transmission System, U.S. Patent 5,019,768 (May 28, 1991).

## *Section II: Role of Oceanographic and Geochemical Methods to Provide Evidence for Global Climatic Change*

Chapter 6

# CLIMATE AND THE MARINE RECORD: SCIENTIFIC RESULTS OF THE OCEAN DRILLING PROGRAM

**P. D. Rabinowitz and J. G. Baldauf**

## TABLE OF CONTENTS



0-8493-4419-0/93/$0.00 + $.50
© 1993 by CRC Press, Inc.

# I. INTRODUCTION

The Ocean Drilling Program (ODP) is a long-term internationally sponsored program of basic research, with Texas A&M University as the Science Operator. The program, which commenced field operations in January 1985, is directed toward studies of the evolution of the solid earth and its environment through the recovery of core samples from beneath the floors of the world's oceans. It is successor to the Deep Sea Drilling Project (DSDP) which operated out of Scripps Institution of Oceanography between 1968 and 1983 and utilized *D/V Glomar Challenger* for its field operations.

The principal facility for ODP is the drillship *SEDCO/BP 471*, also known as *JOIDES Resolution* (Figure 1). Cruises aboard *JOIDES Resolution* are approximately 8 weeks long, carrying a scientific and technical complement of 50 people, plus a crew of about 60 people. Each cruise addresses a particular set of earth science objectives in a particular geographic region (Rabinowitz et al. 1991). *JOIDES Resolution* was built in 1978 as a conventional dynamically positioned drillship, and was converted in late 1984 for scientific ocean drilling (Foss, 1985). A major element in the conversion was construction of a seven-story 12,000 $ft^2$ laboratory structure located forward of the derrick on the starboard side. Within this structure are state-of-the-art instruments for study of sedimentology, physical properties, paleomagnetics, paleontology, and petrology of sediments and rock (Rabinowitz 1986). There are well-equipped laboratories for inorganic and organic geochemical analyses and for analyses by X-ray diffraction and X-ray fluorescence. The scientific tasks are also supported by a photographic laboratory, an electronics repair shop, the capability for gathering and interpreting underway geophysical data, a global positioning navigation system, and a central computer processing unit including 50 microcomputers distributed throughout the laboratory and refrigerated core storage spaces. These facilities allow the international scientific party of 25 to 30 scientists on each cruise to conduct a comprehensive shipboard analysis of the physical and chemical properties of collected core material.

Because of its size and displacement (Table 1), *JOIDES Resolution* has proven itself to be a remarkably stable drilling platform for work in high latitudes, especially important for paleoenvironmental studies (Foss et al. 1988). We have collected samples beneath the deep sea floor (see Figure 2 for generalized locations) in the Norwegian Sea (Leg 104), Baffin Bay and Labrador Sea (Leg 105), Weddell Sea and sub-Antarctic South Atlantic Ocean (Legs 113 and 114), and Prydz Bay and sub-Antarctic Indian Ocean (Legs 119 and 120).

The ship's drilling capabilities were put to severe tests during these cruises. For example, on Leg 120, operating in the southern high latitudes during the months of March and April was a challenge. Nevertheless, the objectives of the cruise were met with less than 2 days lost waiting on weather, despite high winds (average 20 to 40 kn and gusts of 50 to 70 kn) and heavy seas while operating at relatively shallow water depths (about 1 to 2 km). In 7 years of drilling operations, the *JOIDES Resolution* has been unable to retrieve cores as a result of weather for only about 5 days — a remarkable achievement considering the areas of operation. Furthermore, the ship has the world's largest heave compensator, a device that acts as a hydraulic shock absorber in order to minimize the vertical motion of the drillship relative to the sea floor. This is essential for collecting high-quality, relatively undisturbed cores.

On three of the voyages (Legs 105, 113, 119), ice was a factor and support vessels were used for ice management and escort duties. For the Baffin Bay drilling of Leg 105, the Canadian registered support vessel *M/V Chester* reported on ice conditions in the operating area. This vessel did not have capabilities for moving or deflecting icebergs or growlers. For our Antarctic research, we used the Danish vessel *Maersk Master* for ice-management duties. This 83 m, 14,900-ton vessel not only reported ice conditions, but had capabilities to tow or otherwise deflect icebergs (Figure 3). These capabilities have allowed for much of the significant scientific results that will be discussed below.

**FIGURE 1.** *JOIDES Resolution (SEDCO/BP 471),* drillship for the ODP. The vessel is 470′ × 70′, and has a displacement of 16,596 ton. The derrick towers 200 ft above the waterline. The ship was built in Halifax, Nova Scotia, in 1978, and is among the top worldwide dynamically positioned drillships. The rig was converted to core in up to 27,000 ft of water; about 12,000 ft$^2$ of laboratory space was added. The vessel is ice-strengthened and has high latitude capabilities. It is shown above operating in Baffin Bay.

**TABLE 1**
**Characteristics of the Drillship *JOIDES Resolution***

| | |
|---|---|
| Principal dimensions | |
| Overall length | ft: 470 |
| Molded beam | ft: 70 |
| Draft | ft: 24 |
| Total displacement, short ton | 18,636 |
| Power systems | |
| Main: 5 16-cylinder turbo-charged | |
| Diesels: 2 16-cylinder diesels | |
| Total hp | 18,757 |
| Total main propulsion, hp | 9,000 |
| Total thrusters, hp | 9,600 |

The scientific objectives of ODP are determined by a group of international scientists called Joint Oceanographic Institutions for Deep Earth Sampling (JOIDES). The planning structure includes four thematic panels (lithosphere, ocean history, sedimentary geochemistry and physical properties, and tectonics), and five service panels (information handling, downhole measurements, safety and pollution prevention, site survey, and shipboard measurements), as well as various detailed planning groups. Recommendations from these panels

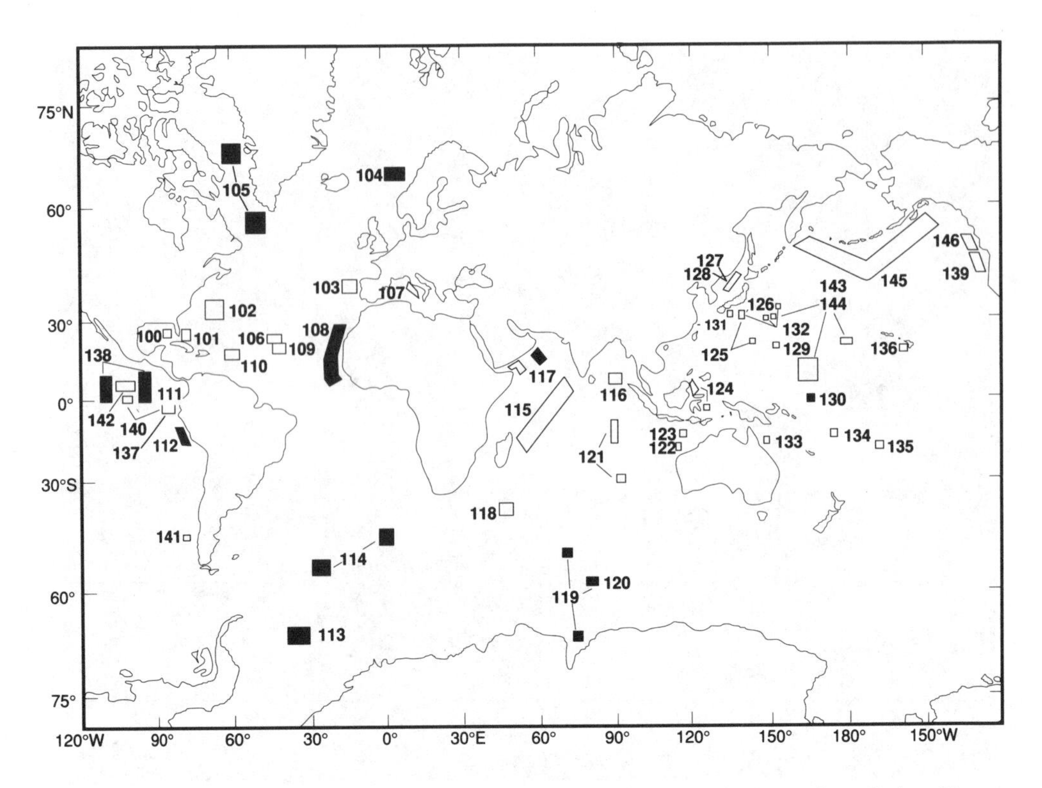


**FIGURE 2.** The location of scientific expeditions during the first 7 years of field operations. Darkened boxes represent regions where primarily paleo-oceanographic or paleoclimatic issues were addressed (see text for discussion).

**FIGURE 3.** The vessel, *Maersk Master,* supported the *JOIDES Resolution* during scientific expeditions to the southern high latitudes (Legs 113 and 119). This ice-support vessel was responsible for ice management, including monitoring and diverting icebergs during coring operations by the *JOIDES Resolution*.

are addressed by several committees (PCOM, EXCOM, TEDCOM, and BCOM), which advise on scientific, technological, and financial directions of the program. Recommendations as to the important paleoceanographic and climatic objectives such as those described in this paper are reviewed and prioritized by the JOIDES Ocean History Panel.

Specific ODP scientific objectives are determined by JOIDES and are based on the long-range objectives set forth by the international oceanographic community at large at major conferences on scientific drilling (COSOD, 1981; COSOD II, 1987). Thus far, many of the scientific objectives defined at COSOD (1981) have been addressed, especially those relating to the tectonic evolution of passive and active continental margins, the nature of oceanic crust, the origin and evolution of marine sedimentary sequences, and causes of long-term changes in the Earth's atmosphere, oceans, cryosphere, biosphere, and magnetic field. COSOD II was held to review and redefine scientific objectives. These new objectives include investigating changes over time in the global environment, mantle/crust interactions, fluid circulation in the crust and global geochemical budget, stress and deformation of the lithosphere, and evolutionary processes in oceanic communities.

## II. CLIMATE RESEARCH METHODS

Our understanding of Earth's climate history is partially derived from analysis of proxy indicators from the marine geological record. These proxies, including, among others, isotope geochemistry, fauna and flora, ice rafting detritus (IRD), eolian input, and clay mineralogy, provide both evidence of climate variability at various temporal scales, and evidence of specific events, such as the onset of glaciation, volcanism, and desertification of continental regions, plus changes in ocean circulation, ocean productivity, and wind regimes (see Barron et al. and Berger et al. 1989).

**Isotope geochemistry** — Oxygen isotopes ($^{16}O$ and $^{18}O$) are commonly used as indicators of global ice volumes and paleotemperatures. The ratio of $^{18}O/^{16}O$ in the world's oceans is measured from the isotopic composition of calcium carbonate tests of organisms (generally foraminifera). This method provides an estimate of paleotemperatures, temperature gradients, and ice volume. Several factors, including diagenetic alteration from recrystallization, severe dissolution, and species effects, influence and often bias the isotope data sets.

**Fauna and flora** — The oceanic biota is used to document changes in oceanographic circulation and to infer changes in climate. Changes in the abundance of key species with known biogeographic distributions or modern affinities are used as indicators for changes in environmental conditions, and often reflect oceanographic response to climate change.

**Clay mineralogy** — The relative proportions of the clay component of marine sediments (e.g., chlorite, montmorillonite, kaolinite, and illite) provide information on the sources and transport paths of solids from the continents to the oceans. As such, this data provide information about weathering and climatic conditions existing on the adjacent landmasses. For example, chlorite is presently abundant in the polar regions and is considered to result from mechanical (glacially derived), rather than chemical, weathering, montmorillonite ''smectite'' is indicative of volcanic regimes and results from the alteration of volcanic ash and the erosion of solid subaerial rocks, illite typically represents river-borne solids to the marine sediments, and kaolinite is indicative of extensive chemical weathering typically found in the low latitudes.

**Ice rafting detritus (IRD)** — The IRD provides evidence of glaciers calving at sea level, and has been used by numerous workers to determine the geographic extent of icebergs in the geological past. IRD is one of the primary marine indicators used by workers (see Barron et al. 1991a) to deduce the glacial history of the polar regions. However, this data set may present a biased reconstruction, as it is often unclear if the lack of IRD in a specific

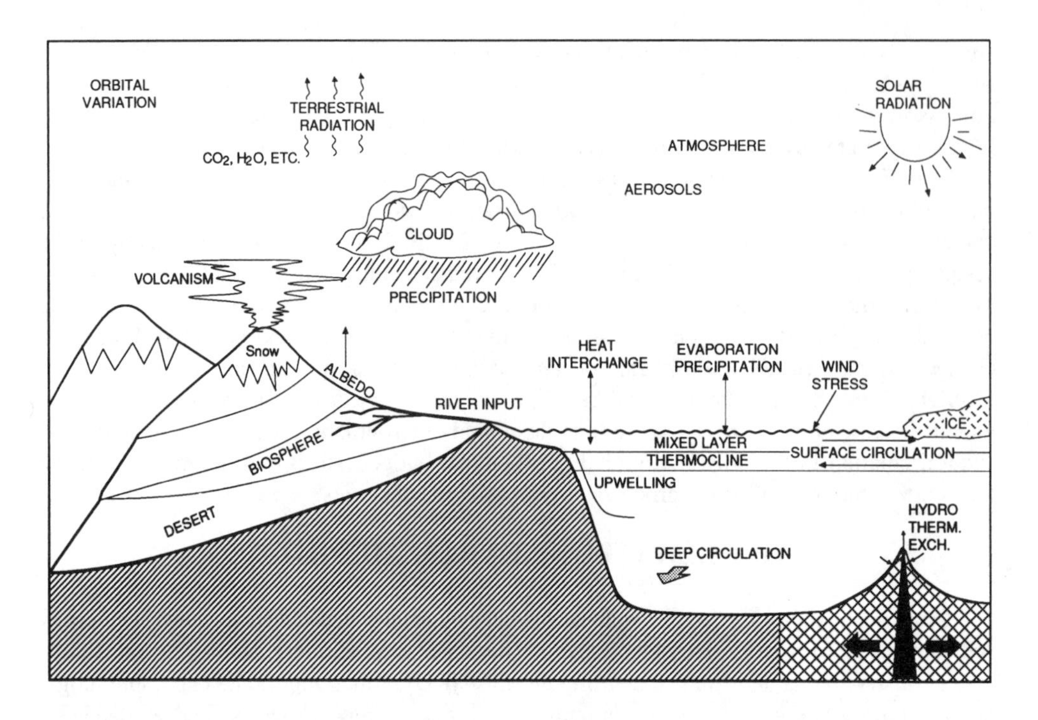


**FIGURE 4.** Schematic representation of the components of the Earth's Climate System. Knowledge of interactions and variability of each component is vital for understanding short- and long-term climatic variation and future climate.

sedimentary sequence reflects a decrease in continental ice sheets or a change in the path of icebergs.

**Eolian input** — Pollen, freshwater diatoms, and opal phytoliths are often used as indicators of wind intensities and source areas. This eolian component of the sediment record also provides clues as to the aridity/humidity cycle of the adjacent continental landmass.

These data sets integrated with terrestrial data sets allow the identification of the primary components of the Earth's climate system (Figure 4), however, they provide only partial insight into the interaction and feedback mechanisms between these components. As such, little is known about the driving mechanism(s), required thresholds, response time, or phase relationship for portions of the climate system.

Our knowledge of the climate system is enhanced by the interactive process of developing general circulation models (GCM) and statistical-dynamical models (SDM) based on known variables, and then ground-truthing and testing these models based on additional proxy data sets obtained from the geological record. The use of proxy data sets in part biases the climatic models in that these data sets are often, limited (based on one or two indicators per site), sparse (few data sets globally distributed), and represent a minimal time interval. As such, the use of these data sets has resulted in conflicting models concerning the mode and pathways of the variables affecting climatic trends.

To remedy this situation, near-continuous sequences containing numerous proxy data sets are required from a global array of sites positioned in climatically sensitive regions and covering a wide range of latitude and water depths to define oceanic fronts, upwelling zones, sea surface temperature (SST) gradients, and changes in oceanic properties with paleodepth. This requirement has long been recognized by the science community and has been the pursuit of paleoclimatologists and paleoceanographers during the last decades.

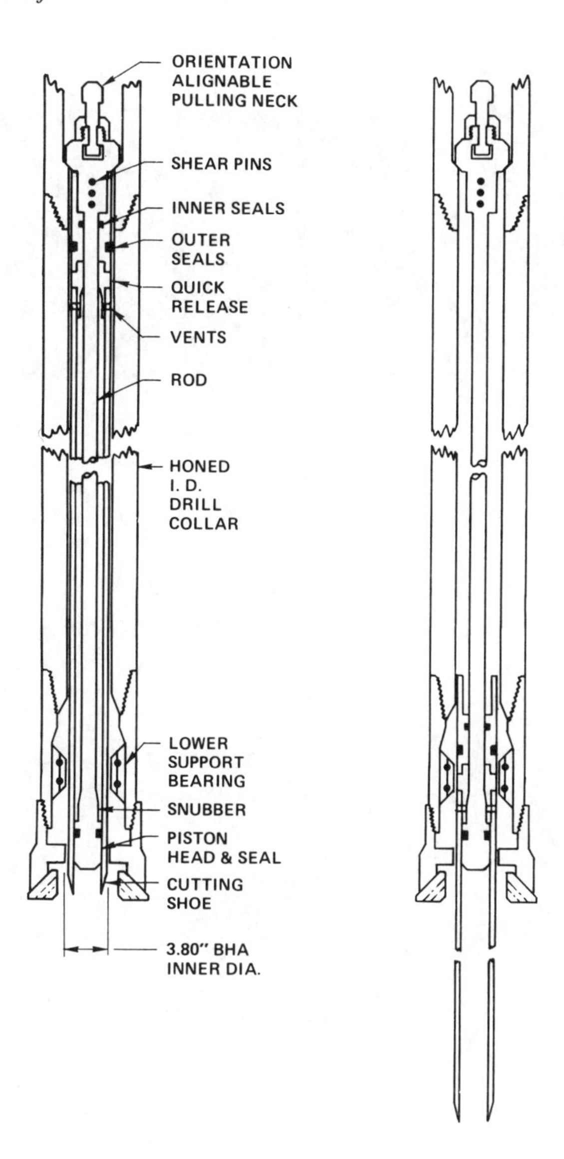


**FIGURE 5.** Schematics of the Advanced Piston Corer (APC).

Initial attempts at deciphering the marine climatic record were completed, using short piston cores (generally <15 m in length) from the world's ocean basins (e.g., Ruddiman and McIntyre 1976). The scientific results were limited by many factors, including incomplete recovery, minimal amount of geological time represented (generally <1.0 Ma), and often by the core quality. Even with these limitations, advances have been made in understanding the near-recent climate system, such as the climatic reconstruction of SST, sea ice, ice sheets, and sea level completed by CLIMAP (1976) for the last glacial maximum.

Technological advances in scientific ocean drilling have allowed for high-quality continuous core retrieval, thereby eliminating many of the previous problems. For example, the Advanced (Hydraulic) Piston Corer (APC; see Figure 5) was developed to overcome problems related to core disturbance in soft sea floor sediments (Storms 1983; Huey 1984). Here, the severe disturbance on unlithified sediments caused by rotary drilling is eliminated by nonrotary methods (Figure 6). Pressurized water drives the core barrel, which acts as a

STANDARD ROTARY CORE
HOLE: 479 CORED INTERVAL: 90.0–107.5 m

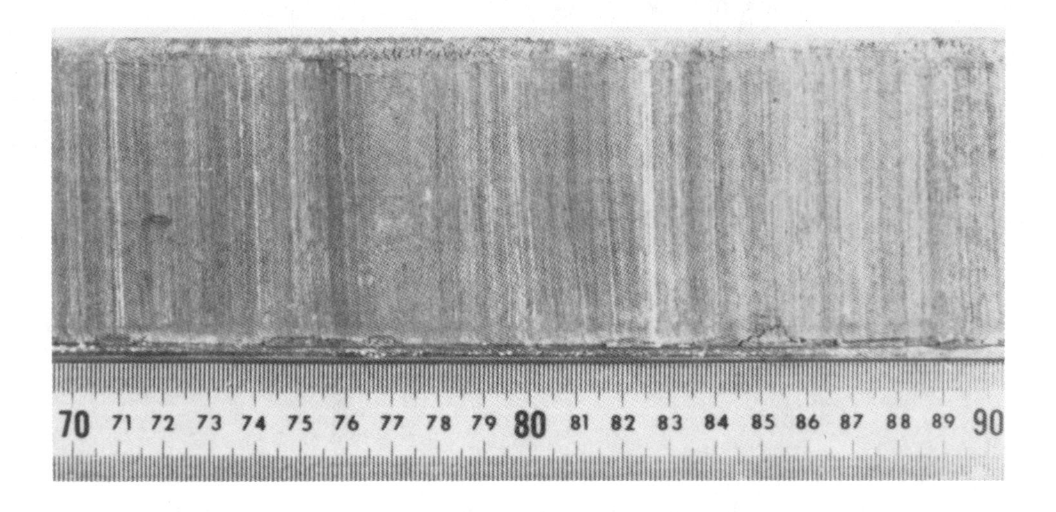


HYDRAULIC PISTON CORE
HOLE: 480 CORED INTERVAL: 95–99.5 m

**FIGURE 6.** Comparison of coring results using the Rotary Core Barrel (RCB) and the Advanced Piston Corer (APC).

high-speed hydraulic arm ~30 ft ahead of the bit at a rate of 10 to 20 ft/s. The hydraulic piston coring effectively separates the coring process from the ship's heave and allows relatively undisturbed 2½ in. diameter cores to be recovered. This technique is successful in the upper ~300 m of the sedimentary sequence. Below this depth, the lithified nature of the sediment generally reduces the core quality.

The Extended Core Barrel (XCB; see Figure 7) has been developed for high quality core collection in more lithified sediments, especially for areas where lithologies alternate

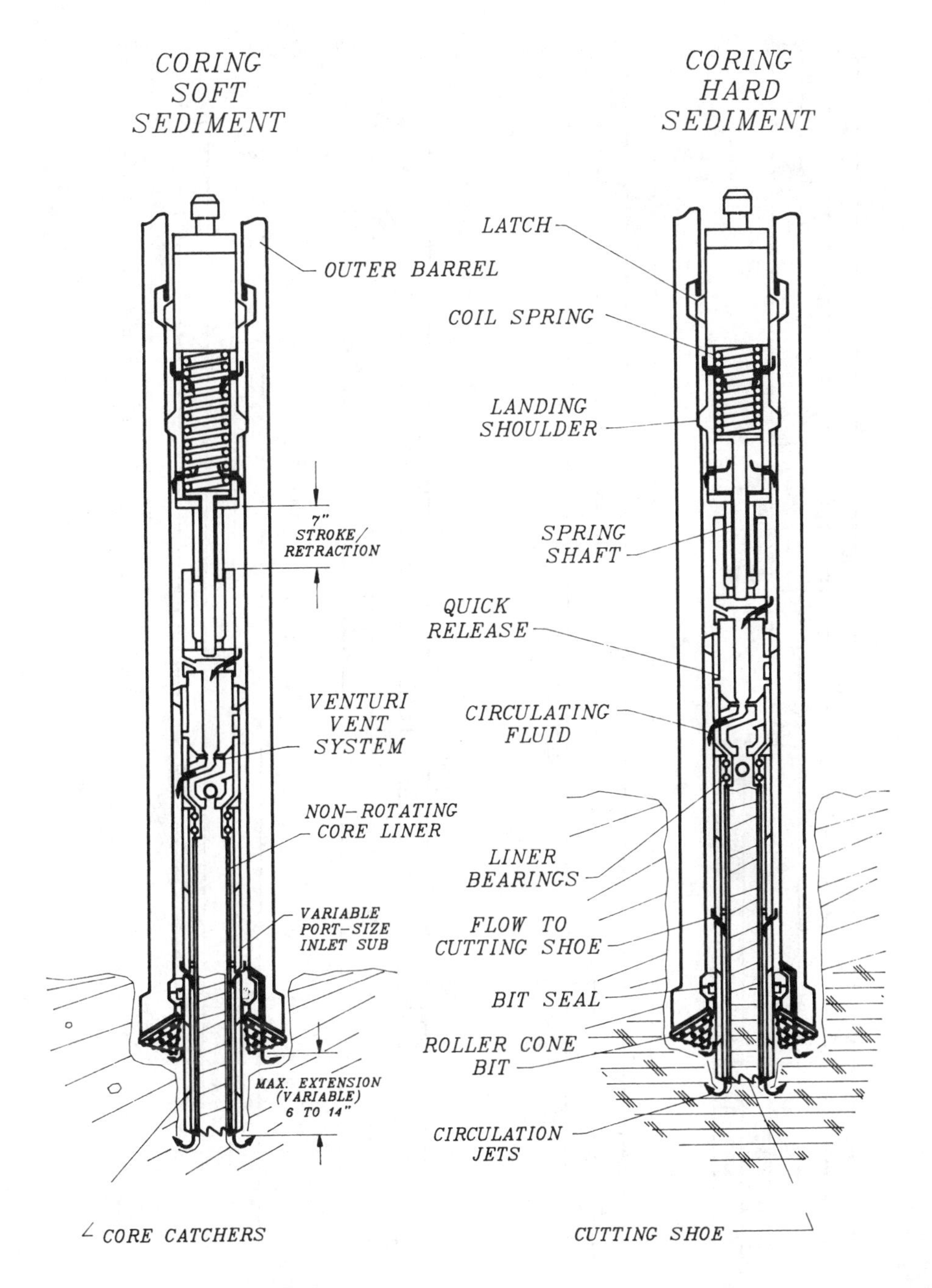


**FIGURE 7.** Schematic of the Extended Core Barrel (XCB).

between hard and soft beds (Cameron 1984). The core barrel with the XCB system extends ~6 in. below the bit in soft formations, and thus protects the sediments from circulating fluids; in harder formations, the core barrel retracts to the bit and the more lithified sediments are cut primarily by the roller cones as in standard rotary coring. The XCB is interchangeable with the APC so that both can be used without tripping the drill string.

The use of these tools, in conjunction with a drilling strategy that consists of multiple holes cored at each site and by using new high resolution correlation techniques (see

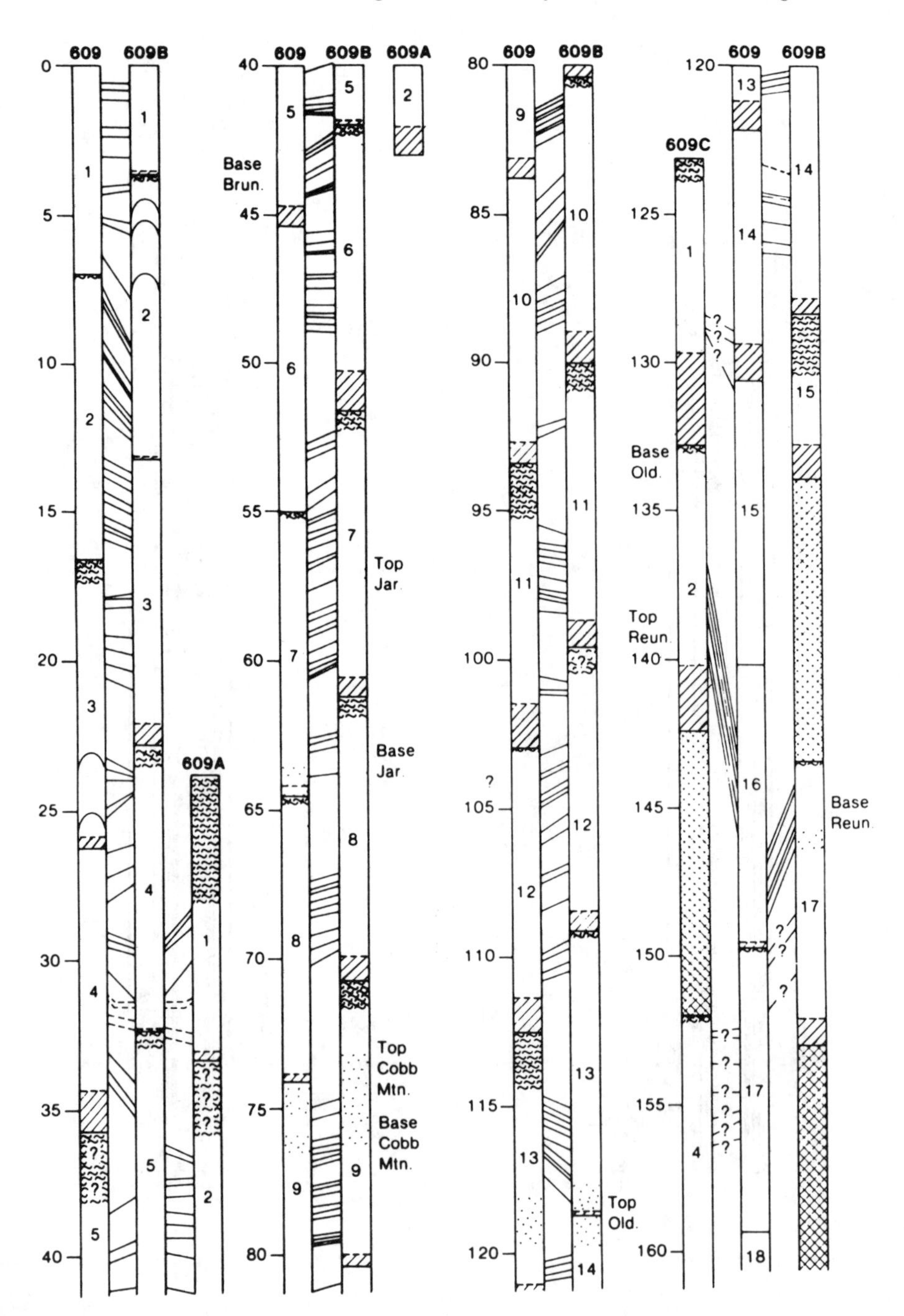


**FIGURE 8.** Correlation diagram for upper Pliocene and Pleistocene interval of four holes at Site 609 (after Ruddiman et al. 1987). Correlations are based on visual comparisons of color layering and stratigraphic position of paleomagnetic transition, e.g., see approximate 8 m depth offset associated with Reunion (Reun.) event associated in core 609C-2 at 137 m, and 609-16 at 145 m.

Ruddiman et al. 1984; Pisias et al. (in press) provides the means to recover near-continuous stratigraphic sedimentary records of high core quality representing the last 3 and occasionally last 10 million years. Multiple cores are required to compensate for coring difficulties which include under-recovery of sediment, deformed or disturbed sediment, repeated sequences, or gaps in the sequence (Figure 8).

Paleo-oceanographers, paleoclimatologists, and climatic modelers working in concert have used constituents of the sedimentary record derived using the above techniques and procedures (including the terrigenous, eolian, and biogenic proxy components) to determine changes in ocean circulation and climate in the geological past. In general, they have demonstrated that the Earth's global ocean-climate system has evolved from the Eocene, a time characterized by warm global temperatures with weak latitudinal gradients, to the late Cenozoic, a time characterized by strong latitudinal and vertical gradients, oceanic fronts, and cold high latitude surface water masses (Shackleton and Boersma 1981). This step in climatic evolution coincides with the development of cold polar regions and the interaction between surface and deep water circulation between the low and high latitudes.

## III. SCIENTIFIC OBJECTIVES OF THE OCEAN DRILLING PROGRAM

High resolution sedimentary sequences are required from a global array of APC sites to assess climatic conditions for the last 25 million years. To achieve this goal, two efforts are underway. These include scientific expeditions to the high latitudes, including the southern high latitudes consisting of the Weddell and Ross Seas, the Kerguelen Plateau region and Prydz Bay, East Antarctica, plus the northern high latitudes region consisting of the Arctic Ocean, Bering, Chukchi, Norwegian-Greenland Seas and Baffin Bay and, the low latitude oceans, including zones of upwelling and high productivity in the Pacific, Atlantic, and Indian Oceans, plus large oceanic plateaus such as the Ontong Java Plateau. Some of these expeditions have been completed and will be discussed later (see Figure 2). Others are planned for the future.

### A. SOUTHERN HIGH LATITUDE

The Southern Ocean and the adjacent Antarctic continent have played a critical role in the development and evolution of Earth's climate. Today, Antarctica influences global climate through numerous pathways including the albedo effect of the ice surface, the influence on ocean and atmospheric circulation and latitudinal position of the polar front, SST and global temperature gradients, and formation of Antarctic intermediate and bottom water. Four ODP scientific expeditions (Legs 113, 114, 119, and 120) have been completed in the southern high latitude in order to unravel the complex oceanographic and climatic history which is essential to understanding global climatic history. Specific, paleoclimatic or paleo-oceanographic objectives were to provide information that would aid in resolving the following scientific issues:

1. The timing and the continuity of Antarctic glaciation has been the center of controversy during the last decade, resulting from numerous yet differing interpretations of the available proxy records, primarily oxygen isotopes (Savin et al. 1975; Shackleton and Kennett 1975; Matthews and Poore 1980; Prentice and Matthews 1988; Shackleton 1986; Miller et al. 1987; and Webb et al. 1984; Harwood and Webb 1989, among numerous others). Barron et al. (1991a) summarized these interpretations to include four hypothesis: (1) that minimal ice was present on Antarctica in the earliest Oligocene (Shackleton 1986); (2) that significant ice was present on the continent in the earliest Oligocene, but disappeared during parts of the Oligocene and Miocene (Miller et al. 1987); (3) that ice volume on the continent has been continually equal to or larger than modern volumes (Prentice and Matthew 1984); and (4) multiple periods of glaciation have occurred (Webb et al. 1984; Harwood and Webb 1989). Therefore, scientific drilling objectives focused on establishing the evolution from preglacial to glacial continental climate, including the timing, magnitude, and extent of Antarctic

glaciation; changes in ice volume, ice drainage patterns between east and west Antarctica, and the extent of glacial erosion.

2. The opening of the Drake Passage between South America and West Antarctica strongly influenced Cenozoic climate because it allowed the complete development of the Antarctic circumpolar current (ACC) and the Polar Frontal Zone (PFZ) bounded by the sub-Antarctic divergence to the north and the Polar Front to the south. The PFZ is characterized by the formation of Antarctic intermediate water. The PFZ acts as a barrier to organisms, as is evident by the modern distribution of biosiliceous sediments which occur to the south and not immediatetly to the north of the PFZ (see Lisitzin 1972 and Baldauf and Barron 1990). Although the significance of the PFZ is understood for the modern oceans, little is known about the evolution of this system, including the timing of onset, the latitudinal migration during the geological past, the ability of this system to act as a surficial barrier, the relationship between variation in this system, and glacial variability on the Antarctic continent.
3. Antarctic bottom water (AABW) influences the geochemistry of the world's oceans and atmosphere as it exists as bottom water throughout the Southern Hemisphere and well north of the equator into the Northern Hemisphere, thereby this watermass ventilates most of the world's oceans. In addition to documenting the timing and water volume generated, scientific objectives also focused on determining the paleoenvironmental evolution of the critical passageways or gateways acting as conduits for this dense bottom water.

## B. NORTHERN HIGH LATITUDE

The northern high latitudes also play a critical role in the global climate system as it interacts with the global system by the formation of permanent and seasonal ice cover, transfer of heat to the atmosphere by ice-albedo, and by deep water formation. The Arctic is an important contributor of deep water which flows south through the Fram Straits into the Norwegian-Greenland Sea. After mixing with waters in the Norwegian-Greenland Sea, it continues to flow south across the Greenland-Scotland Ridge into the Atlantic ocean and the other world oceans.

The inaccessibility of the Arctic and adjacent seas has hampered our ability to understand the sensitivity of the northern high latitudes forcing mechanisms and the response of this region to changes in the low or southern high latitudes. As a result, it is unclear what role the northern high latitudes played in the transformation from warm Eocene climates to the late Cenozoic climates. Results from DSDP Leg 19 (Bering Sea (Creager et al. 1973) and Leg 38 (Norwegian-Greenland Sea (Talwani et al. 1976) provide only minimal information to enhance our understanding of these regions, as holes were generally noncontinuously cored and were completed using the rotary rather than piston coring system. An adventurous program for future drilling has been proposed by various members of the scientific community for drilling in the Bering, Chukchi, Norwegian-Greenland Sea, Baffin Bay, and the Arctic Ocean. To date, ODP has complete scientific expeditions to two of these regions, the Norwegian-Greenland Sea (Leg 104) and Baffin Bay (Leg 105). Specific paleoclimatic and paleoceanographic scientific objectives addressed during these cruises include

1. The onset of Northern Hemisphere glaciation commenced in the early Pliocene. Most proxy indicators examined suggest an age of about 2.5 Ma (Shackleton et al. 1984), although some evidence suggests an age of about 5.0 Ma (see Srivastava et al. [1987] for discussion). In addition to timing, a better understanding is required concerning the forcing mechanisms, critical pathways, and factors responsible for Northern Hemisphere glacial development. Two models for Northern Hemisphere glaciation have been proposed based on the use and interpretation of different proxy indicators. Rud-

diman and McIntyre (1981) using data based on oxygen isotopes from deep sea sediments, suggest that the tempo of Northern Hemisphere glaciation consists of a slow buildup of ice sheets, followed by rapid ice-sheet disintegration triggered by iceberg calving. Andrews et al. (1983) using organic data from continental deposits, suggest that the glacial tempo consists of a very rapid buildup of ice sheets and an equally rapid disintegration. In addition to the mode and mechanism of glaciation, it is currently unclear as to what impact the resulting climate asymmetry between the Northern and Southern Hemisphere had on forcing a change in the climate trend. Therefore, scientific objectives for northern high latitude drilling included determining the timing and frequency of climatic oscillations, glacial and interglacial cycles, and the cause and mechanism(s) of glaciation.

2. The Norwegian-Greenland Sea is one of the conduits for Arctic water to communicate with the world's oceans by the overflow of mixed Arctic-Norwegian-Greenland Sea water into the North Atlantic, forming North Atlantic Deep Water (NADW). In exchange for this water, the Norwegian-Greenland Sea receives warm-saline surface water from the North Atlantic. Evidence for NADW formation based on drift deposition, benthic foraminifera, and isotope data from the North Atlantic (Miller and Tucholke 1983; Dickson and Kidd 1987; Berggren and Schnitker 1983; Thiede and Eldholm 1983; Eldholm and Thiede 1980; Miller et al. 1987; Woodruff and Savin 1989) has resulted in debate over the timing and onset of this exchange, with ages for this event ranging from pre-, mid-, and late Miocene, to early Pliocene. In addition to the overall timing, it is unclear as to what impact climate variability has on this exchange and the formation of NADW. Specific scientific objectives incorporated documenting the evolution of this oceanographic gateway, including the chemical and physical signature of this watermass and its variability through time.
3. The oceanography of the Norwegian-Greenland Sea is complicated by the inflow of cold, dense water from the Arctic and warm-saline water from the North Atlantic. As such, sharp thermal, hydrologic, chemical, and biological gradients exist today. Of particular interest to the scientific community is the timing and variation of these gradients during the late Cenozoic, and the impact these gradients have on the paleoenvironment of the Norwegian-Greenland Sea and the history of bottom water formation.

## C. LOW LATITUDE

The subtropical and equatorial regions figure prominently in Earth's oceanographic and climate history. This, in part, results from the equatorial circulation response to forcing from higher latitudes, specifically the response to variation in ice volume and sea ice, latitudinal thermal gradients, and the strength of the polar cells. However, in addition to this high latitude forcing, numerous local forcing factors also influence the low latitudes. These include trade winds and their influence on oceanic upwelling and productivity, continental aridity and humidity cycles, and tectonic events (such as opening and closing of gateways). The low-latitude region is especially significant in unraveling Earth's climate history, as it is in these regions that the calcareous biota is generally well preserved, allowing detailed benthic and planktonic isotope data sets to be generated.

Numerous cruises have been completed in this region in an effort to complete the global array of low-latitude APC sites. These include Leg 108 (northwest Africa); Leg 112 (Peru-Chile); Leg 117 (Oman); Leg 130 (Ontong, Java); and Leg 138 (eastern equatorial Pacific). Scientific problems addressed during these cruises are discussed below:

1. It is unclear as to the mechanisms responsible for changes in the low latitude SST signals during the geological past. Although the majority of workers suggests that such

changes are a direct response to variation in ice volume and ice-sheet growth or decay, other factors (such as variations in monsoonal- or trade-wind circulation, or variation in $CO_2$ effects) could play an equally important role (see Ruddiman et al. 1988; 1989). Of particular interest is the tempo and mode of tropical variability associated with periods of prominent changes in ice volume in an effort to determine the low latitude SST response.

2. Surface productivity of the world's oceans plays a critical role in the $CO_2$ cycle, and hence in Earth's climate cycle. Regions of high surface water productivity in the low latitudes are associated with upwelling cells such as those associated with eastern boundary currents, monsoonal circulation, or areas of oceanic divergence. Of particular importance is to determine the history of upwelling both in defining the latitudinal persistence of the upwelling cells during varying climates, and in assessing its importance in the broader climatic context of variations in $CO_2$ budget and deposition of organic rich sediments. Also important is to determine the biological and sedimentary response to fluctuations in intensity and source of upwelling. Additional scientific objectives focused on the response of upwelling cells and the associated biota to variation in the monsoonal circulation system.
3. Vertical oceanographic gradients and their linkage to climatic parameters is poorly understood. Improvement in our knowledge of deep water paleoceanography requires in part sediment sequences containing a calcareous biota allowing the generation of oxygen isotope data sets. Specific issues pertaining to the deep oceans include global geochemistry and variation with time as to the chemical constituents and deep water properties, the role $CO_2$ dissolution plays in the overall $CO_2$ cycle, dissolved oxygen cycle, flux rates, and ocean gradients.
4. Changes in atmospheric circulation, as monitored by proxies such as clay mineralogy, wind-blown diatoms, phytoliths, and pollen, provide evidence as to climatic variability of adjacent continental landmasses. By determining changes in these indicators in the geological past one can provide critical information concerning aridity/humidity cycles, desertification of continental regions, history of the intertropical convergence, and the history of the southern and northern trade winds.

## IV. SCIENTIFIC RESULTS

### A. SOUTHERN HIGH LATITUDE

#### 1. Leg 113

Twenty-two holes were drilled at nine sites (Sites 689 to 697; Figures 9A,B) and 1944 m of core recovered in the Weddell Sea in order to study the evolution of circum-Antarctic watermasses and to examine the history of glaciation on the Antarctic continent and relate this to ice volume. Dr. Peter F. Barker (University of Birmingham, U.K.) and Dr. James P. Kennett (University of Rhode Island) were Co-Chief Scientists of this cruise. Dr. Suzanne O'Connell was the Texas A&M University Staff Scientist. (See Barker et al. 1988 and 1990 for detailed discussion of cruise results.)

**Significant Paleoenvironmental Results**

Climatic conditions on East Antarctica were generally warm and semi-arid until the Eocene, with ice buildup and glacial conditions commencing by the early Oligocene. Evidence for this climatic deterioration is based on changes in the sediment composition recovered from Maud Rise, and on the transition from biocalcareous to biosiliceous sediments. The Eocene sediment is dominated by smectite typical of warm conditions, and the Oligocene sediment is dominated by illite typical of physical weathering processes, suggesting the presence of glacial or preglacial continental conditions. Ice-rafted debris is rare on Maud Rise sections, but does occur in Oligocene age sediment.

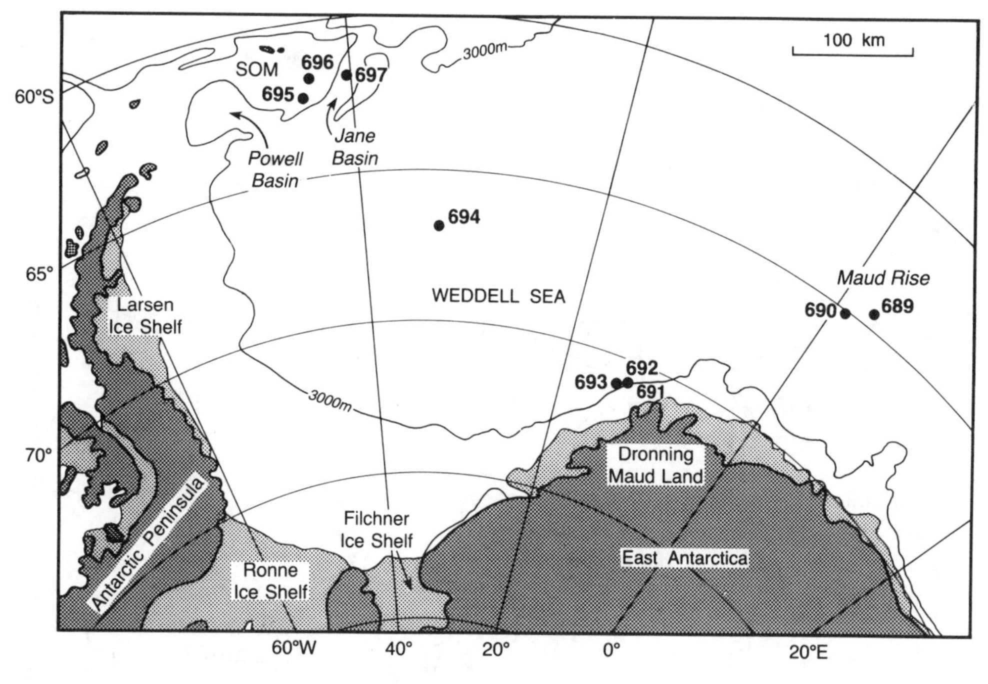

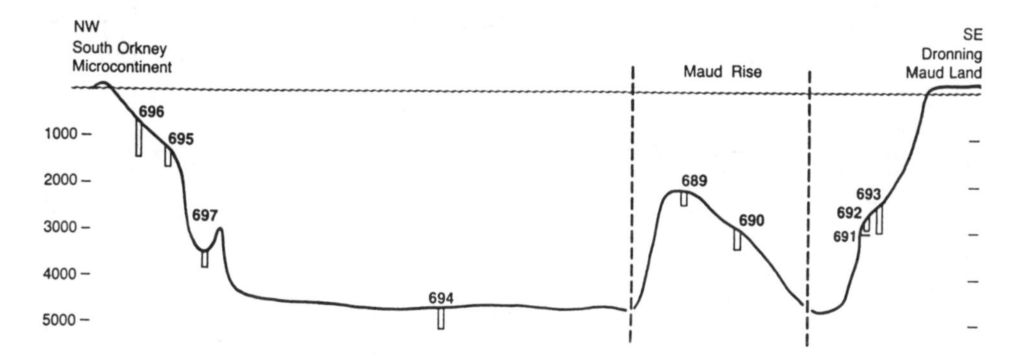


**FIGURE 9.** Location of sites occupied during ODP Leg 113 (top); schematic cross-section showing coring results of the Leg 113 sites (bottom).

The West Antarctica ice sheet underwent major expansion during the middle Miocene, indicated by an increase in ice-rafted debris in the lower/upper Miocene and the replacement of smectite by both illite and chlorite. This is interpreted by the scientific party to suggest the occurrence of physical weathering resulting from climatic cooling. If this interpretation is correct, it also suggests that climatic cooling and glacial conditions developed much later on West Antarctica (late Miocene) than on East Antarctica (early Oligocene).

A dramatic increase in sedimentation rate approximating the Miocene/Pliocene boundary results from an increased terrigenous and biosiliceous input, and indicates an intensification of cooling and establishment of true polar-type glaciation. This increase also suggests that the West Antarctic ice sheet may have become stable at this time.

## 2. Leg 114

Twelve holes were drilled at seven sites (Sites 698 to 704; Figure 10) and 2300 m of core recovered in the sub-Antarctic South Atlantic in order to address tectonic objectives

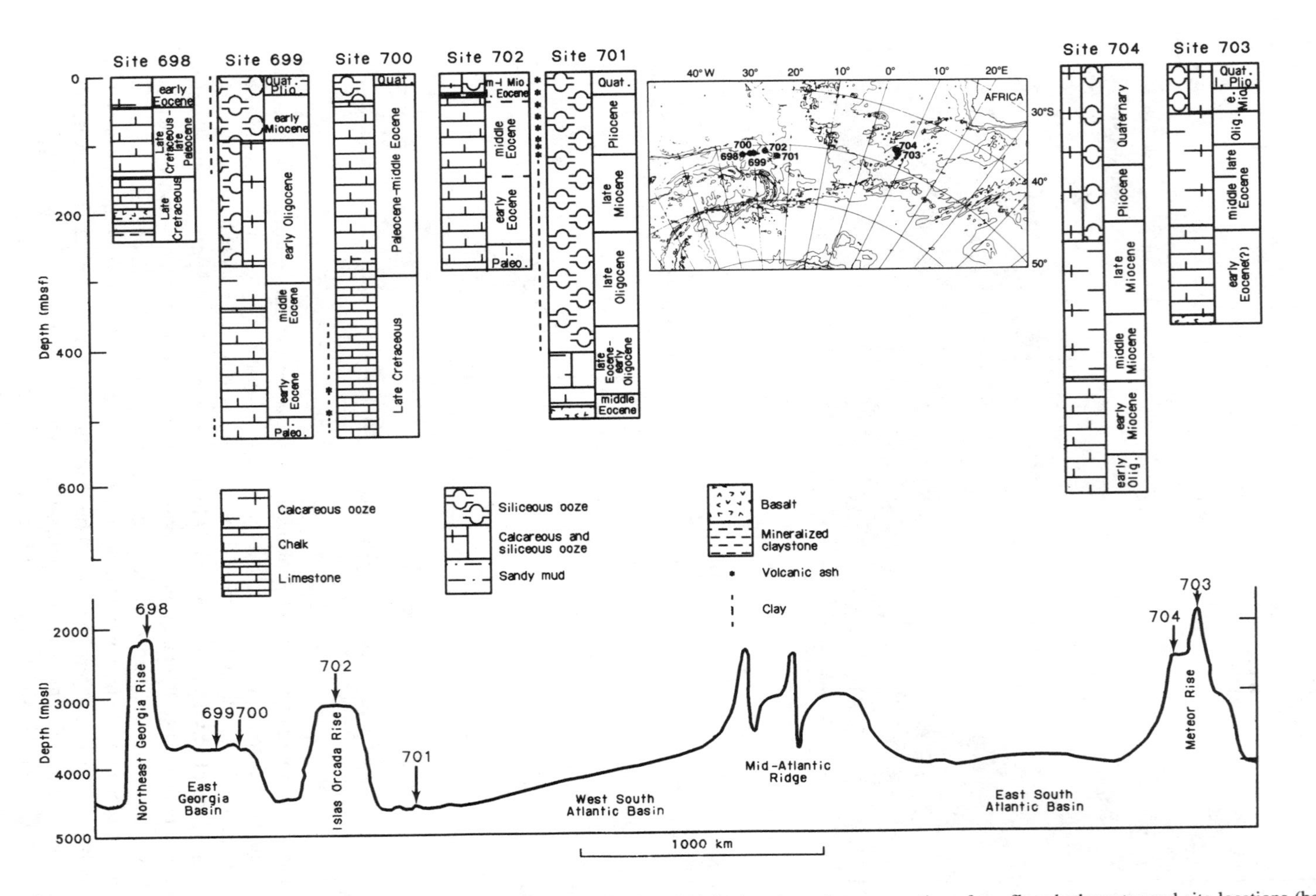


**FIGURE 10.** Location and lithologic columns of sites occupied during ODP Leg 114 (top); schematic cross-section of sea-floor bathymetry and site locations (bottom).

relating to ages, nature of basement, and subsidence histories of the morphological features, as well as to investigate the teleconnective passageways to oceanic circulation within the Atlantic sector of the Southern Ocean. Dr. Paul F. Ciesielski (University of Florida) and Dr. Yngve Kristoffersen (University of Bergen, Norway) were Co-Chief Scientists of this cruise. Dr. Brad Clement was the Texas A&M University Staff Scientist. (See Ciesielski et al. 1988 and 1991 for detailed discussion of cruise results.)

**Significant Paleoenvironmental Results**

The first evidence of increased vigor of the deep-water communication between the Weddell and South Atlantic basins is represented by a hiatus lasting 2 to 5 million years along the eastern foot of the Islas Orcadas Rise. This hiatus was generated by increased bottom currents at the end of the middle Eocene.

Sedimentation shows persistent bottom-current control throughout the section commencing in the late Oligocene. Age equivalent sediments also became more clay-rich, which shows an increased supply of terrigenous material to the Weddell Basin from the surrounding landmasses during the Oligocene sea-level low stand. The clay was probably brought north in the benthic boundary layer of the emerging Antarctic bottom-water communication with the South Atlantic.

An early-to-late-Neogene hiatus foreshadowed the northward advance of the zone of biosiliceous deposition around Antarctica. The intensification of circum-Antarctic circulation was related to the opening of the Drake Passage nearly 23.5 Ma. This event is recorded by hiatuses at all Leg 114 sites (Site 704 was least affected). In part, the missing interval appears to have been caused by multiple erosional and nondepositional events. A greater time interval is represented in the hiatuses of the shallower sites, indicating higher current velocities or longer erosion times during initial stages of the passage opening.

**3. Legs 119 and 120**

Twenty-two holes were drilled at eleven sites (Sites 736 to 746; Figure 11) and over 2100 m of core were recovered on the Kerguelen Plateau and in Prydz Bay, East Antarctica on Leg 119. Twelve holes were drilled at five sites (Sites 747 to 751) and over 1000 m of core were recovered on Leg 120 on the Central Kerguelen Plateau. These two cruises completed a latitudinal transect in the Southern Ocean between Kerguelen Island (49°S) and Prydz Bay, East Antarctica (67°S) a region critical to the understanding Cenozoic climate and oceanographic development. The Kerguelen Plateau lies south of the present-day Polar Front (Antarctic Convergence), and beneath the main flow of the Antarctic Circumpolar Current. Sampling on the Kerguelen Plateau documents the development and evolution of these two oceanographic features, which have a major effect on global climate and surface-water circulation. Dr. John Barron (U.S. Geological Survey, Menlo Park) and Dr. Birger Larsen (Technical University of Denmark) were the Co-Chief Scientists of Leg 119. Dr. Jack Baldauf was the Texas A&M University Staff Scientist. Dr. Roland Schlich (Institut de Physique du Globe Laboratoire de Geophysique Marine, France) and Dr. Sherwood W. Wise, Jr. (Florida State University) were the Co-Chief Scientists for Leg 120. Dr. Amanda Palmer was the Texas A&M University Staff Scientist. (See Barron et al. 1989; 1991b and Schlich et al. 1989 and in press for detailed discussion of cruise results.)

**Significant Paleoenvironmental Results**

The sedimentary sequence recovered from Prydz Bay records major changes in the depositional environment. The oldest preglacial sequence above acoustic basement consists of "red bed" type sediments of continental origin, and suggests deposition within proximal reaches of a river system. Climatic conditions were probably warm and characterized by seasonal rainfall. Although the paucity of microfossils inhibits age determination, these red beds may be equivalent to deposits in the Lambert Graben of possible Permian or later age.

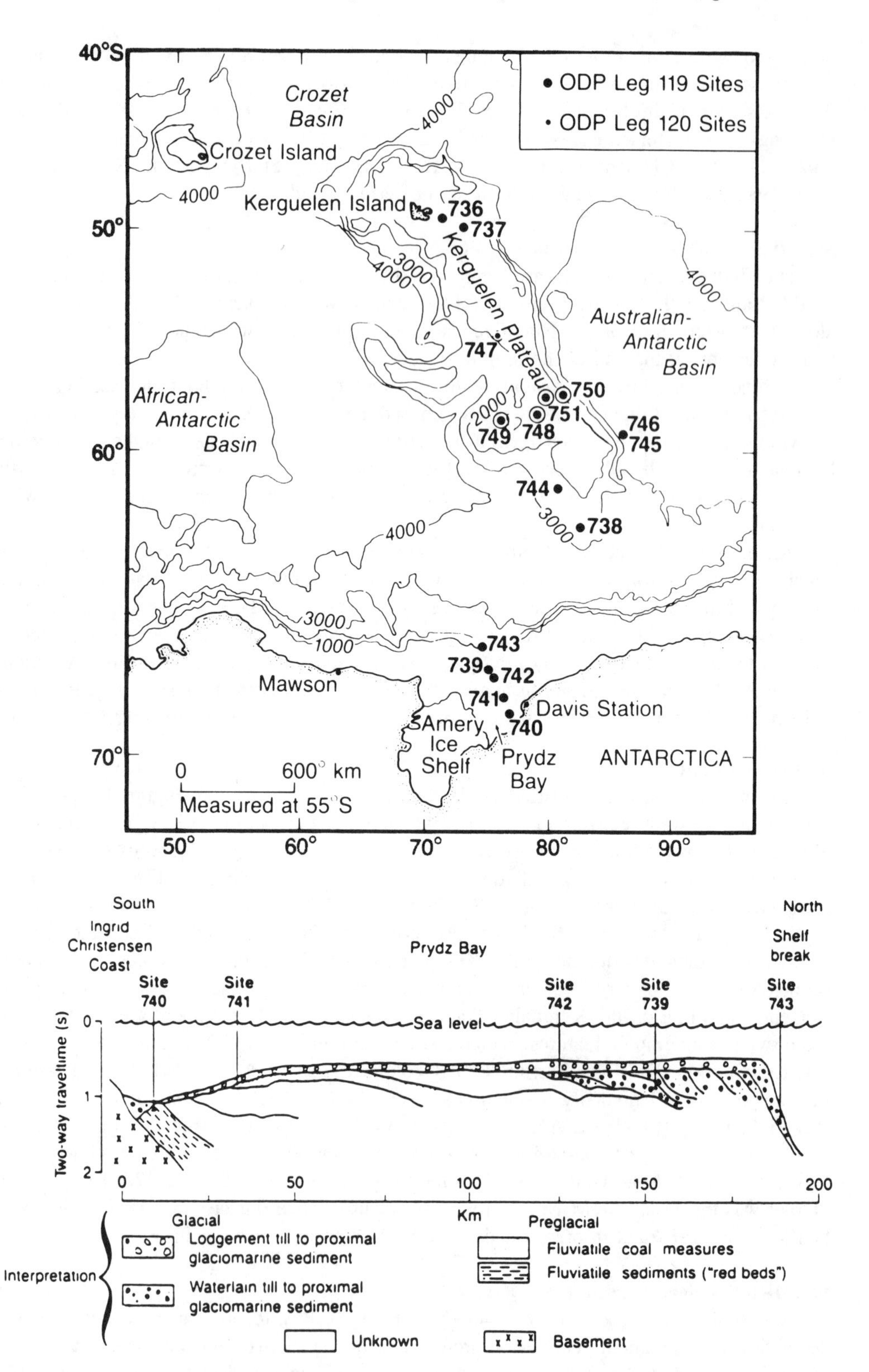


**FIGURE 11.** Location of sites occupied during ODP Legs 119 and 120 (top); schematic cross-section showing the coring results of ODP Leg 119 to Prydz Bay, East Antarctica (bottom).

Changes in the sediment composition indicate that Antarctic glaciation proceeded in phases with the presence of significant late/middle-late Eocene glaciation in East Antarctica (~42 Ma), and the presence of a continental-size ice sheet in East Antarctica during the earliest Oligocene.

The sedimentological evidence indicates that the outer limit of the ice front during the earliest Oligocene was beyond that of the present day by at least 140 km. This suggests that early Oligocene glaciation recorded from the Ross Sea in West Antarctica was more than a local event, and extended beyond the Trans-Antarctic mountains.

The occurrence of marine diatom-rich sediment for intervals of the upper Miocene, lower Pliocene, and upper Pliocene indicates fluctuation in the extent of the ice sheet and the waxing and waning of glaciers across the Prydz Bay shelf during the latter part of the late Miocene and Pliocene.

Sediments on the Kerguelen Plateau record glacial-interglacial cycles during the last 10 million years. Transition from calcareous to siliceous deposition commenced during the earliest Oligocene, and was completed during the late Miocene. Increased deposition of gneissic/granitic ice-rafted debris of probable Antarctic derivation took place during the late Pliocene to Holocene, signaling intensification of glaciation in Antarctica.

Recovered isotopic and biogenic data provided critical information, and with further shore-based analysis, should allow determining the oceanographic response to climatic deterioration and the latitude fluctuation of the polar front.

## B. NORTHERN HIGH LATITUDE

### 1. Leg 104

Eight holes were drilled at three sites (Sites 642 to 644; Figure 12) and over 1650 m of core were collected in order to study the evolution of the passive continental margin of the Vøring Plateau, and to complete a paleo-oceanographic transect with particular objectives relating to the Norwegian current as a continuation of the Gulf Stream system into the polar basins, the initiation and variability of Northern Hemisphere glaciation, and the Cenozoic evolution of polar flora and fauna. Co-Chief Scientists were Dr. Jorn Thiede (Geologisch-Paleontologisches Institute in Kiel, Federal Republic of Germany) and Dr. Olav Eldholm (University of Oslo, Norway). Dr. Elliott Taylor was the Texas A&M University Staff Scientist. (See Eldholm et al. 1987 and 1989 for detailed discussion of cruise results.)

#### Significant Paleoenvironmental Results

The subaerial central Vøring Plateau was covered by dense vegetation growing in the moist and damp climate, generating lateritic soils on the volcanic floor during the early Paleogene. During the Eocene and Oligocene, the warm climate persisted, though less moist than previously. During the latest early Miocene, the climate changed from warm to temperate with a progressive cooling to temperate conditions during the middle-to-late Miocene. From the latest Miocene through the Pliocene/Pleistocene, the climate progressively entered its glacial mode. During the past 2.6 million years the region was characterized by alternating glacial/interglacial climate episodes.

The Norwegian surface waters stayed relatively warm to temperate until the middle Miocene when intensive cooling began. However, based on ice rafting, the first indicators of cold waters and intermittent ice cover appeared 5.5 Ma. Since that time the surface waters have remained cool to temperate with major ice cover occurring at ~2.6 and ~1.2 Ma.

### 2. Leg 105

Twelve holes were drilled at three sites (Sites 645 to 647; Figure 13) and 1884 m of core were recovered in Baffin Bay and the Labrador Sea in order to study their early tectonic history and the paleoceanographic history of the North Atlantic and its connection to the

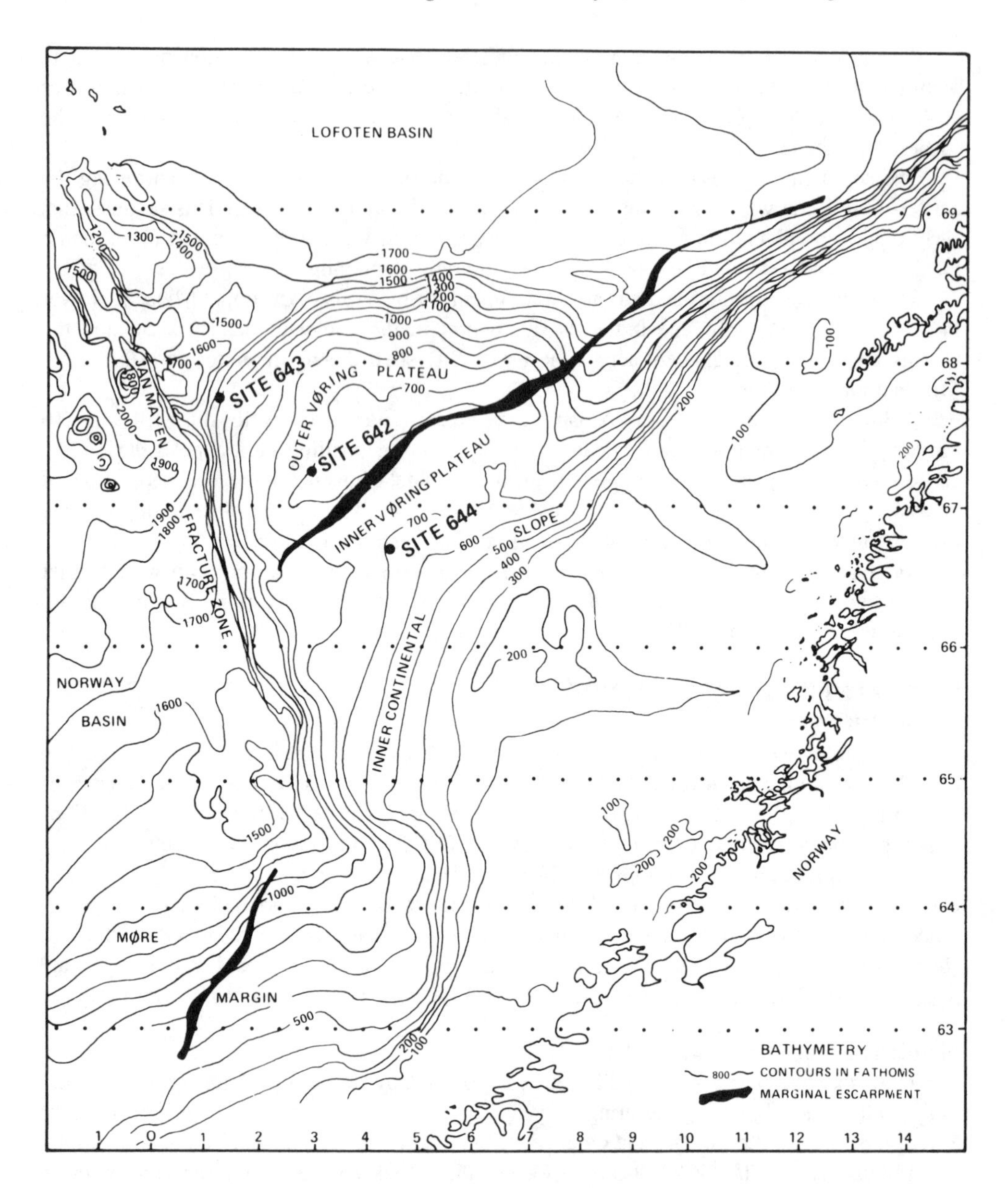


**FIGURE 12.** Location of sites occupied during ODP Leg 104.

Arctic Ocean. Co-Chief Scientists were Dr. Shiri Srivastava (Bedford Institute of Oceanography, Canada) and Dr. Michael Arthur (University of Rhode Island). Dr. Brad Clement was the Texas A&M University Staff Scientist. (See Srivastava et al. 1987 and 1989 for detailed discussion of cruise results.)

## Significant Paleoenvironmental Results

Climatic deterioration and southward transport of Arctic watermasses in Baffin Bay began in the middle Miocene or earlier. Significant cooling of surface waters began as early as the late Eocene.

The building of major sedimentary drifts resulting from intensified deep-water circulation began in the middle-to-late Miocene in the Labrador Sea.

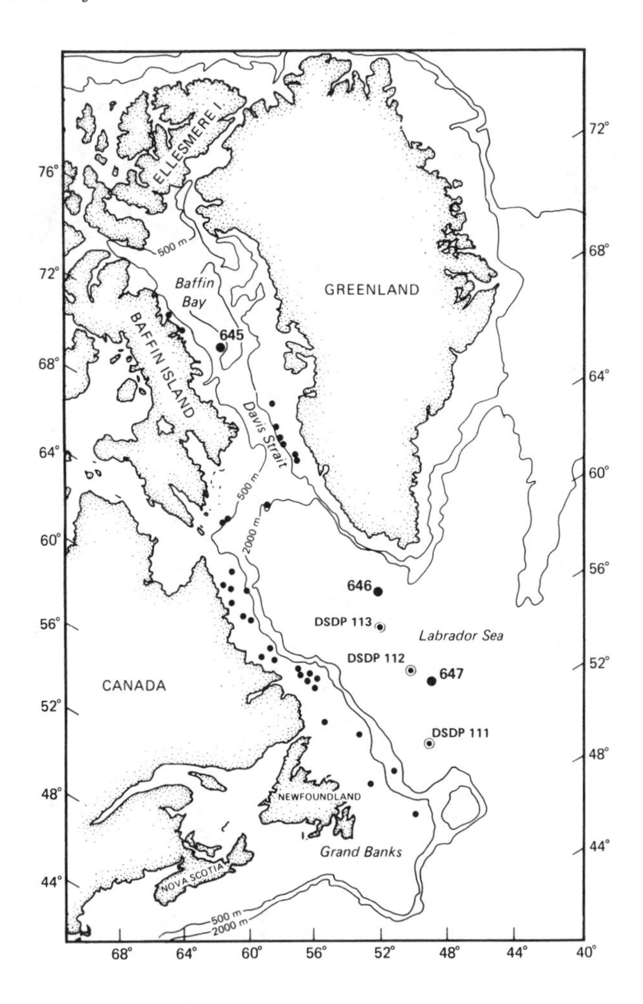


**FIGURE 13.** Location of sites occupied during ODP Leg 105.

Dropstones and other coarse sediment record the onset of ice rafting 2.5 Ma, and possibly as early as 3.4 Ma. The beginning of major glacial activity in Baffin Bay may have preceded ice rafting in the North Atlantic by up to a million years. Isolated pebbles and granules in strata as old as late Miocene may indicate that seasonal sea ice, at least, was present as early as 8 Ma.

In Baffin Bay, ~1 m cycles of alternating dark grey and light grey-brown calcareous silt, clay, or mud in the upper part of the Pleistocene section reveal glacial/interglacial changes. Each cycle spans about 8,000 years, a much shorter period than the previous estimate of 41,000 years based on studies of short piston cores from the region and predicted by the orbital-forcing theory. This suggests that sea-ice melting and advance and retreat of glacial ice on the margins of Baffin Bay fluctuated more rapidly than previously thought.

## C. LOW LATITUDE

### 1. Leg 108

Twenty-seven holes were drilled at twelve sites (Sites 657 to 668; Figure 14) and ~4000 m of core were recovered off Northwest Africa in order to obtain paleoclimatic data on oceanic and atmospheric circulation patterns, ice volume, and sea-level changes, and to provide detailed studies of Neogene biostratigraphy. Co-Chief Scientists for the cruise were

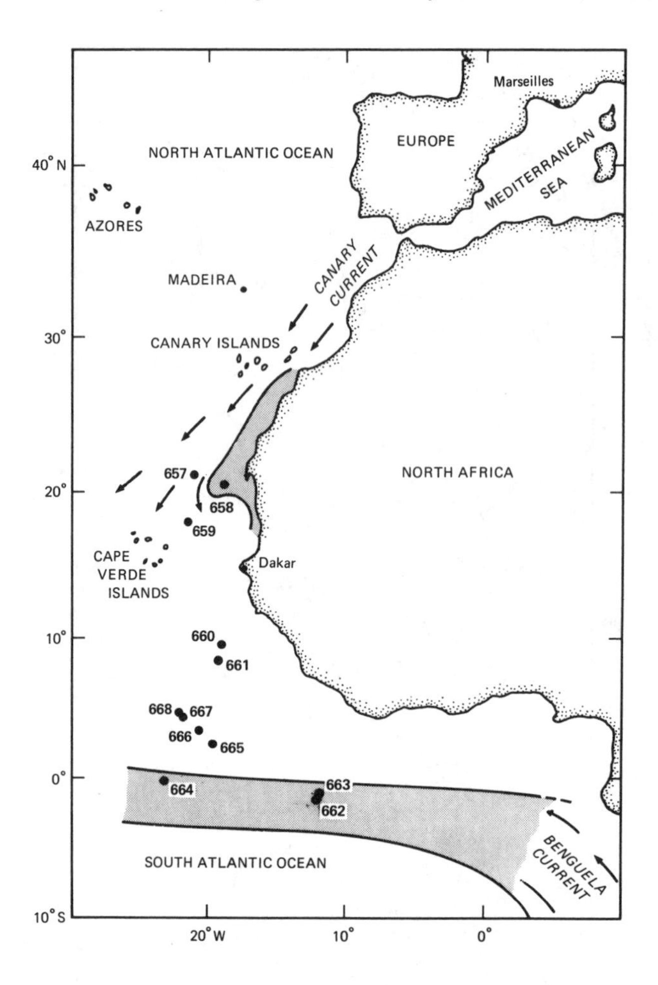


**FIGURE 14.** Location of sites occupied during ODP Leg 108.

Dr. William Ruddiman (Lamont-Doherty Geological Observatory of Columbia University) and Dr. Michael Sarnthein (University of Kiel, Federal Republic of Germany). Dr. Jack Baldauf was the Texas A&M University Staff Scientist. (See Ruddiman et al. 1988 and 1989 for detailed discussion of cruise results.)

## Significant Paleoenvironmental Results

P-wave velocity and magnetic susceptibility signals which contain orbital-scale rhythms that carry much of the key paleoclimatic response of these regions were measured at intervals less than 3 cm in the cores. Such signals lead to a better understanding of the linkages between polar components of the climate system, such as the buildup of ice sheets, the formation of sea ice, and low-latitude climate change.

Between 2.5 and 3.0 Ma, coastal upwelling and south equatorial divergence substantially intensified, continuing to about 500,000 years ago. This result is based on increased amount of diatoms and organic carbon in the sediment during this time interval.

The presence of diatoms — cold-water siliceous microfossils — in the sediments indicates that during the past 3 million years, stronger eastern-boundary currents in both hemispheres moved cold surface water toward the equator. Higher proportions of clay, silt, and freshwater diatoms from lake basins also suggest a higher quantity of wind-borne detritus, which may have resulted from more frequent and severe arid conditions in Africa.

Sediment cycles rich in calcium carbonate occurred earlier than 2.5 to 3 million years ago. A concomitant lack of biogenic opal, freshwater diatoms, and land-derived silt and clay suggests that oceanic productivity of both equatorial divergence and coastal upwelling was much lower at that time. Sedimentation rates strongly increased about 4.5 Ma at sites in water depths of less than 4000 m and about 4 Ma at deeper sites. The change in the calcium compensation at different depths reflects a gradual but major displacement of deep-water masses.

### 2. Leg 112

Twenty-seven holes were drilled at ten sites (Sites 679 to 688; Figure 15) and 2666 m of core were recovered from the Peruvian continental margin, in order to address paleo-oceanographic objectives including, amongst others, the initiation of the Humboldt current, variations of upwelling phenomena, the occurrences of "El Niño," the formation of organic-rich muds, and the dolomitization history of the classic Peru coastal upwelling regime. Dr. Roland von Huene (U.S. Geological Survey, Menlo Park, CA) and Dr. Erwin Suess (Oregon State University) were Co-Chief Scientists for this expedition. Dr. Kay-Christian Emeis was the Texas A&M University Staff Scientist. (See Suess et al. 1988a and 1990 for detailed discussion of cruise results.)

#### Significant Paleoenvironmental Results

The record of paleoenvironmental changes and the diagenetic process of the forearc basins have been shown to be closely linked to the tectonic history of the margin. For example, adjacent Lima and Trujillo basins were at about the same water depth until middle Miocene time, when they were affected differently by deep crustal tectonism. As a result, the position of coastal upwelling centers along the margin has not been stationary, but instead has varied unpredictably with the varied responses of each basin to vertical movement.

The detailed sedimentary records of late Miocene, Pliocene, and Pleistocene-Holocene upwelling obtained by drilling along the transects do not go back to the beginning of the coastal upwelling history along the Peru margin, because of the very large subsidence of the upper slope before middle Miocene time. However, the occurrence of upper Miocene sediment in upwelling facies that were recovered 3820 m water depth, about 150 km offshore, shows that the critical depth zone of coastal upwelling migrated landward in response to margin subsidence and the fluctuation of sea level. The laminated diatomites recovered there are lithified equivalents of shallow-water sediments from centers of coastal upwelling and are correlative with sediments exposed in the Pisco Formation onshore.

Most of the sediment deposited on the slope and in the accreted complex (Site 685) has been produced in the shifting coastal zone characterized by upwelling and intense currents. Once deposited in that zone, sediment has been reworked, transported by slumping, and redeposited on the continental slope in much deeper water.

A supersaturated brine discovered in the pore waters in all shelf sites suggests that a saline brine underlines the upper 300 m of sediments along the entire shelf. The brine replenishes the supply of dissolved sulfate, enhancing bacterial degradation of sedimentary organic matter and providing dissolved calcium and magnesium for diagenetic dolomite and calcite formations.

### 3. Leg 117

Twenty-five holes were drilled at twelve sites (Sites 720 to 731; Figure 16) in three operational areas and 4367 m of core were collected in order to investigate global climate and ocean-atmosphere interactions over the last 10 million years. The general operational region has been under the influence of high biological productivity and high accumulation rates in one of the most fertile sectors of the world's oceans — the area of monsoonal

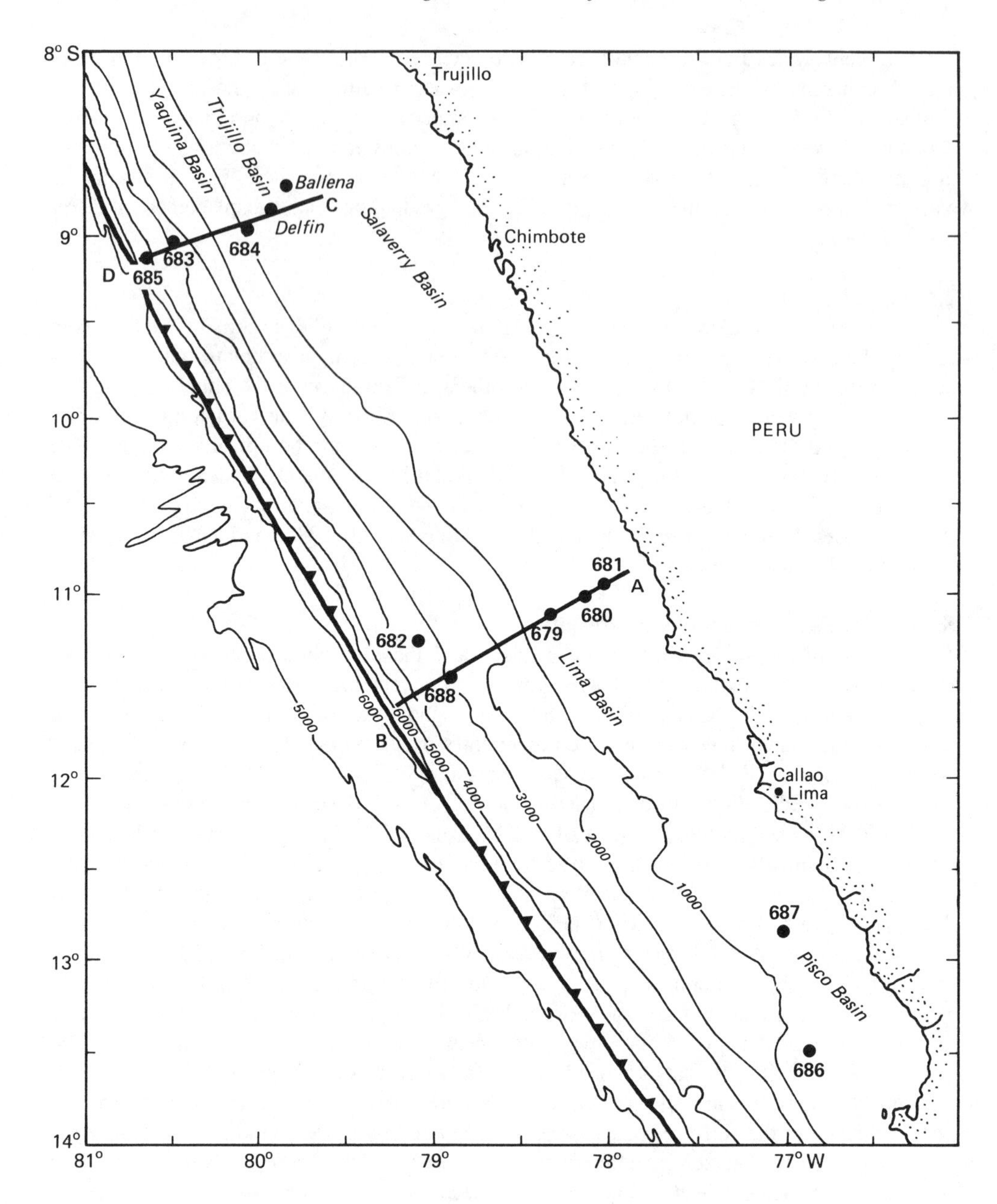


**FIGURE 15.** Location of sites occupied during ODP Leg 112.

upwelling off Arabia. Dr. Warren L. Press (Brown University, Rhode Island) and Dr. Nobuaki Niitsuma (Shizuoka University, Japan) were the Co-Chief Scientists of this cruise. Dr. Kay-Christian Emeis was the Texas A&M University Staff Scientist. (See Prell 1989 and 1991 for detailed discussion of cruise results.)

## Significant Paleoenvironmental Results

Monsoonal upwelling may have been active since the middle Miocene, and the oxygen-minimum zone may have fluctuated in oxygen content during this time.

The high-productivity zone of upwelling waters and the extent and location of the oxygen-minimum zone have been the most important factors for the distribution of sediment facies

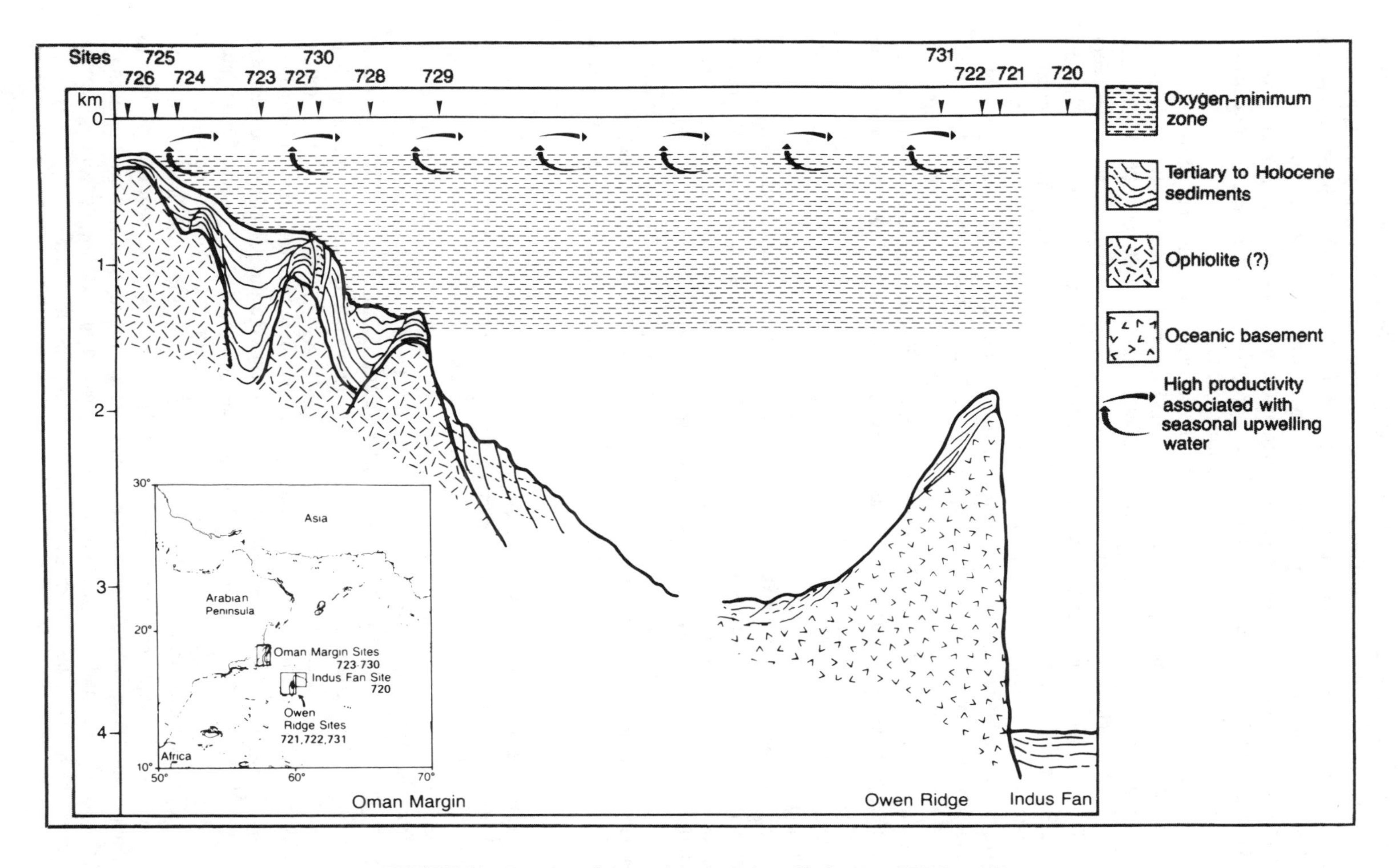


**FIGURE 16.** Location of sites and cross-section of bathymetry ODP Leg 117.

on the Oman margin. The latter is controlled by watermass circulation, oxygen consumption by bacterial activity, and the extent of biological production in the upwelling areas. High organic carbon content, coupled with intervals of primary varve-like laminations, are preserved in upper Pliocene sediments recovered at most sites. They are interpreted as an expression of an intensified oxygen-minimum zone, which episodically and effectively inhibited bioturbation during this time. The laminated, organic-rich interval may record a signal of significantly different oceanographic — and possibly atmospheric — circulation in the Northwest Indian Ocean from about 2.5 to 1.5 Ma.

A pronounced cyclicity of sediment properties is observed, consistent across all sites. Patterns of sediment color, physical properties, and magnetic susceptibility of sediments from the late Miocene to the present reveal distinct periodicities (of 400, 100, 41, 23, and 19 k.y.) that match those of the Milankovitch mechanism. These sedimentary cycles seem to reflect variations in primary productivity, preservation of carbon and carbonate, and the proportion of eolian detritus. The magnetic susceptibility varies together with, and probably originates in, the eolian detritus (as indicated by the abundance of quartz).

Climatic signals recognized in the cores provide independent evidence of the variability in monsoonal intensity. Magnetic susceptibility data vary here directly with terrigenous quartz, which is thought to be eolian. Thus, the magnetic material that causes susceptibility variations may be part of the windblown dust from the Arabian peninsula or from the Iran-Makran region. The susceptibility signal thus may be directly related to wind direction and intensity.

### 4. Leg 130

Leg 130 recovered over 4800 m of core from 16 holes at 5 sites (Sites 803 to 807; Figure 17) drilled along a depth transect on the Ontong Java Plateau that was designed to gain new insight into the evolution of global ocean dynamics and climate during the past 25 million years, and the origin and tectonic history of the world's largest oceanic plateau. Dr. Loren Kroenke (University of Hawaii) and Dr. Wolfgang Berger (University of Bremen, Germany) were the Co-Chief Scientists of this cruise. Dr. Thomas Janecek was the Texas A&M University Staff Scientist. (See Kronke et al. 1991, and in press for detailed discussion of cruise results.)

### Significant Paleoenvironmental Results

Periods of low sedimentation are dominant in the late-early to early-middle Miocene (20 to 15 Ma) and are characterized by hiatuses at some sites, and by condensed sections at others. The onset of the period of low accumulation coincides with a marked change in the carbon isotope composition of the ocean, which is the result of an increased difference in productivity between the deep sea and coastal waters.

Period of high sedimentation centered in the late Miocene to earliest Pliocene (7 Ma) displays some of the highest rates ever recorded in open-ocean pelagic sediments (about 50 m/million years). The timing of this period suggests that these high rates may be associated with the closing of the Tethys seaway (which once connected the Atlantic and the Pacific from the Mediterranean region to the Himalayas) and with important phases of mountain-building, leading to an increased supply to the ocean of continent-derived materials.

A large number of acoustic reflectors are synchronous and are associated in time with important paleoceanographic events. Many reflectors line up with sudden changes in carbonate accumulation. Some reflectors are strongly related to diagenesis, or are enhanced by diagenesis. Others mark the position of hiatuses, which in turn line up with condensed sections, emphasizing the importance of carbonate dissolution pulses.

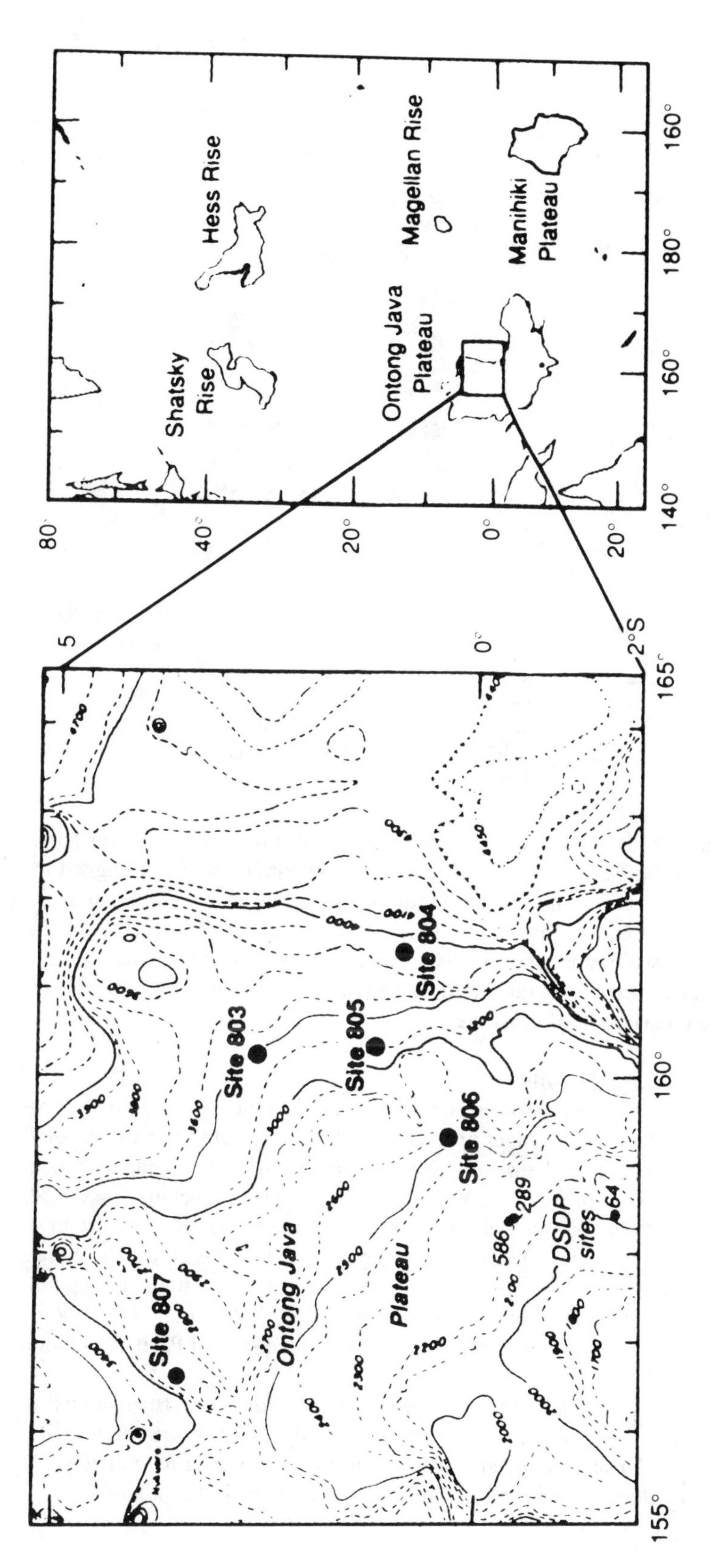


**FIGURE 17.** Location of sites occupied during ODP Leg 130.

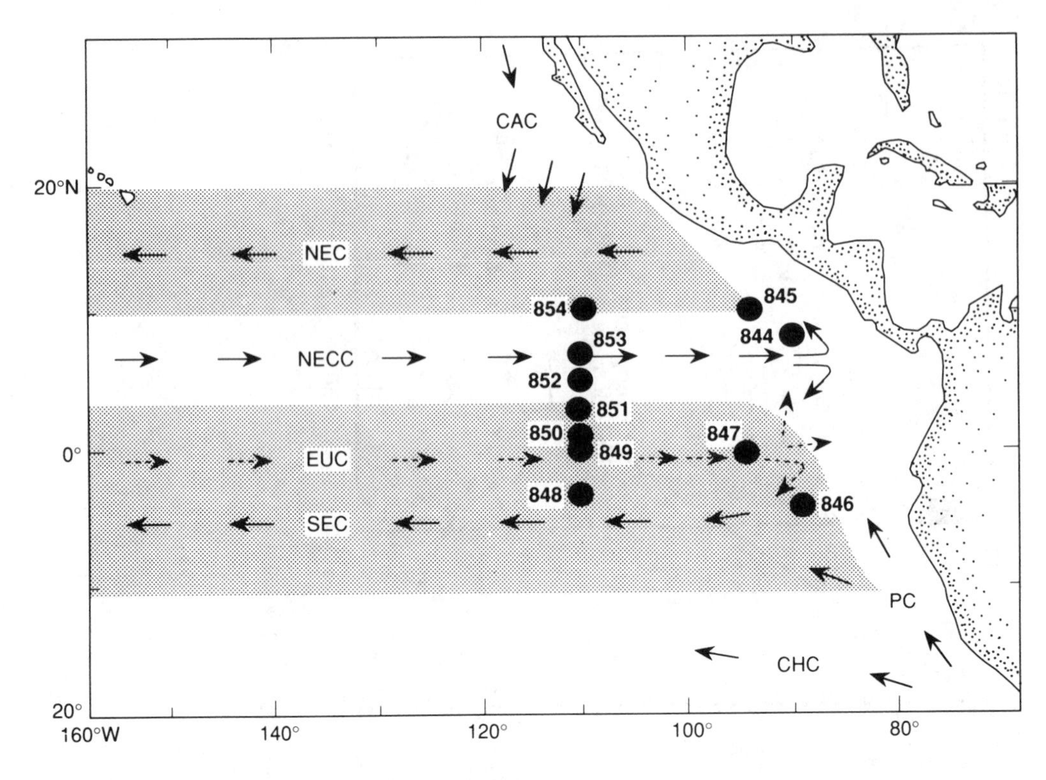


**FIGURE 18.** Location of sites occupied during ODP Leg 138.

## 5. Leg 138

Forty-two holes were drilled at eleven sites (Sites 844 to 854; Figure 18) and over 5500 m of core recovered in the East Equatorial Pacific in order to continue studies designed to examine the evolution of global climate change during the late Cenozoic through high resolution studies of tropical ocean sediments. Dr. Larry Mayer (Dalhousie University, Nova Scotia) and Dr. Nick Pisias (Oregon State University) were Co-Chief Scientists for this cruise. Dr. Thomas Janecek was the Texas A&M University Staff Scientist. (See Pisias et al. [in press] for detailed discussion of cruise results.)

### Significant Paleoenvironmental Results

The broad pattern of temporal variation in sediment accumulation appears to be basin-wide ''events,'' typically reflected by changes in calcium carbonate content (probably related to dissolution), or as intervals rich in the diatom *Thalassiothrix longissima*, some of which contain mm-scale laminated beds. In the modern ocean, *T. longissima* is generally associated with enhanced upwelling and high productivity; the presence of these intervals here are thus indicators of major productivity events. These events may be synchronous across the equatorial Pacific and are imaged with seismic profiling techniques. The broad pattern of temporal variations in sedimentation and accumulation rates is consistent with that found in previous drilling expeditions in the central equatorial Pacific, the western equatorial Pacific, and the equatorial Indian Ocean.

Superimposed on the long-term temporal changes are higher frequency fluctuations in the ratio of sedimentary components, typically carbonate to silica (which are reflected in the continuous core logs of density, color, and susceptibility), that show a periodicity that is consistent with Milankovitch climate-forcing induced by variations in Earth's orbit.

The near-continuous density, susceptibility, and color reflectance core logs collected on this Leg 138 demonstrate that these detailed records, extending into the late Neogene, can be correlated over thousands of km.

## V. CONCLUSIONS

The ODP has recovered near-continuous stratigraphic sequences from drill sites which are globally arranged and which partially complete a spatial and temporal framework for determining Earth's climatic history. Analysis of the sedimentary sequences provides critical information concerning the timing, extent, and stability of Earth's cryosphere, development of the modern-day upwelling regimes (and thus wind regimes), and the relationship between the low and high latitude components of the climate system. These sequences provide the only means to test current and future climatic models. In addition, the scientific results identify limitations or gaps in our understanding of the climate system, thereby providing insight into the future requirements necessary to further advance our understanding of the Earth's oceans and atmosphere.

Scientific data obtained by ocean drilling have dramatically improved the constraints concerned with the timing and development of Earth's cryosphere. Specific results indicate that an ice sheet existed on east Antarctica by 36 Ma, and possibly as earlier as 42 Ma, although the question of size and stability of this ice sheet is still in question. Development of the west Antarctic ice sheet commenced significantly later, occurring by 5 Ma and possibly 10 Ma. This western ice sheet has been a stable and permanent feature since about 4.8 Ma.

Glacial conditions in the northern high latitudes coincide more closely with that of western Antarctic than that of eastern Antarctica. Scientific results suggest that the onset of glaciation occurred at about 2.5 Ma in the Labrador Sea, 2.9 Ma in the Norwegian Sea, and as early as 3.4 Ma in Baffin Bay. As such, the duration of the cryosphere asymmetry is constrained, although detailed information as to the cause and effects of this asymmetry is still required.

Current results also provide evidence for the occurrence of a broad-scale linkage, at least for the last 3 million years, between the polar components of the climate system (ice sheets, sea ice, and polar ocean) and the low-latitude ocean atmosphere components (surface ocean, upwelling, wind circulation, and land climate). Additional data, however, are required to be collected and interpreted prior to determining the specific forcing or response mechanisms present. Preliminary results indicate that low-latitude components, specifically development of the upwelling systems and surface productivity, commence in the eastern equatorial Atlantic at about 3.5 Ma, and in the Oman Region at about 10 Ma. Evidence from the Arabian Sea suggests that this regional productivity may in part be associated with development of the monsoonal system and the uplift of the Himalayas.

Recent scientific results have also identified limitations or gaps in our understanding of Earth's climate system, and provide the impetus for further scientific research. Specific objectives include improving our understanding of (1) the short- and long-term ocean-climate variability including the cause and effect; (2) the stability and characteristics of Cenozoic ice sheets; (3) the role of the Arctic Ocean as a component of the climate system; (4) the variability of Cenozoic ice volume and its relationship to ice-sheet advances and retreats; (5) the evolution and variability of atmospheric circulation; and (6) the relationship and variability of biological productivity and biogeochemical cycles.

## ACKNOWLEDGMENTS

The ODP JOIDES Institutions, which give scientific advice to the program, are the Hawaii Institute of Geophysics, Univeristy of Hawaii; Lamont-Doherty Geological Observatory of Columbia University; Scripps Institution of Oceanography, University of California at San Diego; Rosenstiel School of Marine and Atmospheric Science, University of Miami; College of Oceanography, Oregon State University; Graduate School of Oceanography, University of Rhode Island; Department of Oceanography, Texas A&M University; Institute for Geophysics, University of Texas at Austin; College of Ocean and Fishery Sciences, University of Washington; Woods Hole Oceanographic Institution; The Department of Energy, Mines, and Resources, Earth Sciences Sector, Canada/Bureau of Mineral Resources, Geology and Geophysics, Australia; the European Science Foundation Consortium for the Ocean Drilling Program (Belgium, Denmark, Finland, Iceland, Italy, Greece, The Netherlands, Norway, Spain, Sweden, Switzerland, and Turkey); Bundesanstalt für Geowissenschaften und Rohstoffe, Federal Republic of Germany; Institut Francais de Recherche pour l'Exploitation de la Mer (IFREMER), France; University of Tokyo, Ocean Research Institute, Japan; the Natural Environment Research Council, U.K.; and the Academy of Sciences, U.S.S.R.

The program is funded by the Joint Oceanographic Institutions, Inc., a nonprofit corporation composed of the ten U.S. JOIDES members which, in turn, is funded by the U.S. National Science Foundation (NSF), an independent federal agency. NSF receives contributions from the non-U.S. members. The contributions of the large members of scientists, engineers, technical support, and administrators at TAMU, L-DGO, JOI, JOIDES, NSF, SEDCO/Forex, and the international community who are responsible for ensuring the success of ODP are greatly appreciated.

## REFERENCES

Andrews, J. T., Shilts, W. W., and Miller, G. H., 1983. Multiple deglaciations of the Hudson Bay Lowlands, Canada since deposition of the Missinaibi (last interglacial?) formation. *Quat. Res. (N.Y)* 19:18–37.

Baldauf, J. G. and Barron, J. A., 1990. Evolution of biosiliceous sedimentation patterns for the Eocene through Quaternary: Paleoceanographic response to polar cooling, in Bleil, U., and Theide, J. (Eds.), *NATO ASI Ser.*, Kluwer Academic, Dordrecht, 575–607.

Barron, J. A., Baldauf, J. G., and Larsen, B., 1991a. Evidence for late Eocene to early Oligocene Antarctic glaciation and observations on late Neogene glacial history of Antarctica: results from Leg 119, in Barron, J. A., Larsen, B. et al., *Proc. ODP, Sci. Results,* 119, (Ocean Drilling Program), College Station, TX, 869–894.

Barron, J., B. Larsen, J. G. Baldauf, C. Alibert, S. P. Berkowitz, J.-P. Caulet, S. R. Chambers, A. K. Cooper, R. Cranston, W. U. Dorn, W. U. Ehrmann, R. Fox, G. Fryxell, M. J. Hambrey, B. T. Huber, C. J., Jenkins, S.-H. Kang, B. H. Keating, K. W. Mehl, I. Noh, G. Ollier, A. Pittenger, H. Sakai, C. J. Schroder, A. Solheim, D. Stockwell, H. Thierstein, B. Tocher, B. Turner, and W. Wei, 1989. *Proc. ODP, Initial Reports* 119, (Ocean Drilling Program), College Station, TX, 942 pp.

Barron, J., B. Larsen, J. G. Baldauf, C. Alibert, S. P. Berkowitz, J.-P. Caulet, S. R. Chambers, A. K. Cooper, R. Cranston, W. U. Dorn, W. U. Ehrmann, R. Fox, G. Fryxell, M. J. Hambrey, B. T. Huber, C. J. Jenkins, S.-H. Kang, B. H. Keating, K. W. Mehl, I. Noh, G. Ollier, A. Pittenger, H. Sakai, C. J. Schroder, A. Solheim, D. Stockwell, H. Thierstein, B. Tocher, B. Turner, and W. Wei, 1991b. *Proc. ODP, Sci. Results,* 119, (Ocean Drilling Program), College Station, TX, 1003 pp.

Barker, P. F., J. P. Kennett, S. O'Connell, S. Berkowitz, W. R. Bryant, L. H. Burckle, P. K. Egeberg, D. K. Futterer, R. E. Gersonde, X. Golovchenko, N. Hamilton, L. Lawver, D. B. Lazasrus, M. Lonsdale, M. Mohr, T. Nagao, C. P. G. Pereira, C. J. Pudsey, C. M. Robert, E. Schandl, V. Spiess, L. D. Stott, E. Thomas, K. F. M. Thompson, and S. W. Wise, Jr., 1988. *Proc. ODP, Initial Reports* 113, (Ocean Drilling Program), College Station, TX, 785 pp.

Barker, P. F., J. P. Kennett, S. O'Connell, S. Berkowitz, W. R. Bryant, L. H. Burckle, P. K. Egeberg, D. K. Futterer, R. E. Gersonde, X. Golovchenko, N. Hamilton, L. Lawver, D. B. Lazasrus, M. Lonsdale, B. Mohr, T. Nagao, C. P. G. Pereira, C. J. Pudsey, C. M. Robert, E. Schandl, V. Spiess, L. D. Stott, E. Thomas, K. F. M. Thompson, and S. W. Wise, Jr., 1990. *Proc. ODP, Sci. Results,* 113, (Ocean Drilling Program), College Station, TX, 1033 pp.

Berger, A., Schneider, S., and Duplessy, J., Cl., 1989. *Climate and Geo-Sciences, a Challenge for Science and Society in the 21st Century.* Kluwer Academic, 723p.

Berggren, W., and Schnitker, D., 1983. Cenozoic marine environments in the North Atlantic and Norwegian-Greenland Sea. In Bott, M. H. P., Saxov, S., Talwani, M., and Thiede, J. (Eds.) *Structure and Development of the Greenland-Scotland Ridge.* Plenum Press, New York, 495–548.

Cameron, D. H., 1984. Design and operation of an extended core barrel, Deep Sea Drilling Project, Tech. Report No. 20, Scripps Institution of Oceanography, University of California at San Diego, 217 pp.

Ciesielski, P. F., Y. Kristoffersen, B. M. Clement, J.-P. Blangy, R. Bourrouilh, J. Crux, J. Fenner, P. Froelich, E. Hailwood, D. Hodell, M. Katz, H.-Y. Ling, J. Mienert, D. Mueller, J. Mwenifumbo, D. C. Nobes, M. Nocchi, D. A. Warnke, and F. Westall, 1988. *Proc. ODP, Init. Rep.* 114, (Ocean Drilling Program), College Station, TX, 815 pp.

Ciesielski, P. F., Y. Kristoffersen, B. M. Clement, J.-P. Blangy, R. Bourrouilh, J. Crux, J. Fenner, P. Froelich, E. Hailwood, D. Hodell, M. Katz, H.-Y. Ling, J. Mienert, D. Mueller, J. Mwenifumbo, D. C. Nobes, M. Nocchi, D. A. Warnke, and F. Westall, 1991. *Proc. ODP, Sci. Results,* 114, (Ocean Drilling Program), College Station, TX, 826 pp.

CLIMAP Project Members, 1976. The surface of the ice-age Earth. *Science,* 191:1131–1137.

COSOD, 1981. Report of the Conference on Scientific Ocean Drilling, Joint Oceanographic Institutions, Inc., Washington, D.C., 112 pp.

COSOD II, 1987. Report of the Second Conference on Scientific Ocean Drilling (COSOD II), Joint Oceanographic Institutions, Inc., Washington, D.C., 142 pp.

Creager, J. S., Scholl, D. W., et al., 1973, Initial Reports DSDP 19, Washington, U.S. Govt. Printing Office, 913p.

Dickson, R. R. and Kidd, R. B., 1987. Deep circulation in the southern Rockall Trough-The oceanographic setting of Site 610. in Ruddiman, W. F., Kidd, R. B., and Thomas, E., et al., *Initial Reports* DSDP, Washington, U.S. Govt. Printing Office, 1061–1074.

Eldholm, O. and Thiede, J., 1980. Cenozoic continental separation between Europe and Greenland. *Palaeogeog., Palaeoclimatol., Paleocol.,* 30:243–259.

Eldholm, O., J. Thiede, E. Taylor, K. Bjorklund, U. Bleil, P. Ciesielski, A. Desprairies, D. Donally, C. Froget, R. Goll, R. Henrich, E. Jansen, L. Krissek, K. Kvenvolden, A. LeHuray, D. Love, P. Lysne, T. McDonald, P. Mudie, L. Osterman, L. Parson, J. D. Phillips, A. Pittenger, G. Qvale, G. Schoenharting, and L. Viereck, 1987. *Proc. ODP, Initial Reports* 104, (Ocean Drilling Program), College Station, TX, 783 pp.

Eldholm, O., J. Thiede, E. Taylor, K. Bjorklund, U. Bleil, P. Ciesielski, A. Desprairies, D. Donally, C. Froget, R. Goll, R. Henrich, E. Jansen, L. Krissek, K. Kvenvolden, A. LeHuray, D. Love, P. Lysne, T. McDonald, P. Mudie, L. Osterman, L. Parson, J. D. Phillips, A. Pittenger, G. Qvale, G. Schoenharting, and L. Viereck, 1989. *Proc. ODP, Sci. Results,* 104, (Ocean Drilling Program), College Station, TX, 1141 pp.

Foss, G. N., 1985. The Ocean Drilling Program II: JOIDES RESOLUTION, scientific drillship of the '80s, *Proc. Marine Tech. Soc. "Ocean Engineering and the Environment"* 1:124–132.

Foss, G. N., P. D. Rabinowitz, B. W. Harding, C. Hanson, and L. Hayes, 1988. Scientific Ocean Drilling in Ice-Laden Regions, in *Proc. 9th Intl. Symp. on Ice,* H. Saeki and K. Hirayama, Eds., Vol. 1, p. 674–685.

Harwood, D. and Webb, P., 1989. Early Pliocene deglaciation of the Antarctic ice sheet and late Pliocene onset of bipolar glaciation, *Eos,* 71:538.

Huey, D. P., 1984. Design and operation of an advanced hydraulic piston corer. Deep Sea Drilling Project Tech. Report No. 21, Scripps Institution of Oceanography, University of California at San Diego, 269 pp.

Kroenke, L., W. Berger, T. Janecek, J. Backman, F. Bassinot, R. Corfield, M. Delaney, R. Hagen, E. Jansen, L. Krissek, C. Lange, M. Leckie, I. Lind, J. Mahoney, J. Marsters, L. Mayer, D. Mosher, R. Musgrave, M. Prentice, J. Resig, H. Schmidt, R. Stax, M. Storey, T. Takayama, K. Takahashi, J. Tarduno, R. Wilkens, and G. Wu, 1991. *Proc. ODP, Initial Reports,* 130, (Ocean Drilling Program), College Station, TX, 1240 pp.

Kroenke, L., W. Berger, T. Janecek, J. Backman, F. Bassinot, R. Corfield, M. Delaney, R. Hagen, E. Jansen, L. Krissek, C. Lange, M. Leckie, I. Lind, J. Mahoney, J. Marsters, L. Mayer, D. Mosher, R. Musgrave, M. Prentice, J. Resig, H. Schmidt, R. Stax, M. Storey, T. Takayama, K. Takahashi, J. Tarduno, R. Wilkens, and G. Wu. *Proc. ODP, Sci. Results,* 130, (Ocean Drilling Program), College Station, TX, in press.

Lisitzin, A. P., 1972. Sedimentation in the world ocean. Society of Economic Paleontologists and Mineralogists, Special Publication 17, Tulsa, Oklahoma, 218p.

Matthews, R. K. and Poore, R. Z., 1980. Tertiary 180 record and glacio-eustatic sea-level fluctuations. Geology, 8:501–504.

Miller, K. G., Fairbanks, R. G., and Mountain, G. S., 1987. Tertiary oxygen isotope synthesis, sea-level history, and continental margin erosion. Paleoceanography, 2:1–19.

Miller, K. G. and Tucholke, B. E., 1983. Development of Cenozoic abyssal circulation south of the Greenland Scotland ridge, in Bott, M. H. P., Saxov, S., Talwani, M., and Thiede, J., Eds., *Structure and Development of the Greenland-Scotland Ridge.* Plenum Press, New York, 549–589.

Pisias, N., L. Mayer, T. Janecek, J. Baldauf, S. Bloomer, K. Dadey, K.-C. Emeis, J. Farrell, J.-A. Flores, E. Galimov, T. Hagelberg, P. Holler, S. Hovan, M. Iwai, A. Kemp, D. C. Kim, G. Klinkhammer, M. Leinen, S. Levi, M. Levitan, M. Lyle, A. K. Mackillop, L. Meynadier, A. Mix, T. Moore, I. Raffi, C. Ravelo, D. Schneider, N. Shackleton, J.-P. Valet, and E. Vincent, *Proc. ODP, Initial Reports,* 138, (Ocean Drilling Program), College Station, TX, in press.

Prell, W. L., N. Niitsuma, K. Emeis, D. Anderson, R. Barnes, R. A. Bilak, J. Bloemendal, C. J. Bray, W. H. Busch, S. C. Clemens, P. Debrabant, P. de Menocal, H. L. Ten Haven, A. Hayashida, J. O. R. Hermelin, R. Jarrard, A. N. K. al-Thobbah, L. A. Krissek, D. Kroon, D. M. Murray, C. Nigrini, T. F. Pedersen, W. Ricken, G. B. Shimmield, S. A. Spaulding, Z. K. S. al-Sulaiman, T. Takayama, and G. P. Weedon, 1989. *Proc. ODP, Initial Reports,* 117, (Ocean Drilling Program), College Station, TX, 1236 pp.

Prell, W. L., N. Niitsuma, K. Emeis, D. Anderson, R. Barnes, R. A. Bilak, J. Bloemendal, C. J. Bray, W. H. Busch, S. C. Clemens, P. Debrabant, P. de Menocal, H. L. Ten Haven, A. Hayashida, J. O. R. Hermelin, R. Jarrard, A. N. K. al-Thobbah, L. A. Krissek, D. Kroon, D. M. Murray, C. Nigrini, T. F. Pedersen, W. Ricken, G. B. Shimmield, S. A. Spaulding, Z. K. S. al-Sulaiman, T. Takayama, and G. P. Weedon, 1991. *Proc. ODP, Sci. Results,* 117, (Ocean Drilling Program), College Station, TX, 638 pp.

Prentice, M. L. and Matthews, R. K., 1988. Cenozoic ice volume history: development of a composite oxygen isotope record. *Geology* 17:963–966.

Rabinowitz, P. D., 1986. Science Ocean Drilling, *J. Ocean Sci. Engin.,* Vol. 10, 3 and 4:353–384.

Rabinowitz, P. D., T. J. G. Francis, J. G. Baldauf, B. W. Harding, R. G. McPherson, R. B. Merrill, A. W. Meyer, and R. E. Olivas, 1991. The Ocean Drilling Program: Results from the Sixth Year of Field Operations, in *Proc. Offshore Technology Conf.,* OTC 6506, p. 55–74.

Ruddiman, W., M. Sarnthein, J. Baldauf, J. Backman, J. Bloemendal, W. Curry, P. Farrimond, J.-C. Faugeres, T. Janecek, Y. Katsura, H. Manivit, J. Mazzullo, J. Mienert, E. Pokras, M. Raymo, P. Schultheiss, R. Stein, L. Tauxe, J.-P. Valet, P. Weaver, and H. Yasuda, 1988. *Proc. ODP, Initial Reports,* 108, (Ocean Drilling Program), College Station, TX, 1073 pp.

Ruddiman, W., M. Sarnthein, J. Baldauf, J. Backman, J. Bloemendal, W. Curry, P. Farrimond, J.-C. Faugeres, T. Janecek, Y. Katsura, H. Manivit, J. Mazzullo, J. Mienert, E. Pokras, M. Raymo, P. Schultheiss, R. Stein, L. Tauxe, J.-P. Valet, P. Weaver, and H. Yasuda, 1989. *Proc. ODP, Sci. Results,* 108, (Ocean Drilling Program), College Station, TX, 519 pp.

Ruddiman, W. F. and McIntyre, A., 1976. Northeast Atlantic paleoclimatic changes over the past 600,000 years. In Cline, R. M. and Hays, J. D., Eds., Investigations of Late Quaternary paleo-oceanography and paleoclimatology. *GSA, Memoir* 145, 111–146.

Ruddiman, W. F. and McIntyre, A., 1981. Oceanic mechanisms for the amplifications of the 23,000 yr. ice-volume cycle, *Science,* 212:617–627.

Ruddiman, W. F., Cameron, D., and Clement, B. D., 1984. Sediment disturbance and correlation of offset holes, in Ruddiman, W. F., Kidd, R. B., and Thomas, E., et al., *Initial Reports, DSDP, Washington U.S. Govt. Printing Office,* 615–634.

Savin, S. M., Douglas, R. G., and Stehli, F. G., 1975. Tertiary marine paleotemperatures. *Geol. Soc. Am., Bull.,* 86:1499–1510.

Schlich, R., S. W. Wise, A. A. Palmer, M.-P. Aubry, W. A. Berggren, P. R. Bitschene, N. A. Blackburn, J. Breza, M. Coffin, D. M. Harwood, F. Heider, M. A. Holmes, W. R. Howard, H. Inokuchi, K. R. Kelts, D. Lazarus, A. Mackensen, T. Maruyama, M. Munschy, E. Pratson, P. G. Quilty, F. Rack, V. J. M. Salters, J. H. Sevigny, M. Storey, A. Takemura, D. Watkins, H. Whitechurch, and J. C. Zachos, 1989. *Proc. ODP, Initial Reports,* 120, (Ocean Drilling Program), College Station, TX, 648 pp.

Schlich, R., S. W. Wise, A. A. Palmer, M.-P. Aubry, W. A. Berggren, P. R. Bitschene, N. A. Blackburn, J. Breza, M. Coffin, D. M. Harwood, F. Heider, M. A. Holmes, W. R. Howard, H. Inokuchi, K. R. Kelts, D. Lazarus, A. Mackensen, T. Maruyama, M. Munschy, E. Pratson, P. G. Quilty, F. Rack, V. J. M. Salters, J. H. Sevigny, M. Storey, A. Takemura, D. Watkins, H. Whitechurch, and J. C. Zachos, 1989. *Proc. ODP, Sci. Results,* 120, (Ocean Drilling Program), College Station, TX, in press.

Shackleton, N. J., 1986. Paleogene stable isotope events, *Palaeogeogr., Palaeoclimatol., Palaeoecol.,* 57:91–102.

Shackleton, N. J. and Boersma, A., 1981. The climate of the Eocene ocean. *J. Geol. Soc.* London, 138:153–157.

Shackleton, N. J. and Kennett, J., 1975. Paleotemperature history of the Cenozoic and the initiation of Antarctic glaciation: oxygen and carbon isotope analysis in DSDP Sites 277, 279, 281, in Kennett, J. P., Houtz, R. E., et al., *Initial Reports,* DSDP, 29, Washington U.S. Govt. Printing Office, 743–755.

Shackleton, N. J., Backman, J., Zimmerman, H., Kent, D. V., Hall, M. A., Roberts, D. G., Schnitker, D., Baldauf, J. G., Despraoroes, A., Homrighausen, R., Huddlestun, P., Keene, J. B., Kaltenback, A. J., Smith, J. W., 1984. Oxygen isotope calibration of the onset of ice-rafting and history of glaciation in the North Atlantic region. *Nature,* 307:620–623.

Srivastava, S., M. Arthur, B. Clement, A. Aksu, J. Baldauf, G. Bohrman, W. Busch, T. Cederberg, M. Cremer, K. Dadey, A. De Vernal, J. Firth, F. Hall, M. Head, R. Hiscott, R. Jarrard, M. Kaminsky, D. Lazarus, A.-L. Monjanel, O. Nielsen, R. Stein, F. Thiebault, J. Zachos, and H. Zimmerman, 1987. *Proc. ODP, Initial Reports,* 105, (Ocean Drilling Program), College Station, TX, 917 pp.

Srivastava, S., M. Arthur, B. Clement, A. Aksu, J. Baldauf, G. Bohrman, W. Busch, T. Cederberg, M. Cremer, K. Dadey, A. De Vernal, J. Firth, F. Hall, M. Head, R. Hiscott, R. Jarrard, M. Kaminsky, D. Lazarus, A.-L. Monjanel, O. Nielsen, R. Stein, F. Thiebault, J. Zachos, and H. Zimmerman, 1989. *Proc. ODP, Sci. Results,* 105, (Ocean Drilling Program), College Station, TX, 1038 pp.

Storms, M. A., W. Nugent, and D. H. Cameron, 1983. Design and operations of hydraulic piston corer. Deep Sea Drilling Project, Tech. Report No. 12, Scripps Institution of Oceanography, University of California at San Diego, 203 pp.

Suess, E., R. von Huene, K.-C. Emeis, J. Bourgois, J. del C. Cruzado, P. De Wever, G. Eglinton, R. Garrison, M. Greenberg, E. Herrera, P. Hill, M. Ibaraki, M. Kastner, A. E. S. Kemp, K. Kvenvolden, R. Langridge, N. Lindsley-Griffin, R. McCabe, J. Marsters, E. Martini, L. Ocola, J. Resig, A. W. Sanchez, H. Schrader, T. M. Thornburg, G. Wefer, and M. Yamano, 1988. *Proc. ODP, Init. Rpts,* 112, (Ocean Drilling Program), College Station, TX, 1015 pp.

Suess, E., R. von Huene, K.-C. Emeis, J. Bourgois, J. del C. Cruzado, P. De Wever, G. Eglinton, R. Garrison, M. Greenberg, E. Herrera, P. Hill, M. Ibaraki, M. Kastner, A. E. S. Kemp, K. Kvenvolden, R. Langridge, N. Lindsley-Griffin, R. McCabe, J. Marsters, E. Martini, L. Ocola, J. Resig, A. W. Sanchez, H. Schrader, T. M. Thornburg, G. Wefer, and M. Yamano, 1990. *Proc. ODP, Sci. Results,* 112, (Ocean Drilling Program), College Station, TX, 738 pp.

Talwani, M. and Udinstev, G., et al., 1976. *Initial Reports* DSDP, 38 Washington, U.S. Govt. Printing Office, 1256p.

Thiede, J. Eldholm, O., 1983. Speculations about the paleodepth of the Greeland Scotland Ridge-new methods and concepts. In Bott, M. H. P., Saxov, S., Talwani, M., and Thiede, J., Eds., *Structure and Development of the Greenland-Scotland Ridge,* Plenum Press, New York, 445–456.

Webb, P. N., Harwood, D., McKelvey, B. C., Mercer, J. N., and Stott, L. D., 1984. Cenozoic marine sedimentation and ice-volume variation on the East Antarctic craton, *Geology,* 12:287–291.

Woodruff, F. and Savin, S. M., 1989. Miocene deepwater oceanography, *Paleoceanography* 4(1):87–140.

Chapter 7

# THE CLIMATE CONTINUUM: AN OVERVIEW OF GLOBAL WARMING AND COOLING THROUGHOUT THE HISTORY OF PLANET EARTH

**Anthony D. Socci**

## TABLE OF CONTENTS



0-8493-4419-0/93/$0.00 + $.50
© 1993 by CRC Press, Inc.

# I. INTRODUCTION

This paper is intended to constitute a qualitative model or assessment of climate controls and feedbacks, an attempt to sort out the controls, feedbacks, and preconditions, as well as their sequence, the sum of which is suggested to define climate for any segment of Earth history, and to identify and clarify the circumstances under which climate might be driven to change. It is an attempt to understand the substance, timing, periodicity, rate, and circumstance of global warming and global cooling events, the suggested end-members of a spectrum of climate conditions housed within the historic records (i.e., ice, rock, sediment, and anthropogenic) of this planet's climate. The absolute timing and duration of major climatic changes (i.e., transitions from global warming to global cooling and vice versa), and in particular, the sequence or oder in which climate-forcing feedbacks among Earth systems occur, are critically important to understanding climate and climate change.

From the standpoint of decision- and policymakers, initiating policy for future generations predicated on complex, computer-generated predictions — general circulation models (GCMs) — of future climate scenarios, is undoubtedly unsettling at best. I suspect that much of this anxiety and fear arises because all too often, complex computer models such as GCMs, give the nonscientist, nonmodeler, and even other scientists, a sense of witnessing an illusion which has no frame of reference in one's sphere of reality. Furthermore, even assuming one gets beyond this threshold of mystery, one is then often left with the impression that such models are untestable, that in the end, one is relegated to gauge or test the accuracy and reliability of such model predictions by waiting or marking time, as our societal future unfolds.

From my perspective, however, the above anxieties are not necessarily well founded in that we do indeed know a great deal more about the history of climate and climate change than might otherwise appear to be the case at first glance, while on the other hand, computer models of future climate are philosophically reasonable and sound ways to approach the future. Each of us, in our daily lives, goes through the same decision-making processes as do computer models, instinctively, quickly, and efficiently, much more so than computers.

In light of the above, I wish to emphasize that there is presently approximately 4.1 billion years of decipherable climate history bearing upon this planet and how climate operates in general (Lovelock and Whitfield, 1982). Much of this information is particularly useful in addressing the extremes of climate, global warming, and global cooling, on a variety of timescales and frequencies (Crowley, 1983; Pollack, 1979; Harland, 1981; Kutzbach, 1976; Mitchell, 1976; Worsley, et al., 1986; Worsley and Nance, 1989; Shackleton and Imbrie, 1990; Brouwer, 1983), virtually all of which can be used to constrain, calibrate, and test the accuracy and reliability, to a considerable extent, of current and future generations of computer-generated climate models or predictions.

Climate modelers, decision- and policymakers, and people in general are becoming increasingly aware of the significance of the historic records of this planet's climate (e.g., housed in ice cores, deep-sea sediment cores, caves, rocks, and trees, as well as in anthropogenic climate records) in assisting in better understanding our present and future environmental and climatic plight, while providing a context or backdrop for grasping the nature and scope of our present and future climate prospects. Hence, our understanding of climates past and present is our best hope of understanding the future of this planet's climate, and refining the GCMs which we will likely come to base many decisions upon. It is much like preventive medicine, or an insurance policy of sorts to ideally warn society of possible disasters in-the-making, while providing possible stratagems aimed at resolution.

Complex computer models of climate (or anything) are ideally, typically built upon a qualitative vision or concept of what climate is and how it operates and changes in time, based upon this planet's sole record of its climate history, the stratigraphic record in the broadest sense of the word. Thus armed with an initial hypothesis or vision, as suggested above, one can then write mathematical expressions which serve to define or translate, into computer language, the systems or parts of this vision which one initially believes to be responsible for climate and climate change. Each mathematical expression which is designed to explain some part of the climate-driving system can then be tested against reality/observation, in this instance, the stratigraphic record.

One can therefore test, in piecemeal fashion, how reliable or accurate each equation is in describing increments of reality (climate past and present in this case). In this manner one can construct a complex, computer-generated model, layer by layer, by testing each component of that model, in turn, until one has honed what might be referred to as a fully integrated model which will ideally and reliably predict reality past and present, and, therefore, future climate changes in all of their complexity.

The final goal of this paper is to bring to light, for broader consumption, the full spectrum of this planet's climate history and the suggested root causes of, and responses to, historic climate changes, including the significance of the rates and frequencies of those changes. Armed with such knowledge, one should have a clearer notion of the significance and possible impact of impending climate changes (i.e., global warming) and a context in which to understand these changes as part of planetary continuum of changes, as opposed to temporally isolated and unrelated natural occurrences in which the past has no relation to the present and future, and vice versa.

## II. ORIGIN OF A TWO END-MEMBER GLOBAL CLIMATE SYSTEM

Subsequent to this planet's earliest record of global cooling, in the guise of the first evidence of glacial ice on continents at approximately 2.7 to 2.5 billion years ago (Harland, 1981; 1983; Crowell, 1982; Frakes, 1979), global climate has oscillated between two extremes or end-members of climate. In addition, there is no evidence that the oceans have every boiled away, at least not in total, and invariably associated with the impact of large asteroids during the first 1 billion years of Earth history (4.5 to 3.5 billion years ago) (Sleep et al., 1989; Chyba, 1989; Kasting, 1990). There is also no evidence that the oceans have every frozen solid. Thus, the Earth's climate has remained remarkably well buffered (not too cold and not too warm) for at least the last 2.7 billion years, and probably as far back as 4 billion years ago (Lovelock and Whitfield, 1982).

Moreover, this initial global cooling event (and hence, all subsequent global cooling events in Earth history) constituted a feedback to the evolution of the atmosphere and the origin and evolution of life (the biosphere), the first signs of which are recorded in rocks 3.8 billion years old. Thus, this time marks the beginning of the transition from a primitive, volcanically regulated atmosphere dominated by high levels of carbon dioxide ($CO_2$), carbon monoxide (CO), ammonia ($NH_3$), nitrogen (N), and methane ($CH_4$), to an atmosphere regulated by life, the consequence of which was an increasingly oxygen ($O_2$)-rich atmosphere at the expense of atmospheric $CO_2$, in response to the evolution of the biosphere (Cloud, 1983). As life evolved and removed more and more $CO_2$ from the atmosphere, the surface temperature of the Earth dropped accordingly, consistent with $CO_2$ greenhouse properties, to the point where ice began to form (i.e., glaciers) for the first time, 2.7 billion years ago. Earth's atmosphere, and thus climate, has since been continually regulated by life-processes, plate movements, heat flow within the Earth, and chemical alteration (metamorphism and weathering) of the Earth's crust (Fischer, 1984a; 1984b; Worseley et al., 1986; Walker et al., 1981; Berner, 1990; Cloud, 1968; 1972; 1983; Budyko and Ronov, 1979; Lovelock and Whitfield, 1982; Holland, 1984; Kasting and Ackerman, 1986; Holland et al., 1986; Lovelock, 1987; Towe, 1990; Wolk, 1987; Schwartzman and Volk, 1989.

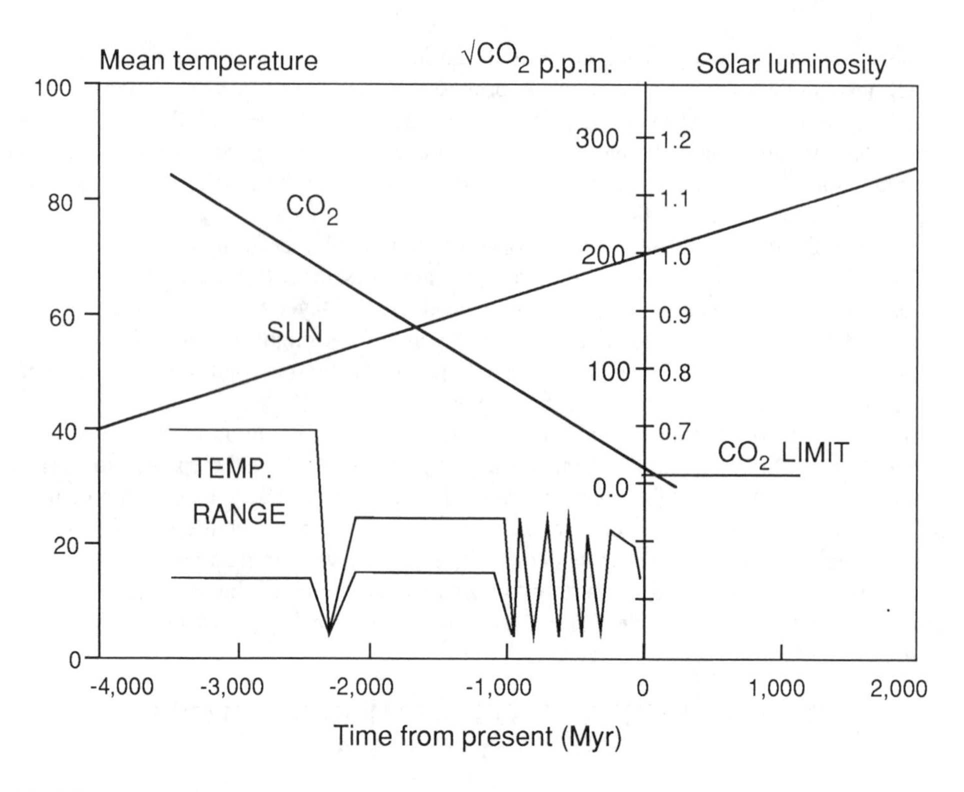


**FIGURE 1.** Modulation of climate resulting from the coincidence of increasing solar luminosity and depletion of atmospheric $CO_2$ over time. Solar luminosity relative to its present value which has been set at 1. $CO_2$ pressures shown are those suggested to have been required to compensate for increased solar luminosity and the resultant buffering of climate within temperature limits conducive for life. (From Lovelock, J. E. and Whitfield, M., *Nature*, 296, 562, 1982. With permission.).

This biotically induced transition from a more reduced (less oxygenated, lifeless (?) atmosphere, to a more oxidized, life and chemically regulated atmosphere, also appeared to serve (and continues to do so) as a first-order precondition to all subsequent long-term global cooling events in the history of this planet.

Furthermore, since the time of the first appearance of life in the form of prokaryotes (one-celled organisms with no cell walls) approximately 3.8 billion years ago or earlier (Cloud, 1983; Sleep, et al., 1989; Kasting, 1990), the average surface temperature of this planet has hovered in the vicinity of 25°C (Lovelock and Whitfield, 1982; Kasting and Ackerman, 1986) despite a 25 to 30% increase in solar luminosity (Lovelock and Margulis, 1974; Sagan and Mullen, 1972; Lovelock and Whitfield, 1982; Endal and Schatten, 1982; Wigley and Brimblecombe, 1981; Walker et al., 1981; Newkirk, 1980; 1983; Lovelock, 1987; Owen et al., 1979; Henderson-Sellers and Schwartz, 1979; Newman and Rood, 1977; Kasting, 1990) which should have resulted in elevated earth-surface temperatures above the threshold of life (terrestrial life?), throughout the history of life (Lovelock and Whitfield, 1982; see also Volk, 1989; 1987; Figure 1).

Life within the vicinity of hydrothermal vents at mid-ocean ridges might have had a higher threshold for life, even in the presence of high atmospheric $CO_2$ partial pressures and, therefore, high surface temperatures (approximately 85°C) during the early history of the planet (Kasting, 1990). However, this would not account for, nor explain, 4 billion years of well buffered, globally averaged earth-surface temperatures in the vicinity of 25°C, paradoxically, in the face of increasing solar luminosity.

Consequently, maintenance of a two end-member system of climate required, and continues to require, a plausible, candidate, planetary temperature-regulating or buffering mechanism (a cooling mechanism) in the face of heating resulting from increasing solar luminosity. One such mechanism is the continuous and/or incremental removal and subsequent burial of atmospheric carbon over time (i.e., rise in the partial pressure of atmospheric oxygen over time — Lovelock and Margulis, 1974; Lovelock and Whitfield, 1982; Walker et al., 1981), possibly by way of abrupt, evolution-driven improvements in the efficiency of scavenging and recycling biolimiting nutrients within successive, evolutionary groups of organisms, i.e., biotic radiations (Worsley et al., 1986; Worsley and Nance, 1989; see Figure 2). Such efficiency would likely result in biotically-driven, abrupt increases in the ratio of $C_{org}$ (the Earth's record of buried organic or unoxygenated carbon) to P (phosphorus — the essential life-limiting nutrient [Broecker and Peng, 1982]) over time (i.e., biotically-driven, abrupt removal and burial of small quantities of atmospheric carbon over time; Figure 2), with possible climatic implications given that any removal of $CO_2$ from the atmosphere results in cooling due to $CO_2$ greenhouse properties.

## III. GLOBAL WARMING AND GLOBAL COOLING DURING THE LAST 750 MILLION YEARS

Global climate cooling events (typically marked by geochemical signatures/osotopic proxies for $CO_2$, and evidence for the former presence of glacial ice) older than 750 million to 1 billion years in age, are generally suspect in that they remain, for the most part, poorly documented, poorly physically and stratigraphically resolved, and poorly constrained with regard to their absolute age and absolute geographic position at the time of occurrence (GARP, 1975; Williams, 1975; Frakes, 1979; Crowell, 1982; Crowley, 1983; Kasting, 1987). In fact, many of the Earth's colder climate warming and cooling events may have been local as opposed to global in extent, not unlike the climate cooling event 450 to 500 million years ago (the Ordovician glaciation), whereby cooling (and the appearance of glacial ice) was largely confined to Africa and South America (Frakes, 1979; Crowell, 1982) as a result of Gondwana (a large continental landmass in the Southern Hemisphere) having drifted over the South Pole at that time (Scotese et al., 1979; Van der Voo, 1988).

Further complications surrounding resolution of climate, and other events recorded within the older portions of the Earth arise due to: (1) unresolved questions related to motion and viscous mixing within and between the Earth's mantle and core resulting, in part, in significant literal movement or drift of the Earth's magnetic poles (true polar wander) over time (Anderson, 1981; 1982; 1984; 1989; Goldreich and Toomre, 1969; Donn, 1989) rendering many attempts at establishing geographic coordinates of former locations of continents through the use of magnetism essentially, spatially meaningless, and (2) changes in the volume and area of continents with time. The latter point is steeped in concerns regarding the timing of the origin and evolution of granitic/continential crust (the first vestiges of which are suggested to have begun to evolve from the mantle 2.5 billion years ago, reaching their present volume roughly 600 million years ago — see Hargraves, 1976; Goodwin, 1981) given that continental glaciers (frequently the sole evidence of this planet's most ancient climate record) cannot become "continental" in magnitude without a substrate (e.g., continental crust) for the ice to accumulate upon and spread.

Continental glaciation by itself is generally universally regarded as an earmark to climate cooling on a global scale. Consequently, in the absence of continents upon which to house and uphold "continent-size" ice sheets, a worldwide global cooling event might otherwise have left no record of glacial ice buildup, and thus, no record at all of its occurrence.

At approximately 750 million years before present, perhaps as early as 600 million years ago, continental crust cooled sufficiently to have become welded to the underlying mantle

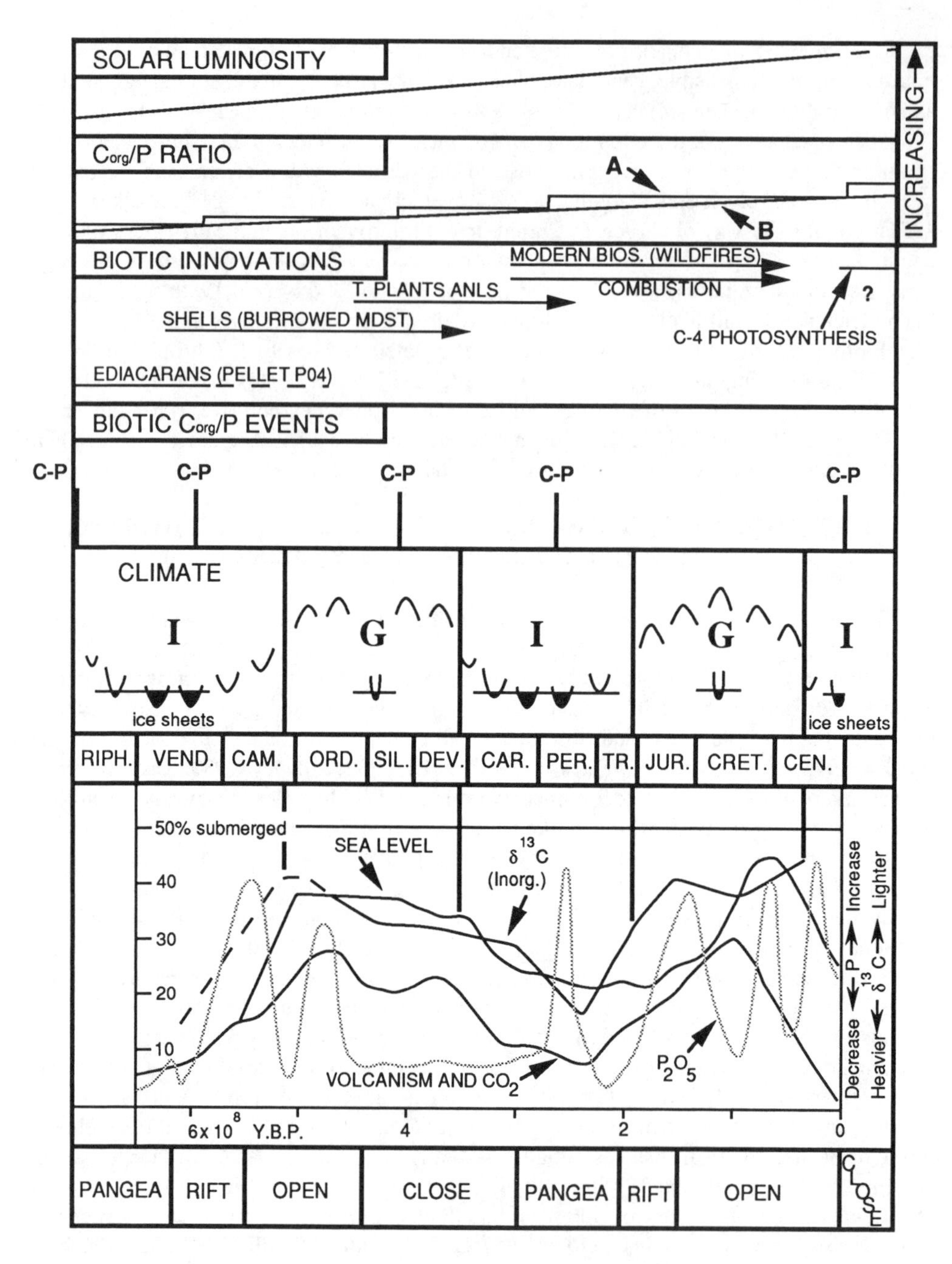


**FIGURE 2.** Earth's suggested tectoclimatic, biogeochemical cycles. From the top down, solar luminosity is increasing to the right. $C_{org}$/P ratio is a proxy for biotic, nutrient efficiency. Increased nutrient efficiency with each new radiation, in the face of increasing solar luminosity, results in buffered earth-surface temperatures (modified from Worsley and Nance, 1989). Operation of the super continent cycle (mantle convection and changes in the shape of the geoid) results in the periodic assembly and disassembly of continents, feedbacks to which are: nested, long- and short-term icehouse/greenhouse climates, long- and short-term fluxes in sea level, fluxes in atmospheric $CO_2$ due to changes in the rate of crustal weathering and volcanism. Storage of P on continental shelves, as well as shifts in the redox state of organic and inorganic carbon, are additional feedbacks to the supercontinent cycle. Sources of information include: Worsley et al., (1984; 1985), Fisher, (1984a; 1984b), Frakes and Francis, (1988), Berner, (1987; 1990); Vail et al., (1977), Gaffin, (1987), Veizer et al., (1980), Engle and Engle, 1964), Cook and Shergold, (1984), and Bond et al., (1984; adjustment for the suggested timing of Vendian rifting). (Adapted from Worsley, T., Nance, D., and Moody, J., *Marine Geology,* 58, 377, 1984. With permission.)

(Hargraves, 1976) thus ushering in a transition to brittle, "Wilson-style" plate tectonics. Prior to this transition in tectonic style from an older, thinner, hotter, more viscous crust, to a thicker, more brittle crust, global warming and global cooling events were not only difficult to identify with certainty, but were more likely aperiodic or quasi-periodic prior to 750 million years ago, due to the then higher temperature of the Earth and the suggested increased circulation of viscous, molten rock at depth within the Earth's mantle (Hargraves, 1976). Other factors affecting the transition from aperiodic to more periodic global warming and cooling events include the volume and evolution of continental crust over time (see below and also Hargraves, 1976; Goodwin, 1981).

## A. GLOBAL COOLING EVENTS OF LONG DURATION

### 1. 700 to 570 Million Years Ago

Perhaps the most paradoxical aspect of this global cooling event (otherwise referred to as the late Proterozoic glacial epoch of continental glaciation) — see Figures 2 and 3), is that with the exception of the work of Bond et al. (1984) the continents are suggested to have been more or less clustered into a large landmass situated at low- to mid-latitudes (Piper, 1976; 1982; Scotese et al., 1979; Morel and Irving, 1978; Morris and Roy, 1977; Chumakov and Elston, 1989; Moores, 1991; Dalziel, 1991). Explanations for the onset and demise of this global cooling event include (1) a change in the Earth's obliquity of the ecliptic (axial tilt) from its present orientation of 23.5 degrees from a vertical line drawn perpendicular to an imaginary plane housing the orbital track of the earth with respect to the sun and other planets within this solar system, to somewhere between 54 and 126 degrees, as a consequence of rotation of the galactic plane with a periodicity of 2.5 billion years (the poles become periodically equatorial, while the equator periodically becomes polar at two points — refer to Williams, 1972; 1973; 1975), (2) changes in the rate of the Earth's rotation (Hunt, 1979), (3) passage of the sun through clouds of interstellar matter (McCrea, 1975), and (4) plate assembly and an associated, tectonically-driven increase in weathering of continental crust, a long-term decrease in atmospheric $CO_2$, and a relative lowering of sea level (Fischer, 1984b; Worsley et al., 1986; Kasting, 1987).

### 2. 340 to 220 Million Years Ago

Peculiar to this period of global cooling is the buildup and confinement of continental glaciers (the Permo-Carboniferous glacial epoch — Figure 3) to a large continental landmass, Gondwana (composed of Africa, Antarctica, Australia, India, Madagascar, New Zealand, Japan, and the Phillippine Islands — Crowell, 1978; Powell et al., 1980; Ross and Ross, 1985; Veevers and Powell, 1987), situated at mid-latitudes in the Southern Hemisphere (Powell and Veevers, 1987). At the same time, but apparently confined to the low latitudes of the Northern Hemisphere, lay another large continental landmass, Laurasia (largely composed of Europe, North America, and Greenland). Consequently, at this time, large portions of Laurasia were periodically inundated in response to frequent sea-level changes. The partial result of which was the intermittent growth and burial of large tracts of swamp surrounded by tropical plants and trees. This is not to suggest that Darwinian evolution and opportunism played insignificant roles in the emergence of these trees and plants.

Thus, Laurasia came to acquire a thick succession of coal deposits, presently the vast bulk of the Earth's coal resources, the thickness and stacking of which appear to have resulted from Laurasia's tropical location at this time, as well as frequent oscillations in sea level worldwide, on the order of $\pm 100$ m. The frequency of sea-level changes/stacking of bands of coal (on the order of 10,000 to 100,000 years in duration), suggests that the position of sea level, worldwide (thus climate and coal), was most likely controlled by orbitally forced changes in the distribution of solar radiation over the surface of the Earth (Ross and Ross, 1985; Heckel, 1986; Veevers and Powell, 1987; Klein and Willard, 1989; Gamundi, 1989).

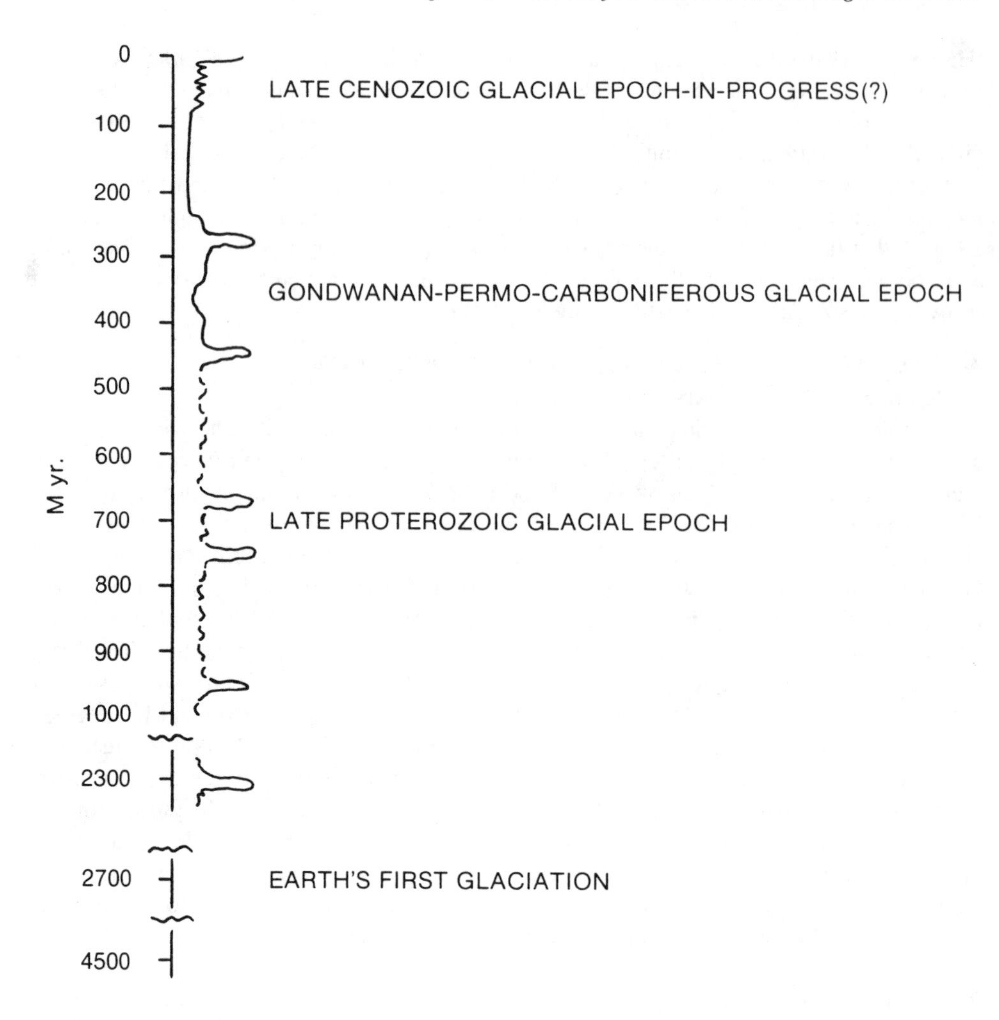


**FIGURE 3.** Chronology of glacial/nonglacial epochs or long-term icehouse/greenhouse climates. (From Crowley, T. J., *Rev. Geophys. Space Phys.*, 21, 835, 1983. With permission.) (see also Figure 2).

Sea-level oscillations associated with this global climate cooling event/interval of time, are similar in scale and rate to global sea-level oscillations associated with the Earth's most recent (and probably ongoing) long-term global cooling event (absenting anthropogenic effects for the moment), otherwise known as the late Cenozoic glacial epoch (see Figure 3), the onset of which appears to have occurred 40 million years ago (Hayes et al., 1976 and Figure 2).

Due in part to the seemingly peculiar and ostensibly contradictory fact that this glacier-bound, Southern-Hemispheric landmass known as Gondwana was situated at mid-latitudes during the Permo-Carboniferous. Powell et al., (1980) and Powell and Veevers (1987) concluded that, in the absence of having been in a polar region, continental glaciation was probably triggered at this time by scattered mountain-building/increased continental elevation resulting from multiple, closely-spaced (in time) continental collisions (closure of multiple ocean[?] basins, or portions of basins) during the formation of the Gondwanan landmass. Thus Powell and Veevers (1987) concluded that long-term global climate cooling at this time (with implications for other times) was tectonic in origin.

Crowell (1982), Fischer (1984b), and Worsley et al. (1984; 1986) have invoked plate assembly and an associated, long-term (on the scale of 100 million years) tectonically-influenced decrease in atmospheric $CO_2$, probably tied to weathering of continental (silicate) crust, and a worldwide lowering of sea level, as the mechanism for the onset of global climate cooling 340 to 220 million years ago.

**3. 40 Million Years Ago to Present and Beyond (?)**

With regard to the Earth's most recent long-term global cooling event in progress (Worsley et al., 1986; Fischer, 1984a), the last 4 million years of which are commonly referred to as the "ice ages", Worsley et al. (1986), Fisher (1984a; 1984b), and Berner (1990) have offered that the climate triggering mechanism for this long-term global cooling event considered to be still in progress, is plate assembly and an associated, long-term (and in progress), tectonically driven increase in the weathering of continental crust, reduction in the volume of $CO_2$ in the atmosphere, a feedback to weathering, and a worldwide lowering of sea level (see Figure 2).

Detailed climate records obtained from deep-sea sediment cores, ice cores, rocks, and tree rings, most of which are temporally confined to the latter portion of this most recent, ongoing, long-term global cooling event, clearly document the existence of numerous global cooling and global warming events (otherwise known as worldwide continental glaciation and deglaciation) superimposed upon/embedded within this longer-term climate cooling event in progress. Furthermore, these overprinted, short-term global warming and global cooling events occur and disappear at Milankovitch-band periodicities or frequencies (on a scale of 10, 000 or fewer years to 100,000 years, by definition) reflective of having been initiated by orbitally driven changes in the distribution of incoming solar radiation (and moisture) over the surface of the Earth (Mesolella et al., 1969; Broecker and van Donk, 1970; Chappel, 1974; Hays et al., 1976; Birchfield et al., 1981; Harland, 1981; Crowell, 1982; Pollack, 1982; Pollard, 1983; Imbrie et al., 1984; Broecker, 1988; Imbrie, 1987; COHMAP Members, 1988; Overpeck et al., 1989).

Perhaps even more compelling is the abundance and variety of natural events and systems, many of which are not climate per se, but constitute a suite of phenomenologically or genetically linked feedbacks, in this case, to Milankovitch-band orbital forcing, at least in part, as described above. Orbitally forced climate (and weather) and related natural phenomena, in addition to the growth and decay of continental ice sheets, include such things as: monsoonal systems (Kutzbach and Otto-Bliesner, 1982; Rossignol-Strick, 1983; Prell, 1984; Kutzbach and Guetter, 1984), hurricanes, storms, and the direction and strength of wind patterns (Janecek and Rea, 1984; Hobgood and Cerveny, 1988), salinity (Thunell et al., 1988; Thunell and Williams, 1989), paleosols and desert rainfall (Spaulding and Graumlich, 1986; Goodfriend and Magaritz, 1988), oceanic dimethylsulfide (Legrand et al., 1991), the distribution of loess (Beget and Hawkins, 1989), productivity and oceanic circulation (Sarnthein et al., 1987; Boyle, 1988a; Lyle, 1988; Mix, 1989), lake levels and evaporite sedimentation (Anderson, 1984), the flux of particles to the deep ocean (Pokras and Mix, 1987; Lyle, 1988; Nair et al., 1989), fluxes in non-anthropogenic levels of atmospheric $CO_2$, $SO_4$, and $CH_4$ (Shackleton et al., 1983; Pisias and Shackleton, 1984; Sarmiento and Toggweiler, 1984; Siegenthaler and Wenk, 1984; Lorius et al., 1985; Oeschger et al., 1985; Barnola et al., 1987; Bates et al., 1987; Charlson et al., 1987; Genthon et al., 1987; Boyle, 1988b; Raynaud et al., 1988; Stauffer et al., 1988; Caldeira, 1989; Mix, 1989) and volumetric changes in levels of sulfates and nitrates (Legrand and Delmas, 1984). All of the above, however, are arguably tied to climate and climate change in very fundamental ways, most of which constitute genetically linked feedbacks to earth systems and processes, a single outcome of which is climate.

Present as well within this most recent long-term period of global cooling is evidence for abrupt changes in climate at frequencies significantly higher than 19,000 years (Siegenthaler and Wenk, 1984; Overpeck et al., 1989; Thompson et al., 1986; 1988; 1989; Imbrie, 1987; Legrand and Delmas, 1987; McKenzie and Eberli, 1987; Thompson and Mosley-Thompson, 1987), yet attributed nonetheless, at least in part to nonlinear (very high frequency) feedbacks to Milankovitch-band, orbitally-forced climate changes (Broecker, 1987; Imbrie, 1987; Overpeck et al., 1989; Sarmiento and Toggweiler, 1984) and/or solar variability (Friis-Christensen and Lassen, 1991), and natural variations of unknown or little understood origin).

The instigation of this most recent long-term global climate cooling event in progress, not to be confused with the record of high-frequency, Milankovitch-band global warming and global cooling events (glacials and interglacials) embedded within, overprinted upon and/or part and parcel of this long-term cooling, has been ascribed to: (1) tectonics and tectonically induced fluxes in atmospheric $CO_2$, weathering, and sea level, as described above (Fischer, 1984b; Worsley et al., 1984); (2) tectonic and thermally induced changes in continental elevation leading to changes in regional atmospheric pressure cells (Ruddiman and Kutzbach, 1989); (3) deep, mantle overturning resulting in continental drifting and true polar wander/geomagnetic excursions (Rampino, 1979; Donn, 1989); and (4) evolution of angiosperms and deciduousness in the late Mesozoic, resulting in increased rates of continental crustal weathering and, thus, increased, long-term drawdown of atmospheric $CO_2$ (Volk, 1989a).

## B. GLOBAL WARMING EVENTS OF LONG DURATION

Interspersed between the aforementioned long-term period of global cooling are at least three longer-term periods of global warming: (1) an unknown but implicitly long period of global warming prior to 700 million years ago; (2) 230 million years of global warming 570 to 340 million years ago; and (3) 180 million years of global warming 220 to 40 million years ago (Fischer, 1982; 1984a; 1984b; Worsley et al., 1986; 1985; 1984; refer to Figure 2), in addition to an unprecedented antropogenically driven climate warming presently in the making (Hansen and Lacis, 1990). This anticipated climate warming would be entirely consistent with greenhouse theory and observations of planetary systems. Furthermore, the duration of antropogenic, greenhouse warming may likely exceed 300,000 years, conceivably preempting at least three anticipated (according to the Milankovitch theory of orbitally forced warming and cooling) orbitally forced thus, short-term, global cooling events normally characterized by thickening and extensive spread of continental glaciers toward the equator (see Kasting and Walker, 1991).

As is shown in Figure 2, with the exception of anthropogenically induced global warming on the horizon, each period of long-term global warming (warming event predating 700 million years ago is not shown) is: (1) a gradual increase in atmospheric $CO_2$ from volcanic sources; (2) significant worldwide rise in sea-level; and (3) breakup and dispersal of exceptionally large continental landmasses otherwise known as "supercontinents" (Worsley et al., 1984; Fischer, 1982; 1984a; 1984b; Berner, 1990).

Anthropogenic climate modulation, largely paced by the onset and evolution of industrialization, principally involves emissions of greenhouse gases (warming influence) and cloud-forming gases and particles, largely emissions of sulfur compounds and particulate carbon from the burning of forests and combustion (cooling influence due to the ability of clouds to reflect incoming radiation back into space — Charlson et al., 1987; Crutzen and Andreae, 1990). The present concentration of $CO_2$ in the atmosphere is 355 ppmv (parts per million by volume), over 100 ppmv is due to anthropogenic emissions since the beginning of the industrial revolution (Lashov and Turpak, 1989). By comparison, the amount of $CO_2$ gas exchanged between the atmosphere and the ocean during the natural transition from the Earth's most recent worldwide continental glaciation (short-term global cooling event), to the period of warming which followed (that which we are presently in, minus the effects of anthropogenically introduced greenhouse gases), is approximately 85 ppmv (Neftel et al., 1985; 1988; Siegenthaler and Wenk, 1984; Sarmiento and Toggweiler, 1984; Gammon et al., 1985; Stauffer et al., 1985; Boyle, 1988b; Broecker and Peng, 1989; Lashov and Turpak, 1989). In other words, the anthropogenic contribution of $CO_2$ to the atmosphere is in excess of the volume of $CO_2$ which nature calls upon to go from one extreme of climate to the other.

Anthropogenic input of $CO_2$ into the atmosphere may already have resulted in a 0.5 to 0.75°C increase in the global average surface temperature of the Earth, with an additional increase in temperature of 3 to 5°C anticipated by the year 2100 (Lashov and Tirpak, 1989).

# IV. CONTINUUM OF CLIMATE CONTROLS AND FEEDBACKS AMONG EARTH AND PLANETARY SYSTEMS

The following represents a qualitative model for the Earth's continuum of climate preconditions, controls and feedbacks, the salient aspects of which are shown in Figure 4, in the order shown.

1. To a large extent, climate and climate change are outcomes or feedbacks to the origin and evolution of life itself, as Lovelock (1987), Lovelock and Margulis (1974) and Cloud (1983) have suggested. As life evolved carbon was sequestered from the atmosphere leaving the atmosphere with a higher partial pressure of oxygen relative to $CO_2$ (Kasting, 1987). Furthermore, given that the Earth's surface temperatures have remained confined to within a very narrow range (for at least the last 4 billion years) in the face of increasing solar luminosity which should have rendered the planet much hotter than it is, or has been, and possibly devoid of life itself (Lovelock and Whitfield, 1982; Sagan and Mullen, 1972; Kasting and Ackerman, 1986; Walker et al., 1981; Newman and Rood, 1977), thus apparently necessitating a planetary temperature-regulating mechanism (in this instance a likely biogeochemical mechanism, poorly defined at present) such as a continual removal of carbon from the atmosphere, resulting in relatively stabilized Earth surface temperatures throughout time.

   Thus, I submit that one of the consequences of the "Dim-Young Sun" paradox described above (Sagan and Mullen, 1972), and the evolution of photosynthetic life, was that this suggested continual incremental removal of carbon from the atmosphere also served as a first-order precondition (not the actual climate trigger or control) to the origin and evolution of a climate system characterized by two extremes or end-members, global warming and global cooling (growth of continental glaciers — see Figure 4). Thus, the first appearance of glaciers 2.7 billion years ago serves as the beginning of our present two end-member system of climate, which is not to say that nature cannot exceed those internal limits presently defined by feedbacks among Earth systems (e.g., Venus and its runaway greenhouse atmosphere).
2. As life was evolving, so too was the planet cooling and segregating into compositionally and densiometrically distinct layers of rock, and molten rock, surrounding a core. However, in terms of the production and preservation of a faithful, continuous record of climate controls and feedbacks, the evolution of continents, the timing of the appearance of continents (continental crust in general), and style of crustal movement and deformation seem to have been eminently important. Without continents to sit on, for example, there may well be no evidence of global cooling in the form of continental glaciers. Furthermore, prior to the time when the continental crust became welded to

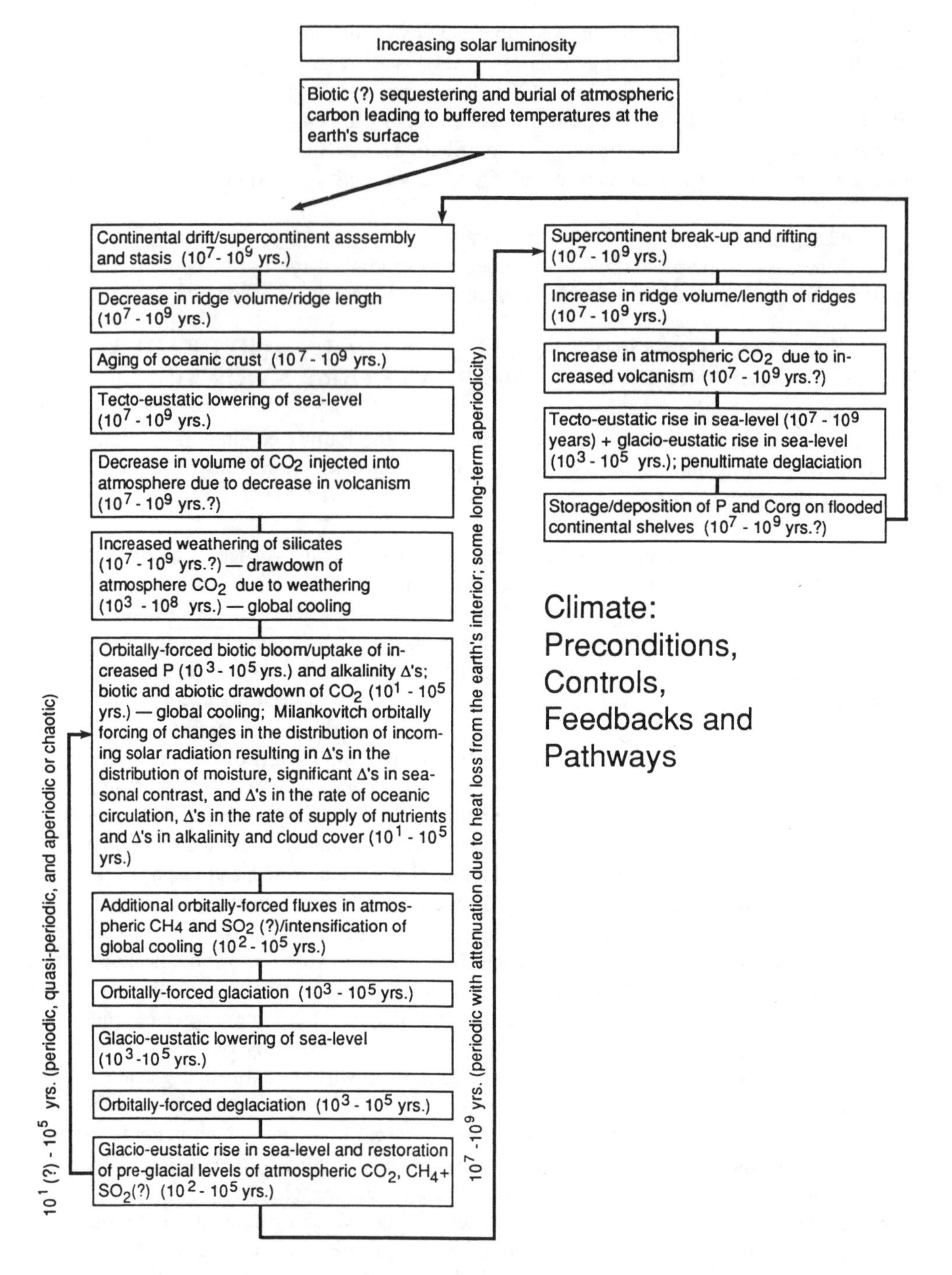


**FIGURE 4.** Synthesis of climate-forcing and climate-modulating feedbacks among the Earth's tectonic, biogeochemical, and astronomical systems, in the presence of life and increasing solar luminosity. The pathways and periodicities and/or rates of feedbacks are also shown (see text for discussion).

the mantle below, roughly 3.5 billion years ago (Hargraves, 1976) the Earth's crust was likely to have been extremely thin, with vigorous mantle convection below rendering the surface crust readily amenable to bending, folding, and breaking, only to subside and be remelted in the hot, viscous material below, much like the fate of the thin, brittle crust which forms atop lava flows as they move downslope.

Consequently, it may not have been until the continental crust literally became welded to the underlying mantle, after sufficient cooling, that the above, choatic, "viscous-style" of plate tectonics evolved into a more brittle, less frenetic style of plate tectonics known as "Wilson-style" tectonics, which have been prevalent on Earth since roughly 600 to 700 million years ago (Hargraves, 1976). Hence, the most preserved and preservable record of this planet's ancient climate is arguably best preserved in rocks younger than 700 million years old. However, this is not to say that there are not vestiges of climate history scattered through the breadth of this planet's ancient record.

3. From the perspective of this more brittle style of tectonics, subsequent preconditions, controls and feedbacks to long- and short-term intervals of global warming and global cooling are as follows.
4. Assembly of existing continents into one or two large landmasses.
5. Assembly of continents leads to a reduction in the volume of ridges present within the world ocean.
6. Since ocean ridges are young and, therefore, thermally elevated portions of ocean crust by definition, as ridge volume decreases during continental assembly, the ocean floor ages.
7. Since young ocean floor is hot and elevated, old ocean crust is chilled and contracted; and being heavier, ultimately sinks into the mantle.
8. As old ocean floor sinks, by definition, sea level drops worldwide. As sea level drops, the thermally insulating effect of water on adjacent continents is reduced.
9. As sea level gradually drops at tectonically defined rates (on a scale of tens of millions of years to hundreds of millions of years) more continental area becomes exposed to weathering at the margins of continents, thus increased weathering due to an increase in weatherable area.
10. Weathering of silicate crust utilizes available atmospheric $CO_2$. Increased weathering results in increased sequestering of atmospheric carbon, all of which results in global cooling at the Earth's surface.
11. Given the approximate establishment of the above tectonically driven events (at long-term rates of tens to hundreds of millions of years in duration), only then are the conditions met for the subsequent growth and decay of continental glaciers (high-frequency global warming and global cooling events), at orbitally-driven rates and frequencies (a scale of 10,000 or fewer years to 100,000 or more years). Thus, I submit, items 2 through 10 constitute a requisite second-order precondition to the subsequent, high-frequency growth and decay of continental glaciers, proxies for short-term global warming and global cooling events in Earth history.
12. At this point, orbital-forcing of changes in the distribution of solar radiation (and implicitly moisture) over the surface of the Earth, at Milankovitch-band rates, brings about further changes (feedbacks to orbitally driven climate-forcing) in oceanic circulation rates, thus, changes in ocean alkalinity and therefore, the amount of dissolved $CO_2$ present in the upper portion of the ocean which, in turn results in exchanges of $CO_2$ with the atmosphere on the order of 90 ppmv (the apparent difference between an ice age global cooling and an interglacial global warming).
13. Exchanges of methane ($CH_4$) between the atmosphere and various candidate, but ill-defined sources, such as $CH_4$ tied up in ice-clathrates in regions of permafrost and $CH_4$ tied up in deep-sea sediments as a result of pressure from the overlying sediment and water column. Since $CH_4$ and $CO_2$ are greenhouse gases the above feedbacks have implications for long-term as well as short-term global warming and global cooling events at Milankovitch-band rates.

14. At this point, just postdating orbitally forced changes in solar insolation, in addition to feedbacks to orbital forcing involving the exchange of greenhouse gases between the ocean, the biosphere, and undefined or vaguely defined sources, continental glaciers grow and decay at Milankovitch-band rates. However, since glacial growth requires more time, by a factor of roughly 10:1 (Charlson et al., 1987; Broecker and van Donk, 1970), than the melting or decay of ice sheets, the aggregate sum of time over which these short-term cycles of global warming and global cooling occur (ice growth and decay as well), are dominated by the buildup of glaciers and are, therefore, predominantly aggregately characterized by long-term global cooling, as opposed to global warming (long-term since I am referring to the sum of all such short-term warming and cooling events).
15. As ice sheets build sea level drops somewhat precipitously, driven by orbitally defined rates. Conversely, as ice sheets melt sea-level rises even more precipitously. The sum of these orbitally driven events are overprinted, or are in addition to the tectonically driven events cited above and below.
16. While all of the above is taking place continents continue to drift at tectonic rates. Thus, at some point, the "supercontinent(s)" which first set in motion (subsequent to the first appearance of glaciers on the planet, and in the backdrop of the "Dim-Young Sun paradox") this network of climate preconditions, feedbacks and controls, now thermally domes and cracks, while the fragments of continents subsequently drift away from the site of doming.
17. During supercontinent rifting there is an increase in the volume of ocean ridges present, by definition. The sites of rupture are implicit ridges. Since ocean ridges are young, hot, and therefore, thermally elevated ocean crust, the ocean crust is said to be "young globally". Since young ocean crust rides higher than older, cooler, more contracted, and thus heavier ocean crust, sea-level is elevated at tectonic rates.
18. A long-term global rise in sea level gradually leads to the flooding of the margins of continents, thus a decrease in continental weathering (reduced continental area exposed) resulting in reduced amounts of $CO_2$ being stripped from the atmosphere through weathering processes.
19. In addition, during continental breakup there is an increase of volcanically derived $CO_2$ injected into the atmosphere. The source of this $CO_2$ is from the metamorphism (literal cooking) of ocean crust laden with carbonate rocks which house the carbon weathered from the surface of the adjacent continental crust. This ocean crust is subducted (forcibly driven below) around the margins of this supercontinent while it was assembled, and during the initial phases of its breakup. Supercontinent stagnation and breakup are, by definition, times of increased volcanic activity, thus increased input of $CO_2$ into the atmosphere (Anderson, 1989).
20. The net climatic effect of items 16 through 19 is the establishment of a long period of global warming generally on the order of 200 million years in duration. At the end of this period of time, dispersed continents will likely reassemble again, as they appear to be in the process of doing at present. It should be noted, however, that during tectonically defined episodes of global warming there is virtually no record of high-frequency, short-term global warming and global cooling events (growth and decay of continental glaciers), except, conceivably within polar regions, nested within these long-term warming events. The tectonically induced warming at these times (tectonic conditions) seems to dwarf, and largely override, the influence of orbital forcing on the rapid growth and decay of continental ice sheets at these times. Thus, high-frequency, Milankovitch-band global warming and global cooling events seem to occur only when sea level is tectonically lowered, and atmospheric $CO_2$ (and other greenhouse gases) are tectonically lowered as well — times of supercontinent assembly and stasis.

This is not to say, however, that orbital forcing of less seemingly dramatic aspects of climate and weather, as well as a host of Earth systems in general, does not take place at this time. To a large extent, orbital forcing of natural systems continues to operate independent of the Earth's tectonic system and vice versa, while in some ways these systems are clearly genetically coupled by intermediate linkages or feedbacks, since orbital forcing of global warming and global-cooling events (short term) appears to be indirect.

# V. DISCUSSION

## A. TECTONISM AND CLIMATE

As shown in Figure 2, for the past 700 million years, and conceivably for as long as the last 2.8 billion years, coincident with the first appearance of granitic continents and the initiation of an oscillatory (quasi-cyclic?) process of continental assembly and disassembly (Engel and Engel, 1964; Worsley et al., 1986; Windley, 1984; Goodwin, 1981; Sutton, 1963), long-term global cooling events have coincided with times of continent assembly, with quasi-periodicity of 300 to 600 million years (Fischer, 1982; 1984a; 1984b; Worsley et al., 1984). Within the past 700 million years there have been two episodes of long-term global cooling, thus, two episodes of supercontinent assembly, and a third episode of long-term global cooling in-the-making, coincident with a third supercontinent in-the-making, separated in time by two long-term global warming events (Fischer, 1982, 1984a; Worsley et al., 1984; refer to Figure 2). Each long-term cooling event also appears to have had an abrupt beginning (see Kasting, 1987).

Furthermore, at higher resolution, long-term global climate cooling events are more accurately, long periods of short-term, high-frequency (10,000 years or less to 100,000s years), Milankovitch-band episodes of global warming and global cooling (Allen, 1973; Garp, 1975; McCrea, 1975; Crowell, 1978). During the most recent period of long-term global cooling in-the-making, for example, continental glaciers (periods of short-term global cooling) were slow in building, yet required relatively little time to melt or recede, by as much as a factor of 10:1 (Charlson et al., 1987; Broecker and von Donak, 1970), to perhaps as little as a factor of roughly 4:1 (Ruddiman and MacIntyre, 1982). Thus long-term global cooling events are characterized by the predominance (time wise) of high-frequency global cooling events, with a far smaller proportion of time marked by higher-frequency global warming events.

Furthermore, the periodic assembly and dispersal of continents govern: (1) the build-up conduction and escape of heat from the interior of the earth; (2) mantle convection; (3) shape of the geoid (Earth itself) hence, changes in the orientation of the Earth (and the Earth's exterior with respect to its interior and vice versa) with respect to the Earth's spin axis; and (4) true polar wander, all of which seem quasi-periodic (Anderson, 1981; 1982; 1984; 1989; Vogt, 1972; Goldreich and Toomre, 1969; Fisher, 1974; Donn, 1989; Worsley et al., 1984; Fischer, 1984a). Thus, long-term global warming and global cooling events, and indirectly, short-term global warming and cooling events, are feedbacks to the dissipation of heat energy outward from the Earth's interior.

### 1. Cycle of Supercontinent Assembly and Disassembly

The following represents a synthesis and an analysis of the work of Olsen et al. (1990), Anderson (1981; 1982; 1984; 1989), Goldreich and Toomre (1969), Fisher (1974), Worsley et al. (1984), and Donn (1989).

Once assembled and stationary, supercontinents become heated by retarding the flow of heat emanating from the earth's mantle, creating so-called "geoid highs" (zones of anomalously high heat flow from deep within the mantle), in addition to inducing upwelling of partially melted mantle material from deep within the Earth's mantle. This results in a reorientation of the mantle with respect to the Earth's spin axis (drift of the Earth's magnetic poles — true polar wander), such that the excess mass (stagnant supercontinent) is ultimately pulled apart (stressed to failure) and gravitationally driven towards the equator (a geoid low — cool zones within the mantle) to become part of the new equatorial bulge relative to the poles of the geoid's spin axis. With cooling and subsidence over time, the former geoid high eventually becomes a geoid low.

Stated otherwise, as a supercontinent retards subcrustal heat flow by remaining assembled and stationary, it heats the mantle below, drawing heat to itself (although itself a poor conductor of heat) and thus becoming thermally domed/physically elevated/thermally expanded, preferentially at its center. However, during this process the margins of the supercontinent, as well as the upper mantle in the vicinity of these margins, remain cooler as a consequence of subduction of cold ocean floor surrounding supercontinents.

Thermal doming/elevation/thermal expansion of the supercontinent continues until the supercontinent develops rifts and breaks into a number of smaller continents which subsequently, gravitationally disperse outward from the site of thermal doming/geoid high/anomalously high heat flow, toward geoid lows (downwelling sites/cooler mantle). While hot and elevated, continental fragments of the now-dispersed supercontinent initially drift away from the geoid high at relatively high rates. In time, as the geoid high cools, as do the initially hot, thermally elevated/expanded fragments of a now-dispersed supercontinent, the rate of plate motion decreases.

In time, as these cold slivers of a supercontinent disperse and gather at the equator (i.e., become part of the new equatorial bulge of the most recent geoid shape), downwelling ceases, heat flow from the mantle becomes retarded causing the lower mantle to heat up, leading to the formation of a new geoid high, and thus starting another cycle of continent dispersal and assembly, with a periodicity in the range of 300 to 600 million years or more (Fischer, 1984a; 1984b; Worsley et al., 1984).

Thus, coalescing blocks of continental crust induce a long-term cycle of convective overturn within the mantle (zones of upwelling and downwelling) resulting in the periodic assembly and disassembly of supercontinents. Consequently, given the fact that the outer surface of the earth, as well as the core and mantle, are free to rotate with respect to each other and to the Earth's spin axis (drifting plates, true polar wander, and changes in the shape of the geoid) there may be no means whatsoever of fixing the location, or former location, of rocks housing climatic (and other) information, within any sphere of reference such as a system of longitude and latitude.

## 2. Supercontinent Cycle and Cycle of Global Warming and Global Cooling

Taking a different perspective of the supercontinent cycle as described in the previous section, recall that while supercontinents remain assembled and stationary, they become hot and therefore, thicken/increase in elevation relative to the position of sea level within the surrounding world ocean. Consequently, while supercontinents rift, cool, contract and thermally subside (become lower in elevation relative to sea level) ocean ridges are created in their wake.

As Russell (1968), Sclater et al. (1971) and Parsons and Sclater (1977) have shown, newly formed, young ocean crust at mid-ocean ridges is hot and thermally elevated. In time, as ocean floor moves away from the mid-ocean axis (site and source of new ocean crust) and ages, it cools, contracts, and therefore lowers in elevation below sea level (Olson et al., 1990). Thus, an increase in the volume of mid-ocean ridges, or an increase in the number of ridges, or an increase in the rate of production of mid-ocean ridges/increase in the rate of plate motion will result in a tectonically induced rise in sea-level (Pitman, 1978; Hays and Pitman, 1973; Worsley et al., 1984). Conversely, a decrease in any of the above will result in a relative fall in sea level.

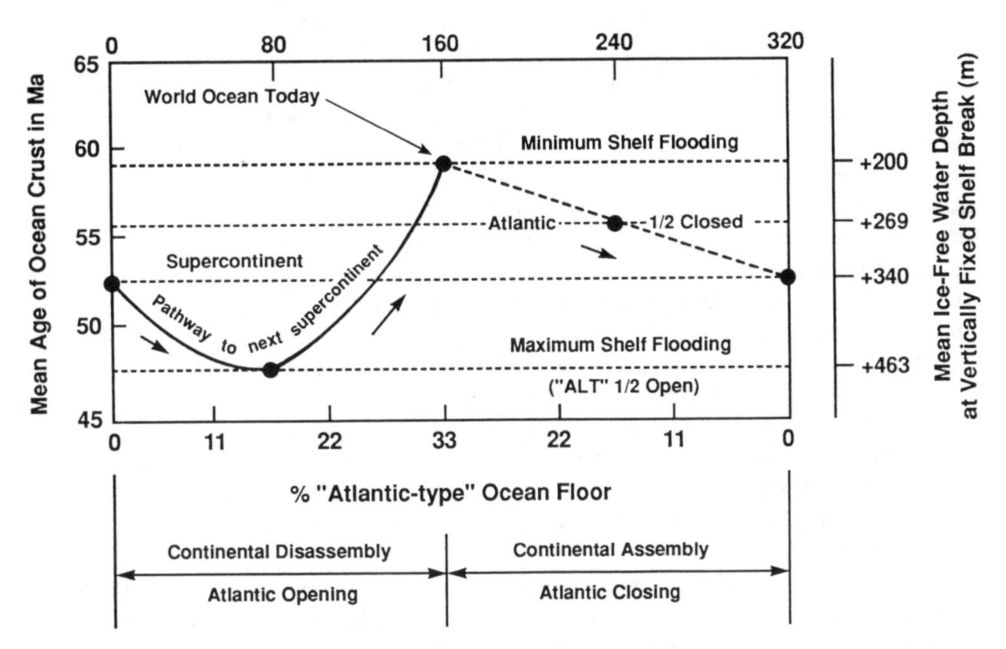


**FIGURE 5.** Response of sea level and ocean crust to the aging and/or younging of ocean floor, as a feedback to the operation of the supercontinent cycle. Since young ocean floor at the sites of rifts is hot, it becomes thermally elevated, causing an initial rise in sea level as a new (Atlantic) Ocean basin begins to form when a supercontinent begins to break apart. During the initial stages of continental dispersal, there is more young, hot crust than there is old, cold ocean crust, on a worldwide basis, and so the average age of the ocean floor decreases (ocean crust youngs — maximum shelf flooding). In time, as this new ''Atlantic'' Ocean widens in response to continued dispersal of continental fragments, the average age of ocean crust increases. Since older crust is colder and more contracted, sea level drops to a point of minimal shelf flooding. However, at this point, continued continental dispersal forces this Atlantic Ocean basin to start closing as the dispersed continental fragments are now beginning to converge again. As this Atlantic Ocean eventually closes, the average age of ocean floor decreases (youngs), forcing sea level to rise once again to a position of maximal shelf flooding. At this point, the continents have converged once again to form a supercontinent surrounded by a single ocean. Insofar as the operation of the supercontinent cycle forces changes in the rates of weathering as a function of sea level, and insofar as these same processes determine how much carbon is being exchanged between sediments/rocks and the atmosphere, these processes serve to regulate and modulate climate in the long-term (after Worsley et al., [1987] and Berger and Winterer, [1974]). (From Worsley, T., Nance, D., and Moody, J., *Marine Geology,* 58, 381, 1984. With permission.)

Consequently, as indicated by Berger and Winterer (1974) and later by Worsley et al. (1984) and Gaffin (1987), when a supercontinent disperses and an Atlantic-type ocean emerges, the average age of ocean crust becomes younger and, therefore, sea level rises (Figure 5) flooding the margins of continents. In time, however, as this new ocean basin becomes wider, the average age of ocean crust increases and sea level drops (Figure 5). And finally, as this Atlantic-type ocean begins to close as continents reassemble to form another supercontinent, the average age of ocean crust becomes younger again, while sea level slowly rises, again flooding the margins of a newly emergent supercontinent (Figure 5). Supercontinents, therefore, sit high in elevation relative to sea level. Maximally dispersed fragments of supercontinents tend to have their margins flooded with seawater to an approximate depth of 200 m. During supercontinent reassembly, continental margins become flooded to an approximate depth of 340 m (Figure 5).

Thus on a time scale of tens to hundreds of millions of years, the assembly and disassembly of supercontinents bring about long-term fluxes in sea level and therefore, rates of weathering of continental crust, as determined by the area of continental crust exposed to the atmosphere in the presence of water at any one time. Insofar as the weathering of calcium-silicate rocks serves as a sink for atmospheric $CO_2$ (Berner, 1990; Walker et al., 1981; Volk, 1987; 1989a; 1989b), lowered sea level and therefore, higher rates of weathering (increased area of denuded continent), and therefore, a volumetric decrease in atmospheric $CO_2$ relative to atmospheric oxygen (and thus long-term global cooling) are all genetically linked feedbacks to supercontinent assembly and stasis (stagnation; Fischer, 1984a; 1984b; Worsley et al., 1984; see Figures 2 and 4). Tectonic lowering of sea-level worldwide, further serves to positively reinforce long-term climate cooling due to an increase in planetary albedo resulting from increased continental surface area (Bates et al., 1987), and a reduction in the warming effect of seawater (thermal capacity of seawater) as sea level drops and becomes confined to a smaller area.

On the other hand however, as stationary supercontinents becomes sufficiently hot due to the buildup of heat at their base, supercontinental crust begins to melt in places, producing volcanos at the surface which serve to inject greenhouse gases from the mantle below (mostly $CO_2$) to the atmosphere with great efficiency (see Olson et al., 1990). Essentially all of this volcanically derived $CO_2$ results from the metamorphism (cooking) and melting of carbonate rocks sitting atop descending oceanic crust surrounding the margins of a supercontinent (Berner and Lasaga, 1989).

This, in turn, results in a gradual, long-term buildup of atmospheric $CO_2$ (and undoubtedly other greenhouse gases), thus ushering in long-term global warming on a scale of 100s of millions of years in duration (Fischer, 1982; 1984b; Worsley et al., 1984; see Figures 2 and 4). Other sources of atmospheric $CO_2$ stem from recrystallization and outgasing during thermal metamorphism associated with the heating of supercontinents while stationary (Berner, 1990; Berner and Lasaga, 1989).

On the other hand, supercontinent fragmentation and dispersal, to the point of maximal dispersal as described above (halfway between a geoid high and a geoid low), bring about a tectonically driven rise in sea-level globally, and the flooding of continental margins (see Figure 2). Since flood margins reduce continental area and thus planetary albedo (Bates et al., 1987), in addition to increasing the effect of the heat capacity of seawater on continents and continental climate, all of these feedbacks serve to positively reinforce climate warming during supercontinent dispersal.

Thus, long-term global climate warming events serve as feedbacks to the disassembly and dispersal of supercontinents, while long-term global climate cooling events serve as feedbacks to the formation and stagnation of supercontinents (see Figure 2).

## 3. Other Feedbacks to the Supercontinent Cycle

Long-term cycles of supercontinent assembly and disassembly also influence or govern long-term stable isotope cycles, many of which are directly and indirectly linked to metabolic processes and climate.

For example, supercontinent assembly and stasis (extended periods of global cooling in which atmospheric abundance of $CO_2$ is low relative to the abundance of atmospheric oxygen — the atmosphere is said to be oxidative at such times [Fischer, 1984a]) bringing about the physical and chemical conditions favorable for high rates of accumulation of plant matter in the form of organic carbon (carbon in the reduced, oxygen-deficient state). Since photosynthetic (life) processes preferentially sequester isotopically light carbon ($CO_2$ preferentially made from $C_{12}$-bearing $CO_2$, as opposed to $C_{13}$-bearing $CO_2$) during periods of high rates of accumulation of organic carbon/reduced carbon (e.g., 650 to 550 million years ago, and 325 to 225 million years ago; refer to Figure 2), freshly/newly deposited carbonate rocks on the continental shelves of the world (oxidized carbon) are said to be isotopically enriched in the heavier isotope of carbon (i.e., contain large amounts of $C_{13}$ as opposed to $C_{12}$; see

Figure 2). Furthermore, oxidation of organic carbon in an $O_2$-rich atmosphere could result in an injection of $CO_2$ into the atmosphere, followed by greenhouse warming.

Consider the following additional example of the operation of tectonically modulated, life-mediated (biogeochemical), and geochemical cycles. When sea level is tectonically low during times of continental assembly and stagnation, previously stored nutrients on the world's continental shelves are now available to be reworked and physically moved to the deeper parts of the ocean basin, thus adding volumetrically more nutrients to the ocean. This is not to be confused with deposition and storage of nutrients on the continental shelves of the world when they are maximally flooded by sea level rising and the flooding of shelves (see Figure 2).

More nutrients (e.g., phosphorous) result in increased biomass and, thus, the drawdown of additional atmospheric $CO_2$, leaving the atmosphere more oxidizing and the Earth's sedimentary environments more reducing. Under these conditions, as stated above, since there would be an increase in plant biomass (preferential uptake of the light isotope of carbon, $C_{12}$, by definition), the oceans would be left enriched in the heavier isotope of carbon, $C_{13}$, available for the formation of carbonate rocks — the discarded shells (not the organic, fleshy tissue) of oceanic organisms (see $C_{13}$ curve, Figure 2).

Thus, during supercontinent disassembly and dispersal carbonate rocks are said to be isotopically light (for an in-depth discussion of stable isotope cycles as a function of supercontinent cycles, see Worsley et al. [1985]).

High stands of sea level favor the storage and preservation of organic/reduced (no oxygen present) carbon and nutrients on the continental shelves of the world (in the form of phosphorites; Cook and Shergold, 1984; Arthur and Jenkyns, 1981) weathered from exposed silicate, continental crust (refer to Figure 2). In other words, when sea level is high (a time when continental shelves are implicitly, maximally flooded), the accumulation of nutrients on continental shelves increases due to the increased storage area and preservation potential. Thus, the rate of nutrient and carbon deposition, and preservation, increases as a function of increased area of available storage space.

Also, any increase in the reduction of carbon (the stripping of oxygen from $CO_2$, leaving the atmosphere more enriched in oxygen) is counterbalanced in nature by the oxidation of sulfur compounds like pyrite, thus adjusting for any changes in the composition of the atmosphere produced by an increase in reduced carbon, as in this example. Thus, the $C_{13}$ and the $S_{34}$ curves are inversely related (when $C_{13}$ is large, $S_{34}$ is small, and vice versa) throughout Earth history, and life-mediated as well (Veizer et al., 1980; refer to the $C_{13}$ curve in Figure 2; see also Worsley et al., 1986).

Finally, in addition to the life-mediated biogeochemical cycles described above, continental crust is observed to have a higher ratio of $strontium_{87}$ to $strontium_{86}$ $Sr_{87}/Sr_{86}$) than the Earth's mantle and ocean crust. Thus, the strontium composition of seawater at any one time is defined by the source or sources from which the strontium is derived. Consequently, for example, during times when continents are assembled (see Figure 2) sea level is low, thus, more continental area is exposed to weathering at this time. Hence, the composition of seawater at these times is tectonically biased toward higher ratios of $Sr_{87}/Sr_{86}$, an outcome of the delivery of more continental material to the ocean, hence, higher values of $Sr_{87}/Sr_{86}$, by definition, when there is more area of continental crust available for weathering, due, in this instance, to tectonic circumstances.

I would point out that Figure 2 contains only those biogeochemical cycles (or feedbacks) which are most directly involved in climate and climate change.

## B. ORBITAL FORCING OF CLIMATE

### 1. Orbital Forcing and the Supercontinent Cycle: The Substance of Long-Term Global Warming and Global Cooling

If the last 50 million years of Earth history represent the most recent supercontinent in-the-making (Worsley et al., 1984; Fischer, 1984a; 1984b; Socci, 1992), and therefore, the most recent long-term global cooling in-the-making (Ruddiman and Kutzbach, 1989; Raymo et al., 1988; Volk, 1989a), it seems reasonable to conclude that the suggested controls and feedbacks to global warming and global cooling associated with this planet's most recent supercontinent in-the-making, should be the same feedbacks and controls on climate that were associated with the older records of supercontinent assembly and disassembly. Yet, as I have stated elsewhere in this chapter, virtually all of the available evidence bearing upon the substance, duration and mechanism(s) responsible for global warming and global cooling during the Earth's most recent candidate global cooling event of long duration indicate that these climatic end-members occurred periodically, at Milankovitch-band and higher frequencies (10,000 year or less to 100,000 years), and were therefore most likely triggered by changes in the distribution of solar radiation and moisture over the surface of the Earth (Berger, 1988; Imbrie et al., 1984; Broecker, 1966; 1987; Hays et al., 1976; Imbrie, 1987; Imbrie and Imbrie, 1980; COHMAP MEMBERS, 1988; Overpeck et al., 1989; Kominz and Pisias, 1979).

In fact, evidence for short-term worldwide fluctuations in sea level, and therefore, short-term episodes of global warming and cooling (at Milankovitch-band frequencies) has been found in rocks scattered throughout the geologic record, dating back to 2.2 billion years before present (Grotzinger, 1986), the approximate time of the Earth's initial glaciation 2.5 to 2.7 billion years ago (see Figure 3).

Consequently, long-term global cooling events, although feedbacks to supercontinent formation, tectonically lowered sea level, and increased rates of continental weathering, are suggested to be long periods of high frequency, Milankovitch-band global warming, and global cooling events triggered by orbitally forced changes in the distribution of solar radiation and moisture (Figure 4; see also Crowley, 1983; Saltzman and Maasch, 1991). Given that the Earth's most recent ice sheets built sluggishly, yet melted rapidly, by as much as a factor of 10:1 (Charlson et al., 1987; Broecker and van Donk, 1970), high-frequency global cooling events may have accounted for the bulk of the duration of long-term global cooling events on the order of 100s of millions of years in duration.

On the other hand, Milankovitch-band frequencies have also been discovered within rocks deposited during times of maximal continental dispersal (periods of long-term global warming — Arthur et al., 1984; Herbert and Fischer, 1986; Laferriere et al., 1987; de Boer and Wonders, 1984; Olsen, 1990; Goldhammer et al., 1987; House, 1985; Barron et al., 1985; Olsen, 1986; Weedon, 1989; Cotillon, 1987; Fischer, 1964; Van Houten, 1962; Fischer, 1980; refer to Figure 6). Consequently, I suggest that when sea level and atmospheric $CO_2$ were tectonically high, resulting in global warming, orbitally-forced changes in the distribution of solar radiation and moisture could not override the warming effects of high levels of tectonically-emplaced, atmospheric $CO_2$, and trigger high-frequency continental glaciation and deglaciation (short-term warming and cooling events). Thus times of maximal continental dispersal were devoid of continental glaciation, for the most part, despite evidence of orbital forcing (see Figures 2 and 6).

Orbital forcing did, however, continue at these (and other) times of maximal continental dispersal, to produce Milankovitch-band, high-frequency changes in rates of oceanic up-welling, degree of deep-ocean oxygenation, velocity of surface currents, surface productivity, carbonate productivity, redox conditions, seasonal contrast, and pole-to-equator temperature gradients (Arthur et al., 1984; Hebert and Fischer, 1986; Arthur et al., 1986).

| Climate | Age of Strata (Myr.) * | Observed Milankovitch Band Frequencies | Source |
|---|---|---|---|
| Ice-house Climate | 0 | PRESENT | |
| | 40 | | |
| Greenhouse Climate | 66 | CENOZOIC | |
| | 76 | 21-41k | Laferriere et al (1987) |
| | 100 | 20k | de Boer and Wonders (1984) |
| | 100 | 20-100k | Barron et al (1985) |
| | 100 | 20k, 41k, 100k | Herbert and Fischer (1986) |
| | 134 | 10-26k(?) | Cotillion (1987) |
| | 175 | 21k, 41k, 100k | Weedon (1989) |
| Icehouse Climate | 195 | | |
| | 210 | 25k, 44k, 100k 133k, 400k | Olsen (1986) |
| | 218 | 20k, 100k | Fischer (1964) |
| | 220 | 21k | Van Houten (1962) |
| | 230 | 21k, 100k | Goldhammer et al (1987) |
| | 240 | 41k | House (1985) |
| | 260 | 20k, 100k | Anderson (1982) |
| | 300 - 275 | none reported | Gamundi (1989) |
| | 320 - 285 | 40-120k, 235-400k | Heckel (1986) |
| | 325 | 20, 40k, 400k | Weller (1930) |
| | 360 - 255 | 23.5 to 39.3k | Veevers and Powell (1987) |
| Green-house Climate | 360 - 320 | none reported | dv. Klein and Willard (1989) |
| | 370 | 20k, 110k | Olsen (1990) |
| Icehouse Climate | 560 | | |
| | 570 | none reported? | Monninger (1979) |
| | 570 - 0 | 21k, 41k, 100k 413k | Algeo and Wilkinson (1988) |
| ? | 750? | | |
| | 1900 | 20k, 100k | Grotzinger (1986) |
| | 2300 - 2700 | FIRST CONTINENTAL GLACIATION | |

**FIGURE 6.** Reported occurrences of Milankovitch-band frequencies within the pre-Cenozoic stratigraphic record. Approximate absolute ages assigned on the basis of reported stratigraphic age and/or position (chronology of icehouse/greenhouse climates from Fischer, [1984a] and Berner, [1990]; see also Figures 2 and 3).

## 2. Changes in Orbital-Forcing Frequencies Over Time

Milankovitch orbital periodicities of precession (21,000 years) and obliquity (41,000 years) are suggested to have been smaller in the past due to the evolution of the Earth-moon system (Figure 7) resulting in a decrease in the rate of the Earth's rotation and an increase in the moon's orbital velocity, thus, shorter days in the early history of the Earth and therefore, higher frequency obliquity and precession components of orbital forcing. The 100,000 year eccentricity cycle of the Earth's path (shape of the orbit) around the sun is unaffected by

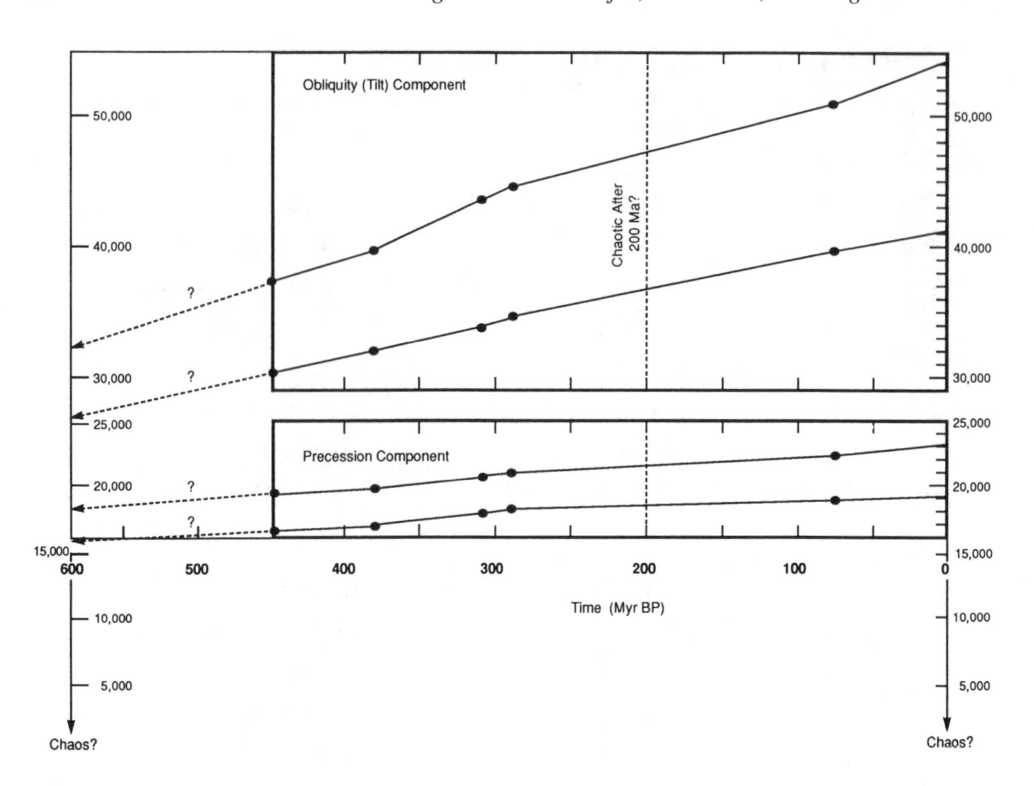


**FIGURE 7.** Suggested changes in the periodicities of the presessional and tilt components of Milankovitch-band, climate forcing, orbital phenomena back in time. Former increased rates of planetary rotation due to the interaction between the Earth and the moon may have resulted in a decrease in the periodicities of these orbital components, more so for higher periodicities, thus the convergence of near the convergence of the periodicities of precession and tilt in the vicinity of 20,000 years, 600 million years ago (modified after Berger et al., [1989]). Suggested chaos within the solar system prior to 200 million years ago may have affected the periodicity of these components of orbital forcing back through time. (From Laskar, J., *Nature,* 338, 238, 1989. With permission.).

the rate of rotation and therefore, is unlikely to have changed over time (Algeo and Wilkinson, 1988; Walker and Zahnle, 1986; Berger et al., 1989). As is shown in Figure 7, the effect upon the longer periods are much greater, such that at approximately 600 million years ago, the period of precession was roughly 17,000 years, while that of obliquity was 28,000 years (Walker and Zahnle, 1986). The chaotic behavior of the solar system may also have implications for Milankovitch-band periodicities within rocks older than 200 million years (Laskar, 1989). However, evidence from the geologic record, even within rocks as old as 2.2 billion years (Grotzinger, 1986), suggests that Milankovitch-band frequencies have remained within the same order of magnitude over this increment of time (Berger et al., 1989).

### 3. Orbitally Forced, High-Frequency Global Warming and Global Cooling: Lessons from the Most Recent Long-Term Cooling Event In Progress

Orbital forcing of climate and climate change (global cooling and warming events of short duration) during the Earth's most recent long-term global cooling even in progress, is well established. Kominz and Pisias, 1979) observed that orbital forcing of climate seems to be confined to the tile (obliquity) and precession components alone, with both components, however, accounting for less than 25% of the observed linear variance in ice volume changes (a proxy for climate change; Kominz and Pisias, 1979). The eccentricity component of orbital-forcing of solar insolation and climate (90,000 year component related to the shape of the Earth's orbit around the sun) accounted for none of the linear variance in ice volume changes associated within the Earth's most recent ice ages (Kominz and Pisias, 1979).

However, with respect to the existence of a near-100,000-year oscillation of climate, beginning at about 800,000 years ago, Saltzman and Maasch (1991) have suggested that this near-100,000-year climate oscillation was borne out of (1) 21,000- and 41,000-year, Milankovitch-band changes in solar insolation, (2) changes in the volume of atmospheric $CO_2$, including tectonically-induced changes, (3) changes in global ice mass, (4) changes in mean ocean temperature, and (5) changes in mean ocean-surface temperature at 65°N. The onset of this 100,000-year periodicity is suggested to have begun by a long-term, tectonically driven (assembly or near-assembly of continents [?] involving increased weathering of continental crust) lowering of atmospheric $CO_2$ to roughly 275 ppmv. At this extremely low level of $CO_2$ the climate system is suggested to be vulnerable to instabilities triggered by a further drop in surface temperature due to orbitally forced changes in insolation, a further decrease in $CO_2$, or an increase in ice mass, all positive feedbacks to the tectonically driven lowering (climate cooling) of atmosphere $CO_2$.

Thus, 770,000 years ago, the suggested convergence of these positive feedbacks to tectonic (long-term) climate cooling resulted in the initiation of a near-100,000-year oscillation in climate. According to Saltzman and Maasch (1991), only when this suite of variables brings about a response from the ocean whereby some $CO_2$ is transferred from the ocean to the atmosphere, does this unstable climate cycle terminate (amounting to a negative feedback to cooling), thus setting in motion the beginning of the next 100,000-year cycle.

Many of the overall climate-driving, or climate-modulating feedbacks to orbitally forced changes in the distribution of solar radiation are suggested to be nonlinear (''normal'', see Imbrie, 1987) responses/feedbacks to orbital forcing, many of which occur at frequencies higher than 19,000 years, not the least of which is the termination of glacial episodes/short-term global warming (Kominz, 1974; Kominz and Pisias, 1979; Imbrie, 1987; Pisias et al., 1975; Pisias and Shackleton, 1984; Shackleton and Pisias, 1985; Siegenthaler and Wenk, 1984; Sarmiento and Toggweilter, 1984; Boyle, 1988b; 1990; Broecker et al., 1985; Jones et al., 1987; Knox and McElroy, 1984; Berger et al., 1987; Broecker and Peng, 1989; Overpeck et al., 1989; Neftel et al., 1982; 1985; Stauffer et al., 1988; Delmas et al., 1980). The asymmetry of the glacial-interglacial cycles is further evidence of the nonlinearity of orbitally forced climate changes and orbital forcing itself.

Despite years of investigation, one of the most fundamental and compelling questions still remains largely unanswered. How, for example, do orbitally forced changes in the distribution of solar insolation at the Earth's surface (not a volumetric change necessarily in solar radiation) result in sweeping climate changes globally? What also are the pathways and rates associated with changes in solar insolation, and specifically, how is climate change carried out? This is, after all, the nuts-and-bolts of understanding orbital-forcing of climate.

In addition to ice growth/decay itself (short-term global warming and cooling), other nonlinear, high-frequency (tens of years to 100,000 years) feedbacks to orbital forcing (and to orbitally-forced ice growth/decay — proxy for climate) include fluxes in salinity (higher during glacials; Thunell and Williams, 1989; Thunell et al., 1988; Broecker, 1982a; 1982b), atmospheric methane (lower during glacials; Stauffer et al., 1988; Raynaud et al., 1988), atmospheric $CO_2$ (lower during glacials; Druffel and Benavides, 1986; Delmas et al., 1980; Neftel et al., 1982; 1985; 1988; Shackleton et al., 1983; Pisias and Shackleton, 1984; Shackleton and Pisias, 1985; Lorius et al., 1985; Jouzel et al., 1987; Barnola et al., 1987; Genthon et al., 1987), and organic carbon burial rates and productivity (higher during glacials; Nair et al., 1989; Lyle, 1988; Mix, 1989; Broecker, 1982a; 1982b; Boyle, 1988a; 1988b; Knox and McElroy, 1984; Martin and Fitzwater, 1988.)

Furthermore, worldwide continental glaciation (short-term global cooling) results in increased planetary albedo and, therefore, further cooling as sea level drops. Increased snow and ice cover during glacials also serve to positively reinforce global cooling (Knox and McElroy, 1984). During interglacials (short intervals of global warming), reduced albedo results in amplification of climate warming. Cloud albedo effects are more complex and largely unknown, especially in the Earth's ancient climate record. Clouds are therefore discussed at more length below.

## C. ORBITAL FORCING, TECTONICS, $CO_2$, AND GLOBAL CLIMATE CHANGE

One of the most significant pieces of evidence bearing on the actual controls on climate and climate change (the sequence of climate controls and feedbacks, at least with respect to the Earth's most recent ice ages) comes from the work of Pisias and Shackleton (1984) and Shackleton and Pisias (1985). Using a highly resolved (time wise) deep sea Shackleton and Pisias indicated that the drawdown of atmospheric $CO_2$ (thus climate cooling) during the last ice age, lagged orbital forcing of solar insolation, but preceded changes in ice volume during the onset of the Earth's last worldwide glaciation (episode of short-term global cooling).

They further concluded that the observed changes in atmospheric $CO_2$ constituted evidence for nonlinear feedbacks to orbital forcing of climate, and that $CO_2$ did not merely serve as a feedback to climate in progress (triggered by some other, more fundamental climate-driving mechanism), but rather, was part of the climate-forcing process itself, a partner with orbitally-forced changes in the distribution of solar insolation, despite the fact that these fluxes in atmospheric $CO_2$ occurred at rates significantly higher (thus nonlinear) than Milankovitch-band rates.

Their conclusions also supported the notion, at least in large part, that orbital forcing account for $<25\%$ of the linear variance associated with ice volume changes/climate changes as reported by Kominz and Pisias (1979). General circulation model calculations (Manabe and Bryan, 1985; Rind et al., 1989) further indicate that orbital forcing alone is too weak a mechanism to alter climate significantly (force a transition from an interglacial period of global warming to a glacial period of global cooling), in the absence of feedbacks from greenhouse gases. The model assumes, however, that orbital-forcing phenomena are sufficiently well known, which is hardly the case.

Furthermore, given that fluxes in atmospheric $CO_2$ cannot be a direct response to orbital forcing (orbital forcing cannot directly change the volume of greenhouse gases in the atmosphere), nutrient fluxes, and therefore fluxes in oceanic biomass, became the obvious, but not necessarily the sole, candidate for intermediate feedbacks linking orbital forcing and atmospheric and oceanic $CO_2$, capable of inducing periodic, very high frequency fluxes in atmospheric $CO_2$, as well as dissolved, oceanic $CO_2$ during glacials and interglacials (short periods of global cooling and global warming).

These rapid (100 to 10,000 years), short-term fluxes in atmospheric $CO_2$ reported by Pisias and Shackleton (1984) and Shackleton and Pisias (1985) were later independently confirmed by analyses of atmospheric gases trapped within ice cores records spanning the last 0.25 million years of earth history (Stauffer et al., 1985; Neftel et al., 1982; 1985; 1988; Delmas et al., 1980; Barnola et al., 1987; Genthon et al., 1987; Lorius et al., 1985; Jouzel et al., 1987).

However, with respect to the transition from the most recent glacial to interglacial period (transition from cooling to warming), recently acquired, Antarctic ice-core records indicate that changes in ice volume predated a rise in atmospheric $CO_2$ by approximately 200 to 1200 years (Neftel et al., 1988). With the exception of the climate-triggering role of orbital forcing, this represents an apparent reversal of the sequence of feedbacks reported by Pisias and Shackleton (1984) and Shackleton and Pisias (1985), regarding the transition from an interglacial warming to a glacial cooling.

It would still appear to be the case, however, that the suggested order or sequence of feedbacks involving the transition from a glacial to an interglacial climate, or vice versa, is still within the error associated with the absolute calibration (time wise) of the sediment and ice-core layers in question. Thus, verification of the order or sequence of feedbacks to climate-driving forces or controls will still have to be more firmly established with even more demonstrably greater time resolution. The asymmetry and mirror-like aspects of the sequence of feedbacks now suggested to be associated with the transition from a glacial to an interglacial climate, and vice versa, are entirely plausible, even in the absence of replications of these efforts. A very recent, highly chronologically resolved ice-core record of a glacial/interglacial transition approximately 110,000 years before present (see Sowers and Bender, 1991) seems to favor the sequence of climate-changing feedbacks suggested by Pisias and Shackleton (1984) and Shackleton and Pisias (1985), with $CO_2$ changes leading changes in temperature and ice volume.

Consequently, in light of these new deep sea and ice-core records, many existing ocean-atmosphere climate models (such as those of Broecker (1982a; 1982b) which sought to account for the depletion of atmospheric $CO_2$ during glacials by way of a volumetric increase in nutrients, eroded from previously deposited, nutrient-rich organic sediments on the continental shelves of the world oceans) had to be abandoned, owing to the need to call upon sea-level changes (begging the question of climate controls in the first place since changes in sea level imply climate changes already in progress) and sluggish rates of erosion which could not have accommodated the delivery of additional nutrients to the oceans at rates sufficiently high enough to account for the extremely rapid, nonlinear rates of changes in atmospheric $CO_2$ recorded in ice cores, on the order of a couple of thousand years (see Siegenthaler and Wenk [1984] for a more complete discussion of Broecker's model). Thus, the following suite of ocean circulation-rate-dependent models, requiring no volumetric changes in nutrient delivery to the oceans, were borne out of attempts to address candidate mechanisms (and feedbacks) which might have led to the rapid fluxes in atmospheric $CO_2$ between glacial and interglacial climates of the most recent ice ages, as observed in ice cores.

Knox and McElroy (1984) have suggested that orbitally forced changes in the rate of oceanic circulation, as well as orbitally forced changes in the availability of incident light (orbitally forced fluxes in cloud cover?) at high, northern latitudes, could result in higher-than-average, incident, winter light levels (due to cloud cover?) at a time critical for bottom-water formation. The net effect of this would lead to decreased biological activity due to the lack of incident light, a resultant increase in preformed nutrients due to underutilization, and an eventual exchange of $CO_2$ between the ocean and the atmosphere due to nutrient underutilization and, hence, orbitally forced global warming. Without nutrient uptake first, oceanic phytoplankton could not acquire biomass, thus depleting the atmosphere of some volume of $CO_2$, the result of which might lead to climate cooling. The inverse of Knox and McElroy's (1984) model would presumably bring about orbitally forced glaciation or at least some climate cooling.

Martin and Fitzwater (1988) have observed that the lack of iron in high-latitude oceans inhibits complete utilization of available nutrients there. Iron, in this case, is also suggested to be biolimiting. Thus when iron and/or iron-rich sediments (in the form of windblow dust) become available, preferentially during cold, dry, glacial climates, all of the available nutrients in high-latitude oceans are completely depleted, resulting in a drawdown of atmospheric $CO_2$ and amplification of global cooling in progress. This mechanism for relatively rapid, climate modulation is implicitly driven by orbital forcing insofar as orbital forcing appears to trigger growth and decay of continental ice sheets in the first place.

Boyle (1988b) proposed a model whereby changes in the rate of oceanic circulation alone result in a chemical restructuring of the oceans such that dissolved $CO_2$ and nutrients are concentrated in deeper water (transferred from intermediate to deeper water) when the rate of oceanic circulation is increased (i.e., during a transition from an interglacial to a glacial climate), presumably in response to orbitally forced changes in the rate of atmospheric and, hence, oceanic circulation. An increase in ther rate of circulation, for example, results in an increase in the probability that organic matter will escape degradation by oxidation as it settles through the water column, thereby increasing the probability that it will accumulate on the ocean floor, relatively intact in the form of reduced carbon. Furthermore, increased circulation increases the likelihood as well, of nutrient accumulation on the sea floor (Figure 8). A decrease in the rate of circulation should have the opposite effect.

Furthermore, as this zone of concentrated nutrients and metabolic $CO_2$ is driven downward, it decreases the carbonate ion concentration of the deep ocean, resulting in higher rates of carbonate dissolution, thus depleting a volume of dissolved $CO_2$ in the ocean in proportion to the amount of carbonate dissolved at depth, via what appears to be subsea ''acid rain'' in a remarkably well-buffered system. In response to this, the alkalinity of the ocean rises and a volume of $CO_2$ is transferred from the atmosphere to the upper ocean (Figure 9), in proportion to the amount of carbonate dissolution at depth, resulting at least in part, in global cooling/glaciation.

The suggested response time for the alkalinity and ocean changes described is approximately 2500 to 6000 years (Boyle, 1988b), with the $CO_2$ change (at least the initial atmosphere-to-ocean transfer) having occurred within several hundred years. Thus, a decrease in the rate of oceanic circulation would result in an upward shift in the zone of concentrated nutrients and dissolved $CO_2$, presumably resulting in global warming. This mechanism accounts for approximately one half of the flux in atmospheric $CO_2$ associated with the last glacial to interglacial transition, about 45 ppmv (refer to Figure 9).

However, the rate and climate implications regarding the reversal of this mechanism are not at all clear. For example, when ocean circulation presumably returns to its slower rate of circulation and this zone of $CO_2$ and nutrient enrichment rises to its original height, in response to the now less-rapid circulation, what happens if carbonate sediment, deposited higher up on the continental shelves when the zone of $CO_2$ enrichment was lower in the water column, comes in contact with this acidic zone of $CO_2$ as it rises to its former level? Will this result in a cooling spike much like the Youger Dryas event which lasted 70 years or less (Broecker and Denton, 1989)? What effect do rates of dissolution versus rates of carbonate sedimentation have on the system's so-called return to its original sluggish state? Are these rates similar? Why should the return to condition A, from condition B, for example (refer to Figure 9 A to C), be symmetrical with respect to time, chemistry, $CO_2$ and climate?

Broecker and Peng's (1989) polar alkalinity model incorporates, and builds upon, the model of Boyle (1988b), as well as rate-induced changes (of undefined origin but conceivably, implicitly orbitally driven) in the periodic(?) shut-down of North Atlantic Deep Water (NADW) circulation, at rates on the order of 20 to 70 years (see also Broecker et al., 1985; Berger et al., 1987; Nisbet, 1990). The suggested increase in ocean alkalinity resulting from the occasional shut-down of NADW, in addition to the increase in alkalinity due to a circulation-rate-dependent, chemical restructuring of the ocean (the model of Boyle [1988b]), collectively account for a net transfer of $CO_2$ from the atmosphere to the ocean, approximately equivalent in volume to the total flux in atmospheric $CO_2$ (on the order of 100 to 1000s of years) associated with the last interglacial to glacial transition, as observed within ice cores records, in this instance depletion of atmospheric $CO_2$ on the order of 70 to 110 ppmv (Barnola et al., 1987).

The viability of Broecker and Peng's (1989) alkalinity model remains somewhat enigmatic, largely because a shutdown in NADW circulation requires a melting event, followed immediately by a rapid discharge of fresh water in the vicinity of where the NADW originates, through salinity-driven subsidence of ocean water (e.g., the Younger Dryas climate event;

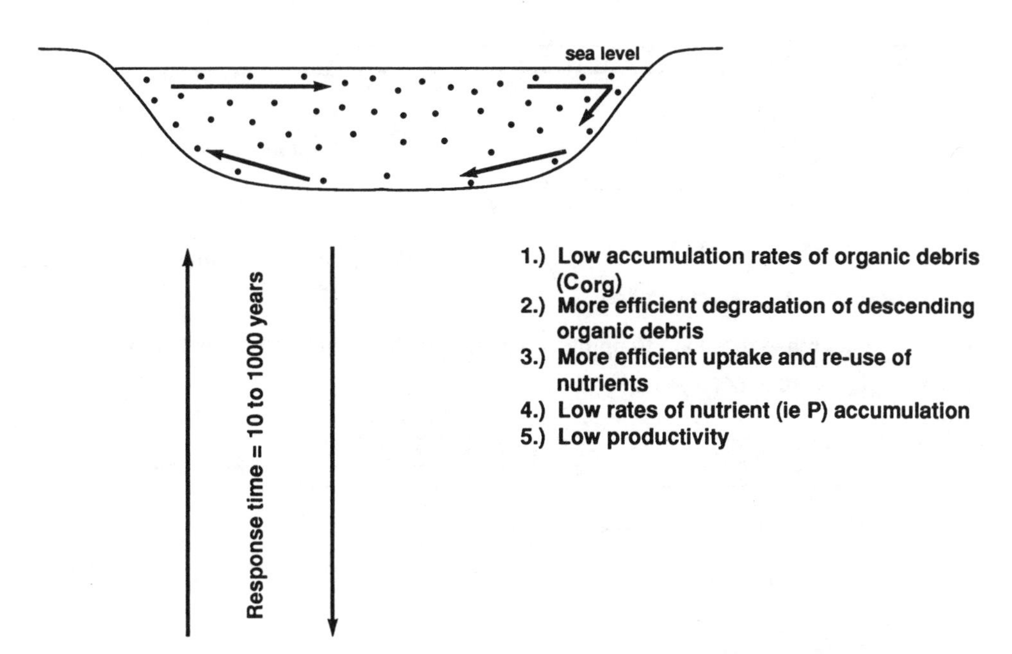


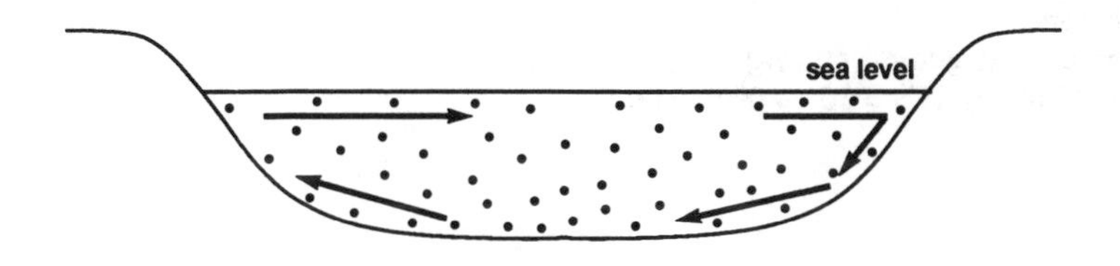


1.) Shift of nutrients and $CO_2$ to greater depths
2.) Higher organic debris ($C_{org}$) accumulation rates
3.) Less efficient degradation of descending organic debris
4.) Higher nutrient accumulation rates
5.) Less efficient uptake and re-use of nutrients
6.) High productivity

**FIGURE 8.** Ocean circulation-rate-driven or rate-dependent phenoma, conceivably orbitally forced in the first place by orbitally driven changes in wind velocity. Such rate-driven changes can lead to an exchange of atmospheric carbon between the atmosphere and the ocean, resulting in high-frequency, nonlinear glaciation and deglaciation (global cooling and global warming; see Figure 9). (A) Slow or low rates of oceanic circulation result in more efficient use of nutrients and degradation of descending organic matter, therefore, a lower probability of accumulation of nutrients and organic carbon at the bottom; (B) faster oceanic circulation results in less efficient use of nutrients at the surface and a higher probability, therefore, or nutrient and organic carbon accumulation at the bottom.

**A. Steady-state ocean circulation**

**Interglacial greenhouse climate/atmospheric $CO_2$ = 280 ppmV**

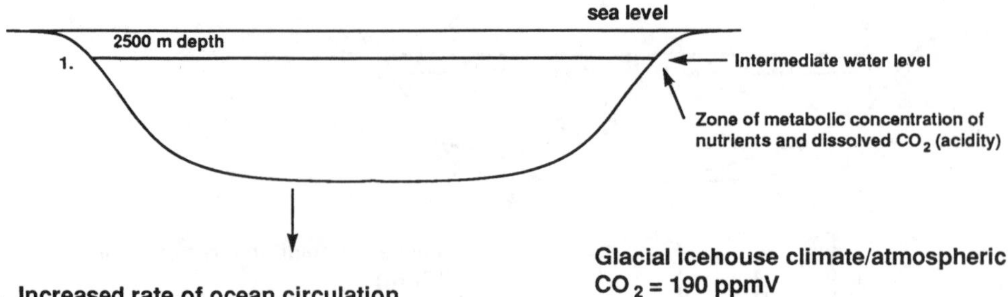


**Glacial icehouse climate/atmospheric $CO_2$ = 190 ppmV**

**B. Increased rate of ocean circulation**

**$pCO_{2_{I-G}}$ = decrease by 90 to 110 ppmV; 45 ppmV due to Δ's in NADW circulation**

**Time $\Delta_{pCO_2}$ = 10 to 6000 years**

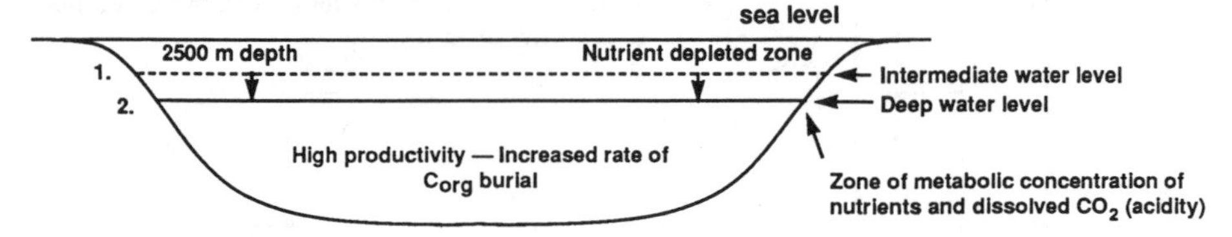


1.) Increased carbonate dissolution
2.) Decrease in carbonate ion concentration
3.) Increase in alkalinity
4.) Transfer of $CO_2$ from atmosphere to ocean
5.) Δ circulation rates due to orbitally-forced Δ's in solar radiation
6.) Lag time between chemical restructuring of oceans, $CO_2$ and alkalinity Δ's = 10 to 6000 years

**Interglacial greenhouse climate/atmospheric $CO_2$ = 280 ppmV**

**C. Decrease in rate of ocean circulation to steady-state condition in A**

**$pCO_{2_{G-I}}$ = increase by 90 to 110 ppmV; 45 ppmV due to Δ's in NADW circulation**

**Time $\Delta_{pCO}$ = 10 to 6000 years**

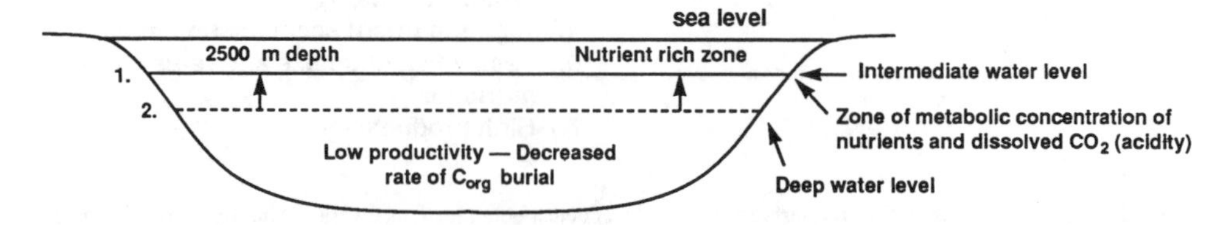


1.) Decrease in carbonate dissolution to steady-state condition of A
2.) Increase in concentration of carbonate ions to steady-state value of A
3.) Decrease in alkalinity to steady-state value of A
4.) Transfer of $CO_2$ from ocean to atmosphere, to steady-state level of A
5.) Δ circulation rates due to orbitally-forced Δ's in solar radiation
6.) Lag time between chemical restructuring of oceans, $CO_2$ and Δ's = 10 to 6000 yrs.

FIGURE 9.

Broecker and Denton, 1989). Shutdown of NADW circulation (and thus, temporary stratification of the oceans?) is suggested to result in a change in the alkalinity of the oceans in the vicinity of Australia, thus leading to a transfer, from the atmosphere to the ocean, of roughly 45 ppmv $CO_2$, one half the total volume of $CO_2$ observed to have been sequestered from the atmosphere during the last interglacial to glacial transition (warming to cooling), according to ice-core records (Barnola et al., 1987). The other half of the observed loss of atmospheric $CO_2$ during this transition, is suggested by Broecker and Peng to have resulted from the chemical restructuring of the oceans, Boyle's (1988b) model.

If Boyle's model requires an increase in the rate of ocean circulation, in order for $CO_2$ to undergo a transfer from the atmosphere to the ocean, how can Boyle's model apply at a time when shutdown of the NADW (the other half of the observed interglacial-to-glacial loss of atmospheric $CO_2$) necessitates, or results from, a slowing of oceanic circulation, perhaps even stratification of the oceans? Perhaps the required flux in atmospheric $CO_2$ is a complex function of changes in wind-driven circulation and thermohaline circulation operating simultaneously, and at times, in the same direction, thus amplifying a climate outcome at times.

Should Boyle's chemical-restructuring model, and Broecker's alkalinity model, prove to be incompatible, as may well be the case, could it be possible that a difference of 45 ppmv $CO_2$, not necessarily a difference of 70 to 110 ppmv $CO_2$, represents enough greenhouse (or reverse greenhouse) forcing to bring about a glacial to interglacial transition, and vice versa with additional feedbacks to climate change itself? The Younger Dryas climate event would seem to indicate that this might indeed be so.

It is entirely plausible as well that abrupt climate changes, on a scale of 10s of years (e.g., such as the Younger Dryas, or Younger Dryas-like events), may be the result of abrupt, sporadic, yet in some instances, volumetrically important greenhouse gas exchanges between the atmosphere, sediments (mainly oceanic slope sediments), and oceans, involving $CH_4$ principally, and/or lesser amounts of $CO_2$ (MacDonald, 1990; Saltzman and Maasch, 1988; Nisbet, 1990). $CH_4$ clathrates in permafrost in tundra might also respond to warming by the release of pressure, via the melting of ice, possibly resulting in the release of $CH_4$ gas trapped below the surface (Nisbet, 1990), thus amplification of warming. The $CH_4$ clathrate reserve would be built again during a subsequent rise in sea level.

Large reservoirs of $CH_4$ are also associated with pock-marked areas of continental slopes where sediment slumping occasionally results in an immediate decrease in overburden pressure, followed by a sudden release of previously trapped $CH_4$ gas (MacDonald, 1990). Furthermore, there is an increase in the incidence of slumping on continental slopes when sea level is low. Nisbet (1990) has suggested, for example, that the Younger Dryas event

**FIGURE 9.** Circulation-rate-driven climate feedbacks suggested to have led to the onset and demise of the Earth's most recent ice ages (short-term global warming and cooling; based upon the constant-nutrient volume model proposed by Boyle, [1988a] and the alkalinity models of Broecker and Peng, [1989], and Broecker and Denton, [1990]. (A) Interglacial steady steady, sluggish, oceanic circulation with development of an $O_2$ minimum, $CO_2$ and nutrient maximum at about 2500 m depth; (B) increased rates of oceanic circulation result in a lowering of the $O_2$ minimum, $CO_2$ and nutrient maximum zone, a decrease in the concentration of carbonate ions, an increase in alkalinity, an increase in the dissolution of carbonates within this zone of high $CO_2$ resulting in a transfer of $CO_2$ from the atmosphere to the oceans (by approximately 45 ppmv + approximately 45 ppmv from a change in the circulation of the North Atlantic Deep Water [NADW] system) hence, high-frequency, nonlinear glaciation and global cooling several 1000 years later; and (C) decreased rates of oceanic circulation during a glacial climate result in a chemical restructuring of the oceans and a return to the sluggish, steady-state conditions in A. Slower circulation brings about a rise in the zone of $O_2$ minimum and $CO_2$ and nutrient maximum, resulting in an increase in the concentration of carbonate ions to the former level in A, a decrease in alkalinity and a decrease in carbonate dissolution to the steady-state values in A, and a transfer of $CO_2$ from the ocean to the atmosphere (by approximately 45 ppmv + 45 ppmv from a change in the circulation of the NADW) to the steady-stage value in A, hence, high-frequency, nonlinear deglaciation and global warming within a few 1000 years, far less than in part B.

(and many other abrupt events), and the warming which followed, may have been related to the release of $CH_4$ associated with slope failure. Evidence of a $CH_4$ spike in ice cores, consistent with the timing of the Younger Dryas event (and others like it), may offer an alternative explanation for the Younger Dryas event, and many others like it (see Nisbet 1991).

Clearly then, although these coupled ocean-atmosphere-sediment models are far from established, they have profound implications over the full range of this planet's past and future climate history, in that sweeping climate changes (i.e., the exchange of atmospheric greenhouse gases) appear to have taken place at rates which were previously unimagined, by something as seemingly uneventful as mere changes in the rate of air and water circulation, in the absence of volumetric fluxes in nutrients entering the world ocean.

I would also point out that insofar as each of the above models requires changes in circulation rates in order to bring about an exchange of gases and hence climate, orbital forcing is called upon as the fundamental rate-driving, climate-triggering mechanism, implicitly or otherwise. Internal natural oscillations of ocean thermohaline circulation are also conceivable, independent of orbital forcing.

Furthermore, while most of the models cited above are oceanic circulation-rate-dependent, increased circulation rates also bring about increased rates of nutrient delivery from intermediate water depths to the ocean surface, the result of which would presumably spike phytoplankton growth somewhat, which in turn would lead to a minor additional draw down of atmospheric $CO_2$, a slight positive feedback to climate cooling.

## D. ELEVATION, TECTONICS, AND CLIMATE

As an alternative to the climate feedbacks and controls suggested herein (see Figure 4), particularly with regard to the long-term global cooling event in progress (beginning roughly 40 million years ago) Ruddiman and Kutzbach (1989) have suggested that this present long-term global cooling event in progress was triggered by an increase in the elevation of the Asian plateform resulting from the collision of India with Asia, as well as an increase in the elevation of southwestern North America due to heating and thermal expansion. In the case of the Asian platform, Ruddiman and Kutzbach have offered that the collision-induced elevation of the Tibetan platform has resulted in the disruption of high and low pressure cells, the jet stream, and monsoons, the net effect of which was to usher in global climate cooling roughly 40 million years ago.

Likewise, Powell and Veevers (1987) have explained the long-term global cooling event which began roughly 340 million years ago and ended 220 million years ago (refer to Figure 2), as resulting from multiple continenal collisions during the formation of a large continental landmass (otherwise known as the Gondwana supercontinent) in the Southern Hemisphere mid-latitude. As continents collided the collisional boundaries would become elevated and thus become cooler, acquiring alpine glaciers which coalesced to become continental-sized glaciers.

From the perspective of a continuum of climate controls and feedbacks, as suggested herein (see Figures 2 and 4, and also Fischer 1984a; 1984b; Worsley et al., 1984), I would argue that each of these long-term global cooling events was more likely to have been a feedback to the assembly and standstill of a former supercontinent in the latter case, and a supercontinent in the making(?) in the former case.

Recall that in the process of supercontinent formation, individual continents assemble at geoid lows and, thereafter, remain relatively stationary. While relatively stationary, geoid lows (cool regions within the Earth's mantle) become geoid highs as continents draw heat unto themselves and effectively act as radiators, resisting the passage of heat from below, thus heating the mantle below as well. As continents become heated from below, they thicken and elevate or dome (Anderson, 1989).

Additionally, in the Earth's present plate configuration, sea level is at its lowest elevation relative to continents (see Figures 2 and 5). Consequently, one might anticipate increased weathering (see Volk 1987) as a feedback to minimal shelf flooding (increased area of weatherable continent crust), resulting in a gradual, long-term draw down of atmospheric $CO_2$ and thus, global cooling.

I would also point out that there is no evidence, to my knowledge, that supercontinent assembly triggers worldwide, high-frequency continental glaciation and deglaciation (short-term global warming and global cooling events). There is abundant evidence, on the other hand, for significant, glacio-eustatic fluxes in sea level (at Milankovitch-band frequencies — orbitally forced), during the late last 40 million years of Earth history, and between 340 million years ago and 220 million years ago (Veevers and Powell, 1987; Klein and Willard, 1989; Heckel, 1986; Gonzalez; 1990; see Figure 6). The model of Veevers and Powell (1987) as originally presented, does not accommodate the record of Milankovitch-band fluctuations in sea level as recorded in coal sequences in North America, Europe, and China.

Finally, the elevation model of Ruddiman and Kutzbach (1989) provides no means of exchanging atmospheric greenhouse gases such as $CO_2$, nor does it accommodate or explain any worldwide sea-level responses to the increased elevation, all of which would presumably have a more global influence on climate than disruption of high and low pressure cells, the jet stream, and the monsoonal system over the Tibetan platform; nor can it explain Milankovitch-band glacials and interglacials throughout the last 40 million years.

## E. LIFE, CARBON, AND CLIMATE

The notion of climate regulation by way of carbon redox (organic/reduced and inorganic/oxidized species of carbon) over time is novel and intriguing. Worsley and Nance (1989) have posed that since post-Archean time, $CO_2$ degassing rates and the growth of continents have remained stabilized. In other words, any $CO_2$ sequestered from the atmosphere by way of weathering of calcium-silicate rocks (continental crust) is eventually returned to the atmosphere by volcanism and metamorphism associated with heating, melting, and recrystallization of carbonate rocks (the Earth's sink for all carbon produced during the weathering of continental crust [Berner and Lasaga, 1989]) during subduction associated with the operation of the supercontinental cycle.

Insofar as Earth surface temperatures appear to have remained remarkably constant over the last 4 billion years (Lovelock and Whitfield, 1982; see Figure 1), despite a 30% increase in solar luminosity, the carbonate-silicate system just described (tectonically driven carbon cycle involving weathering of continental crust, deposition of weathering-derived carbon in the form of carbonates, and rerelease of that carbon back to the atmosphere during subduction and metamorphism associated with the supercontinent cycle) cannot therefore have been responsible for holding earth-surface temperatures constant in the face of a 30% increase in solar luminosity over the last 4.6 billion years. Had this been otherwise, a 30% increase in solar luminosity should have resulted in significantly elevated earth-surface temperatures in the vicinity of 70°C, possibly in excess of the threshold for terrestrial life.

Worsley and Nance (1989) have therefore proposed that life itself (reduced/organic carbon) has held the surface temperature of the planet relatively constant, at lease since 3.5 billion years ago. However, since earth-surface temperatures appear to have remained constant (with a range of approximately $\pm 15$°C) throughout most of this planet's history, the rate of weathering has also remained relatively constant insofar as weathering is temperature-dependent (Walker et al., 1981).

Furthermore, since the volume of living matter (organic carbon) is limited by the availability of the biolimiting nutrient phosphorous (and others), whose only source is continental crust, given that weathering rates have remained constant due to constant earth-surface temperatures, there is no apparent mechanism which might have led to an increase in the supply of biolimiting nutrients, over the history of this planet, which would have led, in turn, to an increase in the biomass and an associated draw down in atmospheric $CO_2$, global cooling, conceivably glaciation, and continued planetary temperature regulation in the face of increasing solar luminosity over time.

Worsley and Nance (1989) further contend that resolution of the apparent paradox lies in the planet's isotopic records of oxidized carbon (carbonates) and reduced carbon (biomass), which suggests that through evolutionary processes of natural selection, the biosphere has become more enzyme-efficient over time (more efficient at recycling or reusing the biolimiting nutrient phosphorous). Thus, each newly evolved biota or radiation (e.g., ediacarans/soft-bodied organisms were followed [outcompeted/naturally selected against] in time by the appearance/radiation of shelled organisms with a capacity for burrowing into the substrate in search of nutrients; refer to Figure 2, C to P events) was more efficient at recycling or reusing biolimiting nutrients than the preceding population/radiation.

Consequently, in a world with a finite volume of biolimiting nutrients, increased biotic efficiency involving the reuse or recycling of nutrients by the biosphere, in lieu of a volumetric increase in the supply of nutrients to the biosphere, will nonetheless permit expansion of the biosphere and, therefore, an increase in the ratio of organic carbon to phosphorous over time (see Figure 2). This in turn might result (in the absence of anthropogenic influences) in a long-term trend or tendency towards global cooling and global cooling events/glacial epochs, and/or a higher frequency of glacial epochs/global cooling events, and, therefore, increased climate sensitivity, over time, whereby a slight adjustment in the composition of the atmosphere might result in rapid shifts in climate extremes.

If Worsley and Nance (1989) are correct, this model has important climatic implications, as well as implications regarding the survival or survivability of the biosphere. One such specific implication is that in the face of increasing levels of atmospheric $O_2$ over the last 3.8 billion years, the very by-product toxins of photosynthetic life itself, life has evolved in response to its own toxicity by sequestering more carbon and thereby continually adjusting the composition of the atmosphere and climate, only to continue to evolve, possibly at accelerated rates, once again in response to its own toxins, by yet again sequestering additional atmospheric carbon, ultimately rendering the atmosphere ignitive.

Ignition of the atmosphere would intuitively lead to biomass burning and the injection of particulate carbon and sulfur into the atmosphere as candidate cloud condensation nuclei (Charleson et al., 1987; Crutzen and Andreae, (1990), thus, perhaps, resulting in some climate cooling, at least locally, if not globally, at least for a while.

Insofar as Cloud (1983) has suggested that the evolution of life (proxy for climate and the composition of the atmosphere) from prokaryotic cells to eukaryotic cells involved in the formation and subsequent thickening of a cell wall surrounding the nucleus where energy (ATP) is made, perhaps the storage of carbon is an evolutionary buffer to increased levels of $O_2$, in response to photosynthetically increased levels of atmosphere $O_2$, which is highly reactive to reduced compounds (lacking oxygen) and, therefore, life-threatening.

Smith (1976) on the other hand, has stated that the evolution of $C_4$ photosynthesis (as opposed to the presently more-prevalent but less-efficient $C_3$ photosynthesis) was a response to a higher oxygen and lower $CO_2$-rich atmosphere (climate/global cooling) during the last 40 million years of Earth history. Given that $C_4$ photosynthesis is a more-efficient photosynthetic process (less energy is required to metabolize an equivalent volume of nutrients) in that carbon, not P, is conserved, this might in fact lead to no changes in the ratio of organic carbon to P over time, or possibly a decrease in the ratio of organic C to P over time (as suggested by Fischer, [1984a; 1984b]), contrary to the conclusions of Worsley and Nance (1989).

In fact, $C_4$ photosynthesis appears to have been alternately, naturally selected for and against throughout the last 40 million years, (Smith, 1976) in response to the redox state of the atmosphere, itself a feedback to evolution and biotic regulation of the atmosphere, hence, climate. Volk (1989a) has independently raised analogous questions regarding the feedbacks and/or controls on climate cooling, and the drawdown of atmospheric $CO_2$, as a consequence of the rise of angiosperms and evolution of deciduousness during the interval 100 million years ago to 65 million years ago, resulting in increased (more efficient) biotic weathering, a mechanism for more efficient removal of $CO_2$ from the atmosphere, at increased rates as well, thus resulting in long-term climate cooling.

Assuming that life can tolerate temperatures conceivably as high as 105°C (e.g., in mid-ocean ridge settings; see Kasting, 1987), why would life evolve, by conserving P while depleting C, particularly if it has an apparent large tolerance for heat in the first place? After all, life evolved in an anoxic (no oxygen), and, therefore, presumably hotter, $CO_2$-rich atmosphere, out of necessity (oxygen was poison to life at this time; see Cloud, 1983).

In addition, if the biota can, in fact, increase rates of weathering, by as much as a factor of 300 according to Lovelock and Whitfield (1982) and Schwartzman and Volk (1989), why again would life conserve nutrients which are biolimiting, but biodriving at the same time, in that evolutionary survival involves the sustenance of life itself through nourishment?

## F. BIOTIC RADIATIONS, ORGANIC CARBON, AND ICEHOUSE CLIMATES

As I have stated elsewhere (Socci, 1992), the contention on the part of Worsley and Nance (1989) that biotic radiations (see Figure 2), representing evolutionary improvements in the efficiency of the biosphere to recycle and reuse biolimiting nutrients, have been responsible for triggering long-term global climate cooling in Earth history, does not appear to be consistent with at least some crucial portions of the Earth's stratigraphic record of carbon (the record of carbon laid down somewhere between 800 and 560 million years ago).

This carbon record and interval of time in the history of the planet are crucial in at least two significant ways: (1) it represents the time of the long-term global cooling event between 700 million years ago and 570 million years ago; and (2) it represents the time over which two biotic radiations evolved, ediacarans and shelled organisms (see Figure 2). Arguably, the most highly, chronologically resolved record of carbon over this interval of time is that of Knoll et al. (1986) from Svalbard and East Greenland (Figure 10).

There are a number of negative carbon excursions (isotopically light excursions representing sequestering of $C_{12}$ in preference over $C_{13}$, hence, biomass expansion or more efficient biotic sequestering of carbon) spanning this interval of Earth history, yet each carbon excursion is approximately of the same magnitude as every other excursion, suggestion multiple glaciations/deglaciations (periods of warming and cooling). Yet there are no obvious correlations between isotopically light carbon excursions and the presence and stratigraphic position of documented glacial markers (implicit markers of local, regional, and in global cooling).

Secondly, in instances where there are glacial events/rocks arguably coincident in position and, therefore, time (with isotopically light carbon excursions) the glacial event(s) in question appears to have predated the carbon depletion event which, if glaciations were driven by biotic radiations, should have been a prerequisite for glaciation rather than an aftermath to glaciation.

Furthermore, this very same isotopically negative carbon spike/event, presumably representing draw down of atmospheric $CO_2$, coincident with a glacial event(s), is later restored to its preexcursion level presumably indicating restoration of preexcursion, higher partial pressures of atmospheric $CO_2$ (global warming?), yet glaciation persisted (climate cooling), as indicated by the rocks themselves, rather than having been terminated by the increase in

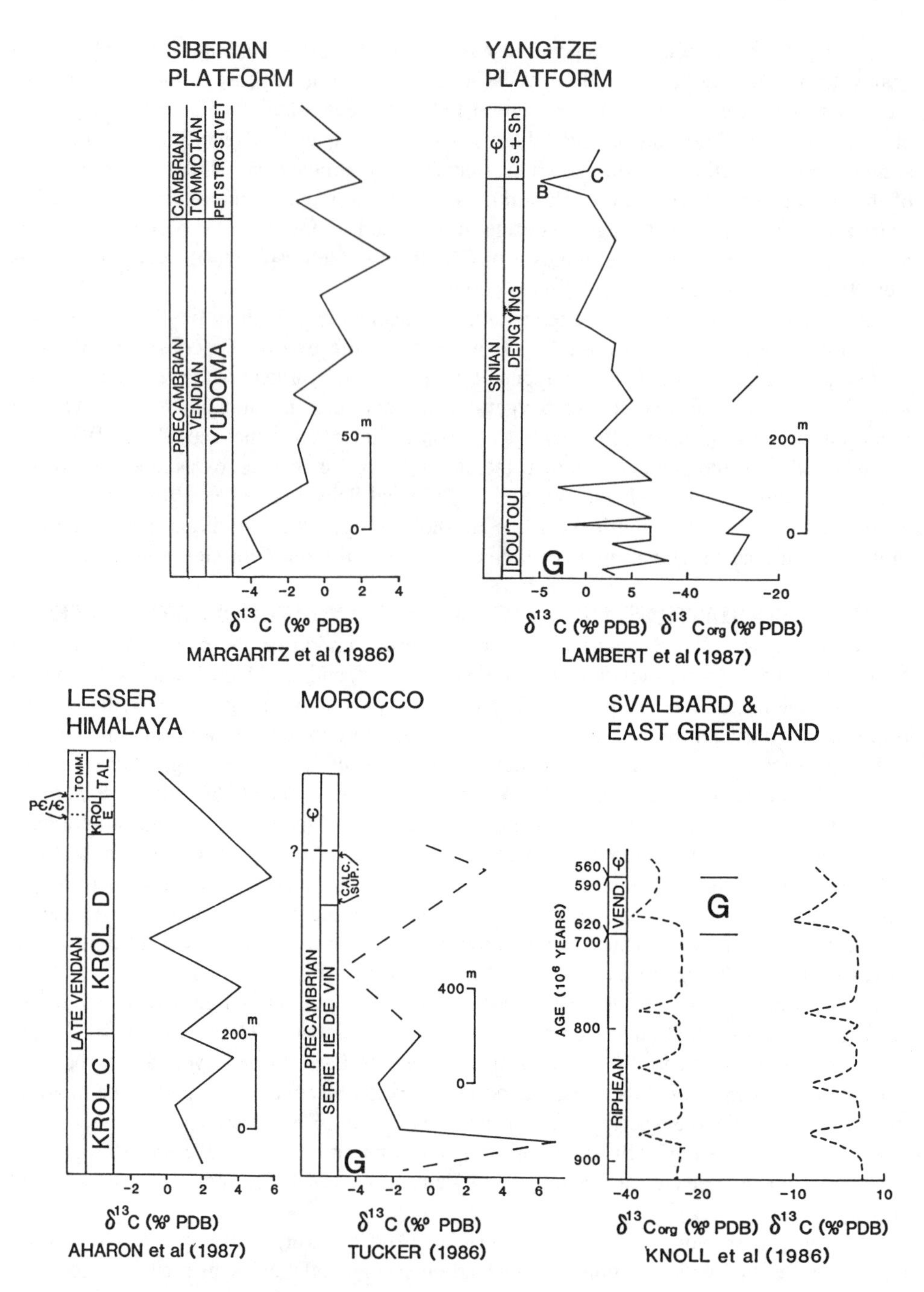


**FIGURE 10.** Carbon isotopic records of the time interval, 800 to 500 million years before present. An absolute chronology has been approximated only for the data of Knoll et al., (1986). **G** indicates presence of glacial climatic indicators at those times or horizons shown, independent of the carbon isotopic data. Note that physical evidence of glaciation in the Svalbard and Greenland section predates a drop in atmospheric $CO_2$ (climate cooling?) and postdates and continues on through an increase? in atmospheric $CO_2$ (excursions toward lighter organic carbon isotopic values). Note also that the magnitude of the carbon shifts from Svalbard and East Greenland, are roughly, internally equivalent, as is the case for carbon data from Morocco and the Yangtze Platform, but are not always associated with climate change. (From Aharon, P., Schidlowski, M., and Singh, I., *Nature,* 327, 700, 1987; Knoll, A., Hayes, J., Kaufman, A., Swett, K., and Lambert, I., *Nature,* 321, 834, 1986; Lambert, I., Walter, M., Wenlong, Y., Songnian, L., and Guogan, M., *Nature,* 325, 141, 1987; Margaritz, M., Holser, W., and Kirschvink, J., *Nature,* 320, 259, 1986; and Tucker, M., *Nature,* 319, 49, 1986. With permission.)

atmospheric $CO_2$, suggesting that worldwide continental glaciers are not implicitly forced by biotic radiations, nor carbon alone, for that matter.

Less well calibrated records of carbon from Morocco (Tucker, 1986) and the Yangtze Platform (Lambert et al., 1987; refer to Figure 10), approximately spanning the same interval of time, exhibit patterns of carbon records similar to those cited above. Moreover, some biotic radiations occurred during intervals of Earth history devoid of any record of glaciation whatsoever (Dao-Yi et al., 1986; Perch-Nielsen et al., 1980; Holser and Margaritz, 1985).

However, the lack of chronologic resolution of these changes so far back in time, pose serious problems to resolution of these and other questions relating to life and climate. This lack of chronologic resolution may also result in an incomplete and inaccurate picture of the actual sequence of events having occurred.

## G. CLOUD CONDENSATION EFFECTS, CLOUD ALBEDO, AND DIMETHYLSULFIDE

Bates et al. (1987), Charlson et al. (1987; 1989), Shaw (1983), and Lovelock et al. (1972) first proposed that most species of phytoplankton in the ocean excrete dimethylsulphide (DMS) which escapes into the atmosphere where it becomes oxidized, to form a sulphate and methane sulphonate (MSA) aerosol, which in turn serves as cloud-condensation nuclei (CCN). Thus, the biosphere may exert yet another influence on climate through cloud production in the open ocean, resulting in periodic(?) cooling at a variety of possible scales of operation and impact.

Particulate carbon released during biomass burning, as is presently the case in the instance of the burning of tropical forests, also serves as cloud-forming nuclei (Crutzen and Andreae, 1990), analagous to the role of sulfur compounds in forming CCN. On a somewhat longer timescale, a biotically regulated, just slightly more oxidative (carbon-depleted) atmosphere than we have at present (see Lovelock 1987) could easily become ignitive, leading to the natural occurrence of widespread wildfires resulting in, among other things, more cloud-seeding, hence, more clouds, thus greater cloud albedo (reflectivity), and thus, increased, or a more widespread distribution of climate-cooling phenomena of presently indeterminate duration and extent or scale.

This is a particularly real possibility in the long run (several 10s of million years down the road), insofar as organic carbon burial may, in fact, be responsible for the Earth having maintained relatively constant, globally averaged surface temperatures in the face of a 30% increase in solar luminosity since the birth of the planet (Lovelock and Whitfield, 1982; Worsley and Nance, 1989). However, I wish to underscore the fact that the burning of biomass/forests (anthropogenic or as a result of natural wildfires due to an ignitive, $O_2$-rich atmosphere) produces clouds in terrestrial regions, while cloud formation by way of DMS is an oceanic phenomenon, thus extending the potential area influenced by the albedo effect of clouds. Furthermore, even though biomass burning results in the production of greenhouse gases as well (Crutzen and Andreae, 1990), in the presence of clouds, any potential greenhouse phenomena in the pipe-line might go undetected for a presently indeterminate period of time due to oceanic thermal inertia, in part (slow response to and regulation of climate by the slow warming of the ocean).

Insofar as cloud albedo (in the case of sulfur) appears to be responsive to CCN density, Wigley (1989), Bates et al. (1987), and Charlson et al. (1987) have proposed that climate may be biotically regulated by phytoplankton, and therefore, serve as a negative feedback to temperature and sunlight, through the emission of DMS, a gaseous precursor of CCN. In places where the ocean is warmest, most saline, and most highly illuminated, the production of DMS (and hence, CCN and cloud albedo) is high, resulting in a reduction in incoming solar radiation and climate cooling (Bates et al., 1987; Charlson et al., 1987).

In the context of glacials and interglacials (short term global warming and global cooling events) it is difficult to evaluate the effects of DMS-related cloud albedo because Milankovitch-type orbital perturbations bring about a change in the distribution of solar radiation over the surface of the earth, which may simply result in the shifting of sites of high vs. low incoming solar radiation with no net change, globally, in the density of CCN and hence cloud albedo. The result in this instance might be no climate regulation via the DMS derived from phytoplankton (see also Schwartz, 1988).

If however, in keeping with the arguments set forth by Rampino and Volk (1988) regarding climatic feedbacks during the K/T (Cretaceous/Tertiary = 65 million years before present) transition, productivity and hence, biomass were to increase during glacials, one might then expect a greater output of DMS and hence, greater cloud albedo, which would serve as a positive feedback to climate cooling, contrary to what one might otherwise have anticipated.

If, on the other hand, DMS production is not directly linked to changes in productivity and hence, changes in biomass, as suggested by Charlson et al. (1987), then orbitally forced climate cooling and the alleged latitudinal shrinking of the tropics (regions of high temperature and high incoming solar radiation) during glacials (see Shackleton, 1977), would likely result in reduced production of DMS and hence, reduced CCN and cloud albedo, and increased warming. In this instance a reduction in cloud albedo would serve as a negative feedback (warming) to orbitally forced cooling.

Alternatively, if DMS production is primarily required for osmotic fluid regulation within phytoplankton, in the face of changing oceanic salinities (Caldeira, 1989), as opposed to serving primarily as a biotic mechanism for climate modulation and ''altruism'' as originally suggested by Charlson et al. (1987), then during glacials when the salinity increases (Thunell and Williams, 1989; Thunell et al., 1988; Broecker, 1982a; 1982b) phytoplankton should require more DMS for osmoregulation. Consequently, upon breakage of the phytoplanktonic tests and the release of DMS to the atmosphere, the net effect should be an increase in cloud albedo and hence, a positive feedback to glaciation, presuming of course that there has been no reduction in biomass which might lead to an entirely different outcome.

Recent ice-core records of atmospheric sulfur and DMS (Legrand et al., 1991) spanning the last 160,000 years of glaciation and deglaciation (short-term global warming and cooling) indicate that DMS production was significantly higher during periods of glaciation or short-term global cooling (additional climate cooling over the glacial oceans? — a positive feedback to global cooling), consistent with Caldeira's (1989) account of the use of sulfur by phytoplankton for osmoregulation, which would be enhanced in a higher-salinity ocean such as during a period of continental glaciation/global cooling (see Thunell and Williams, 1989; Thunell et al., 1988; Broecker, 1982a; 1982b). Thus, it would presently appear that the biotic production of DMS over the oceans is itself a feedback to climate insofar as changes in ocean salinity are feedbacks to climate. Yet insofar as DMS seeds clouds, the biosphere can also be said to be exerting its own influence on climate, thus a feedback to climate.

Finally, contrary to much of the above, with the exception of the work of Caldeira (1989), in attempting to test the hypothesis of Charlson et al. (1987) by investigating the climatic response to the present input of anthropogenic $SO_2$, Schwartz (1988) has determined that temperature records for the past 100 years yield no evidence of any changes in cloud albedo and hence, no evidence of any feedback on climate. However, Schwartz's conclusions appear to have been vigorously contested (see Henderson-Sellers and McGuffie, 1989; Charlson et al., 1987; Gavin et al., 1989; Ghan et al., 1989).

Ironically, however, assuming that there is a relationship between cloud albedo and DMS, efforts to reduce present anthropogenic levels of atmospheric $SO_2$ might serve to accelerate the rate of greenhouse gas-induced global warming since the two gases have an inverse on climate (Wigley, 1989; 1991). However, although these two gases have an inverse influence on climate, sulfur (insofar as it leads to the production of clouds which, in turn, reflect sunlight and thus cool that which lies immediately below) preempts for a time, or temporarily masks, the warming effects of any greenhouse gases, locally and/or regionally. Sulfur dioxide and DMS have no greenhouse properties, unlike $CO_2$, $CH_4$, and others. Furthermore, Charlson et al. (1991) have recently shown that any climate cooling effects of sulfur are regional or local at best, and will not therefore cancel any global warming due to the presence of greenhouse gases. Thus, sulfur compounds might have a temporary cooling effect upon climate in places, but is unlikely to have a global influence. Furthermore, sulfur and DMS have short residence times in the atmosphere, particularly when compared to the residence times of the Earth's principal greenhouse gases.

Finally, given that the life cycle of oceanic phytoplankton (and plankton in general) is on the order of days, the longer-term sulfur and DMS fluxes observed in ice-core records (Legrand et al., 1991) are not readily, nor altogether understandable. It's climatic role clearly requires a great deal more investigation and thinking. The reported trends in DMS from the ice-core records cited above only presently make any sense climatically and biologically, if the additional reservoir of DMS in the glacial atmosphere represents a reservoir of sulfur compounds once present, and therefore, part of the oceanic biosphere prior to the onset of global cooling and glaciation.

It is tempting to suggest that the high levels of DMS in the glacial atmosphere represent that portion of the oceanic biosphere lost during the transition from global warming to global cooling. Why DMS would remain in the atmosphere for so long is still, however, not clear. Perhaps in the absence or near absence of water vapor/precipitation in the glacial atmosphere (much of which has presumably gone into the production of continental glaciers over 2 mi thick), sulfur compounds and DMS have a longer residence time. This would also have implications regarding the issue of whether or not oceanic productivity increases or decreases in a glacial ocean (time of global cooling). The argument cited above would implicitly argue for reduced primary productivity during times of continental glaciation (short-term global cooling).

In addition, high levels of DMS in the glacial atmosphere might arguably reflect a lack of cloud formation at this time, particularly insofar as water vapor might be more scarce in a cooler world.

## VI. CONCLUSIONS

Climate and climate change are not discrete, unconnected happenstance in Earth history, but rather are a continuum of preconditions, controls, and feedbacks to the interactions among the astronomical (orbitally forced changes in the distribution of solar isolation), tectonic (supercontinent cycles driven by mantle convection and heat flow, and in turn, driving mantle convection), and chemical and biogeochemical systems (linked abiotic and biotic cycling of nutrients, other metabolically important isotopes such as sulfur and oxygen, and nonmetabolic isotopes such as strontium), which collectively govern the sum and substance of this planet and its mode(s) of operation. Nearly all of these systems are periodic, occasionally chaotic (having suggested degrees of chaotic behavior which itself may be periodic as well), while some systems also attenuate over time, thus altering their periodicity, perhaps in a predictable manner (e.g., supercontinent cycle and long-term global warming and cooling) since heat is continually lost to the atmosphere from the interior of the Earth, while the volume of radioactive elements present within the Earth (the source of much of this heat) decreases over time.

Long-term global cooling and global warming events appear to be responses or feedbacks to a suite of tectonic events, many of which result in an exchange of atmospheric $CO_2$, at tectonic rates on the order of 10s of millions to 100s of millions of years.

Short-term, Milankovitch-band global warming and cooling events (synonymous with glaciation and deglaciation) occur at times of supercontinent assembly (time of long-term global cooling), when sea level and atmospheric $CO_2$ are already tectonically low. During supercontinent breakup and dispersal (times of circumstances of long-term global warming), orbital forcing of short-term global warming and cooling events (high-frequency glacials and interglacials) seems to be damped by tectonically induced high levels of atmospheric $CO_2$, except perhaps in polar regions and/or highly elevated regions, thus resulting in significantly higher, globally averaged, Earth surface temperatures as in the Cretaceous when temperatures (at least according to some estimates (Budyko and Ronov, 1979), were thought to have been 10 to 14°C higher than at present.

This is not to say, however, that orbital forcing at Milankovitch-band rates was not in operation at these times, or that orbital forcing ceased to have an influence on earth systems and processes, only one of which is the growth and decay of continental glaciers (see Herbert and Fischer, 1986).

Orbital-forcing of climate is a nonlinear process, often involving a multiplicity of linked mechanisms which appear to serve as feedbacks to the original forcing. Many of these intermediate feedback mechanisms remain unknown in detail. However, since most responses to the original orbital forcing of climate occur at rates far in excess of Milankovitch-band, orbitally defined rates, seemingly by orders of magnitude (e.g., $CO_2$ variations between a transition from global cooling to global warming, as well as changes in ocean circulation rates and changes in the chemical structuring of oceans [see Boyle, 1988a; 1988b; Siegenthaler and Wenk, 1984]), orbitally forced climate events are nonlinear. Thus, orbitally forced climate events often seem to play a major role in climate change even on the scale of a human lifetime, decades to centuries, in addition to as opposed to scales on the order of 10,000 to 100,000 years, the rates of orbital forcing of solar radiation, without intermediate feedbacks which ultimately translates orbital forcing (the flow of energy set in motion) into climate and other natural phenomena.

The composition of the atmosphere and therefore climate, is influenced and regulated by life which in turn is influenced by climate itself (is a feedback to climate in some instances [Smith, 1976]), and the composition of the atmosphere at any one time in the history of this planet, and presumably, it's future as well.

A relatively constant surface temperature has arguably been maintained throughout the life of this planet, in the face of a 30% increase in solar luminosity (Sagan and Mullen's "Dim-Young Sun" paradox), conceivably by way of an incremental removal of carbon from the atmosphere by the biosphere. Should this trend continue, as appears imminent, the planet may become more ignitive and prone to wildfires, in the presence of a slightly more oxygen-rich atmosphere (see Lovelock, 1987). This may be particularly worrisome in the long-term (hundreds of thousands to millions of years from now), given that the Earth's atmosphere presently contains only 0.035% $CO_2$ by volume, relative to other gases present (Lovelock, 1987; see also Figure 1).

In light of all of the above, in addition to the fact that human kind has almost assuredly already programmed into the pipeline (passed along to the future generations of human and other life systems), so to speak, conceivably as much as, or more than 300,000 years of human-induced climate change (global warming), with respect to $CO_2$ alone (refer to Kasting and Walker, 1991), in addition to the compound environmental degradation and resource-depletion observable virtually everywhere on this planet, in particular, the life-modulating resources of carbon and phosphorous.

In addition, there are at least several well-documented cases of rapid climate changes on this planet (see Broecker and Denton, 1989; Siegenthaler and Wenk, 1984; Sarmiento and Toggweiler, 1984; Boyle, 1988b), having taken place on a timescale of tens to hundreds to thousands of years, resulting in wholescale, sweeping changes in climate, resources and life itself, the impetus for migration, for example. In each of these case histories of climate change involving nearly instantaneous changes in natural systems, the system or systems in question appeared to approach certain natural thresholds of behavior, setting in motion an immediate response or reorganization, changing the world considerably, in the process. In addition, despite recent sweeping advances in understanding the Earth's record of climate, there is still much uncertainty as to how systems work and interact with other systems, in detail, on very fundamental levels.

Consequently, in my estimation, this society should take immediate, concrete measures to: (1) conserve and recycle resources; (2) cease charging the atmosphere with carbon dioxide and other greenhouse gases and pollutants, especially those which lead to depletion of the ozone layer; (3) embark on a vigorous program to changeover from a carbon-based system of energy to a non-carbon-based system of renewable energy; (4) restore presently degraded elements of our environment; and (5) implement a national and global environmental/resource management program for sustaining resources and life itself, in the long- and short-term.

For the moment, society has shortsightedly and previously unknowingly perhaps, embarked upon what can only be described as the single most dangerous, uncontrolled experiment in the history of human civilization, for which there is no historic counterpart.

## ACKNOWLEDGMENTS

I wish to acknowledge the following individuals for their comments, criticisms, suggestions, insights, and willingness on many occasions to engage in lengthy discussions which were invariably stimulating: John Crowell, Larry Frakes, Eric Barron, Poppe de Boer, Peter Worsley, Andrew Knoll, Bruce Smith, Nicholas Eyles, Alfred Morner, and many others too numerous to name, but whose assistance is not forgotten.

Lastly, I wish to give credit to the general body of knowledge which has been passed along from generation to generation, without which the task at hand would have been impossible.

## REFERENCES

Aharon, P., Schidlowski, M., and Singh, I. B., Chronostratigraphic markers in the end-Precambrian carbon isotope record of the Lesser Himalaya, *Nature (London)*, 327, 699–702, 1987.

Algeo, T. J. and Wilkinson, B. H., Periodicity of mesoscale Phanerozoic sedimentary cycles and the role of Milankovitch orbital modulation, *J. Geol.*, 96, 312–322, 1988.

Allen, C. W., *Astrophysical Quantities*, 3rd ed., University of London, Athlone Press, 1973, 112–117.

Anderson, D. L. *Theory of the Earth*, Blackwell Scientific, Oxford, 1989, 239–257.

Anderson, D. L., The Earth as a planet: paradigms and paradoxes, *Science*, 223, 347–355, 1984.

Anderson, D. L., Hotspots, polar wander, Mesozoic convection and the geoid, *Nature (London)*, 297, 391–393, 1982.

Anderson, D. L., Hotspots, basalts, and the evolution of the mantle, *Science*, 213, 83–89, 1981.

Anderson, R. Y., Orbital forcing of evaporite sedimentation, in *Milankovitch and Climate*, Part 1, Berger, A., Imbrie, J., Hays, J., Kukla, G., and Saltzman, B., Eds., D. Reidel Publishing, Boston, MA, 1984, 147–162.

Anderson, R., A long record from the Permian, *J. Geophys. Res.*, 87, 7285–7294, 1982.

Arthur, M. A., The carbon cycle-controls on atmospheric CO-2 and climate in the geologic past, in *Climate in Earth History*, Geophysics Study Committee, Eds., National Acadmy of Sciences, Washington, D.C., 1982, 55–67.

Arthur, M. A. and Jenkyns, H. C., Phosphorites and paleoceanography, in *Ocean Geochemical Cycles*, Berger, W. H., Ed., Special Suppl., *Oceanol. Acta*, 4, 83–96, 1981.

Arthur, M. A., Dean, W. E., Bottjer, D., and Scholle, P. A., Rhythmic bedding in Mesozoic-Cenozoic pelagic carbonate sequences: the primary and disgenetic origin of Milankovitch-like cycles, in *Milankovitch and Climate*, Part I, Berger, A., Imbrie, J., Hays, J., Kukla, G., and Saltzman, B., Eds., Reidel Publishing, Boston, MA, 1984, 191–222.

Arthur, M. A. and Garrison, R. E., Eds., Special section on Milankovitch cycles through geologic time, *Paleoceanography*, 1, 369–586, 1986.

Barnola, J. M., Raynaud, Y. S., Korotkevich, Y. S., and Lorius, C., Vostok ice core provides 160,000-year record of atmospheric CO-2, *Nature (London)*, 329, 408–414, 1987.

Barron, E. J., Arthur, M. A., and Kauffman, E. G., Cretaceous rhythmic bedding sequences: a plausible link between orbital variations and climate, *Earth Planet. Sci. Lett.*, 72, 327–340, 1985.

Bates, T. S., Charlson, R. J., and Gammon, R. H., Evidence for the climatic role of marine biogenic sulphur, *Nature (London)*, 329, 319–321, 1987.

Beget, J. E. and Hawkins, D. B., Influence of orbital parameters on Pleistocene loess deposition in central Alaska, *Nature (London)*, 337, 151–153, 1989.

Berger, A., Milankovitch theory and climate, *Rev. Geophys.*, 26, 624–657, 1988.

Berger, W. H. and Winterer, E. L., Plate stratigraphy and the fluctuating carbonate line, in *Pelagic Sediments: On Land and Under the Sea*, Spec. Publ. Int. Assoc. Sedimentol., Hsu, K. J. and Jenkyns, H. C., Eds., Blackwell Scientific, Oxford, 1974, 11–98.

Berger, A., Loutre, M. F., and Dehant, V., Pre-quaternary Milankovitch frequencies, *Nature (London)*, 342, 133, 1989.

Berger, W. H., Burke, S., and Vincent, E., Glacial-holocene transition: climate pulsations and sporadic shutdown of NADW production, in *Abrupt Climatic Change*, Berger, W. H. and Labeyrie, L. D., D. Reidel Publishing, Boston, MA, 1987, 279–297.

Berner, R. A., Atmospheric carbon dioxide levels over Phanerozoic time, *Science*, 249, 1382–1386, 1990.

Berner, R. A., Models for carbon and sulfur cycles and atmospheric oxygen: application to Paleozoic geologic history, *Am. J. Sci.*, 287, 177–196, 1987.

Berner, R. A. and Lasaga, A. C., Modeling the geochemical carbon cycle, *Sci. Am.*, 260, p. 74–81, 1989.

Birchfield, G. E., Weertman, J., and Lunde, A. T., A paleoclimate model of northern hemisphere ice sheets, *Quat. Res. (NY)*, 15, 126–142, 1981.

Bond, G. C., Nickeson, P. A., and Kominz, M. A., Breakup of a supercontinent between 625 Ma and 555 Ma: new evidence and implications for continental histories, *Earth Planet. Sci. Lett.*, 70, 325–345, 1984.

Boyle, E. A., Quaternary deepweater oceanography, *Science*, 249, 863–870, 1990.

Boyle, E. A., Vertical oceanic nutrient fractionation and glacial/interglacial $CO_2$ cycles, *Nature (London)*, 331, 55–56, 1988a.

Boyle, E. A., The role of vertical chemical fractionation in controlling late Quaternary atmospheric carbon dioxide, *J. Geophys. Res.*, 93, 701–715, 1988b.

Broecker, W. S., Terminations, in *Milankovitch And Climate*, Part 2, Berger, A., Imbrie, J., Hays, J., Kukla, G., and Saltzman, B., Eds., D. Reidel Publishing, Boston, MA, 1984, 687–698.

Broecker, W. S., Glacial to interglacial changes in ocean chemistry, *Prog. Oceanogr.*, 11, 151–197, 1982a.

Broecker, W. S., Ocean chemistry during glacial time, *Geochim. Cosmochim. Acta*, 46, 1689–1705, 1982b.

Broecker, W. S., Absolute dating in the astonomical theory of glaciation, *Science*, 151, 299–304, 1966.

Broecker, W. S. and Denton, G. H., The role of ocean-atmosphere reorganizations in glacial cycles, *Geochim. Cosmochim. Acta*, 53, 2465–2501, 1989.

Broecker, W. S. and Peng, T. H., *Tracers in the Sea*, Columbia University, New York, 1982, 690p.

Broecker, W. S. and Peng, T. H., The cause of the glacial to interglacial atmospheric $CO_2$ change: a polar alkalinity hypothesis, *Global Biogeochem. Cycles*, 3, 215–239, 1989.

Broecker, W. S., Peteet, D. M., and Rind, D., Does the ocean-atmosphere have more than one stable mode of operation?, *Nature (London)*, 315, 21–26, 1985.

Broecker, W. S., Thurber, D. L., Goddard, J., Teh-lung, Ku, Matthews, R. K., and Mesolella, K. J., Milankovitch hypothesis supported by precise dating of coral reefs and deep-deep sediments, *Science*, 159, 297–300, 1987.

Broecker, W. S. and van Donk, J., Insolation changes, ice volumes, and the O-18 record in deep-sea cores, *Rev. Geophys. Space Phys.*, 8, 169–197, 1970.

Brouwer, A., Global controls in the Phanerozoic sedimentary record, *Stratigr. Newsl.*, 12, 166–174, 1983.

Budyko, M. I. and Ronov, A. B., Chemical evolution of the atmosphere in the Phanerozoic, *Geochem. Int.* (transl. from *Geokhimiya*, 5), 1–9, 643–653, 1979.

Caldeira, K., Evolutionary pressures on planktonic production of atmospheric sulphur, *Nature (London)*, 337, 732–734, 1979.

Chappell, J., Relationships between sea levels, O-18 variations and orbital perturbations, during the past 250,000 years, *Nature (London)*, 252, 199–202, 1974.

Charlson, R. J., Langner, J., Rodhe, H., Leovy, C. B., and Warren, S. G., Perturbation of the northern hemisphere radiative balance by backscattering from the anthropogenic sulfate aerosols, *Tellus*, 43(AB), 152–163, 1991.

Charlson, R. J., Lovelock, J. E., Andreae, M. O., and Warren, S. G., Sulphate aerosols and climate, *Nature (London)*, 340, 437–438, 1989.

Charlson, R. J., Lovelock, J. E., Andreae, M. O., and Warren, S. G., Oceanic phytoplankton, atmospheric sulfur, cloud albedo and climate, *Nature (London)*, 326, 655–661, 1987.

Chumakov, N. M. and Elston, D. P., The paradox of Late Proterozoic glaciations at low latitudes, *Episodes*, 12, 115–120, 1989.

Chyba, C. F., Impact delivery and erosion of planetary oceans in the early inner solar system, *Nature (London)*, 343, 129–133, 1990.

Cloud, P. E., Atmospheric and hydrospheric evolution on the primitive Earth, *Science*, 160, 729–736, 1968.

Cloud, P. E., A working model of the primitive earth, *Am. J. Sci.*, 272, 537–548, 1972.

Cloud, P. E., The biosphere, *Sci. Am.*, 249, 176–189, 1983.

COHMAP Members, Climatic changes of the last 18,000 years: observations and model simulations, *Science*, 241, p. 1043–1052, 1988.

Cook, P. J. and Shergold, J. H., Phosphorous, phosphorites and skeletal evolution at the Precambrian-Cambrian boundary, *Nature (London)*, 308, 231–236, 1984.

Cotillon, P., Bed-scale cyclicity of pelagic Cretaceous successions as a result of world-wide control, *Mar. Geol.*, 78, 109–123, 1987.

Crowell, J. C., Continental glaciation through geologic times, in *Climate in Earth History*, Geophysics Study Committee, Eds., National Academy of Sciences, Washington, D.C., 1982, 77–82.

Crowell, J. C., Gondwanan glaciation, cyclothems, continental positioning, and climate change, *Am. J. Sci.*, 278, 1345–1372, 1978.

Crowley, T. J., The geologic record of climate change, *Rev. Geophys. Space Phys.*, 21, 828–877, 1983.

Crutzen, P. J. and Andreae, M. O., Biomass burning in the tropics: impact on atmospheric chemistry and biogeochemical cycles, *Science*, 250, 1669–1677, 1990.

Dalziel, I. W. D., Pacific margins of Laurentia and East Antarctica-Australia as a conjugate rift pair: evidence and implications for an Eocambrian supercontinent, *Geology*, 19, 598–601, 1991.

Dao-Yi, X., Zheng, Y., Qin-Wen, Z., Zhi-Da, S., Yi-Yin, S., and Lian-Fang, Y., Significance of a C-13 anomaly near the Devonian/Carboniferous boundary at the Muhua section, South China, *Nature (London)*, 321, 854–855, 1986.

deBoer, P. L. and Wonders, A. A., Astronomically induced rhythmic bedding in Cretaceous pelagic sediments near Moria, (Italy), in *Milankovitch and Climate*, Part 1, Berger, A, Imbrie, J., Hays, J., Kukla, G., and Saltzman, B., D. Reidel, Publishing, Boston, MA, 1984, 177–190.

Delmas, R. J., Ascencio, J., and Legrand, M., Polar ice evidence that atmospheric CO-2 20,000 yr BP was 50% of present, *Nature (London)*, 284, 155–157, 1980.

Donn, W. L., Paleoclimate and polar wander, *Palaeogeogr. Palaeoclimatol. Palaeoecol.*, 71, 225–236, 1989.

Druffel, E. R. M. and Benavides, L. M., Input of excess CO-2 to the surface ocean based on C-13/C-12 ratios in a banded Jamaican sclerosponge, *Nature (London)*, 321, 58–61, 1986.

Endal, A. S. and Schatten, K. H., The faint young sun-climate paradox: continental influences, *J. Geophys. Res.*, 87, 7295–7302, 1982.

Engel, A. E. J. and Engel, C. B., Continental accretion and the evolution of North America, in *Advancing Frontiers in Geology and Geophysics*, Subramanian, A. P. and Balakrishna, S., Eds., Indian Geophysical Union, Delhi, 1964, 17–37e.

Fischer, A. G., The two Phanerozoic supercycles, in *Catastrophes in Earth History: The New Uniformitarianism*, Berggren, W. A. and Van Couvering, J. A., Eds., Princeton University Press, Princeton, NJ, 129–150, 1984a.

Fischer, A. G., Biological innovations and the sedimentary record, in *Patterns of Change in Earth Evolution*, Holland, H. D. and Trendall, A. F., Eds., Springer-Verlag, New York, 1984b, 145–157.

Fischer, A. G., Long-term climatic oscillations recorded in stratigraphy, in *Climate in Earth History*, Geophysics Study Committee, Eds., National Academy of Sciences, Washington, D.C., 1982, 97–104.

Fischer, A. G., Gilbert-Bedding rhythms and geochronology, Geol. Soc. Am. Spec. Pap., 183, 93–104, 1980.

Fischer, A. G., The Lofer Cyclothems of the Alpine Triassic, *Kans. Geol. Surv. Bull.*, 169, 107–149, 1964.

Fisher, D., Some more remarks on polar wander, *J. Geophys. Res.*, 79, 4041–4045, 1974.

Frakes, L. A., *Climates Through Geologic Time*, Elsevier, Amsterdam, 1979, 1–310.

Frakes, L. A. and Francis, J. E., A guide to Phanerozoic cold polar climates from high-latitude ice-rafting in the Cretaceous, *Nature (London)*, 333, 547–549, 1988.

Friis-Christensen, E. and Lassen, K., Length of the solar cycle: an indicator of solar activity closely associated with climate, *Science*, 254, 698–700, 1991.

Gaffin, S., Ridge volume dependence on seafloor generation rate and inversion using long term sea level change, *Am. J. Sci.*, 287, 596–611, 1987.

Gammon, R. H., Sundquist, E. T., and Fraser, P. J., History of carbon dioxide in the atmosphere, in *Atmospheric Carbon Dioxide and the Global Carbon Cycle*, Trabalka, J. R., Ed., Office of Energy Research, U.S. Department of Energy, Washington, D.C., 1985, 25–62.

Gamundi, O. L., Postglacial transgressions in late Paleozoic basins of western Argentina: a record of glacioeustatic sea level rise, *Palaeogeogr. Palaeoclimatol. Palaeoecol.*, 71, 257–270, 1989.

GARP, *Understanding Climatic Change: A Program for Action*, National Academy of Sciences, Washington, D.C., 1975.

Gavin, J., Kukla, G., and Karl. T., Sulphate aerosols and climate, *Nature (London)*, 340, 438, 1989.

Genthon, C., Barnola, J. M., Raynaud, D., Lorius, C., Jouzel, J., Barkov, N. I., Korotkevich, Y. S., and Kotlyakov, V. M., Vostok ice core: climatic response to CO-2 and orbital forcing changes over the last climatic cycle, *Nature (London)*, 329, 414–418, 1987.

Ghan, S. J., Penner, J. E., and Taylor, K. E., Sulphate aerosols and climate, *Nature (London)*, 340, 438, 1989.

Goldhammer, R., Dunn, P., and Hardie, L., High frequency glacio-eustatic sealevel oscillations with Milankovitch characteristics recorded in middle Triassic platform carbonates in northern Italy, *Am. J. Sci.*, 287, 853–892, 1987.

Goldreich, P. and Toomre, A., Some remarks on polar wandering, *J. Geophys. Res.*, 74, 2555–2567, 1969.

Goodfriend, G. A. and Magaritz, M., Paleosols and late Pleistocene rainfall fluctuations in the Negev Desert, *Nature (London)*, 332, 144–146, 1988.

Goodwin, A. M., Precambrian perspectives, *Science*, 213, 55–61, 1981.

Goodwin, P. W. and Anderson, D. L., Punctuated aggradational cycles: a general hypothesis of episodic stratigraphic accumulation, *J. Geol.*, 93, 515–534, 1985.

Gonzalez, C. R., Development of the late Paleozoic glaciations of the South American Gondwana in western Argentina, *Palaeogeogr. Palaeoclimatol. Palaeoecol.*, 79, 275–287, 1990.

Grotzinger, J. P., Upward-shallowing platform cycles: a response to 2.2 billion years of low-amplitude, high-frequency (Milankovitch band) sea level oscillations, *Paleoceanography*, 1, 403–416, 1986.

Hansen, J. E. and Lacis, A. A., Sun and dust versus greenhouse gases: an assessment of their relative roles in global climate change, *Nature (London)*, 346, 713–719, 1990.

Hargraves, R. B., Precambrian geologic history, *Science*, 193, 363–371, 1976.

Harland, W. B., The Proterozoic glacial record, in *Proterozoic Geology*, Select. Pap. Int. Proterozoic Symp., Medaris, L. G., Jr., Ed., Geological Society of America Memoir, 161, 279–288, 1983.

Harland, W. B., Chronology of Earth's glacial and tectonic record, *J. Geol. Soc. London*, 138, 197–203, 1981.

Hays, J. D. and Pittman, W. C., III, Lithospheric plate motion, sea-level changes and climatic and ecological consequences, *Nature (London)*, 246, 18–22, 1973.

Hays, J. D., Imbrie, J., and Shackleton, N. J., Variations in the Earth's orbit: pacemaker of the ice ages, *Science*, 194, 1121–1132, 1976.

Heckel, P. H., Sea-level curve for Pennsylvanian eustatic marine transgressive-regressive depositional cycles along midcontinent outcrop belt, North America, *Geology*, 14, 330–334, 1986.

Henderson-Sellers, A. and McGuffie, K., Sulphate aerosols and climate, *Nature (London)*, 340, 436–437, 1989.

Henderson-Sellers, A. and Schwartz, A., Chemical evolution and ammonia in the early Earth's atmosphere, *Nature (London)*, 287, 526–528, 1980.

Herbert, T. D. and Fischer, A. G., Milankovitch climatic origin of mid-Cretaceous black shale rhythms in central Italy, *Nature (London)*, 321, 739–743, 1986.

Hobgood, J. S. and Cerveny, R. S., Ice-age hurricanes and tropical storms, *Nature (London)*, 333, 243–245, 1988.

Holland, H. D., *The Chemical Evolution of the Atmosphere and Oceans*, Princeton University Press, Princeton, NJ, 1874, 29–127 & 441–551.

Holland, H. D., Lazar, B., and McCaffrey, M., Evolution of the atmosphere and oceans, *Science*, 320, 27–33, 1986.

Holser, W. T. and Magaritz, M., The Late Permian carbon isotope anomaly in the Bellerophon Basin, Carnic and Dolomite Alps, *Jahrb. Geol. Bundesanst. Wein*, 128, 75–82, 1985.

House, M. R., A new approach to an absolute timescale from measurements of orbital cycles and sedimentary microrhythms, *Nature (London)*, 315, 721–725, 1985.

Hunt, B. G., The effects of the past variations of the Earth's rotation rate on climate, *Nature (London)*, 281, 188–191, 1979.

Imbrie, J., Abrupt terminations of late Pleistocene ice ages: a simple Milankovitch explanation, in *Abrupt Climatic Change*, Berger, W. H. and Labeyrie, L. D., Eds., D. Reidel Publishing, Boston, MA, 1987, 365–368.

Imbrie, J. and Imbrie, J., Modeling the climatic response to orbital variations, *Science*, 207, 943–953, 1980.

Imbrie, J., Hays, J. D., Martinson, D. G., McIntyre, A., Mix, A. C., Morley, J. J., Pisias, N. G., Prell, W. L., and Shackleton, N. J., The orbital theory of Pleistocene climate: support from a revised chronology of the marine O-18 record, in *Milankovitch and Climate*, Part 1, Berger, A., Imbrie, J., Hays, J., Kukla, G., and Saltzman, B., Eds., D. Reidel Publishing, Boston, MA, 1984, 269–305.

Janecek, T. R. and Rea, D. K., Pleistocene fluctuations in northern hemisphere tradewinds and westerlies, in *Milankovitch and Climate*, Part 1, Berger, A., Imbrie, J., Hays, J., Kukla, G., and Saltzman, B., Eds., D. Reidel Publishing, Boston, MA, 1984, 331–347.

Jones, P. D., Wigley, T. M. L., and Raper, S. C. B., The rapidity of $CO_2$-induced climatic change: observations, model results and paleoclimatic implications, in *Abrupt Climatic Change*, Berger, W. H. and Labeyrie, L. D., Eds., D. Reidel Publishing, Boston, MA, 1987, 47–55.

Jouzel, J., Lorius, C., Petit, J. R., Genthon, C., Barkov, N. I., Kotlyakov, V. M., and Petrov, V. M., Vostok ice core: a continuous isotope temperature record over the last climatic cycle (160,000 years) *Nature (London)*, 329, 403–408, 1987.

Kasting, J. F., Impacts and the origin of life, *Earth and Miner. Sci.*, 59, 37–42, 1990.

Kasting, J. F., Theoretical constraints on oxygen and carbon dioxide concentrations in the Precambrian atmosphere, *Precambrian Res.*, 34, 205–229, 1987.

Kasting, J. F. and Ackerman, T. P., Climatic consequences of very high carbon dioxide levels in the Earth's early atmosphere, *Science*, 234, 1383–1391, 1986.

Kasting, J. F. and Walker, J. C. G., The geochemical carbon cycle and the uptake of fossil fuel CO-2, *Am. Phys. Soc. Symp. Global Warm.*, in press.

Klein, G. D. and Willard, D. A., Origin of the Pennsylvanian coal-bearing cyclothems of North America, *Geology*, 17, p. 152–155, 1989.

Knoll, A. H., Hayes, J. M., Kaufman, A. J., Swett, K., and Lambert, I. B., Secular variation in carbon isotope ratios of Upper Proterozoic successions of Svalbard and East Greenland, *Nature (London)*, 321, 832–838, 1986.

Knox, F. and McElroy, M., Changes in atmospheric CO-2: influence of the marine biota at high latitude, *J. Geophys. Res.*, 89, 4629–4637, 1984.

Kominz, M. A., Development of a Detailed Chronology and Determination of the Major Element Geochemistry of Equatorial Pacific Core V 28-238, M.S. thesis, University of Rhode Island School of Oceanography, Kingston, RI, 1974.

Kominz, M. A. and Pisias, N. G., Pleistocene climate: deterministic or stochastic, *Science*, 204, 171–173, 1979.

Kutzbach, J. E., The nature of climate and climatic variations, *Quat. Res. (NY)*, 6, 471–480, 1976.

Kutzbach, J. E. and Guetter, P. J., The sensitivity of monsoon climates to orbital parameter changes for 9,000 years BP: experiments with the NCAR general circulation model, in *Milankovitch and Climate*, Part 2, Berger, A., Imbrie, J., Hays, J., Kukla, G., and Saltzman, B., Eds., D. Reidel Publishing, Boston, MA, 1984, 801–820.

Kutzbach, J. E. and Otto-Bliesner, B. L., The sensitivity of the African-Asian monsoonal climate to orbital parameter changes for 9,000 years BP in a low resolution general circulation model, *J. Atmos. Sci.*, 39, 1177–1188, 1982.

Laferriere, A. P., Hattin, D. E., and Archer, A. W., Effects of climate, tectonics, and sea-level changes on bedding patterns in the Niobrara Formation (Upper Cretaceous), U.S. western interior, *Geology*, 15, 233–236, 1987.

Lambert, I. B., Walter, M. R., Wenlong, Y., Songnian, L., and Guogan, M., Palaeoenvironment and carbon isotope stratigraphy of Upper Proterozoic carbonates of the Yangtze Platform, *Nature (London)*, 325, 140–142, 1987.

Lashov, D. A. and Tirpak, D. A., Eds., The greenhouse gases, in *Policy Options for Stabilizing Global Climate*, Draft Rep. Congr., Exec. Summ., p. I-8, I-13, 1989.

Laskar, J., A numerical experiment on the chaotic behavior of the solar system, *Nature (London)*, 338, 237–238, 1989.

Legrand, M., Feniet-Saigne, C., Saltzman, E. S., Germain, C., Barkov, N. I., and Petrov, V. N., Ice-core record of oceanic dimethylsulphide during the last climate cycle, *Nature (London)*, 350, 144–146, 1991.

Legrand, M. and Delmas, R. J., Environmental changes during last deglaciation inferred from chemical analysis of the Dome C ice core, in *Abrupt Climatic Change*, Berger, W. H. and Labeyrie, L. D., Eds., D. Reidel Publishing, Boston, MA, 1987, 247–259.

Legrand, M. and Delmas, R. J., The ionic balance of Antarctic snow: a 10 yr. detailed record, *Atmos. Environ.*, 18, 1867–1876, 1984.

Lorius, C., Jouzel, J., Ritz, C., Merlivat, L., Barkov, N. I., Korotkevich, Y. S., and Kotlyakov, V. M., A 150,000-year climatic record from Antarctic ice, *Nature (London)*, 316, 591–596, 1985.

Lovelock, J. E., Gaia, in *A New Look at Life on Earth,* Oxford University Press, New York, 1987, 1–157.
Lovelock, J. E. and Whitfield, M., Life span of the biosphere, *Nature (London),* 296, 561–563, 1982.
Lovelock, J. E. and Margulis, L., Atmospheric homeostasis by and for the biosphere: the gaia hypothesis, *Tellus,* 26, 2–10, 1974.
Lovelock, J. E., Maggs, J., and Rasmussen, R. A., Atmospheric dimethylsulphide and the natural sulfur cycle, *Nature (London),* 237, 452–453, 1972.
Lyle, M., Climatically forced organic carbon burial in equatorial Atlantic and Pacific Oceans, *Nature (London),* 335, 529–532, 1988.
MacDonald, G. D., Role of methane clathrates in past and future climates, *Climat. Change,* 16, 247–281, 1990.
Margaritz, M., Holser, W. T., and Kirschvink, J. L., Carbon-isotope events across the Precambrian/Cambrian boundary on the Siberian Platform, *Nature (London),* 320, 258–259, 1986.
Manabe, S. and Bryan, K., CO-2-induced change in a coupled ocean-atmosphere model and its paleoclimatic implications, *J. Geophys. Res.,* 90, 11,689–11,707, 1985.
Martin, J. H. and Fitzwater, S. E., Iron deficiency limits phytoplankton growth in the north-east Pacific subarctic, *Nature (London),* 331, 341–343, 1988.
McCrea, W. H., Ice ages and the galaxy, *Nature (London),* 255, 607–609, 1975.
McKenzie, J. A. and Eberli, G. P., *Abrupt Climatic Change,* Berger, W. H. and Labeyrie, L. D., Eds., D. Reidel Publishing, Boston, MA, 1987, 127–136.
Mesolella, K. J., Matthews, R. K., Broecker, W. S., and Thurber, D. L., The astronomical theory of climatic change: Barbados data, *J. Geol.,* 77, 250–274, 1969.
Mitchell, J. M., An overview of climatic variability and its causal mechanisms, *Quat. Res. (NY),* 6, 481–493, 1976.
Mix, A. C., Influence of productivity variations on long-term atmospheric $CO_2$, *Nature (London),* 337, 541–546, 1989.
Moores, E. M., Southwest U.S.-East Antarctica (SWEAT) connection: a hypothesis, *Geology,* 19, 425–428, 1991.
Monninger, W., The section of Tiout (Precambrian/Cambrian Boundary Beds, Anti-Atlas, Morocco): An Environmental Model, Doctoral dissertation, Julius-Maximilians-Universitat, Wurzburg, Germany, 1979, 197.
Morel, P. and Irving, E., Tentative paleocontinental maps for the early Phanerozoic and Proterozoic, *J. Geol.,* 86, 535–561, 1978.
Morris, W. A. and Roy, J. L., Discovery of the Hadrynian polar track and further study of the Grenville problem, *Nature (London),* 266, 689–692, 1977.
Nair, R. R., Ittekkot, V., Manganini, S. J., Ramaswamy, V., Haake, B., Degens, E. T., Desai, B. N., and Honjo, S., Increased particle flux to the deep ocean related to monsoons, *Nature (London),* 338, 749–751, 1989.
Neftel, A., Moore, E., Oeschger, H., and Stauffer, B., Evidence from polar ice cores for the increase in atmospheric $CO_2$ in the past two centuries, *Nature (London),* 315, 45–47, 1985.
Neftel, A., Oeschger, T., Staffelbach, T., and Stauffer, B., $CO_2$ record in the Byrd ice core, 50,000-5,000 years BP, *Nature (London),* 331, 609–611, 1988.
Neftel, A., Oeschger, H., Schwander, J., Stauffer, B., and Zumbrunn, R., Ice core sample measurements give atmospheric $CO_2$ content during the past 40,000 yr, *Nature (London),* 295, 220–223, 1982.
Newkirk, G., Jr., Variations in solar luminosity, *Annu. Rev. Astron.,*21, 429–467, 1983.
Newkirk, G., Jr., Solar variability on time scales of $10^5$ to $10^6$ years, in *The Ancient Sun, Fossil Record in the Earth, Moon and Meteorites,* Pepen, R. O. et al., Eds., Pergamon Press, Elmsford, NY; *Geochim. Cosmochim. Acta Suppl.,* 1980, 293–320.
Newman, M. J. and Rood, R. T., Implications of solar evolution for the Earth's early atmosphere, *Science,* 198, 1035–1037, 1977.
Ninkovich, D. and Donn, W. L., Explosive Cenozoic volcanism and climatic implications, *Science,* 194, 899–906, 1976.
Nisbet, E. G., The end of the Ice age, *Can. J. Earth Sci.,* 27, 148–157, 1990.
Oeschger, H., Stauffer, B., Finkel, R., and Langway, C. C., Variations of the $CO_2$ concentration of occluded air and of anions and dust in polar ice core, in *Carbon Dioxide and the Carbon Cycle, Archean to Present,* Sundquist, E. T. and Broecker, W. S., Eds., Geophys. Monogr. Ser., Am. Geophys. Union, Washington, D.C., 1985, 132–142.
Olsen, H., Astronomical forcing of meandering river behavior: Milankovitch cycles in the Devonian of East Greenland, *Palaeogeogr. Palaeoclimatol. Palaeoecol.,* 79, 99–115, 1990.
Olsen, P. E., A 40-million-year record of early Mesozoic orbital climatic forcing, *Science,* 234, 842–848, 1986.
Olson, P., Silver, P. G., and Carlson, R. W., The large-scale structure of convection in the Earth's mangle, *Nature (London),* 344, 209–215, 1990.
Overpeck, J. T., Peterson, L. C., Kipp, N., Imbrie, J., and Rind, D., Climatic change in the circum-North Atlantic region during the last deglaciation, *Nature (London),* 338, 553–557, 1989.
Owen, T., Cess, R. D., and Ramanathan, V., Enhanced $CO_2$ greenhouse to compensate for reduced solar luminosity on early Earth, *Nature (London),* 277, 640–642, 1979.

Parsons, B. and Sclater, J. G., An analysis of the variation of ocean floor bathymetry and heat flow with age, *J. Geophys. Res.*, 82, 803–827, 1977.

Perch-Nielsen, K., McKenzie, J., and Quiziang, He., Biostratigraphy and isotope stratigraphy and the "catastrophic" extinction of calcareous nannoplankton at the Cretaceous/Tertiary boundary, in *Geological Implications of Impacts of Large Asteroids and Comets on the Earth,* Silver, L. T., Ed., Geol. Soc. Am. Spec. Pap. 190, 1980, 353–371.

Piper, J. D. A., The Precambrian paleomagnetic record: the case for the Proterozoic supercontinent, *Earth Planet. Sci. Lett.*, 59, 61–89, 1982.

Piper, J. D. A., Palaeomagnetic evidence for a Proterozoic supercontinent, *Philos. Trans. R. Soc. London,* 280, 469–490, 1976.

Pisias, N. G. and Shackleton, N. J., Modeling the global climate response to orbital forcing and atmospheric carbon dioxide changes, *Nature (London),* 310, 757–759, 1984.

Pisias, N. G., Heath, G. R., and Moore, T. C., Lag times for oceanic responses to climatic change, *Nature (London),* 256, 716–717, 1975.

Pittmann, W. C., III, Relationship between eustasy and stratigraphic sequences of passive margins, *Geol. Soc. Am. Bull.*, 89, 1389–1403, 1978.

Pokras, E. M. and Mix, A. C., Earth's precession cycle and Quaternary climatic change in tropical Africa, *Nature (London),* 326, 486–488, 1987.

Pollack, J. B., Solar, astronomical, and atmospheric effects on climate, in *Climate in Earth History,* Geophysics Study Committee, Ed., National Academy of Sciences, Washington, D.C., 1982, 68–76.

Pollack, J. B., Climatic change on the terrestrial planets, *Icarus,* 37, 479–553, 1979.

Pollard, D., A coupled climate-ice sheet model applied to the Quarternary ice ages, *J. Geophys. Res.*, 88, 7705–7718, 1983.

Powell, C.McA., Johnson, B. D., and Veevers, J. J., A revised fit of east and west Gondwanaland, *Tectonophysics,* 63, 13–29, 1980.

Powell, C.McA. and Veevers, J. J., Namurian uplift in Australia and South America triggered the main Gondwanan glaciation, *Nature (London),* 326, 177–179, 1987.

Prell, W. L., Monsoonal climate of the Arabian Sea during the late Quaternary: a reponse to changing solar radiation, in *Milankovitch and Climate,* Part 1, Berger, A., Imbrie, J., Hays, J., Kukla, G., and Saltzman, B., Eds., D. Reidel Publishing, Boston, MA, 1984, 349–366.

Rampino, M. R., Possible relationships between changes in ice volume, geomagnetic excursions, and the eccentricity of the Earth's orbit, *Geology,* 7, 584–587, 1979.

Rampino, M. R. and Volk, T., Mass extinctions, atmospheric sulphur and climatic warming at the K/T boundary, *Nature (London),* 332, 63–66, 1988.

Raymo, M. E., Ruddiman, W. F., Froelich, P. N., Influence of late Cenozoic mountain building on ocean geochemical cycles, *Geology,* 16, 649–654, 1988.

Raynaud, D., Chappelaz, J., Barnola, J. M., Korotkevich, Y. S., and Lorius, C., Climatic and CH-4 cycle implications of glacial-interglacial CH-4 change in the Vostok ice core, *Nature (London),* 333, 655–657, 1988.

Rind, D., Peteet, D., and Kukla, G., Can Milankovitch orbital variations initiate the growth of ice sheets in a general circulation model?, *J. Geophys. Res.*, 94, 12,851–12,871, 1989.

Ross, C. A. and Ross, J. R. P., Carboniferous and Early Permian biogeography, *Geology,* 13, 27–30, 1985.

Rossignol-Strick, M., African monsoons, an immediate climate response to orbital insolation, *Nature (London),* 304, 46–49, 1983.

Ruddiman, W. F. and Kutzbach, J. E., Forcing of late Cenozoic northern hemisphere climate by plateau uplift in southern Asia and the American West, *J. Geophys. Res.*, 94, 18409–18427, 1989.

Ruddiman, W. F. and McIntyre, A., Severity and speed of northern hemisphere glaciation pulses: the limiting case?, *Geol. Soc. Am. Bull.*, 93, 1273–1279, 1982.

Russell, K. L., Oceanic ridges and eustatic changes in sea level, *Nature (London),* 218, 861–862, 1968.

Sagan, C. and Mullen, G., Earth and Mars: evolution of atmospheres and surface temperatures, *Science,* 177, 52–56, 1972.

Saltzman, B. and Maasch, K. A., A first-order global model of late Cenozoic climate change, *Clim. Dyn.*, 5, 201–210, 1991.

Saltzman, B. and Maasch, K. A., Carbon cycle instability as a cause of the Late Pleistocene Ice Age oscillations: modeling the asymmetric response, *Global Biogeochem. Cycles,* 2, 177–185, 1988.

Sarmiento, J. L. and Toggweiler, J. R., A new model for the role of the oceans in determining atmospheric $_{p}CO_2$, *Nature (London),* 308, 621–624, 1984.

Sarnthein, M., Winn, K., and Zahn, R., Paleoproductivity of oceanic upwelling and the effect on atmospheric CO-2 and climate change during deglaciation times, in *Abrupt Climatic Change,* Berger, W. H. and Labeyrie, L. D., Eds., D. Reidel Publishing, Boston, MA, 1987, 311–337.

Schwartz, S. E., Are global cloud albedo and climate controlled by marine phytoplankton?, *Nature (London),* 336, 441–445, 1988.

Schwartzman, D. W. and Volk, T., Biotic enhancement of weathering and the habitability of Earth, *Nature (London)*, 340, 457–460, 1989.

Sclater, J. G., Anderson, R. N., and Bell, M. L., The elevation of ridges and the evolution of the central eastern Pacific, *J. Geophys. Res.*, 76, 7883–7915, 1971.

Scotese, C. R., Bambach, R. K., Barton, C., Van der Voo, R., and Ziegler, A. M., Paleozoic base maps, *J. Geol.*, 87, 217–277, 1979.

Shackleton, N. J., Carbon-13 in Uvigerina: tropical rainforest history and the equatorial Pacific carbonate dissolution cycles, in *The Fate of Fossil Fuel CO-2 in the Ocean,* Andersen, N. R. and Malahoff, A., Eds., Plenum Press, New York, 1977, 401–427.

Shackleton, N. J. and Imbrie, J., The O spectrum of oceanic deep water over a five-decade band, *Climat. Change*, 16, 217–230, 1990.

Shackleton, N. J. and Pisias, N. G., Atmospheric carbon dioxide orbital forcing, and climate, in *Carbon Dioxide and the Carbon Cycle, Archean to Present,* Sundquist, E. T. and Broecker, W. S., Eds., Geophys. Monogr. Ser. Am. Geophys. Union, Washington, D.C., 1985, 313–318.

Shackleton, N. J., Hall, M. A., Line, J., and Shuxi, C., Carbon isotope data in core V19-30 confirm reduced carbon dioxide concentration in the ice age atmosphere, *Nature (London),* 306, 319–322, 1983.

Shaw, G. E., Bio-controlled thermostasis involving the sulfur cycle, *Climat. Change,* 5, 297–303, 1983.

Siegenthaler, U. and Wenk, Th., Rapid atmospheric CO-2 variations and ocean circulation, *Nature (London),* 308, 624–626, 1984.

Sleep, N. H., Zahnle, K. J., Kasting, J. F., and Morowitz, H. J., Annihilation of ecosystems by large asteroid impacts on the early Earth, *Nature (London),* 342, 139–142, 1989.

Smith, B. N., Evolution of C-4 photosynthesis in response to changes in carbon and oxygen concentrations in the atmosphere through time, *BioSystems,* 8, 24–32, 1976.

Socci, A. D., Climate, glaciation and deglaciation, controls, pathways, feedbacks, rates and frequencies. *Mod. Geol.*, 16, 279–316, 1992.

Sowers, T. and Bender, M., The $S^{18}O$ of atmospheric $CO_2$ from air inclusions in the Vostok ice core: timing of $CO_2$ and ice volume changes during the penultimate deglaciation, *Paleoceanography,* 6, 679–696, 1991.

Spalding, W. G. and Graumlich, L. J., The last pluvial climatic episodes in the deserts of southwestern North America, *Nature (London),* 320, 441–445, 1986.

Stauffer, B., Lochbronner, E., Oeschger, H., and Schwander, J., Methane concentration in the glacial atmosphere was only half that of the preindustrial Holocene, *Nature (London),* 332, 812–814, 1988.

Stauffer, B., Fischer, G., Neftel, A., and Oeschger, H., Increase of atmospheric methane recorded in Antarctic ice core, *Science,* 229, 1386–1388, 1985.

Sutton, J., Long-term cycles in the evolution of the continents, *Nature (London),* 198, 731–735, 1963.

Thompson, L. G. and Mosley-Thompson, E., Evidence of abrupt climatic change during the last 1,500 years recorded in ice cores from the tropical Quelccaya ice cap, Peru, in *Abrupt Climatic Change,* Berger, W. H. and Labeyrie, L. D., Eds., D. Reidel Publishing, Boston, MA, 1987, 99–110.

Thompson, L. G., Xiaoling, W., Mosley-Thompson, E., and Zichu, X., Climatic records from the Dundee ice cap, China, *Ann. Glaciol.*, 10, 1–5, 1988.

Thompson, L. G., Mosley-Thompson, E., Dansgaard, W., and Grootes, P. M., The Little Ice Age as recorded in the stratigraphy of the tropical Quelccaya ice cap, *Science,* 234, 361–364, 1986.

Thompson, L. G., Mosley-Thompson, E., Davis, M. E., Bolzan, J. F., Dai, J., Yao, T., Klein, L., Gundestrup, N., Tao, T., Wu, X., and Xie, Z., Glacial stage ice-core records from the subtropical Dundee ice cap, China, *Ann. Glaciol.*, 14, 288–297, 1989.

Thunell, R. C. and Williams, D. F., Glacial-Holocene salinity changes in the Mediterranean Sea: hydrographic and depositional effects, *Science,* 338, 493–496, 1989.

Thunell, R. C., Locke, S. M., and Williams, D. F., Glacio-eustatic sea-level control on Red Sea salinity, *Nature (London),* 334, 601–604, 1988.

Towe, K. M., Aerobic respiration in the Archean?, *Nature (London),* 348, 54–56, 1990.

Tucker, M., Carbon isotope excursions in Precambrian/Cambrian boundary beds, Morocco, *Nature (London),* 319, 48–50, 1986.

Vail, P. R., Mitchum, R. M., Jr., and Thompson, S., III, Global cycles of relative changes in sea level, in *Seismic Stratigraphy — Applications to Hydrocarbon Exploration,* Payton, C. E., Ed., Am. Assoc. Petrol. Geol. Mem., 26, 83–97, 1977.

Van der Voo, R., Paleozoic paleogeography of North America, Gondwana, and intervening displaced terranes: comparisons of paleomagnetism with paleoclimatology and biogeographical patterns, *Geol. Soc. Am. Bull.*, 100, 311–324, 1988.

Van Houten, F. B., Cyclic sedimentation and the origin of analcime-rich Upper Triassic Lockatong Formation, west-central New Jersey and adjacent Pennsylvania, *Am. J. Sci.*, 260, 561–576, 1962.

Veevers, J. J. and Powell, C.McA., Late Paleozoic glacial episodes in Gondwanaland reflected in transgressive-regressive depositional sequences in Euramerica, *Geol. Soc. Am. Bull.*, 98, 475–487, 1987.

Veizer, J., Holser, W. T., and Wilgus, C. K., Correlation of C-13/C-12 and S-34/S-32 secular variations, *Geochim. Cosmochim. Acta,* 44, 579–587, 1980.

Vogt, P. R., Evidence for global synchronism in mantle plume convection, and possible significance for Geology, *Nature (London),* 240, 338–342, 1972.

Volk, T., Rise of angiosperms as a factor in a long-term climatic cooling, *Geology,* 17, 107–110, 1989a.

Volk, T., Sensitivity of climate and atmospheric CO-2 to deep-ocean and shallow-ocean carbonate burial, *Nature (London),* 337, 637–640, 1989b.

Volk, T., Feedbacks between weathering and atmospheric $CO_2$ over the last 100 million years, *Am. J. Sci.,* 287, 763–779, 1987.

Walker, J. C. G. and Zahnle, K. J., Lunar nodal tide and distance to the moon during the Precambrian, *Nature (London),* 320, 600–602, 1986.

Walker, J. C. G., Hays, P. B., and Kasting, J. F., A negative feedback mechanism for the long-term stabilization of Earth's surface temperature, *J. Geophys. Res.,* 86, 9776–9782, 1981.

Weedon, G. P., The detection and illustration of regular sedimentary cycles using Walsh power spectra and filtering, with examples of the Lias of Switzerland, *J. Geol. Soc. London,* 146, 133–144, 1989.

Weller, M. J., Cyclic sedimentation of the Pennsylvanian period and its significance, *J. Geol.,* 38, 97–135, 1930.

Wigley, T. M. L., Could reducing fossil-fuel emissions cause global warming?, *Nature (London),* 349, 503–506, 1991.

Wigley, T. M. L., Possible climate change due to $SO_2$-derived cloud condensation nuclei, *Nature (London),* 339, 365–367, 1989.

Wigley, T. M. L. and Brimblecombe, P., Carbon dioxide, ammonia and the origin of life, *Nature (London),* 291, 213–216, 1981.

Williams, G. E., Late Precambrian glacial climate and the Earth's obliquity, *Geol. Mag.,* 112, 441–544, 1975.

Williams, G. E., Geotectonic cycles, lunar evolution, and the dynamics of the Earth-Moon system, *Mod. Geol.,* 4, 159–183, 1973.

Williams, G. E., Geological evidence relating to the origin and secular rotation of the solar system, *Mod. Geol.,* 3, 165–181, 1972.

Windley, B. F., *The Evolving Continents,* 2nd ed., John Wiley & Sons, New York, 1984, 1–385.

Wollin, G., Ericson, D. B., and Ryan, B. F., Magnetism of the earth and climatic changes, *Earth Planet. Sci. Lett.,* 12, 175–183, 1971.

Worsley, T. R., Moody, J. B., and Nance, R. D., Proterozoic to recent tuning of biogeochemical cycles, in *Carbon Dioxide and the Carbon Cycle: Archean to Present,* Sundquist, E. T. and Broecker, W. H., Eds., Geophys. Monogr. Ser. Am. Geophys. Union, Washington, D.C., 1985, 561–572.

Worsley, T. R. and Nance, R. D., Carbon redox and climate control through earth history: a speculative reconstruction, *Palaeogeogr. Palaeoclimatol. Palaeoecol.,* 75, 259–282, 1989.

Worsley, T. R., Nance, R. D., and Moody, J. B., Tectonic cycles and the history of the Earth's biogeochemical and paleoceanographic record, *Paleoceanography,* 1, 233–263, 1986.

Worsley, T. R., Nance, R. D., and Moody, J. B., Global tectonics and eustasy for the past 2 billion years, *Mar. Geol.,* 58, 373–400, 1984.

Chapter 8

# ANTHROPOGENIC FOSSIL CARBON SOURCES OF ATMOSPHERIC METHANE

**Timothy R. Barber and William M. Sackett**

## TABLE OF CONTENTS



0-8493-4419-0/93/$0.00 + $.50
© 1993 by CRC Press, Inc.

# I. INTRODUCTION

Measurements of atmospheric methane, prior to 1970, indicated some variability (Bainbridge and Heidt, 1966; Cavanaugh et al., 1969; Swinnerton et al., 1969; Lamontagne et al., 1971; Behar et al., 1972), but it was generally accepted that differences represented short-term fluctuations about a steady state value (Ehhalt, 1974). Recent systematic measurements using high resolution gas chromatographic techniques have detected a 0.7 to 1.9% annual increase from 1968 to 1981 (Graedel and McRae, 1980; Rasmussen and Khalil, 1981; Fraser et al., 1981; Khalil and Rasmussen, 1983). Measurements from 1983 to 1986 indicated an increase between 0.78 to 1.1% per year (Steele et al., 1987; Blake and Rowland, 1988). Khalil et al. (1989) reported a rate slightly $<1\%$ per year during 1960 to 1980, when the data were corrected for differences in interlaboratory calibration. The well-documented increase in atmospheric methane is an important environmental problem. Methane takes part in several important tropospheric and stratospheric reactions and is an effective greenhouse gas which could contribute significantly to global warming (Ramanathan, 1980; Dickinson and Cicerone, 1986; Hansen et al., 1989).

Analysis of air trapped in polar ice cores indicates that the concentration of $CH_4$ has been increasing about 1% per year for the last 100 to 150 years (Craig and Chou, 1982; Rasmussen and Khalil, 1984). The present concentration of 1.75 ppmv is twice as high as it was before anthropogenic activities significantly affected the global methane budget. The doubling of atmospheric methane, however, is not unprecedented in Earth's history. Reported values during the last glacial period ($\approx$17,000 years ago) are only 0.35 ppmv (Stauffer et al., 1988; Raynaund et al., 1988). Nevertheless, the present concentration and rate of increase in atmospheric methane surpass anything in recorded history.

The timing of the methane increase suggests anthropogenic activities have perturbed the natural methane cycle (Khalil and Rasmussen, 1983; Bolle et al., 1986). Human related activities may produce 60 to 70% of the total amount of $CH_4$ released to the atmosphere (Figure 1). Anthropogenic sources include: microbially mediated production from rice paddies, domestic livestock, and landfills, incomplete combustion of recently fixed organic matter (biomass burning), and fossil $CH_4$ mostly derived from coal and natural gas exploitation.

Radiocarbon ($^{14}C$) data suggest that about 80% of atmospheric methane is of recent biological origin (Wahlen et al., 1989; Manning et al., 1990; Quay et al., 1991). The remaining 20% possesses a fossil signature. Microbial production from ancient peat deposits, destabilization of methane hydrates at high northern latitudes (Kvenvolden, 1988), and release of primordial methane (Gold, 1979; Welhan and Craig, 1979) have been qualitatively identified as fossil sources; however, they are currently considered relatively minor sources (<5 Tg/yr) (Tyler, 1991).

This study estimates the emission of methane from several anthropogenic fossil sources (AFS). Potential AFS include: coal mining, loss of natural gas from offshore production platforms and pipelines, production of methane from petrogenic materials exposed to sunlight, losses associated with incomplete combustion of fossil fuels, and other processes related to the use of fossil fuels. Annual production is calculated by multiplying the total amount of fossil precursor produced or distributed and an empirical emission factor. Emission rates are crude estimates, based on incomplete data sets, numerous model simplifications, and several unjustified assumptions. Our study is not a rigorous treatment of AFS, but an attempt to illustrate the relationship between human population, demand for fossil fuels, and methane emissions. The release of nonmethane light hydrocarbons (NMLHC) from anthropogenic fossil sources is also reported.

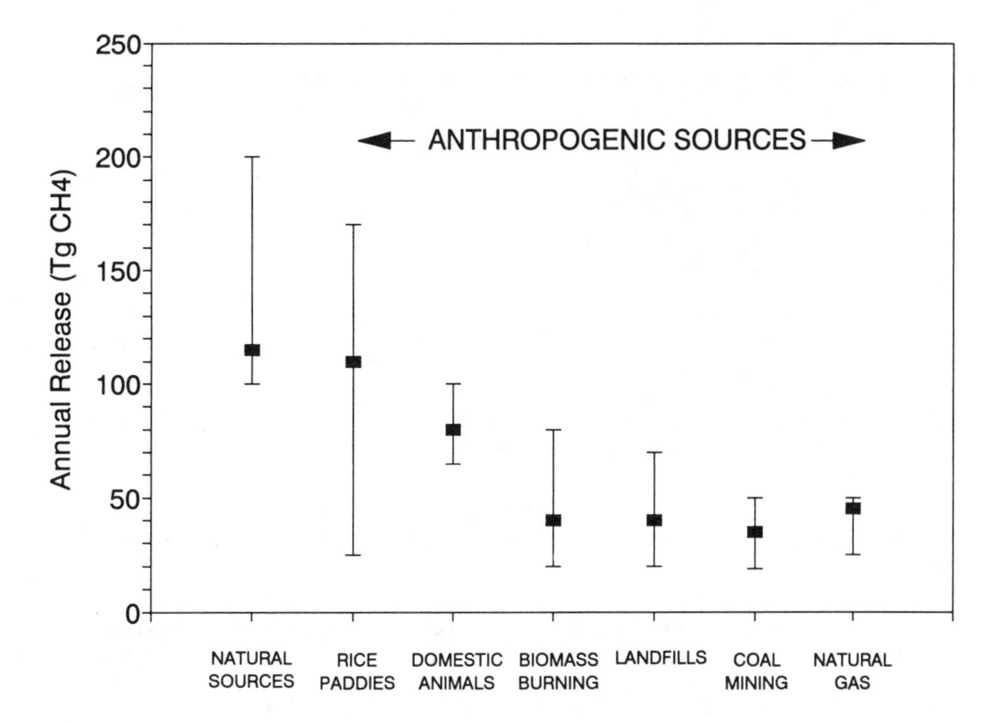


**FIGURE 1.** Annual methane release rates for identified sources (Cicerone and Oremland, 1988). Uncertainty expressed as the range.

## II. GLOBAL METHANE CYCLE

The total atmospheric reservoir of methane (N) is 4990 Tg, corresponding to a globally averaged atmospheric concentration of 1.75 ppmv. It is increasing about 45 Tg/yr (Cicerone and Oremland, 1988). The globally averaged residence time ($\tau$) is defined as

$$\tau = N/(dN/dt) \qquad (1)$$

where dN/dt is the steady state source flux of methane. The annual release of methane to the atmosphere is constrained fairly well by independent estimates of the atmospheric residence time from two-dimensional transport models (Ehhalt, 1978; Khalil and Rasmussen, 1983), anthropogenic halocarbon data (Mayer et al., 1982; Prinn et al., 1987), and controlled experiments (Vaghjiani and Ravishankara, 1991). Recent estimates have converged on 500 Tg (Cicerone and Oremland; 1988; Fung et al., 1991), assuming a residence time of 10 years. A summary of the global methane cycle data is given in Table 1.

Fossil sources contribute about 20% of the annual release of 500 Tg (Fung et al., 1991). Recent attempts to allocate the flux among various fossil sources are listed in Table 2. Despite a 20-year interval in publication dates, no temporal trend is observed in the magnitude of the AFS (Table 2); however, the amount of $CH_4$ given off should be proportional to the amount of the fossil fuel being produced and distributed. The demand for fossil fuels increases as a function of human population and level of industrialization. Figure 2 illustrates the relationship between the increase in source flux of $CH_4$ (relative to 1950) and human population. A strong linear relationship is suggested ($r^2 = 0.997$, for $\tau = 10$ years). As the population increased from 2.5 billion in 1950 to 5.3 billion in 1990, the annual net flux increased by 135 Tg. Bolle et al. (1986) suggested that increased usage of natural gas and coal resulted in an increasing emission of $CH_4$ from 24 Tg/yr in 1950, to 70 Tg/yr in 1980.

**TABLE 1**
**Summary of Data for Atmospheric Methane Cycle[a]**

| Variable | Value |
|---|---|
| Average tropospheric concentration | 1.75 ppmv |
| Rate of change | 0.9%/yr |
| Total atmospheric burden | 4990 Tg |
| Residence time | 10 yr |
| Steady state source | 500 Tg |
| Known methane sinks | |
| Reaction with OH | 450 Tg (90%) |
| Microbial oxidation | 5–50 Tg (up to 10%) |
| Escape to stratosphere | 25–50 Tg (<10%) |

[a] Table adapted from Cicerone and Oremland, 1988.

**TABLE 2**
**Anthropogenic Fossil Sources of $CH_4$ and their Global Production Rate[a]**

| Sources | Hitchcock and Wechsler (1972) | Sheppard et al. (1982) | Senum and Gaffney (1985) | Cicerone and Oremland (1988) | Barns and Edmunds (1990) | Tyler (1991) |
|---|---|---|---|---|---|---|
| Coal mining | 7.9–27.7 | 7.9–27.7[b] | 9.5–25.9 | 25–45 | 25 | 10–35 |
| Natural gas | | | 7–21 | 25–50 | | |
| Venting and flaring | | 30 | | | 21–26.2 | 15–30 |
| Pipeline losses | | 20 | | | 16.1 | 10–20 |
| Industrial | 7–21 | 7–21[b] | | | | 5–25 |
| Automobiles | 0.5 | 0.5[b] | 0.5 | | | 0.5 |
| Volcanos | 0.2 | 0.2[b] | 0.3 | | 1.3 | 0.5 |
| $CH_4$ hydrates | | | | 0–100 | | 2–4 |

[a] Values reported in Tg $CH_4$/year.
[b] Estimated by Hitchcock and Wechsler (1972), as cited in Ehhalt (1974).

The data of Bolle et al. (1986) imply one third of the increase in $CH_4$ concentration is explained by an increase in AFS.

# III. FOSSIL SOURCES

We consider three main categories of anthropogenic fossil sources of atmospheric methane: coal mining, natural gas, and petroleum. Natural gas is on average 90% $CH_4$ and is often produced during petroleum and coal formation. The gross production of fossil fuels in 1988 was about 6000 Tg (oil equivalent); therefore, losses during production, distribution, and use of these fuels may be a significant source of atmospheric methane.

## A. COAL MINING

Methane is formed during coalification and is trapped in the coal-containing rock strata. During the mining operation, the mines are ventilated to reduce their likelihood of explosion due to the buildup of explosive levels of methane. While technologies exist to harvest this often commercial quantity of natural gas, it is usually not exploited due to short-term economic pressure.

Early evaluations of the coal mining source were plagued with faulty assumptions and incomplete analysis. A recent review by Boyer et al. (1990) on the emission of methane

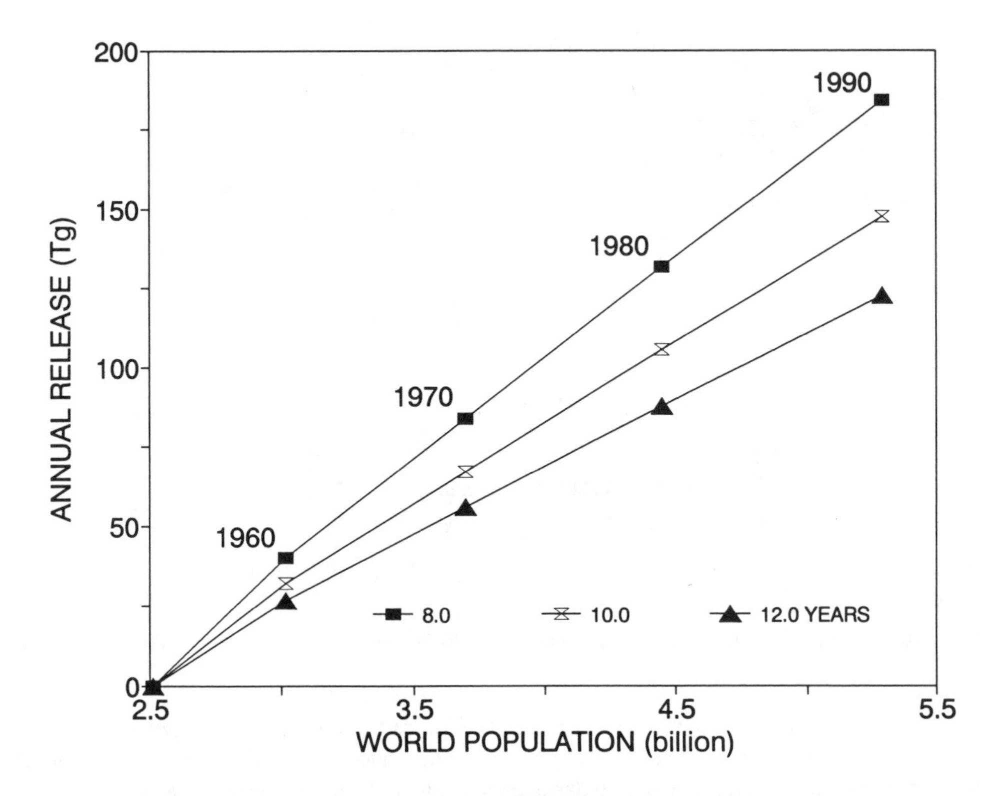


**FIGURE 2.** Growth of human population and increase of annual release rate of $CH_4$ from 1950. Plot constructed using Equation 1, and assuming atmospheric lifetimes of 8, 10, and 12 years.

from coal mines is the most comprehensive and reliable data set to date. Boyer et al. (1990) considers all types of coal: anthracite (hard), lignite (brown), and bituminous (soft). Also, surface and deep mining activities were treated separately, with deep mining operations releasing more methane. A first order approximation of country-by-country methane emissions distinguishes between developed and developing nations. Over 90% of the total methane released from coal mining is accounted for by 25 countries (Boyer et al., 1990).

## B. NATURAL GAS

Fossil natural gas, mainly $CH_4$, is derived from cracking of organic matter at high temperature and pressure in processes associated with petroleum formation and physical isolation of biogenic gas for at least 50,000 yr. The reduction of $CO_2$ by $H_2$ within the Earth's mantle and primordial $CH_4$ is not considered in this study.

The three main inputs are underwater venting of waste gas from offshore production platforms, losses associated with production facilities, and transmission losses during distribution. Offshore petroleum operations are responsible for increasing the NMLHC content of the surface water on the Louisiana Shelf several orders of magnitude over normal background values (Sackett and Brooks, 1975; Brooks, 1976; Brooks et al., 1977). The amount of natural gas vented underwater or burned in a flare from offshore petroleum production facilities is highly variable (Barns and Edmunds, 1990). The market price of natural gas often determines if recovery is economically feasible. If not, venting and flaring are thought to be safe methods of disposal. Also, the maturity of a country's petroleum industry is an important factor (Barns and Edmunds, 1990). As the supporting infrastructure develops, uses for previously wasted gas are often found.

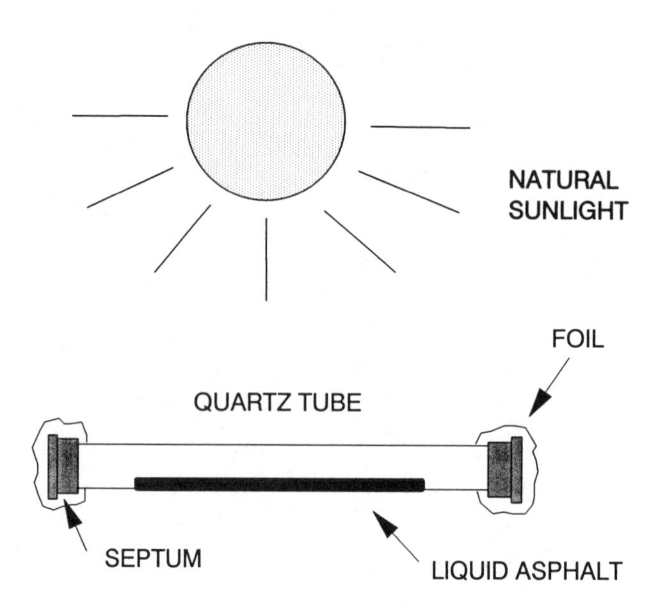


**FIGURE 3.** Schematic of reaction tube used to investigate the photodegradation of asphalt under ambient light conditions.

Transmission losses generally refer to the difference between the amount of natural gas supplied to a system and that delivered by the system. Unaccounted for gas sources include: venting of lines associated with maintenance operations, transmission pipeline leakage, losses from safety valves during system upset, and leaks around joints and valves (Barns and Edmunds, 1990). Blake et al. (1984) have suggested natural gas losses from low pressure distribution lines in urban centers are significant on a global scale, perhaps accounting for 20% of the observed increase in atmospheric methane. Rowland et al. (1990) reported leakage from pipelines in non-U.S. cities is significantly higher than that from U.S. cities.

## C. PETROLEUM

Two sources of $CH_4$ from oil-related processes are examined: photodegradation of petrochemicals and losses associated with incomplete combustion of fossil fuels.

Preliminary experiments conducted in our laboratory (Figure 3) provide strong evidence that the photodecomposition of asphalt yields significant quantities of methane (Figure 4). Asphalt was chosen for our experiments because it is used extensively for road surfaces and roofing materials, both ideally suited for exposure to the sun's radiant energy. These preliminary experiments demonstrate the potential for photochemical production of low molecular weight hydrocarbons from synthetic organic materials (Sackett and Barber, 1988; Sackett et al., 1988).

The petrochemical industry has devoted enormous resources to the study of the decomposition (i.e., wear and abrasion) of petroleum-derived materials in sunlight. Most studies have concentrated on the deterioration of materials, and not on the production of quasi-stable, relatively nontoxic entities like methane, ethane, ethylene, etc. In an extensive review, Ranby and Rabek (1975) reported that light hydrocarbons were produced from the photodecomposition of several polymers (polypropylene, polymethylacrylate, polystyrene, polydienes, polyacrylonitrile, etc). Almost any product may be formed during irradiation, and nearly every possible simple bond rupture and rearrangement may take place (Calvert and Pitts, 1966).

Hydrocarbon loss by automobiles occur via three pathways: release of incomplete combustion products, crankcase blowby, and fuel tank and carburetor evaporation. Recent efforts

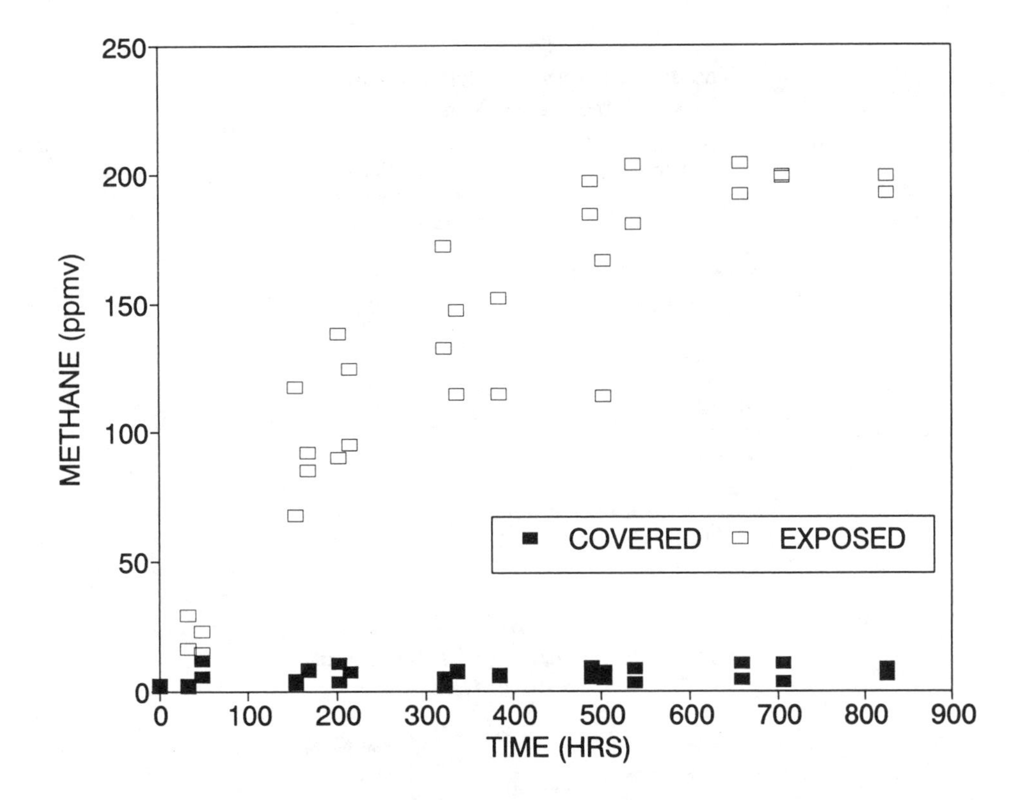


**FIGURE 4.** Enclosed atmospheric $CH_4$ concentration vs. time deployed.

have focused on the reduction of hydrocarbon emissions from motor vehicles in the U.S. due to chronic urban air pollution episodes. There are several important variables in estimating the emission factors from autos: exhaust temperature, air/fuel ratio, combustion efficiency, presence of pollution control devices, driving habits, and equipment maintenance (Barns and Edmunds, 1990).

The amount of $CH_4$ released from other fossil fuel combustion processes (excluding automobiles) is usually assumed to be negligible (Barns and Edmunds, 1990). However, Hansen et al. (1989) have correlated atmospheric $CH_4$ concentrations with carbon black, an indicator of incomplete combustion of carbonaceous materials, suggesting $CH_4$ emission from fossil fuel combustion processes is important on a local scale. Therefore, we include estimates of the amount of $CH_4$ released from incomplete combustion of coal, oil, and natural gas.

## IV. ESTIMATES OF EMISSION FACTORS

Estimates of the emission factors are difficult to derive. Clearly, the $CH_4$ flux from every coal mine, production platform, pipeline, automobile, and road surface cannot be measured. A number of assumptions must be made to derive the following production rates from AFS. Our model divides the world into seven geopolitical regions. Emission factors are assigned from literature values, usually based on a detailed analysis for a specific year. Emission rates are combined with fossil fuel production and consumption databases to estimate the annual $CH_4$ flux. The emission rates are assumed to be constant with time. There is no *a priori* reason to make this assumption. In fact, it is most likely that emission factors vary depending on geographical heterogeneity, evolving technologies, and supply and demand for fossil fuel reserves (Barns and Edmunds, 1990; Boyer et al., 1990).

**TABLE 3**
**Emission Factors for Methane Emissions from Coal Mining[a]**

| Region/country | Coal Mining Emission Factors ($Tg\ CH_4/PJ$)[b,c] |
|---|---|
| North America | $3.44 \times 10^{-4}$ |
| South America | $3.91 \times 10^{-4}$ |
| Europe | $4.93 \times 10^{-4}$ |
| Asia | $7.22 \times 10^{-4}$ |
| U.S.S.R. | $5.35 \times 10^{-4}$ |
| Africa | $5.39 \times 10^{-4}$ |
| Oceania | $2.83 \times 10^{-4}$ |
| Weighted average | $3.77 \times 10^{-4}$ |

[a] Table adapted from Boyer et al., 1990.
[b] A heating value of $28.8 \times 10^9$ J/ton of coal was assumed.
[c] 1 PJ = $10^{15}$ J.

**TABLE 4**
**Sources Associated with the Use of Natural Gas**

| Sources | Emission factors (Tg/PJ)[a] |
|---|---|
| Production | $9.60 \times 10^{-5}$ |
| Venting | $1.72 \times 10^{-4}$ |
| Transmission | $2.59 \times 10^{-4}$ |

[a] Emission factors provided by Barns and Edmunds (1990), assuming a heating value of 39.02 $\times 10^6$ J/m$^3$ gas.

The emission factors for coal mining (Table 3) were estimated by dividing the Boyer et al. (1990) source flux by the total amount of coal produced in that country in 1986 (Boyer et al., 1990). The amount of $CH_4$ emitted into the atmosphere in 1986 was estimated between 33 and 64 Tg methane. It is expected that the emission factor associated with coal mining will increase in the future because surface mines are dwindling and further exploitation of methane-rich deep mines is necessary to offset demand (Boyer et al., 1990).

Our model segments the natural gas flux into three components: venting, production losses, and transmission losses. Emission factors for natural gas have been proposed by Barns and Edmunds (1990); Table 4). The venting emission factor assumes 25% of the vented and flared gas is released to the atmosphere (Hamilton, 1977). The total natural gas source from 1950 to 1990, for the identified regions, is plotted in Figure 5.

Methane emitted from asphalt surfaces exposed to light depends on production rates, surface area of exposed material, and the lifetime of the material in the natural solar environment. The global source of $CH_4$ from asphalt was estimated at 0.01 Tg/yr by Tyler et al. (1990). The proposed emission factor for asphalt is estimated in a noncomprehensive way by assuming 3% of the crude oil is used in asphalt production (Table 5; OGJ Reports, 1986).

The emission factors for the incomplete combustion of fossil fuels at power plants were obtained from Daniel (1968), assuming methane makes up 2% of the total hydrocarbons (Table 5; Altshuller et al., 1966).

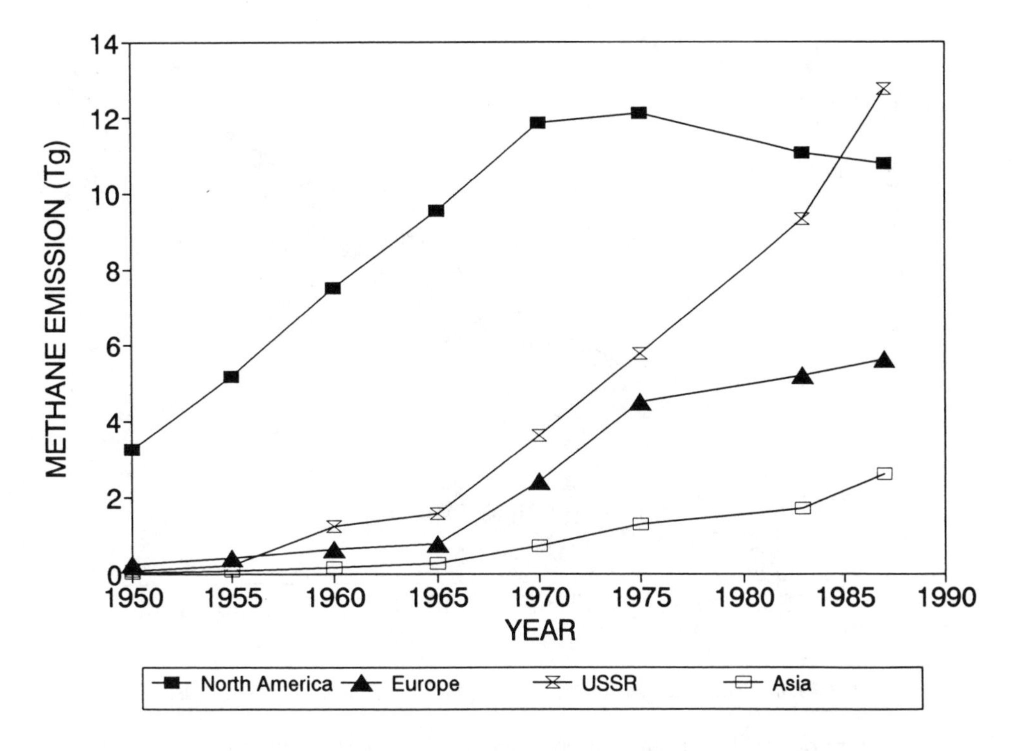


**FIGURE 5.** Loss of methane associated with the use of natural gas from 1950 to 1990.

**TABLE 5**
**Estimated Emission Factors for Sources Associated with the Production and Use of Oil**

| Source | Emission factors ($Tg\ CH_4/PJ$) |
|---|---|
| Photochemical-asphalt[a] | $1.77 \times 10^{-6}$ |
| Incomplete combustion[b] | |
| Coal | $3.47 \times 10^{-7}$ |
| Oil | $2.07 \times 10^{-7}$ |
| Natural gas | N[c] |
| Automobiles[d] | $3.29 - 5.76 \times 10^{-5}$ |

[a] Global source flux from asphalt (Tyler et al., 1990) divided by total production of asphalt for 1988, assuming 3.0% of crude oil is used for asphalt production.
[b] Emission factors provided for fossil fuel combustion at power plants (Daniel, 1976), assuming 2% of hydrocarbon emission is methane.
[c] N = negligible.
[d] Assuming fuel rating of 20 mpg, a heating value of $5.07 \times 10^9$ J/bbl (Barns and Edmunds, 1990).

Barns and Edmunds (1990) proposed emission factors for automobiles (see Table 5). Charcoal canisters and evaporative loss control systems have been effective in reducing hydrocarbon emissions in the U.S. Automobile emmissions from developing nations are larger than those from the U.S.; however, the total flux from motor vehicles is relatively small compared to coal and natural gas.

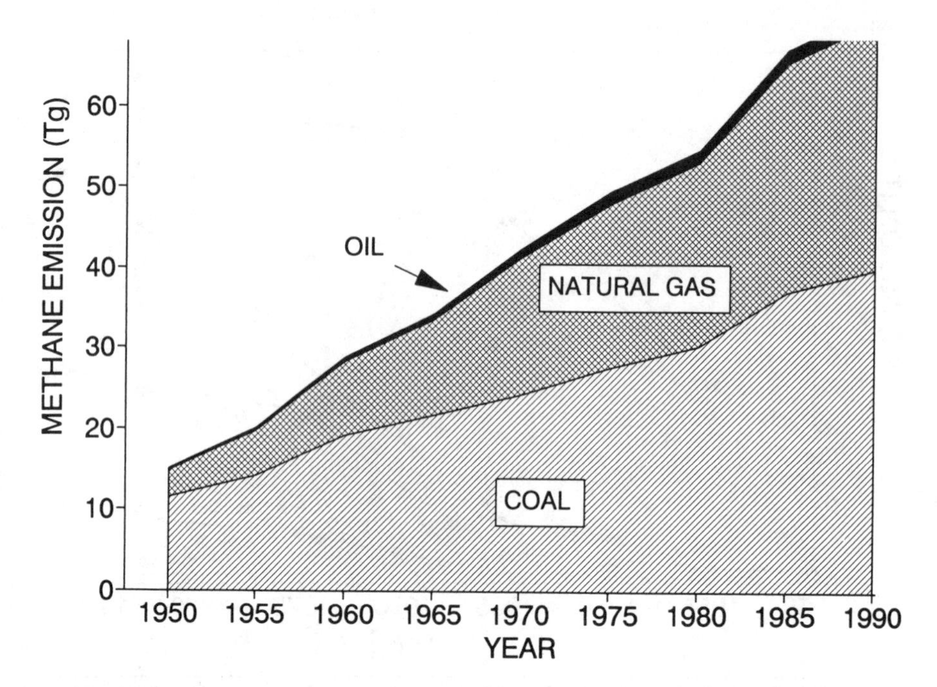


**FIGURE 6.** Estimated trend of $CH_4$ emission rates from anthropogenic fossil sources.

The record of $CH_4$ flux from anthropogenic fossil sources (Figure 6) was estimated from the emission factors in Tables 3, 4, and 5, and regional fossil fuel production and consumption statistics (OGJ Reports, 1986; Crabbe and McBride, 1978; United Nations, 1987; 1988; World Resources Institute, 1990).

## V. OTHER LIGHT HYDROCARBONS

Ethane, propane, and other NMLHC are greenhouse gases (Ramanathan et al., 1985; Ramanathan, 1988) and play important roles in the carbon and nitrogen cycle in the troposphere and lower stratosphere (Crutzen, 1979; Singh and Hanst, 1981; Kanakidou et al., 1991). The oxidation of these precursors forms photochemically active aldehydes, ketones, organic acids, and organic nitrates (Crutzen, 1979; Hanst and Gay, 1983). Owing to their high reactivity with Cl and OH radicals, NMLHC may compete with $CH_4$ for available hydroxyl radicals, and serve as important sinks for Cl atoms in the upper troposphere and lower stratosphere (Chameides and Cicerone, 1978; Rudolph et al., 1981).

The primary processes for the production of light hydrocarbons in the environment are: bacterial degradation of organic matter in anoxic sediments yielding principally $CH_4$ (Claypool and Kaplan, 1974), and thermal alteration of organic matter associated with fossil fuel genesis (Schoell, 1983). Cracking of organic matter, either thermal or catalytic, produces a wide spectrum of hydrocarbons. The ratio of $CH_4$ to higher hydrocarbons is usually $<100$ (Bernard et al., 1978; Bernard, 1979; Schoell, 1983); therefore, venting of petrogenic gas from coal mines, offshore production platforms, and transmission pipelines contributes to the atmospheric burden of $C_2$+ hydrocarbons. Assuming a $CH_4$ source of 70 Tg, an additional 2.5 Tg of carbon would be emitted into the atmosphere, or 15% of the estimated steady state source flux of ethane, 16 Tg, (Kanakidou et al., 1991).

Our photodegradation experiments indicate $C_2$ to $C_4$ hydrocarbons, both saturated and unsaturated, are photochemically produced from asphalt (Figure 7). The total amount of carbon emitted to the atmosphere from asphalt surfaces may be a factor of 4 or larger when NMLHC are included.

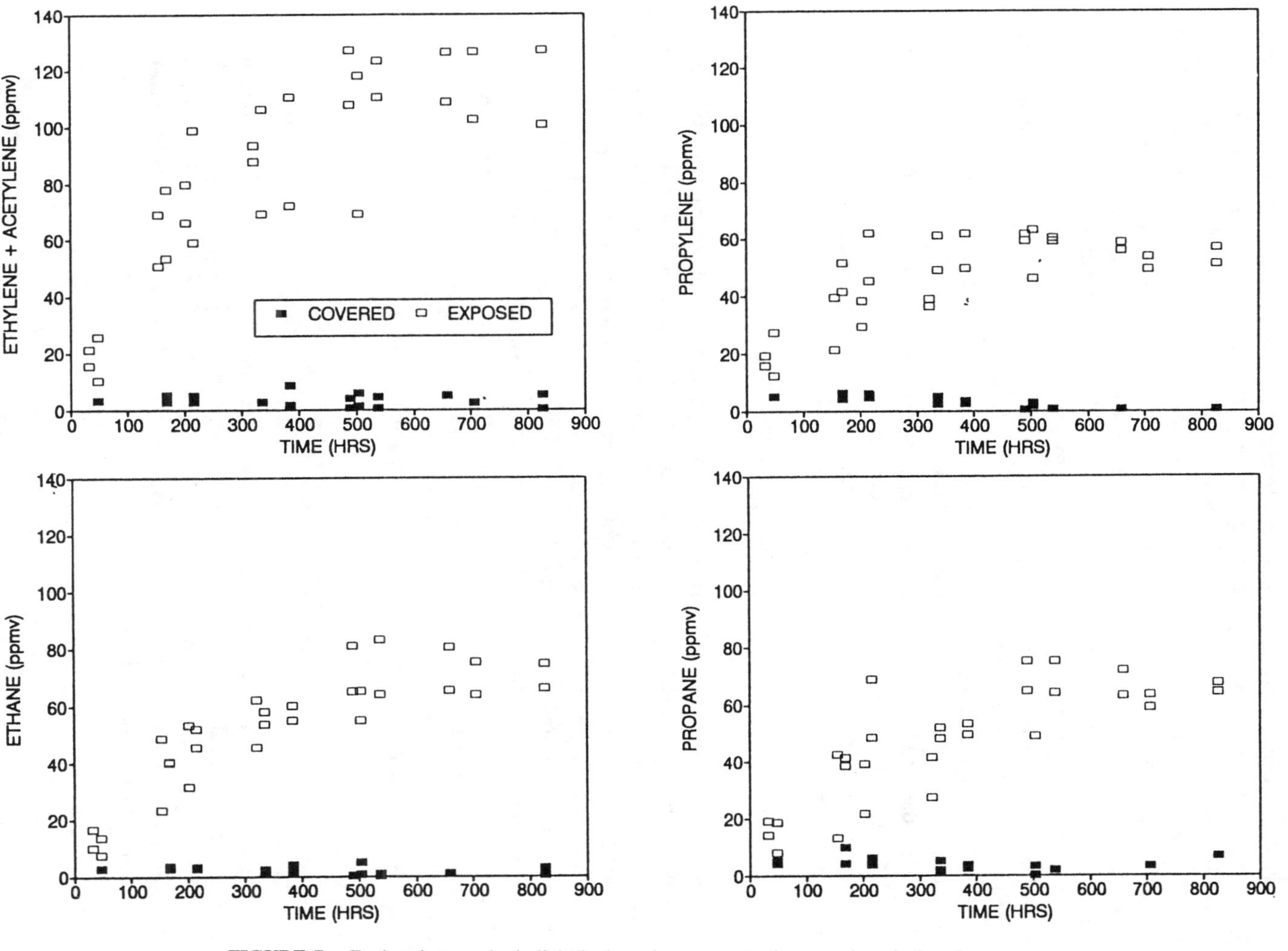


**FIGURE 7.** Enclosed atmospheric light hydrocarbon concentrations vs. time deployed.

Another approach in estimating the amount of $CH_4$ and NMLHC emitted to the atmosphere from the photodecomposition of asphalt and other petrochemicals each year is to consider the total amount of C–H bonds produced during the manufacture of these materials from crude oil. In 1987 the crude oil feed to refineries was about $1.46 \times 10^{10}$ barrels/year, with about 10% being converted to various petrochemicals (i.e., paints, rubber, and plastics). Using the appropriate conversion factors, this amounts to about $21 \times 10^{10}$ kg/year. Assuming a density of 0.9 g/cm$^3$ for petroleum and 85% carbon yield, $17.8 \times 10^{13}$ g of petrochemicals on a per carbon basis are produced each year. This is equivalent to $1.5 \times 10^{13}$ mol carbon/year. If each carbon is bonded to a hydrogen, then there are the same numbers of C–H bonds produced each year and introduced into the environment. As methane has 4 C–H bonds, we think in terms of there being $1.5 \times 4 \times 10^{13}$ methane equivalents available for emission to the atmosphere by photochemical processes. On a per gram basis this is 60 Tg of methane. The residence time of asphalt, rubber, plastics and other petrochemicals may be estimated at 10 years or more so that there may be a pool of about 600 Tg of methane precursor carbon that may interact with solar radiation to give C–H compounds that absorb IR radiation and enhance the greenhouse effect.

During the operation of a motor vehicle, a wide variety of saturated and unsaturated NMLHC are produced through incomplete combustion, and then emitted to the atmosphere via the exhaust (Daniel, 1968). Only 2% of the hydrocarbons released by automobiles are methane (Hitchcock and Wechsler, 1972).

## VI. CONCLUSIONS

The increase in atmospheric $CH_4$ is still only explained in a qualitative way. One class of sources, anthropogenic fossil sources, can be estimated based on production and use of fossil fuel precursors. Emission from AFS has been estimated for the period between 1950 and 1990. As the energy requirements for an expanding society grow, the exploitation of fossil fuel and the concurrent release of $CH_4$ and other light hydrocarbons to the atmosphere increases. The total fossil fuel increased from 15 Tg/yr in 1950 to 70 Tg/yr in 1990. Also, significant NMLHC are emitted to the atmosphere. We estimate an additional 5 to 10 Tg of carbon is released from AFS as $C_2$ to $C_4$ hydrocarbons. Anthropogenic fossil sources are especially troublesome because they release previously buried carbon back into the atmosphere. Because the emission of anthropogenic fossil sources is directly linked to human population and activities, it may be useful to evaluate mitigation strategies aimed at slowing or limiting this source of atmospheric methane.

## ACKNOWLEDGMENTS

We thank Eliot Atlas for his assistance and insight into the photochemical production of methane. We are also grateful to Melisa Reiter for conducting the majority of the asphalt incubation experiments. This work was supported by the NASA Upper Atmosphere Research Program through grant NAGW 836. Additional support to WMS was provided by the National Science Foundation via OCE-90-15580.

# REFERENCES

Altshuller, A. P., Ortman, G. C., Saltzman, B. E., and Neiligan, R. E., Continuous monitoring of methane and other hydrocarbons in urban atmospheres, *J. Air. Pollut. Contr. Assoc.*, 16, 87–91, 1986.

Bainbridge, A. E. and Heidt, L. E., Measurements of methane in the troposphere and lower stratosphere, *Tellus,* 18(2), 221–224, 1966.

Barns, D. W. and Edmunds, J. A., An evaluation of the relationship between the production and use of energy and atmospheric methane emissions, *Carbon Dioxide Res. Progr.*, Rep. TR047, U.S. Department of Energy, Washington, D.C., 1990.

Behar, J. V., Zafonte, L., Cameron, R. E., and Morelli, F. A., Hydrocarbons in air samples from Antarctic dry valley drilling sites, *J. Antarct.*, 7,94–96, 1972.

Bernard, B. B., Methane in marine sediments, *Deep-Sea Res.*, 26A, 429–443, 1979.

Bernard, B. B., Brooks, J. M., and Sackett, W. M., Light hydrocarbons in recent Texas continental shelf and slope sediments, *J. Geophys. Res.*, 92, 2181–2187, 1978.

Blake, D. R. and Rowland, F. S., Continuing worldwide increase in tropospheric methane, 1978 to 1987, *Science,* 239, 1129–1131, 1988.

Blake, D., Woo, V., Tyler, S., and Rowland, F., Methane concentrations and source strengths in urbane locations, *Geophys. Res. Lett.*, 11, 1211–1214, 1984.

Bolle, H. J., Seiler, W., and Bolin, B., Other greenhouse gases and aerosols assessing their role for atmospheric radiative transfer, in *The Greenhouse Effect, Climatic Change, and Ecosystems,* John Wiley & Sons, Chichester, England, 1986, 157–203.

Boyer, C. M., Kelafant, J. R., Kuuskraa, V. A., Manger, K. C., and Kruger, D., Methane Emissions from Coal Mining: Opportunities for Reduction, *ANR-445,* U.S. Environmental Protection Agency, Washington, D.C., 1990.

Brooks, J. M., The flux of light hydrocarbons into the Gulf of Mexico via runoff, in *Marine Pollution Transfer,* Windom, H. L. and Duce, R. A., Eds., Lexington Books, Lexington, MA, 1976, 185–200.

Brooks, J. M., Bernard, B. B., and Sackett, W. M., Input of low molecular weight hydrocarbons from petroleum operations into the Gulf of Mexico, in *Fate and Effects of Petroleum Hydrocarbons in Marine Ecosystems and Organisms,* Wolfe, D. A., Ed., Pergamon Press, Elmsford, NY, 1977, 373–384.

Calvert, J. G. and Pitts, J. N., *Photochemistry,* John Wiley & Sons, New York, 1966, 899 pp.

Cavanaugh, L. A., Schadt, C. F., and Robinson, E., Atmospheric hydrocarbon and carbon monoxide measurements at Point Barrow, Alaska, *Environ. Sci. Technol.*, 3, 251–257, 1969.

Chameides, W. L. and Cicerone, R. J., Effects of non-methane hydrocarbons in the atmosphere, *J. Geophys. Res.*, 83, 947–952, 1978.

Cicerone, R. J. and Oremland, R. S., Biogeochemical aspects of atmospheric methane, *Global Biogeochem. Cycles,* 2(4), 299–327, 1988.

Claypool, G. E. and Kaplan, I. R., The origin and distribution of methane in marine sediment, in *Natural Gases in Marine Sediments,* Kaplan, I. R., Ed., Plenum Press, New York, 1974, 99–139.

Crabbe, D. and McBride, R., *The World Energy Book,* Nichols Publishing, New York, 1978.

Craig, H. and Chou, C. C., Methane: the record in polar ice cores, *Geophys. Res. Lett.*, 9(11), 1221–1224, 1982.

Crutzen, P. J., The role of NO and $NO_2$ in the chemistry of the troposphere and stratosphere, *Annu. Rev. Earth Planet. Sci.*, 7, 443–472, 1979.

Daniel, W., Engine variable effects on exhaust hydrocarbon composition — a single engine study with propane as the fuel, *SAE Trans.*, 76, 774–795, 1968.

Dickinson, R. E. and Cicerone, R. J., Future global warming from atomspheric trace gases, *Nature (London),* 319(9), 109–118, 1986.

Ehhalt, D. H., The atmospheric cycle of methane, *Tellus,* 26, 58–70, 1974.

Ehhalt, D. H., The $CH_4$ concentration over the ocean and its possible variation with latitude, *Tellus,* 30, 169–176, 1978.

Fraser, P. J., Khalil, M. A., Rasmussen, R. A., and Crawford, A. J., Trends of atmospheric methane in the southern hemisphere, *Geophys. Res. Lett.*, 8(10), 1063–1066, 1981.

Fung, I., John, J., Lerner, J., Matthews, E., Prather, M., Steele, L., and Fraser, P., Three-dimensional model synthesis of the global methane cycle, *J. Geophys. Res.*, 96, 13033–13065, 1991.

Gold, T. J., Terrestrial sources of carbon and earthquake outgassing, *J. Pet. Geol.*, 1, 3–19, 1979.

Graedel, T. E. and McRae, J. E., On the possible increase of the atmospheric methane and carbon monoxide concentrations during the last decade, *Geophys. Res. Lett.*, 7(11), 977–979, 1980.

Hamilton, R. E., Present and predicted levels of consumption of fossil fuels, in *Global Chemical Cycles and Their Alteration by Man,* Stumm, W., Ed., Dahlen Konferenzen, Abakon Verlags-gesellschaft, Berlin, 1977, 155–164.

Hansen, A. D. A., Conway, T. J., Steele, L. P., Bodhaine, B. A., Thoning, K. W., Tans, P., and Novakov, T., Correlations among combustion effluent species at Barrow, Alaska: aerosol black carbon, carbon dioxide, and methane, *J. Atmos. Chem.*, 9, 283–299, 1989.

Hansen, J., Lacis, A., and Prather, M., Greenhouse effect of chlorofluorocarbons and other trace gases, *J. Geophys. Res.*, 94, 16417–16421, 1989.

Hanst, P. L. and Gay, B. W., Atmospheric oxidation of hydrocarbons: formation of hydroperoxides and peroxyacids, *Atmos. Environ.*, 17, 2259–2265, 1983.

Hitchcock, D. and Wechsler, A., Biogeological cycling of atmospheric trace gases, *Final Report, NASW-2128*, 1972, 117–154.

Kanakidou, M., Singh, H., Valentin, K., and Crutzen, P., A two-dimensional study of ethane and propane oxidation in the troposphere, *J. Geophys. Res.*, 96, 15395–15413, 1991.

Khalil, M. A. K. and Rasmussen, R. A., Sources, sinks, and seasonal cycle of atmospheric methane, *J. Geophys. Res.*, 88, 5131–5144, 1983.

Khalil, M. A. K., Rasmussen, R. A., and Shearer, M. J., Trends of atmospheric methane during the 1960s and 1970s, *J. Geophys. Res.*, 94, 18279–18288, 1989.

Kvenvolden, K. A., Methane hydrates and global climate, *Global Biogeochem. Cycles*, 2, 221–229, 1988.

Lamontagne, R. A., Swinnerton, J. W., and Linnenbom, V. J., Nonequilibrium of carbon monoxide and methane at the air-sea interface, *J. Geophys. Res.*, 76, 5117–5121, 1971.

Manning, M. R., Lowe, D. C., Melhuish, W. H., Sparks, R. J., Wallace, G., and Brenninkmeijer, C. A. M., The use of radiocarbon measurements in atmospheric studies, *Radiocarbon*, 32, 1990.

Mayer, E. W., Blake, D. R., Tyler, S. C., Makide, Y., Montague, D. C., and Rowland, F. S., Methane: interhenispheric concentration gradient and atmospheric residence time, *Proc. Natl. Acad. Sci. USA*, 79, 1366–1370, 1982.

OGJ Reports, Worldwide report, *Oil & Gas Journal*, 84, 33–103, 1986.

Prinn, R., Cunnold, D., Rasmussen, R., Simmons, P., Alyea, F., Crawford, A., Fraser, P., and Rosen, R., Atmospheric trends in methylchloroform and the global average for the hydroxyl radical, *Science*, 238, 945–950, 1987.

Quay, P. D., King, S. L., Stutsman, J., Wilbur, D. O., Steele, L. P., Fung, I., Gammond, R. H., Brown, T. A., Farwell, G. W., Grootes, P. M., and Schmidt, F. H., Carbon isotopic composition of atmospheric $CH_4$: fossil and biomass burning source strengths, *Global Biogeochem. Cycles*, 5, 25–47, 1991.

Ramanathan, V., Climatic effects of anthropogenic trace gases, in *International Energy Climate*, Bach, W., Pankrath, J., Williams, J., Eds., D. Reidel, 1980, 269–280.

Ramanathan, V., The radiative and climatic consequences of the changing atmospheric composition of trace gases, in *The Changing Atmosphere*, Rowland, F. and Isaken, I., Eds., 1988, 159–186.

Ramanathan, V., Cicerone, R. J., Singh, H. B., and Kiehl, J. T., Trace gas trends and their potential role in climate change, *J. Geophys. Res.*, 90, 5547–5566, 1985.

Ranby, B. G. and Rabek, J. F., *Photodegradation, Photo-oxidation, and Photostabilization of Polymers*, John Wiley & Sons, New York, 1975, 573 pp.

Rasmussen, R. A. and Khalil, M. A. K., Atmospheric methane (CH4): trends and seasonal cycles, *J. Geophys. Res.*, 86, 9826–9832, 1981.

Rasmussen, R. A. and Khalil, M. A. K., Atmospheric methane in the recent and ancient atmospheres: concentrations, trends, and interhemispheric gradient, *J. Geophys. Res.*, 89, 11599–11605, 1984.

Raynaud, D., Chappellaz, J., Barnola, J. M., Korotkevich, Y. S., and Lorius, C., Climatic and $CH_4$ cycle implications of glacial–interglacial $CH_4$ change in the Vostok ice core, *Nature (London)*, 333, 655–657, 1988.

Rowland, F. S., Harriss, R. P., and Blake, D. R., Methane in cities, *Nature (London)*, 347, 432–433, 1990.

Rudolph, J., Ehhalt, D. H., and Tonnissen, A., Vertical profiles of ethane and propane in the stratosphere, *J. Geophys. Res.*, 86, 7276–7272, 1981.

Sackett, W. M. and Barber, T. R., Fossil carbon sources of atmospheric methane, *Nature (London)*, 334, 201, 1988.

Sackett, W. M., Barber, T. R., and Atlas, E. L., Fossil carbon sources of atmospheric methane, *Eos*, 69, 1079, 1988.

Sackett, W. M. and Brooks, J. M., Origin and distributions of low molecular weight hydrocarbons in Gulf of Mexico coastal waters, ACS Symp. Ser., 18, 211–230, 1975.

Schoell, M., Genetic characterization of natural gases, *Am. Assoc. Pet. Geol. Bull.*, 67, 2225–2238, 1983.

Senum, G. I. and Gaffney, J. S., A reexamination of the tropospheric methane cycle: geophysical implications, in *The Carbon Cycle and Atmospheric $CO_2$: Natural Variations Archean to Present*, Geophys. Monogr. Ser., Vol. 32, Sundquist, E. T. and Broecker, W. S., Eds., AGU, Washington, D.C., 1985, 61–69.

Sheppard, J. C., Westberg, H., Hopper, J. F., and Ganesan, K., Inventory of global methane sources and their production rates, *J. Geophys. Res.*, 87, 1305–1312, 1982.

Singh, H. B. and Hanst, P. L., Peroxyacetyl nitrate (PAN) in the unpolluted atmosphere: an important reservoir for nitrogen oxides, *Geophys. Res. Lett.*, 8, 941–944, 1981.

Stauffer, B., Lochbronner, E., Oeschger, H., and Schwander, J., Methane concentration in the glacial atmosphere was only half that of the preindustrial Holocene, *Nature (London)*, 322, 812–814, 1988.

Steele, L. P., Fraser, P. J., Rasmussen, R. A., Khalil, M. A. K., Conway, T. J., Craword, A. J., Gammon, R. H., Masarie, K. A., and Thoning, K. W., The global distribution of methane in the troposphere, *J. Atmos. Chem.*, 5, 125–171, 1987.

Swinnerton, J. W., Linnenbom, V. J., and Cheek, C. H., Distribution of methane and carbon monoxide between the atmosphere and natural waters, *Environ. Sci. Technol.*, 3, 836–838, 1969.

Tyler, S. C., The global methane cycle, in *Microbial Production and Consumption of Greenhouse Gases: Methane, Nitrogen Oxides, and Halomethanes*, Rogers, J. E. and Whitman, W. B., Eds., American Society of Microbiology, Washington D.C., 1991.

Tyler, S. C., Lowe, D. C., Dlugokenecky, E., Zimmerman, P. R., and Cicerone, R. J., Methane and carbon monoxide emissions from asphalt pavement: measurements and estimates of their importance to global budgets, *J. Geophys. Res.*, 95, 14007–14014, 1990.

United Nations, *Environmental Data Report*, United Nations, New York, 1987.

United Nations, *1986 Yearbook of Energy Statistics*, United Nations, New York, 1988.

Veghjiani, G. and Ravishankara, A., New measurement of the rate coefficient for the reaction of OH with methane, *Nature (London)*, 350, 406–409, 1991.

Wahlen, M., Tanaka, N., Henry, R., Deck, B., Zeglen, J., Vogel, J. S., Southon, J., Shemesh, A., Fairbanks, R., and Broecker, W., Carbon-14 in methane sources and in atmospheric methane: the contribution from fossil carbon, *Science*, 245, 286–290, 1989.

Welhan, J. A. and Craig, H., Methane and hydrogen in East Pacific rise hydrothermal fluids, *Geophys. Res. Lett.*, 6, 829–831, 1979.

World Resources Institute, *World Resources 1990-91*, World Resources Institute, Oxford University Press, Oxford, 1990.

Chapter 9

# ELEVATED SEA SURFACE TEMPERATURES CORRELATE WITH CARIBBEAN CORAL REEF BLEACHING

**Thomas J. Goreau, Raymond L. Hayes, Jenifer W. Clark, Daniel J. Basta, and Craig N. Robertson**

## TABLE OF CONTENTS



0-8493-4419-0/93/$0.00+$.50
© 1993 by CRC Press, Inc.

# I. INTRODUCTION

Mass coral reef bleaching began to occur in the 1980s across the Caribbean, Indian Ocean, and Pacific Coral Reef Provinces (Fankboner and Reid 1981; Lasker et al. 1984; Fisk and Done 1985; Harriott 1985; Jaap 1985; Oliver 1985; Hoegh-Guldberg et al., 1987; Nishihira 1987; Williams et al., 1987; Brown and Suharsono 1990; Gates 1990; Goreau 1990a,b; Williams and Bunkley-Williams, 1990; Lang et al. in press). Coral bleaching takes place when reefbuilding corals expel intracellular symbiotic algae (Mayor 1918; Yonge and Nicholls 1931) on which they rely for most of their nutrition and energy for skeletal deposition (Goreau and Goreau 1959, 1960). Corals are unable to flee stress, and can only respond to environmental extremes by bleaching or by dying. A large number of factors can experimentally induce bleaching in the laboratory, including temperatures that are too high (Mayor 1918; Yonge and Nicholls 1930; Hoegh-Guldberg and Smith 1989; Glynn and D'Croz 1990) or too low (Steen and Muscatine 1987), light levels that are too high (Dustan 1982; Hoegh-Guldberg and Smith 1989; Glynn and D'Croz 1990) or too low (Yonge and Nichols 1930) salinity that is too high (Reimer 1971) or too low (Goreau 1964), and excessive turbidity (Trench, 1986). Although bleaching had been observed in the field before the 1980s, all such events were confined to corals in shallow water subject to clearly identifiable and locally confined stresses, such as muddy freshwater plumes from rivers swollen by hurricane rains (Yonge and Nicholls 1930; Goreau 1964), or locally high temperatures caused by poor water circulation (Jaap, 1985; Lasker et al., 1984; Mayor 1918; Yonge and Nicholls 1930; Glynn 1984). In most cases the stress was brief and recovery was rapid (within weeks), but in one case the stress was long and mortality was nearly complete (Glynn 1984).

In contrast to previous bleaching events, those of the late 1980s affected unprecedented numbers of individual colonies and species of coral. Bleaching was observed at the greatest depths to which symbiotic corals are found, and appeared over large areas free from any obvious stress other than elevated temperature (Williams et al. 1987). Bleached corals show altered pigment concentrations (Kleppel et al. 1989), disruption of function at the cellular and biochemical levels (Glynn et al. 1985; Hayes 1988), reduced photosynthesis (Porter et al. 1989) and failure to grow a skeleton (Goreau and Macfarlane 1990a) or to deposit annual dense skeletal bands (Goreau and Dodge unpublished data). Corals which survive bleaching may take up to ten months to regain normal pigmentation and growth, but reduced symbiotic algae causes starvation which weakens their ability to complete with seaweed overgrowth (Goreau and Macfarlane 1990b).

Numerous observers of recent mass coral reef bleaching events have suggested that elevated water temperatures were a likely environmental cause, however, there has remained a lack of adequately documented seawater temperatures from coral reef sites. In this paper, biweekly National Oceanic and Atmospheric Administration (NOAA) satellite-derived Ocean Features Analysis (OFA) is examined for seven sites in the Caribbean Coral Reef Province where bleaching was reported. These analyses show that bleaching events have invariably accompanied the highest extremes of water temperature in the late 1980s.

# II. METHODS: PROCESSING AND CALIBRATION OF SATELLITE TEMPERATURE DATA

The seven locations which were selected as our data set are indicated geographically in Figure 1. High resolution satellite data were used, with a pixel size of 4 km. Polar orbiting Tiros-N/NOAA (Kidwell 1986) weather satellites with Advanced Very High Resolution Radiometer sensors (AVHRR) provided these data. The AVHRR features a cross-track scanning system with three channels which record in the thermal IR spectral range. The

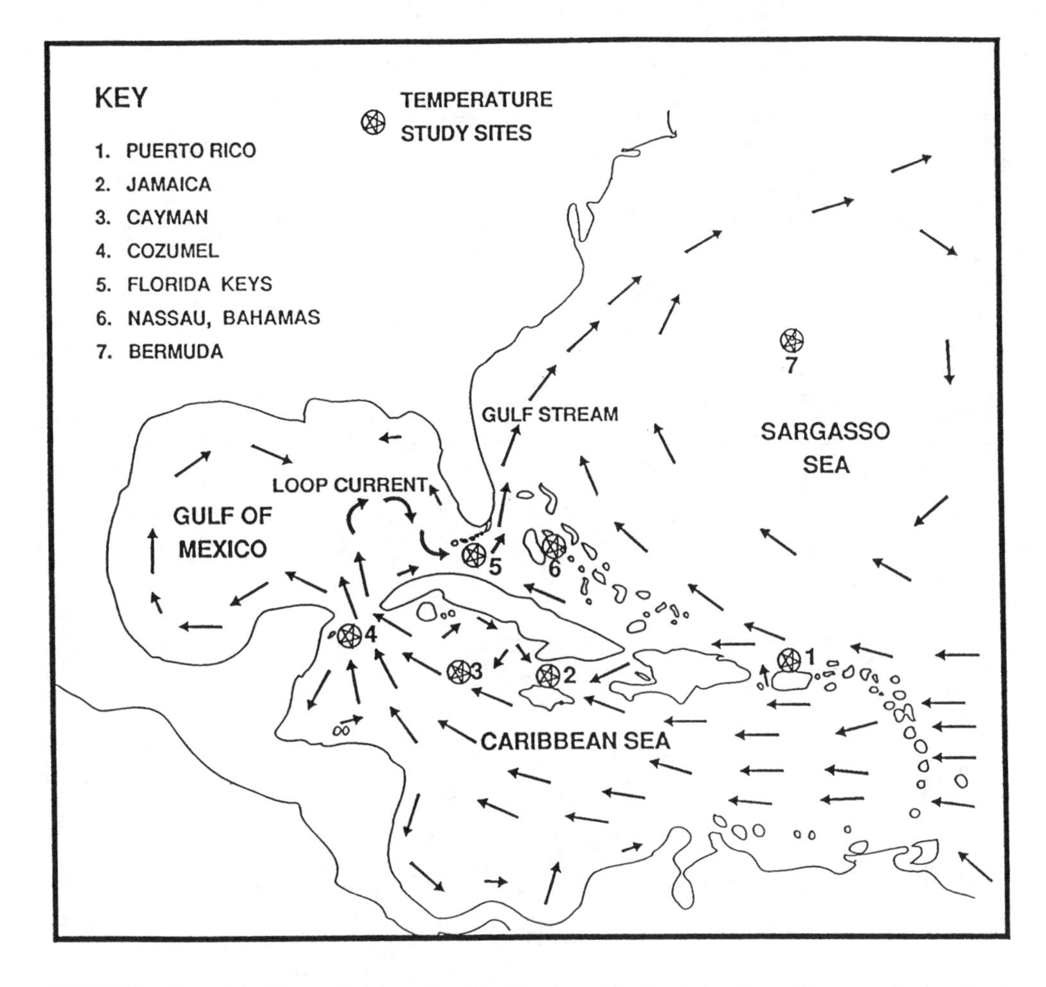


**FIGURE 1.** Map of the Greater Caribbean Coral Reef Province, showing the locations of the seven sites monitored and major regional surface current flows according to Sverdup et al., 1942.

southern boundary of those data was 18°N, passing through central Jamaica and Puerto Rico, and covering most of the region's reefs where bleaching has been intensively observed (see Figure 2). This range did not include sites such as Belize, Curacao, Colombia, Barbados, and the Lesser Antilles, where bleaching has also been reported.

Satellite data provide global coverage of sea surface temperature (SST). The U.S. Department of Commerce's NOAA has satellites equipped with high resolution IR sensors. Polar-orbiting IR satellite imagery is the primary data source for generating the OFA. These IR sensors detect sea surface brightness which is analyzed to generate sea surface thermal structures of the OFA. The South Panel OFA (Figure 2) is generated twice a week. The bulk of SST values on the chart are from thermal IR satellite data, with a spatial resolution of 4 km, supplemented by much spatially sparser SST measurements from buoys, expendable bathythermographs (XBTs), and ships in areas where cloud cover prevents direct surface observations. Because the final data set contains numbers from a variety of sources, it is referred to as ''blended''. Almost all readings published in the OFA have been made by the same oceanographer (J. W. Clark). Sea surface without cloud cover is selected for enumeration to avoid artifactual reduction in SST. Presence of even small and thin clouds in the pixel would bias the data toward lower temperatures. However, this represents a minor problem in the Caribbean since well developed convection patterns in the lower atmosphere

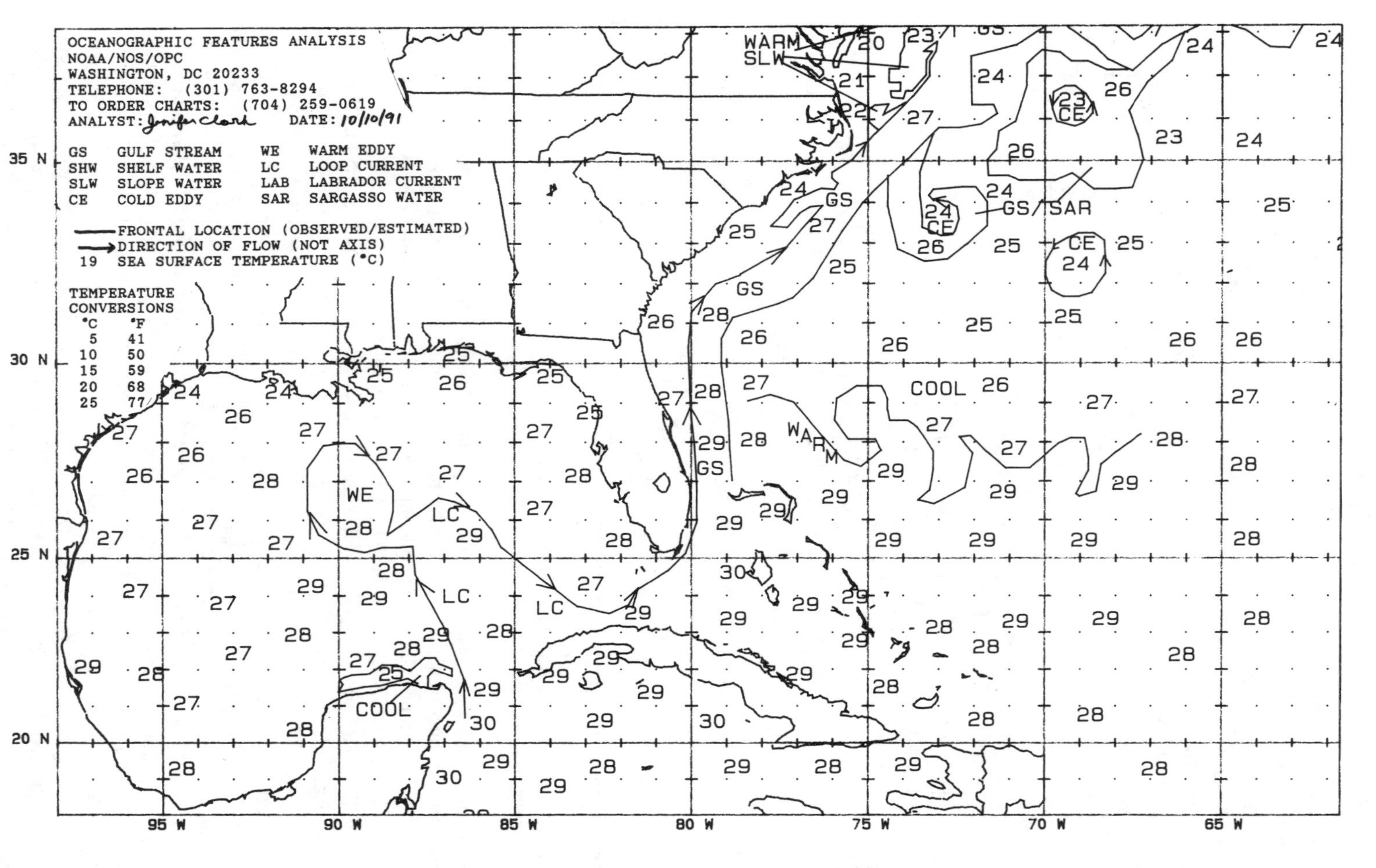


**FIGURE 2.** South Panel, Ocean Feature Analysis (OFA).

provide a fine scale mosaic of cloudy and cloud-free areas in regions of ascending and descending air, respectively. Cloud distributions are usually such that cloud-free areas can be found fairly close to any given site, except during relatively rare cases of large scale frontal cloud systems which typically occur in winter.

To construct the SST at each site, the measured values nearest to each location were averaged. The data used could consist of a single measurement immediately offshore, or the average of up to four measurements within several hundred kilometers of the site, weighted by hand according to distance. This means that the values reported have in effect been averaged over a spatial scale of a few hundred kilometers, although on very cloudy days, or in the case of Puerto Rico, where data is relatively sparse compared to other sites, the average scale may be larger. Averaging done independently by different observers (T. J. Goreau and R. L. Hayes) produced monthly average values which are within 0.09 ± 0.18°C (n = 14).

Monthly mean values from the OFA as described above were calibrated for the north coast of Jamaica against *in situ* measurements made at 3 m depth on the fore reef at the Discovery Bay Marine Laboratory (Discovery Bay, Jamaica) from 1985 to 1990, showing that:

$$T_{Sat} = 1.998 + 0.920 \times T_{Meas} \quad R = 0.897 \quad P = <0.001 \quad n = 65$$

where R is the correlation coefficient, P is the probability that the variables are uncorrelated, and n is the number of measurements. Satellite derived values were lower than measured water temperatures by 0.08 degrees at 26.00°C, and by 0.40 degrees at 30.00°C. Comparison of OFA data and monthly mean water temperatures measured on the west coast of Grand Cayman at 10 m depth from May to October 1990 showed that the two correlated strongly (R = 0.9634, P = 0.002, n = 6), but satellite values were higher than *in situ* values by a statistically insignificant (P = 0.060) amount, 0.177 ± 0.502°C. Differences between satellite and *in situ* data at higher temperatures in Jamaica could be caused by increased surface evaporation and atmospheric water vapor as well as sea surface layer effects. Cloud contamination, which is more likely near Jamaica than near Cayman due to topographically-induced convection, could bias satellite SST estimates to lower values when ocean temperatures are high. Although differences between satellite and *in situ* records are small, additional data from the OFA data set should be examined to see if it may offer clues for evaluating greenhouse feedbacks caused by water vapor.

Local heating of coastal waters where bleaching occurs is a function of the extent of shallow coastal banks and of circulation between shelf waters and open ocean. Shallow banks are nonexistent at Jamaica, Cayman, and Puerto Rico sites, small at Cozumel and Bermuda, but extensive at Florida and Bahama sites. Exchange between shelf waters and ocean waters is a rapidly changing function of fluctuating winds and currents, so corrections for inshore-offshore differences would be both site specific and temporally variable. Jamaica is unique among Caribbean islands in being a high, wet, limestone island, with numerous nearshore groundwater springs with water temperatures as low as 25.7°C, which tend to cool Jamaican nearshore waters. As water temperature records were only available to us from Jamaica and Cayman, corrections were not applied and could not be verified at other sites. Only detailed year-round measurements at each site would allow accurate estimation of local correction factors. In their absence, uncorrected open-ocean satellite values have been used in our analyses, even though they may be slightly different than the temperatures to which reef corals are actually exposed.

Temperatures for the whole Caribbean are also available as maps at 4 to 6 degrees (about 400 to 600 km) spatial mean resolution (Reynolds, personal communication) published in NOAA's Oceanographic Monthly Summaries (OMS). These low resolution data were

**TABLE 1**
**Regression of OMS vs. OFA Data Sets**

| Location | A(°C) | B | R | P |
|---|---|---|---|---|
| Puerto Rico | −0.03159 | 0.29528 | 0.423 | 0.052 |
| Jamaica | −0.08399 | 0.27908 | 0.602 | 0.003 |
| Cayman | −0.06039 | 0.29241 | 0.561 | 0.007 |
| Cozumel | +0.02282 | 0.33613 | 0.516 | 0.014 |
| Florida | −0.11022 | 0.35211 | 0.510 | 0.015 |
| Nassau | +0.09246 | 0.04585 | 0.069 | 0.760 |
| Bermuda | +0.41257 | 0.58513 | 0.708 | <0.001 |

*Note:* Temperature deviations from long-term averages were taken from NOAA's OMS and regressed against values taken from the OFA data set from January 1989 to October 1990 at each site (n = 22). **A** is the y intercept in degrees Celcius, **B** is the slope of the regression, **R** is the correlation coefficient, and **P** is the probability that the two data sets are unrelated. If both data sets agreed perfectly, A would be 0, B would be 1, R would be 1, and P would be 0. All regressions have slopes <1, indicating that the low resolution OMS data set underestimates positive temperature anomalies compared to the high resolution OFA data.

examined by Atwood and co-workers, to conclude that there were no unusual sea temperature anomalies in the Caribbean during the 1987 bleaching event (Atwood et al. 1988). Other researchers, using satellite SST data, reported that positive temperature anomalies were associated with coral bleaching in the Caribbean in 1987 (McCormack and Strong 1990; Strong and McCormack 1991). The low resolution OMS data were not used in our study because details of monthly SST anomalies did not correspond closely with *in situ* seawater temperature measurements, whereas the OFA data correlated highly with observed water temperatures. Table 1 shows regressions of OMS temperature anomalies against anomalies determined from the OFA data set. These results show that the OMS data considerably underestimate large positive temperature anomalies recorded by OFA data and in *in situ* measurements. A possible explanation is cloud contamination in the much larger pixels of the low resolution OMS data.

## III. SITE LOCATIONS AND HYDROGRAPHIC FEATURES

Water entering the Caribbean is predominantly supplied by the westward Guyana current, driven by trade winds through passages between the Lesser Antilles (Figure 1). A small amount of water enters via the Mona Passage between Puerto Rico and the Dominican Republic and via the Windward Passage between Haiti and Cuba. Surface water is transported westward through most of the Caribbean, veering northwestward and then northward towards the only exit, the Yucatan Channel. There the outflow is squeezed into a powerful northward jet flow into the Eastern Gulf of Mexico, which curves around 180 degrees to form the Loop Current. The rate of water flow out of the Caribbean determines the strength of the current, its northern extent, and amount of mixing with water from the Gulf of Mexico. The current then flows eastward between Florida and Cuba, before turning north between Florida and the Bahamas to form the Florida Current. The flow follows the edge of the continental shelf to Cape Hatteras, where it turns northeastward as the Gulf Stream. The

Gulf Stream is subsequently influenced by admixture with North American coastal shelf waters and the Sargasso Sea as it passes north of Bermuda towards northwestern Europe. The weather of eastern North America and Europe is influenced by the position and strength of the Gulf Stream.

Eddies form in geographically protected embayments on the fringes of the Caribbean: (1) in the southwestern Caribbean between Colombia, Panama, Costa Rica, and Nicaragua; (2) in the western Caribbean between Hondruas and Belize; (3) in the shelf waters of Southern Cuba; (4) in the Gulf of Mexico; and (5) also in the central Caribbean between Haiti, Jamaica, Cayman, and Cuba. Surface waters in eddies have a prolonged local residence time, and may gain additional heat from solar radiation. This is counteracted by localized sporadic upwelling along leeward shores, heavy rainfall along the Colombian, Panamanian, Costa Rican, Honduran, and Belizean coasts, and seasonally strong winter cooling along the northern fringes of the Gulf of Mexico. Fringing reef growth in mainland coastal waters is often poor, with reefs found offshore (e.g., the Belize Barrier Reef, the Bay Islands, San Andres, and Providencia), or in locally protected coastal sites. The region between Cayman, Cuba, Jamaica, and Haiti lies in the lee of the mountains of Hispaniola, and warms more than central Caribbean waters when winds are calm, especially in the Bay of Gonave between the Massif de la Hotte and the Massif du Nord of western Haiti and in the Gulf of Batabano off southern Cuba.

The three northernmost sites lies outside the trade-wind belt during winter, when they are exposed to continentally derived cyclonic air masses, but the strength of their major ocean currents is primarily influenced by volume and strength of Caribbean heat loss. Most of the rest of the region is subjected to uniform trade winds from the east year-round, but during winter, cold continental air masses from North America cause occasional frontal systems to penetrate the region. Distinctive hydrographic features of each site are

1. Northern Puerto Rico — There is no coastal shelf and the bottom drops steeply into the Puerto Rico Trench, the deepest waters in the North Atlantic. Currents are predominantly westward, and representative of entering water pushed into the Caribbean Sea by the Trade Winds. Reef development in Puerto Rico is best in the southwest, but is poor and declining due to high sediment loads from erosion (Acevedo et al. 1989).
2. North-Central Jamaica — There is no coastal shelf and the bottom drops steeply into the Cayman Trench, the deepest waters in the Caribbean. Currents are predominantly westward. Because Northern Jamaica is mostly limestone, there is little surface drainage or erosion, and extremely well-developed fringing reefs extend all along the North Coast.
3. Grand Cayman — The Cayman Islands lie atop a narrow ridge, with the Cayman Trench to the south and a deep basin to the north. There is virtually no terrestrial runoff from these flat and dry islands. Currents and winds are very similar to the north coast of Jamaica. Cayman lies directly in the path of Caribbean surface waters flowing towards the Yucatan Channel.
4. Eastern Cozumel — Cozumel lies on the western side of the Yucatan Channel, and offshore currents are predominantly northward. Cozumel waters are an admixture of central Caribbean outflow with entrained waters which have been trapped in the gyre between Yucatan, Belize, and Honduras, and in the southwestern Caribbean eddy.
5. Mid-Florida Keys — Florida reefs lie on an extensive submarine bank, and reefs are bathed by the edge of the Florida Current. When the Loop Current is strong, the flow axis tends to lie further offshore, exposing reefs to water moving between the Keys from Florida Bay. Water circulates poorly on the shallow West Florida Shelf and in

the Northern Gulf of Mexico. Calm conditions intensify local heating in summer, and cold fronts cause intense winter cooling. Reef conditions have deteriorated recently due to climatic extremes, boat damage, and eutrophication from septic tanks (LaPointe et al., 1990) and coral cover is declining (Porter, personal communication).

6. New Providence (Nassau), Bahamas — Water masses are influenced by the Florida Current to the west, and by the westward flowing Antillean current driven by trade winds and the Sargasso Sea to the east. Very shallow and extensive banks limit water exchange and cause seasonal extremes of cooling and heating during calm weather. Reefs are limited to narrow fringes at bank edges. Temperatures were evaluated for an area about 12 km south of Nassau, near the boundary of the Bahama banks and the deep water Tongue of the Ocean passage, at a site where bleaching has been recorded (Hayes, personal communication).
7. Bermuda — The extreme northern limit of coral reef growth, Bermuda is a largely submerged seamount atoll. A small shallow shelf causes some local heating. Waters do not get as warm here as at the other sites, since solar radiation is less intense and the dominant influence is interior waters of the North Atlantic Sargasso Sea Gyre. Conditions are near the lower tolerance limit of corals (ca. 18°C) during passage of North-American winter air masses.

Most sites monitored represent coastal waters of islands that have active private or governmental marine laboratories that are conducting coral reef research. The University of Puerto Rico at Mayaguez, Department of Marine Sciences, maintains a marine laboratory off La Parguera in the southwestern portion of the island. For over 30 years, research activities have been conducted at that facility. The University of the West Indies has operated the Discovery Bay Marine Laboratory and the Port Royal Marine Laboratory since the early 1950s. These laboratories were the sites for the pioneering reef studies using SCUBA initiated by the late Professor T. F. Goreau and colleagues (Goreau 1959; Goreau and Goreau 1973; Goreau et al. 1979). In the Cayman Islands, the governmental Natural Resources Unit has conducted coral reef research and monitoring of the Marine Park system for several years. Active sport diving enterprises on Grand Cayman Island, Nassau, and Cozumel have been developed during the past 20 years. The Florida Department of Natural Resources, the Looe Key National Marine Sanctuary, and the Key Largo National Marine Sanctuary have monitored reef conditions during the past 20 years. The Bermuda Biological Laboratory is one of the oldest research facilities in the western North Atlantic, and has been the site of offshore oceanographic expeditions and shore-based reef studies since the 1960s (Logan, 1988). There are no marine laboratories at Cozumel or Nassau.

## IV. RESULTS AND DISCUSSION: SPATIO-TEMPORAL PATTERNS OF TEMPERATURE AND BLEACHING

Figure 3a shows mean monthly SST at the seven sites from May 1980 to October 1990. These data were obtained by computing monthly means of over 1000 measurements at each site from the charts, typically around 8 or 9 biweekly measurements per month. Tabular display of these numerical data are given in Table 2. All existing chart observations were used. Figure 3b shows the monthly temperature variability.

Table 3 shows the mean water temperature at each site during the 1980 baseline period (May 1980 to April 1990). Annual temperature variability at different sites, measured by the standard deviation (S.D.) of monthly average temperature, decreases with rising mean temperature (MT):

$$\text{S.D.} = 11.880 - 0.383 \times \text{MT} \quad R = 0.946 \quad P = <0.001 \quad n = 7$$

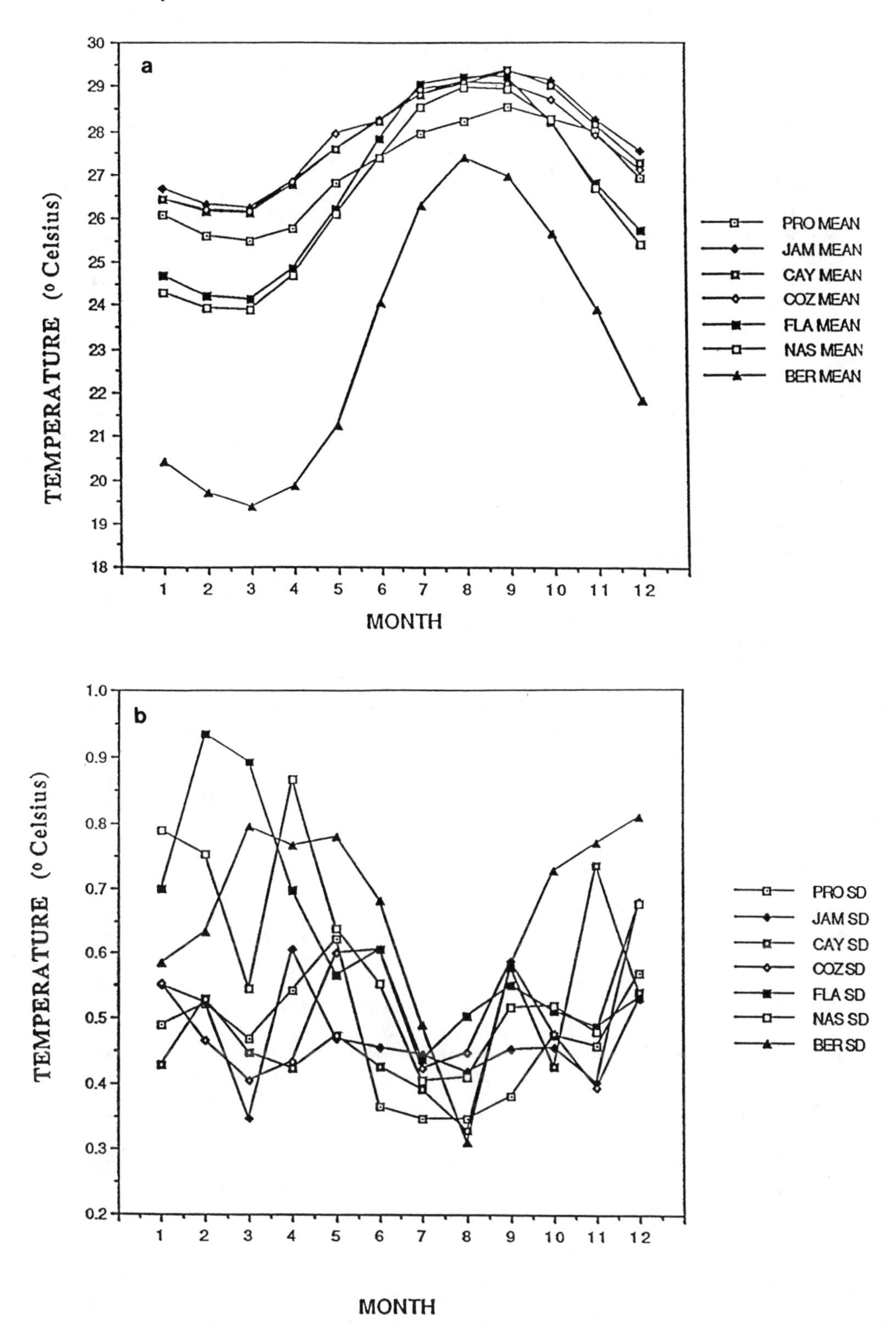


**FIGURE 3a and b.** (a) Mean annual water temperatures at each site during the 1980s; (b) monthly temperature standard deviations (S.D.) at each site show minimum variability in summer. January is month 1 and December is month 12.

**TABLE 2**
**Caribbean Monthly Average SST (May 1980 to October 1990)**

| | Month | Puerto Rico | Jamaica | Cayman | Cozumel | Florida | Nassau | Bermuda |
|---|---|---|---|---|---|---|---|---|
| 1 | MAY 80 | 28.000 | 28.000 | 28.000 | 28.000 | 27.000 | 26.000 | 22.000 |
| 2 | JUN 80 | 27.329 | 28.429 | 28.429 | 27.929 | 28.571 | 26.800 | 22.857 |
| 3 | JUL 80 | 28.150 | 29.200 | 29.200 | 29.400 | 29.250 | 28.590 | 26.300 |
| 4 | AUG 80 | 28.222 | 29.033 | 29.167 | 29.500 | 29.333 | 28.889 | 26.889 |
| 5 | SEP 80 | 28.333 | 29.500 | 29.500 | 29.611 | 29.556 | 28.589 | 26.889 |
| 6 | OCT 80 | 28.944 | 29.556 | 29.556 | 29.067 | 28.611 | 28.022 | 26.000 |
| 7 | NOV 80 | 28.375 | 28.400 | 28.188 | 28.050 | 27.038 | 27.050 | 23.625 |
| 8 | DEC 80 | 27.056 | 27.167 | 27.111 | 27.044 | 25.967 | 25.444 | 20.778 |
| 9 | JAN 81 | 25.667 | 25.889 | 25.889 | 25.611 | 24.389 | 23.222 | 20.056 |
| 10 | FEB 81 | 25.562 | 25.587 | 25.300 | 25.500 | 23.625 | 23.375 | 18.812 |
| 11 | MAR 81 | 24.944 | 25.900 | 25.800 | 25.556 | 23.833 | 23.600 | 18.178 |
| 12 | APR 81 | 24.911 | 26.556 | 26.489 | 26.422 | 25.167 | 24.489 | 19.389 |
| 13 | MAY 81 | 26.250 | 28.125 | 28.125 | 27.938 | 26.375 | 26.075 | 21.000 |
| 14 | JUN 81 | 27.444 | 28.778 | 28.722 | 29.000 | 28.222 | 27.778 | 24.556 |
| 15 | JUL 81 | 28.056 | 28.833 | 28.778 | 29.778 | 29.500 | 28.778 | 26.889 |
| 16 | AUG 81 | 28.714 | 29.286 | 29.257 | 29.929 | 29.643 | 28.743 | 27.643 |
| 17 | SEP 81 | 29.056 | 29.322 | 29.611 | 29.700 | 29.178 | 28.256 | 26.444 |
| 18 | OCT 81 | 28.778 | 29.222 | 29.222 | 29.044 | 28.278 | 27.867 | 25.333 |
| 19 | NOV 81 | 27.788 | 28.337 | 28.512 | 28.462 | 26.875 | 26.562 | 22.750 |
| 20 | DEC 81 | 26.700 | 27.180 | 26.810 | 26.700 | 25.700 | 24.800 | 20.750 |
| 21 | JAN 82 | 26.062 | 26.663 | 26.562 | 26.375 | 24.062 | 24.625 | 19.500 |
| 22 | FEB 82 | 25.188 | 26.113 | 26.012 | 26.375 | 24.062 | 23.562 | 19.062 |
| 23 | MAR 82 | 25.500 | 26.233 | 26.133 | 26.422 | 25.056 | 24.878 | 19.111 |
| 24 | APR 82 | 25.322 | 26.933 | 26.556 | 27.222 | 25.722 | 25.133 | 19.500 |
| 25 | MAY 82 | 26.000 | 27.538 | 27.375 | 28.062 | 25.750 | 25.875 | 19.438 |
| 26 | JUN 82 | 26.833 | 28.089 | 28.022 | 27.478 | 27.000 | 27.156 | 23.611 |
| 27 | JUL 82 | 27.778 | 29.000 | 28.933 | 28.511 | 28.944 | 28.656 | 26.622 |
| 28 | AUG 82 | 28.322 | 28.722 | 28.944 | 29.333 | 29.522 | 28.789 | 27.667 |
| 29 | SEP 82 | 28.583 | 29.217 | 29.217 | 29.167 | 29.217 | 28.883 | 26.867 |
| 30 | OCT 82 | 28.188 | 28.788 | 28.562 | 29.250 | 28.188 | 28.488 | 25.663 |
| 31 | NOV 82 | 27.650 | 27.900 | 28.137 | 27.500 | 26.188 | 26.000 | 23.650 |
| 32 | DEC 82 | 26.756 | 27.056 | 26.744 | 26.778 | 26.056 | 24.978 | 21.889 |
| 33 | JAN 83 | 25.825 | 25.962 | 26.050 | 26.913 | 25.062 | 24.288 | 20.637 |
| 34 | FEB 83 | 25.312 | 26.512 | 26.500 | 26.450 | 24.625 | 23.775 | 19.538 |
| 35 | MAR 83 | 26.150 | 26.160 | 26.020 | 26.240 | 23.200 | 23.070 | 18.840 |
| 36 | APR 83 | 26.100 | 26.850 | 26.575 | 26.500 | 24.188 | 23.688 | 19.750 |
| 37 | MAY 83 | 27.500 | 26.889 | 27.200 | 29.111 | 25.833 | 25.633 | 20.822 |
| 38 | JUN 83 | 27.911 | 28.278 | 28.078 | 28.933 | 27.833 | 27.400 | 24.444 |
| 39 | JUL 83 | 28.300 | 28.913 | 29.012 | 29.238 | 28.762 | 28.250 | 25.800 |
| 40 | AUG 83 | 28.867 | 29.556 | 29.400 | 29.411 | 29.411 | 28.556 | 27.411 |
| 41 | SEP 83 | 28.133 | 29.267 | 29.089 | 28.922 | 29.489 | 28.967 | 26.689 |
| 42 | OCT 83 | 27.938 | 28.475 | 28.750 | 28.462 | 27.562 | 28.250 | 26.038 |
| 43 | NOV 83 | 28.200 | 28.550 | 28.038 | 27.938 | 26.150 | 26.462 | 23.000 |
| 44 | DEC 83 | 27.311 | 27.767 | 27.511 | 27.578 | 24.678 | 26.078 | 21.811 |
| 45 | JAN 84 | 26.486 | 27.233 | 27.056 | 26.659 | 23.839 | 25.439 | 20.706 |
| 46 | FEB 84 | 25.660 | 26.700 | 26.600 | 25.740 | 23.000 | 24.800 | 19.600 |
| 47 | MAR 84 | 25.389 | 26.256 | 26.000 | 26.267 | 22.500 | 23.778 | 19.422 |
| 48 | APR 84 | 26.500 | 26.725 | 26.462 | 26.688 | 24.000 | 24.762 | 19.413 |
| 49 | MAY 84 | 26.220 | 27.200 | 27.070 | 27.160 | 25.300 | 25.630 | 21.330 |
| 50 | JUN 84 | 27.371 | 27.686 | 27.643 | 27.443 | 26.857 | 26.471 | 23.357 |
| 51 | JUL 84 | 27.944 | 28.267 | 28.467 | 28.711 | 28.233 | 27.867 | 25.667 |
| 52 | AUG 84 | 27.889 | 28.567 | 28.567 | 28.500 | 28.356 | 28.500 | 27.100 |
| 53 | SEP 84 | 28.175 | 28.762 | 28.637 | 28.400 | 28.438 | 28.500 | 25.925 |
| 54 | OCT 84 | 27.589 | 28.822 | 28.444 | 28.389 | 27.333 | 27.256 | 24.489 |
| 55 | NOV 84 | 27.062 | 27.438 | 26.200 | 27.087 | 26.087 | 25.887 | 23.350 |

**TABLE 2 (continued)**
**Caribbean Monthly Average SST (May 1980 to October 1990)**

| | Month | Puerto Rico | Jamaica | Cayman | Cozumel | Florida | Nassau | Bermuda |
|---|---|---|---|---|---|---|---|---|
| 56 | DEC 84 | 25.586 | 26.529 | 26.357 | 25.929 | 25.143 | 25.071 | 21.443 |
| 57 | JAN 85 | 25.278 | 26.033 | 25.878 | 25.522 | 23.556 | 23.189 | 19.578 |
| 58 | FEB 85 | 25.288 | 25.562 | 25.462 | 25.828 | 22.438 | 22.938 | 19.125 |
| 59 | MAR 85 | 25.750 | 26.087 | 25.775 | 26.000 | 23.562 | 24.337 | 18.950 |
| 60 | APR 85 | 25.933 | 26.456 | 26.511 | 26.800 | 24.811 | 24.078 | 19.122 |
| 61 | MAY 85 | 26.475 | 27.575 | 27.750 | 27.913 | 26.525 | 25.750 | 21.375 |
| 62 | JUN 85 | 27.587 | 28.550 | 28.225 | 28.325 | 28.438 | 28.000 | 24.312 |
| 63 | JUL 85 | 28.188 | 29.038 | 28.625 | 28.562 | 28.562 | 28.188 | 25.775 |
| 64 | AUG 85 | 27.833 | 28.489 | 28.956 | 28.700 | 28.189 | 29.233 | 27.367 |
| 65 | SEP 85 | 28.500 | 29.400 | 29.613 | 28.700 | 28.450 | 28.475 | 27.275 |
| 66 | OCT 85 | 28.020 | 28.890 | 28.720 | 28.310 | 28.130 | 28.000 | 25.850 |
| 67 | NOV 85 | 27.629 | 28.129 | 28.086 | 28.229 | 26.671 | 26.757 | 24.386 |
| 68 | DEC 85 | 26.867 | 27.411 | 27.344 | 27.144 | 25.833 | 25.389 | 23.011 |
| 69 | JAN 86 | 25.667 | 26.633 | 26.411 | 26.378 | 25.222 | 23.944 | 21.278 |
| 70 | FEB 86 | 25.125 | 25.875 | 25.975 | 25.850 | 24.625 | 23.375 | 20.800 |
| 71 | MAR 86 | 24.913 | 25.762 | 25.788 | 25.863 | 23.913 | 23.375 | 19.913 |
| 72 | APR 86 | 25.711 | 26.389 | 26.111 | 25.978 | 24.389 | 23.644 | 20.289 |
| 73 | MAY 86 | 26.978 | 26.889 | 26.678 | 26.889 | 25.889 | 25.389 | 21.278 |
| 74 | JUN 86 | 27.286 | 27.671 | 27.729 | 28.029 | 27.657 | 26.886 | 23.929 |
| 75 | JUL 86 | 27.380 | 28.360 | 28.370 | 28.520 | 28.990 | 28.450 | 26.070 |
| 76 | AUG 86 | 28.188 | 28.587 | 28.712 | 28.850 | 29.000 | 28.837 | 27.413 |
| 77 | SEP 86 | 28.411 | 28.656 | 28.522 | 28.178 | 28.689 | 29.144 | 26.722 |
| 78 | OCT 86 | 28.000 | 28.720 | 28.670 | 28.050 | 28.020 | 28.100 | 24.620 |
| 79 | NOV 86 | 28.317 | 28.100 | 28.183 | 28.150 | 27.167 | 26.583 | 24.817 |
| 80 | DEC 86 | 27.250 | 28.450 | 28.087 | 27.337 | 26.288 | 26.875 | 22.788 |
| 81 | JAN 87 | 26.663 | 27.188 | 26.725 | 26.800 | 25.688 | 25.500 | 20.663 |
| 82 | FEB 87 | 26.786 | 26.571 | 26.700 | 26.571 | 24.429 | 24.657 | 19.686 |
| 83 | MAR 87 | 26.222 | 26.756 | 26.822 | 26.478 | 24.556 | 24.144 | 19.333 |
| 84 | APR 87 | 26.222 | 27.556 | 27.278 | 27.000 | 24.578 | 24.144 | 19.922 |
| 85 | MAY 87 | 27.000 | 28.100 | 28.000 | 28.000 | 26.438 | 26.188 | 22.062 |
| 86 | JUN 87 | 27.556 | 29.033 | 29.067 | 29.000 | 28.444 | 27.600 | 24.167 |
| 87 | JUL 87 | 28.444 | 29.678 | 29.656 | 29.111 | 29.722 | 29.367 | 27.000 |
| 88 | AUG 87 | 28.087 | 29.200 | 29.337 | 28.663 | 30.062 | 29.650 | 26.938 |
| 89 | SEP 87 | 29.322 | 29.578 | 29.711 | 30.000 | 30.078 | 29.989 | 27.489 |
| 90 | OCT 87 | 28.933 | 29.567 | 29.344 | 29.278 | 28.056 | 29.089 | 26.378 |
| 91 | NOV 87 | 28.600 | 28.157 | 28.929 | 27.743 | 27.214 | 27.071 | 24.786 |
| 92 | DEC 87 | 27.800 | 28.633 | 27.883 | 27.833 | 25.667 | 24.917 | 22.750 |
| 93 | JAN 88 | 26.725 | 27.350 | 27.012 | 27.312 | 24.750 | 24.062 | 20.250 |
| 94 | FEB 88 | 26.143 | 27.057 | 26.886 | 26.971 | 24.714 | 23.271 | 20.000 |
| 95 | MAR 88 | 25.712 | 26.675 | 26.712 | 26.812 | 24.312 | 23.375 | 19.300 |
| 96 | APR 88 | 26.250 | 27.050 | 27.363 | 27.387 | 24.125 | 24.800 | 19.163 |
| 97 | MAY 88 | 26.889 | 27.744 | 27.811 | 27.444 | 25.778 | 26.578 | 21.167 |
| 98 | JUN 88 | 27.756 | 28.389 | 28.211 | 27.989 | 27.667 | 27.589 | 24.067 |
| 99 | JUL 88 | 27.688 | 28.288 | 28.600 | 28.875 | 29.188 | 28.625 | 26.000 |
| 100 | AUG 88 | 27.856 | 29.333 | 29.244 | 28.900 | 29.033 | 28.867 | 27.822 |
| 101 | SEP 88 | 28.300 | 29.178 | 29.078 | 28.689 | 29.111 | 29.111 | 28.056 |
| 102 | OCT 88 | 28.200 | 29.600 | 29.562 | 28.065 | 28.775 | 28.788 | 26.750 |
| 103 | NOV 88 | 28.250 | 28.788 | 28.663 | 27.837 | 27.413 | 27.312 | 24.812 |
| 104 | DEC 88 | 26.867 | 27.022 | 27.167 | 27.278 | 25.556 | 24.700 | 21.311 |
| 105 | JAN 89 | 25.867 | 26.722 | 26.167 | 26.522 | 25.222 | 24.700 | 20.533 |
| 106 | FEB 89 | 25.188 | 26.125 | 25.900 | 26.000 | 25.150 | 24.500 | 20.500 |
| 107 | MAR 89 | 25.289 | 25.967 | 25.522 | 25.556 | 24.944 | 24.256 | 21.178 |
| 108 | APR 89 | 25.643 | 27.071 | 26.971 | 26.857 | 25.786 | 26.357 | 21.700 |
| 109 | MAY 89 | 26.500 | 27.920 | 27.740 | 28.200 | 27.000 | 27.620 | 22.060 |
| 110 | JUN 89 | 26.720 | 27.840 | 28.180 | 28.000 | 27.400 | 28.200 | 25.300 |

**TABLE 2 (continued)**
**Caribbean Monthly Average SST (May 1980 to October 1990)**

| | Month | Puerto Rico | Jamaica | Cayman | Cozumel | Florida | Nassau | Bermuda |
|---|---|---|---|---|---|---|---|---|
| 111 | JUL 89 | 27.512 | 28.725 | 28.550 | 28.775 | 29.188 | 28.750 | 26.700 |
| 112 | AUG 89 | 28.100 | 29.600 | 29.650 | 29.030 | 29.270 | 29.660 | 27.480 |
| 113 | SEP 89 | 28.429 | 30.300 | 30.514 | 29.071 | 29.786 | 29.429 | 27.000 |
| 114 | OCT 89 | 27.886 | 29.771 | 29.329 | 28.900 | 29.000 | 28.586 | 25.214 |
| 115 | NOV 89 | 27.925 | 28.725 | 28.300 | 28.137 | 27.062 | 27.137 | 23.725 |
| 116 | DEC 89 | 27.000 | 28.050 | 27.688 | 27.288 | 26.488 | 25.688 | 21.750 |
| 117 | JAN 90 | 26.467 | 27.078 | 26.578 | 26.711 | 25.133 | 24.167 | 21.067 |
| 118 | FEB 90 | 25.663 | 26.875 | 26.375 | 26.562 | 25.375 | 25.062 | 20.050 |
| 119 | MAR 90 | 25.022 | 26.667 | 26.611 | 26.300 | 25.333 | 24.000 | 19.756 |
| 120 | APR 90 | 25.000 | 27.667 | 27.233 | 27.267 | 25.667 | 25.667 | 20.200 |
| 121 | MAY 90 | 26.000 | 27.810 | 27.790 | 27.910 | 27.290 | 26.740 | 21.230 |
| 122 | JUN 90 | 27.000 | 28.780 | 28.560 | 28.600 | 28.940 | 28.410 | 22.280 |
| 123 | JUL 90 | 27.790 | 29.220 | 28.660 | 28.770 | 29.670 | 29.000 | 26.060 |
| 124 | AUG 90 | 28.250 | 29.840 | 30.050 | 29.230 | 30.380 | 30.400 | 27.540 |
| 125 | SEP 90 | 28.780 | 31.060 | 30.780 | 30.700 | 30.410 | 30.950 | 27.310 |
| 126 | OCT 90 | 29.110 | 30.660 | 30.020 | 30.120 | 29.300 | 29.440 | 25.980 |

**TABLE 3**
**Mean Decadal Temperatures (May 1980 to April 1990)**

| Location | MT | S.D. | Minbl | Tmin | Tmax |
|---|---|---|---|---|---|
| Puerto Rico | 27.17 | 1.17 | 29.1 | 24.91 | 29.32 |
| Jamaica | 27.84 | 1.17 | 29.6 | 25.57 | 30.30 |
| Cayman | 27.75 | 1.24 | 29.6 | 25.30 | 30.51 |
| Cozumel | 27.71 | 1.18 | 29.6 | 25.50 | 30.00 |
| Florida | 26.67 | 1.99 | 29.6 | 22.44 | 30.08 |
| Nassau | 26.42 | 1.99 | 29.3 | 22.94 | 29.99 |
| Bermuda | 23.05 | 2.95 | 27.6 | 18.18 | 28.06 |

*Note:* Mean temperatures (MT) and standard deviations (S.D.) are computed from monthly average values from May 1980 to April 1990, Minbl is the highest monthly average temperature at which bleaching was not observed, and is an estimate of the minimum bleaching temperature, Tmin and Tmax are the lowest and highest, respectively, monthly average temperatures during the baseline period.

Increasing seasonal variability at colder sites is largely due to lower minimum temperatures: the highest and lowest monthly mean values during the baseline period are both proportional to mean temperature, but the lowest values increase over three times faster with increasing mean temperature as do the highest values:

$$\text{Tmax} = 17.743 + 0.451 \times \text{MT} \quad R = 0.976 \quad P = <0.001 \quad n = 7$$

$$\text{Tmin} = -18.049 + 1.561 \times \text{MT} \quad R = 0.976 \quad P = <0.001 \quad n = 7 \quad \text{(ue 3)}$$

where Tmax and Tmin are the highest and lowest monthly average temperatures during the baseline period, respectively.

Patterns of absolute temperature maxima and of bleaching covary in both space and time:

1. Puerto Rico (Figure 4a) had widespread mass bleaching only in 1987 (Williams and Bunkley-Williams 1990; Williams et al. 1987) and 1990 (Goenaga 1991), the warmest years of the decade. Isolated local bleaching was reported in other years (Avecedo and Goenaga 1986; Goenaga and Canals 1979) but these were largely in shallow reefs downstream from river mouths, after hurricane rains caused heavy runoff and erosion.
2. Jamaica (Figure 4b) had mass bleaching in 1987 (Gates 1990; Goreau 1990a,b; Goreau and Macfarlane, 1990), 1989 (Goreau 1990a,b), and 1990, the three warmest years in the satellite SST record. The 1987 maximum temperatures were only slightly above previous maxima, but were unique in being unusually prolonged. Most previous maxima lasted only a month, but that of 1987 was prolonged for 4 months. Reef-water temperatures reached 30°C in July, and remained at those levels until December (Gates 1990). Mass bleaching began in mid-July and reached a maximum in December. Corals then gradually recovered pigmentation, with around 5% being still visibly pale in May 1988 (Goreau and Macfarlane 1990a). Temperatures in 1988 never reached above 29.5°C. The water cooled immediately following Hurricane Gilbert, the strongest hurricane measured in the Caribbean, whose eye passed over Jamaica, Cayman, and Cozumel on September 13 to 15, prior to the historical October maximum. In 1989, reef water temperatures reached 30°C by early August (Goreau 1990a,b). Bleaching was first observed in early October, and around 80% of all corals along the north coast bleached. In 1990, temperatures were even higher, and bleaching was more intense. Only minor bleaching was found on the south coast of the island, where water was observed to be noticeably cooler than the north coast. Most of Jamaica's population, almost all industry and manufacturing, and the bulk of its pollution is on the south coast, reducing the possibility of local pollution as a cause of bleaching. Such widespread mass bleaching events had never been seen in Jamaica since coral reef research began in 1951. Isolated local bleaching occurred in Jamaica in 1963 (Goreau, 1964) and 1988 (Goreau, 1990a,b) but only in shallow reefs directly affected by muddy freshwater plumes of rivers exceptionally swollen by Hurricanes Flora and Gilbert, respectively.
3. Cayman (Figure 4c) had mass bleaching in 1987, 1989, and 1990 (Hayes 1988; Hayes and Bush 1989 and 1990; Hayes, personal communication). The exceptionally warm patterns in these years are almost identical to Jamaica, with 1987 being most prolonged and 1990 hottest. Bleaching was most intense in 1990 and more intense in 1987 than in 1989, in contrast to Jamaica.
4. Cozumel (Figure 4d) bleaching was reported only in 1987 (Williams and Bunkley-Williams 1990) and 1990 (R. Sammon, personal communication). The water temperature pattern was very different from Jamaica or Cayman: 1989 was not unusually warm, and high temperatures also occurred in the early 1980s. The temperature difference with Cayman implies that central Caribbean waters are significantly mixed with waters from the Belize-Honduras and Colombia-Nicaragua embayments at Cozumel, and that either those waters were warm in the early 1980s but cooled at the end of the decade, or that Cozumel has been exposed to a greater proportion of water which has rapidly traversed the entire Caribbean. The difference may lie either in changes of cloudiness, sunshine, and rainfall in the southwestern Caribbean, or changes in ocean current flows reducing water residence time. These alternative possibilities are not distinguishable from the data available in this study, but might be tested by more detailed ocean current measurements, or by satellite altimetry. Water temperatures in 1981 were nearly as warm as in 1987, so it is possible that bleaching could have occurred then. No known reports from that period are available. Observations of bleaching at that time could strongly the hypothesis that bleaching is caused by elevated

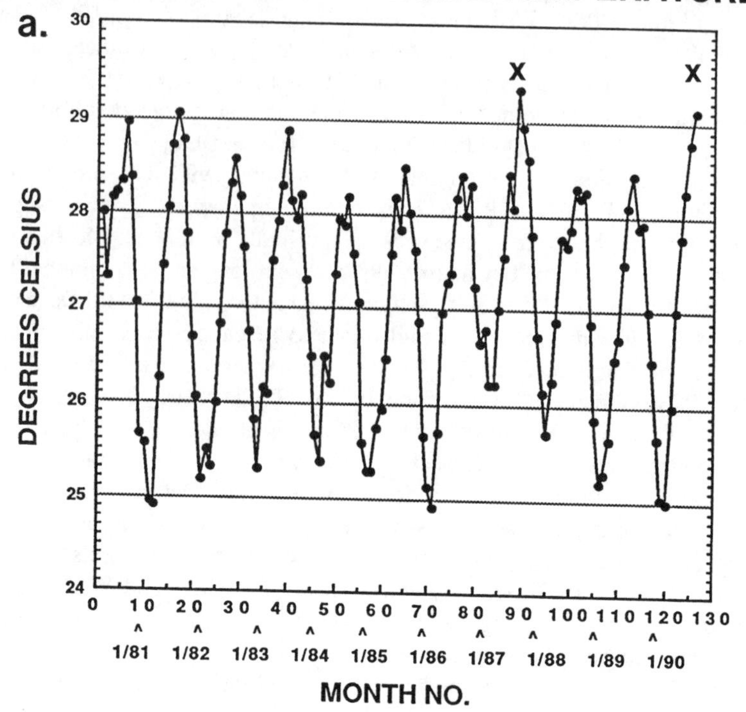


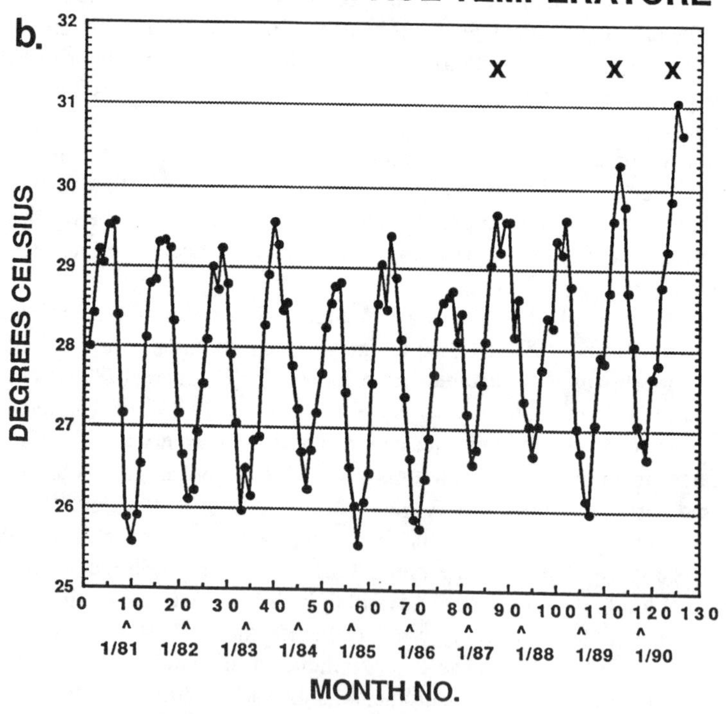


**FIGURE 4a to g.** Mean monthly SST at each site from May 1980 to October 1990 from the NOAA satellite-borne AVHRR. They have not been corrected for nearshore effects. Observed mass coral-bleaching events are marked x.

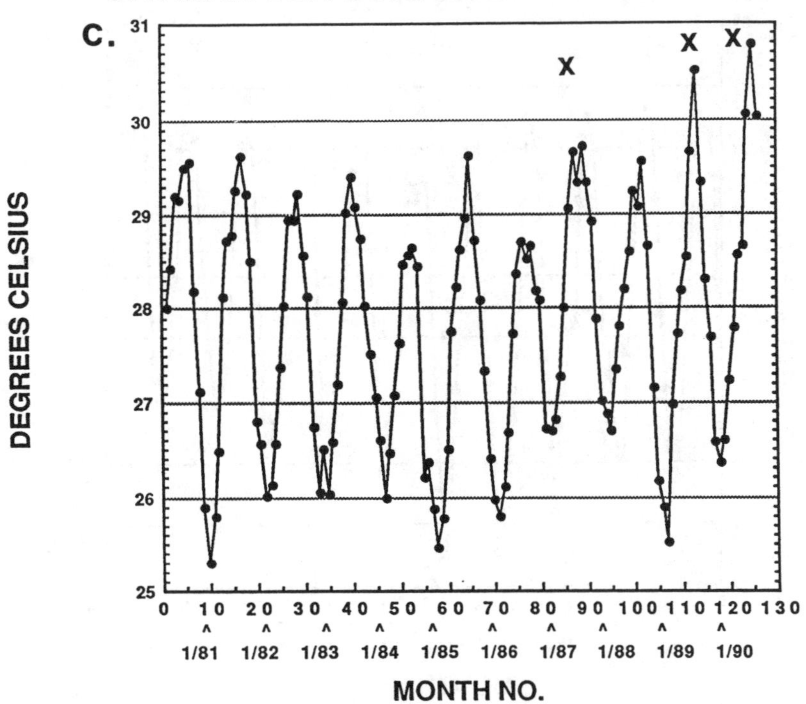


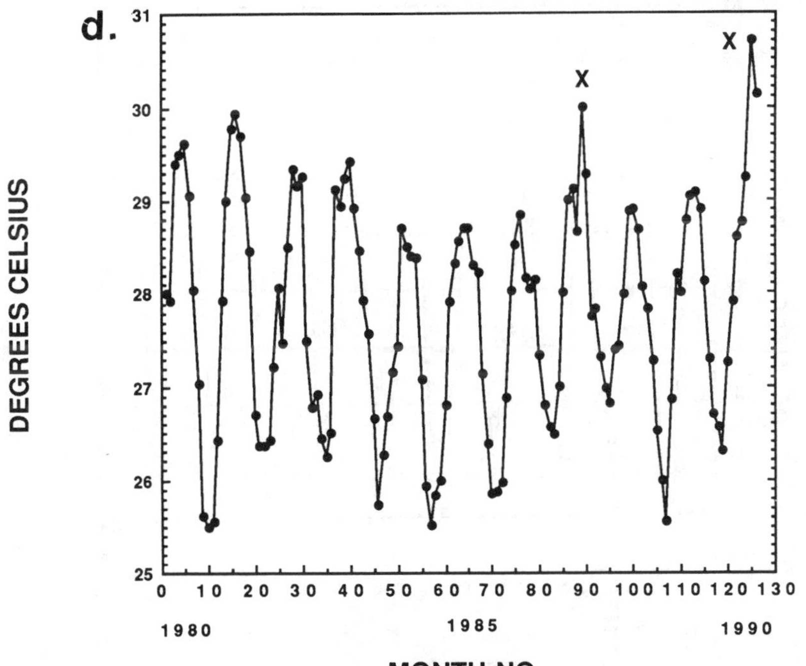


**FIGURE 4 (continued).**

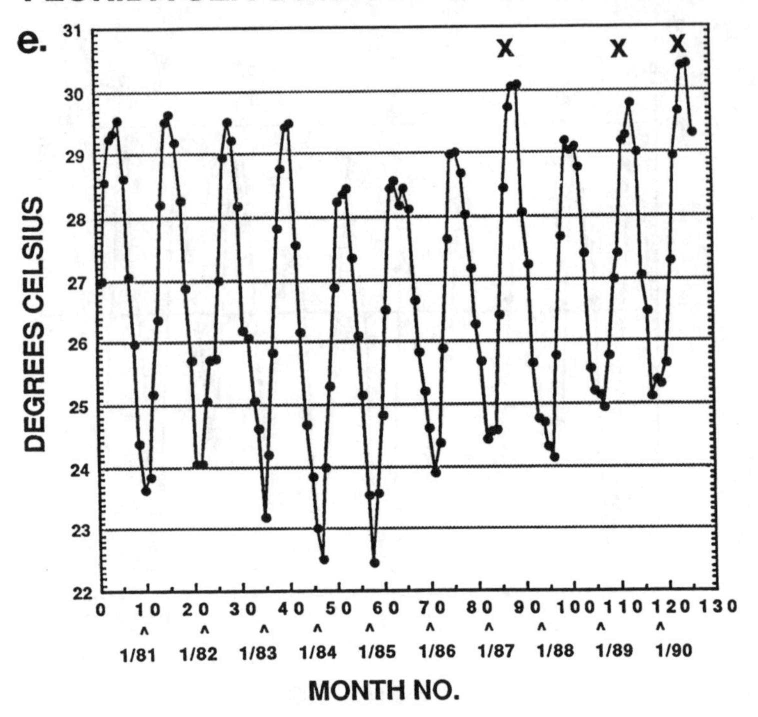


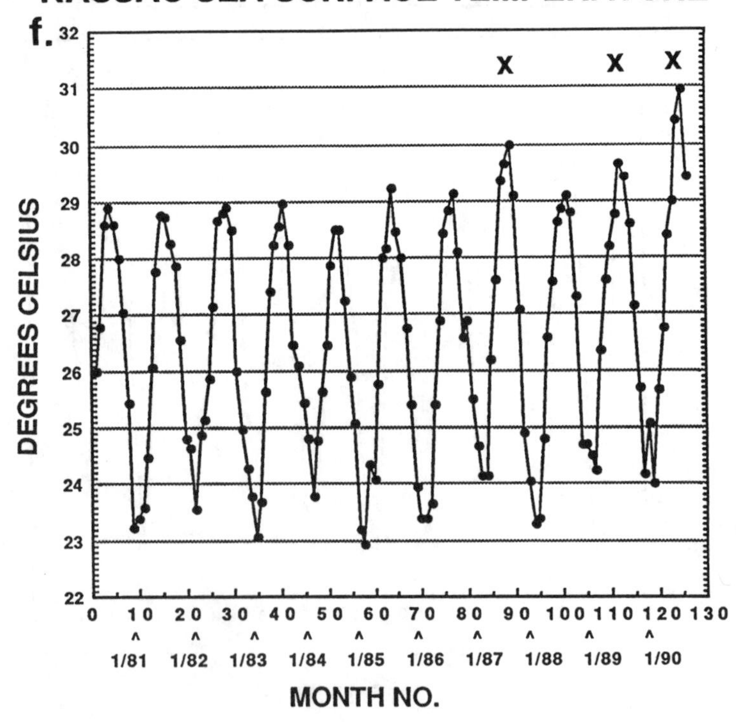


FIGURE 4 (continued).

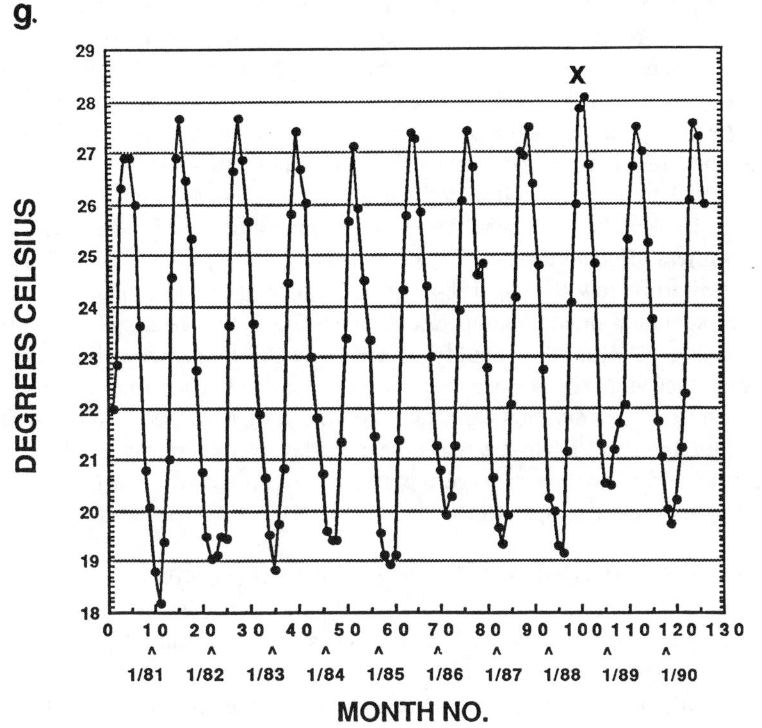


**FIGURE 4g.**

temperatures, but its absence would imply that corals have become more sensitive temperature.

5. Florida (Figure 4e) mass bleaching was most severe in 1987 (Causey 1988) 1989 and 1990 (B. Causey, personal communication), the three hottest years in the record. The 1987 bleaching was more intense than the 1989 event, which was brief because an early cold front rapidly cooled down the water. Bleaching in 1990 caused severe coral mortality. Localized bleaching was reported in 1918 (Mayor 1918), 1979 (Jaap 1979), and 1983 (Jaap 1985). The first was confined to a shallow lagoon on the Dry Tortugas, heated during calm, cloudless, sunny weather, and the second was very local in extent. The 1983 event was also local in extent. Although that year did not have an exceptional maximum monthly mean value, it included a short hot spell with one of the highest single daily water temperature value measured in shallow Florida Bay, just to the north. Fluctuations of the Florida Current could have caused highly local warming patterns around flow channels between the Keys (Jaap 1979, 1985).
6. Nassau (Figure 4f) had mass bleaching reported in 1987 (Lang), 1989 (Hayes and personal communication), and 1990 — the three hottest years in the record. 1989 was cooler, and bleaching less severe than 1987 and 1990, consistent with water temperatures.
7. Bermuda (Figure 4g) shows a different temperature pattern, with exceptional warmth occurring only in 1988. This was the first year in which mass bleaching was reported (Cook et al. 1990). Bleaching in Bermuda did not coincide with bleaching at Caribbean sites in 1987.

The spatial and temporal pattern of mass bleaching correlates very closely with the hottest water temperatures during the satellite record. Maximal temperatures and bleaching intensities in 1990 were the highest in the record at most sites. Obvious abnormalities in SST or bleaching were not seen during the strongest El Niño recorded in 1982 to 1983, or the weaker one in 1986 to 1987, and warm conditions were found in the Caribbean in 1989 despite cold water temperatures in the eastern Pacific. No correction was made for the effects of El Chichon, the Mexican volcano which put large amounts of sulfur aerosols into the atmosphere in 1982, reducing satellite-derived SST with a contribution from stratosphere aerosols (Strong 1984). The Caribbean lies directly upwind of the Mexican Cordillera, and would have been less affected than the Eastern Pacific, where SST measured from satellite records are known to be low (Strong 1984, 1989; Robock 1989; Reynolds et al. 1989).

The lowest maximum monthly temperature coincident with bleaching (see Table 3) was estimated by taking the warmest monthly maximum water temperature from the satellite record for which bleaching was known *not* to occur, Minbl. Because bleaching is likely to be sensitive to magnitude, duration, and rate of change of temperature (Hoegh-Guldberg and Smith 1989; Glynn and D'Croz 1990) higher values over shorter intervals could result in bleaching. Minimum monthly average temperatures required for bleaching are strongly correlated with mean water temperatures:

$$\text{Minbl} = 18.118 + 0.41591 \times \text{MT} \quad R = 0.953 \quad P = <0.001 \quad n = 7$$

Bleaching in Bermuda, the coldest site, took place at a temperature 1.5°C lower than at Jamaica, Cayman, Cozumel, and Florida. This implies that corals and/or their symbiotic algae physiologically or genetically adapt to local temperatures, so there is no single critical threshold temperature for bleaching across the entire geographical range of each species. However, all sites exceeded their local thresholds in the 1987 to 1990 period to an unprecedented amount.

Our temperature data suggest that coral reef bleaching is a response to exceptionally warm water temperatures. These data can also be examined to determine whether mean water temperatures rose over the period. Trends should be most readily detectable at sites with the smallest annual range, particularly Jamaica and Cayman. Figure 5a to g shows the linear regressions to the detrended monthly temperature values from May 1980 through October 1990 at each site, using May 1980 to April 1990 as a baseline period. Table 4 shows the statistical parameters of each trend. Linear regression fits to changes in SST deviations from mean values are presented. A is the y intercept in degree Celsius; B is the slope of the regression, the mean rate of change of temperature over time, in degrees Celsius per month; the next column converts these values into average warming rates in degrees Celsius per decade over the 1980s; P is the probability that this regression could result from chance; and the final column shows the uncorrected F ratios. These values should be divided by a factor to correct for a reduced number of degrees of freedom caused by "memory" lags in SST (Sciremammano 1979). This was estimated to be about 2 months or less from transient relaxation time scales. This correction did not change the significance of the temperature trends. Out of the seven sites examined, five showed significant increasing temperature trends during the 1980s. The trends in Puerto Rico and Cozumel are small and not statistically significant, but the increases at the five other sites are highly significant ($P < 0.001$).

Figure 6a to 6g shows the standard deviation (S.D.) of the temperature variance each month over the entire period. There is no statistically significant change in the monthly temperature variability at Bermuda or Puerto Rico, a barely significant rise at Nassau, and strongly significant decreases in monthly variability at Jamaica, Cayman, Cozumel, and

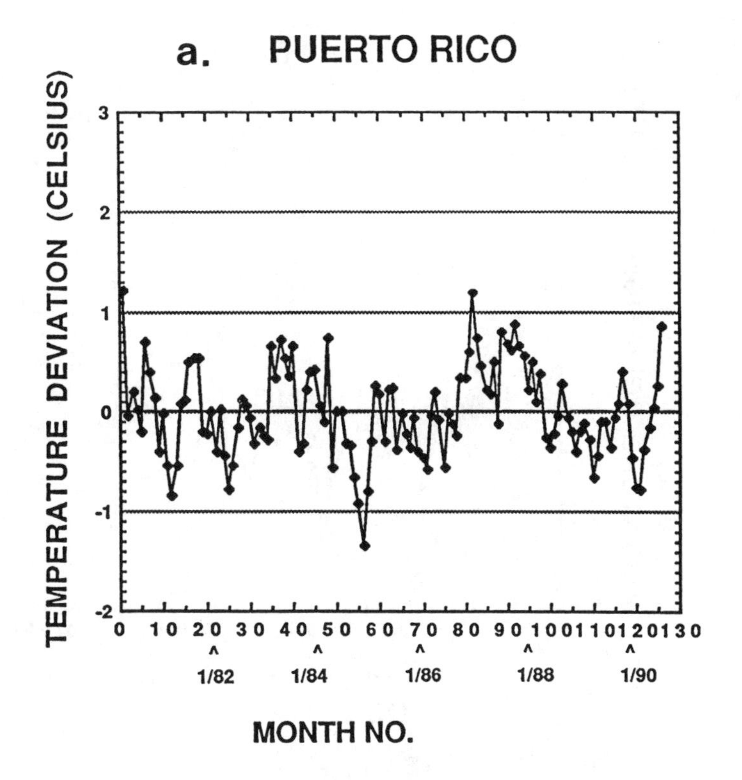


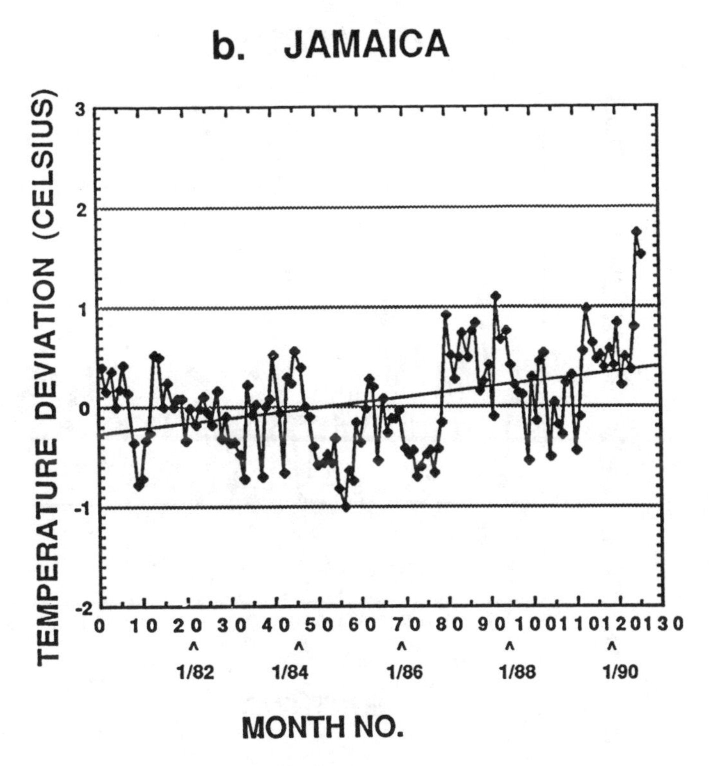


**FIGURE 5a to g.** Linear regression of detrended temperature records, whose parameters are given in Table 4.

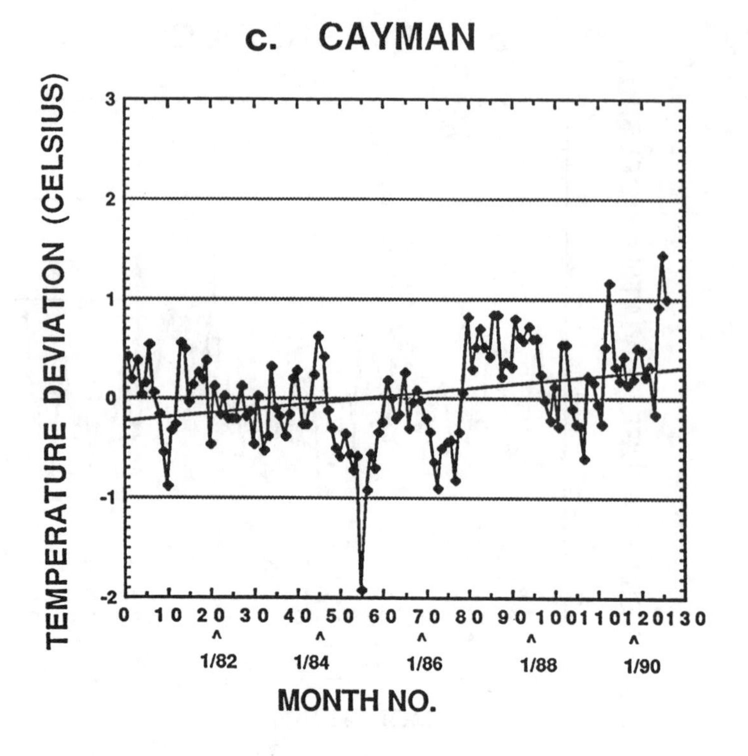


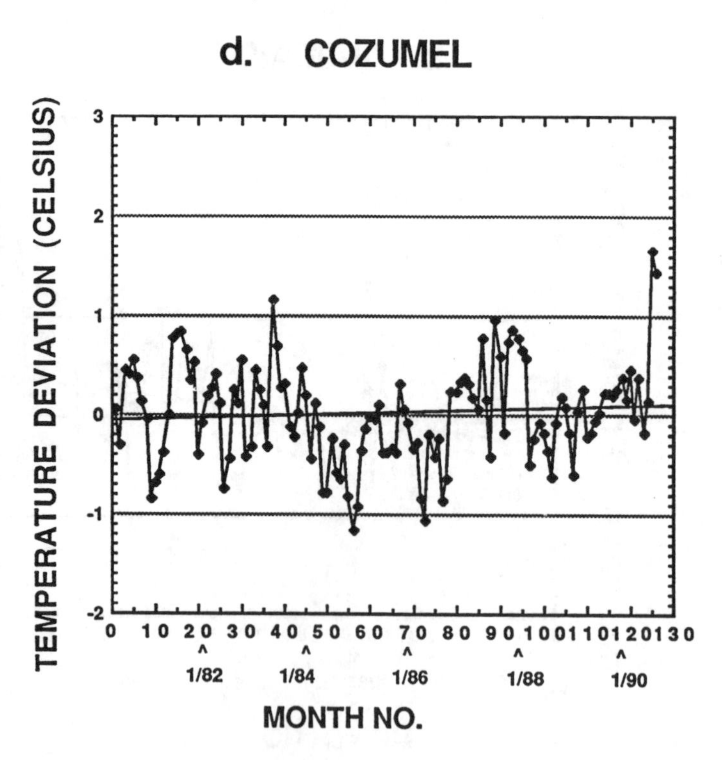


FIGURE 5 (continued).

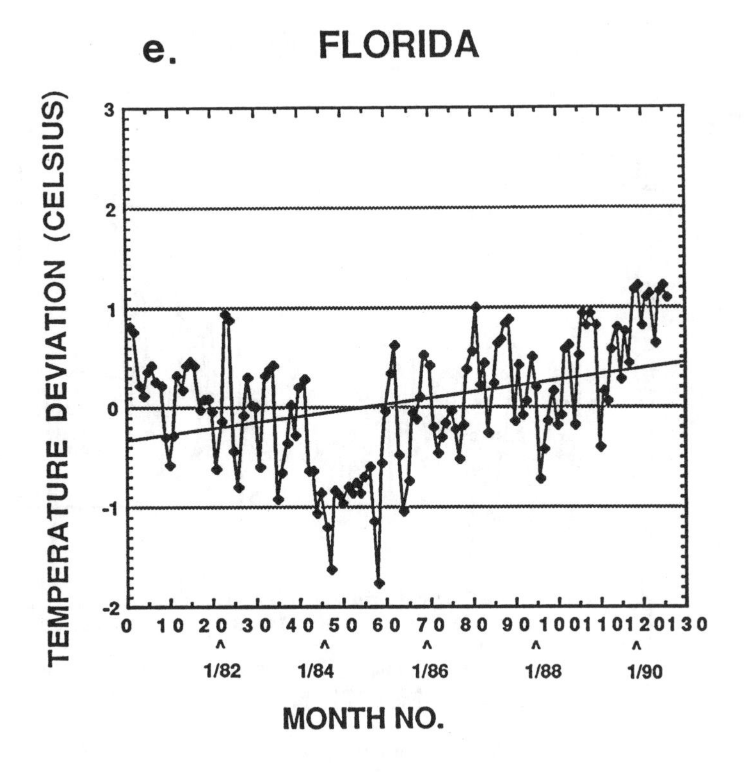


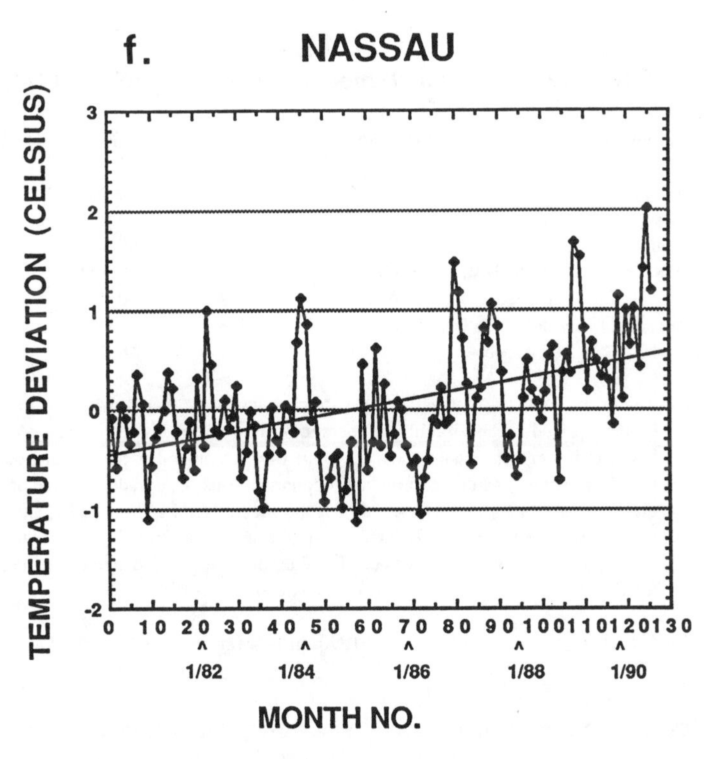


FIGURE 5 (continued).

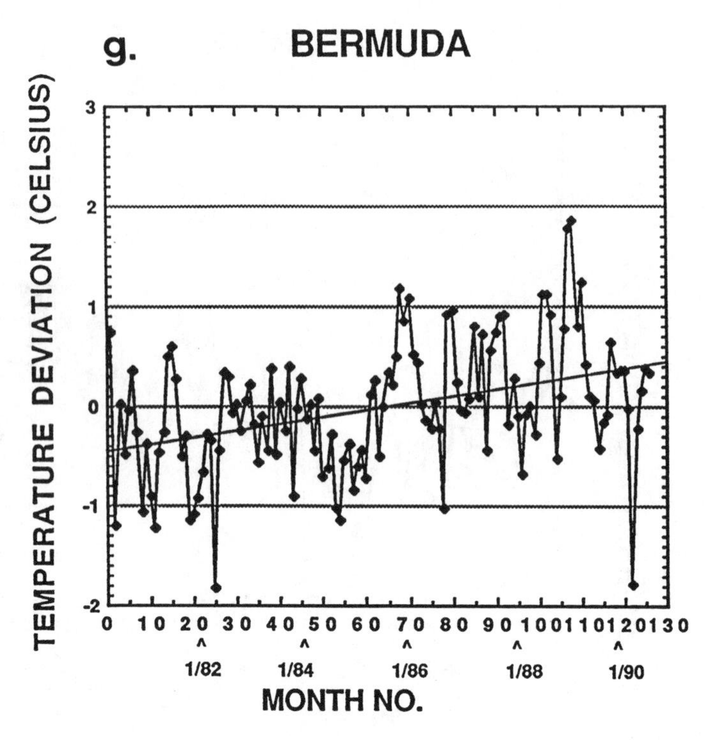


FIGURE 5g.

**TABLE 4**
**Linear Regressions to Temperature Trends (1980 to 1990)**

| Location | A(°C) | B(°C/mo) | (°C/dec) | P | F ratio |
|---|---|---|---|---|---|
| Puerto Rico | 0.00229 | −0.00005 | −0.006 | 0.967 | 0.00168 |
| Jamaica | −0.28097 | 0.00519 | 0.623 | <0.001 | 21.21659 |
| Cayman | −0.23071 | 0.00410 | 0.492 | <0.001 | 12.91092 |
| Cozumel | −0.05107 | 0.00116 | 0.139 | 0.352 | 0.87426 |
| Florida | −0.34227 | 0.00614 | 0.737 | <0.001 | 17.36845 |
| Nassau | −0.45303 | 0.00798 | 0.958 | <0.001 | 33.93642 |
| Bermuda | −0.45386 | 0.00701 | 0.841 | <0.001 | 22.76517 |

*Note:* Regression analysis of the detrended temperature data, derived by substracting the baseline-period mean monthly temperatures from the measured values. **A** is the y intercept in degrees Celsius, **B** is the slope of the temporal trend, in units of degrees Celsius per month and then in units of degrees Celsius per decade. **P** is the probability that there is no change with time, and **F** is the uncorrected F ratio. F values and numbers of degrees of freedom were divided by two to account for serial dependence (''memory effects''), and compared to critical F values for the reduced number of degrees of freedom using an F table. Significance of trends were unchanged.

[a] Values are from May 1980 to October 1990, using May 1980 to April 1990 as a baseline period.

Florida. These data suggest that the trend in increasing mean temperature is not a statistical artifact of increasing variability (Table 5), since variability has decreased at most sites.

Figure 7 shows the temperature trends month-by-month at each site. Positive values indicate that temperatures in that month have risen over the period. At all sites the increase

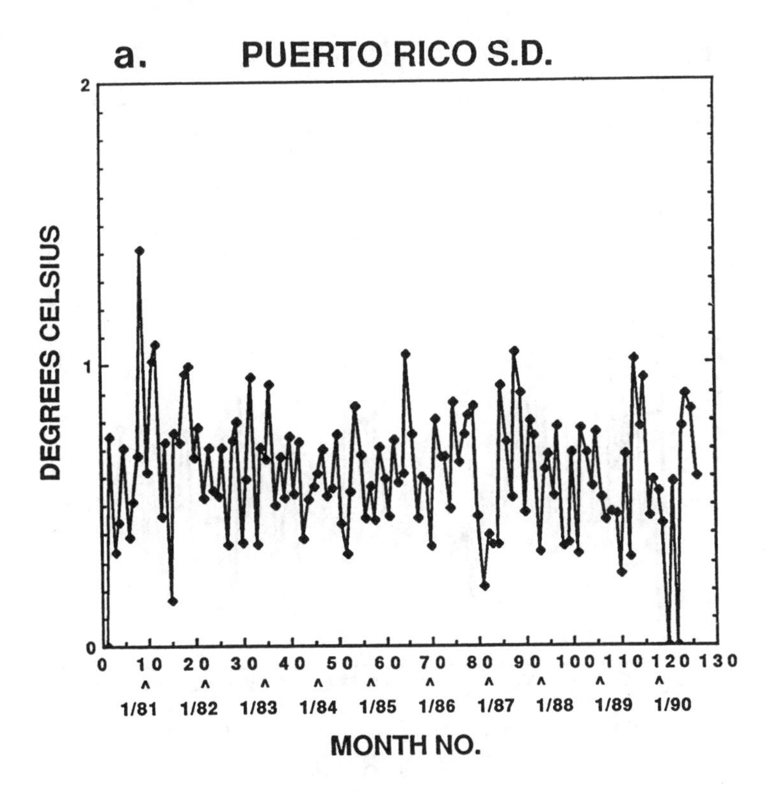


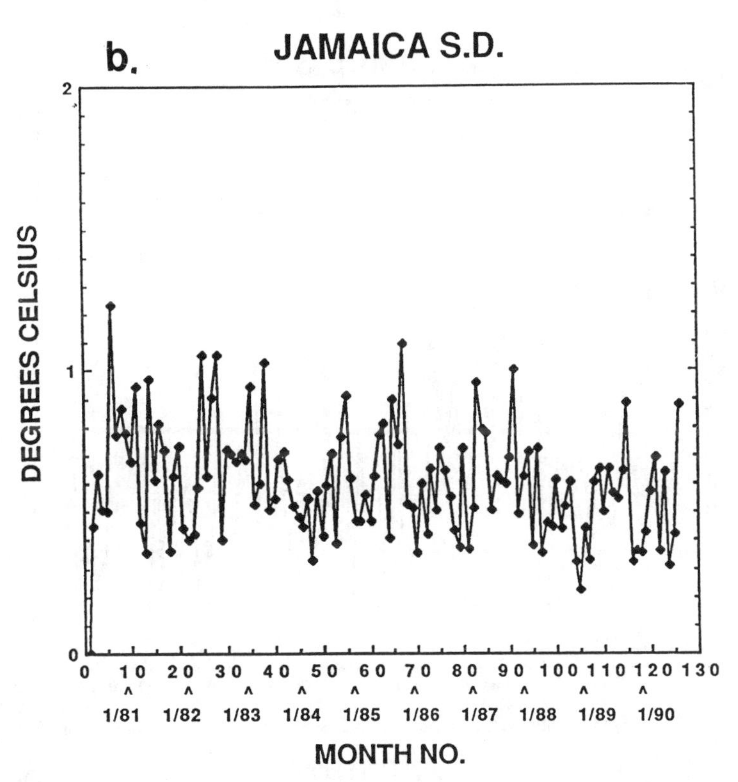


**FIGURE 6a to g.** Within-month temperature variance at each site. The value plotted is the standard deviation (S.D.) of all measurements made each month — a measure of the day-to-day temperature variability.

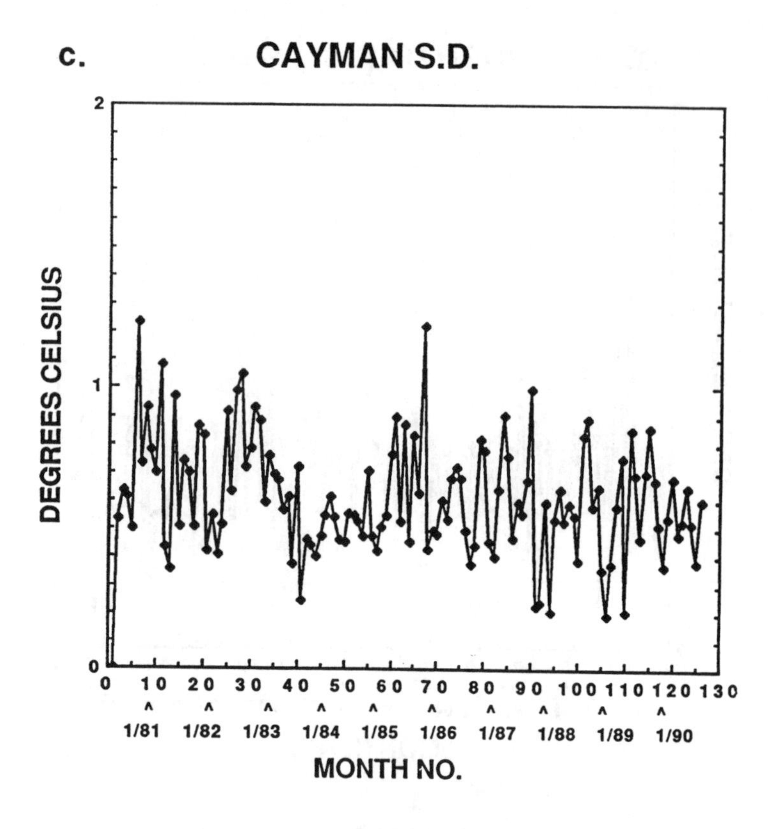


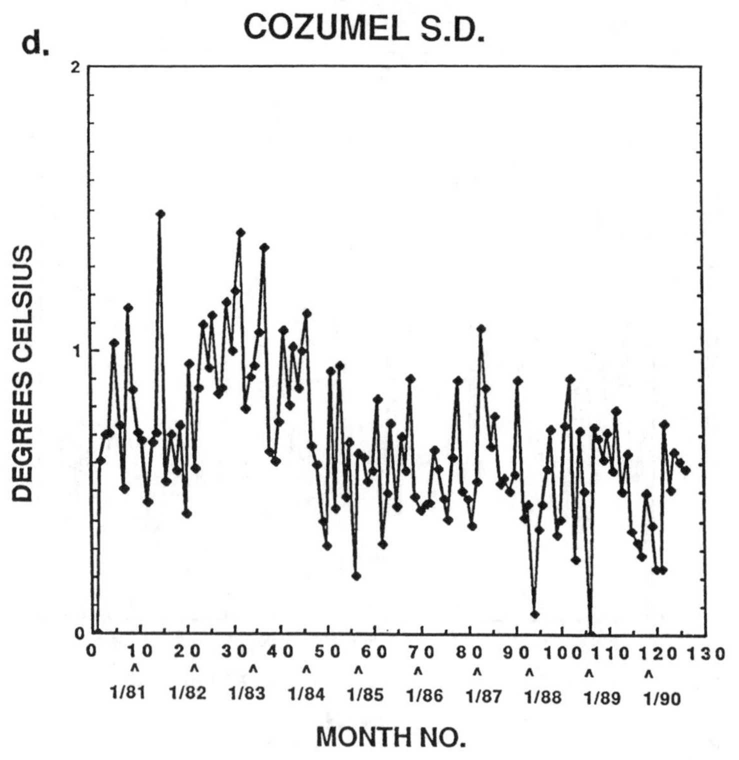


**FIGURE 6 (continued).**

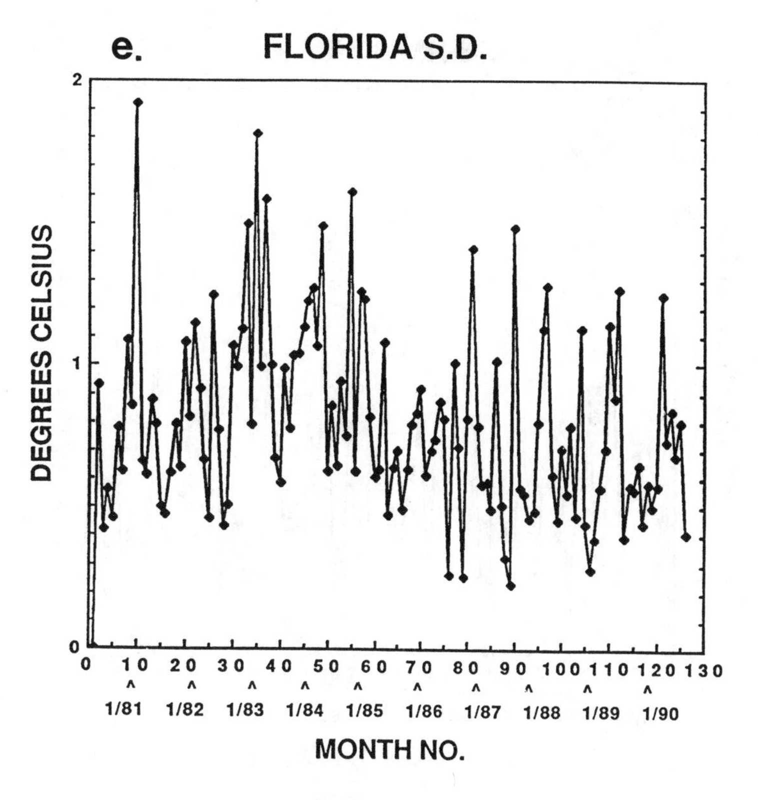


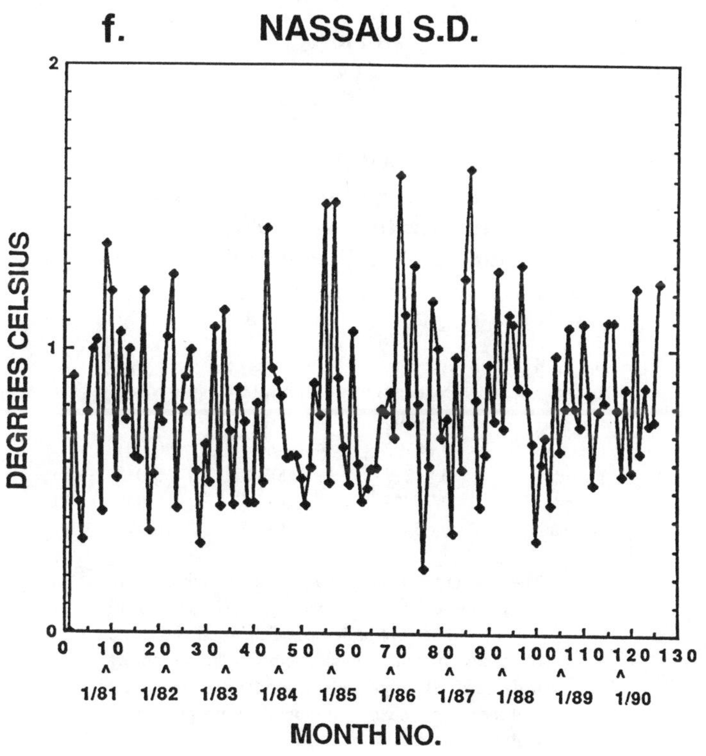


FIGURE 6 (continued).

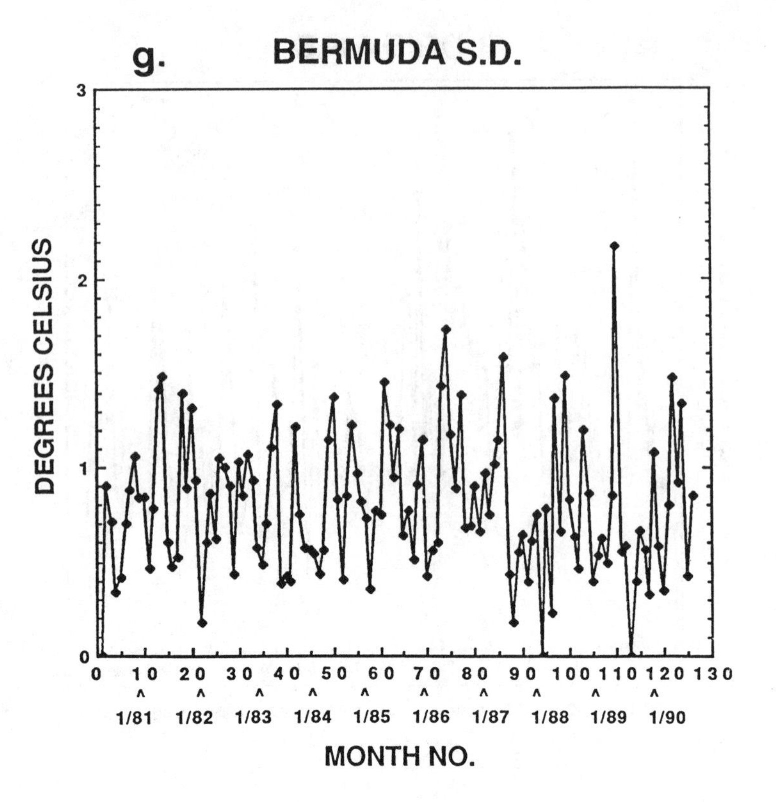


FIGURE 6g.

**TABLE 5**
**Linear Regressions to Within-Month Temperature Variability (June 1980 to October 1990)**

| Location | A(°C) | B(°C/mo) | P |
|---|---|---|---|
| Puerto Rico | 0.67785 | −0.00090 | 0.080 |
| Jamaica | 0.69986 | −0.00146 | 0.002 |
| Cayman | 0.69692 | −0.00134 | 0.008 |
| Cozumel | 0.88792 | −0.00344 | <0.001 |
| Florida | 0.93297 | −0.00208 | 0.010 |
| Nassau | 0.75967 | +0.00075 | 0.046 |
| Bermuda | 0.83446 | −0.00049 | 0.595 |

*Note:* The standard deviation of within-month temperature variability for each month was regressed against time. **A** is the y intercept in degrees Celsius, **B** is the rate of change of within-month temperature variability in degrees Celsius per month, and **P** is the probability that there is no change in monthly temperature variability.

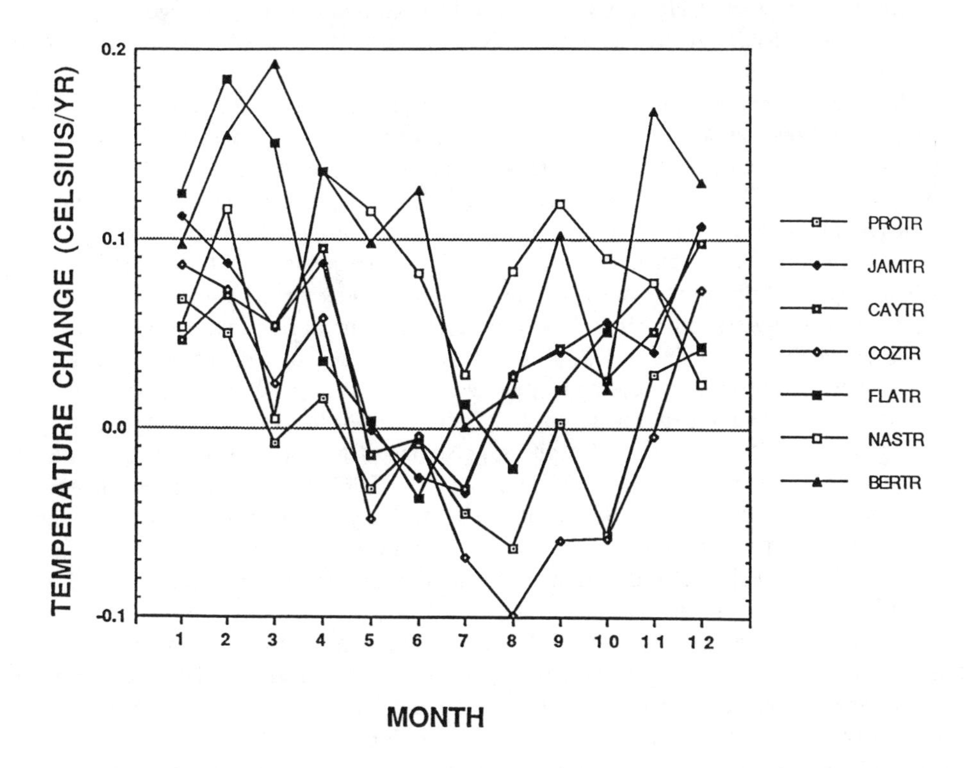


**FIGURE 7.** Monthly trends in temperature change at each site. The temperature rate of change for each month at each site is shown. At all sites the increase in temperature is greater in winter than in summer. January is month 1 and December is month 12.

in temperature is greater in winter than in summer. This implies that seasonal temperature variability is decreasing because minimum temperatures are rising faster than maximum temperatures. Although maximum temperatures are correlated with bleaching, rising minimum temperatures are reducing the length of the recovery period between bleaching events. Puerto Rico and Cozumel show a summer cooling trend.

## V. CONCLUSIONS

The exact spatial and temporal synchronies seen between bleaching and elevated water temperature provide strong support for the view that high water temperatures have been correlating recently with mass bleaching. No diseases or pathogenic organisms have been found (Hayes 1988; Glynn et al. 1985) and other factors known to induce localized bleaching are absent. An alternative hypothesis (that pollution is lowering the resistance of corals) is contradicted by intense bleaching in sites far from obvious sources of pollution. For a pollutant to cause this pattern, it would have to be ubiquitous and uniformly distributed in tropical waters, and have a strongly temperature dependent biological effect.

The fact that the high water temperatures of the late 1980s and mass coral bleaching have not been seen before in the Greater Caribbean during nearly 40 years of continuous study of coral reef ecosystems implies that temperatures have only recently exceeded tolerance limits of the corals. Reports of increased bleaching in the Indo-Pacific region in the

1980s may suggest that these are global and not regional events. Temperature data reported from Hawaii (Jokiel and Coles 1990) show that the maximum monthly average temperatures have risen by 0.764°C per decade between 1978 and 1989 ($P = 0.021$), a similar trend to that seen in the Caribbean.

The temperature data which we present in this paper cover a limited period. If bleaching events are part of a natural long-term cycle they must have a longer recurrence period than the observational record. Detailed studies of coral banding from 1918 to 1983 in eastern Florida corals reveal precise year-to-year correlations of growth rates with hydrological changes in the Everglades (Goreau et al. 1988). Cessation of growth during bleaching would cause missing years in the banding pattern. Exact temporal matches between coral banding and hydrological chronologies imply that there are no missing years between 1918 and 1983 and, hence, that massive bleaching did not occur in Florida during that period.

Our data indicate that since 1985, tropical SST has exceeded 30°C for extended periods. Heat transfer and evaporative cooling by release of excess heat to the atmosphere (Ellsaesser 1990) or cloud formation (Ramanathan and Collins 1991) have not prevented SST from reaching 31 to 32°C on occasion. Despite suggestions that the rate of warming would not be expected to exceed 1°C/20 days, (Ellsaesser 1984, 1990), we have observed changes of 1°C within 7 to 10 days. Local SST is affected by changes in current and wind patterns, changes in vertical mixing profiles, changes in cloudiness or rainfall, and changes in sea surface height, as well as global warming. Further data are needed to separate each of these as contributing factors to the changes we report.

The two coldest years, 1984, and 1986, fall in the middle of the record, and so do not bias the decadal trend. This pattern may explain why Spencer and Christie found no significant trend in their satellite microwave sounding unit (MSU) data from the 1980s (Spencer and Christy, 1990), the warmest decade on record (Jones and Wigley 1990; Hansen and Lebedeff 1987). MSU data are primarily sensitive to temperatures in the middle troposphere (channel 2) and lower stratosphere (channel 4). Since the tropical atmosphere is generally heated from below by the ocean, changes at the lower troposphere should preceed changes in the upper troposphere or stratosphere.

Only 10 years of observations from a small part of the Earth's surface cannot confirm long term global trends, and only time will tell if they are short term or continuing. Mass regional bleaching in the 1980s may be accentuated in tropical surface waters by evaporative increases in total humidity, which amplify warming by a local greenhouse effect (Flohn and Kapala 1989; Ramanathan et al. 1989a,b). Organisms in tropical habitats live at temperatures much closer to their upper tolerance limits than those elsewhere, and are negatively affected by much smaller absolute temperature increases than those in cooler habitats. Corals are highly sensitive to extreme maxima in water temperature, which become more statistically frequent as mean temperature rises (Hansen and Lebedeff 1987).

The increased level of mortality seen in 1990 implies that if bleaching continues many Caribbean reef corals will not survive. Reduced coral growth could eventually cause serious economic losses in terms of fisheries reductions, species losses, lack of water clarity and limestone sand production in tourist beach resorts, and loss of shoreline protection from hurricane waves and rising sea level (Williams and Bunkley-Williams 1990; Goreau 1990a,b; Williams et al. 1987). Coral's exquisite biological sensitivity may make it a more reliable natural indicator of rising temperature than our best available technology. Coral reef bleaching may be providing an early visible signal of global warming.

Caribbean water temperatures during 1990 reached the highest levels yet measured, according to both field measurements and the AVHRR satellite data (Hayes 1991). Coral reefs are the most species-rich oceanic ecosystem, and are major sites of ocean limestone burial. Their productivity will be severely affected if bleaching continues to inhibit coral growth.

Because tropical surface waters represent a major heat storage in the Earth's climate system, and a major driving force of winds and currents which affect global weather and climate, priority should be placed on increasing the network of systematic direct water temperature measurements in the tropics. Such a program could be most economically carried out by funding the required equipment and personnel in marine laboratories near coral reefs, and increased use of the information which the high resolution OFA data contains.

## NOTE ADDED IN PROOF

Update since this chapter was submitted: (1) 1991 — Water temperatures reached near maximum levels and strong mass bleaching occurred at all seven sites and most of the Caribbean. Elevated coral mortality was seen. (2) 1992 — Through August press time of this book, temperatures were lower, close to long-term averages, and only very minor coral bleaching was reported.

## ACKNOWLEDGMENTS

We thank Cy Macfarlane, Ruth Gates, and Ian Sandeman for providing some of the thermometer records of sea temperatures from Jamaica, and Phillippe Bush for the data from Cayman. We thank Audrey Allen, Marshall Hayes, Steve Orzack, Nora Goreau, Richard Reynolds, Kevin McCarthy, Mark Waters, Catherine Woody, Jim Lynch, Alan Strong, and Kilho Park for assistance with data, discussions, and review of the manuscript. Special thanks go to Peter Goreau for preparing Figure 1. Statistics were calculated using Statworks 1.2 software on a Macintosh computer. Partial support for this investigation was provided by a HURD grant from Howard University to RLH.

## REFERENCES

Acevedo, R. and Goenaga, C., 1986, Note on a coral bleaching after a chronic flooding in southwestern Puerto Rico, *Carib. J. Sci.*, 22, 225.

Acevedo, R., Morelock, J., and Olivieri, R., 1989, Modification of coral reef zonation by terrigenous sediment stress, *Palaios,* 4, 92–100.

Atwood, D. K., Sylvester, J. C., Corredor, J. E., Morell, J. M., Mendez, A., Nodal, W. J., Huss, B. E., and Foltz, C., 1988, Sea surface temperature anomalies for the Caribbean, Gulf of Mexico, Florida reef track, and Bahamas considered in light of the 1987 regional coral bleaching event, *Proc. Assoc. Mar. Labs. Caribbean,* 21, 47.

Brown, B. and Suharsono, 1990, Damage and recovery of coral reefs affected by El Nino related sea water warming in the Thousand Islands, Indonesia, *Coral Reefs,* 8, 163–170.

Causey, B., 1988, Observations of environmental conditions preceeding the coral bleaching event of 1987, *Proc. Assoc. Island Mar. Labs. Caribbean,* 21, 48.

Causey, B., unpublished data.

Cook, C., Logan, A., Ward, J., Luckhurst, B., and Berg, C., 1990, Elevated temperatures and bleaching on a high latitude coral reef: the 1988 Bermuda event, *Coral Reefs,* 9, 45–49.

Dustan, P., 1982, Depth dependent photoadaptation by zooxanthellae of the reef coral *Montastrea annularis, Mar. Biol.,* 68, 253–264.

Ellsaesser, H. W., 1984, The climatic effect of $CO_2$: a different view, *Atmosph. Envir.,* 18, 431–434.

Ellsaesser, H. W., 1990. Oceanic role in terrestrial climate: a commentary, in *The Ocean in Human Affairs,* pp. 117–134, S. F. Singer, Ed. Paragon House, New York.

Fankboner, P. V. and Reid, R. G. B., 1981. Mass expulsion of zooxanthellae by heat stressed reef corals: a source of food for giant clams, *Experientia* 37, 251–252.

Fisk, D. A. and Done, T. J., 1985. Taxonomic and bathymetric patterns of bleaching, Myrmidon Reef, *Proc. 5th Intern. Coral Reef Symp.* 6, 149–154.

Flohn, H. and Kapala, A., 1989. Changes of tropical sea-air interaction processes over a 30-year period, *Nature* 338, 244–246.

Gates, R., 1990. Seawater temperature and sublethal coral bleaching in Jamaica, *Coral Reefs* 8, 192–197.

Glynn, P., 1983. Extensive "bleaching" and death of reef corals on the Pacific coast of Panama, *Environ. Conserv.* 10, 149–154.

Glynn, P., 1984. Widespread coral mortality and the 1982/1983 El Niño warming event, *Environ. Conserv.* 11, 133–146.

Glynn, P. and D'Croz, L., 1990. Experimental evidence for high temperature stress as the cause of El Niño-coincident mortality, *Coral Reefs* 8, 181–191.

Glynn, P. W., Peters, E. C., and Muscatine, L., 1985. Coral tissue microstructure and necrosis: relation to catastrophic coral mortality in Panama, *Dis. Aquat. Org.*, 1, 29–37.

Goenaga, C. and Canals, M., 1990. Island-wide coral bleaching in Puerto Rico: 1990, *Carib. J. Sci.* 26, 171–175.

Goenaga, C. and Canals, M., 1979. Relacion de mortandad masiva de *Millepora complanata* (Cnidaria, Anthozoa) con alta pluviosidad y escorrentia del Rio Fajardo en Cayo Ahogado, Fajardo, *Symp. Nat. Res. San Juan, Puerto Rico* 6, 84–95.

Goreau, T. F., 1959. The ecology of Jamaican reefs. I. Species composition and zonation, *Ecology* 40, 67–90.

Goreau, T. F., 1964. Mass expulsion of zooxanthellae from Jamaican reef communities after Hurricane Flora, *Science* 145, 383–386.

Goreau, T. F. and Goreau, N. I., 1959. The physiology of skeleton formation in corals. II. Calcium deposition by hermatypic corals under various conditions in the reef, *Biol. Bull.* 117, 239–250.

Goreau, T. F. and Goreau, N. I., 1960. Distribution of labelled carbon in reef building corals with and without zooxanthellae, *Science* 131, 668–669.

Goreau, T. F. and Goreau, N. I., 1973. The ecology of Jamaican reefs. II. Geomorphology, zonation, and sedimentary phases, *Bull. Mar. Sci.* 23, 399–464.

Goreau, T. F., Goreau, N. I., and Goreau, T. J., 1979. Corals and Coral Reefs, *Sci. Amer.* 241, 124–136.

Goreau, T. J., 1990a. Coral bleaching in Jamaica, *Nature* 343, 417.

Goreau, T. J., 1990b. Coral bleaching, *New Scientist,* (Jan. 13, 1990b) p. 76.

Goreau, T. J. and Dodge, R. E., unpublished data.

Goreau, T. J., Dodge, R. E., Goreau, P., and Dunham, J., 1988. Coral fluorescence records Everglandes hydrology, 1918–1983, *Proc. Assoc. Island Mar. Labs. Caribbean* 21, 43.

Goreau, T. J. and Macfarlane, A. H., 1990a. Reduced growth rate of *Montastrea annularis* following the 1987–1988 bleaching event, *Coral Reefs* 8, 211–216.

Goreau, T. J. and Macfarlane, A. H., 1990b. Mass bleaching, temperature, and Hurricane Gilbert: effects on Jamaican coral growth, 1987–1990, *Proc. Assoc. Mar. Labs. Caribbean* 23, 141.

Hansen, J. and Lebedeff, S. J., 1987. Global trends of measured surface temperature, *Geophys. Res.* 92, 13,345–13,372.

Harriott, V. J., 1985. Mortality rates of scleractinian corals before and during a mass bleaching event, *Mar. Ecol. Prog. Ser.* 21, 81–88.

Hayes, R. L., 1988. Histological and histochemical comparisons of bleached and normal tissues from *Agaricia lamarckii, Proc. Assoc. Island Mar. Labs. Caribbean* 21, 12.

Hayes, R. L., 1991. Testimony presented to the U.S. House of Representatives, Committee on Health and the Environment, February, 1991.

Hayes, R., unpublished data.

Hayes, R. and Bush, P., 1990. Microscopic observations of recovery in the reef-building scleractinian coral *Montastrea annularis,* after bleaching on a Cayman reef, *Coral Reefs* 8, 203–209.

Hoegh-Guldberg, O., McCloskey, L. R., and Muscatine, L., 1987. Expulsion of zooxanthellae by symbiotic cnidarians from the Red Sea, *Coral Reefs* 5, 201–204.

Hoegh-Guldberg, O. and Smith, G. J., 1989. Light, salinity, and temperature and the population density, metabolism, and export of zooxanthellae from *Stylophora pistillata* and *Seriatopora hystrix, J. Exper. Mar. Biol. Ecol.* 129, 279–303.

Jaap, W., 1979. Observation on zooxanthellae expulsion at Middle Sambo Reef, Florida Keys, *Bull. Mar. Sci.* 29, 414–422.

Jaap, W., 1985. An epidemic zooxanthellae expulsion during 1983 in the lower Florida Keys coral reefs: hyperthermic etiology, *Proc. 5th Intern. Coral Reef Symp.* 6, 143–148.

Jokiel, P. L. and Coles, S. L., 1990. Resposne of Hawaiian and other Indo-Pacific corals to elevated temperatures, *Coral Reefs* 8, 155–162.

Jones, P. and Wigley, T. M. L., 1990. Global warming trends, *Scientific American* 263, 84–91.

Kidwell, K. B., 1986. *NOAA Polar Orbital Data Users Guide.*

Kleppel, G. S., Dodge, R. E., and Reese, C. J., 1989. Changes in pigmentation associated with the bleaching of stony corals, *Limnol. Oceanogr.* 34, 1331–1335.

Lang, J., Wicklund, R., and Dill, R., in press. Unusual depth-related bleaching of reef corals near Lee Stocking Island, Bahamas, *Proc. 6th Intern. Coral Reef Symp.*

LaPointe, B., O'Connell, J., and Garrett, G., 1990. Nutrient couplings between on-site sewage disposal systems, groundwaters, and nearshore surface waters of the Florida Keys, *Biogeochemistry* 10, 289–307.

Lasker, H., Peters, E., and Coffroth, M. A., 1984. Bleaching of reef coelenterates in the San Blas Islands, Panama, *Coral Reefs* 3, 183–190.

Logan, A., 1988. *Holocene Reefs of Bermuda, Sedimenta* X, 62pp.

Mayor, A. G., 1918. Toxic effects due to high temperature, *Carnegie Inst. Wash. Pap. Mar. Biol.* 12, 175–178.

McCormack, R. C. and Strong, A. E., 1990. Correlation between sea surface temperature trends and Caribbean coral bleaching events, *Eos* 71, 104.

Nishihira, M., 1987. Natural and human interference with the coral reef and coastal communities in Okinawa, *Galaxea* 6, 311–324.

Oliver, J., 1985. Recurrent seasonal bleaching and mortality of corals on the Great Barrier Reef, *Proc. 5th Intern. Coral Reef Symp.* 4, 201–206.

Porter, J., Fitt, W., Spero, H., Rogers, C., and White, M., 1989. Bleaching in reef corals: physiological and stable isotopic responses, *Proc. Nat. Acad. Sci.* 86: 9342–9346.

Porter, J., personal communication; unpublished data.

Ramanathan, V., Barkstrom, B. R., and Harrison, E. F., 1989. Climate and the Earth's radiation budget, *Physics Today* 42, 22–32.

Ramanathan, V., Cess, R. D., Harrison, E. F., Minnis, P., Barkstrom, B. R., Ahmad, E., and Hartmann, D., 1989. Cloud-radiative forcing and climate: results from the Earth Radiation Budget Experiment, *Science* 342, 57–63.

Ramanathan, V. and Collins, W., 1991. Thermodynamic regulation of ocean warming by cirrus clouds deduced from observations of the 1987 El Niño, *Nature* 351, 27–32.

Reimer, A., 1971. Observations on the relationship between several species of tropical zoanthids (Zoanthideea, Coelenterata) and their zooxanthellae, *J. Exper. Mar. Biol. Ecol.* 7, 207–217.

Reynolds, R. W., Folland, C. K., and Parker, D. E., 1989. Biases in satellite-derived sea-surface-temperature data, *Nature* 341, 728–731.

Reynolds, R. W., personal communication.

Robock, A., 1989. Satellite data contamination, *Nature* 341, 695.

Sammon, R., personal communication, 1990.

Sciremammano, F., 1979. A suggestion for the presentation of correlations and their significance levels, *J. Phys. Oceanogr.* 9, 1273–1276.

Spencer, R. and Christy, J., 1990. Precise monitoring of global temperature trends from satellites, *Science* 247, 1558–1562.

Steen, R. G. and Muscatine, L., 1987. How temperature evokes rapid exocytosis of symbiotic algae by a sea anemone, *Biol. Bull.* 172, 245–263.

Strong, A. E., 1984. Monitoring El Chichon aerosol distribution using NOAA-7 satellite AVHRR sea surface temperature observations, *Geofisica Int.* 23, 129–141.

Strong, A. E., 1989. Greater global warming revealed by satellite-derived sea-surface-temperature trends, *Nature* 338, 642–645.

Strong, A. E., 1989. Satellite data contamination, *Nature* 341, 695.

Strong, A. E. and McCormack, R. C., 1991. Coral bleaching and sea surface temperatures, *Proc. Oceanogr. Soc. Conf.*, St. Petersburg, FL.

Sverdup, H., Johnson, M., and Fleming, R., 1942. *The Oceans: their Physics, Chemistry, and General Biology*, Prentice-Hall, Englewood Cliffs, N.J.

Trench, R., 1986. Dinoflagellates in non-parasitic symbioses, in *Biology of Dinoflagellates*, F. R. J. Taylor, Ed., Blackwell, Oxford, 530–570.

Williams, E. H., Testimony presented to the U.S. House of Representatives, Committee on Health and the Environment, February, 1991.

Williams, E. and Bunkley-Williams, L., 1990. The world-wide coral reef bleaching cycle and related sources of coral mortality, *Atoll Res. Bull.* 335, 1–71.

Williams, E., Goenaga, C., and Vicente, V., 1987. Mass bleaching on Atlantic coral reefs, *Science* 238, 877–878.

Yonge, C. M. and Nicholls, A. G., 1931. Studies on the physiology of corals, 5, *Great Barrier Reef Expedition Scientific Papers* 1, 135–176; 177–211.

Chapter 10

# GEOCHEMICAL METHODS TO EVALUATE METHANE CONCENTRATIONS IN SOILS AND SEDIMENTS

**Ronald C. Pflaum**

## TABLE OF CONTENTS



0-8493-4419-0/93/$0.00 + $.50
© 1993 by CRC Press, Inc.

# I. INTRODUCTION

The study of methane as it relates to global warming has addressed the relative importance of several natural and anthropogenic sources including wetlands, rice paddies, and cattle production. Additional inputs of fossil fuel methane may result as a consequence of oil and gas production, and from the seepage of natural gas from marine and onshore reservoirs. Increased temperature conditions may also cause releases of methane currently held in soil and shallow marine sediments. Releases of methane from soil and sediment, along with the input from wetlands and seepage of natural gas, represent inputs which are not subject to control by regulation or legislation. Any measures proposed to stem the increase in the atmosphere methane content should consider these factors carefully in order to determine the length of time necessary to evaluate the methods effectiveness. The application of a technique developed for surface geochemical prospecting for oil and natural gas, which measures the concentrations of low-molecular weight hydrocarbons (LMWH) bound in soils and sediments, may allow investigators to determine more accurately the methane inventory for a given sample and the probability of that methane being released by altered climatic conditions. The following is a brief description of the historical development of the "adsorbed gas" or "bound gas" method, a detailed description of the technique, and a discussion of how modifications of the technique may provide data which allow the design of more accurate models for the effects of global warming.

# II. HISTORICAL DEVELOPMENT

Most of the research involving bound methane and other LMWH has been linked to geochemical exploration for petroleum. The historical development of the technique must be viewed in light of this goal. Further, in petroleum exploration studies it is vital to distinguish between methane of biogenic and thermogenic origin (Bernard et al., 1978; Claypool and Kvenvolden, 1983; Pflaum, 1989). For the purposes of this discussion, however, such a distinction is unnecessary and would require needless additional analytical effort.

The common method to determine the methane content of a sediment or soil sample is based on measurement of the gases partitioned into the headspace of an enclosed space or container (Bernard et al., 1978). Headspace techniques are very useful and form the first step in determining the total methane inventory of a sample. Among the earliest direct applications of headspace techniques to soil analysis, used first in Germany, was the "interstitial air" method, which was performed by sealing off a shallow bore hole for several days and measuring the methane concentration in the hole (Laubmeyer, 1933). The interstitial air method suffered from the low sensitivity of the combustion techniques used to quantify the bulk concentration of LMWH prior to the advent of gas chromatography methods, and early experiments suggested that analysis of bound (adsorbed and occluded) hydrocarbons would yield a greater concentration of LMWH than the analysis of interstitial air. Bound gas techniques are also superior to the technique of sealing off a bore hole because sampling is faster and it is possible to collect samples in marshy areas where the soil contains enough water to fill the hole. A procedure to release hydrocarbons from soil was developed (Horvitz, 1939), and later improved with the addition of an acidification step (Horvitz, 1954). Many investigators have questioned the rationale for the acid extraction technique. Horvitz mentions that acidification of the soil samples under vacuum was an improvement adopted in the 1940s to decompose carbonate minerals. He states that, from the time of the earliest attempts to improve on the techniques of Laubmeyer, carbonates were thought to play an important role in bound gas distribution (Horvitz, 1954). Two patents from the early 1970s also claim that carbonate minerals are the key elements in the mechanism for binding the LMWH to the sediments (Thompson, 1970; Thompson et al., 1974). For the purpose of determining the concentration of methane which may be released from soils or sediments by a warming of the environment, it may not be desirable to release those LMWH occluded in carbonate minerals. Fortunately, the method is sufficiently flexible to allow the investigator to choose the temperature and chemical treatment appropriate to his or her particular study.

## III. METHODS

The methods are detailed here as they would be used to maximize the release of LMWH for geochemical prospecting. Specialized modifications can be made to optimize the methods for other types of studies.

Samples to be analyzed for free or bound LMWH are taken in duplicate and each is placed in a No. 2 size tin-alloy can having a volume of approximately 600 ml. For free gas analysis in soils, 100 ml of helium purged, distilled water is added to improve sample breakup during agitation. Before the cans are sealed, the headspace is purged with nitrogen or helium to reduce background hydrocarbon levels and to impede the activities of aerobic organisms. Samples should fill 70 to 80% of the volume of the can. Cans are sealed with a mechanical canner which rolls and crimps the edges of the can and lid to form a gas-tight seal. Samples are stored in a freezer maintained at $-20°C$ until analyzed.

The analysis scheme for LMWH consists of two steps: (1) removal of the free or bound LMWH from the sediment and (2) quantification of the hydrocarbons by gas chromatography. For geochemical prospecting there is a third step, the isolation and combustion of the methane to $CO_2$ and the subsequent measurement of the $^{13}C/^{12}C$ ratio by isotope ratio mass spectrometry (Pflaum, 1989).

Concentrations of headspace LMWH are measured on gases partitioned into the headspace following warming of the sample containers to room temperature and agitation on a high-speed shaker. This technique results in a very favorable partitioning of the gases into the headspace. Each can is weighed to determine the mass of the sample, subtracting out the weight of any water added, and is then placed in a metal frame having a special fitting which pierces the can and simultaneously seals the hole in such a way that aliquots of headspace gas can be withdrawn through a septum with a gas-tight syringe. Alternatively, headspace gases can be swept from the container and isolated with a dual septa, purge and trap system (Bernard et al., 1978). Following removal of the headspace gas, the container is removed from the frame, opened, and the volume of the headspace determined by measuring the amount of water necessary to fill the can.

Bound gas analysis is performed on samples of undried bulk soil or sediment weighing from 60 to 100 g. Use of larger samples is possible by increasing the overall volume of the extraction line (Figure 1). The bound LMWH are released by placing the sample in a vacuum extraction line patterned after the one described in Horvitz (1972) and incorporating several modifications similar to those described by Wesson and Armstrong (1975). The sample is placed in reservoir D (Figure 1). The 1000 ml round bottom flask C is filled with 200 to 300 ml of a 50% solution (w/v) of potassium hydroxide (KOH). The system is closed and then evacuated with a rotary vacuum pump. A 50% solution (w/v) of phosphoric acid ($H_3PO_4$) is added to the sample slowly, 10 to 25 ml at a time, from reservoir A to avoid rapid $CO_2$ generation from the decomposition of carbonates. Once wetted with the acid, the sample is agitated by a magnetic stirrer to improve contact between the sample and the viscous $H_3PO_4$ solution. The reservoir C is agitated by a second magnetic stirrer to assist in the uptake of evolved $CO_2$ by the KOH. Phosphoric acid is added until the sample no longer generates $CO_2$ (i.e., ceases to bubble). A phosphoric acid excess of 25 ml is then added and a heating

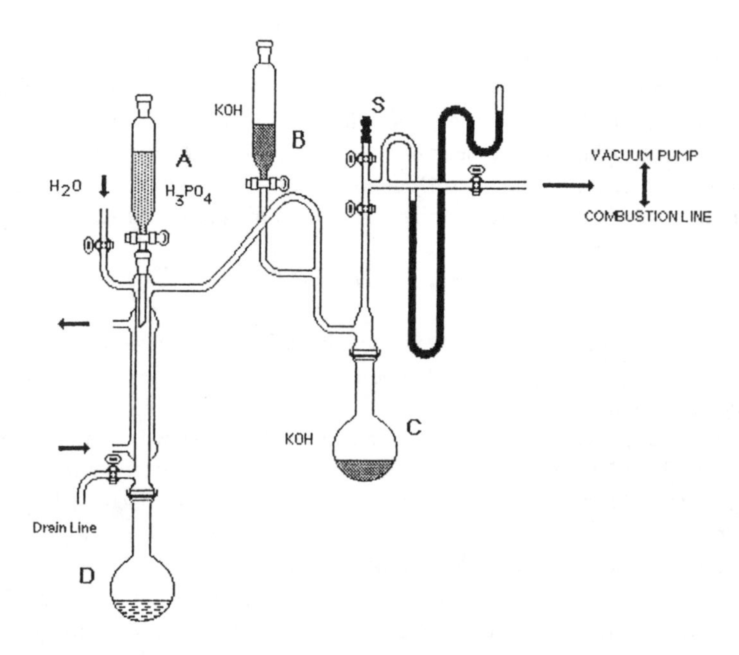


**FIGURE 1.** Apparatus for the release of bound gases from soil or sediment. Modified from Horvitz, (1972).

mantle is placed between the same reservoir and the magnetic stirrer. The sample is heated to approximately 100°C, or whatever temperature is desired, and the temperature held there for 30 min (total time 45 to 50 min). The heating mantle and stirrers are then switched off and the system is filled with distilled water to concentrate the gases released into a smaller volume. Distilled water is used because the process of heat distillation effectively degases the water so that it does not contribute to the volume of gases released. The water is added slowly over a period of 10 to 15 min to fill the system uniformly and avoid trapping the released gases in small pockets against the walls of the extraction line. The water will stop flowing into the system when the back pressure becomes equal to the hydrostatic head on a water reservoir located approximately 2 ft (0.6 m) above the $H_2O$ inlet valve. A pressurized faucet should not be used to supply water to the system as the pressure may be enough to cause the system components to separate. Once the released gas is concentrated, its volume is measured manometrically and an aliquot for molecular analysis is withdrawn through the septum located at S. Molecular analysis of LMWH is typically performed on a gas chromatograph equipped with a packed column, a flame-ionization detector (FID), and an electronic integrator.

## IV. DISCUSSION

The method described above allows the investigator to vary the temperature at which bound hydrocarbons are released and to choose the type and concentration of the acid used, or to use no acid at all. Table 1 shows a portion of the data collected on replicate analyses of a typical marine sediment from the East Breaks region of the Gulf of Mexico. In this series of experiments different temperatures, heating times, and concentrations of phosphoric acid were used to evaluate the optimal parameters of the method for geochemical prospecting (Pflaum, 1989). It can be readily observed that relatively little methane was released from this particular sediment without the acidification step. Whether acidification is needed for release of significant concentrations of methane from all samples is unclear.

**TABLE 1**
**Bound Gas Extraction Parameter Matrix — East Breaks Sediment**

| $H_3PO_4$ (%) | Temp (°C) | Time (min) | Methane (ml/100g dry sediment) |
|---|---|---|---|
| 0 | 100 | 30 | 0.20 |
| 0 | 150 | 30 | 0.14 |
| 10 | 25 | 30 | 3.14 |
| 10 | 50 | 30 | 4.32 |
| 10 | 50 | 120 | 4.53 |
| 10 | 100 | 30 | 3.53 |
| 50 | 100 | 30 | 4.39 |
| 50 | 100 | 30 | 4.40 |
| 50 | 100 | 30 | 4.15 |
| 50 | 25 | 30 | 4.40 |
| 50 | 100 | 120 | 3.76 |
| 85 | 150 | 30 | 4.16 |

Most published geochemical surveys which differentiate between methane and other LMWH and report values for both free and bound gases have been done on marine sediments. In marine sediments, where either free or bound methane may dominate, free methane is usually found in higher concentrations in anoxic environments, while bound methane is higher in oxic sediments. In the surface sediments of the anoxic Orca Basin (Gulf of Mexico) the free methane averages 80.5 ml/100 g (dry weight), while the bound methane averages 4.5 ml/100 g (Pflaum, 1989). In the similarly anoxic Santa Barbara Channel, Horvitz (1986) found free methane concentrations ranging from $<2 \times 10^{-5}$ to $>6$ ml/100g (wet weight), while bound methane in the same samples ranged from $<150$ to $>350$ ppb by weight. In the oxic sediments of the Pigmy Basin (Gulf of Mexico) free methane averages 0.1 ml/100 g and bound methane averages 2.4 ml/100 g (Pflaum, 1989). Typical arable soils will probably have methane distributions similar to oxic sediments, while wetlands and rice paddies will more closely mirror anoxic sediments.

While it is most probable that releases of methane to the atmosphere due to climatic changes will be greatest from soils and sediments with high concentrations of free methane, it is still useful to quantify the total methane inventory. An understanding of the distribution between free and bound methane and the conditions under which either can be released to the environment will increase our knowledge of what limits can be placed on the ability to regulate the increase of atmospheric methane.

## REFERENCES

Bernard, B. B., Brooks, J. M., and Sackett, W. M., Light hydrocarbons in recent Texas continental shelf and slope sediments, *J. Geophys. Res.*, 83, 4053–4061, 1978.

Claypool, G. E. and Kvenvolden, K. A., Methane and other hydrocarbon gases in marine sediment, *Annu. Rev. Earth Planet. Sci.*, 11, 299–327, 1983.

Horvitz, L., On geochemical prospecting I., *Geophysics*, 4, 210–225, 1939.

Horvitz, L., Near-surface hydrocarbons and petroleum accumulation at depth, *Min. Eng. Rev.*, 1205–1209, 1954.

Horvitz, L., Vegetation and geochemical prospecting for petroleum, *Am. Assoc. Pet. Geol. Bull.*, 56, 925–940, 1972.

Horvitz, L., Hydrocarbon geochemical exploration after fifty years, in *Unconventional Methods in Exploration for Petroleum and Natural Gas IV,* Davidson, M. J., Ed., Southern Methodist University, Dallas, 1986, 147–161.
Laubmeyer, G., A new geophysical prospecting method, *Z. Petrol.*, 29, 1–4, 1933.
Pflaum, R. C., Gaseous Hydrocarbons Bound in Marine Sediments, Ph.D. dissertation, Texas A&M University, College Station, 1989.
Thompson, R. R., Extraction of Hydrocarbon Gases from Earth Samples, U.S. Patent 3539299, 1970.
Thompson, R. R., Duschatko, R. W., and Nash, A. J., Critical carbonate minerals in geochemical prospecting, U.S. Patent 3801281, 1974.
Wesson, T. C. and Armstrong, F. E., The determination of $C_1$-$C_4$ hydrocarbons adsorbed on soils, BERC/RI-75/13 USERDA, Bartlesville, OK, 1975.

Chapter 11

# POSSIBLE REDUCTION OF ATMOSPHERIC $CO_2$ BY IRON FERTILIZATION IN THE ANTARCTIC OCEAN

**Tsung-Hung Peng**

## TABLE OF CONTENTS



This chapter was completed under the auspices of the U.S. Government and is therefore in the public domain.

# I. INTRODUCTION

The rising atmospheric $CO_2$ concentration resulting from fossil fuel combustion and terrestrial ecosystem disturbances poses a potential threat to our living environment by causing undesirable climate changes as a consequence of greenhouse warming. Because of this concern, consideration is being given to the possibility that the power of the ocean's biological carbon pump could be artificially strengthened. The organic material formed in surface water by photosynthesis in the photic zone takes up about 130 carbon atoms per phosphorus atom. Thus, the effect of biological activity is to reduce the amount of total $CO_2$ ($\Sigma CO_2$) in the surface water. The $CO_2$ partial pressure ($pCO_2$) in the surface ocean water is influenced, in turn, by the extent to which photosynthesis reduces the $\Sigma CO_2$ content of the water. The lowered $pCO_2$ in surface water enhances the transfer of $CO_2$ from the atmosphere into the ocean through a gas exchange process across the sea-air interface. The magnitude of $\Sigma CO_2$ reduction in seawater is controlled by the efficiency with which the limiting nutrients $PO_4$ and $NO_3$ are used. In temperate and tropical oceans the utilization efficiency is high, and hence $pCO_2$ reduction is near maximum. By contrast, in the polar oceans the utilization efficiency is low, leaving plenty of nutrients unused. Therefore, if a way can be found to increase the efficiency of nutrient utilization in these waters, their $pCO_2$, and in turn that for the atmosphere, could be reduced.

Martin and co-workers[1-6] have proposed iron fertilization in the Antarctic Ocean as a potential means of enhancing the biological carbon pump for drawing down $pCO_2$ in this region, thereby absorbing more $CO_2$ from the atmosphere to reduce the rising atmospheric $CO_2$. They have shown in field bottle incubation experiments that plant growth rates in waters collected from polar regions can be accelerated through the addition of trace amounts of dissolved iron. Iron is an essential micronutrient required for the metabolism of all forms of life. Its primary function is in cytochrome formation. Because iron is one of the most particle reactive elements, its concentration in the sea is very low, with the lowest values occurring in regions like the Antarctic, which are most remote from the continental sources. The lack of iron in this region is considered by them to be one of the main reasons that the limiting nutrients are inefficiently used and, hence, are readily available for biological consumption if iron can be properly introduced.

Some objections have been raised about Martin's iron fertilization hypothesis.[7-11] The increase in productivity in the bottle incubation experiment after the addition of a trace amount of iron could also be interpreted as a result of the elimination of predators in the bottle. The reduced grazing activity in the bottle leads to the increased accumulation of photosynthetic products. This condition would imply that no real increase in productivity in the ocean is to be expected. Even if these objections are invalid, most biologists remain skeptical that iron alone limits plant production in the Antarctic. For example, winter darkness and the extreme turbulence are known to be important. The likely result of iron fertilization will be a modest reduction in the concentration of $NO_3$ and $PO_4$ rather than a complete removal of these nutrients.

Assuming the iron fertilization works perfectly in a hypothetical case in which water circulation does not exist in the Antarctic Ocean, only an amount of $CO_2$ equivalent to that removed as a result of the initial fertilization would be sequestered from the atmosphere. In such a situation the atmosphere and surface oceans of Antarctic and non-Antarctic regions would rapidly reach a new equilibrium, leaving the atmosphere with a $CO_2$ content only slightly lower than what it was before iron fertilization was instituted. In order to achieve a significant reduction of the atmosphere's $CO_2$ content, the surface waters of the Antarctic must be frequently replaced from below. Thus, the critical issue in addition to iron availability, is the rate of vertical mixing in the Antarctic Ocean.

Simulations using box-advection-diffusion ocean models have been made to estimate the effect of iron fertilization in the Antarctic on reducing the atmospheric $CO_2$ content. As Peng and Broecker[12] pointed out, even if iron fertilization did strip the nutrients, because of dynamic constraints only a modest reduction in atmospheric $pCO_2$ would be achieved. They concluded that the dynamical limitations alone would allow no more than a 10% reduction in atmospheric $CO_2$ if the Antarctic Ocean is fertilized with iron for 100 years. The main reason is that the volume of seawater available in the Antarctic Ocean for storage of excess $CO_2$ from the atmosphere is too small.

Joos et al.[13,14] have run model simulations of the possible effects of iron fertilization in the Antarctic Ocean on $CO_2$ content of the atmosphere. Their model is a high-latitude exchange and interior diffusion advection four-box model, calibrated with bomb-$^{14}C$ distribution and verified by freon distribution in the Antarctic. Under a standard condition for their model parameters, they obtained a reduction of 107 μatm for the business-as-usual scenario of $CO_2$ emission after 100 years of successful iron fertilization. They have also conducted numerous sensitivity tests and found that the most important factors affecting the magnitude of $CO_2$ reduction are the area of fertilization and the amount of future $CO_2$ emissions.

Sarmiento and Orr[15] have taken a step further in simulating the depletion of nutrients in the Southern Ocean by using a three-dimensional ocean general circulation model (GCM). In contrast to one-dimensional box-advection-diffusion models used in early simulations, the GCM-based model of the ocean carbon cycle incorporates geographical and bottom topographical details of the ocean and dynamic details of ocean circulation. Results of three-dimensional simulation should be a more realistic representation of what would happen to the ocean-atmosphere system when iron is purposely added to the Southern Ocean. Their results indicate that the additional uptake of $CO_2$ from the atmosphere in the first 100 years after the iron fertilization would lower the atmospheric $pCO_2$ by about 46 to 85 μatm. Their global mean of 72 μatm reduction is compatible with 64 μatm reduction in Peng and Broecker's[16] box model simulation if the business-as-usual scenario of anthropogenic carbon release over the next century is used.

In this paper, possible reduction of atmospheric $CO_2$ by iron fertilization in the Antarctic Ocean is reviewed on the basis of the box-advection-diffusion model of Peng and Broecker.[12] A more detailed description of the effects of iron fertilization on atmospheric $CO_2$ partial pressure can be obtained from publications by Peng and Broecker,[12,16] Peng et al.,[17] and Joos et al.[13,14]

## II. MODEL OF ANTARCTIC AND NON-ANTARCTIC OCEANS

The box model of global ocean designed to simulate the redistribution of carbon between the atmosphere and ocean is made of two separate water columns containing vertical advection and diffusion, one representing Antarctic Ocean and the other representing non-Antarctic oceans. The division between Antarctic and non-Antarctic oceans is based on the distribution of nutrients and anthropogenic tracers in the Antarctic. Measurements made as part of the Geochemical Ocean Sections Study (GEOSECS) program (Figure 1 for station locations) provide such basic data. The geographic domain in which surface waters contain appreciable amounts of unused $NO_3$ and $PO_4$ can thus be identified. As can be seen from the map in Figure 1, the ambient $PO_4$ level for Antarctic surface water (during the summer months) of 1.6 μmol $kg^{-1}$ drops off rapidly between 50°S and 40°S. The latitude of 45°S is adopted to be the northern boundary of the region where iron fertilization has potential. The total area south of this boundary represents 16.8% of the global ocean.

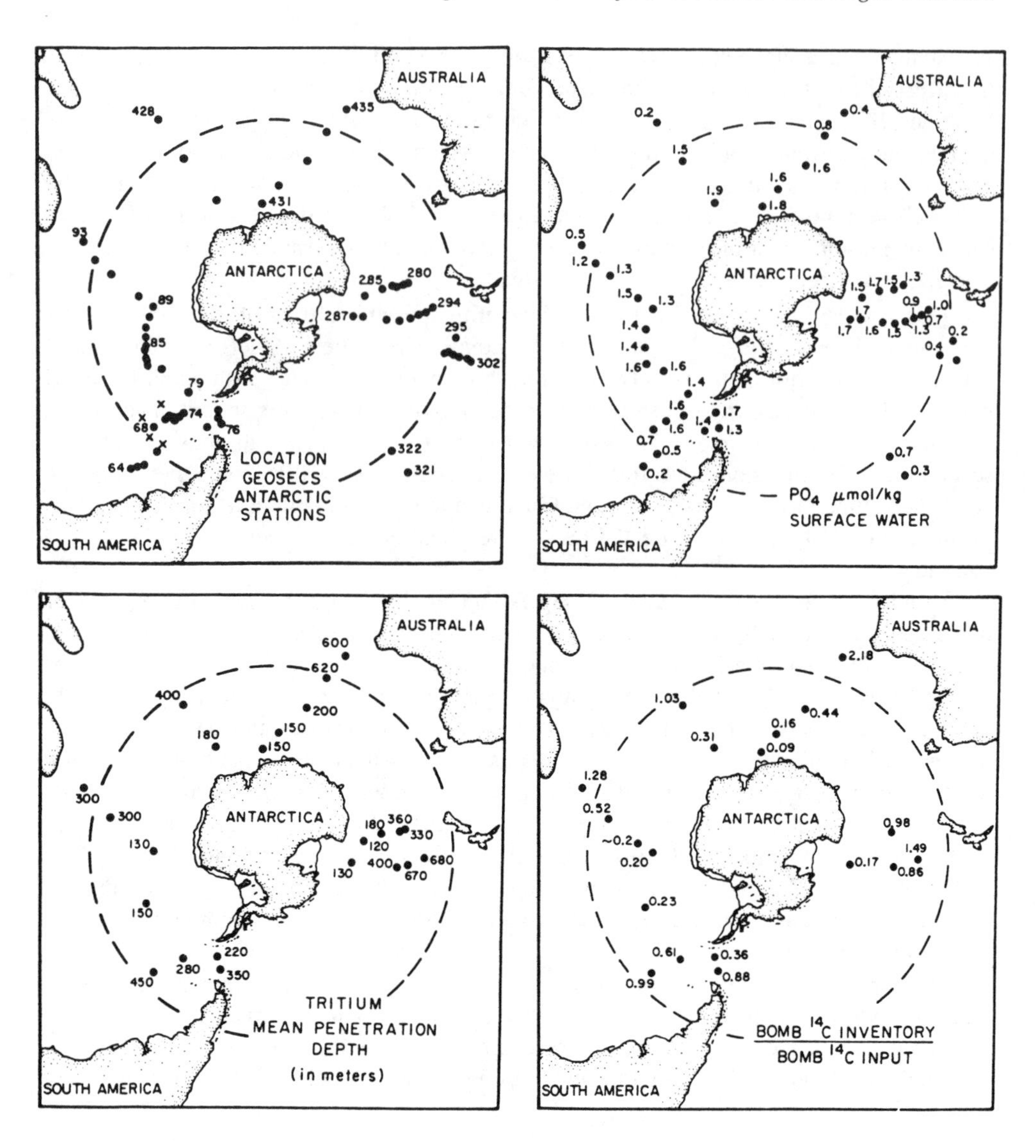


**FIGURE 1.** The upper left panel is a map showing the location of the GEOSECS Antarctic stations (black dots) and four SAVE stations (crosses). The other three maps show the surface water $PO_4$ concentrations at these stations, the mean penetration depth of tritium at the time of the GEOSECS surveys (see Reference 26), and the ratio of the water column inventory of nuclear testing radiocarbon to the input of nuclear testing radiocarbon (see Reference 18). The dashed circle at 45°S marks the latitude where, on the average, surface $PO_4$ reaches one half its ambient Antarctic concentration.

The results of the $^3H$ and $^{14}C$ measurements for these stations are also summarized in Figure 1. The tritium[26] results allow an estimate to be made of the extent of downward mixing into the thermocline on the time scale of one decade (i.e., the time between the tritium delivery and the GEOSECS surveys). As can be seen from the summary in Figure 1, this depth ranges from as little as 150 m in the poleward Antarctic to as much as 600 m at 45°S. The other important source of information comes from the $^{14}C$ results. Broecker et al.[18] used measurements on prenuclear samples to estimate the prenuclear $\Delta^{14}C$ values for surface waters at the GEOSECS stations at which $^{14}C$ measurements were made. The $^3H$ measurements allow an estimate to be made of the depth to which significant amounts of bomb $^{14}C$ had penetrated at the time of GEOSECS surveys. With these two end points and a knowledge of the shape of the $^{3}H$ profile, the prenuclear $^{14}C$ profile for each station can be established. The area between the measured profile and the prenuclear profile provides an estimate of the water column burden of excess $^{14}C$. Broecker et al.[18] showed that these excesses have a distinct geographic pattern (Figure 2). Higher-than-average inventories are found in the temperate regions of the ocean and in the northern Atlantic. Lower than average inventories are found in the equatorial zone, the northern Pacific, and the Southern Ocean. These authors point out that the areas of high inventory correspond to regions of downwelling, and the areas of low inventory correspond to regions of upwelling. Further, they attribute this correspondence to lateral transport of bomb testing $^{14}C$ from regions of upwelling to regions of downwelling. Experiments conducted in the Geophysical Fluid Dynamics Laboratory ocean GCM confirm that such transports can explain the inventory pattern.[19]

Broecker et al.[18] compared the inventory at any given station with the net amount of bomb $^{14}C$ invading that station from the atmosphere. This allows the magnitude of the excess or deficiency to be quantified (Figure 3). Their calculations neglect the dependence of the $CO_2$ exchange rate on wind speed. Were wind speed to be taken into account, the magnitude of the deficiencies for the Antarctic stations would be increased. The reason is that the wind speed, and therefore the $CO_2$ exchange rate over the Antarctic, is higher than the global average. At all latitudes in the Antarctic, less bomb $^{14}C$ is present than what originally entered the ocean. The deficiencies range from a very small percentage at 45°S, to as much as 90% of the total input closer to the Antarctic continent (Figure 1).

Together the $^{3}H$ penetration depths and the bomb-$^{14}C$ deficiencies allowed Broecker et al.[18] to obtain estimates of the average upwelling and downwelling velocities for various regions of the ocean. The values they obtained are summarized in Table 1. For the Antarctic they required an average vertical eddy diffusivity of 3 $cm^2\ s^{-1}$ and upwelling velocities ranging from 9 m/year (for the Indian sector) to 31 m/year (for the Atlantic sector). The vertical mixing in the Antarctic derived from these tracer distributions is in agreement with traditional thinking that deep ocean water upwells in this region. A portion of this upwelled water moves beneath the sea-ice fringe, where its density is increased by brine release, causing it to sink back into the deep sea and form Antarctic bottom water. The remainder moves to the north. Part of this is converted to Antarctic intermediate water, which penetrates northward into the Atlantic, Pacific, and Indian Oceans at a depth of about 1000 m. The remainder is mixed into the temperate surface water.

Based on these tracer distributions and the lateral transport mechanism of Broecker et al.[18] a box model of Antarctic and non-Antarctic oceans is constructed. It involves two side-by-side, box-diffusion columns by Oeschger et al.[20] They are linked together by an overlying atmosphere and underlying deep sea (Figure 4). One column represents the Antarctic and the other the non-Antarctic region of the ocean. Each column is capped by a 75-m thick mixed layer. These mixed layers are underlain by 2000 m thick diffusive zones. Beneath the diffusive zone is a single well-mixed deep reservoir. The area and volume of the Antarctic column are taken to be 10% of the ocean total. The area of the Antarctic is not set at 16.8% because iron fertilization could draw the surface water $pCO_2$ only during those months when sunlight is plentiful. In the light-poor austral winter months the $pCO_2$ would be driven back toward its prefertilization value by vertical mixing in the upper water column.

Consistent with the distribution of bomb-radiocarbon, the vertical eddy diffusivity in the Antarctic column is set at 3 $cm^2\ s^{-1}$, and in the non-Antarctic column at 1 $cm^2\ s^{-1}$. The upwelling flux in the Antarctic column is set at 17.4 sverdrups (equivalent to the upwelling rate of 15.2 m/year if the Antarctic surface area is $3.6 \times 10^{13}\ m^2$). Because there are no firm means to determine the fate of the upwelled water, two limiting scenarios are adopted. In the first, all the water is transferred entirely to the surface of the non-Antarctic column. In the second it is transferred entirely to the deep reservoir.

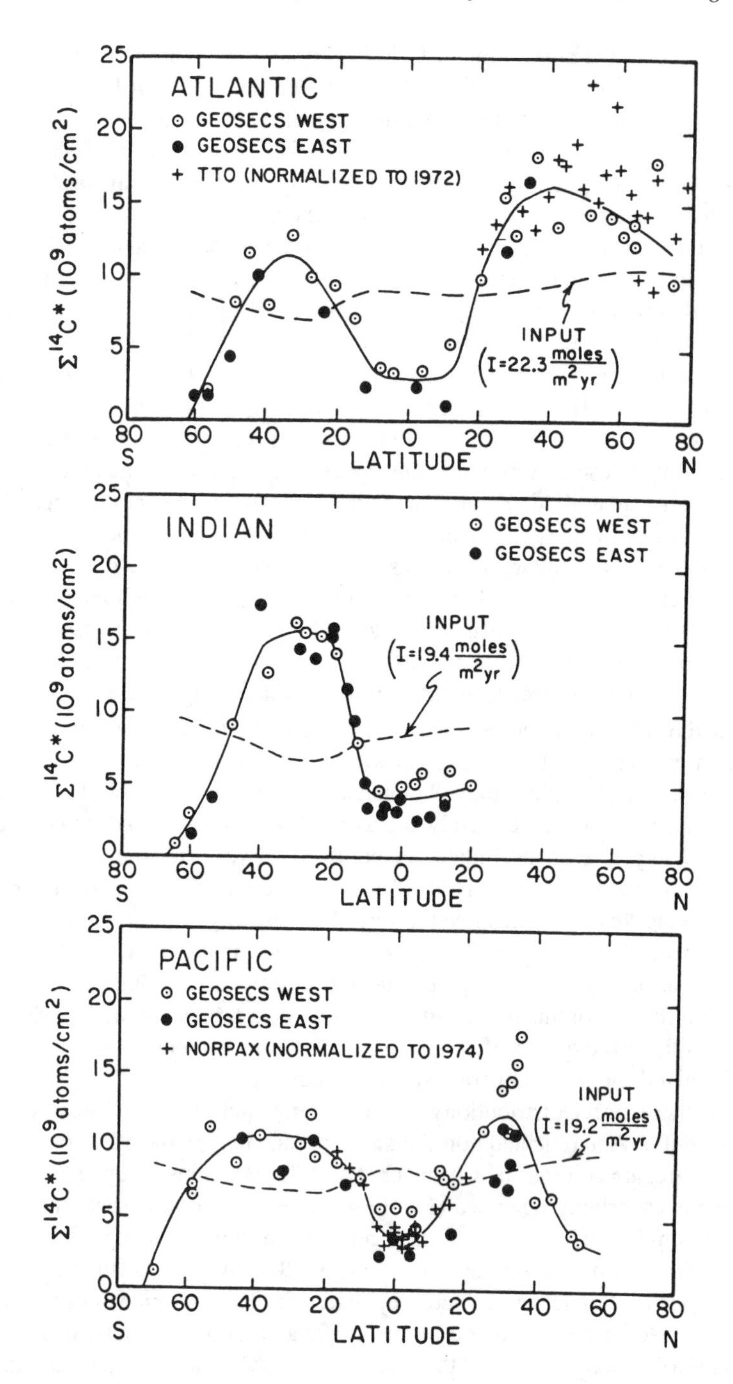


**FIGURE 2.** Bomb-testing $^{14}C$ inventories as measured during the GEOSECS, NORPAX, and TTO programs (as summarized by Broecker in Reference 18. The dashed lines show the amount expected at the time of the GEOSECS program if there were no lateral transport from one zone to other. "I" represents $CO_2$ invasion rate.

The model circulation starts with a steady state. The residence times for $PO_4$ with respect to biological removal from the surface reservoirs are set to yield 1.6 μmol kg$^{-1}$ $PO_4$ in the surface water above the Antarctic column, and near zero $PO_4$ in the surface waters of the non-Antarctic column. The regeneration function for falling organic debris is set to yield $PO_4$ vs. depth profiles similar to the observed (it should be pointed out that the choice of

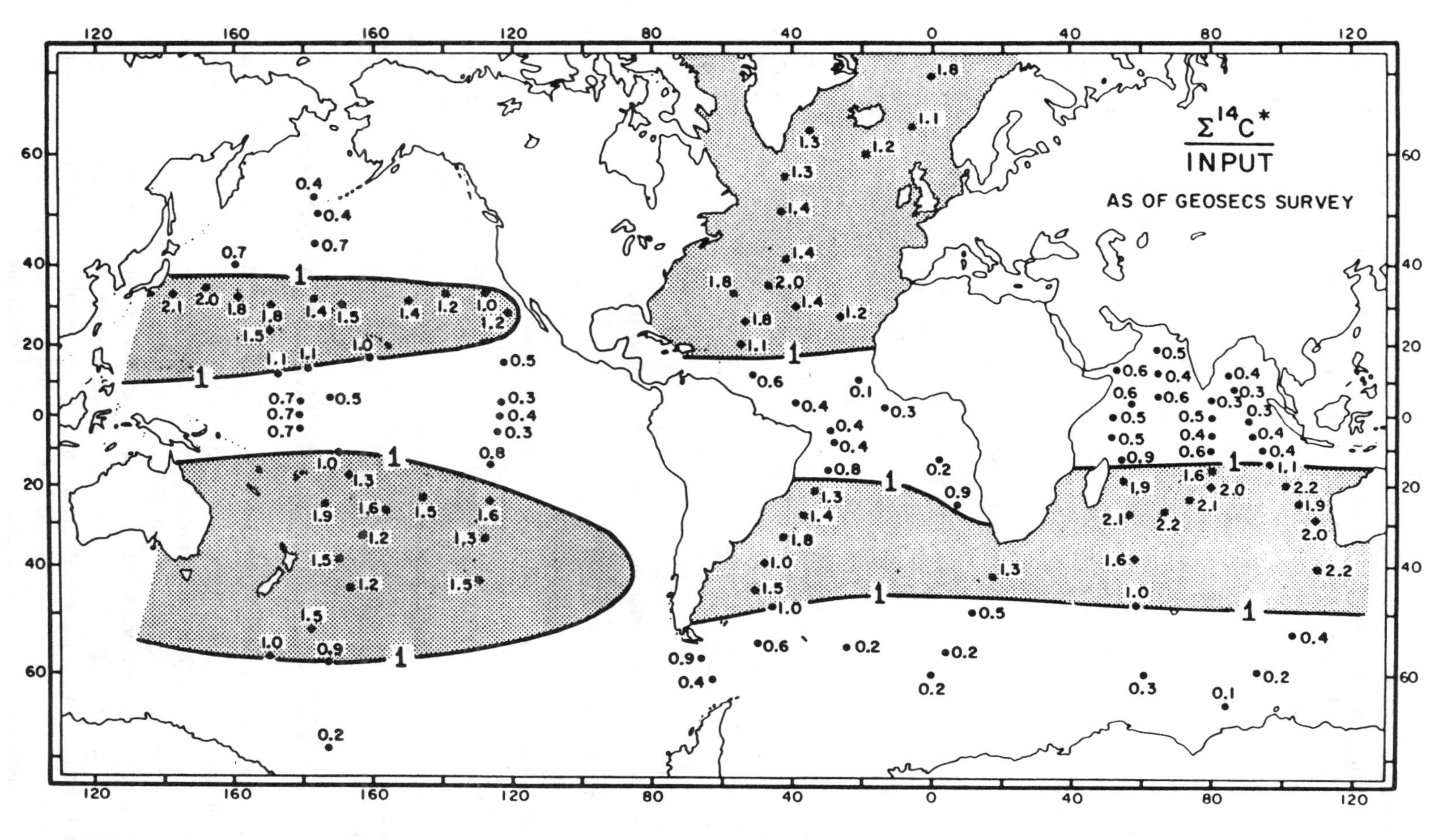


**FIGURE 3.** Map showing the ratios of observed nuclear testing radiocarbon inventories to those calculated if there were no lateral transport (see Reference 18). The stippled areas represent regions with the observed $^{14}C$ inventories in excess of estimated input by gas exchange process alone.

**TABLE 1**
**Comparison of Surface Water Δ¹⁴C Excesses and Water-Column Inventories of Bomb-Testing Radiocarbon Generated by the Model of Broecker et al. (1985) With Those Observed During the GEOSECS Program**

| Latitude belt | Area $10^{12}m^2$ | W m/yr | Flux sverdrups | K $cm^2/s$ | $z^a$ m | $\Delta^{14}C-\Delta^{14}C°$ ‰ | Bomb-$^{14}C$ input, $10^{26}$ atoms | Bomb-$^{14}C$ inventory, $10^{26}$ atoms |
|---|---|---|---|---|---|---|---|---|
| | | | **Atlantic (I = 22.3 mol/m²/yr)** | | | | | |
| 80°N–40°N | 18.6 | −8.5 | 5 | 9.9 | 710 | 130 (120) | 20 | 26 (26) |
| 40°N–20°N | 15.8 | −30.0 | 15 | 0.5 | 470 | 215 (195) | 14 | 24 (23) |
| 20°N–20°S | 26.7 | +21.2 | 18 | 1.0 | 190 | 146 (150) | 25 | 11 (10) |
| 20°S–45°S | 18.4 | −22.3 | 13 | 1.0 | 400 | 185 (170) | 14 | 20 (18) |
| 45°S–80°S | 15.1 | +31.2 | 15 | 3.0 | 270 | 100 (100) | 14 | 6 (7) |
| | | | **Indian (I = 19.4 mol/m²/yr)** | | | | | |
| 25°N–15°S | 27.0 | +15.2 | 13 | 1.0 | 215 | 130 (160) | 26 | 11 (12) |
| 15°S–45°S | 29.8 | −20.1 | 19 | 1.0 | 510 | 190 (180) | 23 | 43 (42) |
| 45°S–70°S | 20.7 | +9.1 | 6 | 3.0 | 410 | 115 (100) | 19 | 14 (13) |
| | | | **Pacific (I = 19.2 mol/m²/yr)** | | | | | |
| 65°N–40°N | 15.1 | +10.4 | 5 | 1.5 | 260 | 152 (145) | 13 | 9 (8) |
| 40°N–15°N | 35.0 | −10.8 | 12 | 1.0 | 365 | 206 (210) | 30 | 39 (37) |
| 15°N–10°S | 50.0 | +16.4 | 26 | 1.0 | 255 | 134 (150) | 44 | 25 (25) |
| 10°S–55°S | 63.0 | −12.5 | 25 | 2.0 | 410 | 166 (170) | 48 | 64 (60) |
| 55°S–80°S | 13.8 | +13.7 | 6 | 3.0 | 335 | 111 (100) | 11 | 9 (7) |

*Note:* The bomb-$^{14}C$ excesses for surface water (in $\Delta^{14}C$ units) and the observed water column inventories are given in parentheses for comparison. "I" represents $CO_2$ invasions rate.

[a] Mean penetration depth, i.e., K/W.

this respiration function has no influence on the result of the calculation of the atmospheric $CO_2$ response to iron fertilization). The atom ratio of carbon to phosphorus in the organic matter falling from the surface mixed layer is 130. The $\Sigma CO_2$/alkalinity ratio in the model ocean is then adjusted to yield an atmospheric $pCO_2$ of 280 μatm. The computation of $pCO_2$ in seawater is based on equations of carbonate chemistry.[21,22] The gas exchange rate between atmosphere and both surface mixed layers is taken to be 20 $mol \cdot m^{-2} \cdot yr^{-1}$ for a $pCO_2$ of 280 μatm.

## III. EFFECTS ON PREINDUSTRIAL ATMOSPHERE

To simulate a totally successful iron fertilization, the preindustrial steady state is perturbed by a great decrease in the residence time with respect to the biological removal of $PO_4$ from Antarctic surface water, bringing its $PO_4$ content to near zero. In this simulation the iron fertilization is continuously carried out for 100 years. The $PO_4$ content of the surface Antarctic water is maintained at zero over this period. The evolutions of the vertical distributions of $PO_4$ and $\Sigma CO_2$ in the Antarctic column are shown in Figure 5. As can be seen, while the water column integral of $PO_4$ remains unchanged, a bulge of excess $\Sigma CO_2$ appears. This bulge represents the $CO_2$ transferred from the atmosphere and the non-Antarctic column to the Antarctic column. The time trends for the $CO_2$ partial pressure in the two ocean-

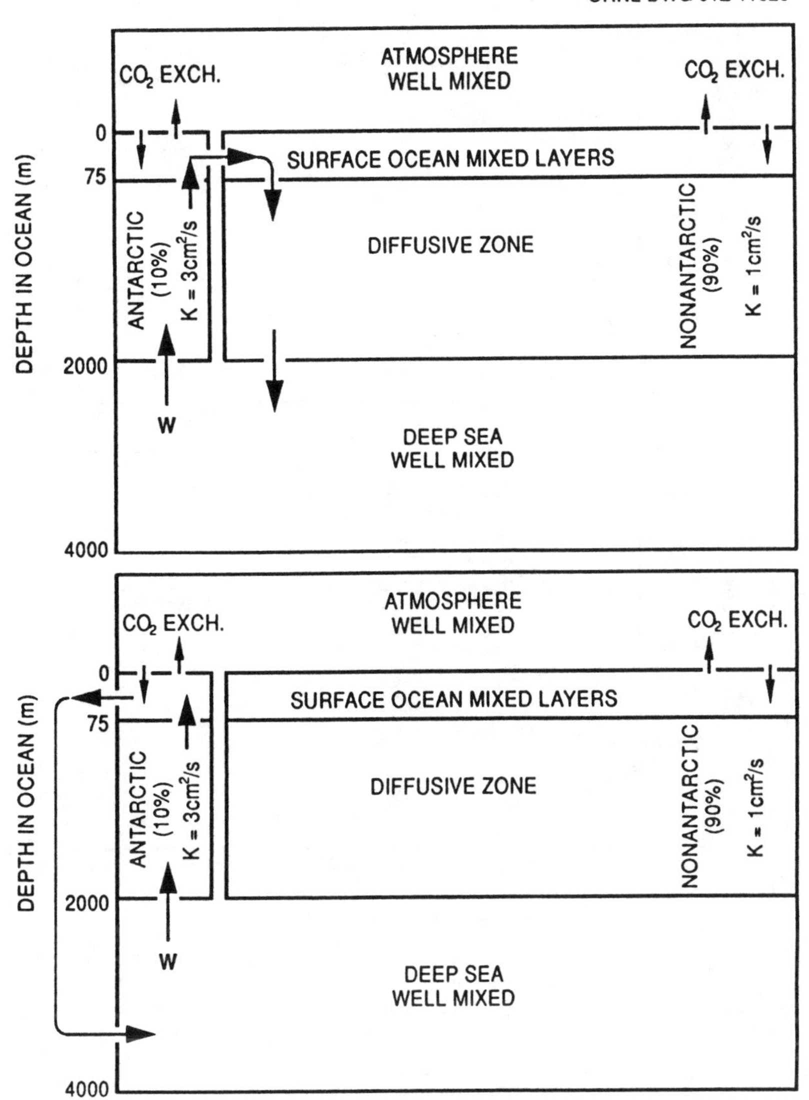


**FIGURE 4.** Linked vertical advection-diffusion model used to evaluate the response to iron fertilization of Antarctic surface waters. The upper panel shows the case in which the water upwelled in the Antarctic is transported laterally to the non-Antarctic surface ocean. The lower panel shows the case in which this upwelled water is converted to deep water and transferred directly to the model's deep reservoir.

surface layers and in the atmosphere are shown in Figure 6. In the lateral transport scenario, the atmospheric $CO_2$ content drops ever more slowly as the century progresses, reaching an asymptote of about 15 μatm lower than the initial value. In the deep transport scenario, the decrease continues reaching about 34 μatm after one century. The reason for the difference is that in one case the surface water from the Antarctic is transferred to the surface of the non-Antarctic region (allowing the excess $CO_2$ to reenter the atmosphere) while in the other this water is removed to the deep sea, isolating it from the atmosphere.

The question naturally arises concerning what would happen if at some point iron fertilization were terminated. As shown in Figure 6, the answer is that any reduction in

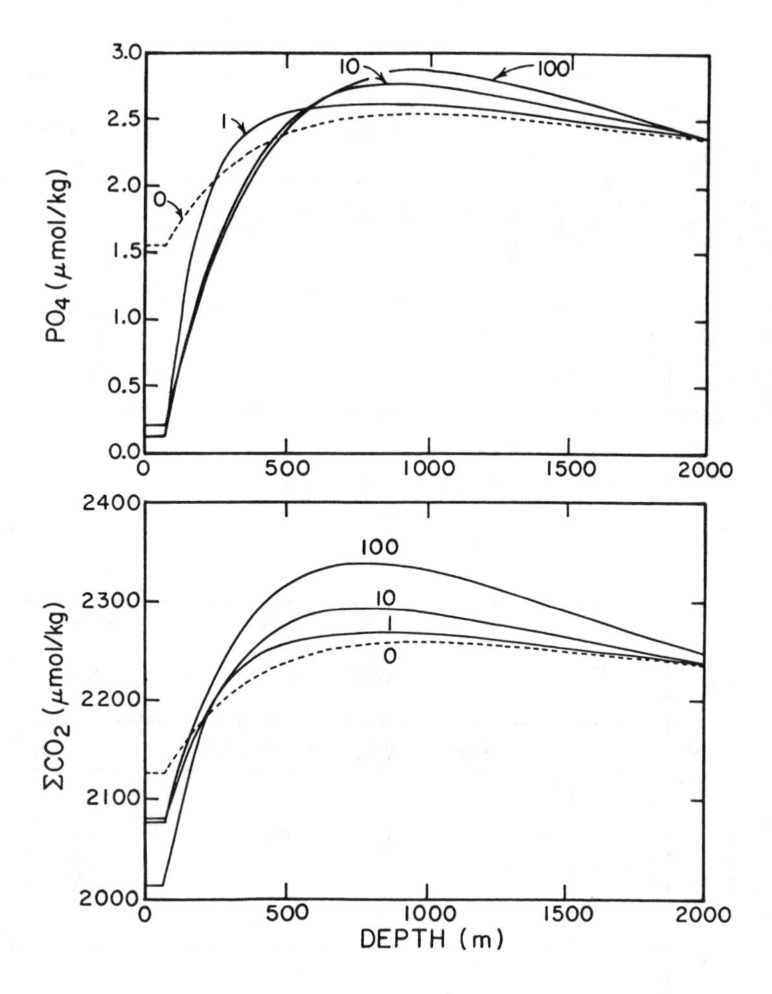


**FIGURE 5.** Vertical distribution of $PO_4$ and $\Sigma CO_2$ in the Antarctic column before the onset of fertilization (dashed line) and 1, 10, and 100 years after the onset of fertilization for the scenario in which water is upwelled at the rate of 17.4 sverdrups and transferred to the deep sea.

atmospheric $CO_2$ content accomplished by the fertilization would be lost on more or less the same time scale as it was gained.

# IV. EFFECTS ON ANTHROPOGENICALLY AFFECTED ATMOSPHERE

To test the effects of iron fertilization on an anthropogenically affected atmosphere, the excess $CO_2$ is introduced into the atmosphere starting in 1800 and continuing through 1990. The atmospheric $pCO_2$ is computed according to the transient distribution of carbon in the whole ocean-atmosphere carbon system. The input function of fossil fuel-production is based on a recent estimate.[23] The release of $CO_2$ from the perturbed terrestrial ecosystem is derived from deconvolution[24] of the time history of atmospheric $pCO_2$ obtained from $pCO_2$ measurements of air bubbles in ice cores[25] and of air samples.[26] The release scenario of a business-as-usual case for the next century is taken from an Intergovernmental Panel on Climate Change (IPCC) report.[27] The $CO_2$ emission between 1800 and 2100 is shown in Figure 7. Before anthropogenic $CO_2$ is introduced, a steady state of this ocean-atmosphere model with $pCO_2$ of 280 μatm is the same as described earlier. Shown in Figure 8 is the

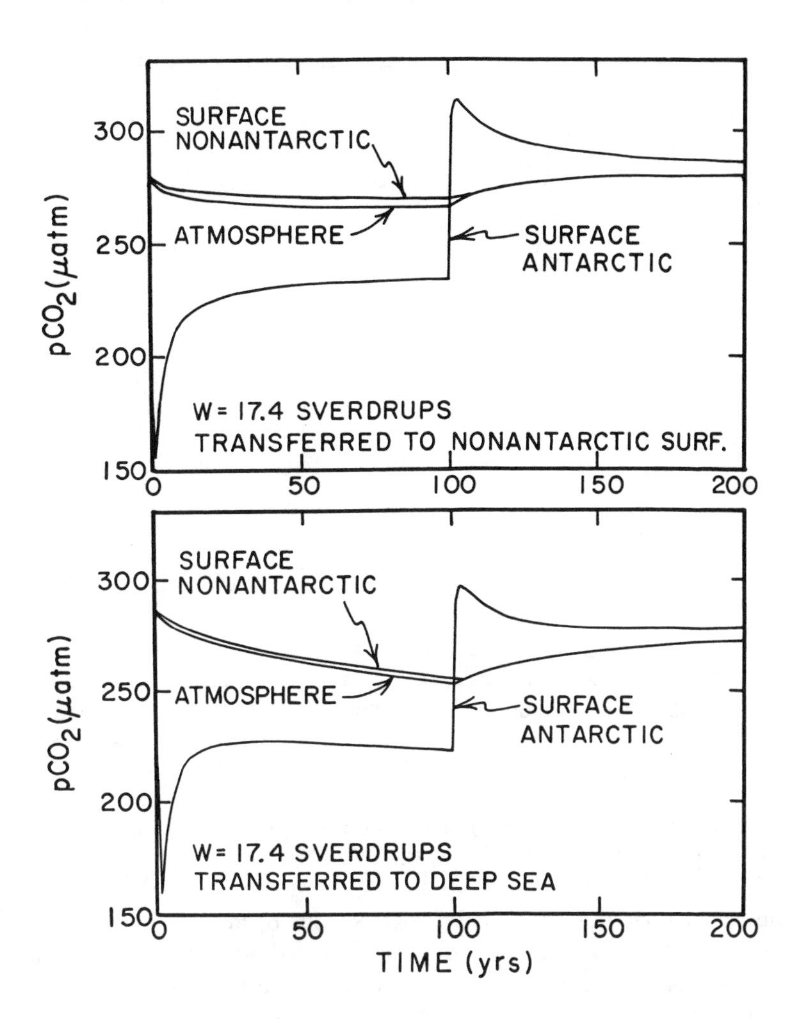


**FIGURE 6.** Model runs for the preindustrial ocean-atmosphere system with an Antarctic upwelling flux of 17.4 sverdrups. The upper panel shows the case in which the upwelled water is transferred to the surface of the non-Antarctic ocean; the lower panel shows case in which the upwelled water is transferred to the deep sea. In each case, totally successful iron fertilization is conducted for 100 years and then stopped. For the steady state conditions that precede the onset of fertilization, the $pCO_2$ of 280 μatm for Antarctic surface waters is nearly identical to that for the atmosphere.

atmospheric $pCO_2$ for the next century resulting from the IPCC $CO_2$ release scenario without iron fertilization in the Antarctic Ocean. The upwelling flux of 17.4 sverdrups is used, and the upwelled water is transferred to the deep reservoir. The result of a successful iron fertilization to reduce the atmospheric $pCO_2$ under such dynamic conditions is also shown as a dashed curve (Figure 8) for comparison. The difference between these two curves represents the net effect of iron fertilization.

To explore the possible reduction in atmospheric $CO_2$ content under various oceanographic conditions during the iron fertilization period, a series of sensitivity calculations are performed. When such model calculations are made, the iron fertilization is assumed to work perfectly in the Antarctic. Results of these computations are given below.

**Upwelling flux** — Upwelling flux is considered to be the most important factor in determining the possible reduction of atmospheric $CO_2$[12] in the event of iron fertilization in the Antarctic. To evaluate its possible effects on reducing atmospheric $CO_2$, the Antarctic upwelling flux is changed by twofold and by one half of the standard value used in the

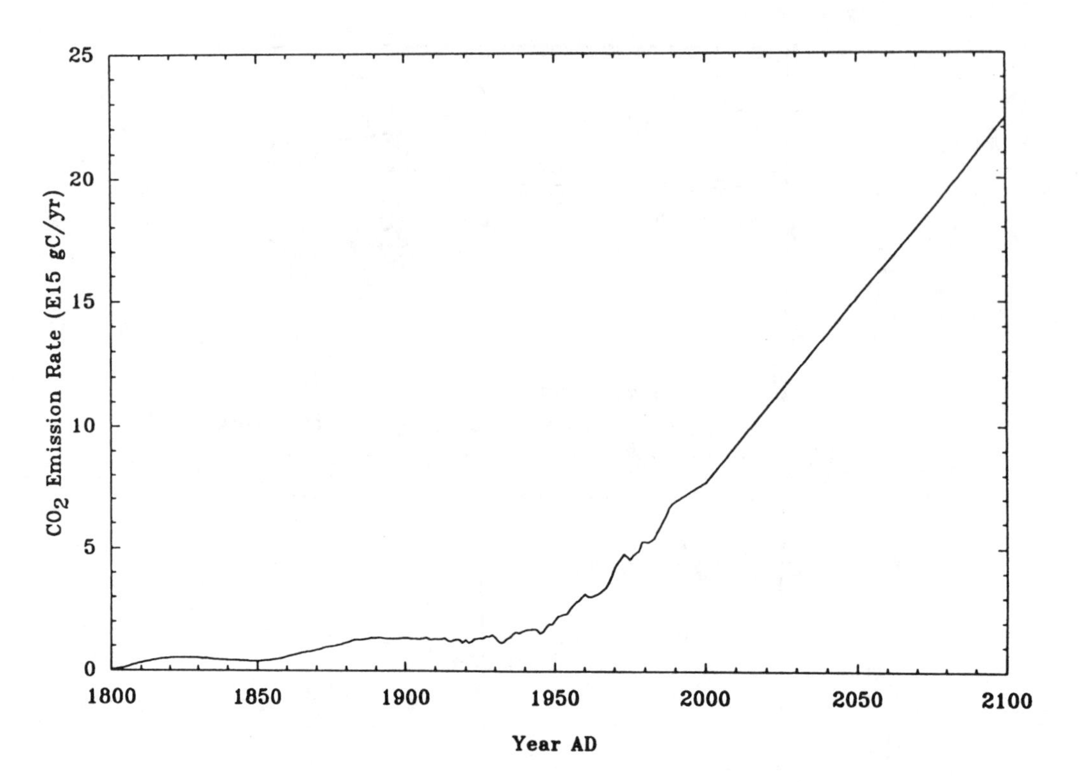


**FIGURE 7.** Time history of $CO_2$ emission used as model input function. The business-as-usual $CO_2$ release scenario for the period between 1991 and 2100 is adopted from IPCC report.

steady-state calculations (i.e., 17.4 sverdrops). Results of these simulations are shown in Figure 9. In the case of the lateral transport of upwelled water to the surface layer of the non-Antarctic region, the reduction is not sensitive to the upwelling fluxes. However, in the case of deep transport, the reduction increases with upwelling fluxes. The deep transport of Antarctic surface water effectively carries away excess carbon to the deep and thus enhances the uptake of $CO_2$ from the atmosphere. Shown in the lower panel of Figure 9 is the percentage of reduction with respect to atmospheric $CO_2$ content. Although the value of absolute reduction increases as the atmospheric $CO_2$ rises in response to increasing $CO_2$ emission, the percentage of reduction reaches a nearly constant value after an initial rapid increase for each dynamical condition. For lateral transport cases, the reduction is about 3%. For deep transport cases it increases from about 6% for 8.7 sverdrups upwelling flux, to about 12% for an upwelling flux of 34.8 sverdrups. A standard case with 17.4 sverdrups (best estimate according to tracer distribution) yields an 8% reduction. Thus, if after 100 years of business-as-usual $CO_2$ emission (1991 to 2090), the atmospheric $CO_2$ partial pressure were 800 μatm, the reduction resulting from continuous iron fertilization during the same period with unreasonably high upwelling flux would be about 96 μatm, and that for the best case would be 64 μatm.

**The size of Antarctic surface area** — The size of Antarctic surface area is considered to be another major factor in determining the effect of iron fertilization.[13] Figure 10 shows the estimated reduction if the Antarctic surface area is taken to be 16% of the ocean total instead of 10% as used in standard computations. A reduction of 71 μatm after 100 years of fertilization is estimated, which is not significantly different from a reduction of 64 μatm in the standard case. The reason for such a small effect is that the upwelling flux of 17.4

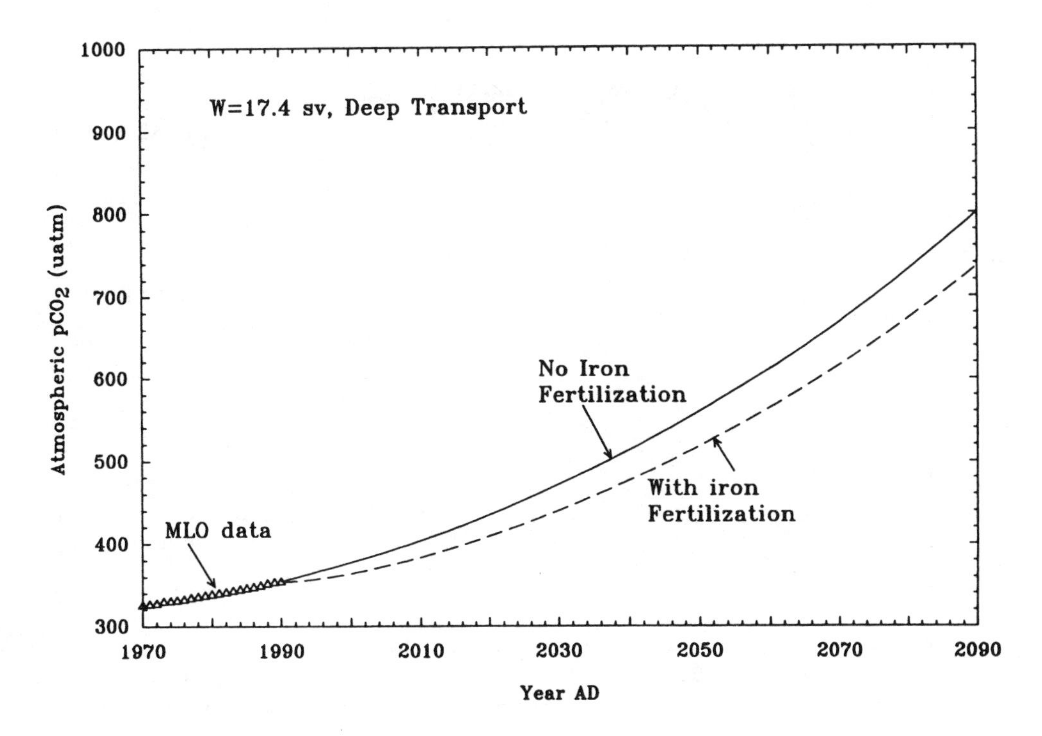


**FIGURE 8.** Atmospheric $pCO_2$ resulting from the IPCC business-as-usual $CO_2$ emission scenario for the period between 1991 and 2090. The solid curve is the predicted atmospheric $pCO_2$ without iron fertilization, and the dashed curve is the one with iron fertilization in the Antarctic Ocean. Standard dynamic conditions of 17.4 sverdrups (sv) upwelling flux and deep transport of the upwelled water are used for simulation.

sverdrups is kept the same in spite of the increased surface area. The total upwelling flux remains constant because it was this flux rather than an upwelling rate that was constrained by Broecker et al.[18] by their analysis of the bomb-$^{14}C$ data set. As a result, the upwelling rate drops from 15.2 to 9.5 m $yr^{-1}$. If the upwelling rate of 15.2 m $yr^{-1}$ does not change (or the upwelling flux increases proportionally), the increase in surface area to 16% would cause a reduction of 99 μatm after 100 years of fertilization. This value is comparable with the estimate of 107 μatm obtained by Joos et al.[14] using the same percentage of surface area in their four-box ocean model.

**Vertical mixing** — A standard vertical diffusivity of 3 $cm^2\ s^{-1}$ in the Antarctic column gives a reduction of 64 μatm. As shown in Figure 11, a reduction of 75 μatm is obtained for a doubled vertical diffusivity (6 $cm^2\ s^{-1}$) and a 58 μatm reduction for a value of 1 $cm^2$ $s^{-1}$. Thus, we see that the vertical diffusivity is not as important as the upwelling flux in determining the reduction in the atmospheric $CO_2$ content.

**Gas exchange** — Gas exchange across the sea-air interface provides a linkage between the atmosphere and the ocean. Changes in the gas exchange rate over the Antarctic Ocean should have a direct effect on reducing atmospheric $CO_2$ during the fertilization period. However, as shown in Figure 12, only an excess of 4 μatm reduction (i.e., 68 vs. 64 μatm) is obtained if the gas exchange rate over the Antarctic is doubled to 40 $mol{\cdot}m^{-2}{\cdot}yr^{-1}$ (at $pCO_2$ = 280 μatm) during the period 1991 to 2090. If the exchange rate is cut in half, the difference in reduction is about 5 μatm. The $CO_2$ exchange rate of 20 $mol{\cdot}m^{-2}{\cdot}yr^{-1}$ (at $pCO_2$ = 280 μatm) used in the best case is fast enough to keep the ocean-atmosphere system at approximate equilibrium. Hence, the $CO_2$ exchange rate is not an important limiting factor in reducing atmospheric $CO_2$ during Antarctic iron fertilization.

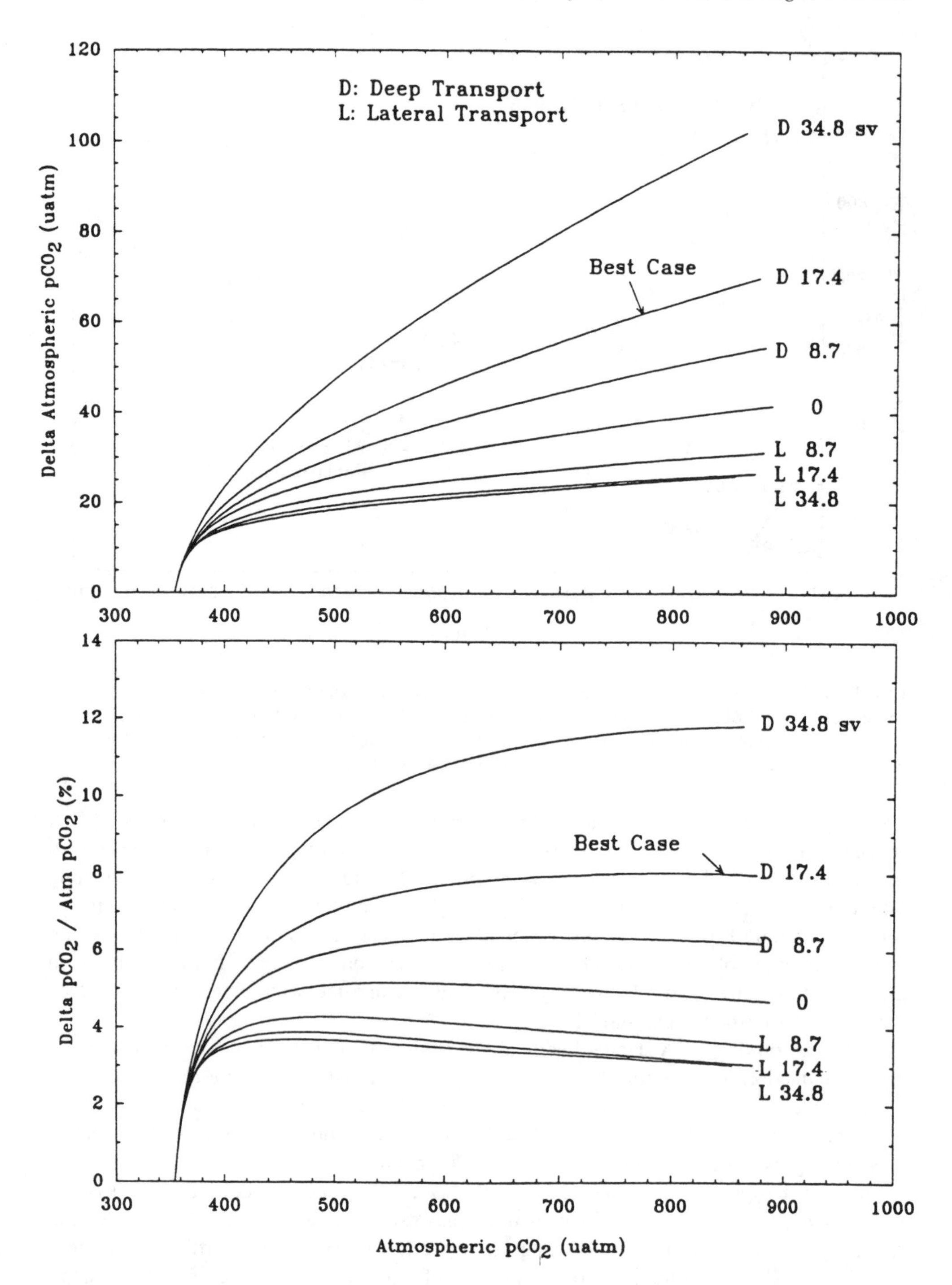


**FIGURE 9.** Difference in partial pressure of $CO_2$ in the atmosphere with and without iron fertilization in the Antarctic Ocean as a function of atmospheric $pCO_2$ for the period between 1991 and 2090. The Antarctic upwelling flux and its fate are the main model variables. The IPCC business-as-usual $CO_2$ emission scenario is used. The magnitude of $pCO_2$ reduction is plotted against the atmospheric $pCO_2$ in the upper panel, whereas the percentage of reduction is plotted in the lower panel; (sv = sverdrups).

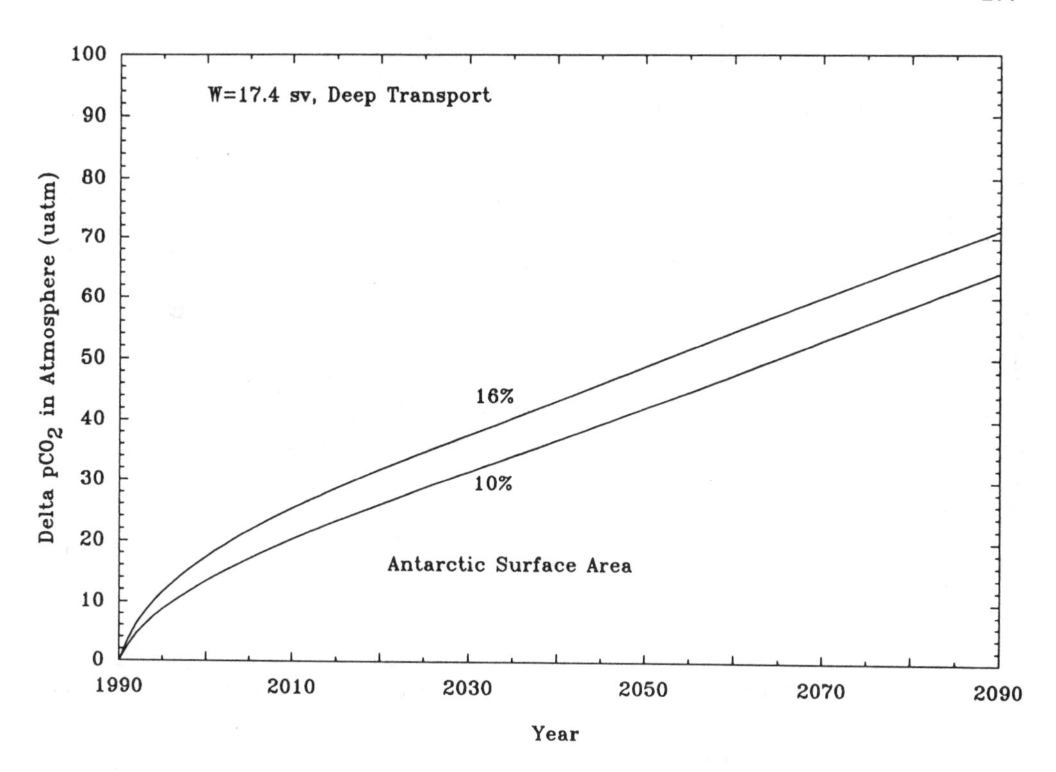


**FIGURE 10.** Effect of difference in fertilized Antarctic surface area on the reduction of atmospheric $pCO_2$. Deep transport of 17.4 sverdrups (sv) of upwelled water remains the same when the fertilized surface area increases from 10 to 16%.

**Biological flux removal** — In the model simulation the biological flux is completely reoxidized, and the carbon and nutrients are returned to inorganic dissolved form. It has been suggested that if organic particulate were to fall directly to the sea floor and become buried in the sediment, fertilization would be more effective in reducing the atmospheric $CO_2$. However, as shown in Figure 13, even with 75% direct removal, the effect is insignificant (less than 3 $\mu$atm extra reduction). The reason for this is simple: if the $PO_4$ is held at zero in surface waters, it would not even matter whether oxidation occurred, let alone at what depth it occurred. This is a subtle but extremely important point, having to do with the fact that $PO_4$ and $\Sigma CO_2$ changes are all tied together by the Redfield ratio. The small change shown in Figure 13 results because the $PO_4$ content of surface waters does not reduce to exactly zero. If this were the case, there would have been no change at all in the extent of $CO_2$ reduction.

**Seasonality** — Light availability certainly limits plant production in the Antarctic, especially during the winter months. The impact of this limitation was evaluated by turning off the effect of iron fertilization for a number of months each year, while the model's mixing rates remained unchanged throughout the year. The period of totally successful fertilization is set at 8, 4, and 2 months. As shown in Figure 14, 4 winter months of no productivity cause only a small reduction in the long-term $CO_2$ drawdown. Even when the period of fertilization is reduced to only 2 months per year, two-thirds of the atmospheric $CO_2$ drawdown achieved for the full-year scenario occurs. It is tempting to conclude from this that were iron added for only one-sixth of the year, two-thirds of the maximum possible atmospheric drawdown would be achieved. Caution is needed in this regard; this result depends very strongly on the ratio of the $PO_4$ drawdown time to the water replacement time

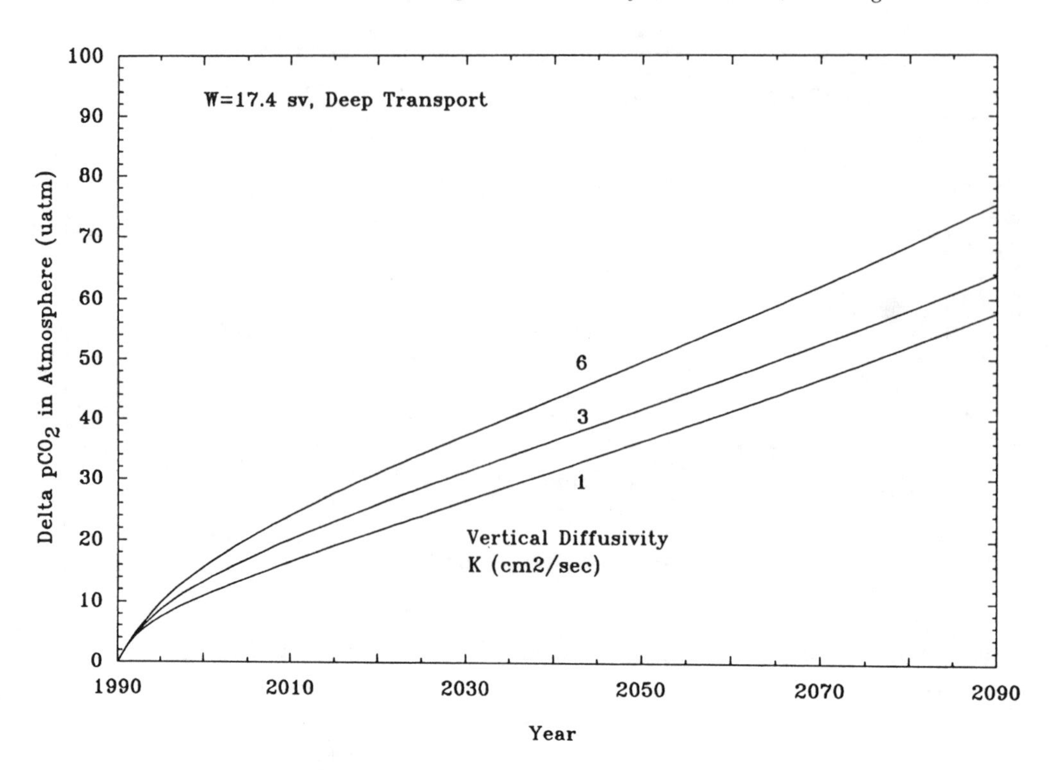


**FIGURE 11.** Effect of difference in vertical diffusivity of Antarctic column on the reduction of atmospheric $pCO_2$ if iron fertilization is carried out over the next 100 years. The standard case has a vertical diffusivity of 3 $cm^2 s^{-1}$; (sv = sverdrups).

for the mixed layer. Were a less favorable ratio to be adopted, the turn off would have more nearly a proportional impact. The choice of 1 month for the phosphate drawdown time is just a guess. Because the surface water replacement time is chosen to match the vertical distribution of tritium about one decade after the cessation of large-scale bomb testing, it has little bearing on the actual rate of water exchange between the mixed layer and the underlying thermocline. Furthermore, the actual vertical exchange has a strong seasonality. Winter cooling thickens the mixed layer, whereas summer warming and sea-ice melting thins it. The interplay of changing light availability and vertical mixing with the timing of iron addition could be used to optimize the amount of atmospheric $CO_2$ drawdown.

**Iron fertilization efficiency** — Efficiency of iron fertilization is assumed to be 100% effective in removing nutrients from the surface waters of the Antarctic Ocean in all cases studied in the model. In light of the possibility that iron fertilization may not be efficient enough to use up nutrients in the surface water, the magnitude of $pCO_2$ reduction under such less-than-perfect conditions needs to be estimated. Since the iron fertilization is simulated by a time constant for the removal of nutrients, changes in such a time constant can be used to simulate the efficiency of iron fertilization. For very efficient iron fertilization, as demonstrated earlier, a nutrient removal time constant of 0.1 year is used. Arbitrary time constants of 1.0 and 1.9 years are chosen for comparison. As shown in the upper panel in Figure 15, the Antarctic surface water $PO_4$ concentration reaches a constant value after about ten years of iron fertilization under less-efficient conditions. The magnitude of $pCO_2$ reduction is drastically diminished as a result of inefficiency in iron fertilization, as shown in the lower panel of Figure 15. As expected, the unused nutrient concentrations are high (0.85 to 1.2 $\mu mol\ kg^{-1}$) under less-efficient conditions, even with continuous iron fertilization.

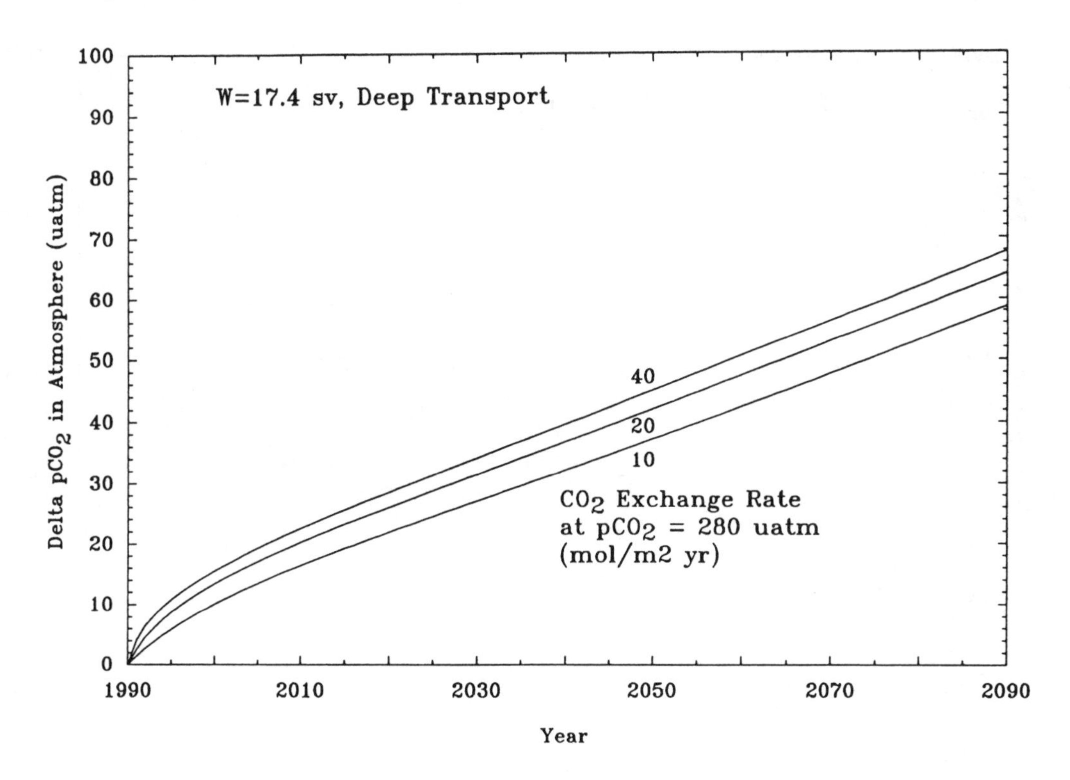


**FIGURE 12.** Effect of $CO_2$ exchange rate over the Antarctic Ocean on the reduction of atmospheric $pCO_2$ if iron fertilization is carried out over the next 100 years. The standard case has a $CO_2$ exchange rate of 20 $mol{\cdot}m^{-2}{\cdot}yr^{-1}$ at $pCO_2$ of 280 μatm; (sv = sverdrups).

However, the efficient iron fertilization case used for all other model calculations results in a surface $PO_4$ concentration of 0.1 μmol $kg^{-1}$ after a few years of iron fertilization.

# V. OXYGEN CONSUMPTION

An excessive amount of oxygen is demanded in the subsurface water to reoxidize the enhanced flux of falling organic particulate resulting from a successful iron fertilization. One critical consideration in this regard is the possibility that anoxic conditions might occur in the Antarctic Ocean, which would have an adverse effect on the marine ecosystem. Unfortunately, the model parameters set up for nutrient and bomb-$^{14}C$ distribution are unable to reproduce a depth profile of oxygen distribution consistent with the observed data. However, the difference in $O_2$ profiles before and after 100 years of iron fertilization in the Antarctic can be calculated. Such $O_2$ profiles represent the consumption of $O_2$ resulting from iron fertilization. In the simulation of oxygen distribution, $O_2$ in the atmosphere exchanges with the surface water, and $O_2$ is generated in the mixed layer during photosynthesis processes according to the Redfield ratio of 175 ($O_2$:P). The same ratio is also used in the subsurface water where $O_2$ is consumed in the respiration processes of organic flux.

Figure 16 shows the distribution of $O_2$ consumption for the best case scenario after 100 years of fertilization. Under this model condition, a maximum depletion of about 133 μmol $kg^{-1}$ takes place at a depth of 600 m. For waters deeper than 2000 m the $O_2$ depletion reaches only about 18 μmol $kg^{-1}$. To predict a possible $O_2$ distribution after 100 years of iron fertilization, this $O_2$ consumption is subtracted from the observed $O_2$ distribution in the Antarctic. In Figure 16, $O_2$ distributions in GEOSECS Atlantic stations 78, 82, and 89 (near

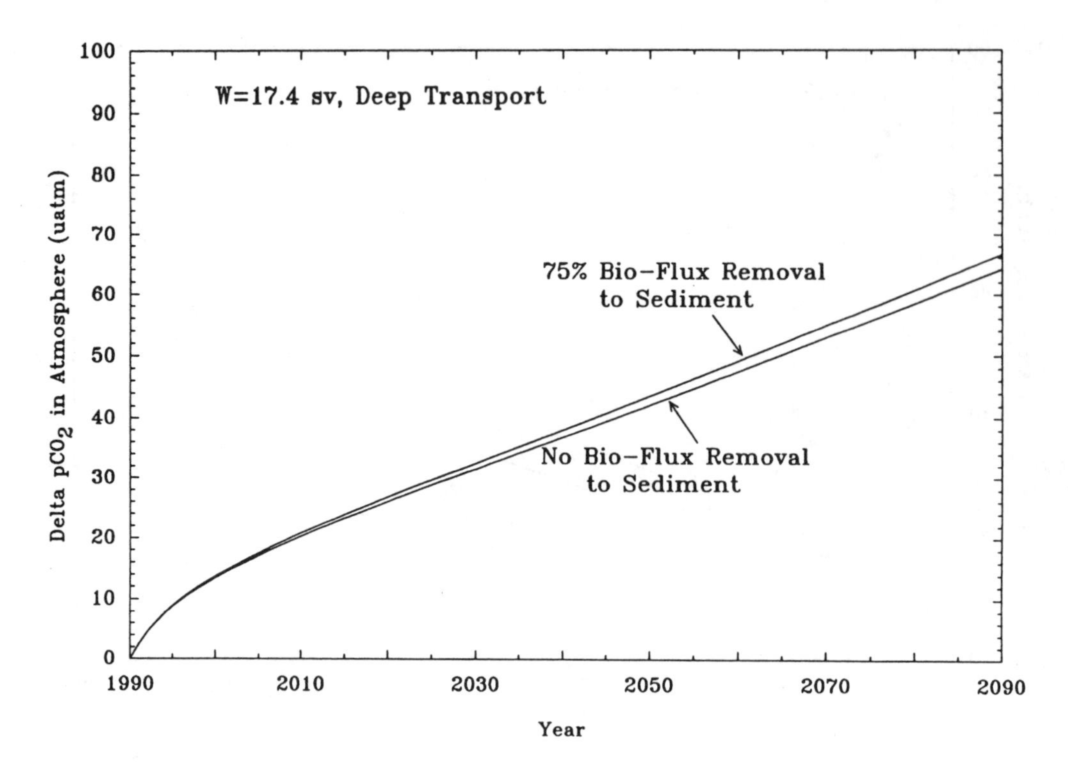


**FIGURE 13.** Effect of direct removal to the sediment of biological flux in the Antarctic Ocean on the reduction of atmospheric $pCO_2$ if iron fertilization is carried out over the next 100 years. No direct removal of biological flux to the sediment in the standard case; (sv = sverdrups).

60°S) are illustrated. The predicted $O_2$ reduction does not lead to anoxic conditions. It should be noted, however, that $O_2$ consumption depends on the reoxidation function below the surface mixed layer. In the standard case, this reoxidation function extends to the well-mixed deep box to produce a reasonable $PO_4$ distribution in the water column. To produce a maximum effect of iron fertilization on $O_2$ distribution, the reoxidation process can be limited to the upper 2000 m of the water column. Shown in the lower panel of Figure 16 is the $O_2$ consumption resulting from such a scenario, where 6% of organic flux is reoxidized right below the mixed layer, and this percentage of reoxidation is decreased exponentially with a half-depth of 275 m. The maximum consumption in 100 years of iron fertilization is almost 500 $\mu$mol $kg^{-1}$ at 500 m. The predicted $O_2$ distribution shows anoxic condition in the depth range of 200 to 1300 m. However, if, as most believe, the extent of nutrient drawdown through iron fertilization would not be anywhere near as great as assumed here, neither would the extent of $O_2$ drawdown.

## VI. CONCLUSIONS

J. Martin's idea of iron fertilization in the nutrient-rich polar ocean as a means of enhancing the biological pump and, hence, of significantly reducing the atmospheric $CO_2$ appears at first to be an appealing alternative for easing the potential problem of the ever-increasing concentration of greenhouse warming gas in the atmosphere. However, simulations using box models of ocean calibrated with tracer distribution indicate that a reduction of no more than 10% of the atmospheric $CO_2$ partial pressure is predicted if the Antarctic Ocean is perfectly fertilized with iron over the next 100 years. To obtain such results also

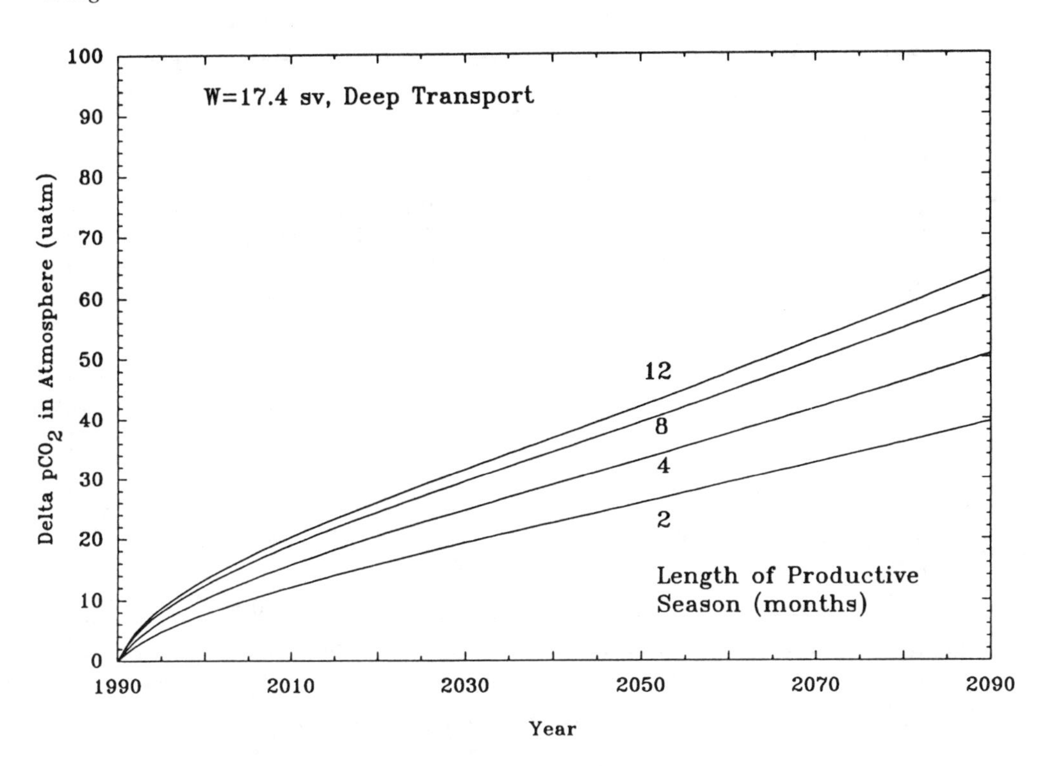


**FIGURE 14.** Reduction of $pCO_2$ in the atmosphere resulted from seasonal iron fertilization for the length of 12, 8, 4, and 2 months. The standard case assumes a 12-month productive season. A two-thirds total reduction in the atmosphere can be achieved by one-sixth yearly fertilization with iron in the Antarctic Ocean. Deep transport of the upwelled water (17.4 sv [sverdrups]) is assumed for all these cases.

requires that iron be continuously added throughout the fertilization period. As soon as the fertilization is terminated, the atmospheric $CO_2$ partial pressure would rise on more or less the same time scale as it was lowered.

Sensitivity tests of model simulations with respect to $pCO_2$ in the atmosphere under the influence of anthropogenic $CO_2$ emission over the next century indicate that the major dynamic factors limiting the oceanic uptake of $CO_2$ in response to a perfect iron fertilization are the rate of upwelling in the Antarctic and the fate of this upwelled water. For lateral transport cases, the net reduction of atmospheric $pCO_2$ is about 3% of atmospheric level and is insensitive to upwelling fluxes. For deep transport cases, however, such reduction increases from about 6% for low upwelling flux of 8.7 sverdrups to about 12% for high flux of 34.8 sverdrups. The best case with 17.4 sverdrups yields an 8% reduction, which is equivalent to a reduction of 64 μatm if the atmosphere $pCO_2$ reaches 800 μatm in the next century, resulting from anthropogenic $CO_2$ emission according to a business-as-usual scenario.

Such other dynamic factors as the air-sea $CO_2$ exchange rate and vertical diffusivity in the upper water column are of secondary importance. The area adopted for the Antarctic is also of secondary importance in this box model. The influence of the dark winter months in which phytoplankton growth cannot take place depends on the ratio of the response time for $PO_4$ removal from surface water, to the response time for replacement of surface mixed layer water by upwelling or vertical mixing.

The extent of $CO_2$ drawdown in the thermocline relies strongly on the depth dependence of respiration. Because this dependence is not known, it is not possible to properly evaluate the extent of anoxia generated by a successful iron fertilization.

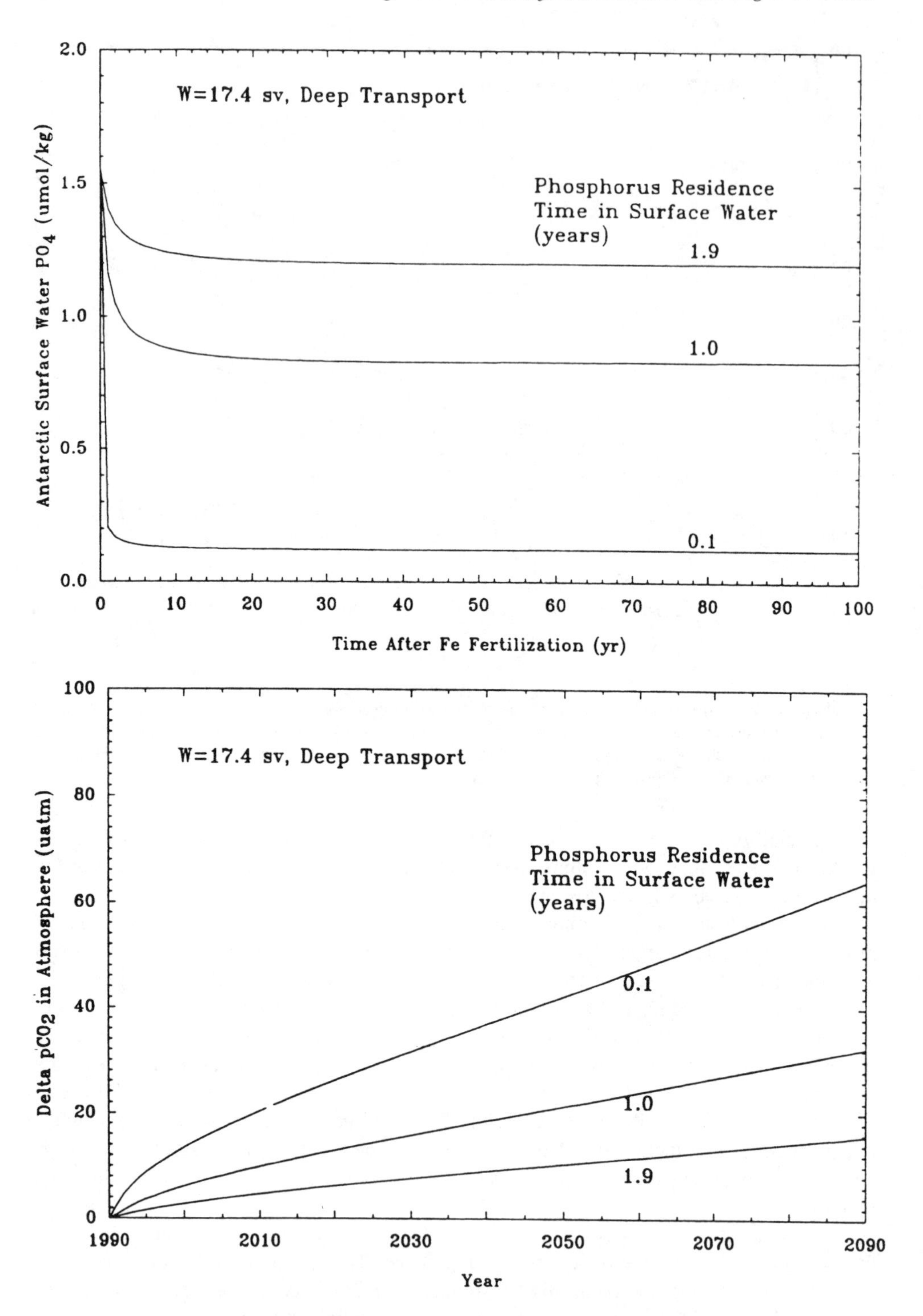


**FIGURE 15.** Effect of iron-fertilization efficiency on the reduction of atmospheric $pCO_2$ in the next 100 years. In the upper panel, surface water $PO_4$ concentration is plotted as a function of fertilization efficiency, which is represented by the mean residence time of nutrients in surface mixed layer of 0.1, 1.0, and 1.9 years. In the lower panel, the corresponding atmospheric $pCO_2$ reductions are plotted for the next century. The mean residence time of 0.1 year is used to represent the most efficient iron fertilization for all cases studied in this paper; (sv = sverdrups).

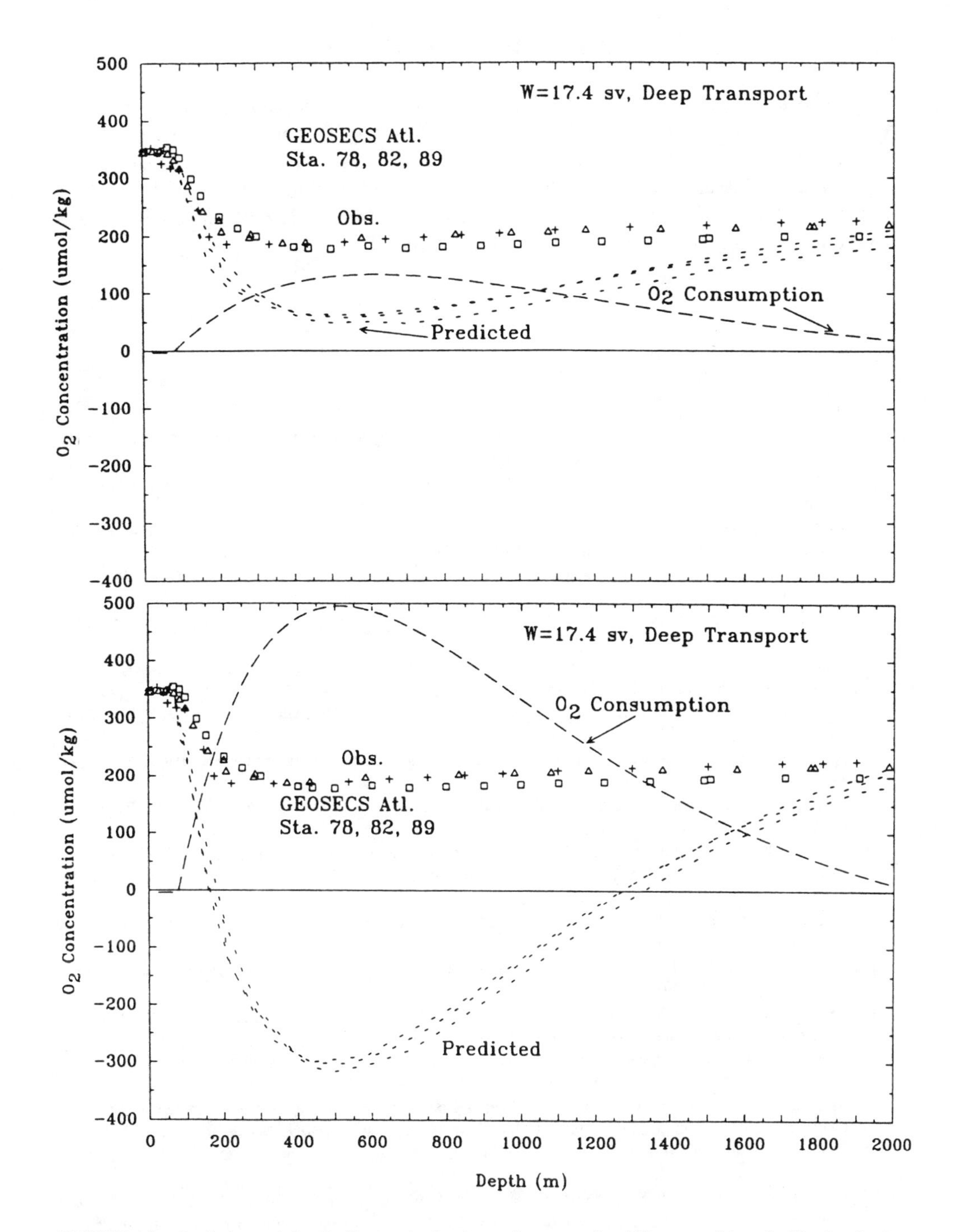


**FIGURE 16.** Predictions of $O_2$ distribution in the Antarctic water after 100 years of iron fertilization in the Antarctic Ocean. Model $O_2$ consumption is shown as dashed lines. The predicted curves are obtained by subtracting $O_2$ consumption value from the observed GEOSECS data at stations 78, 82, 89. In the upper panel, a standard reoxidation function for falling organic flux is used. In the lower panel, the organic flux is assumed to reoxidize completely in the upper 2000 m of the water column. In the latter case, a serious anoxic condition would result in the depth range of 200 to 1300 m; (sv = sverdrups).

## ACKNOWLEDGMENTS

Thanks to W. M. Post and A. W. King of Environmental Sciences Division, Oak Ridge National Laboratory, for their review of this manuscript. This research is sponsored by Carbon Dioxide Research Program, Environmental Sciences Division, Office of Health and Environmental Research, U.S. Department of Energy, under contract DE-AC05-840R21400 with Martin Marietta Energy Systems, Inc., Publication No. 3953, Environmental Sciences Division, Oak Ridge National Laboratory, Oak Ridge, Tennessee.

## REFERENCES

1. Martin, J. H. and Fitzwater, S. E., Iron deficiency limits phytoplankton growth in the northeast Pacific subarctic, *Nature (London),* 331, 341–343, 1988.
2. Martin, J. H. and Gordon, R. M., Northeast Pacific iron distributions in relation to phytoplankton productivity, *Deep-Sea Res.,* 35, 177–196, 1988.
3. Martin, J. H., Gordon, R. M., Fitzwater, S., and Broenkow, W. W., VERTEX: phytoplankton/iron studies in the Gulf of Alaska, *Deep-Sea Res.,* 36, 649–680, 1989.
4. Martin, J. H., Gordon, R. M., and Fitzwater, S., Iron in Antarctic waters, *Nature (London),* 345, 156–158, 1990a.
5. Martin, J. H., Fitzwater, S. E., and Gordon, R. M., Iron deficiency limits phytoplankton growth in Antarctic waters, *Global Biogeochem. Cycles,* 4, 5–12, 1990b.
6. Martin, J. H., Glacial-interglacial $CO_2$ change: the iron hypothesis, *Paleoceanography,* 5, 1–13, 1990.
7. DeBaar, H. J., Buma, A. G. J., Nolting, R. F., Cadee, G. C., Jacques, G., and Treguer, P. J., On iron limitation of the southern ocean: experimental observations in the Weddell and Scotia Seas, *Mar. Ecol. Prog. Ser.,* 65, 105–122, 1990.
8. Banse, K., Does iron really limit phytoplankton production in the offshore subarctic Pacific?, *Limnol. Oceanogr.,* 35, 772–775, 1990.
9. Banse, K., Iron availability, nitrate uptake, and exportable new production in the subarctic Pacific, *J. Geophys. Res.,* 96, 741–748, 1991.
10. Dugdale, R. C. and Wilkerson, F. P., Regional prospectives in global new production, in *Oceanologie Actualite et Prospective,* Denis, M. M., Ed., Centre d'Oceanologie de Marseille, Marseille, France, 1989, 289.
11. Dugdale, R. C. and Wilkerson, F. P., Iron addition experiments in the Antarctic: a reanalysis, *Global Biogeochem. Cycles,* 4, 13–19, 1990.
12. Peng, T.-H. and Broecker, W. S., Dynamic limitations on the Antarctic iron fertilization strategy, *Nature (London),* 349, 227–229, 1991.
13. Joos, F., Sarmiento, J. L., and Siegenthaler, U., Potential enhancement of oceanic $CO_2$ uptake by iron fertilization of the Southern Ocean, *Nature (London),* 349, 772–775, 1991a.
14. Joos, F., Siegenthaler, U., and Sarmiento, J. L., Possible effects of iron fertilization in the Southern Ocean on atmospheric $CO_2$ concentration, *Global Biogeochem. Cycles,* 5, 135–150, 1991b.
15. Sarmiento, J. L. and Orr, J. C., Three dimensional ocean model simulations of the impact of Southern Ocean nutrient depletion on atmospheric $CO_2$ and ocean chemistry, *Limnol. Oceanogr.,* in press, 1992.
16. Peng, T.-H. and Broecker, W. S., Factors limiting the reduction of atmospheric $CO_2$ by iron fertilization, *Limnol. Oceanogr.,* 1992.
17. Peng, T.-H., Broecker, W. S., and Östlung, H. G., Dynamic constraints on $CO_2$ uptake by an iron-fertilized Antarctic, in *Modeling the Earth System,* Ojima, D., Ed., UCAR/OIES, Boulder, CO, Global Change Institute, Vol. 3, 77–106, 1992.
18. Broecker, W. S., Peng, T.-H., Östlung, G., and Stuiver, M., The distribution of bomb radiocarbon in the ocean, *J. Geophys. Res.,* 90, 6953–6970, 1985.
19. Toggweiler, J. R., Dixon, K., and Bryan, K., Simulations of Radiocarbon in a coarse-resolution world ocean model 2. Distributions of bomb-produced carbon 14, *J. Geophys. Res.,* 94, 8243–8264, 1989.
20. Oeschger, H., Siegenthaler, U., Schotterer, U., and Gugelman, A., A box diffusion model to study the carbon dioxide exchange in nature, *Tellus,* 27, 168–192, 1975.

21. Peng, T.-H., Takahashi, T., Broecker, W. S., and Olafsson, J., Seasonal variability of carbon dioxide, nutrients and oxygen in the northern North Atlantic surface water: observations and a model, *Tellus,* 39B, 439–458, 1987.
22. Takahashi, T., Broecker, W. S., Bainbridge, A. E., and Weiss, R. F., Carbonate chemistry of the Atlantic, Pacific, and Indian Oceans: the results of the GEOSECS expeditions, 1972–1978. Technical Report No. 1, CU-1-80, Lamont-Doherty Geological Observatory, Palisades, N.Y.
23. Marland, G., Global $CO_2$ emissions, in *Trends '90: A Compendium of Data on Global Change,* Boden, T. A., Kanciruk, P., and Farrel, M. P., Eds., ORNL/CDIAC-36, Carbon Dioxide Information Analysis Center, Oak Ridge National Laboratory, Oak Ridge, TN, 1990, 92.
24. Peng, T.-H., Oceanic $CO_2$ uptake and future atmospheric $CO_2$ concentrations, in *Air-Water Mass Transfer* Selected Papers from the 2nd Int. Symp. on Gas Transfer at the Water Surfaces, Wilhelms, S. C. and Gulliver, J. S., Eds., ASCE, New York, 1991, 618–636.
25. Neftel, A., Moor, E., Oeschger, H., and Stauffer, B., Evidence from polar ice cores for the increase in atmospheric $CO_2$ in the past two centuries, *Nature (London),* 315, 45–47, 1985.
26. Keeling, C. D., Bacastow, R. B., Carter, A. F., Piper, S. C., Whorf, T. P., Heimann, M., Mook, W. G., and Roeloffzen, H., A three-dimensional model of atmospheric $CO_2$ transport based on observed winds. I. Analysis of observational data, *Geophys. Monogr. 55,* Am. Geophys. Union, 1989, 165–231.
27. Houghton, J. T., Jenkins, G. J., and Ephraums, J. J., *Climate Change: The IPCC Scientific Assessment,* Cambridge University Press, Cambridge, 329–341, 1990.
28. Broecker, W. S., Peng, T.-H., and Östlund, G., The distribution of bomb tritium in the ocean, *J. Geophys. Res.,* 91, 14331–14344, 1986.

Chapter 12

# LAND-WATER INTERACTIONS AND CLIMATE CHANGE: A FISH-EYE VIEW

**Penelope Firth**

## TABLE OF CONTENTS



0-8493-4419-0/93/$0.00 + $.50
© 1993 by CRC Press, Inc.

# I. INTRODUCTION

The fish-eye lens is an interesting tool of the photographer's trade. It gives the viewer a broader field of view than an ordinary lens, but the picture is distorted around the edges. This chapter explores the area of land-water interactions and climate change much like a fish-eye lens might show a waterfall. You can tell which direction the water is moving, but you're not sure how wide the fall is and whether or not there are rocks at the bottom.

In recent decades we have begun to understand how important land-water interactions are to the integrity of freshwater ecosystems. Functional freshwater ecosystems, and the resources they provide, are, in turn, critical to the water resources that provide for the existence and welfare of all human populations. The link between climate and land-water interactions is a direct one, mediated by precipitation and temperature. Climate change will thus have an immediate and far-reaching implication for freshwater systems. Most scientists believe changes are likely,[1] but we do not know when they will become apparent, what the regional effects will be, or how much warming, cooling, raining, and baking we will face.

# II. CLIMATE CHANGE

The Earth's climate ordinarily changes on a time scale of thousands of years. Our current climate only seems permanent because the climate of historic times has been relatively similar to the present. Examination of deep ice cores has provided extensive records indicating that climate and the composition of the atmosphere have tracked each other through prehistoric time. This is significant because human activities have caused, and are continuing to cause, great changes in the composition of the atmosphere. Global climate change is the likely outcome.

## A. THE GREENHOUSE EFFECT AND EARTH'S CLIMATE

The greenhouse effect is caused by certain atmospheric trace gases trapping heat near the Earth. Without these greenhouse gases, Earth's temperature would be 33°C lower,[2] and the habitability of our planet would be greatly compromised.

Carbon dioxide ($CO_2$), which has increased by about 25% since 1850, is responsible for about half of the present and projected effective warming. The other trace gases are much less abundant than $CO_2$, but they last longer, are complementary (i.e., they trap different thermal spectra), and they are many times as effective. The annual growth rates of these gases result in very large projected concentrations in the short term. Between 1975 and 1985, the natural gases increased by less than 5%, while the synthetic gases (CFCs) increased by more than 100%.[3]

## B. GLOBAL WARMING?

The global mean surface air temperature has increased by 0.3 to 0.6°C over the last century. The size of this warming is consistent with predictions of climate models, but it is also of the same magnitude as natural climate variability. The unequivocal detection of the enhanced greenhouse effect from observations is not likely for a decade or more.[1]

The Intergovernmental Panel on Climate Change (IPCC) predicts, based on current models, that a 3°C increase in global mean temperature is likely before the end of the next century. To put this in perspective, there has only been about 5°C of global warming since the last ice age 18,000 years ago. In ecological terms, the post-ice-age warming caused some forests to move more than 500 miles to the north.[4]

### C. FEEDBACKS, UNCERTAINTIES, AND PROJECTIONS

Positive feedbacks from warming global temperatures, such as increased respiration and methanogenesis, are of great concern. Biotic feedbacks such as these are among the most neglected mechanisms in the modeling effort.[5] Other uncertainties in the predictions are due to our incomplete understanding of the sources and sinks of greenhouse gases, clouds, oceans, and ice sheets.[1]

The models disagree on many of the details, but there is general agreement that while overall global precipitation may increase, continental interiors may be hotter and drier.[6-8] Increased summer dryness in midcontinents has also been suggested based on historical analogs.[9,10]

It is important to recognize that by adding greenhouse gases to the atmosphere, humans are performing a giant global experiment. If the models are off in one direction, the climate changes could be relatively benign and might occur over a time period of centuries. If the models are off in the other direction, however, in near decades we could see extraordinary climate changes, with ominous consequences for ecological and social systems.

## III. LAND-WATER INTERACTIONS

The lake shore, the river bank, the edge of the marsh, these are what come to mind when we think of the interface between land and water. But land-water interactions are far broader, and more complicated, than this. Land and water are closely coupled through the hydrologic cycle, biogeochemical cycles, and energy flows via sunlight and organic compounds. This coupling lends unique characteristics to each freshwater system, and drives patterns and processes at many spatial and temporal scales. The impacts of human presence may add other dimensions to this coupling.

### A. THE INTEGRATING CATCHMENT

Lakes, streams, and wetlands integrate the terrestrial and atmospheric events taking place in their catchments. A great deal of research has emphasized the importance of the catchment and the riparian zone to freshwater ecosystems.[11-16] Much of this research has emphasized the key role of catchment vegetation, which can control or influence the amount and kind of materials entering the water body.

Catchment vegetation is, of course, directly related to climate, and it is generally agreed that terrestrial plant assemblages will change if populations shift in response to altered weather patterns.[17,18] These changes would certainly have direct effects upon aquatic systems. For example, as the mix of catchment species changed, organic inputs to surface and groundwaters would be expected to change as well. Likewise, extensive tree mortality resulting from a less hospitable environment would be expected to increase the amount of woody debris in streams, and simultaneously increase the amount of light reaching the water.

### B. ANASTOMOSING EFFECTS

Indirect effects of changed catchment vegetation would impact aquatic ecosystems in a variety of ways. At higher temperatures, plants would be expected to have higher transpiration rates, lowering runoff. However, this effect might be balanced somewhat in a higher $CO_2$ environment because plant transpiration would be decreased, resulting in decreased water consumption — and hence, greater runoff.[19] Cloudiness, humidity, and windiness also affect transpiration, and the regional changes that might take place in these factors are as yet unknown.[20]

Higher $CO_2$ levels in the atmosphere would have other important effects on catchment vegetation that could directly affect freshwater systems. Meyer and Pulliam[21] discuss these in detail. In one example, groundwater nutrient content could be altered as a result of a complex set of sequential reactions involving (1) increased allocation of photosynthate to roots in a high $CO_2$ atmosphere, resulting in (2) increased volume of root exudates, that allows the development of (3) increased micorrhizal biomass, and hence, (4) increased rates of phosphorus uptake from the subsurface water.[21] In this scenario, aquatic systems in watersheds thus affected could experience a shift in the base of the food chain away from algae or aquatic plants and toward leaves and other organic matter falling into the water.

Unfortunately, high $CO_2$ levels would also be expected to directly affect the quality of leaf litter: leaf nitrogen content has been found to be lowered under elevated $CO_2$ conditions.[22-24] The quality, or nutrient content, of leaf litter directly influences its decomposition, as well as the feeding and production of the organisms that depend on it.[25,26]

## C. SOILS AND SHORES

Catchment vegetation and soil dynamics may also significantly influence nutrient flux to aquatic systems. Microbial processes in the soil are tightly linked to soil moisture and temperature, as are higher plant processes such as nutrient uptake. Increased decomposition, nitrification, and chemical weathering rates resulting from warmer temperatures could result in increased nutrient release from the catchment. Likewise, accelerated respiration of soil organic matter could decrease the dissolved organic matter than normally leaches from soils and enters aquatic systems.[21]

Specific edge habitats, such as river floodplains and lake shores, mediate land-water interactions at many levels. Alterations in temperature and riparian inundation patterns affect biogeochemical processes in the riparian zone. These processes, in turn, alter the exchange of elements between rivers and their floodplains.[21]

## D. ALTERED HYDROLOGY

The hydrologic cycle that determines the distribution of water resources is particularly sensitive to climate change.[27-29] All evidence suggests that movements in the hydrologic cycle were slower in glacial time, but they are likely to increase with climatic warming.[30] Warmer temperatures would increase evaporation from the oceans, leading to increased precipitation globally, perhaps by as much as 7 to 15%.[6] Regionally, however, average annual precipitation could change by ±20% in a doubled $CO_2$ climate scenario.[31]

In many parts of the world there is a strong inverse relationship between temperature and runoff for a given amount of precipitation: as temperature increases, runoff decreases. On a regional basis, climate controls the quantity and temporal distribution of precipitation, setting an upper limit to runoff over any arbitrary time scale.[32] As discussed above, the local vegetative cover is also extremely important, controlling the rate at which precipitation enters the channel and the rate at which evapotranspiration returns water to the atmosphere. Finally, the underlying geology and topography of the catchment also impact the rate at which water enters the channel, by either overland or subsurface flow. The ability of the drainage basin to sustain surface flow through aquifier discharge during dry periods is critical to the aquatic biological communities.[33]

It is clear that regional shifts in climate will not only alter the quantity of runoff, but its variability and timing as well.[32] This can have very serious ramifications for both human and ecological systems. For example, snowpack is of critical importance in many parts of the world to sustain summer flows in arid-land streams. Not surprisingly, alpine glaciers are thought to be particularly sensitive to significant changes in climate.[40] In a warmer climate, more precipitation will fall as rain, and snowpack will probably be reduced, lowering summer water availability in arid regions.[34] The speed and duration of melting might also change. Earlier and shorter-duration melting can cause high flows accompanied by rapid pH drops. The timing and spatial extent of such acidic spates could have particular significance for vulnerable life history stages of aquatic organisms including algae, insect larvae, and fish.[35]

## E. GEOPHYSICAL ELEMENTS

As Ward et al.[33] discuss, the underlying lithology influences hydrology, composition of weathering products (dissolved and particulate), soil characteristics, and terrestrial vegetation. Within this template, changes in precipitation and temperature impact geophysical attributes of the catchment and can have major implications for land-water interactions.

One of the ubiquitous processes that link land and water is erosion and sedimentation; precipitation is the dominant controlling climatic factor. If an area were to receive less precipitation, soil moisture and vegetative cover would be reduced, and higher sediment yields to streams and lakes would result. In addition, during the period while precipitation declined, any increase in precipitation variability would make increased erosion and sediment yields more likely.[33]

Land and water are also linked by the physical transfer of dissolved materials. The dissolved load of aquatic systems is determined by chemical weathering, biological processes, atmospheric inputs, and evaporation/crystallization.[33] Interestingly, warmer temperatures are not expected to have as great an impact on chemical weathering as changes in pH induced by higher atmospheric $CO_2$ levels. Doubling of atmospheric $CO_2$ levels would decrease the pH of rainwater by about 0.4 U through the production of carbonic acid.[33] Acid precipitation has a relatively great ability to dissolve minerals.

## F. LAKES, STREAMS, AND WETLANDS: SYSTEM DIFFERENCES

Specific types of aquatic ecosystems are known to respond in different ways to the forcing functions of temperature and precipitation change. In the case of lakes, temperature change alone can have major effects on aquatic communities. For example, Schindler et al.[36] examined a 20-year record of climatic, hydrologic, and ecological records for the Experimental Lakes area of northwestern Ontario. This study showed that moderate climatic changes (2°C temperature increase, 3 week increase in ice-free season) have significant effects on the lake ecosystems. These included increased concentrations of most chemicals, increased populations and diversity of phytoplankton, and deeper thermoclines with resulting decreased summer habitats for cold stenotherms (e.g., lake trout, opposum shrimp).

The biota of rivers and streams also respond sensitively to fluctuations in temperature and precipitation that occur with climate change. Poff[32] speculates that the ecological vulnerability of streams to different climate change scenarios will vary, depending on their recent historical hydrologic characteristics. For example, mesic groundwater streams are buffered against flow reductions, but may be relatively susceptible to increased flood frequency and flow variability. Streams that flood seasonally now (e.g., snowmelt streams) may be more vulnerable to nonseasonal flooding than streams that typically flood nonseasonally.[33]

Perhaps the closest land-water interactions occur in wetlands. Regional climate change, particularly precipitation rate, is one of the most important determinants of total freshwater wetland area.[37] Of greater significance than changes in wetland area are potential changes in the species composition of wetland plant communities and the habitat, food, and protection they provide.[38] If climate changes result in altered fluctuation regimes for temperature and water level, wetland stability and community diversity would be adversely impacted.[39]

## IV. CONCLUSIONS

Freshwater systems are ubiquitous on Earth and are critical to humans as well as thousands of other species. They are, however, neither evenly distributed nor abundant. This makes the implications of global climate change for land-water interactions extremely serious. Altered patterns and timing of precipitation, snowmelt, and evapotranspiration will affect catchment characteristics and runoff on a region-specific basis worldwide, altering the nature and timing of critical land-water interactions.

The implications for aquatic life are grave. Why is this important to humans? Because structural and functional changes at the level of the ecosystem can, in turn, impact attributes of great importance to humans, such as fish production, water quality, and waste assimilative capacity.

The fish-eye lens shows us clear evidence for very specific effects of climate change on land-water interactions. Climate change, particularly a global warming, will impact the geophysical and geochemical aspects of the fresh waters on Earth, and thereby change the ecological aspects that are vital to both environmental health and human benefits derived from water resources. However, the complexity of the problem, and the dearth of easy solutions, blur the edges of the picture. We are clearly near a crossroads where scientific information can feed into the social and political decision-making process.

## REFERENCES


1. IPCC (Intergovernmental Panel on Climate Change), Policymakers Summary of the Scientific Assessment of Climate Change, Rep. to IPCC from Working Group 1, Exec. Summ., Intergovernmental Panel on Climate Change, 1990.
2. Abrahamson, D. E., Ed., *The Challenge of Global Warming,* Island Press, Washington, D.C., 1989.
3. Ramananthan, V., The greenhouse theory of climate change: a test by an inadvertent global experiment, *Science,* 240, 293–299, 1988.
4. Kutzbach, J. E., Historical perspectives: climatic changes throughout the millennia, in *Global Change and Our Common Future,* DeFries, R. S. and Malone, T. F., Eds., National Academy Press, Washington, D.C., 1989.
5. Lashoff, D. A., The dynamic greenhouse: feedback processes that can influence global warming, in *Coping with Climate Change,* Topping, J. C., Jr., Ed., The Climate Institute, Washington, D.C., 1989.
6. Bolin, B., Doos, B. R., Jaeger, J., and Warrick, R. A., Eds., *The Greenhouse Effect, Climatic Change, and Ecosystems,* SCOPE 29, John Wiley & Sons, Chichester, England, 1986.
7. Hansen, J., Fung, I., Lacis, A., Rind, D., Lebedeff, S., Ruedy, R., Russell, G., and Stone, P., Global climate changes as forecast by Goddard Institute for Space Studies three-dimensional model, *J. Geophys. Res.,* 93, 9341–9364, 1988.
8. Manabe, S., Changes in soil moisture, in *The Challenge of Global Warming,* Abrahamson, D. E., Ed., Island Press, Washington, D.C., 1989.
9. Jager, J. and Kellogg, W. W., Anomalies in temperature and rainfall during warm Arctic seasons, *Climat. Change,* 5, 39–60, 1983.
10. Wigley, T. M. L., Jones, P. D., and Kelley, P. M., Scenarios for a warm, high $CO_2$ world, *Nature (London),* 283, 17–21, 1980.
11. Hynes, H. B. N., The stream and its valley. *Verh. Int. Vereinig. Theor. Angew. Limnol.,* 19, 1–15, 1975.
12. Odum, W. E. and Prentki, R. T., Analysis of five North American lake ecosystems. IV. Allochthonous carbon inputs, *Verh. Int. Vereinig. Theor. Angew. Limnol.,* 21, 574–580, 1978.
13. Vannote, R. L., Minshall, G. W., Cummins, K. W., Sedell, J. R., and Cushing, C. E., The river continuum concept, *Can. J. Fish. Aquat. Sci.,* 37, 130–137, 1980.
14. Swanson, F. J., Gregory, S. V., Sedell, J. R., and Campbell, A. G., Land-water interactions: the riparian zone, in *Analysis of Coniferous Forest Ecosystems in the Western United States,* Edmonds, R. L., Ed., Hutchinson Ross, Stroudsberg, PA, 1982, 267–291.

15. Minshall, G. W., Cummings, K. W., Peterson, R. C., Cushing, C. E., Bruns, D. A., Sedell, J. R., and Vannote, R. L., Developments in stream ecosystem theory, *Can. J. Fish. Aquat. Sci.*, 42, 1045–1055, 1985.
16. Cummins, K. W., Wilzbach, M. A., Gates, D. M., Perry, J. B., and Taliaferro, W. B., Shredders and riparian vegetation, *BioScience,* 39, 24–30, 1989.
17. Andrasko, K. and Wells, J. B., North American forests during rapid climate change: overview of effects and policy response options, in *Coping with Climate Change,* Topping, J. C., Jr., Ed., The Climate Institute, Washington, D.C., 1989.
18. Davis, M. B., Insights from paleoecology on global change, *Bull. Ecol. Soc. Am.,* 70, 222–228, 1989.
19. Rosenberg, N. J., The increasing $CO_2$ concentration in the atmosphere and its implication on agricultural productivity. I. Effects on photosynthesis, transpiration and water use efficiency, *Climat. Change,* 3, 265–279, 1981.
20. Martin, P. H., Rosenberg, N. J., and McKenney, M. S., Sensitivity of evapotranspiration in a wheat field, a forest and a grassland to changes in climate and direct effects of carbon dioxide, *Climat. Change,* 14, 117–151, 1989.
21. Meyer, J. D. and Pulliam, W. M., Modification of terrestrial-aquatic interactions by a changing climate, in *Global Climate Change and Freshwater Ecosystems,* Firth, P. L. and Fisher, S. G., Eds., Springer-Verlag, New York, 1991, 177–191.
22. Strain, B. R. and Cure, J. D., Direct Effects of Increasing Carbon Dioxide on Vegetation, U.S. Department of Energy, Washington, D.C., 1985.
23. Fajer, E. D., How enriched carbon dioxide environments may alter biotic systems even in the absence of climatic changes, *Conserv. Biol.,* 3, 318–320, 1989.
24. Fajer, E. D., Bowers, M. D., and Bazzaz, F. A., The effects of enriched carbon dioxide atmospheres on plant/insect herbivore interactions, *Science,* 243, 1198–1200, 1989.
25. Cummins, K. W. and Klug, M. J., Feeding ecology of stream invertebrates, *Annu. Rev. Ecol. Syst.,* 10, 147–172, 1979.
26. Webster, J. R. and Benfield, E. F., Vascular plant breakdown in freshwater ecosystems, *Annu. Rev. Ecol. Syst.,* 17, 567–594, 1986.
27. Revelle, R. R. and Waggoner, P. E., Effects of a carbon-dioxide-induced climatic change on water supplies in the Western U.S., in Carbon Dioxide Assessment Committee, Eds., *Changing Climate,* National Academy Press, Washington, D.C., 1983.
28. Wigley, T. M. L. and Jones, P. D., Influences of precipitation changes and direct $CO_2$ effects on streamflow, *Nature (London),* 314, 149–152, 1985.
29. Gleick, P. H., Regional hydrologic consequences of increases in atmospheric $CO_2$ and other trace gases, *Climat. Change,* 10, 137–161, 1987a.
30. Schlesinger, W. H., *Biogeochemistry: An analysis of global change,* Academic Press, San Diego, 1991.
31. Department of Energy (DOE), Multi-laboratory Climate Change Committee, *Energy and Climate Change,* Lewis Publishers, Chelsea, MI, 1990.
32. Poff, N. L., Regional hydrologic response to climate change: an ecological perspective, in *Global Climate Change and Freshwater Ecosystems,* Firth, P. L. and Fisher, S. G., Eds., Springer-Verlag, New York, 1991, 88–115.
33. Ward, A. K., Ward, G. M., Harlin, J., and Donahoe, R., Geological mediation of stream flow and sediment and solute loading to stream ecosystems due to climate change, in *Global Climate Change and Freshwater Ecosystems,* Firth, P. L. and Fisher, S. G., Eds., Springer-Verlag, New York, 1991, 116–142.
34. Gleick, P. H., The development and testing of a water balance model for climate impact assessment: modeling the Sacramento Basin, *Water Resour. Res.,* 23, 1049–1061, 1987b.
35. Resh, V. H., Brown, A. V., Covich, A. P., Gurtz, M. E., Li, H. W., Minshall, G. W., Reice, S. R., Sheldon, A. L., Wallace, J. B., and Wissmar, R. C., The role of disturbance in stream ecology, *J. North Am. Benthol. Soc.,* 7, 433–455, 1988.
36. Schindler, D. W., Beaty, K. G., Fee, E. J., Cruikshank, D. R., DeBruyn, E. R., Findlay, D. L., Linsey, G. A., Shearer, J. A., Stainton, M. P., and Turner, M. A., Effects of climatic warming on lakes of the central boreal forest, *Science,* 250, 967–970, 1990.
37. Mitsh, W. J. and Gosselink, J. G., *Wetlands,* Van Nostrand Reinhold, New York, 1986.
38. Meisner, J. D., Goodier, J. L., Regier, H. A., Shuter, B. J., and Christie, W. J., An assessment of the effects of climate warming on Great Lakes Basin fishes, *J. Great Lakes Res.,* 13, 340–352, 1987.
39. Keddy, P. A. and Reznicek, A. A., Great Lakes vegetation dynamics: the role of fluctuating water levels and seeds, *J. Great Lakes Res.,* 12, 25–36, 1986.
40. Wood, F. B., Global alpine glacier trends, 1960s to 1980s, *Arct. Alp. Res.,* 20, 404–413, 1988.

# *Section III: Global Assessment of Greenhouse Gas Production Including Need for Additional Information*

Chapter 13

# CLIMATE HISTORY DURING THE RECENT GREENHOUSE ENHANCEMENT

**Patrick J. Michaels and Robert C. Balling**

## TABLE OF CONTENTS



0-8493-4419-0/93/$0.00 + $.50
© 1993 by CRC Press, Inc.

# I. INTRODUCTION

Virtually all scientists directly involved in research on climatic change believe that the Earth will undergo some warming as a result of the increase in anthropogenerated emissions that absorb IR radiation, or enhance the ''greenhouse effect''. However, within certain broad limits, how much the world warms is irrelevant. Rather, the critical policy question is how will the world warm, because the real effect will be expressed by its regionality, seasonality, and distribution within the day-night cycle. Viewed in light of these factors, there are now several compelling lines of evidence that indicate the chance of an ecologically or economically disastrous global warming is becoming more remote.

General circulation model (GCM) climate simulations of the mid-1980s predicted a mean global warming of 4.2°C, with winter warming as much as 18°C for the north polar regions (Manabe and Wetherald, 1980; Hansen et al., 1984; Schlesinger, 1984; Mitchell, 1983; Washington and Meehl, 1983). This, along with Congressional testimony of June 23, 1988 that there was a ''high degree of cause and effect'' between current temperatures and human greenhouse alterations (Hansen, 1988) has helped to heighten both public and scientific interest in this problem.

In light of that interest, we examine here the problems of (1) global and hemispheric temperature histories and trace gas concentrations; (2) artificial warming from urban heat islands; (3) high latitude and diurnal temperatures; and (4) more recent climate models.

# II. TRACE GAS CONCENTRATIONS AND TEMPERATURE HISTORIES

While there are several IR-absorbing trace gases that have increased as a result of anthropogeneration, almost all of the current increase in radiative forcing is associated with (in descending order) carbon dioxide ($CO_2$), methane ($CH_4$), nitrous oxide ($N_2O$), and chlorofluorocarbons (CFC). As a result of their IR absorption and the concentration of these species in the lower atmosphere, an increasing but very small fraction of energy that would normally escape to space is redirected towards the Earth's surface, resulting in increased warming. The sensitivity of the current mean surface temperature is thought to be about 1°C for every additional $W/m^2$ of downward-directed radiation; this calculation can be made readily from work by Kiehl and Dickinson (1987).

Continuous instrumental records of $CO_2$ concentration date from the late 1950s at Mauna Loa Observatory, where the 1958 annual average was 315 ppm. Initial ice-core studies gave a background of 270 to 290 ppm, with a most likely value of 279 ppm (Neftel et al., 1985). Another analysis obtained a lower bound of 260 ppm (Raynaud and Barnola, 1985). The Soviet/French work on the long Vostok core seems to corroborate the lower values (Lorius et al., 1987), which if assumed to be 270 ppm, results in an anthropogenerated rise of 31% to date.

The non-$CO_2$ greenhouse gases have also increased considerably. The current climate forcing of $CH_4$ is nearly 40% of the effect of increased $CO_2$ (Craig et al., 1988) and the overall radiative effect of all of the non-$CO_2$ anthropogenerated greenhouse enhancers is approximately an additional 80% of that caused by a change in $CO_2$ alone. Hansen and Lacis (1990) give the current increase in overall forcing as 2 to 2.5 $W/m^2$, so that the current contribution from $CO_2$ alone is approximately 1.5 $W/m^2$. Because of the additional greenhouse enhancers, the current effective $CO_2$ concentration is approximately 420 to 430 ppm, or over 150% of the background range of 260 to 279 ppm. The United Nations Intergovernmental Panel on Climate Change (IPCC) corroborates this by stating that ''Greenhouse gases have increased since preindustrial times...by an amount that is radiatively equivalent

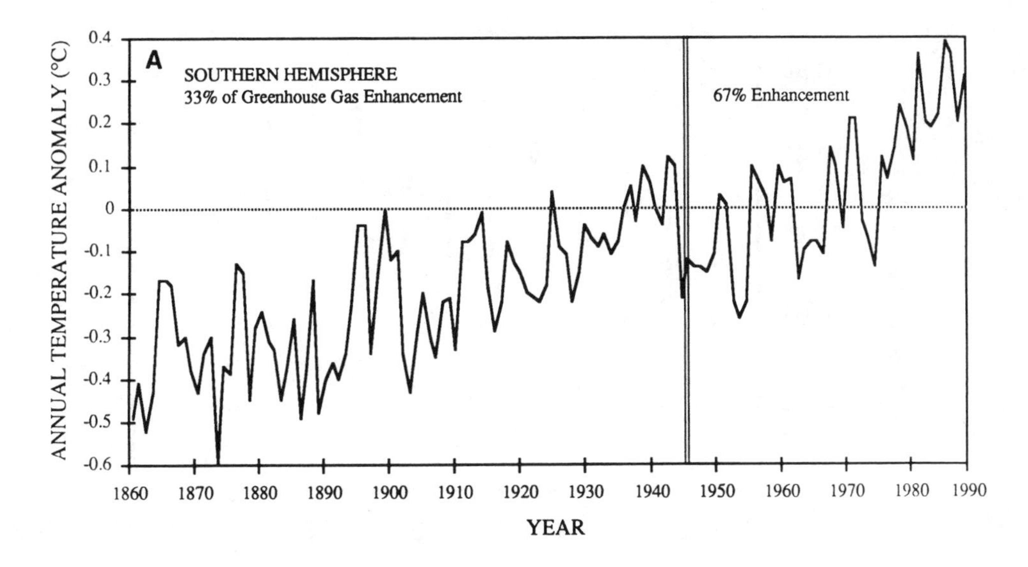


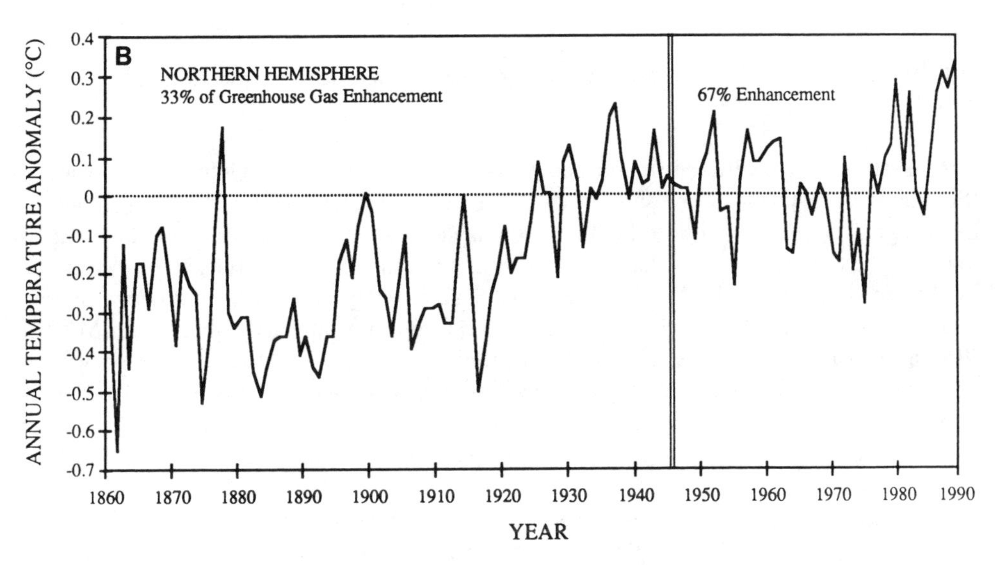


**FIGURE 1.** (A). Land/ocean annual mean surface air temperatures for the Southern Hemisphere through 1990. Updated from the U.S. Department of Energy (1990). (B). Land/ocean annual mean surface air temperatures for the Northern Hemisphere through 1990. Updated from U.S. Department of Energy (1990).

to about a 50% increase in carbon dioxide'' (IPCC, 1990), which gives a current effective value of 390 to 430 ppm. By all calculations, we have already proceeded more than halfway to an effective doubling of the background $CO_2$ concentration.

While the citation is often made that the history of global temperature as ''at least not contradictory to'' (MacCracken and Kukla, 1985) the mid-1980s climate model projections, those models produce an expected equilibrium warming for this trace gas change of over 2.0°C, assuming the climate temperature sensitivity is $1°C/W/m^2$, while a least-squares fit of the updated set of Jones et al. (1990) data of the U.S. Department of Energy (1990) global data gives a rise of $0.45 \pm 0.10°C$ in the last 100 years (Balling and Idso, 1990). Further, Figure 1 demonstrates that virtually all of the warming of the Northern Hemisphere was prior to the major postwar emissions of the trace gases; a linear trend through the data

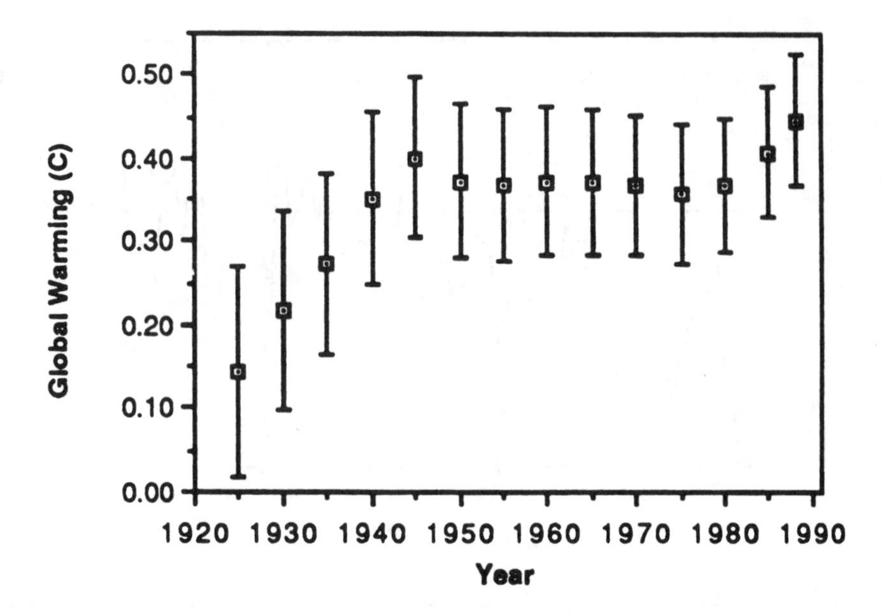


**FIGURE 2.** Least-squares trend in global land-based temperatures, with trends beginning in 1985 (Balling and Idso, 1990).

since 1940 is statistically indistinguishable from zero (p = .05). In addition, the ''water'' (Southern) Hemisphere, which should warm up slower, in fact shows the more ''greenhouse-like'' signal. While most of the Southern Hemisphere stations are on land, the fact that so much of the Southern Hemisphere is water results in more oceanic influence on the record than occurs in the Northern Hemisphere. Least-squares analysis of this record demonstrates, in a statistical sense, that much of the observed warming was already realized prior to 1945 (Balling and Idso, 1990); see Figure 2. At this time the actual $CO_2$ concentration was in the range of 307 ppm and the effective concentration was approximately 322 ppm. This is only one third of the total change in forcing to date, and implies that there has been very little additional warming during the period in which the majority of the radiative forcing has taken place.

Oceanic thermal lag may not in fact account for the difference. Wigley (1987) calculated, using liberal estimates of the lag, that the Hemisphere should still have warmed 1°C (primarily after 1950), and the least-squares trend since then is in fact 0.33°C. Sea surface temperature analyses of Bottomley et al. (1990) and Oort et al. (1989) demonstrate very little lag between that record and land readings. The relationship between deep ocean circulation and climate is very unclear and is thought to be a source of possible climatic ''surprises'' in the future; i.e., sudden warmings that will explain current temperatures.

## III. URBAN HEAT ISLANDS AND LONG-TERM SITE BIAS

Localized warming related to urban buildup has long been recognized as a confounding factor in regional and global temperature analysis. While causes include changes in the surface energy balance, local wind flow, and amount of open sky, precise calculations on the various components of the ''urban effect'' are not available. However, Karl et al. (1988a) have found in the U.S. that the effect is statistically detectable at population levels of 2500 and up, and Balling and Idso (1989) note the effect with population as low as 500. The temperature record is further confounded by the fact that exponential populations increases suggest the urban warming bias should be of the same functional form as enhanced greenhouse forcing, and that it will therefore tend to contaminate the latest years of temperature measurement. In other words, it would be striking if the last years in the record *were not* the warmest, although the least-squares global warming in the Jones and Wigley record since 1970 (which, on a smoothed basis, is the lowest point since 1915) of 0.35°C is certainly greater than any urban warming. The global trend for the last half-century (since 1940) is 0.21°C.

Karl et al. (1988b) have created a de-urbanized "Historical Climate Network" (HCN) of stations that have been checked for instrumental and location changes, and adjusted for population. While this record only applies to the coterminous U.S. and is therefore not necessarily representative of global trends (i.e., it shows a slow but unsteady cooling for much of the last 50 years), it serves as an important check on other global records because they can be compared over this limited region.

Karl et al. (1988a) found an apparent urban warming of 0.10 to 0.15°C in the global temperature records published in Figure 1. Curiously, the bias appears greatest in the U.S. portion of the record. A similar analysis of an analogous Soviet Union record shows no overall urban bias (Jones et al., 1990), although urban stations are in fact cooler than surrounding rural ones from 1953 to 1967; there is no difference between the two through 1980, and then there is some apparent urban warming after 1980. A similar negative urban effect is observed in the 1960s in mainland China (Jones et al., 1990). It therefore seems that the global urban bias could in fact be <0.10°C (a reasonable figure, according to the IPCC report, Chapter 7), but it should be noted that it will be more pronounced at the end of the record, which shows the highest temperatures. Because of the urban effect, it appears that the true global warming in Figure 1 during the last 100 years is 0.35 to 0.40°C.

Urban contamination compromises our estimates of historical temperature change. In an attempt to determine the reliability of the various long-term climate records, Spencer and Christy (1990) correlated satellite-microwave sensed temperatures with Karl et al.'s (1988b) U.S. HCN, and the hemisphere records of Jones et al. (1990) (see U.S. Department of Energy, 1990) and the Hansen (1988) record from his June 1988 congressional testimony. Spencer and Christy (1990) claim an accuracy of 0.02°C in the microwave-sensed temperatures. The HCN is the most correspondent to the satellite-sensed temperatures ($R^2 = 0.86$). The hemispheric records of both Jones et al. (1990) and Hansen (1988) were much less with the satellite data, with $R^2$ values of 0.42 and 0.20 for the Jones et al. (1990) record (Northern and Southern Hemisphere, respectively), and 0.46 and 0.34 for the Hansen (1988) record (updated records supplied as personal communication from John Christy). Spencer and Christy's (1990) temperature record from 1979 through 1990 shows a global trend of only +0.04°C/decade.

We calculated annual departures from the 1979 through 1990 mean period in the Jones et al. (1990) data set and subtracted from them anomalies in the Spencer and Christy (1990) data. When these are plotted as function of time (beginning in 1979), it is apparent that there is a dramatic relative warming of the Jones et al. (1990) data, compared to the satellite, in the Southern Hemisphere (Figure 3); no similar effect occurs in the Northern Hemisphere. The very high correlation between the satellite data and the U.S. HCN suggests that the measurements are accurate and, therefore, that there is a great deal of spurious warming in the recent Jones and Wigley data for the Southern Hemisphere. If this is the case, the true warming in the Southern Hemisphere since 1950 is on the order of 0.2°C, while the Northern Hemisphere warming over the same period is zero.

An additional and more insidious problem may be introduced into long-term land based climate records by a general site bias. Over what was the "developing world" of the late 19th century, the longest-standing records have tended to originate at points of commerce.

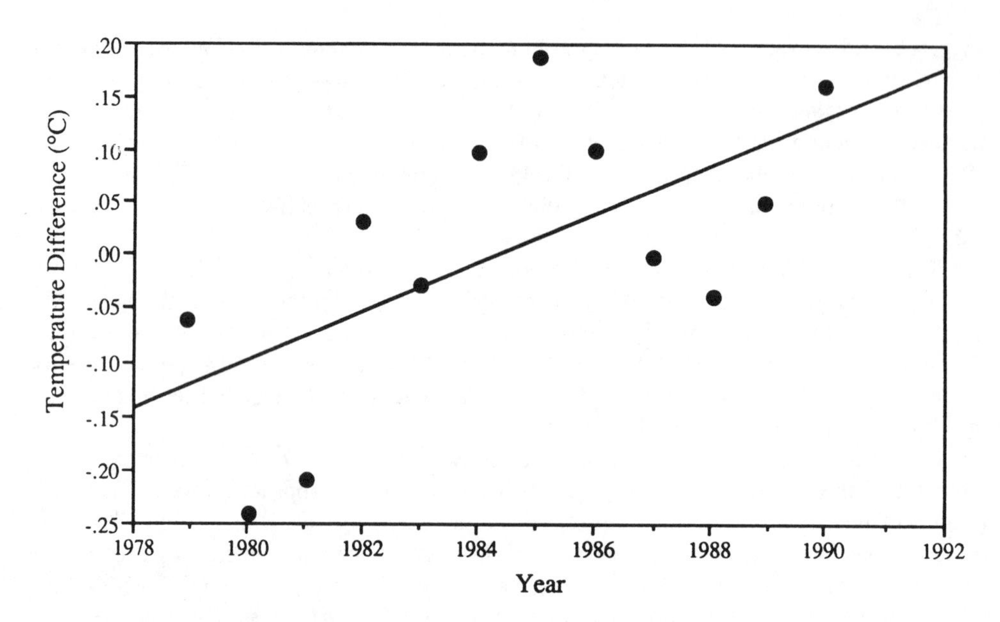


**FIGURE 3.** Annual departures from the 1979 to 1990 mean of Spencer and Christy's (1990) satellite-microwave data, subtracted from annual departures from the 1979 to 1990 mean of the Northern Hemisphere temperature history.

Because of the predominance of water power during that era, it is likely that the longest records are in fact at sites that are preferential for cold air drainage. Such locations may be buffered from nighttime warming.

As an example, the mean elevation of the three climatological divisions in western Virginia is approximately 2000 ft. Of the 22 stations maintaining enough record length to be included in the 1951 to 1980 climatological normals for temperature, the mean elevation is 1400 ft. Out of 22 stations, 13 are located on valley floors, while 3 are on ridgetops. It is doubtless that this set is preferentially sited in regions favoring night-time cold pooling.

Several attempts have been made to compensate both for population and general biases, usually involving temperatures in the free atmosphere. These can be determined from combinations of barometric readings and weather balloon ascents, and therefore cannot suffer from the urban effect on thermometers. Angell (1990), using the atmospheric layer from 850 to 300 mb, found a net warming of 0.24°C in the last 30 years after adjusting for El Niño warming and volcanic cooling. This is slightly less than one half of the least-squares linear trend of 0.42°C in the U.S. Department of Energy (1990) record shown in Figures 1A,B. A Recent extension shows 1990 as the warmest year (by 0.04°C), but this is not replicated in the Spencer and Christy (1990) data. Their data correlated very well with Angell's data up to 1990. Why there has been such a discrepancy in 1 year is unknown.

Michaels (1990) used the depth of the bottom half of the atmosphere over North America back to the beginning of systematic balloon observations in 1948, and a statistical surrogate back to 1885, and concluded that the observed secular variability in surface temperatures calculated from this record was approximately three times larger than that measured from ground-based thermometers. The rise in the first half of the 20th century and the subsequent fall through the mid-1970s was in fact concurrent with that of the ground-based record, although the magnitude was again greater. Notably, the rise in the period 1885 to 1950 calculated by this method (of 1.8°C) does not differ appreciably from the expected regional greenhouse warming calculated by GCMs for the next 50 years. It is unknown whether this magnitude is related to noise in the statistical transfer functions between the cyclone frequency record used to calculate the 500 mb height from 1885 to 1948, or whether it truly reflects free-atmosphere temperature variability.

Weber (1990a) also examined the same layer over the entire Northern Hemisphere since 1950, and found warming in the Northern Hemisphere tropics and subtropics, but cooling in the higher latitudes, with no overall hemispheric warming or cooling between the 1950s and 1980s. This signal is opposite to what would be expected from most climate models that suggest enhanced high latitude warming and minimal tropical warming.

## IV. HIGH LATITUDE AND DAILY TEMPERATURE REGIMES

In the climate simulation of Manabe and Wetherald (1980), which is quite representative of the mid-1980s generation of climate models, the difference between 2 and 4 $\times$ $CO_2$ at high northern latitudes is twice the 8°C difference between 1 and 2 $\times$ $CO_2$, suggesting that for the currently effective 1.5 $\times$ $CO_2$ there should be considerable Arctic warming, even with oceanic thermal lag. Nonetheless, the Ellsaesser et al. (1986) compendium of temperature records indicates a rapid rise in Arctic temperatures prior to the majority of trace gas emissions and concurrent with the rise in the Northern Hemisphere average. In most records the rise is followed by a decline from the 1930s to the mid-1970s of similar proportion; see Figure 4A.

All GCMs suggest that the polar warming will be magnified in winter. Nonetheless, there has been a substantial secular decline in winter temperatures over the Atlantic Arctic since 1920 (Rogers, 1989), and there has been no change in polar night temperatures at the South Pole (Figure 4B) — a station that surely has no urban warming (Sansom, 1989). Kalkstein et al. (1990) have documented that while there has been no net warming of the North American arctic, the coldest air masses, whose mean surface temperatures are approximately $-40°$, have warmed some 2°. Thus, the most severe air mass in North America may be undergoing some mitigation that is consistent with an enhanced greenhouse; nonetheless, the lack of any observed warming in the Antarctic polar night data seems at variance with this finding. In their coupled ocean-atmosphere model, Manabe et al. (1991) project very little Antarctic warming, and even some cooling where cloud cover may increase.

The very negative vision of future climate is not supported by recent studies on daily temperature regimes. Karl et al. (1988a) examined maximum and minimum values from the U.S. HCN and found that the daily range (difference between the two) has declined precipitously since 1950, and is now two standard deviations below the mean for the century. In that record, maximum values have actually declined, while minimum values have risen (Figure 5).

This behavior of the HCN is consistent with an enhanced greenhouse combined with the increases in cloudiness (of 3.5%) and reduced sunshine that have been documented across the coterminous U.S. (Angell, 1990). Weber (1990b) has also documented a decline in sunshine in Germany, and notes that the effect is enhanced in the mountains, which implicates stratocumulus (Sc) — the low altitude cloud type most effective at surface cooling — as the cause. Warren et al. (1988) have also found an increase in global cloudiness (with increases greater in the Northern Hemisphere) at most marine locations around the globe, although the shipboard observations used are subject to substantial observer and scale biases. Nonetheless, the cloud type that shows the most increase is again the low-level Sc, and again in the Northern Hemisphere. The finding of a decline in UV radiation at low elevations (Scotto et al., 1988) in combination with increased values at heights $>$10 km (Bruhl and Crutzen, 1989) is also consistent with an increase in low level cloudiness.

Subtraction of maximum and minimum temperature curves for the other Northern Hemisphere locations in the IPCC report (on observed temperature) gives the same result: a

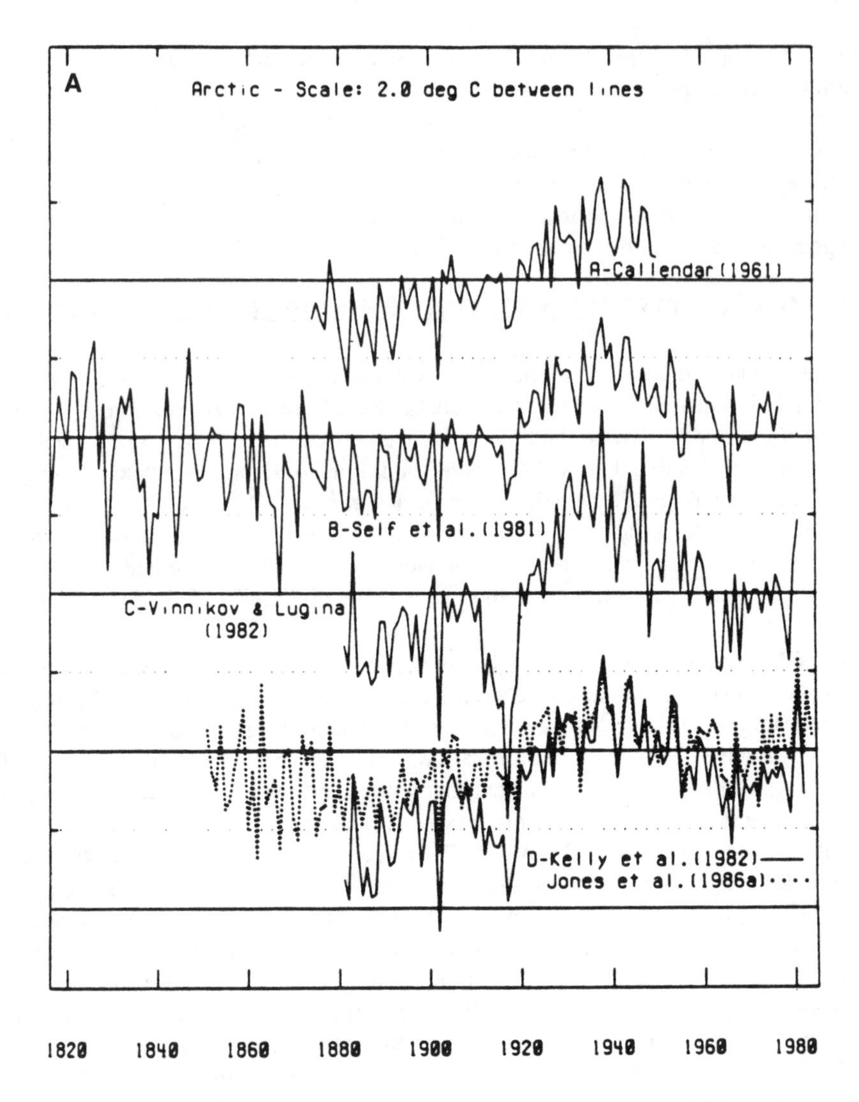


**FIGURE 4.** (A). Arctic temperature records from Ellsaesser et al. (1986). (B). South Polar temperatures from Sansom (1989).

considerable narrowing of the daily range since 1950 that is mainly a result of a rise in night temperatures (Figure 6). Rural data for the Soviet Union exhibit similar behavior (Karl, personal communication).

If anthropogenerated warming takes place primarily at night, the negative vision of future climate is probably wrong. Evaporation rate increases, which are a primary cause of projected increases in drought frequency (Rind et al., 1990) are minimized. The growing season is longer, because that period is primarily determined by night low temperatures. Finally, many plants, including some agriculturally important species, will show enhanced growth with increased moisture efficiency (Sionit et al., 1980) because of the well-known "fertilizer" effect of $CO_2$. While there may be some concern about increasing pest populations with warmer nights, it is a fact that standing biomass increases in a statistically significant fashion as the daily temperature range decreases. Thus, nature has already declared the general vegetation result of atmospheric changes that are already occurring.

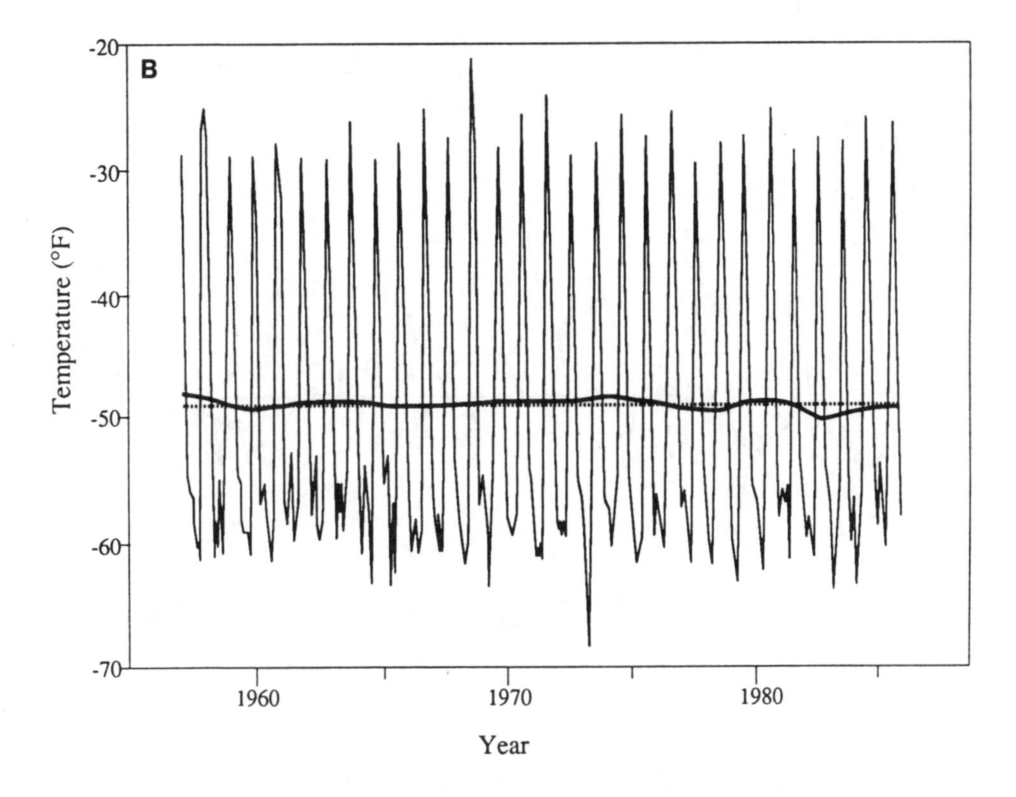


**FIGURE 4 B.**

Night warming also minimizes polar melting because mean temperatures are so far below freezing during winter that the enhanced greenhouse is insufficient to induce melting. Thus, "how" the world has warmed already appears to be beneficial. If this is indeed the expression of the altered greenhouse, the Earth will have to reverse a course it has already embarked upon to get to an enhanced-greenhouse disaster.

Curiously, none of the analyzed Southern Hemisphere temperature ranges are declining, in calculations made from IPCC data. Such an important interhemispheric differential in one of the prime components of the diurnal energy regime could be very important; as a likely cause would be an increase in Northern Hemisphere cloudiness with less change in the Southern Hemisphere. One prime candidate is, therefore, anthropogenerated sulfate, which both reflects solar radiation directly and brightens clouds by serving as condensation nuclei. These particulates are produced in much lower volume in the Southern Hemisphere, and they have a very short residence time (days or weeks), so that very little is advected from the Northern to the Southern Hemispheres.

Idso (1990) recently compared Northern and Southern Hemisphere temperature histories and found a striking difference that appears to be associated with the onset of world industrialization after 1950; an updated version is presented here in Figure 7. Beginning in the mid-1950s, Northern Hemisphere temperatures stopped rising at the rate that characterized the 20th century, while those in the Southern Hemisphere continue to rise nearly at the rate that characterized the first half of this century.

Mayewski et al. (1990) recently demonstrated that the anthropogenerated sulfate load in the Northern Hemisphere atmosphere is now equivalent to the maximum loading from the Tambora volcano, which has been associated with a 1 to 2° cooling of short duration. Their data are presented here in Figure 8.

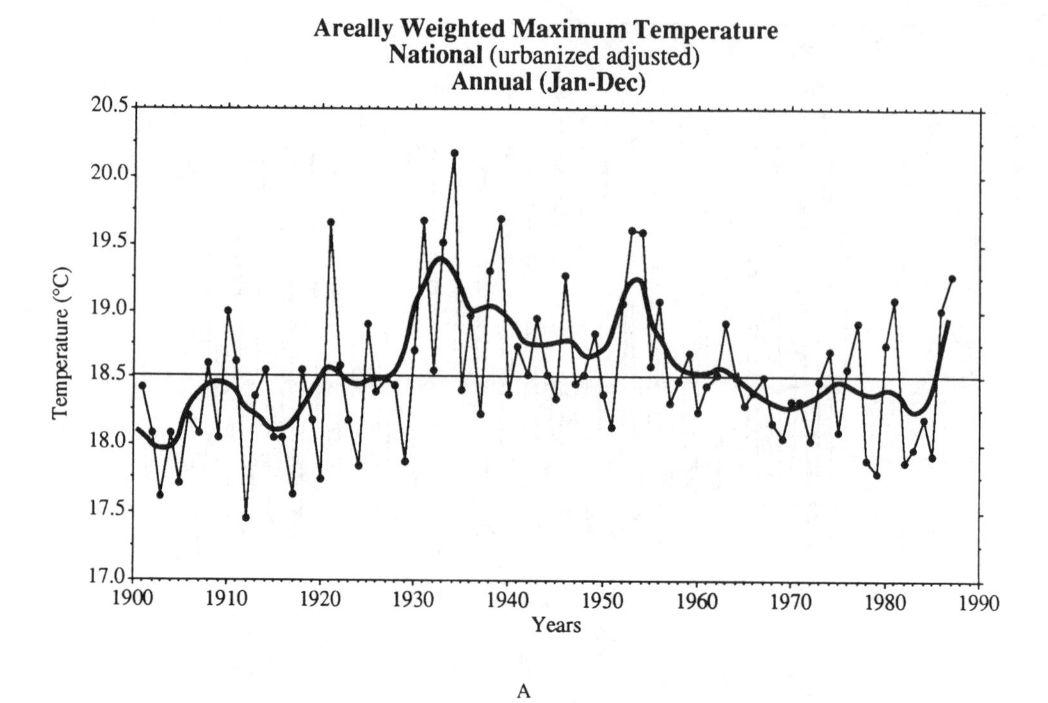


A

**Areally Weighted Minimum Temperature**
**National** (urbanized adjusted)
**Annual (Jan-Dec)**

Temperature (°C)
5.5
5.0
4.5
4.0
3.5
3.0
2.5
1900
1910
1920
1930
1940
1950
1960
1970
1980
1990
Years

B

**FIGURE 5.** (A) Maximum, (B) minimum, and (C) daily temperature range in the U.S. Historical Climate Network (Karl et al., 1988a,b).

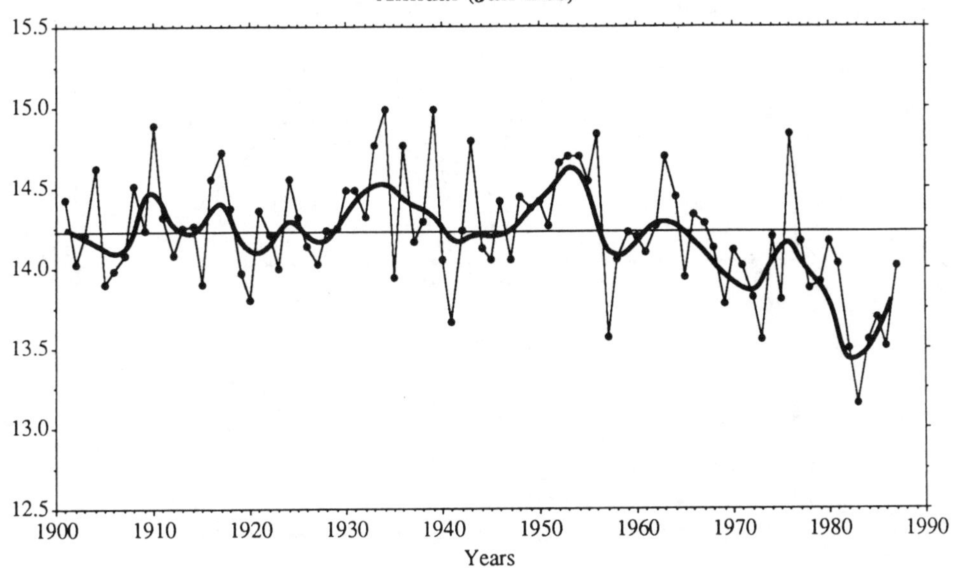


**FIGURE 5 C.**

Hansen and Lacis (1990) and Wigley (1991) both hypothesized substantial radiative effects from industrial aerosol, although the former paper emphasized direct backscattering, while Wigley was more concerned with increased cloud condensation nuclei (CCN) that would tend to enhance low-level cloudiness. In combination, the effects could in fact force net cooling at the current time, rather than warming.

In fact, if low-level cloudiness of industrial origin were increasing in a climatically significant fashion, we should see the following:

1. Night warming from both the increase in greenhouse gases as well as the increase in cloudiness
2. A counteraction of daytime warming because of cloud albedo
2a. A consequent decrease in the daily temperature range
3. The greatest warming (night effect) of clouds should occur on (long) winter nights
4. The greatest cooling (day effect) should occur on (long) summer days
5. The least warming (night effect) should occur on (short) summer nights
6. The least cooling (day effect) should occur on (short) winter days
7. The effects should be concentrated in the industrial (Northern) hemisphere
8. Cloudiness should be enhanced near the CCN source regions of North America and Eurasia

Evidence cited above is supportive of hypotheses 1, 2, and 7.

Four of the likely consequences of a cloud-modified greenhouse warming would seem especially hard to fulfill with random numbers. These are the hypotheses pertaining to what should happen to the day and night temperatures during different seasons.

The U.S. HCN is very suggestive that these difficult hypotheses may be fulfilled, but it only comprises a rather small portion of the globe or Northern Hemisphere. Karl et al.

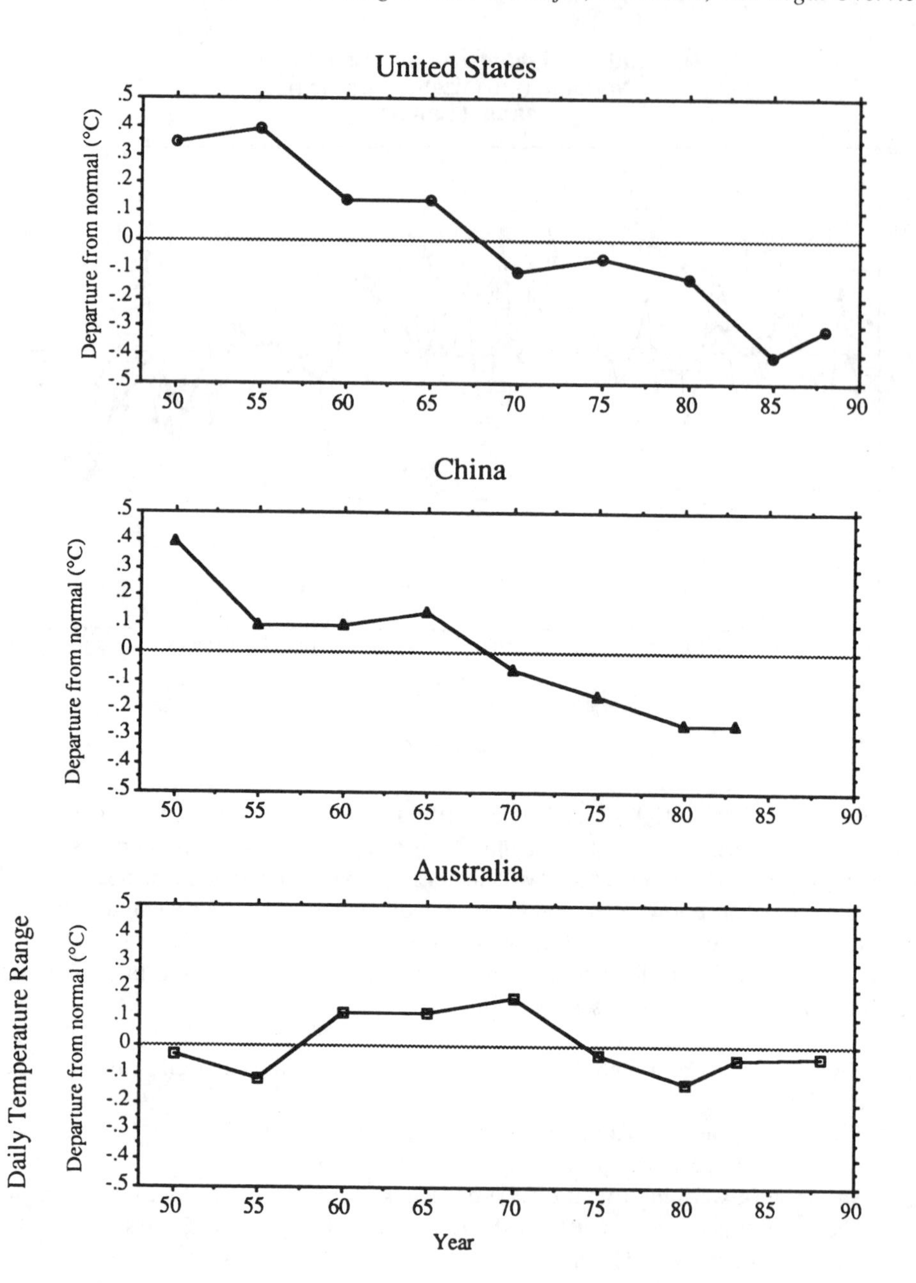


**FIGURE 6.** Mean daily temperature range at 5-year intervals (and including the last year of record) for the U.S., mainland China, and Australia.

(1991) created a HCN-like data set for U.S.S.R. and Continental China. In aggregate, these records, along with the U.S. HCN, now cover 42% of the landmass of the Northern Hemisphere, and include the world's most productive agricultural regions.

Rather than examine their results on a country-by-country basis, it is instructive to aggregate their results into a hemispheric average, as was done in Table 1. National totals were adjusted for the relative area of each country.

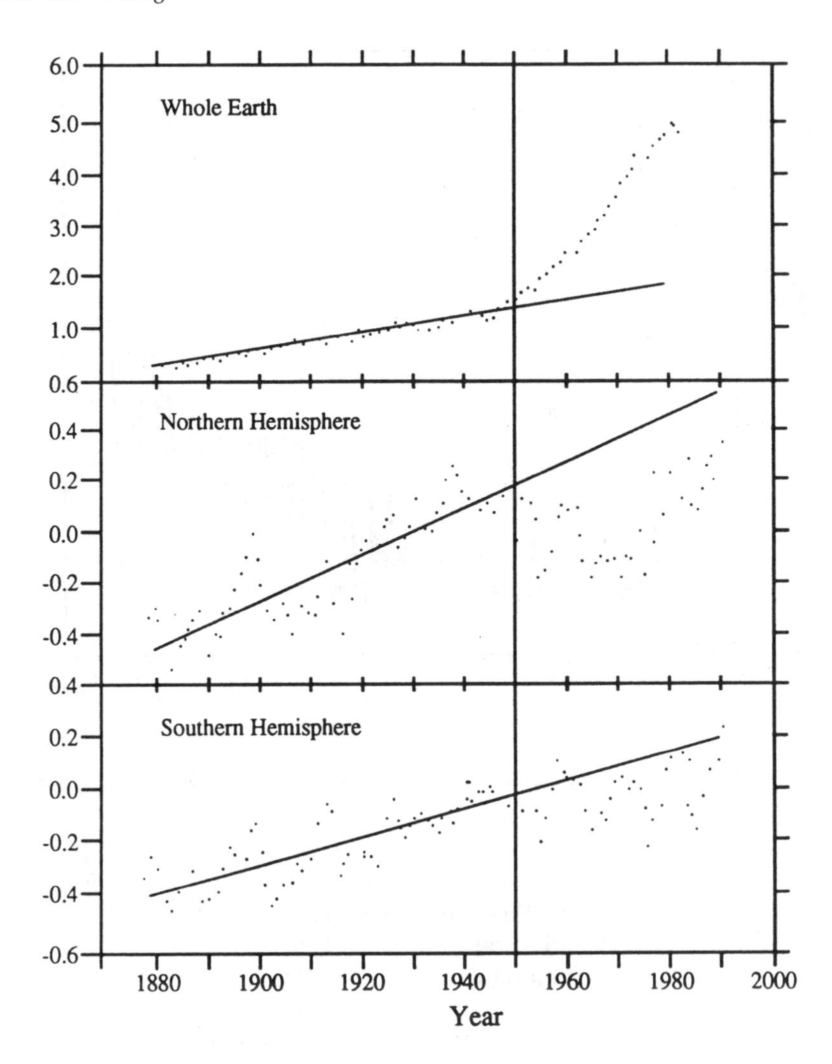


**FIGURE 7.** Global carbon emissions, and Northern and Southern Hemisphere temperatures, originally from Idso (1990), updated through 1990. A least-squares trend has been fit to the data through 1950.

Hypothesis 3. *The greatest warming (night effect) of clouds should occur in winter, when clouds have the greatest length of time to trap heat.* In fact, when this record is aggregated, winter low temperatures show a warming of 1.8°C/100 years, with 3.6°C in the Soviet Union. Russian winter nights give rise to the very cold anticyclones that occasionally reach North America; these air masses now show some warming (of 2.0°C) when they pass over Alaska (Kalkstein et al., 1990), although mean temperatures north of 55% have actually declined over the last half-century (Walsh, 1991) in the region in which GCMs project the most warming to have occurred.

Hypothesis 4. *The greatest cooling (day effect) should occur in the summer, when clouds have the greatest length of time to reflect away radiation and counter greenhouse warming.* Aggregated data demonstrate that summer days actually show a cooling trend, of 0.4°C/100 years. Idso and Balling (1991) demonstrate that summer warming stopped in the U.S. HCN at the time that the increase in $CO_2$ changed dramatically.

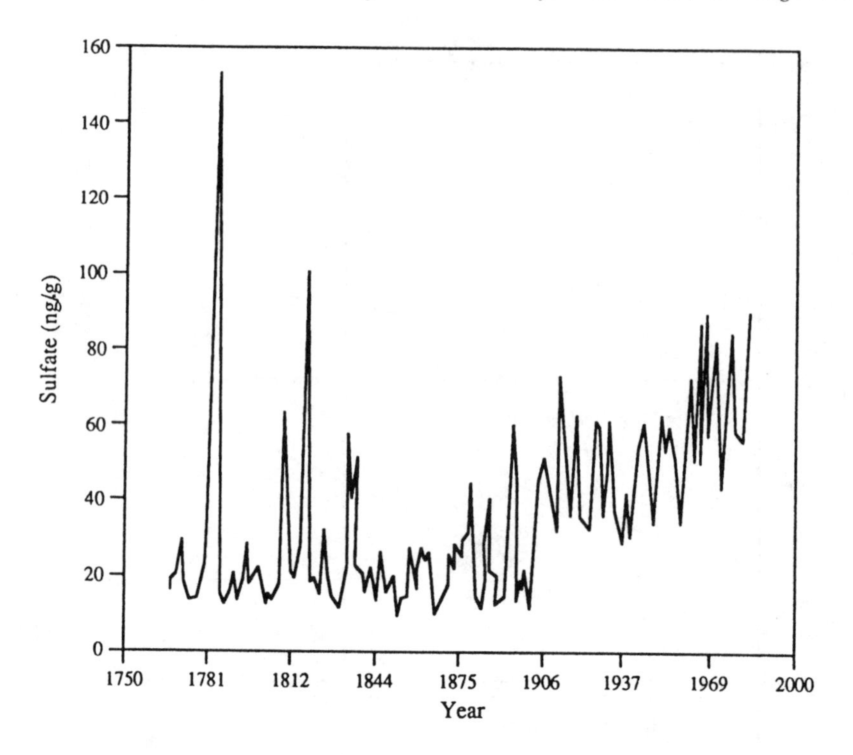


**FIGURE 8.** The history of sulfate aerosol from a Greenland ice core (Mayewski et al., 1990).

## TABLE 1
## Area-Weighted Aggregate U.S., Continental China, and U.S.S.R. Temperature Trends (°C/100 yr)

| Season | Mean max (day) | Mean min (night) |
|---|---|---|
| Winter | *+0.6°* | *+1.8°* |
| Spring | +0.6° | +1.5° |
| Summer | *−0.4°* | *+0.4°* |
| Fall | −0.6° | +0.7° |
| Annual | +0.05° | +1.1° |

*Note:* Each of the four (italicized) hypotheses concerning the day and night breakdown between seasons as modified by cloudiness is supported by the data.

Hypothesis 5. *The least warming (night effect) should be on the short summer nights, when clouds have the least length of time to trap heat.* This is also borne out by the aggregated data.

Hypothesis 6. *The least cooling (day effect) should occur on the short winter days, when clouds have the least length of time to reflect away radiation.* This turns out to be a tie with spring, when results were biased by the lack of an increase in cloudiness in the U.S. record in that season (Angell, 1990), which is the only season that shows no increase.

The observed ratio of night to day warming in this aggregate record is in excess of 10 to 1.

With regard to hypothesis 8, that the effect should be enhanced near the CCN source regions, Cess (1989) presented a limited set of satellite data depicting an increase in brightness in ocean-surface Sc that was heightened near the source regions of Asia and North America. The effect, in which reflectivity was increased by as much as 8%, persisted for 1000s of miles downstream from the source regions. A ''clean'' swath of the South Pacific ocean served as a ''control'', and showed no brightening. Thus, it appears that all eight hypotheses, consistent with an increase in low-level cloudiness and a generally benign greenhouse enhancement, can continue to be entertained by scientists.

## V. MORE RECENT CLIMATE MODELS

An earlier generation of GCMs calculated a mean equilibrium warming for a doubling of atmospheric $CO_2$ of 4.2°C, with maximum warming of as much as 18°C during north polar winter, and less significant warming (approximately 2°C) over tropical oceans (Manabe and Wetherald, 1980; Hansen et al., 1984; Schlesinger, 1984; Mitchell, 1983; Washington and Meehl, 1983); see Figure 9.

Major shortcomings in these calculations included unrealistic ocean dynamics, ocean-atmosphere coupling, and inadequate cloud parameterization; prior to the Earth Radiation Budget Experiment (ERBE) analysis even the sign of the present temperature forcing by the Earth's cloud cover was unknown and debatable. It is now thought to be negative (Ramanathan et al., 1989). These models also suffered from the use of stepwise (instantaneous) doubling of $CO_2$, rather than the low-order exponential increase that occurs in the real world.

With a modified ice-water interaction between clouds, projected warming in the U.K. Meterorological Office model (UKMO) dropped from 5.2 to 1.9°C (Mitchell et al., 1989). With a coupled ocean and climate GCM, the net warming in the National Center for Atmospheric Research (NCAR) model dropped to 1.6°C for 30 years after a shock doubling, compared to 3.7°C in an earlier equilibrium calculation (Washington and Meehl, 1989). Manabe et al. (1991) report a net equilibrium warming for a doubling with a coupled atmosphere-ocean model of 4.0°C, but the combination of a transiet model for time of approximate doubling, and a random background run for the Northern Hemisphere, indicates that the calculation is probably too warm for the current enhancement of one half of a doubling.

The transient response in that model for approximate time of doubling in the Northern Hemisphere is given as 2.8°C. If one superimposes one half of that response onto their random (unperturbed) run for the Hemisphere, and then plots observed vs. trended temperature, it is apparent that this model appears to be too warm; see Figure 10. The mean error over the last 50 years is very nearly 0.5°C, which implies that the transient error at time of doubling will be about 1°C.

Figure 11 details the NCAR calculation with the coupled land and ocean GCM for 30 years after a shock doubling. Areas of warming of greater than 4°C have been dramatically reduced (compare to Figure 9), and in fact there are none in either Hemisphere for June to August.

Because maximum warming in all GCMs is concentrated at high latitudes, use of this Mercator, or related projections, presents a highly distorted view of model results that overemphasizes areas of warming. In terms of actual area of the globe, on an annual basis, the area of >4°C of warming in the illustrated model is <5%. Because almost all of that warming is confined to latitudes higher than 60°C in the Northern Hemisphere, December to February calculations, virtually all of the strong warming is now projected either for twilight or night, which is consistent with what has been observed throughout the Northern Hemisphere.

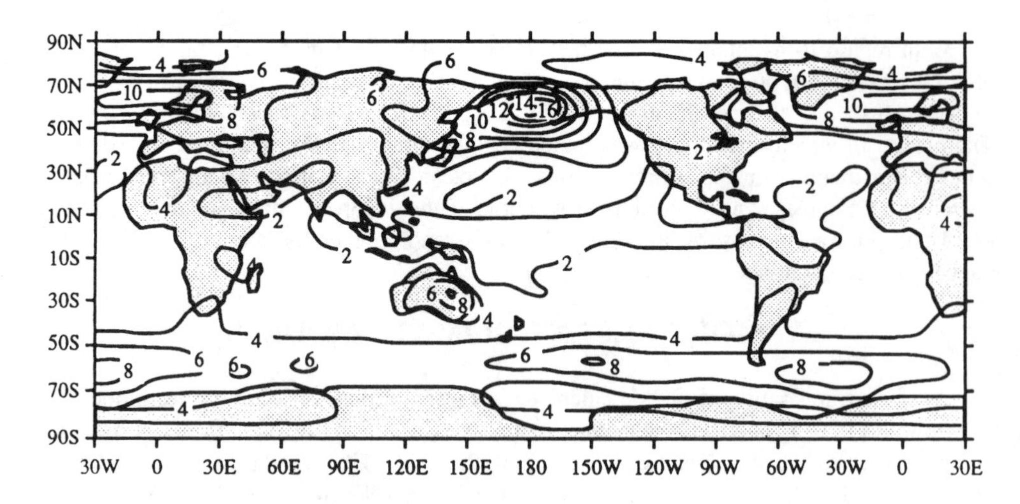


**FIGURE 9.** Winter temperature changes for a doubling of $CO_2$ calculated by Washington and Meehl (1983).

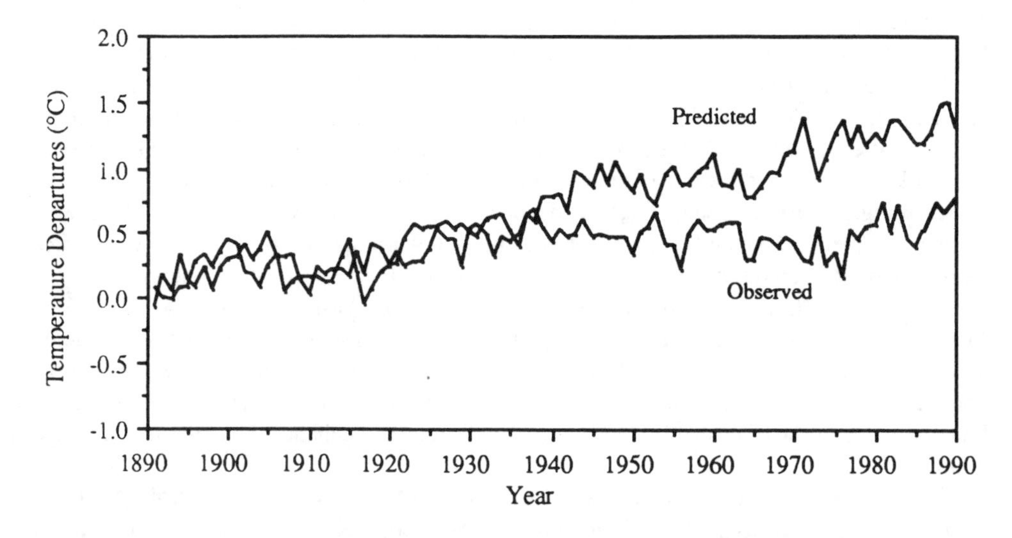


**FIGURE 10.** Comparison of observed northern hemisphere temperature with a combination of the random (nonperturbed) model of Manabe et al. (1991), and one-half of temperature the increase calculated for the time of $Co_2$ doubling.

It is tempting to view the unanimity of these newer models as a sign of reliability, but that is not the case. In a version of the NCAR model that increases greenhouse gases by a realistic 1%/year (as opposed to the "shock doubling" used in the model cited above), calculations based upon its results show that the current global temperature should be 0.7°C above the 1950 mean; in fact the rise has been on the order of 0.3°C. Thus, even the most conservative climate model appears to have substantially overestimated global warming. This version also predicts very unrealistic regional temperatures for the current greenhouse enhancement. Figure 12 details the distribution of December to February temperature in the 5-year average of years 26 to 30 in the 30 year transient run. Roughly speaking, the map should be analogous to a 5-year aggregate in the period 1975 to 1985 (years 26 to 30 beginning in 1950). The most significant projected anomalies are the 2 to 4°C warming of

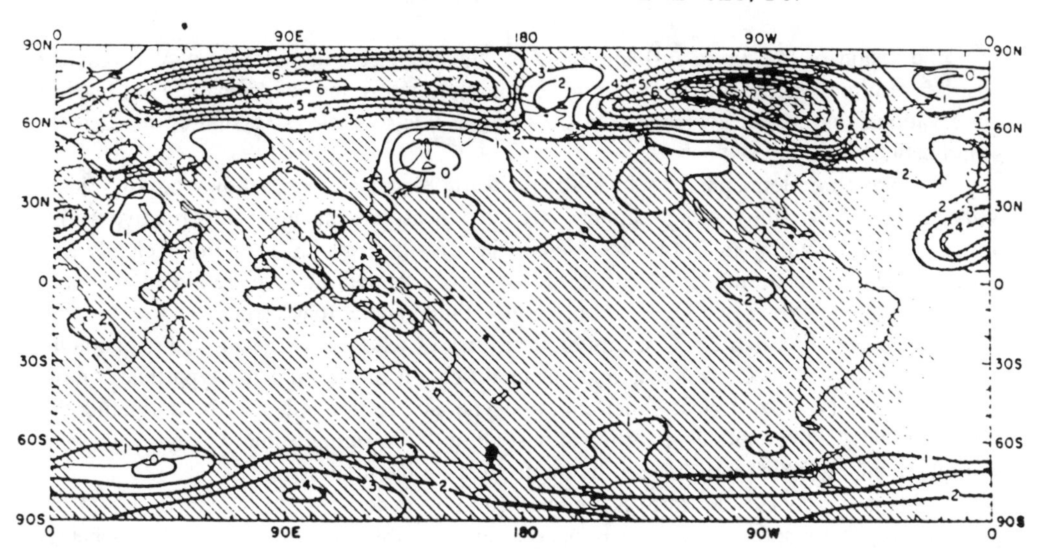


**FIGURE 11.** Winter temperature changes for a 30 years after a shock doubling of $CO_2$ calculated by Washington and Meehl (1989).

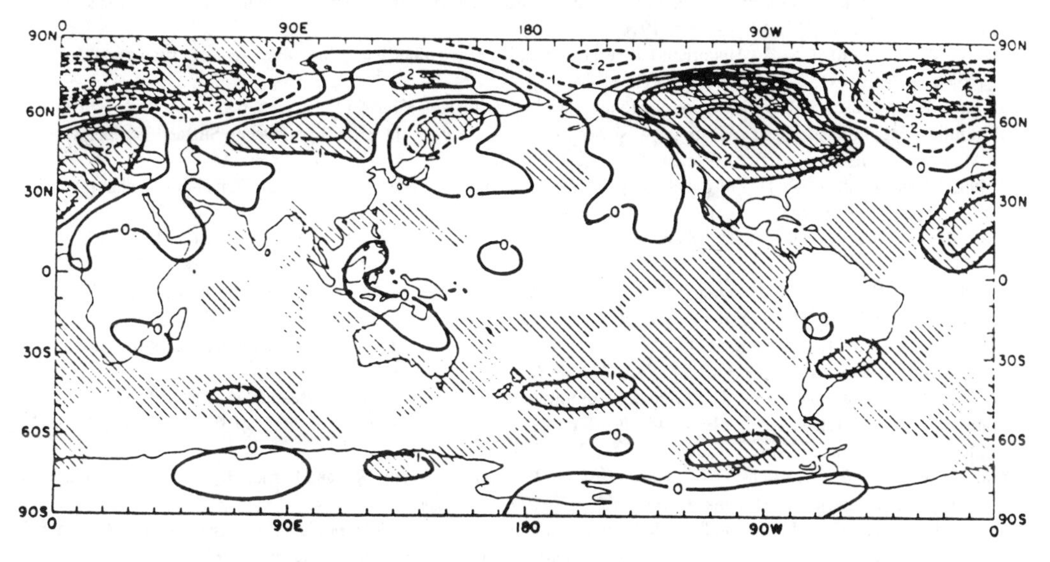


**FIGURE 12.** Calculated regional winter temperature changes from model background in the year 26–30 average of a model that realistically mimics anthropogenerated greenhouse forcing since 1950 (Washington and Meehl, 1989).

the northern half of North America and the 3 to 6°C cooling of the North Atlantic. Neither occurred, although Atlantic temperatures may have dropped some (Rogers, 1989). This projected cold anomaly does not appear in later years of this transient model. On 5-year scales, anomalies of this magnitude over such large areas are very rare.

## VI. CONCLUSION

One vision of the world influenced by an enhanced greenhouse is that of ecological disaster resulting from rising temperature, evaporation rates, and sea level. Several lines of observational and model evidence now suggest that this scenario is becoming increasingly improbable.

The Northern Hemisphere, which should warm first and most, shows no significant change in trend over the last half-century. Repeated measurements now show relative warming at night, which may in fact be beneficial. The amount of global warming is clearly less than it should be according to the earlier GCMs, given the fact that we are already halfway to an effective doubling of $CO_2$. The only high-latitude signal that seems consistent with a greenhouse alteration is a 2.0°C warming of air masses whose average surface temperature is approximately −40°C, which is a slight modification of the air mass type that is most inhospitable in North America. Other anthropogenerated compounds may be mitigating the warming. New climate models partition almost all of the warming of more than 4°C to polar twilight or night, which will have a minimal effect on ice melting.

If this is the course the Earth has embarked upon in response to human insults of the atmosphere, that response is primarily benign, and possibly beneficial. However, whether indeed this is the true response remains to be seen.

## REFERENCES

Angell, J. K., Variation in global tropospheric temperature after adjustment for the El Niño influence, 1958–89, *Geophys. Res. Lett.*, 17, 1093–1096, 1990.

Balling, R. C. and Idso, S. B., Historical temperature trends in the United States and the effect of urban population growth, *J. Geophys. Res.*, 94, 3359–3363, 1989.

Balling, R. and Idso, S. B., 100 years of global warming?, *Environ. Conserv.*, 17, 165, 1990.

Bottomley, M., Folland, C. K., Hsuing, J., Newell, R. E., and Parker, D. E., Global Ocean Surface Temperature Atlas, U.K. Meteorol. Office, Bracknell, U.K., 1990, 20.

Bruhl, C. and Crutzen, P. J., On the disproportionate role of tropospheric ozone as a filter against solar UV-B radiation, *Geophys. Res. Lett.*, 16, 703–709, 1989.

Cess, R. E., Presentation to Department of Energy Research Agenda Workshop, Germantown, MD, April 25, 1989.

Craig, H., Chou, C. C., and Wehlan, J. A., Stevens, C. M., and Engelmeier, A., The isotopic composition of methane in polar ice cores, *Science*, 242, 1535–1539, 1988.

Ellsaesser, H. W., MacCracken, M. C., Walton, J. R., and Grotch, S. L., Global climatic trends as revealed by the recorded data, *Rev. Geophys.*, 24, 745–792, 1986.

Hansen, J. E., Testimony to the U.S. Senate, Committee on Energy and Natural Resources, June 23, 1988.

Hansen, J. E. and Lacis, A. A., Sun and dust versus greenhouse gases: an assessment of their relative roles in global climate change, *Nature (London)*, 346, 713–718, 1990.

Hansen, J. E., Lacis, A., Rind, D., Russell, G., Stone, P., Fung, I., Ruedy, R., and Lerner, J., Climate sensitivity: analysis of feedback mechanisms, *Geophys. Monogr. Ser.*, 29, 130–163, 1984.

(IPCC) Intergovernmental Panel on Climate Change Policymakers Summary of the Scientific Assessment of Climate Change. World Meteorological Organization, United Nations Environment Programme, 1990, 39.

Idso, S. B., Evidence in support of Gaian climate control: hemispheric temperature trends of the past century, *Theor. Appl. Climatol.*, 42, 135–137, 1990.

Idso, S. B. and Balling, R. C., Jr., Surface air temperature response to increasing global industrial productivity: a beneficial greenhouse effect?, *Theor. Appl. Climatol.*, 44, 37–41, 1991.

Jones, P. D., Groisman, P., Coughlan, M., Plummer, N., Wang, W.-C., and Karl, T. R., How large is the urbanization bias in large area-averaged surface air temperature trends?, *Nature (London)*, 347, 169–172, 1990.

Kalkstein, L. S., Dunne, P. C., and Vose, R. S., Detection of climatic change in the western North American arctic using a synoptic climatological approach, *J. Climate*, 3, 1154–1167, 1990.

Karl, T. R., Diaz, H. F., and Kukla, J., Urbanization: its detection and effect in the United States climate record, *J. Climate*, 1, 1099–1123, 1988a.

Karl, T. R., Baldwin, R. G., and Burgin, M. G., Historical Climatology Ser. 4 to 5, National Climatic Data Center, Asheville, NC, 1988b, 107 pp.

Karl, T. R., Kukla, G., Razuvayev, V. N., Changery, M. G., Quayle, R. G., Heim, R. R., Jr., Easterling, D. R., and Fu, C. B., Global warming: evidence for asymmetric diurnal temperature change, *Geophys. Res. Lett.*, 92, 2252–2256, 1991.

Kiehl, J. T. and Dickinson, R. E., A study of the radiative effects of enhanced atmospheric $CO_2$ and $CH_4$ on early earth surface temperatures, *J. Geophys. Res.*, 92, 2991–2998, 1987.

Lorius, C., Jouzel, J., Ritz, C., Merlivat, L., Barkov, N. I., Korotkevich, Y. S., and Kotlyakov, V. M., A 150,000 climatic record from Antarctic ice, *Nature (London)*, 329, 591–596, 1987.

MacCracken, M. C. and Kukla, G. J., Detecting the Climatic Effects of Increasing Carbon Dioxide, U.S. Department of Energy Publ. DOE/ER-1235, 1985, 163–176.

Manabe, S. and Wetherald, R. T., On the distribution of climate change resulting from an increase in the $CO_2$ content of the atmosphere, *J. Atmos. Sci.*, 37, 99–118, 1980.

Manabe, S., Stouffer, R. J., Spelman, M. J., and Bryan, K., Transient responses of a coupled ocean-atmosphere model to gradual changes of atmospheric $CO_2$: Part 1: Annual Mean Response, *J. Climate*, 4, 785–818, 1991.

Mayewski, P. A., Lyons, W. B., Spencer, M. J., Twickler, M. S., Bock, C. F., and Whitlow, S., An ice-core record of atmospheric response to anthropogenic sulphate and nitrate, *Nature (London)*, 346, 554–556, 1990.

Michaels, P. J., The greenhouse effect and global change: review and reappraisal, *Int. J. Environ. Stud.*, 36, 55–71, 1990.

Mitchell, J. F. B., The seasonal response of a general circulation model to changes in $CO_2$ and sea temperature, *Q. J. R. Meteorol. Soc.*, 109, 113–153, 1983.

Mitchell, J. F. B., Senior, C. A., and Ingram, W. H., $CO_2$ and climate: a missing feedback, *Nature (London)*, 341, 132–134, 1989.

Neftel, A., Moor, E., Oeschger, H., and Stauffer, B., Evidence from polar ice cores for the increase in atmospheric $CO_2$ in the past two centuries, *Nature (London)*, 315, 45–47, 1985.

Oort, A. H., Pan, Y. H., Reynods, R. W., and Ropelewski, C., Historical trends in surface temperature over the oceans based on the COADS, *Clim. Dyn.*, 2, 29, 1989.

Ramanathan, V., Cess, R. D., Harrison, E. F., Minnis, P., Barkstrom, G. B. R., Ahmad, E., and Hartmann, D., Cloud-radiative forcing and climate: results from the Earth Radiation Budget Experiment, *Science*, 243, 53–67, 1989.

Raynaud, D. and Barnola, M. J., An Antarctic ice core reveals atmospheric $CO_2$ variations over the past few centuries, *Nature (London)*, 315, 309–311, 1985.

Rind, D., Goldberg, R., Hansen, J., Rosensweig, C., and Ruedy, R., Potential evapotranspiration and the likelihood of future drought, *J. Geophys. Res.*, 95, 9,983–10,004, 1990.

Rogers, J. C., Proc. 13th Annu. Climate Diagnostics Workshop, NOAA; Available from NTIS, 1989.

Sansom, J., Antarctic surface temperature time series, *J. Climate*, 2, 1164–1172, 1989.

Schlesinger, M. E., Climate model simulation of $CO_2$-induced climatic change, *Adv. Geophys.*, 26, 141–235, 1984.

Scotto, J., Cotton, G., Urbach, F., Berger, D., and Fears, F., Biologically-effective ultraviolet radiation: surface measurements in the United States, *Science*, 239, 762–763, 1988.

Sionit, N., Hellmers, H., and Strain, B. R., Growth and yield of wheat under carbon dioxide enrichment and water stress conditions, *Crop. Sci.*, 20, 687–690, 1980.

Spencer, R. W. and Christy, J. R., Precise monitoring of global temperature trends from satellite, *Science*, 247, 1558, 1990.

U.S. Department of Energy, Trends '90, Carbon Dioxide Inf. Analysis Center, Oak Ridge, TN, 1990, 257 pp.

Walsh, J. E., The Arctic as bellweather, *Nature (London)*, 352, 19–20, 1990.

Warren, S. G., Hahn, C. J., London, J., Chervin, R. M., and Jenne, R. L., Global Distribution of Total Cloud Cover and Cloud Type Amounts over the Ocean, U.S. Department of Energy Publ. DOE/ER-0406, 1988, 42 pp + maps.

Washington, W. M. and Meehl, G. A., General circulation model experiments on the climate effects due to a doubling and quadrupling of carbon dioxide concentration, *J. Geophys. Res.*, 88, 6600–6610, 1983.

Washington, W. M. and Meehl, G. A., Climate sensitivity due to increased $CO_2$: experiments with a coupled atmosphere and ocean general circulation model, *Clim. Dyn.*, 2, 1–38, 1989.

Weber, G.-R., Tropospheric temperature anomalies in the northern hemisphere 1977–86, *Int. J. Climatol.*, 10, 3–19, 1990a.

Weber, G.-R., Spatial and temporal variation of sunshine in the Federal Republic of Germany, *Theor. Appl. Climatol.*, 41, 1–9, 1990b.

Wigley, T. M. L., Could reducing fossil-fuel emissions end global warming?, *Nature (London)*, 349, 503–505, 1991.

Wigley, T. M. L., Relative contributions of different trace gases to the greenhouse effect, *Climate Monitor*, 16, 14–28, 1991.

Wigley, T. M. L., Analytical solution for the effect of increasing $CO_2$ on global mean temperature, *Nature (London)*, 315, 649–652, 1987.

Chapter 14

# ENVIRONMENTAL AND ECONOMIC BENEFITS OF NATURAL GAS USE FOR POLLUTION CONTROL

**Partha R. Dey, Eugene E. Berkau, and Karl B. Schnelle**

## TABLE OF CONTENTS



0-8493-4419-0/93/$0.00 + $.50
© 1993 by CRC Press, Inc.

## I. SUMMARY OF RESULTS

One of the primary goals of this research effort was to document and compare the economic and environment benefits of using natural gas for pollution control in boilers, furnaces and internal combustion engines, with conventional control technologies.

The study indicated that replacement of 15% of the coal used in coal-fired boilers employed in the generation of electric power in the U.S., with natural gas, would considerably reduce the emissions of acid rain precursors such as sulfur and nitrogen oxides, and do so in a cost-effective manner. The reductions achieved were also in concordance with the reductions in sulfur dioxide emissions mandated by the new Clean Air Act (CAA) Amendments of 1990.[1]

The combustion of natural gas would also produce less carbon dioxide as compared to the combustion of coal with an equivalent amount of heat content. Carbon dioxide is a "greenhouse gas", i.e., it is believed to play a major role in global warming. Natural gas technology therefore presents a cost-effective step in the eventual mitigation of two of the main environmental problems presently facing us, acid rain, and global warming.

## II. INTRODUCTION

Three criteria air pollutants generated by burning fossil fuels are sulfur oxides, nitrogen oxides ($NO_x$), and particulate matter. The primary sources of such pollution are coal-fired industrial and utility boilers, stationary engines and vehicles. Acid rain, ozone and photochemical smog, and atmospheric temperature increases, also known as "global warming", are products of chemical reactions of these criteria pollutants in the atmosphere. Currently there is a great public and Congressional concern in the U.S. regarding environmental issues such as acid rain, non-attainment of the National Ambient Air Quality Standards (NAAQS) for ozone, and global warming or the "greenhouse effect".[2-5] Pollution abatement work in this area has primarily focused on (1) improving the performance of emission control devices such as scrubbers, baghouses, electrostatic precipitators (ESPs), afterburners, etc; (2) improving the quality of fossil fuels, by operations such as coal-washing and petroleum sweetening — which reduce the sulfur content of the fuel; and (3) by improving energy conversion processes used in stationary internal-combustion engines by modifications designed to burn fuel more effectively and cleanly. These approaches are presently being used by industries to handle the air-pollution problem. While being reasonably effective in complying with emission limits and air quality standards these methods are expensive.

## III. NATURAL GAS FOR POLLUTION CONTROL

A relatively new approach in control of the criteria pollutants generated in coal-fired boilers is cofiring small amounts of natural gas (e.g., 5 to 20% of the total heat input) with coal. Natural gas occurring in the U.S. is comprised of 90 volume percent methane on an average, 0.002 and 0.02 $g/m^3$ of hydrogen sulfide and sulfur, respectively, and no ash. However, the gas available to the consumer after sweetening contains negligible hydrogen sulfide and sulfur. The coals used in the U.S. on the other hand contain 0.2 to as much as 7% (by weight) sulfur, and 8 to 15% (by weight) ash.[6] Consequently, particulate and $SO_2$ reductions are proportional to the amount of gas used. It should be noted that the particulate reductions are not only because of the absence of ash in burning natural gas, but also because combustion of a light hydrocarbon-like methane goes to completion more easily than coal or other fuels. Incomplete combustion of fuels leads to the formation of soot which is the major source of fine particulate matter (less than 10 microns in diameter — PM-10) pollution problems.

Reductions in $NO_x$ emissions resulting from the use of natural gas can range from 30 to 70% because natural gas contains negligible amounts of organic nitrogen, as compared with coal and oil which contains as much as 2% by weight of organic nitrogen.[6] Natural gas has the added benefit of burning without generating any residuals as in the case of oil and coal. Cofiring gas with coal and the use of natural gas in place of other fuels (e.g., diesel fuel in fleet vehicles) are pollution control strategies that have not been studied in detail.

The following is a description of the calculation done to study the effect of replacing 15% of the heating value of coal with an equivalent amount of natural gas (using 1 lb of coal as the basis), on carbon dioxide production. Using 13,890 Btu/lb as an average heating value for coal used in the U.S., and a value of 78% as the carbon content, a pound of coal produces 2.86 lbs of $CO_2$. If 15% of the coal is replaced with natural gas then it means that 15% of 2.86, or 0.42 pounds of $CO_2$ is replaced. Assuming that the composition of natural gas is 100% methane, and using 23,861 Btu/lb as the heat of combustion of methane to carbon dioxide and water,[6] a stoichiometric calculation shows that the amount of natural gas equivalent to 15% of the heat content in 1 lb of coal is 0.087 lbs. This quantity of natural gas would produce 0.24 lbs of $CO_2$. Hence there is approximately a 43% reduction in carbon dioxide production. This is a very significant reduction in $CO_2$ emissions and will have a correspondingly large effect on the global warming phenomenon.

Despite all the advantages of using natural gas to control environmental problems, it is not economical to completely substitute natural gas for coal due to a number of reasons. Among these are the higher price of natural gas based on equivalent heat content, assuring adequate and uninterrupted supplies of natural gas, the absence of adequately sized gas pipelines for the volume of gas that would have to be transported, possible disruption to the coal-mining industry, and retrofit difficulties and costs associated with combustion of a fuel for which the boilers have not been initially designed, i.e., modification costs.

Natural gas reburn (NGR) represents a technology which strikes a compromise by allowing cheap and cost-effective combined $SO_2 + NO_x$ emissions reduction. Field studies with NGR technology used substitution of 15% of the total heat input from coal in coal-fired boilers for a 15% reduction in $SO_2$ emissions and a 50% reduction in $NO_x$ emissions.[7,8] Therefore, the present study used the same amount of gas and $SO_2$ and $NO_x$ emissions reduction in its calculations. The stimultaneous $SO_2$ and $NO_x$ emissions reduction is one of the most attractive features of this technology which has not been extensively studied as an option in control strategies. The present study compares the costs of control strategies using NGR technology with those using only conventional technologies such as scrubbing. It also compares the costs of switching to low sulfur coals before applying additional controls.

## IV. LITERATURE REVIEW

The various sources of $SO_2$, $NO_x$, and PM-10 emissions were classified into stationary and mobile sources. Mobile sources were not studied due to the dearth of reliable data on the use of natural gas as compared to other fuels. The stationary sources were comprised of boilers and stationary engines which could be further categorized into turbines and internal combustion engines. The current federal emission standards[9] for each category of these sources is shown in Table 1. Table 1 shows the emission standards for stationary source types which include utility and industrial boilers, gas turbines, and internal combustion engines.

## VI. STATIONARY SOURCES

Table 2 is a compilation of emission data and Table 3 is a compilation of the corresponding cost data for gas technologies applied to various stationary sources. The emissions reduction

**TABLE 1**
**Stationary Source Emissions Standards**

| | | | Emission Standards | | |
|---|---|---|---|---|---|
| **Source category** | **Description** | **Fuel** | **$SO_2$** | **$NO_2$ lbs/ mm Btu** | **TSP** |
| Utility boilers[a] | >73 MW | Coal, coal/wood residue | 1.2 | 0.7 | 0.10 |
| | | Oil, oil/wood residue | 0.8 | 0.3 | 0.10 |
| | | Gas, gas/wood residue | — | 0.2 | 0.10 |
| | | Mixed fossil fuel | Prorated | Prorated | 0.10 |
| | | Lignite, lignite/wood residue | 1.2 | 0.6 | 0.10 |
| | | | | 0.8[b] | |
| Utility boilers[c] | >73 MW | Solid & solid-derived | 1.2 | 0.5 | 0.03 |
| | | | | 0.8[b] | |
| | | | | 0.6[d] | |
| | | Liquid fuel | 0.8 | 0.3 | 0.03 |
| | | Gaseous fuel | 0.8 | 0.2 | 0.03 |
| Industrial boilers | 25<MW<73 | Coal | 1.2 | 0.7 | 0.05 (CF<10%) |
| | | | | 0.6[d] | 0.10 (CF>10%) |
| | | | | 0.8[b] | 0.20 (5) |
| | | | | 0.5[f] | |
| | | Oil | 0.3 | 0.1[g] | 0.10 |
| | | | | 0.2[h] | |
| | | | | 0.3[i] | |
| | | | | 0.4[j] | |
| | | | | 0.1[g] | |
| | | Natural Gas | — | 0.1[g] | 0.10 |
| | | | | 0.2[h] | |
| | | Municipal solid waste + other | 1.2 | 0.3 | 0.10 (CF< = 10%) |
| | | | | | 0.20[e] |
| | | Wood | — | — | 0.10 (CF>30%) |
| | | | | | 0.20 (CF<30%) |
| Gas Turbines[k] | >2.9 MW | Natural gas, Distillate fuel #2, or other | 150 | 150 | — |
| | | | | 75 (>29 MW) | |
| Internal Combustion Engines[l] | >350 CID/cyl or > = 8 cyl & >240 CID/cyl | Gas | — | 700 | — |
| | >560 CID/cyl or >1500 CID/rotor | Diesel/Dual Fuel | — | 600 | — |

[a] Construction commenced after Aug. 17, 1971.
[b] For lignite mined in ND, SD, and MT, and lignite fired in cyclone unit.
[c] Construction commenced after Sept. 18, 1978.
[d] Lignite, bituminous, anthracite and spreader stoker and fluidized bed combustion.
[e] Built after June 19, 1984 and before Nov. 25, 1986 (CF< = 30%).
[f] Mass feed stoker and coal-derived synthetic fuels.
[g] Distillate oil with low heat release rate (<7000 Btu/hr-sqft).
[h] Distillate oil with high heat release rate (>7000 Btu/hr-sqft) and duet burner used in a combined cycle system with natural gas.
[i] Residual oil with low heat release rate (<7000 Btu/hr-sqft).
[j] Residual oil with high heat release rate (>7000 Btu/hr-sqft) and duet burner used in a combined cycle system.
[k] Emission in ppm; emergency, military, training, firefighting and R&D turbines exempt from $NO_x$ standards.
[l] Emissions in ppm.

and costs associated with using natural gas cofiring and natural gas reburn in boilers were calculated using an EPA cost estimation computer program known as Integrated Air Pollution Control Systems (IAPCS-3) model, that can be used on a PC.[10] For all the other cases, calculations were made using the data available in the literature so that the results could be expressed on a comparable basis.

Most of the emissions and cost data in Tables 2 and 3 were from previous studies and EPA background documents for federal emission standards. However in cases where no specific emission data was available, the EPA AP-42 document on generic air pollution emission sources was used for the purpose.[11] The costs were differentiated with respect to capital costs, and operating and maintenance costs (O & M), which were then levelized on an annual basis. The levelized annual costs were then also expressed in terms of dollars/ton of $SO_2$ + $NO_x$ removed which is a measure of cost-effectiveness of a given technology. In the boiler studies, EPRI guidelines for cost estimation were followed.[12] These capital, O & M, total levelized annual, and cost-effectiveness dollars were then used as the basis of economic comparisons of the given technologies. The following comparisons for the indicated categories can be made from Tables 2 and 3.

### A. BOILERS

Under the category of boilers, natural gas applications comparable with conventional systems which were examined were gas-fired combined cycle, natural gas cofiring and reburn, and natural gas repowering applications.

#### 1. Combined Cycle Electric Power Generation

In the first case, a new natural gas-fired combined cycle unit is compared with a new coal-fired unit having the same power output.[13] Figure 1 illustrates a 240-MW combined cycle system. Natural gas is fed into 2 gas turbines of 80-MW capacity each. These turbines directly drive two generators. The heat from the spent gases from the generators is then recovered in a heat recovery steam generator. The steam thus generated is used to drive another 80-MW steam turbine-generator unit to get additional power. The efficiency of this system is in the range of 45% as compared to a conventional boiler which can only be 30 to 34% efficient. Thus, the fuel requirement of the combined cycle system is reduced to about 80% of that of a coal-fired boiler. Emissions of $SO_2$ and TSP from the combined cycle unit are negligible as compared to the coal-fired unit which requires an expensive flue gas-desulfurization unit and an electrostatic precipitator just to meet emission limits. The $NO_x$ emissions, based on the current $NO_x$ emission limits, from the combined cycle unit are less than 30% of that from the coal-fired plant. Another major advantage of the combined cycle unit over the coal-fired boiler is its relatively low capital cost which is less than 30% of that of the coal-fired boiler in terms of dollars/kilowatt. The operation and maintenance costs of the combined cycle system are only 30% that of the coal-fired unit. The cost-effectiveness of the combined cycle unit, in terms of dollars/ton of $SO_2$ + $NO_x$ removed is approximately 25% of that of the coal-fired unit. It is also possible to bring a combined cycle unit on-line approximately 5 times more rapidly than a coal-fired boiler because of the pre-packaged gas turbine component of the system. Lastly all costs associated with ash and sludge handling for a new coal-fired boiler can be avoided by using the combined cycle system.

#### 2. Natural Gas Cofiring and Reburn

In the natural gas cofiring process up to 20% of the heat input to a coal- or oil-fired boiler is replaced with natural gas. Natural gas has negligible sulfur and ash content and therefore, there is a reduction of $SO_2$ and particulate emissions directly proportional to the

**TABLE 2**
**Compilation of Emissions Using Gas Technologies (Stationary Sources)**

| Equipment type | Technology description | Application | Gas use (%) | Size (MW) | CF | $SO_2$ | $NO_x$ #mmBtu | TSP | HC Total | CO | Ref. |
|---|---|---|---|---|---|---|---|---|---|---|---|
| Boilers | Gas-fired[a] (Combined Cycle) | Energy conv. | 100 | 240 | 65 | — | 0.25 | — | 0.002 | 0.04 | 11,13 |
| | Coal-fired[b] (w/t FGD + ESP) | Energy conv. | — | 240 | 65 | 0.60 | 0.70 | 0.03 | 0.004 | 0.03 | 11,13 |
| | Cofiring[c] (w/t ESP) | Pollution ctrl. | 6 | 570 | 65 | 2.10 | 0.46 | 0.04 | 0.003 | 0.025 | 11,14 |
| | Reburn[d] (w/t ESP) | Pollution ctrl. | 15 | 570 | 65 | 1.90 | 0.31 | 0.04 | 0.003 | 0.025 | ?,15 |
| | Repower(gas)[e] | Peaking | 100 | 80 | 65 | — | 0.41 | — | 0.002 | 0.04 | 11,18 |
| | coal | Peaking | — | 80 | 65 | 0.60 | 0.70 | 0.03 | 0.004 | 0.03 | |
| | gas | Heat recovery | 100 | 198 | 65 | — | 0.30 | — | 0.002 | 0.04 | |
| | coal | Heat recovery | — | 198 | 65 | 0.60 | 0.70 | 0.03 | 0.004 | 0.03 | |
| | gas | Boiler | 100 | 75 | 65 | −0.55 | 0.30 | −0.03 | 0.002 | 0.04 | |
| | coal | Boiler | — | 75 | 65 | 0.62 | 0.70 | 0.03 | 0.004 | 0.03 | |
| Stationary engines | Turbines[f] | Standby (gas) | 100 | 1 | 2 | — | 0.21 | — | 0.04 | 0.17 | 11,19 |
| | | DF-2 | — | 1 | 2 | 0.11 | 0.49 | — | 0.04 | 0.40 | |
| | | Industrial | 100 | 3 | 91 | — | 0.40 | — | 0.04 | 0.03 | |
| | | (gas) DF-2 | — | 3 | 91 | 0.11 | 0.61 | — | 0.04 | 0.15 | |
| | | Utility (gas) | 100 | 66 | 91 | — | 0.50 | — | 0.04 | 0.01 | |
| | | DF-2 | — | 66 | 91 | 0.11 | 0.87 | — | 0.04 | 0.01 | |
| | | Offshore (gas) | 100 | 3 | 91 | — | 0.40 | — | 0.04 | 0.03 | |
| | | DF-2 | — | 3 | 91 | 0.11 | 0.61 | — | 0.04 | 0.15 | |
| | I.C. Engines[g] | Utility (diesel) | — | 3 | 91 | — | 2.17 | — | 0.07 | 1.13 | 11,20 |
| | | Utility (dual) | 95 | 3 | 91 | — | 1.86 | — | 1.11 | 2.20 | |
| | | Oil & Gas Trans. | 100 | 5 | 91 | — | 3.35 | — | 1.03 | 0.36 | |
| | | Oil & Gas Prod. | 100 | 5 | 91 | — | 3.35 | — | 1.03 | 0.36 | |

[a] Costs converted to 1988 $ from 1985 $, using Chemical Engg. (CE) cost indices.

[b] 2% S, 10% ash, 10000 Btu/lb coal, FPD $1.00/mmBtu, costs converted to 1988 $ from 1985 $, using CE cost indices.

[c] 1.4% S, 12% ash, 12500 Btu/lb coal, FPD $1.00/mmBtu, costs converted to 1988 $ from 1986 $ using CE cost indices, 25% $NO_x$ reduction. All costs calculated using IAPCS-3 computer cost model. Baseline emissions (#/mmBtu): $SO_2$ = 2.24, $NO_x$ = 0.62, TSP = 0.04 (w/t ESP).

[d] 1.4% S, 12% ash coal, FPD $1.00/mmBtu, costs converted to 1988 $ from 1986 $ using CE cost indices, 50% $NO_x$ reduction. All costs calculated using IAPCS-3 computer cost model. Baseline emissions (#/mmBtu): $SO_2$ = 2.24, $NO_x$ = 0.62, TSP = 0.04 (w/t ESP).

[e] Power in MW added, 2% S, 10% ash coal, FPD $1.00/mmBtu, costs converted to 1988 $ from 1986 $ using CE cost indices.

[f] Base Fuel — Distillate Fuel Oil #2 (DF-2), 0.1% S. All costs converted to 1988 $ from 1977 $ using CE cost indices. Uncontrolled $NO_x$ emissions (#/mmBtu):

Standby (gas) 0.2 (27% $NO_x$ reduction
(DF-2) 0.8 (27% $NO_x$ reduction)
Industrial (gas) 0.3 (34% $NO_x$ reduction)
(DF-2) 0.5 (34% $NO_x$ reduction)
Utility (gas) 0.5 (40% $NO_x$ reduction)
(DF-2) 0.6 (40% $NO_x$ reduction)
Offshore (gas) 0.3 (34% $NO_x$ reduction)
(DF-2) 0.5 (34% $NO_x$ reduction)

[g] Control costs based on cheapest 40% $NO_x$ reduction technology (as per average $NO_x$ reduction estimated in Reference 14). All costs converted to 1988 $ from 1977 $ using CE cost indices.

**TABLE 3**
**Compilation of Costs for Gas Technologies Applied to Stationary Sources**

| Equipment type | Technology description | Application | Capital MM | $/KW | O&M MM | m/KWH | LEV. MM | ANN. m/KWH | $/TON ($SO_2 + NO_x$) | Ref. |
|---|---|---|---|---|---|---|---|---|---|---|
| Boilers | Gas-fired[a] (Combined Cycle) | Energy conv. | 126 | 526 | 4 | 2.8 | 16.5 | 12.1 | 718 | 11,13 |
| | Coal-fired[b] (w/t FGD + ESP) | Energy conv. | 430 | 1790 | 13 | 8.9 | 55.1 | 40.4 | 2509 | 11,13 |
| | Cofiring[c] (w/t ESP) | Pollution ctrl. | 4 | 9 | 2 | 0.6 | 2.8 | 0.9 | 447 | 11,14 |
| | Reburn[d] (w/t ESP) | Pollution ctrl. | 4 | 9 | 6 | 1.9 | 6.7 | 2.0 | 1361 | (?)15 |
| | Repower(gas)[e] | Peaking | 21 | 263 | 1 | 3.1 | 18.8 | 41.2 | 2661 | 11,18 |
| | coal | Peaking | 114 | 1425 | 4 | 9.8 | 30.1 | 67.8 | 5193 | |
| | gas | Heat recovery | 66 | 333 | 3 | 3.1 | 41.6 | 37.0 | 2510 | |
| | coal | Heat recovery | 282 | 1425 | 11 | 9.8 | 61.8 | 54.8 | 4425 | |
| | gas | Boiler | 34 | 457 | 1 | 3.1 | 18.6 | 43.6 | 2226 | |
| | coal | Boiler | 106 | 1425 | 4 | 9.8 | 19.3 | 68.6 | 4796 | |
| Stationary engines | Turbines[f] | Standby (gas) | 0.07 | 77 | 0.02 | 83.7 | 0.02 | 128 | 1E + 05 | 11,19 |
| | | DF-2 | 0.07 | 77 | 0.02 | 83.7 | 0.02 | 128 | 8E + 04 | |
| | | Industrial (gas) | 0.12 | 42 | 1.20 | 49.8 | 1.21 | 50 | 3E+04 | |
| | | DF-2 | 0.12 | 42 | 1.20 | 49.8 | 1.21 | 50 | 5E+04 | |
| | | Utility (gas) | 3.39 | 52 | 20.5 | 38.9 | 20.8 | 39.6 | 3E+04 | |
| | | DF-2 | 3.39 | 52 | 20.5 | 38.9 | 20.8 | 39.6 | 3E+04 | |
| | | Offshore (gas) | 0.12 | 42 | 1.20 | 49.8 | 1.21 | 50 | 3E+04 | |
| | | DF-2 | 0.12 | 42 | 1.20 | 49.8 | 1.21 | 50 | 4E+04 | |
| | I.C. Engines[g] | Utility (diesel) | 0.22 | 69 | 1.29 | 50.7 | 1.31 | 51.5 | 7915 | 11,20 |
| | | Utility (dual) | 0.22 | 69 | 1.48 | 57.9 | 1.49 | 58.7 | 11018 | |
| | | Oil & Gas Trans. | 0.30 | 67 | 1.58 | 43.9 | 1.60 | 44.8 | 4620 | |
| | | Oil & Gas Prod. | 0.30 | 67 | 1.58 | 43.9 | 1.60 | 44.8 | 4620 | |

[a] Costs converted to 1988 $ from 1985 $, using Chemical Engg. (CE) cost indices.

[b] 2% S, 10% ash, 10000 Btu/lb coal, FPD $1.00/mmBtu, costs converted to 1988 $ from 1985 $, using CE cost indices.

[c] 1.4% S, 12% ash, 12500 Btu/lb coal, FPD $1.00/mmBtu, costs converted to 1988 $ from 1986 $ using CE cost indices, 25% $NO_x$ reduction. All costs calculated using IAPCS-3 computer cost model. Baseline emissions (#/mmBtu): $SO_2$ = 2.24, $NO_x$ = 0.62, TSP = 0.04 (w/t ESP).

[d] 1.4% S, 12% ash coal, FPD $1.00/mmBtu, costs converted to 1988 $ from 1986 $ using CE cost indices, 50% $NO_x$ reduction. All costs calculated using IAPCS-3 computer cost model. Baseline emissions (#/mmBtu): $SO_2$ = 2.24, $NO_x$ = 0.62, TSP = 0.04 (w/t ESP).

[e] Power in MW added, 2% S, 10% ash coal, FPD $1.00/mmBtu, costs converted to 1988 $ from 1986 $ using CE cost indices.

[f] Base Fuel — Distillate Fuel Oil $2 (DF-2), 0.1% S. All costs converted to 1988 $ from 1977 $ using CE cost indices. Uncontrolled $NO_x$ emissions (#/mmBtu):

Standby (gas) 0.2 (27% $NO_x$ reduction
(DF-2) 0.8 (27% $NO_x$ reduction)
Industrial (gas) 0.3 (34% $NO_x$ reduction)
(DF-2) 0.5 (34% $NO_x$ reduction)
Utility (gas) 0.5 (40% $NO_x$ reduction)
(DF-2) 0.6 (40% $NO_x$ reduction)
Offshore (gas) 0.3 (34% $NO_x$ reduction)
(DF-2) 0.5 (34% $NO_x$ reduction)

[g] Control costs based on cheapest 40% $NO_x$ reduction technology (as per average $NO_x$ reduction estimated in Reference 14). All costs converted to 1988 $ from 1977 $ using CE cost indices.

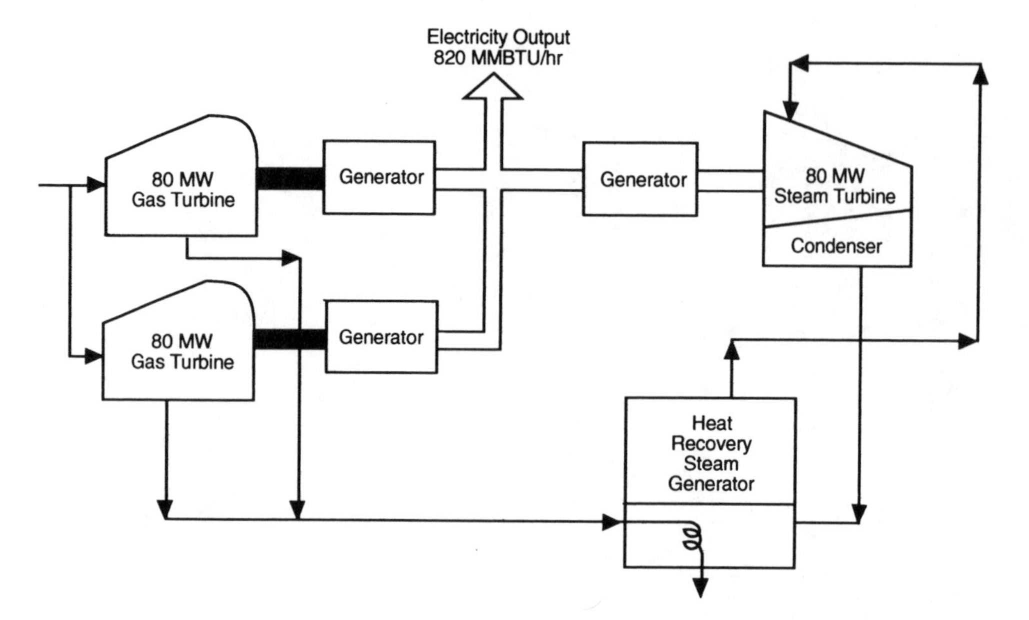


**FIGURE 1.** Illustrative diagram of a 240-MW natural gas fired combined cycle electric powerplant.

percentage of gas used. There is also a considerable, but varying degree of reduction in $NO_x$ emissions. In one study reported in the literature there was a 25% $NO_x$ reduction achieved with a 6% natural gas use in a 570-MW tangentially fired coal-burning boiler.[14] The mechanism of $NO_x$ reduction is outside the scope of this study. An independent study has conducted an economic analysis of cofiring benefits.[15] These benefits include $SO_2$ reduction, $NO_x$ reduction, start-up improvement, availability improvement, coal switching opportunities, operation and maintenance costs reduction, peaking capacity increase, turn-down capability improvement, and boiler efficiency improvement.

Reburning, illustrated in Figure 2 is a specific application of cofiring in which 15% of the boiler heat input is substituted with natural gas. The field data which supports 15% gas use is contained in references 7, 15, and 16.

The amount of natural gas used in the Natural Gas Reburn (NGR) technology has been fixed at 15% for the present study based on the minimum gas use necessary to effect 50% $NO_x$ reduction in field studies mentioned in the cited reference. The natural gas is fired into a secondary combustion region above the primary combustion zone to create a fuel-rich region in which the $NO_x$ formed in the primary zone is reduced to molecular nitrogen. This is followed by a burnout zone where unburnt fuel is combusted to completion. Use of NGR results in a 50% reduction in $NO_x$ emissions and a 15% reduction in $SO_2$ emissions with 15% gas use.

The costs shown in Table 3 were calculated using the IAPCS-3 computer cost estimation model for both the corfiring and reburning options. It should be noted that NGR alone is not capable of reducing the boiler $SO_2$ emissions to the full extent required for permitting purposes and therefore some additional $SO_2$ controls would be necessary. A study on the utility boiler population in West Virginia showed that NGR combined with blending with state-of-the-art washed coal and Duct Sorbent Injection Technology (DSIT) was cheaper than switching to washed coal and using wet Flue Gas Desulfurization (FGD).[17] Thus using NGR for combined $SO_2$ and $NO_x$ reduction along with other technologies can be effective in overall emission reduction strategies for a given area.

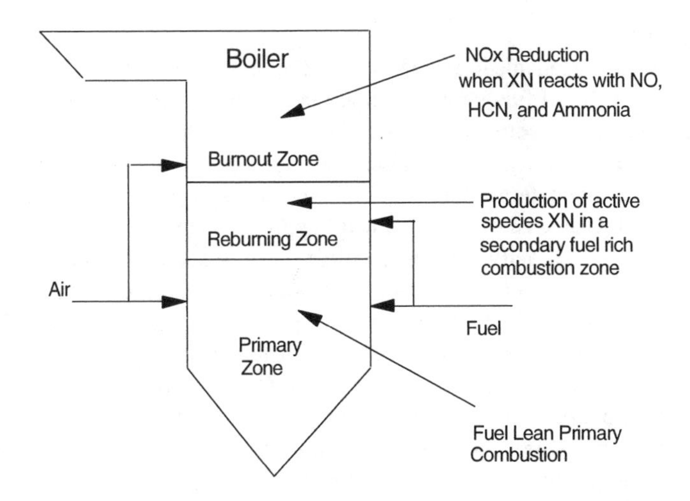


**FIGURE 2.** Schematic of the reburning process.

### 3. Repowering

Repowering is the process by which an electric power generation system output is increased, either during maximum load periods or during its entire operating cycle by introducing additional capacity and/or increasing generating efficiency.[18] The other option available is to replace the entire system with a new and higher capacity system. There were three specific cases of repowering which were studied: (1) peaking turbine repowering, which refers to the addition of a steam turbine and heat recovery unit to an existing gas turbine, with the efficiency improvement allowing the unit to convert from peaking to baseload operation; (2) heat recovery repowering, where an old coal boiler is replaced with a gas turbine and heat recovery unit, leaving the existing steam turbine in place to be operated by steam generated in the heat recovery unit; and (3) boiler repowering, in which the exhaust from a new gas turbine is fed into an existing coal boiler, reducing power requirements by replacing existing forced draft fans and reducing the load on air heaters. Each of these cases was compared with the option of adding new coal-fired boilers, on the basis of cost and emissions.

The $SO_2$ emissions and capital and annual costs for the coal-fired option in each of the above three types of repowering is based on the need for a scrubber in order to comply with $SO_2$ emission standards. The peaking turbine repowering $NO_x$ emission is based on the fact that pre-1982 gas turbines had no $SO_2$ or $NO_x$ federal emission limitations. The heat recovery and boiler repowering $NO_x$ emissions are based on the current gas turbine emission limits. Table 2 shows the $SO_2$ and particulate emission numbers for the boiler repowering gas option as negative because the use of a gas turbine actually reduces the coal requirement of the boiler and therefore reduces $SO_2$ and particulates.

As can be seen from Table 3 the capital and annual costs in each case of repowering are approximately 30% of that of the coal-fired option. Cost-effectiveness in terms of dollars/ton of $SO_2$ + $NO_x$ removed is reduced by 50% when the gas option is used as compared to the coal option.

## B. STATIONARY ENGINES

Stationary engines are categorized into gas turbines and stationary internal combustion engines.

### 1. Gas Turbines

Gas turbines are mechanical devices where the energy from the expanding combustion gases resulting from fossil fuel combustion are used directly to drive the generator shaft. In the case of turbines, gas is compared with distillate fuel oil No. 2 (DF-2) for four applications which include standby service, industrial use, utility power generation, and offshore drilling and transportation applications.[19] It is found that in each case gas use results in a $NO_x$ emission rate which is at least 0.21 lbs/mmBtu (approximately 30%) less than when distillate fuel No. 2 (DF-2) is used and complete elimination of $SO_2$ emissions occurs. In estimating operating and maintenance costs for each fuel option it is assumed that the fuel costs are $2.18/mmBtu for both DF-2 and natural gas. Depending on fuel prices at specific sites the actual capital and O & M costs may be slightly different. The present analysis shows that cost-effectiveness, in terms of dollars/ton of $SO_2$ + $NO_x$ removed, of using gas is less than that of using DF-2 because of negligible sulfur in natural gas. The actual cost-effectiveness numbers are very large when compared to the previous values because the amounts of $NO_x$ removed range from 27 to 40% of the uncontrolled $NO_x$ emissions, while the actual size of the turbines, in terms of megawatts, is much smaller than the previous systems.

### 2. Internal Combustion Engines

Internal combustion (I.C.) engines different from turbines because in these engines the energy from the expanding combustion gases resulting from fossil fuel combustion are used to move pistons which are in turn connected by a crank shaft to the generator. Emissions from I.C. engines using diesel, dual fuel (95% gas, 5% diesel), and I.C. engines used in oil and gas production and transportation, are listed in Table 2 and costs are listed in Table 3.[20] The background document which contains all the I.C. engine cost and emission data used for the present analysis uses a fuel price of $2.50/mmBtu for diesel, $3.00/mmBtu for dual fuel, and $2.00/mmBtu for natural gas used by the oil and gas transport and production industry. Diesel and dual fuel operation of I.C. engines costs 50% more than natural gas engines in terms of cost-effectiveness ($/ton of $SO_2$ + $NO_x$ removed). $NO_x$ emissions listed in Table 2 are based on a 40% reduction of $NO_x$ using retarded firing, manifold cooling and water injection techniques.

In the above literature review $CO_2$ emissions were not specifically addressed. However, it may be reiterated that NGR technology significantly reduces the emission of $CO_2$ based on the lower carbon content per unit of heating value as compared to coal.

## VI. NATURAL GAS AND CLEAN AIR LEGISLATION

### A. HISTORY OF CLEAN AIR LEGISLATION

The present concern over the deleterious effects of acid rain and the focus on ways to mitigate the acid rain problem has a long history behind it. Brown et al. have given a detailed history and background of Clean Air Act (CAA) of 1963, which was the first piece of legislation marking the national awakening of the public to the value of our environment.[21] The CAA of 1963 was followed by the amendments of 1970 and 1977.[22] The above-mentioned amendments to the Clean Air Act set ambient air quality standards for a number of pollutants, with primary standards being designed to protect public health, and secondary standards designed to protect public welfare. Public welfare is a broad concept that includes natural resources, aesthetics, economics, and the general quality of life. These amendments required preparation of State Implementation Plans (SIPs) to be enforced by local agencies at the state level. There was a provision for deadlines by which the SIPs were to be enforced. Except for the non-attainment of the ozone standards, the CAA has been largely successful according to the EPA. In addition to the ozone non-attainment problem, there is another environmental issue in the form of acid rain, which is of great concern. Acid rain is primarily caused by the formation of acidic compounds due to photochemical atmospheric reactions of $SO_2$, $NO_x$ and other acid-precursors. These compounds are then scrubbed from the atmosphere with precipitation but, when this happens, the precipitation in turn becomes acidic in nature (the pH becoming less than 7.0). The acidic precipitation has a deleterious effect on the ecology and flora and fauna and eventually, human health.[23-25] With the passing of years, there have been many experiments and tests on acid rain and a lot of data documenting the effects of acid rain has accumulated.[26,27] The large amounts of $SO_2$ being released by coal-fired utility boilers have largely incriminated utility companies as being the culprits.[28] Most of the research work has been in Canada because the direction of the jet stream across the U.S. is such that the emissions from the midwestern and northeastern U.S. are carried into southeastern Canada.[29,30] An interim report from the National Acid Precipitation Assessment Program (NAPAP) has assessed that power plants contribute up to 65% of the national annual emissions of $SO_2$ and up to 29% of the $NO_x$ emissions.[31] In the light of the circumstances explained above, it is not surprising that $SO_2$ and $NO_x$ emissions from utility coal-fired boilers have come under the scrutiny of legislators. The current legislation is described in the next section.

### B. CLEAN AIR ACT AMENDMENTS OF 1990

The new Clean Air Act (CAA) amendments of 1990, were signed into law by President Bush on November 15, 1990. It required a 10 million ton reduction in $SO_2$ emissions and a 2 million ton reduction in $NO_x$ emission, using 1985 through 1987 as the baseline years.

Phase I of the CAA amendments required electric power generation boilers above 100 MW to restrict $SO_2$ emissions to 2.5 lbs/mmBtu by 1995. Phase II of the CAA amendments takes effect from 2000, and requires all utility units greater than 25 MW to restrict $SO_2$ emissions to 1.2 lbs/mmBtu. Within 18 months of the enactment of the CAA amendments, $NO_x$ emissions from tangentially fired boilers will be restricted to 0.45 lbs/mmBtu, and dry bottom wall-fired boilers to 0.50 lbs/mmBtu. EPA will establish similar $NO_x$ emission limits for wet bottom wall-fired boilers, cyclones, and all other types of boilers, not later than January 1, 1997.

An important feature of the CAA amendments is that it allows emissions trading. It provides for "allowances", an "allowance" being worth one ton of $SO_2$, which are fully marketable. In other words, a utility which controls its $SO_2$ emissions beyond the mandatory limit may sell the additional tonnage to another utility which will then be treated as having achieved the corresponding amount of $SO_2$ reduction.

## VII. THE ROLE OF NATURAL GAS IN COMPLIANCE STRATEGIES

The use of natural gas in compliance strategies was studied on both a statewide and a county level.

### A. STATEWIDE LEVEL

The State of Ohio was chosen for the current study because it is the highest $SO_2$ emitting region in the U.S. and it has a large utility-boiler population.[32] Figure 3 graphically shows the $SO_2$ and $NO_x$ emissions as state totals for 1980.

In order to have a common basis for evaluating the control strategies, a statewide emission limit was calculated on the basis of the current New Source Performance Standards (NSPS) emission limits imposed on all new boilers built after September 1978.[9] The NSPS standards were used because the CAA amendments had not been enacted at the time the present

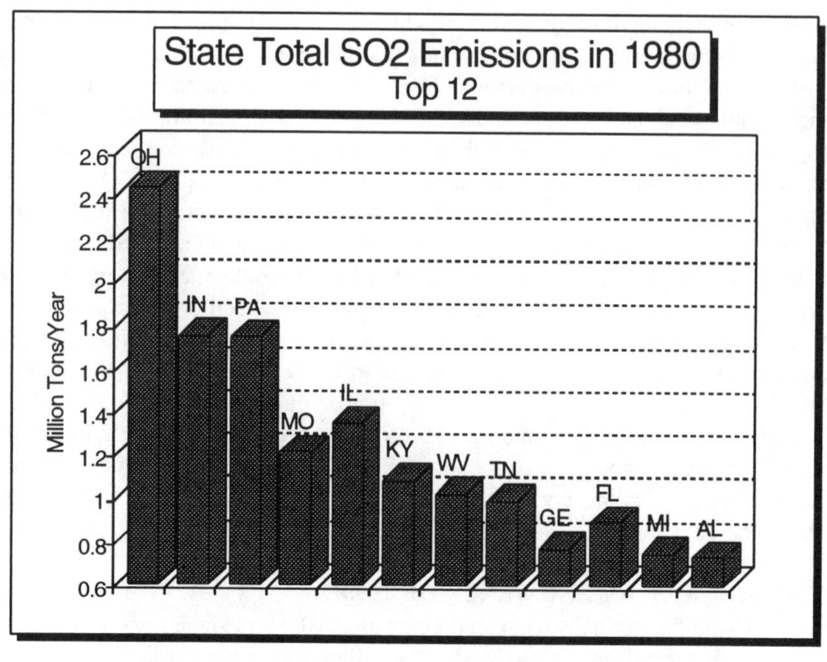


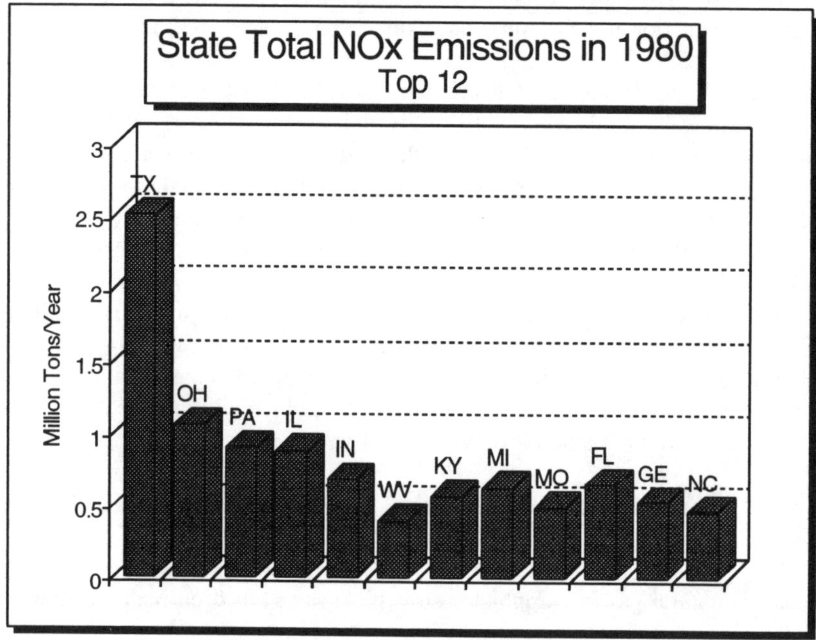


**FIGURE 3.** Sulfur dioxide and nitrogen oxide emissions.

research was conducted. The NSPS standards require an $SO_2$ emission limit of 1.2 lbs/mmBtu and an $NO_x$ emission limit of 0.6 lbs/mmBtu on all types of coal-fired utility boilers except cyclones which have a limit of 0.8 lbs/mmBtu. The NSPS emission limits were applied to all the 101 Ohio utility boilers and the resulting statewide $SO_2$ and $NO_x$ emission reductions were calculated. Using these emission limits led to 70% reduction in $SO_2$ emissions and 32% reduction in $NO_x$ emissions for the entire state. Therefore, these were chosen as the emissions reduction targets for each control strategy that was studied.

**TABLE 4**
**$SO_2$ and $NO_x$ Control Costs for 70% $SO_2$ Reduction and 32% $NO_x$ Reduction**

| No. | Strategy | $SO_2$ Control Costs (M $) | | | $NO_x$ Control Costs (M $) | | |
|---|---|---|---|---|---|---|---|
| | | Capital | Annual | C E | Capital | Annual | C E |
| 7 | NGR + FGD | 3024 | 832 | 539 | 107 | 88 | 567 |
| 8 | NGR + LMB | 2451 | 819 | 532 | 107 | 88 | 567 |
| 6 | LNC + LMB | 2667 | 862 | 562 | 191 | 28 | 181 |
| 5 | LNC + FGD | 3106 | 845 | 552 | 191 | 28 | 181 |
| 3 | LNC + OH2.5 + FGD | 2103 | 753 | 491 | 191 | 28 | 181 |
| 1 | NGR + OH2.5 + FGD | 1864 | 691 | 451 | 107 | 88 | 567 |
| 4 | LNC + OH2.5 + LMB | 1061 | 706 | 460 | 191 | 28 | 181 |
| 2 | NGR + OH2.5 + LMB | 904 | 645 | 419 | 107 | 88 | 567 |

*Note:* C E = Cost effectiveness in dollars/ton.

The control technologies used in the analysis were Natural Gas Reburn (NGR), Low $NO_x$ Combustion (LNC), Limestone Injection Multistage Burners (LIMB), Flue Gas Desulfurization (FGD), and Coal Switching (OH2.5, in-state washed Ohio coal emitting 2.5 lbs/mmBtu $SO_2$).

The above technologies were combined into the following eight strategies:

1. NGR + OH2.5 + FGD
2. NGR + OH2.5 + LIMB
3. LNC + OH2.5 + FGD
4. LNC + OH2.5 + LIMB
5. LNC + FGD
6. LNC + LIMB
7. NGR + FGD
8. NGR + LIMB

The above strategies were chosen in order to make the following comparisons:

1. NGR with LNC
2. FGD with LIMB
3. Switching to washed Ohio coal with no coal switching or washing.

Strategies were ranked on the basis of capital costs, levelized annual costs and the cost-effectiveness ($/ton $SO_2$ + $NO_x$ removed). The main economic assumptions used in the levelization of costs was a 30-year plant life, an annual inflation rate of 6%, and a discount rate of 11%. The costs are expressed in constant 1988 dollars. Tables 4 and 5 show the $SO_2$ and $NO_x$, and the total control costs for each strategy, respectively. The costs are differentiated into capital and levelized annual costs and sorted in the order of descending total annual costs and cost-effectiveness.

These costs were then compared on three levels, coal switching vs. no coal switching, FGD vs. LIMB, and NGR vs. LNC. Table 6 shows the comparison of coal switching vs. no coal switching. Coal switching results in a 46% average savings in capital costs and a 16% average savings in levelized annual costs and the cost effectiveness. Similarly Tables 7 and 8 show the comparison of FGD with LIMB strategies, NGR with LNC strategies, respectively. It was found that LIMB strategies represented a 42% average capital cost

**TABLE 5**
**Total Control Costs for 70% $SO_2$ and 32% $NO_x$ Reduction**

| No. | Strategy | Capital M$ | Annual M$ | C E / $SO_2 + NO_x$ |
|---|---|---|---|---|
| 7 | NGR + FGD | 3131 | 920 | 542 |
| 8 | NGR + LMB | 2558 | 907 | 534 |
| 6 | LNC + LMB | 2858 | 890 | 527 |
| 5 | LNC + FGD | 3297 | 873 | 518 |
| 3 | LNC + OH2.5 + FGD | 2294 | 781 | 462 |
| 1 | NGR + OH2.5 + FGD | 1971 | 779 | 461 |
| 4 | LNC + OH2.5 + LMB | 1252 | 734 | 434 |
| 2 | NGR + OH2.5 + LMB | 1011 | 733 | 433 |

*Note:* C E = Cost effectiveness in dollars/ton

**TABLE 6**
**Advantages of Coal Switching**

| No. | Strategy | Capital M$ | Annual M$ | C E / $SO_2 + NO_x$ |
|---|---|---|---|---|
| 7 | NGR + FGD | 3131 | 920 | 542 |
| 1 | NGR + OH2.5 + FGD | 1971 | 779 | 461 |
| | Difference | 1160 | 141 | 81 |
| | % Savings | 37 | 15 | 15 |
| 8 | NGR + LMB | 2558 | 907 | 534 |
| 2 | NGR + OH2.5 + LMB | 1011 | 733 | 433 |
| | Difference | 1547 | 174 | 101 |
| | % Savings | 60 | 19 | 19 |
| 6 | LNC + LMB | 2858 | 890 | 527 |
| 4 | LNC + OH2.5 + LMB | 1252 | 734 | 434 |
| | Difference | 1606 | 156 | 93 |
| | % Savings | 56 | 18 | 18 |
| 5 | LNC + FGD | 3297 | 873 | 518 |
| 3 | LNC + OH2.5 + FGD | 2294 | 781 | 462 |
| | Difference | 1003 | 92 | 56 |
| | % Savings | 30 | 11 | 11 |
| | Avg. % Savings | 46 | 16 | 16 |

*Note:* C E = Cost effectiveness in dollars/ton

savings and a 3% average levelized annual cost and cost-effectiveness savings over FGD strategies. Lastly, it was found that NGR strategies represented a 12% average capital cost savings and a negative average levelized annual cost and cost-effectiveness savings over LNC strategies. The negative numbers mean that the levelized annual costs of LNC strategies are actually less than the corresponding NGR strategy costs. This could be explained by the fact that the overall emissions reduction target in the State of Ohio was such that it could not be achieved by using LIMB alone where applicable. Therefore, the LIMB strategy costs also include the cost of a very minimal number of FGD units which were added to meet the $SO_2$ and $NO_x$ emissions target.

Although the statewide emission target could not be met by using LIMB alone, it was considerably cheaper to use LIMB and provide additional scrubbers than to use scrubbers alone. The use of NGR made strategies cheaper than those using no gas. In some cases

**TABLE 7**
**Advantages of LIMB Over FGD**

| No. | Strategy | Capital M$ | Annual M$ | C E $SO_2 + NO_x$ |
|---|---|---|---|---|
| 7 | NGR + FGD | 3131 | 920 | 542 |
| 8 | NGR + LMB | 2558 | 907 | 534 |
| | Difference | 573 | 13 | 8 |
| | % Savings | 18 | 1 | 1 |
| 1 | NGR + OH2.5 + FGD | 1971 | 779 | 461 |
| 2 | NGR + OH2.5 + LMB | 1011 | 733 | 433 |
| | Difference | 960 | 46 | 28 |
| | % Savings | 49 | 6 | 6 |
| 5 | LNC + FGD | 3297 | 873 | 518 |
| 6 | LNC + LMB | 2858 | 890 | 527 |
| | Difference | 439 | −17 | −9 |
| | % Savings | 56 | −2 | −2 |
| 3 | LNC + OH2.5 + FGD | 2294 | 781 | 462 |
| 4 | LNC + OH2.5 + LMB | 1252 | 734 | 434 |
| | Difference | 1042 | 47 | 28 |
| | % Savings | 45 | 6 | 6 |
| | Avg. % Savings | 42 | 3 | 3 |

*Note:* C E = Cost effectiveness in dollars/ton

**TABLE 8**
**Advantages of NGR Over LNC**

| No. | Strategy | Capital M$ | Annual M$ | C E $SO_2 + NO_x$ |
|---|---|---|---|---|
| 5 | LNC + FGD | 3297 | 873 | 518 |
| 7 | NGR + FGD | 3131 | 920 | 542 |
| | Difference | 166 | −47 | −24 |
| | % Savings | 5 | −5 | −5 |
| 3 | LNC + OH2.5 + FGD | 2294 | 781 | 462 |
| 1 | NGR + OH2.5 + FGD | 1971 | 779 | 461 |
| | Difference | 323 | 2 | 1 |
| | % Savings | 14 | <1 | <1 |
| 6 | LNC + LMB | 2858 | 890 | 527 |
| 8 | NGR + LMB | 2558 | 907 | 534 |
| | Difference | 300 | −17 | −7 |
| | % Savings | 10 | −2 | −2 |
| 4 | LNC + OH2.5 + LMB | 1252 | 734 | 434 |
| 2 | NGR + OH2.5 + LMB | 1011 | 733 | 433 |
| | Difference | 241 | 1 | 1 |
| | % Savings | 19 | <1 | <1 |
| | Avg. % Savings | 12 | NA | NA |

*Note:* C E = Cost effectiveness in dollars/ton; NA = Not applicable

annual costs using LNC were almost the same as those using NGR. However it would be advisable to use NGR because it can be applied to all types of boilers whereas LNC doesn't work well in tangentially fired boilers, and cannot be used in wet-bottom or cyclone-type boilers. Secondly, NGR has the advantage of lower capital costs. Another argument is that combined $SO_2$ and $NO_x$ reduction is achieved by using NGR whereas LNC has no effect

**TABLE 9**
**Total Control Cost for Compliance With New Legislation (Washington County)**

| No. | Strategy | Capital M$ | Annual M$ | C E / $SO_2 + NO_x$ |
|---|---|---|---|---|
| 1 | FGD + SCR + OH2.5 + LNC | 616 | 150 | 461 |
| 2 | FGD + NGR + OH2.5 | 463 | 129 | 395 |
| 3 | FGD + DSI + OH2.5 + SCR + LNC | 459 | 128 | 393 |
| 4 | FGD + DSI + OH2.5 + NGR | 248 | 96 | 295 |

*Note:* C E = Cost effectiveness in dollars/ton; NGR = Natural gas reburn; OH2.5 = Washed Ohio emitting 2.5 lbs/mmBtu $SO_2$; DSI = Duct sorbent injection of limestone slurry; FGD = Flue gas desulfurization; SCR = Selective catalytic reduction; LNC = Low $NO_x$ combustion

**TABLE 10**
**Total Control Costs for Compliance With New Legislation (Gallia County)**

| No. | Strategy | Capital M$ | Annual M$ | C E / $SO_2 + NO_x$ |
|---|---|---|---|---|
| 1 | DSI + SCR + OH2.5 + LNC | 388 | 164 | 296 |
| 2 | DSI + NGR + OH2.5 | 171 | 139 | 255 |
| 3 | FGD + OH2.5 + SCR + LNC | 835 | 238 | 433 |
| 4 | FGD + DSI + OH2.5 + NGR | 451 | 192 | 352 |

*Note:* C E = Cost effectiveness in dollars/ton; NGR = Natural gas reburn; OH2.5 = Washed Ohio coal emitting 2.5 lbs/mmBtu $SO_2$; DSI = Duct sorbent injection of limestone slurry; FGD = Flue gas desulfurization; SCR = Selective catalytic reduction; LNC = Low $NO_x$ combustion

on $SO_2$ emissions and in some cases there may be difficulty retrofitting LNC to old boilers. The cheapest strategy in terms of cost-effectiveness, coal switching + NGR + LIMB cost 433 \$/ton of $SO_2$ + $NO_x$ removed. Additional sensitivity analysis was also carried out to study the effect of cost-based and technology-based parameters such as fuel price differential (FPD) between natural gas and coal.[33]

## B. COUNTY LEVEL

The above analysis was also done on a county level to provide a cost comparison on a "micro" level as compared to the statewide "macro" level. This analysis was also carried out in the State of Ohio. It was decided to select particular counties where large utilities were located, and to examine the costs of using strategies incorporating natural gas technologies. The three counties selected were Washington, Gallia, and Coshocton. The emission limits applied in each county were 1.2 lbs/mmBtu for $SO_2$ and a 50% reduction in $NO_x$ emissions. These values are the same as the Phase II requirements of the 1990 CAA amendments. The $SO_2$ emissions reduction required were 85, 80, and 63% in Washington, Gallia, and Coshocton counties, respectively.

Tables 9, 10, and 11 list the capital, annual costs, and the cost-effectiveness associated with the strategies selected in Washington, Gallia, and Coshocton counties, respectively. Noteworthy distinctions between the strategies selected here and those for the entire State of Ohio, are that instead of using LIMB, a variation of the same technology known as Duct

**TABLE 11**
**Total Control Costs for Compliance With New Legislation (Coshocton County)**

| No. | Strategy | Capital M$ | Annual M$ | CE / $SO_2 + NO_x$ |
|---|---|---|---|---|
| 1. | FDG + SCR + OH2.5 + LNC | 230 | 63 | 555 |
| 2. | DSI + NGR + OH2.5 | 54 | 26 | 236 |

*Note:* C E = Cost effectiveness in dollars/ton; NGR = Natural gas reburn; OH2.5 = Washed Ohio coal emitting 2.5 lbs/mmBtu $SO_2$; DSI = Duct sorbent injection of limestone slurry; FGD = Flue gas desulfurization; SCR = Selective catalytic reduction; LNC = Low $NO_x$ combustion

Sorbent Injection (DSI), was used. This resulted in 70% $SO_2$ reductions compared to 50% reductions for the LIMB technology at very little additional cost. In certain cases Selective Catalytic Reduction (SCR) technology was used where LNC could not be used. The results were again in favor of using NGR technology in conjunction with coal switching and DSI. Here again additional sensitivity analysis was also carried out to study the effect of cost-based and technology-based parameters.[33]

## VIII. NATURAL GAS SUPPLY LOGISTICS

Our studies indicated that in order to achieve $SO_2$, and $NO_x$ reductions of 70 and 32%, respectively, in Ohio, 93 billion cubic feet (BCF) of natural gas would be required for the natural gas strategies.

The annual average quantity of natural gas used by utilities from 1985 to 1987 was approximately 300 BCF.[34-36] The data also indicated that there was a downturn in gas use from 1985 to 1987. An American Gas Association (AGA) report projected that natural gas use in the USA would increase by 25% between 1988 and 2010.[37] Based on the above data and projections, it can be seen that if the natural gas strategy were to be implemented in Ohio, then the increase in natural gas use would be approximately 30% of the average use in Ohio between 1985 and 1987.

The total average quantity of natural gas use by the utility sector in all of the U.S. during this period was approximately 5 trillion cubic feet (or 5000 BCF).[37] The total average gas use for all consumers, including residental, commercial, industrial, and utilities, during this same period was approximately 17 trillion cubic feet (or 17,000 BCF).[37]

Hence, the increase in natural gas use resulting from gas use for acid rain control strategies, would be less than 2% of the total utility sector gas use in the U.S., and less than 1% of the total gas used by all sectors of the gas market in the U.S.

From the above discussion it can be surmized that the choice of using natural gas strategies is more likely to be influenced by the proximity of a natural gas pipeline. Discussions with Texas Gas Company officials have indicated that the cost of laying new natural gas pipelines may range all the way from $100,000 to $1,000,000/mile of pipeline, depending on the terrain.[38] A GRI study has indicated that of the 27 utility plant sites in Ohio, 13 plants are within close proximity of a major natural gas pipeline.[39] However, the estimation of actual costs that would be incurred in transporting gas to the individual plant sites, merits further study.

## IX. CONCLUSIONS AND RECOMMENDATIONS

Literature data on boiler systems indicated that a combined cycle electric power generation system was superior to a new coal-fired system of comparable size and electrical output.[13] In boiler repowering scenarios, i.e., in situations where boiler steam generation capability was increased, the use of natural gas resulted in a cost-effectiveness 50% less than a comparable coal-based system.[18]

Under the category of stationary engines natural gas turbines had a 30% lower $NO_x$ emissions rate and no $SO_2$ emissions with insignificant cost differences as compared to distillate fuel. Internal combustion engines fuelled with natural gas cost 50% less than diesel or dual fuel engines in terms of cost effectiveness.

Our research estimated the costs of achieving 70% reduction in $SO_2$ emissions, and 32% reduction in $NO_x$ emissions, in the State of Ohio using eight different strategies. These strategies showed that the use of coal switching resulted in cost benefits, limestone injection was more cost effective in strategies than flue gas desulfurization, and the use of natural gas reburn technology with coal switching and limestone injection represented the most cost-effective strategy among those studied. This result holds good until the price differential of natural gas over coal increases to 2.1 $/mmBtu. The same was found to be true on the county level also.

On the basis of the above analysis it can be said that natural gas technologies present a technologically and economically viable alternative to using conventional flue gas scrubber technologies which are much more capital and maintenance-intensive. Natural gas accomplished a dual $SO_2$ and $NO_x$ emissions reduction from coal-fired boilers that is chemistry-based and does not require additional raw materials or equipment other than a fuel burner. Lastly, on the basis of stoichiometric calculations it has been shown that the use of 15% natural gas reduces the emission of $CO_2$ by 43% with a concomitant reduction in the "global warming" phenomenon. If NGR technology becomes widely adopted then the fuel price of natural gas might fall making this technology even more attractive. It is the opinion of the researchers that because of the large supply of natural gas in the market prices of natural gas will not take an upward turn in the immediate future. The present research has demonstrated that a systems approach using a mix of technologies can produce the same emissions reductions as conventional technologies, at a much lower cost, and without large-scale disruption in the coal-mining industry.

## REFERENCES

1. Clean Air Act Amendments, Public Law 101–549, November 15, 1990.
2. "Urban Ozone and the Clean Air Act: Problems and Proposals for Change", Staff Paper from OTA's Assessment of New Clean Air Act Issues, Oceans and Environment Program, Office of Technology Assessment, Washington, D.C., April 1988.
3. Meyer, E. L., "Review of Control Strategies for Ozone and Their Effects on other Environmental Issues", EPA-450/4-86-011, OAQPS-USEPA, November 1986.
4. "The Effects of Oxides of Nitrogen on California Air Quality", Report #TSD-85-01, California Air Resources Board, Technical Support Division, March, 1986.
5. Walker, H. M., "Ten Year Ozone Trends in California and Texas", *JAPCA*, 35(9), 903–912, 1985.
6. Perry, R. H. and Green, D., *Perry's Chemical Engineers' Handbook*, 6th ed., McGraw-Hill, New York, N.Y., 1984, 9–16.
7. Gas Research Institute, *"Gas Reburning-Sorbent Injection (GR-SI): Development Status and Field Evaluation Plan"*, topical report, August 1987.

8. La Fond, J. F. and Chen, S. L., "An Investigation To Define The Physical/Chemical Constraints Which Limit $NO_x$ Emission Reduction Achievable by Reburning", Quarterly report No. 1, prepared for U.S. Department of Energy by Energy and Environmental Research Corporation, Irvine, California, May 1987.
9. 40 CFR 60. Code of Federal Regulations, Title 40, Part 60, *Standards of Performance for New Stationary Sources,* Office of Federal register, Washington, D.C., July 1, 1988.
10. PEI Associates, Inc., *"User's Manual for the Integrated Air Pollution Control System Performance and Costing Model: Version III",* EPA Contract No. 68-02-4284, September, 1988.
11. "Compilation of Air Pollutant Emission Factors", (AP-42), 4th ed., OAQPS, Environmental Protection Agency, RTP, NC, September 1985.
12. Technical Assessment Guide (TAG), Volume I: Electricity Supply — 1986, prepared by the Electric Power Research Institute, EPRI P-4463-SR, December 1986.
13. "Natural gas-fueled combined cycle electric power generation — an attractive option", AGA 1985-9, 1985.
14. Booth, R. C., and Glickert, R. W., "Extended development of gas cofiring to reduce sulfur dioxide and nitric oxide emissions from a tangentially coal-fired utility boiler", Topical report, Environmental System Associates, Pittsburgh PA, (prepared for GRI), August 1988.
15. Gas Research Institute, *"Preliminary Assessment of the Economics of Natural Gas Cofiring in Utility Coal-Fired Boilers",* report by Energy and Environmental Analysis, Inc., Virginia, June, 1988.
16. Bartok, W., Folsom, B. A., Elbl, M., Kurzynske, F. R., and Ritz, H. J., "Gas Reburning-Sorbent Injection for Controlling $SO_2$ and $NO_x$ in Utility Boilers", Environmental Progress, Vol. 9, No. 1, 1990.
17. Berkau, E. E., Glickert, R. W., and Dey, P. R., *"A Strategy for Acid Rain Control in West Virginia Using Natural Gas Cofiring and Clean Coal Technology",* Energy Systems Associates report prepared for Consolidated Natural Gas Service Company, Pittsburgh, PA, June 1989.
18. Wilkinson, P. L. and Hay, N., "Repowering with natural gas — An electricity generation option", AGA 14, 1986.
19. "Standards support and environmental impact statement: Stationary Gas Turbines", EPA-450/2-77-017a, OAQPS, Environmental Protection Agency, RTP, NC, September 1977.
20. "Stationary Internal Combustion Engines: Standards Support and Environmental Impact Statement — Vol. 1", EPA-450/2-78-125a, OAQPS, Environmental Protection Agency, RTP, NC, July 1979.
21. Brown, G. A., Cramer, J. J., and Samela, D., "The Impact of the Proposed Clean Air Act Amendments", Chemical Engineering Progress, 84(12), 1988.
22. Easton, E. B. and O'Donnell, F. J., "The Clean Air Act Amendments of 1977", *JAPCA,* 27(10), pp. 943–47, 1977.
23. Franklin, C. A., Burnett, R. T., Paolini, R. J., and Raizenne, M. E., "Heath Risks from Acid Rain", Environmental Health Perspective, 63, pp. 155–68, 1985.
24. Rennie, P. J., "Evidence for effects on Canadian Forests", Symposium on Air Pollution Effects for Ecosystems, St. Paul, MN, pp. 111–22, 1985.
25. Foster, N. W., "Acid Precipitation and Soil Solution Chemistry within a Maple-Birch Forest in Canada", Forest Ecology Management, 12, pp. 215–31, 1985.
26. Hern, J. A., "Chemical Effects of Simulated Acid Precipitation on Two Canadian Shield Forest Soils", Dissertation Abstract International B, 47(5), 1985.
27. Bertram, H. L., Das, N. C., and Lau, Y. K., "Precipitation Chemistry Measurement in Alberta", Water, Air, Soil Pollution, 30(1–2), 1986.
28. Hidy, G. M., Hansen, D. A., Henry, R. C., Ganesan, K., and Collins, J., "Trends in Historical Acid Precursor Emissions and their Airborne and Precipitation Products", *JAPCA,* 34(4), pp. 333–54, 1984.
29. Whelpdale, D. M. and Bottenheim, J. W., "Recent Canada-USA Transboundary Air Pollution Studies", Studies in Environmental Science, 21, pp. 357–63, 372–3, 1982.
30. Whelpdale, D. M. and Barrie, L. A., "Atmospheric Monitoring Network Operations and Results in Canada", Water, Air, Soil Pollution, 18(*–2–3*), 1982.
31. "NAPAP Interim Assessment: The Causes and Effects of Acidic Deposition, Volume II: Emissions and Control", National Acid Precipitation Assessment Program, 1986.
32. "Acid Rain and Transported Air Pollutants: Implications for Public Policy", Office of Technology Assessment Report, June 1984.
33. Dey, P. R., *"Environmental and Economic Benefits of Natural Gas Use for Pollution Control",* Ph.D. Dissertation, Vanderbilt University, 1990.
34. *Steam Electric Plant Factors, 1986 Edition,* published by the National Coal Association, Washington, D.C.
35. *Steam Electric Plant Factors, 1987 Edition,* published by the National Coal Association, Washington, D.C.
36. *Steam Electric Plant Factors, 1988 Edition,* published by the National Coal Association, Washington, D.C.
37. American Gas Association, "The A.G.A. — TERA 1989 Mid-Year Base Case", Total Energy Resources Analysis (TERA) Model, September 29, 1989.

38. Discussion with Mr. James Allison, Texas Gas Company, August 3, 1990.
39. Emmel, T. E., Rimpo, E. T., Spaite, P. S., and Szabo, M. F., ''*Analysis of Natural Gas-Based Technologies for Control of* $SO_2$ *and* $NO_x$ *from Existing Coal-Fired Boilers*'', draft topical report by Radian Corporation, prepared for Gas Research Institute, January 27, 1987.

Chapter 15

# METHANE EMISSIONS TO THE GLOBAL ATMOSPHERE FROM COAL MINING

**Robert C. Harriss, Terry Bensel, and Denise Blaha**

## TABLE OF CONTENTS



0-8493-4419-0/93/$0.00 + $.50
© 1993 by CRC Press, Inc.

# I. INTRODUCTION

The observed rates of increase in atmospheric concentrations of carbon dioxide ($CO_2$), methane ($CH_4$), nitrous oxide ($N_2O$), and chlorofluorocarbons (CFCs), and their contribution to a potential future warming of the Earth's lower atmosphere, are well documented.[1,2] These data are, however, not sufficient for developing estimates of the contributions of specific human activities as sources of increasing greenhouse gas emissions, or for formulating policies which would regulate sources. In this chapter, we report the results of a detailed assessment of coal mining and use as a source for atmospheric $CH_4$. The results of our study have important implications for the policy process which might be used to develop a strategy for mitigating the global warming potential from $CH_4$ accumulation in the atmosphere.

$CH_4$ is a greenhouse gas which also partially controls the oxidizing capacity of the atmosphere.[3] Ambient air measurements indicate that $CH_4$ is increasing at an annual rate of about 1%, although in recent years the rate of increase has slowed.[4] Future growth in atmospheric $CH_4$ concentrations is likely to contribute more to a greenhouse warming effect than any other gas except $CO_2$.[3] Historical records of atmospheric $CH_4$ derived from polar ice cores indicate that preindustrial concentrations varied over a range of approximately 0.30 to 0.70 ppm, compared to the present average concentration of 1.7 ppm.[5] Contemporary atmospheric $CH_4$ concentrations and the currently observed rates of increase are unprecedented, at least during the past 160,000 years. The strong correlation between increasing atmospheric $CH_4$ and human population growth during the past 150 years supports the contention that human activities are involved in causes for the increase.[6] $CH_4$ is emitted to the atmosphere from flooded soils, ruminant animals, fires, termites, natural gas exploitation, and coal mining. Annual $CH_4$ release from these sources has been estimated to be 400 to $640 \times 10^{12}$ g ($10^{12}$ g = Tg). The 1%/year increase in atmospheric $CH_4$ requires a net increase of approximately 50 Tg/year to the atmosphere. This net increase is thought to reflect the combination of increased emissions and a reduction in the oxidative removal potential of the lower atmosphere.[6]

The annual emission rates for individual sources of atmospheric $CH_4$ are highly uncertain by factors of 2 to 25.[3] If annual $CH_4$ emissions from coal mining are approximately 25 to 45 Tg, as suggested by preliminary estimates,[3] they may represent one of the $CH_4$ sources potentially most amenable to control in any future program to stabilize the composition of the atmosphere.

# II. ESTIMATING METHANE EMISSIONS FROM COAL MINING: A CONSERVATIVE APPROACH

Coal-bed gas, composed primarily of $CH_4$, is a natural product of microbiological and thermogenic processes in coal-forming environments.[7] The coalbed gas is trapped during burial and exists at depth adsorbed on coal particle surfaces and trapped in pore spaces and fractures. The major correlates with the $CH_4$ content in coal are physical properties (rank) and present depth of burial.[7,8] Several independent approaches have been taken to determine quantitative relationships between coal rank, depth of occurrence, and $CH_4$ content.[7-9] This literature is focused on coalbed $CH_4$ as a mine hazard or potential energy source, but also provides potential emission factor data for estimating $CH_4$ loss to the atmosphere as a result of global coal production.

Our study has included a comprehensive synthesis of research by the U.S. Bureau of Mines on $CH_4$ degassing from coal as a mine and transportation safety problem, a compilation of data on the occurrence of $CH_4$ in coal deposits, and a compilation of global coal mining statistics on a country basis.[10] From these data we estimated emission factors for $CH_4$ loss to the atmosphere as a function of coal type and mine depth. These emission factors were used with production and mining statistics for the top 20 coal-producing countries to estimate global $CH_4$ emissions from coal mining. In any year the top 20 coal-producing countries accounted for greater than 95% of total production. Historical emission trends were calculated for the period 1950 to 1988. The data base on production, mining depths, and coal type were updated at 5-year intervals.

Because these emission factor data are derived from data obtained on small sections of cored material, they do not include the ''free'' gas component of coalbed $CH_4$ which is associated with macropores and fracture zones. Thus, our calculated emissions based on losses of adsorbed $CH_4$ are conservative estimates. An approach for including emissions of the heterogeneously distributed and less well-quantified free gas component of $CH_4$ losses from coal mining is discussed later in this chapter. Our rationale for emphasizing the ''conservative'' characteristic of our estimate is to address the public interest in near-term policy actions which reduce global warming potential. Estimates of all global sources, with the possible exception of $CO_2$ from combustion of commercial fossil fuels, are fraught with significant scientific undertainties. One criteria that can be used to develop rational policy initiatives is to be certain that a greenhouse gas source will not go below a certain level as estimates of sources improve. By selecting conservative choices at each decision point in our estimation procedure, we conclude that future estimates of $CH_4$ emissions from coal mining are very likely to be higher than our estimate. If research or mitigation actions are justified by the magnitude of our current estimate, the probability is high that improved estimates in the future will reinforce the course of action.

## III. RESULTS

Methane emissions for 5 of the 38 years studied are listed at decadal intervals (except for the 1988 column) in Table 1 to indicate trends. Since the top ten source countries account for over 90% of the global emissions in any year, we limit the list to these countries. (Data for the remaining source countries are available on request.) For the entire 1950 to 1988 interval, the 13 countries in Table 1 accounted for 91 to 95% of the total $CH_4$ emissions from coal production. Our results indicate that research to refine estimates of emissions, and initial negotiating of international agreements on source reduction strategies, could be productive if only three to ten of the top producers were participants. The factors which determined the $CH_4$ emissions from each country included a complex interaction of variables such as energy demand, estimated coal reserves, fuel prices and market dynamics, energy and environmental policies, and existing infrastructures for use of specific fuels. The geographical patterns of $CH_4$ emissions from coal production have changed with time (Figure 1), and can be expected to be dynamic in future years. The U.S. and Soviet Union have been major coal-related $CH_4$ producers throughout the 38-year study period. China has advanced from tenth place in 1950 to the world's leading coal-related $CH_4$ producer in 1988. These three countries have the world's largest remaining coal reserves.[11] China will play a particularly important role in future attempts to mitigate coal-related sources of atmospheric $CH_4$. Coal supplied 80% of China's total commercial energy supply in 1988. The Soviet Union and the U.S. utilized coal for 22 and 25%, respectively, of their commercial energy.[11] Large domestic coal reserves and the relatively low current market costs compared to alternative energy sources are strong incentives to continue and perhaps expand coal use for energy production.

The relative contribution of a greenhouse gas to the potential for climate change depends on a number of factors including emissions, the relative contribution of each gas to radiative forcing, and the removal rate from the atmosphere.[12] The combined effects of such factors

**TABLE 1**
**Trends in $CH_4$ Emissions Indicated by Annual Emissions at the Beginning of Each Decade[a] (1950 to 1988)**

| | Methane Emissions (Tg) | | | | |
|---|---|---|---|---|---|
| Country | 1950 | 1960 | 1970 | 1980 | 1988 |
| Australia | — | — | 0.26 | 0.45 | 0.50 |
| Belgium | 0.29 | — | — | — | — |
| China | 0.18 | 2.11 | 2.11 | 3.69 | 5.61 |
| Czechoslovakia | 0.19 | 0.27 | 0.29 | 0.29 | 0.27 |
| France | 0.46 | 0.50 | 0.33 | — | — |
| India | — | 0.24 | 0.34 | 0.46 | 0.68 |
| Japan | 0.20 | 0.38 | — | — | — |
| Poland | 0.58 | 0.77 | 1.04 | 1.43 | 1.42 |
| South Africa | — | — | — | 0.48 | 0.73 |
| Soviet Union | 1.74 | 2.78 | 3.47 | 3.91 | 4.13 |
| U.K. | 1.56 | 1.41 | 1.03 | 0.88 | 0.69 |
| U.S. | 2.27 | 1.67 | 2.14 | 2.47 | 2.79 |
| West Germany | 1.18 | 1.52 | 1.24 | 0.93 | 0.84 |
| % of world | 91 | 91 | 91 | 92 | 93 |

*Note:* The top ten producers account for >90% of world $CH_4$ emissions from coal mining in any year. The unit teragrams (Tg) = $10^{12}$ g.

[a] Except for the 1988 column.

determine the time history of the concentration anomalies in the atmosphere which actually produce the greenhouse effect. We determined the emissions history for coal-related $CH_4$ for each country with commercial coal production during the period 1950 to 1988. The cumulative emissions for the top ten producers for the entire study period are illustrated in Figure 2. For pollutant gases with relatively long atmospheric lifetimes, like $CH_4$, cumulative emissions should be used as an index of a country's contribution to potential global environmental change rather than current emissions.[12]

## IV. LIMITATIONS ON ESTIMATING ACTUAL EMISSIONS

To this point, we have used the extensive data on adsorbed $CH_4$ associated with coal to develop a conservative estimate of emissions to the atmosphere. Several factors can enhance $CH_4$ emissions associated with coal production above the amounts calculated from emission factors based only on adsorbed $CH_4$. For example, gas seepage along fracture surfaces in coalbeds, losses from coal left behind to provide structural support in the mine, and the diffusion of $CH_4$ from unmined coalbed materials into the lower pressure zones created by ventilated mine shafts can contribute to total mine mouth emissions.[13,14] However, estimating an upper boundary on potential $CH_4$ emissions associated with coal production is difficult due to limited data on sources and losses of "free $CH_4$" associated with macropores and fractures in coalbeds. The only study we have located which bears directly on this issue was a study of total $CH_4$ emissions from six operating mines in the U.S., the measured total emissions were six to nine times higher than predicted on the basis of the adsorbed $CH_4$ content of the coal.[14] Even these mine mouth measurements may have underestimated total $CH_4$ losses from produced coal. Mine coal is known to continue to emit $CH_4$ during transportation and storage.

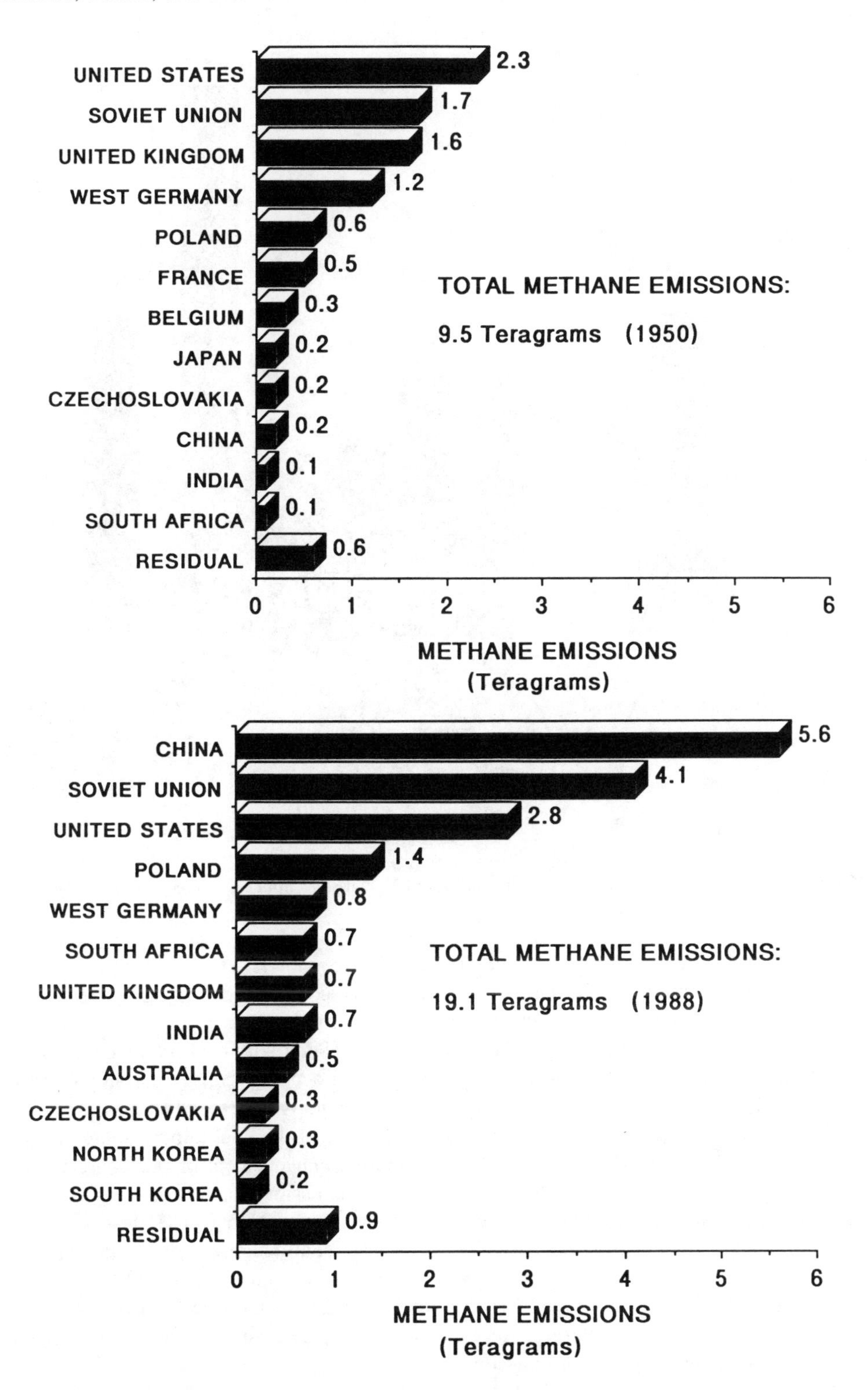


**FIGURE 1.** A comparison of lower limits of $CH_4$ emissions from coal mining for top 12 source countries. (A) 1950; (B) 1988. The residual emissions are a sum of all emissions from 8 additional countries which make up the top 20 $CH_4$ source countries. The top 20 countries account for more than 95% of global $CH_4$ emissions from coal mining in any year.

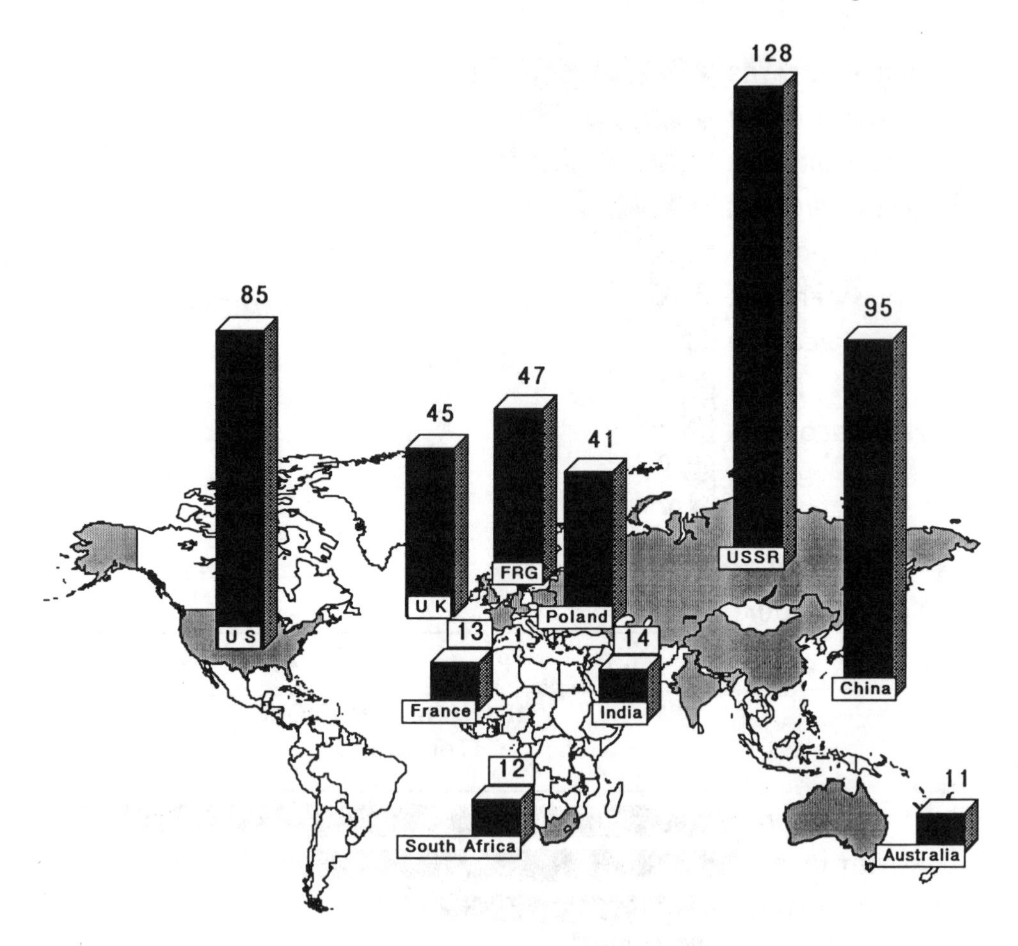


**FIGURE 2.** Comparison of the lower limits of cumulative $CH_4$ emissions from coal mining for the period 1950 to 1988 for the top ten source countries.

Factors such as the very limited data on total mine mouth $CH_4$ emissions, uncertainties associated with the application of data derived from U.S. coal to calculate global estimates, and changing coal mining technologies seriously compromise present efforts to proceed beyond our conservative estimate of $CH_4$ emissions associated with coal production. Large uncertainties in emissions of "free gas" preclude setting accurate constraints on an upper limit for $CH_4$ associated with coal mining. From a qualitative perspective it is unlikely, based on our current understanding of the $CH_4$ budget,[3] that the coal mining source could reach 114 to 171 Tg, which would be a six to ninefold enhancement of our conservative estimate. The few mines studied for total mine mouth emissions have been older, deeper mines with extensive underground tunneling which would exacerbate $CH_4$ losses. If we assume that approximately 12% of all mines are exceptionally "gassy", as is the case in the U.S.,[15] a more reasonable preliminary estimate of "total" emissions would be 30 to 37 Tg. Systematic measurements of total $CH_4$ emissions from a large number of mines, stratified according to geological characteristics and mining techniques used, are urgently needed to establish an improved estimate of emissions.

The ICF Resources, Inc., conducted an independent study for the U.S. Environmental Protection Agency (EPA) which estimated methane emissions from coal mining to be in the order of 45 Tg. However, the ICF calculation used a simplified approach of assuming one emission factor of 27.1 $m^3$/ton for all underground coal production, and a factor of 2.5

$m^3$/ton for surface coal production.[16] This approach does not even attempt to account for the well-documented variability of methane content in underground coal deposits. The EPA is currently attempting to organize a more comprehensive international program to both quantify and mitigate methane emissions from coal mining systems.[16] It is clear even from our conservative emissions estimate that reducing methane emissions from coal mining could contribute significantly to slowing the potential for global climate change.

Our systematic conservative estimates of $CH_4$ losses associated with coal production on a country basis lead to the following conclusions. Total $CH_4$ emissions to the atmosphere from coal production almost certainly exceeded 19 Tg in 1988. For the period 1950 to 1988, 13 countries were primarily responsible for this source of atmospheric $CH_4$ (see Table 1; Figure 2). In 1988, three countries (China, the Soviet Union, and the U.S.) produced 12.5 Tg $CH_4$, 66% of the estimated total source. These results indicate that research to further quantify $CH_4$ emissions related to coal production should emphasize these countries.

Hazard assessment studies by the U.S. Bureau of Mines have determined that as few as 25 of 200 underground mines may account for approximately 55% of the $CH_4$ emitted to the atmosphere by mine ventilation systems.[15] Thus, a reasonable possibility exists that a strategy for significant reductions in coal mining as a source of atmospheric $CH_4$ can be developed by as few as three to five major coal-producing countries, and might impact only a fraction of each country's industry.

In the U.S. 86% of coal consumption is currently associated with electrical power generation. In China and the Soviet Union the end uses of coal are more broadly spread across electrical power generation, industrial, and domestic consumption. Factors such as end uses, costs and availability of alternative fuels, and options for energy conservation will influence future attempts to reduce $CH_4$ emissions from coal mines as part of a strategy for reducing greenhouse gas emissions.

## V. CONCLUSIONS

In summary, a conservative approach to estimating $CH_4$ emissions to the atmosphere from world coal mining indicates a minimum of 19.1 mt in 1988. China, the Soviet Union, and the U.S. produced 66% of total emissions in 1988. For a longer period, 1950 to 1988, 13 countries produced 95% of $CH_4$ emissions from coal mining. The magnitude of this $CH_4$ source is large enough to justify consideration in strategies to stabilize the composition of the atmosphere and reduce global warming potential. Our results suggest that a strategy of selective closing of mines with the highest emission rates in the top three to ten coal-producing countries could make a significant contribution to reducing the potential for future global warming.

## ACKNOWLEDGMENTS

The authors thank Craig Rightmire, Dan Thompson, Whitney Telle, Jeff Levine, Pat Diamond, and John LaScola for their helpful comments and clarifications on the interpretation of data on $CH_4$ associated with coal. Professor Ralph Cicerone provided important comments on the manuscript. This research is supported by grants from the National Aeronautics and Space Administration (NASA).

# ENDNOTES


1. Schneider, S. H., The greenhouse effect: science and policy, *Science,* 243, 771–781, 1989.
2. Dickinson, R. E. and Cicerone, R. J., Future global warming from atmospheric trace gases, *Nature (London),* 319, 109–115, 1986.
3. Cicerone, R. J. and Oremland, R. S., Biogeochemical aspects of atmospheric methane, *Global Biogeochem. Cycles,* 2, 299–327, 1988.
4. Khalil, M. A. K. and Rasmussen, R. A., Atmospheric methane: recent global trends, *Environ. Sci. Technol.,* 24, 549–553, 1990.
5. Chappellaz, J., Barnola, J. M., Raynaud, D., Korotkevich, Y. S., and Lorius, C., Ice-core record of atmospheric methane over the past 160,000 years, *Nature (London),* 345, 127–131, 1990.
6. Khalil, M. A. K. and Rasmussen, R. A., Causes of increasing atmospheric methane: depletion of hydroxyl radicals and the rise of emissions, *Atmos. Environ.,* 19, 397–407, 1985.
7. Rightmire, C. T., *Coalbed Methane Resources of the United States,* Rightmire, C. T., Eddy, W. and Kirr, F., Eds., Am. Assoc. Petrol. Geol. Studies Geol., No. 17, Tulsa, OK, 1984, 1–13.
8. Kim, A. G., Experimental Studies on the Origin and Accumulation of Coalbed Gas, Rep. Investigation 8317, Bureau of Mines, U.S. Department of Interior, Washington, D.C., 1978.
9. Diamond, W. P. and Levine, J. R., Direct Method Determination of the Gas Content of Coal: Procedures and Results, Rep. Investigation 8515, Bureau of Mines, U.S. Department of Interior, Washington, D.C., 1981.
10. Using the data from papers cited in Reference 8, weighted by the predominance of bituminous coal which is approximately 66% of the total coal mined, and by Hilt's Law which correlates increasing rank with increasing depth, we adopted the following emission factors for loss of adsorbed $CH_4$ from mined coal: 2.11 kg$CH_4$/mt (kg/t) for the depth interval from the surface to 0.15 km, 5.27 kg/t for 0.15 to 0.30 km, 7.38 kg/t for 0.30 to 0.45 km, 9.48 kg/t for 0.45 to 0.60 km, and 10.54 kg/t for coal mined from depths greater than 0.60 km. An emission factor of 0.13 kg/t was used for soft coal (lignite), based on data in Schatzel, S. J., Hyman, D. M., Sainato, A., and Lascola, J. C., Methane Contents of Oil Shale from the Piceance Basin, Colorado, Rep. Investigation 9063, Bureau of Mines, U.S. Department of Interior, Washington, D.C., 1987.
11. British Petroleum Corp., *Statistical Review of World Energy,* London, 1989, 36 pp.
12. Rodhe, H., A comparison of the contribution of various gases to the greenhouse effect, *Science,* 248, 1217–1219, 1990.
13. Kissel, F. N., McCulloch, C. M., and Elder, C. H., The Direct Method of Determining Methane Contents of Coalbeds for Ventilation Design, Rep. Investigation 7767, Bureau of Mines, U.S. Department of Interior, Washington, D.C., 1973.
14. Irani, M. C., Thimons, E. D., Bobick, T. G., Deul, M., and Zabetakis, M. G., Methane Emission from U.S. Coal Mines: A Survey, Inf. Circular 8558, Bureau of Mines, U.S. Department of Interior, Washington, D.C., 1972.
15. Grau, R. H. and LaScola, J. C., Methane Emissions from U.S. Coal Mines in 1980, Inf. Circular 8987, Bureau of Mines, U.S. Department of Interior, Washington, D.C., 1984.
16. Kruger, D., U.S. Environmental Protection Agency, letter communication to Harriss, R., June 1991.

Chapter 16

# WHAT TO DO ABOUT GREENHOUSE WARMING: LOOK BEFORE YOU LEAP*

**S. Fred Singer, Roger Revelle, and Chauncey Starr**

## TABLE OF CONTENTS



* From Singer, S. F., Revelle, R., and Starr, C., *Cosmos,* Vol. 1, No. 1, 1991. With permission.

*Editor's Note:* Additional information on R. Revelle's ideas on the topic are available in his article published in *Oceanography,* Vol. 2 No. 2, pp 126–127, 1992. The article is based upon a Revelle session presentation February 16, 1990 at an A.A.A.S. meeting in New Orleans, LA.; the topic was "Climate Change: Scientific Uncertainties and Policy Responses".

# I. INTRODUCTION

Greenhouse warming has emerged as one of the most complex and controversial environmental foreign-policy issues of the 1990s.

It is an environmental issue because carbon dioxide ($CO_2$), generated from the burning of oil, gas, and coal, is thought to enhance (by trapping heat in the atmosphere) the natural greenhouse effect that has kept the planet warm for billions of years. Some scientists predict drastic climatic changes in the 21st Century.

It is a foreign-policy issue because, for a number of reasons, the U.S. has taken a more cautious approach to dealing with $CO_2$ emissions than have many industrialized nations. Wide acceptance of the Montreal Protocol, which limits and rolls back the manufacture of chlorofluorocarbons (CFCs) to protect the ozone layer, has encouraged environmental activists at international conferences the past three years to call for similar controls on $CO_2$ from fossil-fuel burning.

These activists have expressed disappointment with the White House for not supporting immediate action. But should the U.S. assume ''leadership'' in a hastily-conceived campaign that could cripple the global economy, or would it be more prudent to assure first, through scientific research, that the problem is both real and urgent?

We can sum up our conclusions in a simple message: *The scientific base for a greenhouse warming is too uncertain to justify drastic action at this time.*

There is little risk in delaying policy responses to this century-old problem since there is every expectation that scientific understanding will be substantially improved within the next decade. Instead of premature and likely ineffective controls on fuel use that would only slow down and not stop the further growth of $CO_2$, we may prefer to use the same resources — trillions of dollars, by some estimates — to increase our economic and technological resilience so that we can then apply specific remedies as necessary to reduce climate change or to adapt to it.

That is not to say that prudent steps cannot be taken now; indeed, many kinds of energy conservation and efficiency increases make economic sense even without the threat of greenhouse warming.

# II. THE SCIENTIFIC BASE

The scientific base for greenhouse warming (GHW) includes some facts, lots of uncertainty, and just plain lack of knowledge — requiring more observations, better theories, and more extensive calculations. Specifically, there are reliable measurements of the increase in so-called greenhouse gases (GHG) in the Earth's atmosphere, presumably as a result of human activities. There is uncertainty about the strength of sources and sinks for these gases, i.e., their rates of generation and removal. There is major uncertainty and disagreement about whether this increase has caused a change in the climate during the last century. There is also disagreement in the scientific community about predicted future changes as a result of further increases in GHG. The models used to calculate future climate are not yet good enough because the climate-balancing processes are not sufficiently understood, nor are they likely to be good enough until we gain more understanding through observations and experiments.

As a consequence, we cannot be sure whether the next century will bring a warming that is negligible or a warming that is significant. Finally, even if there are global warming and associated climate changes, it is debatable whether the consequences will be good or bad; likely some places on the planet would benefit, some would suffer.

## III. GREENHOUSE GASES

It has been common knowledge for about a century that the burning of fossil fuels would increase the normal atmospheric content of $CO_2$, causing an enhancement of the natural greenhouse effect and a possible warming of the global climate. Advances in spectroscopy in the last century produced evidence that $CO_2$ — and other molecules made up of more than two atoms — absorb IR radiation and thereby would impede the escape of such heat radiation from the Earth's surface. In fact, it is the greenhouse effect from naturally occurring $CO_2$ and water vapor ($H_2O$) that has warmed the Earth's surface for billions of years; withouth the natural greenhouse effect ours would be a frozen planet without life.

Adding in the other trace gases that produce greenhouse effects, we have already gone halfway to an effective GHG doubling — something that cannot be reversed in our lifetime — and, according to the prevailing theory, locked in a temperature increase of about 1.6°C.

Has there been a climate effect caused by the increase of GHG in the last decades? The data are ambiguous, to say the least. Advocates for immediate action profess to see a global warming of about 0.5°C since 1880, and point to record global temperatures in the 1980s, plus the warmest year on record in 1990. Most atmospheric scientists tend to be cautious, however, they call attention to the fact that the greatest temperature increase occurred *before* the major rise in GHG concentration. It was followed by a quarter-century decrease between 1940 and 1965, when concern arose about an approaching ice age! Following a sharp increase during 1975 to 1980, there has been no clear upward trend (Figure 1) during the 1980s, despite some very warm individual years and record GHG increases. Similarly, global atmospheric (rather than surface) temperatures measured by Tiros weather satellites show no trend in the last decade.

K. Hanson, T. Karl, and G. Maul, scientists of the National Oceanic and Atmospheric Administration (NOAA), find no overall warming in the U.S. temperature record, contrary to the global record assembled by J. Hansen, of the National Aeronautics and Space Administration (NASA). Using a technique that eliminates urban "heat islands" and other local distorting effects, the NOAA scientists confirm the temperature rise before 1940, followed, however, by a general decline. Newell and colleagues at the Massachusetts Institute of Technology (MIT) report no substantial change in the global sea-surface temperature in the South Indian Ocean and global network.

Precise measurements of the increase in atmospheric $CO_2$ date to the International Geophysical Year of 1957 to 1958. More recently, it has been discoverd that other GHG, i.e., gases that absorb strongly in the IR have also been increasing — at least partly as a result of human activities. These gases currently produce a greenhouse effect nearly equal to that of $CO_2$, and could soon outdistance it.

**Methane ($CH_4$)** — Is produced in large part by sources that relate to population growth; among these are rice paddies, cattle, landfills, forest fires, coal mines, and oil field operations. Indeed, methane, now 20% of the GHG effect and growing twice as fast as $CO_2$, has more than doubled since preindustrial times; it would soon become the most important GHG if $CO_2$ emissions were to stop.

**Nitrous oxide ($N_2O$)** — has increased by 10% and most likely because of soil bacterial action promoted by the increased use of nitrogen fertilizers.

**Ozone ($O_3$)** — From urban air pollution adds about 10% to the global greenhouse effect. It may decrease in the U.S. as a result of Clean Air legislation, but increase in other parts of the world.

**CFCs** — Manufactured for use in refrigeration, air conditioning, and industrial processes, are making an important contribution but will soon be replaced by less-polluting substitutes.

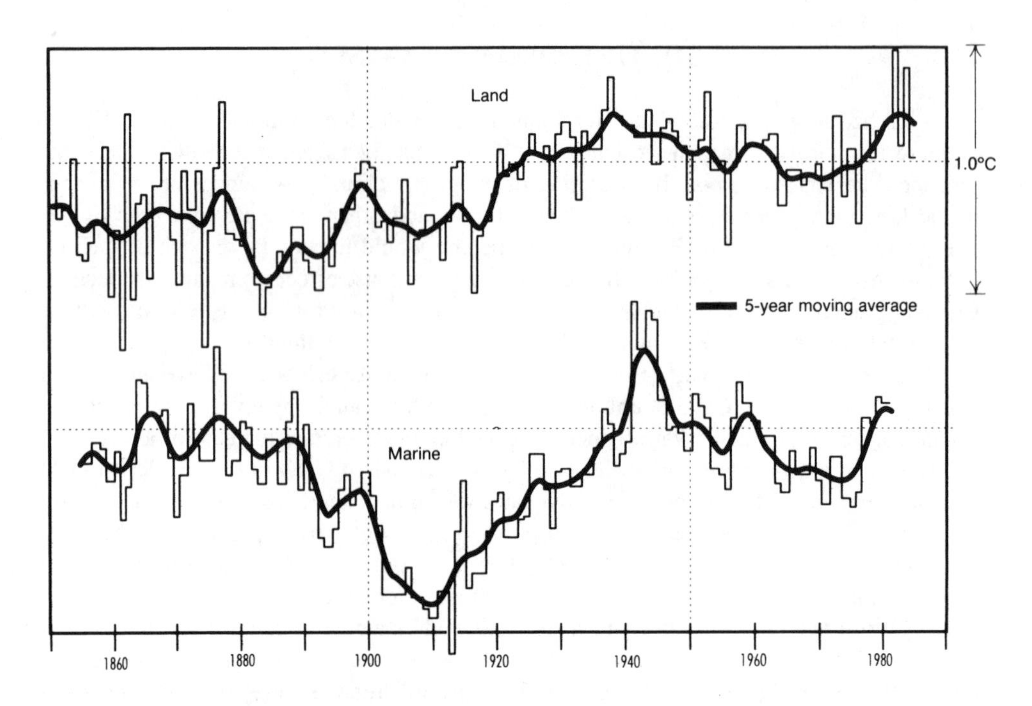


**FIGURE 1.** Yearly average temperatures in the Northern Hemisphere. Note the major increases before 1938, when GHG concentrations were low, and the sustained cooling to 1975. Although temperatures remained near record levels during the 1980s, there was no clear upward trend. Data from Jones, P. D., Wigley, T. M. L., and Wright, P. B., *Nature,* 322, 430, 1986. (From Singer, S. F., Revelle, R., and Starr, C., *Cosmos*, Vol. 1, No. 1, 1991. With permission.)

**Water vapor ($H_2O$)** — Turns out to be the most effective GHG by far. It is not man-made, but is assumed to amplify the warming effects of the gases produced by human activities. We don't really know whether $H_2O$ has increased in the atmosphere or whether it will increase in the future — although that is what all the model calculations assume. Indeed, predictions of future warming depend not only on the amount, but also on the horizontal and especially vertical distribution, of $H_2O$, and on whether it will be in the atmosphere in the form of a gas, or liquid cloud droplets, or ice particles. The current computer models are not complete enough to test these crucial points.

## IV. THE CLIMATE RECORD

The issue now is whether the 25% increase of $CO_2$ in the atmosphere, mainly since World War II, calls for immediate and drastic action to limit and roll back global energy use. Taking account increases of land temperatures in the past century; yet the ocean, because of its much greater heat inertia, should control any atmospheric climate change.

Perhaps most interesting are the NOAA studies that document a relative rise in night temperatures in the U.S. over the last 60 years, while daytime values stayed the same or declined. This is what one would expect from the increase in atmospheric GHG concentration, but its consequences, as University of Virginia climatologist Michaels and others have pointed out, are benign: a longer growing season, fewer frosts, and no increase in soil evaporation.

It is, therefore, fair to say that we haven't seen the huge GHW (of between 0.7 and 2.5°C) expected from the conventional theories. Why not? This scientific puzzle has many suggested solutions:

- The warming has been "soaked up" by the ocean and will appear after a delay of some decades. Plausible — but there is no evidence to support this theory until deep-ocean temperatures are measured on a routine basis, as suggested by Scripps Institution oceanographer W. Munk. Feasibility tests are currently underway, using a sound source at Heard Island and an array of microphones, but data over at least a decade will be needed to provide an answer.
- The warming has been overestimated by the existing models. Meteorologists H. Ellsaesser (Livermore National Laboratory) and R. Lindzen (MIT) propose that the models do not take proper account of tropical convection and thereby overestimate the amplifying effects of water vapor over this important part of the globe. Other atmospheric scientists suggest that the extent of cloudiness may increase as ocean temperatures try to rise and as evaporation increases. Clouds reflect incoming solar radiation; the resultant cooling could offset much of the GHW. Most intriguing has been the suggestion by British researchers that sulfates from smokestacks — the precursors of acid rain — may have played a role in producing an increase in bright stratocumulus clouds.
- The warming exists as predicted, but has been hidden by offsetting climate changes caused by volcanoes, solar variations, or other natural causes as yet unspecified — such as the cooling from an approaching ice age. (Some, like R. Balling of Arizona State University, consider the warming before 1940 to be a recovery from the "Little Ice Age" that prevailed from 1600 to about 1850; if correct, this would imply no net warming at all in the past century due to GHW.) Each hypothesis has vocal proponents — and opponents — in the scientific community, but the jury is out until better data become available.

## V. MATHEMATICAL MODELS

Indeed, there is much to complain about when it comes to predictions of future climate, but there is really no alternative to global climate models. "Models are better than hand-waving," claims S. Scheider of the National Center for Atmospheric Research (NCAR) — but how much better? Half a dozen of these general circulation models (GCM) are now running, mostly in the U.S. Even though they use similar basic atmospheric physics, they give different results. There is general agreement among them that there should be global warming but, with an effective GHG doubling, the calculated average global increase ranges between 1.5 and 4.5°C! These predicted values were unchanged for many years, then crept up and have recently dropped back to the lower end of the range. Just during 1989 some of the modelers cut their predictions in half as they tried to include clouds and ocean currents in a better way. Furthermore, there is serious disagreement among the models on the regional distribution of this warming and on where the increased precipitation will go.

The models are "tuned" to give the right mean temperature and seasonal temperature variation, but they fall short of modeling other important atmospheric processes, such as the poleward transport of energy via ocean currents and atmosphere from its source in the equatorial region. Nor do the models encompass longer-scale processes that involve the deep layers of the oceans or the ice and snow in the Earth's cryosphere, nor fine-scale processes that involve convection, cloud formation, boundary layers, or processes that depend on the Earth's detailed topography.

There are serious disagreements also between model results and the actual experience from the climate record of the past decade, according to Ellsaesser. Existing models retroactively predict a strong warming of the polar regions and the tropical upper atmosphere, and less warming in the Southern Hemisphere than the Northern — all contrary to observations. Yet there is hope that research, including satellite observations and ocean data, will provide many of the answers within this decade. Faster computers will also allow higher spatial resolution and incorporate the detailed and more complicated interactions that are now neglected.

## VI. IMPACTS OF CLIMATE CHANGE

Assume what we regard as the most likely outcome: a modest average warming in the next century — well below the normal year-to-year variation — and mostly at high latitudes and in the winter. Is this necessarily bad? One should recall perhaps that only a decade ago when climate *cooling* was a looming issue, economists of the National Academy of Sciences' National Research Council calculated a huge national cost associated with such cooling. More to the point perhaps, actual climate cooling, experienced during the Little Ice Age or in the famous 1816 New England "year without a summer", caused large agricultural losses and even famines.

If cooling is bad, then warming should be good, it would seem — provided the warming is slow enough so that adjustment is easy and relatively cost free. Even though crop varieties are available that can benefit from higher temperatures with either more or less moisture, the soils themselves may not be able to adjust that quickly. Agriculturalists, like S. Idso of the U.S. Department of Agriculture, and Yale professor W. Reifsnyder, generally expect that with increased atmospheric $CO_2$ — which is, after all, plant food — plants will grow faster and need less water. The warmer night temperatures suggested by P. Michaels, using the data of T. Karl, translate to longer growing seasons and fewer frosts. Increased global precipitation should also be beneficial to plant growth.

Keep in mind also that year-to-year changes at any location are far greater and more rapid than what might be expected from GHW; nature, crops, and people are already adapted to such changes. It is the extreme climate events that cause the great ecological and economic problems: crippling winters, persistent droughts, extreme heat spells, killer hurricanes, etc. There is no indication from modeling or actual experience that such extreme events would become more frequent if GHW becomes appreciable. The exception might be tropical cyclones, which — R. Balling and R. Cerveney argue — would be more frequent but weaker, would cool vast areas of the ocean surface, and increase annual rainfall. In summary, climate models predict that global precipitation should increase by 10 to 15%, and polar temperatures should warm the most, thus reducing the driving force for severe winter-weather conditions.

There is finally the question of sea-level rise as glaciers melt — and fear of catastrophic flooding. The cryosphere certainly contains enough ice to raise sea level by 100 m; and, conversely, during recent ice ages, enough ice accumulated to drop sea level 100 m below the present value. These are extreme possibilities; tidal-gauge records of the past century suggest that sea level has risen modestly, about 0.3 m, but the gauges measure only relative sea level, and many of the gauge locations have dropped because of land subsidence. Besides, the test locations are too highly concentrated geographically, mostly on the U.S. East Coast, to permit global conclusions. The situation will improve greatly, however, in the next few years as precise absolute global data become available from a variety of satellite systems.

In the meantime, satellite radar-altimeters have already given a surprising result. As reported by NASA scientist J. Zwally in *Science*, Greenland ice sheets are gaining in thickness - a net increase in the ice stored in the cryosphere and an inferred drop in sea level — leading to somewhat uncertain predictions about future seal level. Modeling results suggest little warming of the Antarctic Ocean because the heat is convected to deeper levels. It is clearly important to verify these results by other techniques, and to also get more direct data on current sea-level changes.

Summarizing the available evidence, we conclude that even if significant warming were to occur in the next century, the net impact to the entire planet may well be beneficial — with some regions enjoying improved climate and some encountering worse. This would be even more true if the long-anticipated ice age were on its way.

In view of the uncertainties about the degree of warming, and the even greater uncertainty about its possible impact — what should we do? During the time that an expanded research program reduces or eliminates these uncertainties, we can be putting into effect policies and pursuing approaches that make sense even if the greenhouse effect did not exist.

## VII. ENERGY POLICIES

**Conserve energy by discouraging wasteful use globally** — Conservation can best be achieved by pricing rather than by command-and-control methods. If the price can include the external costs that are avoided by the user and loaded onto someone else, this strengthens the argument for proper pricing. The idea is to have the polluter or the beneficiary pay the cost. An example would be peak-pricing for electric power. Yet another example, appropriate to the greenhouse discussion, is to increase the tax on gasoline to make it a true highway-user fee — instead of having most capital and maintenance costs paid by various state taxes, as is done now. Congress has lacked the courage for such a direct approach, preferring instead regulation that is mostly ineffective and produces large indirect costs for the consumer.

**Improve efficiency in energy use** — Energy efficiency should be attainable without much intervention, provided it pays for itself. A good rule of thumb: if it isn't economic, then it probably wastes energy in the process and we shouldn't be doing it. Over-conservation can waste as much energy as under-conservation. (For example, destroying all older cars would certainly raise the fuel efficiency of the fleet, but replacing these cars would consume more energy in their manufacture.) If energy is properly priced, i.e., not subsidized, the job for government is to remove the institutional and other road blocks:

- Provide information to consumers, especially on life-cycle costs of home heating, lighting, refrigerators, and other appliances.
- Encourage — but not force — the turnover and replacement of older, less efficient (and often more polluting) capital equipment: cars, machinery, power plants. Some existing policies that make new equipment too costly go counter to this goal.
- Stimulate the development of more efficient system, such as a combined-cycle power plant or a more efficient internal-combustion engine.

**Use nonfossil-fuel energy sources wherever this makes economic sense** — Nuclear power is competitive now, and in many countries is cheaper than fossil-fuel power — yet it is often opposed on environmental grounds. The problems cited against nuclear energy, such as disposal of spent nuclear fuel, are more political and psychological than technical. To address safety concerns, nuclear engineers are focusing on an ''inherently'' safe reactor. Nuclear energy from fusion rather than from fission may be a longer-term possibility, but the time horizon is uncertain.

Solar energy, and other forms of renewable energy, should also become more competitive as their costs drop and as fossil-fuel prices rise. Solar energy applications are restricted not only by cost; solar energy is both highly variable and very dilute; it takes a football field of solar cells to supply the total energy allocated to the average U.S. household. Wind energy and biomass are other forms of solar energy, competitive in certain applications.

Schemes to extract energy from temperature differences in the ocean have been suggested as inexhaustible sources of nonpolluting hydrogen fuel, once we solve the daunting technical problems.

## VIII. DIRECT INTERVENTIONS

If GHW ever becomes a problem, there are a number of proposals for removing $CO_2$ from the atmosphere. Rebuilding forests is widely talked about, but may not be cost-effective; yet natural expansion of boreal forests — those in high latitude regions — in a warming climate would sequester atmospheric $CO_2$. A novel idea, proposed by California oceanographer J. Martin, is to fertilize the Antarctic Ocean and let plankton growth do the job of converting $CO_2$ into biomaterial. The limiting trace nutrient may be iron, which could be supplied and dispersed economically.

If all else fails, there is always the possibility of putting "venetian blind" satellites into Earth orbit to modulate the amount of sunshine reaching the Earth. These satellites could also generate electric power and beam it to the Earth, as originally suggested by P. Glaser of A. D. Little. Such schemes may sound farfetched, but so did many other futuristic projects in the past — and in the present, like covering the Sahara with solar cells or Australia with trees.

## IX. CONCLUSION

Drastic, precipitous — and, especially, unilateral — steps to delay the putative greenhouse impacts can cost jobs and prosperity and increase the human costs of global poverty, without being effective. Stringent controls enacted now would be economically devastating — particularly for developing countries for whom reduced energy consumption would mean slower rates of economic growth — without being able to greatly delay the growth of GHG in the atmosphere. Yale economist, W. Nordhaus, one of the few who has been trying to deal quantitatively with the economics of the greenhouse effect, has pointed out that ". . . Those who argue for strong measures to slow greenhouse warming have reached their conclusion without any discernible analysis of the costs and benefits . . . ." It would be prudent to complete the ongoing and recently expanded research so that we will know what we are doing before we act. "Look before you leap" may still be good advice.

## APPENDIX A: "ANOTHER ICE AGE COMING?"

Global temperatures have been declining since the dinosaurs roamed the Earth some 70 million years ago. About 2 million years ago, a new "ice age" began — most probably as a result of the drift of the continents and the buildup of mountains. Since that time, the Earth has seen 17 or more cycles of glaciation, interrupted by short (10,000 to 20,000 years) interglacial or warm periods. We are now in such an interglacial interval, the Holocene, that started 10,800 years ago. The onset of the next glacial cycle cannot be very far away.

It is believed that the length of a glaciation cycle, about 100,000 to 120,000 years, is controlled by small changes in the seasonal and latitudinal distribution of solar energy received as a result of changes in the Earth's orbit and spin axis. While the theory can explain the timing, the detailed mechanism is not well understood — especially the sudden transition from full glacial to interglacial warming. Very likely an ocean-atmosphere interaction is triggered and becomes the direct cause of the transition in climate.

The climate record also reveals evidence for major climatic changes on time scales shorter than those for astronomical cycles. During the past millennium, the Earth experienced a ''climate optimum'' around 1100 A.D., when Vikings found Greenland to be green and Vinland (Labrador?) able to support grape growing. The ''Little Ice Age'' found European glaciers advancing well before 1600 and suddenly retreating, starting in 1860. The warming reported in the global temperature record since 1880 may thus simply be the escape from this Little Ice Age rather than our entrance into the human greenhouse.

Chapter 17

# ATMOSPHERIC TRACE GASES: SOURCES, SINKS, AND ROLE IN GLOBAL CHANGE

**M. A. K. Khalil**

## TABLE OF CONTENTS



0-8493-4419-0/93/$0.00 + $.50
© 1993 by CRC Press, Inc.

# I. INTRODUCTION

More than 99.9% of the Earth's (dry) atmosphere is nitrogen ($N_2$), oxygen ($O_2$), and argon (Ar). Carbon dioxide ($CO_2$) makes up about 0.034% of the atmosphere; the rest of the gases exist at exceedingly small concentrations. It is remarkable that such trace amounts of methane ($CH_4$), nitrous oxide ($N_2O$), carbon monoxide (CO), ozone ($O_3$), and exotic fluorocarbons can have a perceptible affect on the Earth's environment. This chapter is about the long-lived atmospheric trace gases that are expected to have the greatest influence on the future climate and global change caused by human activities.

Although there are many environmental problems of global magnitude, only two are directly related to atmospheric pollution. The first is the potential for global warming caused by the buildup of trace gases from various human activities, and the second is the possibility of stratospheric ozone depletion caused by emissions of exotic man-made chlorofluorocarbons (CFCs) and other gases. The major gases involved in both these effects are the same — namely $CO_2$, $CH_4$, $N_2O$, $CCl_3F$, and $CCl_2F_2$. These gases are prototypes for a large number of man-made and natural trace gases that have similar environmental effects, but exist at such low concentrations or are increasing so slowly that they are not likely to have perceptible effects in the future. The five gases ($CO_2$, $CH_4$, $N_2O$, $CCl_3F$, and $CCl_2F_2$) are the most important because concentrations are rising and have already reached levels that may affect the environment. Moreover, the uses of these gases are such as to require continuing or increasing emissions for a long time to come. The CFCs are now controlled by the "Montreal Protocol", which is an international agreement designed to greatly reduce and perhaps even eliminate their production as we enter the 21st century. Because of the important industrial and consumer uses of these compounds, substitutes have to be developed that may bring new environmental and safety concerns. The five gases mentioned here, and some others, will be considered in more detail in this chapter.

Figure 1 shows the connections among the key trace gases and their potential environmental effects. $CO_2$ is likely to be a major contributor to future global warming. $CH_4$, $N_2O$, and the two main CFCs may together have a warming affect comparable to $CO_2$.

Figure 1 also shows that the gases involved in global warming are also effective in the chemistry of stratospheric ozone. The CFCs and $N_2O$ destroy ozone. Global warming at the Earth's surface is likely to cool the upper stratosphere, which will have a net effect of increasing stratospheric ozone concentrations by slowing down chemical reactions. $CH_4$ increases have a complex effect on stratospheric ozone (see Section II.B.).

Hydroxyl (OH) is the key oxidizing radical that removes many man-made and natural trace gases from the atmosphere. Atmospheric concentrations of CO, $CH_4$, $O_3$, and OH are inextricably tied together. The formation of OH requires $O_3$, water vapor, and sunlight; OH is destroyed mostly by reacting with $CH_4$ and CO.

$CH_4$ oxidation initiated by reaction with OH eventually produces CO. Increases of CO from its many sources and $CH_4$ can have an indirect effect on the environment by depleting tropospheric OH radicals and thus, reducing the oxidizing capacity of the Earth's atmosphere. If this happens, many natural and man-made trace gases will accumulate in the atmosphere and affect global climate and atmospheric chemistry. The increase of $CH_4$, for instance, would speed up if OH radicals are depleted. CO and $CH_4$ also produce $O_3$, which is itself potent in global warming.

Methylchloroform ($CH_3CCl_3$) is an industrial degreasing solvent. By itself it has a small effect on global warming or ozone depletion. It is, however, a unique indicator of OH concentrations. For these reasons, the global balances of CO and $CH_3CCl_3$ are included here (see Section III.F).

$O_3$ is an important atmospheric species that has many roles. In urban environments it is a pollutant that can adversely affect human health and is, therefore, monitored and its

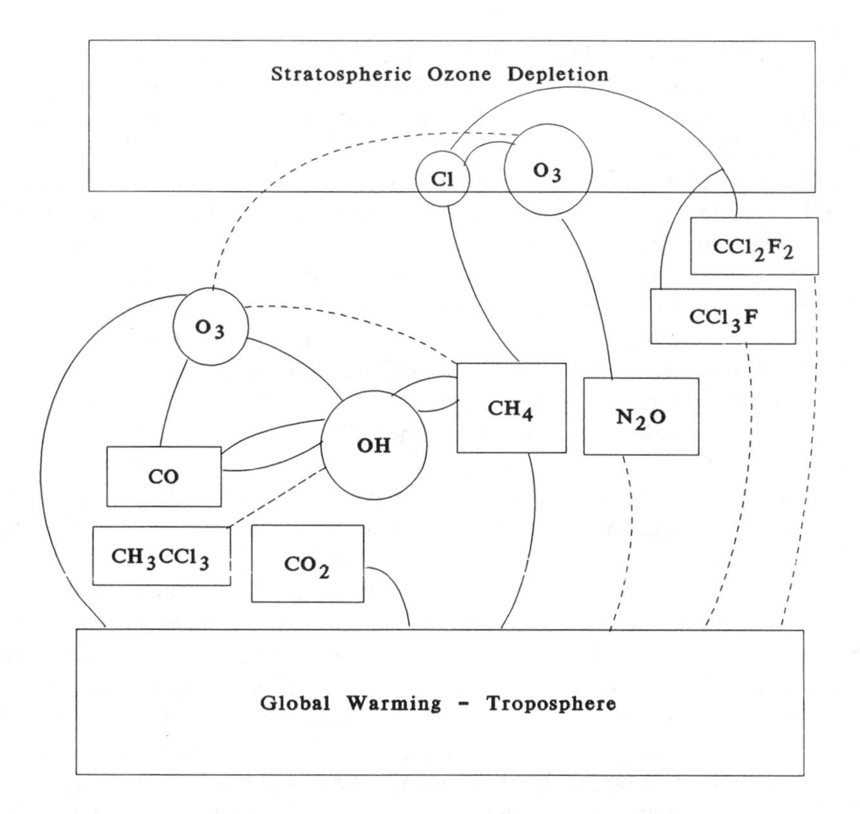


**FIGURE 1.** Atmospheric trace gases that may affect future global climate. The stable gases are in rectangles; the key reactive species in the global atmosphere are shown in circles. The connections between the various gases are shown by connecting lines.

precursors are controlled in industrial countries. In the troposphere, it is needed to make most of the OH radicals that oxidize and remove hundreds of pollutants. It is also a powerful absorber of the Earth's heat and thus contributes to the greenhouse effect, which in its natural manifestation is greatly beneficial to life on Earth. Finally, in the stratosphere, $O_3$ filters out UV radiation and prevents it from reaching the Earth's surface, thus protecting life from the ill affects of high energy solar radiation (see WMO, 1985, 1988, 1989).

Section II discusses the atmospheric behavior of trace gases, specifically the greenhouse effect and the balance of stratospheric $O_3$. Changes in these characteristics of the Earth's atmosphere are the cause of concern about the future of the global environment. Section III is a discussion of the global sources, emission rates, sinks (removal processes), removal rates, atmospheric concentrations, and trends of the key trace gases. Section IV contains comparisons of the trends and concentrations of the gases. It also includes a table that summarizes the current knowledge of atmospheric trace gases believed to be most important in global change. The chapter concludes with Section V. All $\pm$ values quoted are estimates of 90% confidence limits.

## II. ATMOSPHERIC BEHAVIOR OF TRACE GASES

### A. THE GREENHOUSE EFFECT

Energy the Earth receives from the sun is balanced by the energy radiated from the Earth into outer space. The incoming solar energy is in the form of electromagnetic radiation of wavelengths between 0.3 and 4 μm, the outgoing electromagnetic energy radiated by the

Earth is in the range of frequencies from 4 to 100 μm; both functions are essentially described by Planck's black body radiation law (see Fleagle and Businger, 1980; Dickinson and Cicerone, 1986). Atmospheric constituents absorb very little of the incoming radiation — most reaches the lower atmosphere and the Earth's surface where a large fraction is absorbed mostly at the ground while the rest is reflected, mostly by clouds. The energy radiated by the Earth (IR radiation) is readily absorbed by atmospheric constituents. The most important absorbers of the Earth's heat are $H_2O$, $O_3$, and $CO_2$. Other gases such as $CH_4$ and $N_2O$ also add to this natural greenhouse effect. Most of these gases are increasing because of human activities, thus leading to an increase in the heat-absorbing ability of the atmosphere, which will ultimately lead to global warming. Some believe that the accumulations of these gases have already caused a warming ( ~0.6°C) of the Earth over the last century. In addition to these gases, human activities have added exotic CFCs such as $CCl_3F$ (F-11) and $CCl_2F_2$ (F-12) and many other gases to the atmosphere. These CFCs and other gases also contribute to the greenhouse effect (Ramanathan et al., 1985; Wang and Molnar, 1985).

There are two noteworthy factors that complicate the relationship between trace gas increases and global warming. First, each gas absorbs radiation in characteristic bands determined by the quantum mechanical properties of the gas molecules. As the concentration of the gas rises, it absorbs almost all the radiation in its characteristic bands so that further increases of concentration cause no more warming. This is the "saturation effect", which acts to reduce the efficiency of a gas for causing global warming. When the concentration of the gas is very low, increases in concentration lead to proportional increases in the absorption of IR radiation in the characteristic bands; at high concentrations the increase of absorption is essentially a logarithmic function of increasing concentration. Along similar lines, the characteristic absorption bands of one gas may overlap with the bands of another gas. This is called "band overlap" and has the same effect as saturation. For instance, addition of man-made trace gases that absorb IR radiation only in the bands that are already saturated by naturally occurring species such as $H_2O$, $CO_2$, and $O_3$ will not cause substantial global warming. The case of the chlorofluorocarbons F-11 and F-12 is the opposite — they absorb in a region where there is little absorption by any natural gas. This makes the fluorocarbons, molecule for molecule, among the most potent gases for causing global warming. The part of the Earth's radiation spectrum where there is little absorption of IR radiation lies between 8 and 12 μm and is called the "window region".

The second effect is the "thermal inertia" of the oceans. As trace gases increase radiative forcing, the oceans warm very slowly because of the large heat capacity of water. Heat that would warm the surface is therefore taken up to heat the water. This effect slows down and even buffers global warming from increasing concentrations of trace gases. Similarly, if circumstances change and the atmosphere begins to absorb less of the Earth's radiation each year, the warmed up oceans will delay global cooling.

The global climate and warming are affected by many other processes that include transport of energy by the circulations of the atmosphere and oceans, the dynamics of ice sheets, and the properties of clouds. These processes are included in a number of mathematical models that are solved numerically on modern computers. Perhaps the simplest such models are the one-dimensional (vertical) radiative convective models, commonly used for studying the global warming from trace gases. A detailed description and results of one such model are given by MacKay and Khalil (1991).

There are a number of ways to assess the relative importance of trace gases for causing global warming. In all such assessments, $CO_2$ is the largest contributor, followed by $CH_4$. When taken together, the two fluorocarbons (F-11 and F-12), $CH_4$, and $N_2O$ are 50 to 70% of the effect of $CO_2$. There are several factors that complicate the comparison of the relative importance of trace gases in causing global warming. In addition to the physical effects of band overlaps and saturation effects, the ultimate importance of the trace gases also depends on the rates of accumulation and atmospheric lifetimes.

One way to look at the relative importance of trace gases is to consider the global warming effect for each additional mole of the gas added to the atmosphere. On this basis, each molecule of $CH_4$ is about 20 to 30 times more effective in causing global warming than $CO_2$; each molecule of $N_2O$ is 150 to 200 times more effective than $CO_2$; each molecule of F-11 and F-12 is 12,000 and 15,000 times more effective, respectively (based on Lacis et al., 1981; Dickinson and Cicerone, 1986; MacKay and Khalil, 1991). This view is instructive because it shows why the CFCs should be considered in the global warming problem — the very small concentrations and accumulation rates in number of extra molecules added to the atmosphere every year are compensated by the enormous effectiveness of each molecule to trap the Earth's radiation.

Taking into account the accumulation rates provides another comparison. First, the annual increase in radiative forcing ($dQ_i/dt$) from the increasing concentrations of a trace gas "i" is calculated as the product of the global warming from a trace gas ($\partial Q/\partial C_i$) in $W/m^2/ppmv$ at present concentrations and the rate of increase ($dC_i/dt$) in ppmv/yr.

$$dQ_i/dt = (\partial Q/\partial C_i)_{C_p} \cdot dC_i/dt \quad (1)$$

The importance of a trace gas relative to $CO_2$ is then the ratio $R_i$ of this quantity for any gas "i" to $CO_2$, evaluated at present concentration ($C_p$) and rates of change:

$$R_i = (dQ_i/dt)/(dQ_{CO_2}/dt) \quad (2)$$

By this method, each year $CH_4$ adds 30% as much radiative forcing as $CO_2$, $N_2O$ adds 8%, F-11 adds 7%, and F-12 adds 14%, all relative to $CO_2$ (based on the results in Dickinson and Cicerone, 1986). In this calculation, increases of the four trace gases taken together are adding about 60% as much radiative forcing each year as the increases of $CO_2$. Similar conclusions can also be derived from the results in Lacis et al. (1981) — the four trace gases are about 50% as effective as $CO_2$ at present concentrations and trends. Readers may reproduce these calculations by using the model of MacKay and Khalil (1991).

Over the decade of 1970 to 1980, Lacis et al. (1981) estimate that the combined effect of $CH_4$, $N_2O$, F-11, and F-12 was about 70% as large as the effect of $CO_2$, based on equilibrium global warming calculations. Over the last century the combined effect from increases of the major trace gases is about 60% as much as the effect from the increase of $CO_2$ alone (Dickinson and Cicerone, 1986; MacKay and Khalil, 1991). All these calculations suggest the trace gases, particularly $CH_4$, $N_2O$, F-11, and F-12, are together almost as important as $CO_2$ in causing global warming at present and past concentrations and rates of increase (see also Ramanathan et al., 1985).

Finally there is yet another view that accounts for the differences of the atmospheric lifetimes and possible indirect effects and leads to "global warming potentials". In this view, an integrated effect of each gas is compared to the integrated effect of $CO_2$. A mole of $CO_2$ once emitted into the atmosphere causes substantial global warming year after year for many years until most of it is removed from the atmosphere. Releasing a mole of a shorter-lived gas such as $CH_4$ does not continue to cause global warming for as long as a mole of $CO_2$. In this comparison, the longer lifetime makes a gas more potent for global warming compared to a shorter-lived gas of otherwise similar characteristics (Lashof and Ahuja, 1990).

## B. OZONE DEPLETION

In the stratosphere, $O_3$ is formed by reactions of $O_2$, first delineated by Chapman (1930).

$$O_2 + h\nu \rightarrow O(^3P) + O(^3P)$$

$$O + O_2 + M \rightarrow O_3 + M$$

$$O + O_3 \rightarrow O_2 + O_2$$

$$O_3 + h\nu \rightarrow O + O_2$$

Calculations based on the Chapman Equations, however, lead to more $O_3$ than is observed. Much of the remaining chemistry of the $O_3$ layer is controlled by catalytic destruction processes of the following type:

$$X + O_3 \rightarrow XO + O_2$$

$$XO + O \rightarrow X + O_2$$

The net effect of such reactions is that an $O_3$ molecule and an O atom are converted to $O_2$ while the catalyst X is recovered, which goes on to attack another $O_3$. X can be NO, Cl, Br, or H and OH ($HO_2$ also directly reacts with $O_3$ and recovers an OH through the reaction $HO_2 + O_3 \rightarrow OH + HO_2$). X continues to destroy $O_3$ until it is sequestered by some reaction that creates a stable chemical compound that can be transported to the troposphere and removed by rain or another process.

There are many anthropogenic sources of the radical species here designated as X. Substantial amounts of Cl may come from the man-made CFCs, particularly $CCl_3F$ (F-11) and $CCl_2F_2$ (F-12), NO may come from the oxidation of $N_2O$ $O(^1D) + N_2O \rightarrow NO + NO$ and also $O(^3P) + N_2O \rightarrow NO + NO$, and HOx may come from the oxidation of $CH_4$.

$CH_4$ also acts to terminate the catalytic destruction of stratospheric $O_3$ by CFCs through the reaction: $CH_4 + Cl \rightarrow HCL + CH_3$. The HCl formed by this and other similar processes is quite stable and eventually is transported to the troposphere where it is removed by rain. A fuller description of stratospheric $O_3$ chemistry is given by Chamberlain and Hunten (1987).

The upshot of these processes is that man-made emissions of CFCs, including F-11 and F-12, and of $N_2O$, are likely to reduce stratospheric ozone. $CH_4$ emissions also will affect stratospheric $O_3$, including some protection of the $O_3$ layer from the effects of the CFCs. Even increased $CO_2$ can have an indirect affect on stratospheric $O_3$ by cooling the stratosphere (Groves and Tucks, 1979; Haigh and Pyle, 1979). This cooling would slow down the catalytic destruction of $O_3$, thus buffering the effects of increasing CFCs and $N_2O$.

Both global warming and $O_3$ depletion depend on the amounts of the trace gases that accumulate in the atmosphere. The accumulation of gases is governed by rules for the global mass balances, which are discussed in the following section.

## C. SOME RULES FOR GLOBAL BALANCES OF TRACE GASES

**Mass balances** — The main components that determine the atmospheric concentrations of a trace gas are summarized in a mass balance equation:

$$\frac{\partial}{\partial t} C(\underline{x},t) = S(\underline{x},t) - L(\underline{x},t) - \frac{1}{n} \underline{\nabla} \cdot n(C\underline{V} - \overleftrightarrow{K} \cdot \underline{\nabla} C) \tag{3}$$

C is the concentration expressed as the mixing ratio, at a point $\underline{x}$ and time t, S is the emissions of the trace gas from natural or anthropogenic sources emitted either directly into the atmosphere or produced by atmospheric chemical processes, L is the losses from chemical, deposition, or other processes, and the last term is the transport by atmospheric winds ($\underline{V}$) and turbulent processes (K). This equation can be cast into forms that incorporate various degrees of complexity (Khalil and Rasmussen, 1984a). The simplest manifestation of Equation 3 is obtained by taking an average over the whole atmosphere (over $\underline{x}$); then it becomes:

$$dC/dt = S - C/\tau \quad (4)$$

In Equation 4, C is the average mixing ratio, which is the number of molecules of the trace gas in atmosphere divided by the total number of molecules of air in the atmosphere (about $10^{44}$), S is the total emissions of the gas from all sources, man-made or natural (mixing ratio/yr), and $\tau$ is the atmospheric lifetime in years. In the transition from Equation 3 to Equation 4, $L(\underline{x},t)$ is approximated by $C/\tau$, since most processes that remove long-lived trace gases from the atmosphere tend to be proportional to the amount of the gas present. While Equation 4 is easy to use, a considerable amount of information has been lost in the transition from Equation 3.

**Causes of trends** — According to Equation 4, a trace gas can increase only if the emissions exceed the losses persistently over many years. For gases that have natural sources and once were in balance, increases occur either if the sources start increasing or if the lifetime gets longer. Equation 4 also shows that four pieces of information, namely the trend, concentration, lifetime, and emissions, are needed to define the balance of a trace gas in the Earth's atmosphere. This balance is the foundation for understanding the chemical and physical characteristics of a trace gas and its potential for affecting the global environment. We would know very little about a gas if two or more of these pieces of information were missing.

The lifetime is perhaps the single most important number that determines the behavior of a trace gas. It represents the reactivity of a gas in the atmosphere, particularly in the troposphere, and it determines the role of the gas in the global environment. Without an accurate knowledge of the lifetime, it is practically impossible to design a plan to control future concentrations of a trace gas.

**Urban and global pollution** — City dwellers are well aware of urban air pollution characterized by high levels of $O_3$ and irritating smog. Global air pollution, on the other hand, is much more subtle and not yet perceptible. The trace gases involved in global and urban pollution are quite different — almost mutually exclusive. The reason is that short-lived trace gases emitted directly from sources rarely affect the global environment, whereas long-lived gases rarely affect urban air pollution. Gases such as some hydrocarbons react rapidly and may therefore produce $O_3$ and other familiar forms of urban air pollution. The gases are depleted before the winds and mixing processes can move them across the hemispheres.

As a rule then, only long-lived gases have the potential for affecting the global environment. For long-lived trace gases, even modest annual emission rates can lead to large ultimate concentrations in the atmosphere (see Equation 4). The large tropospheric concentrations in turn are needed to significantly cause global warming or upset the balance of stratospheric ozone. The only exceptions are gases that are formed from the chemical reactions in the atmosphere. In such cases, short-lived trace gases may occur at high concentrations far from the sources of man-made pollution. CO is an example of such a gas.

The lifetime also controls the global distributions and atmospheric variability of trace gases. For very short-lived gases ($\tau <$ a week or so), the concentrations are determined by local sources and sinks according to Equation 3, since transport is not an effective mechanism to bring the gas to a particular location (chemistry-controlled, e.g., reactive hydrocarbons, OH). For very long-lived gases, the atmospheric mixing processes cause the gas to become uniformly distributed throughout the atmosphere, with relatively small differences of concentrations from place to place; local sources and sinks have a small effect on the concentrations (transport-controlled, e.g., $N_2O$, F-11, F-12).

**Variability and lifetime** — The variability of trace gas, with some restrictions, is inversely proportional to the atmospheric lifetime. This rule was proposed by Junge, who defined variability as the ratio of standard deviation (SD) of measured concentrations to the mean concentration at locations far from the sources. The rule applies to many other forms of variability. This effect arises principally because the ambient concentrations of long-lived gases are the result of many years of accumulated emissions. Any changes in the emissions, removal, or transport that cause variability are small compared to the amount of the trace gas already present in the air at any given time. In the context of this chapter, the main effect of variability is to mask significant trends in trace gases caused by anthropogenic sources. Therefore, increasing trends of shorter-lived gases are harder to detect and quantify than trends of longer-lived gases. CO and $H_2$ are examples of gases for which increasing trends should exist but are barely observable even after many years of intensive measurements.

**Additive effects** — There is one more rule, easily overlooked, that governs the importance of trace gases in the global environment. Trace gases often individually contribute little to global warming or even to stratospheric $O_3$ depletion. Their collective effects, however, can be substantial. Because of this effect, the individual gases appear to be quite innocuous, but unfortunately there are too many such gases.

The rules discussed here can be used to construct self-consistent global mass balances that tie together the trends, emissions, and lifetimes of $CO_2$ and each of the trace gases $CH_4$, $N_2O$, $CCl_3F$, $CCl_2F_2$, $CH_3CCl_3$, and CO. The various pieces of the mass balances of these trace gases will be discussed next.

## III. GLOBAL BUDGETS: CONCENTRATIONS, TRENDS, EMISSIONS, AND ATMOSPHERIC LIFETIMES

### A. CARBON DIOXIDE

Unlike the other gases discussed here, every year natural processes cause enormous exchanges of $CO_2$ between its various reservoirs. The most significant exchanges over short time scales are between the ocean mixed layer and the atmosphere, and the biota and the atmosphere (see Warneck, 1988; Chamberlain and Hunten, 1987; Wuebbles and Edmonds, 1991).

**Concentrations** — The concentrations of $CO_2$ undergo seasonal variations. The average cycles at Mauna Loa Observatory, Hawaii, and the South Pole station are shown in Figure 2. After this average seasonal cycle is subtracted from the data, the remaining deseasonalized concentrations are shown in Figure 3 (see Keeling et al., 1982; data shown here were achieved at CDIAC (1990) where additional primary references are listed; deseasonalization methods are discussed by Khalil and Rasmussen, 1990b). The concentration of $CO_2$ is rising steadily. At present, the average rate of increase is about 1.63 $\pm$ 0.05 ppmv/yr, or 0.47%/yr calculated from the above-mentioned data. The rate of increase is much faster in recent years compared to earlier times when measurements were first taken, but the rate also varies from year to year (Figure 4).

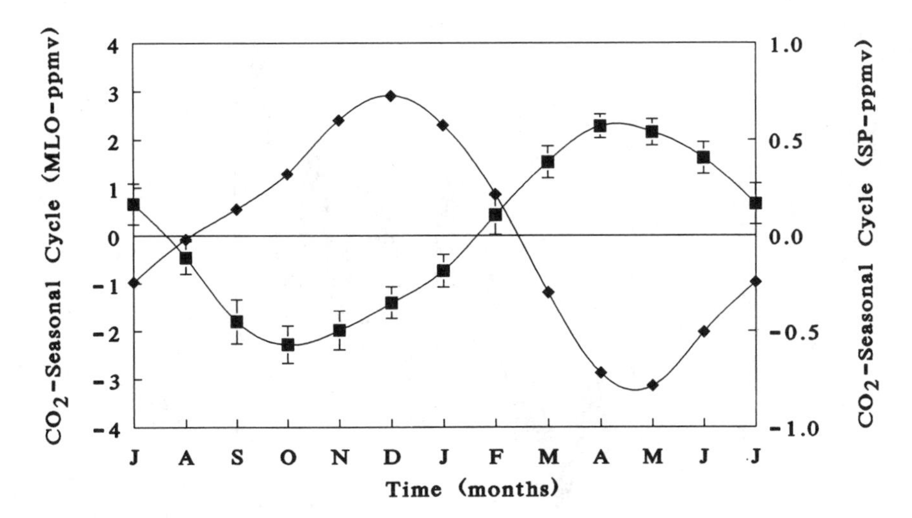


**FIGURE 2.** Seasonal cycles of $CO_2$ at Mauna Loa (MLO), Hawaii (diamonds), and the South Pole (SP) (squares). Note the different vertical scales — the cycle at Mauna Loa has a much higher amplitude.

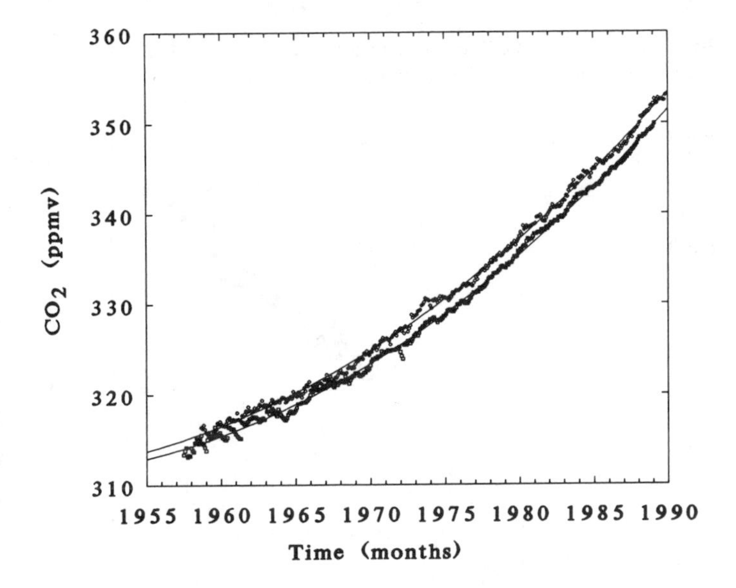


**FIGURE 3.** Deseasonalized monthly concentrations of $CO_2$ at Mauna Loa and the South Pole. Lines through the data are quadratic fits (see Table 4).

**Sources and sinks** — The anthropogenic influences on the $CO_2$ cycle arise from the burning of fossil fuels and global deforestation. Although the annual emissions from these sources are small compared to the natural annual exchanges of $CO_2$ between the active oceanic and biotic reservoirs, the systematic addition of $CO_2$ by human activities is large enough to appear as a clear trend in the atmospheric record.

It is estimated that in 1988 some 22,000 Tg of $CO_2$ were emitted from the combustion of fossil fuels (see Wuebbles and Edmonds, 1991; for earlier estimates see Marland and Rotty, 1984). The amount of $CO_2$ emissions per unit of energy obtained are highest from coal burning and lowest from burning natural gas; intermediate amounts are released from burning oil.

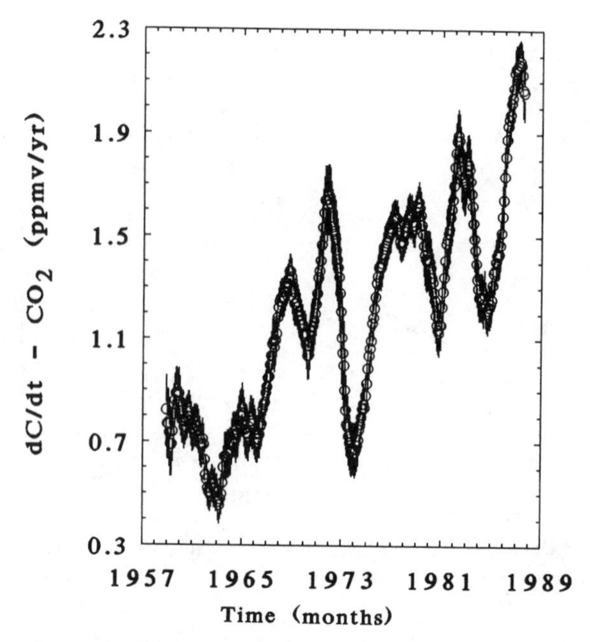


**FIGURE 4.** 3-year moving trends of $CO_2$ concentrations (average of Mauna Loa and South Pole deseasonalized data). $CO_2$ concentrations are increasing at faster rates now than in the past.

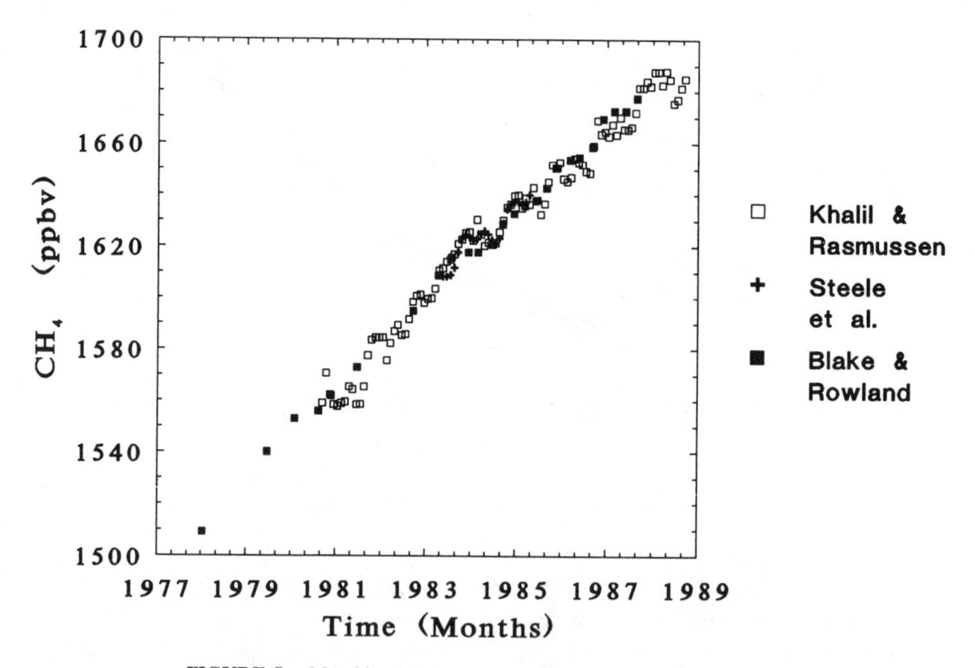


**FIGURE 5.** Monthly global average concentrations of $CH_4$.

Global deforestation is considered to be an important but highly uncertain source of $CO_2$, estimated to be between 4000 and 10,000 Tg $CO_2$/yr (Houghton et al., 1987). Almost all is believed to come from tropical deforestation.

Only about half the amount of $CO_2$ emitted by human activities still remains in the atmosphere. The rest is believed to be taken up by the oceans.

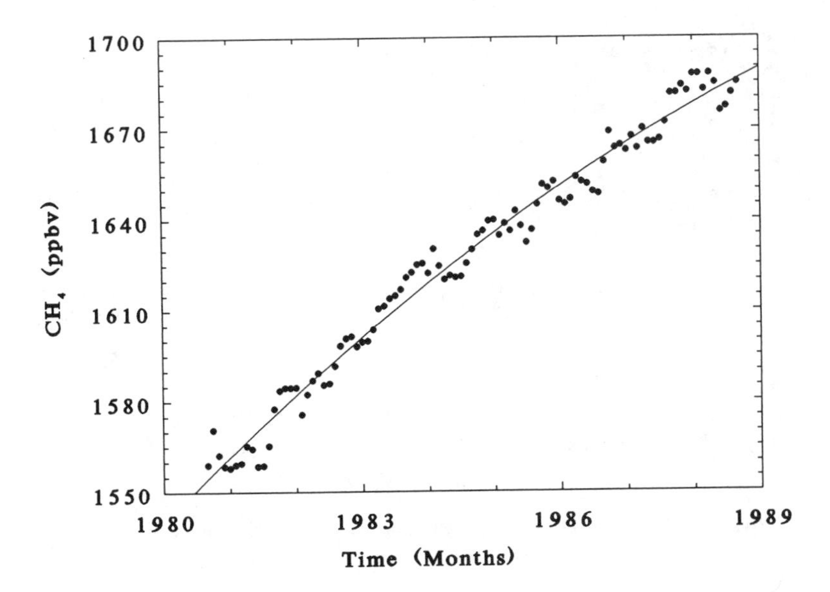


**FIGURE 6.** Monthly global average concentrations of $CH_4$, from Khalil and Rasmussen (1990b), based on a single calibration standard. The line through the data is a quadratic fit (see Table 4).

## B. METHANE

$CH_4$ controls the concentrations of OH in the troposphere, and its oxidation is a source of CO and eventually $CO_2$. It also contributes to the greenhouse effect, both directly and through its production of $O_3$ and $CO_2$. $CH_4$ is increasing so rapidly at present that it is second only to $CO_2$ in affecting global warming, as mentioned earlier. In the stratosphere it terminates the destruction of $O_3$ from Cl atoms and also produces water vapor, which can destroy $O_3$ through the HOx cycle. In the long-term, $CH_4$ is an indicator of the vitality of life on Earth — concentrations are highest during warm interglacial periods, and fall to low levels during ice ages (see also Rambler et al., 1989).

**Concentrations** — The global average atmospheric concentrations over the last decade are shown in Figures 5 and 6. In Figure 6, data are shown from Khalil and Rasmussen (1990b) based on a single constant calibration. The concentrations of $CH_4$ also vary by season and latitude (Khalil and Rasmussen, 1983a). These data show that $CH_4$ is increasing at about 16.4 $\pm$ 0.5 ppbv/yr, or about 1%/yr. The trend appears to be slowing down, as can be seen in Figure 6. A detailed analysis of the trends is shown in Figure 7, in which the 24-month moving trends are plotted. This graph shows a (statistically) significant reduction in the rate of increase in the later part of the data compared to the early measurements (for details see Khalil and Rasmussen, 1990b).

**Sources and sinks** — The major sources of $CH_4$ controlled by human activities are believed to be raising cattle, rice agriculture, natural gas use, venting of gas from drilling for oil, landfills, and biomass combustion. The wetlands are thought to produce the largest natural emissions. Smaller natural sources include forest soils, oceans, lakes, tundra, wild animals, and insects.

$CH_4$ is removed principally by reacting with tropospheric OH radicals: $CH_4 + OH \rightarrow CH_3 + H_2O$. The methyl radical undergoes further oxidation and produces CO and other products.

Among the other processes that remove atmospheric $CH_4$, dry soils are perhaps the most important (see Steudler et al., 1989). Minor chemical reactions and stratospheric processes may remove some $CH_4$ every year.

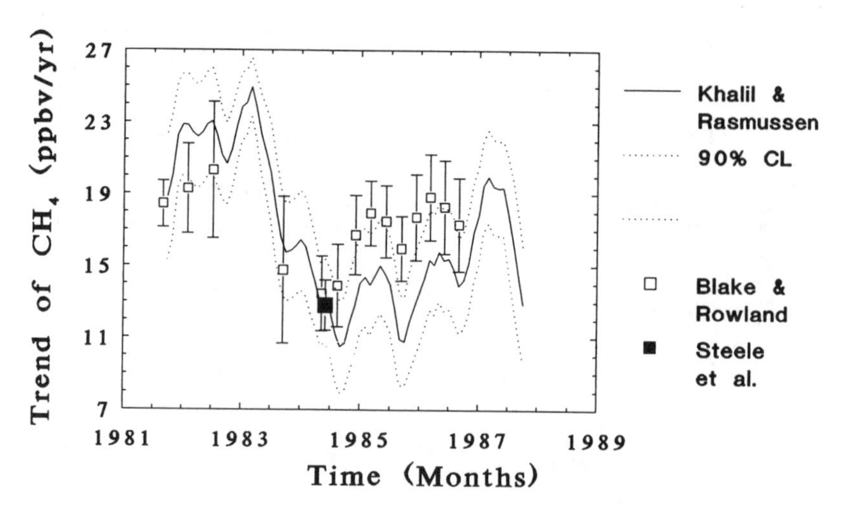


**FIGURE 7.** 2-year moving slopes of the $CH_4$ concentrations. The rate of increase of $CH_4$ is slowing down in recent years.

In recent years, several budgets of $CH_4$ have been proposed (summarized in Table 1). Estimates of emissions from individual sources are less certain than estimates from larger groups of sources such as the total anthropogenic or natural emissions or the total global emissions.

For $CH_4$ there are three major constraints that reduce the uncertainties in the estimated emission rates of the total annual emissions and combined emissions from groups of sources. The global budget of the man-made $CH_3CCl_3$ (Equation 4) can be used to deduce the average concentration of OH (see Section III.F for further discussion). Based on this OH calculation, the total annual loss of $CH_4$ by reactions with OH can be estimated accurately. Knowing the concentrations and the accumulation rate from atmospheric measurements gives the total annual emissions of $CH_4$ from all sources, as shown in the following equation:

$$S(t) \approx \frac{dC(t)}{dt} + (K_{eff}[OH] + \eta_o)C(t) \qquad (5)$$

In Equation 5, $\eta_o C$ are the annual losses of $CH_4$ by removal process other than reactions with OH. $K_{eff}$ is a suitably averaged rate constant for the reaction of $CH_4$ with OH (Khalil and Rasmussen, 1984d). The total emissions based on such calculations are estimated to be around 500 Tg/yr. Data from polar ice cores define the total anthropogenic emissions as the difference between present total emission rates and the emission rates in the atmosphere of several 100 years ago (based on Equation 5). Current calculations suggest that some 350 Tg/yr come from sources affected by human activities (Khalil and Rasmussen, 1990c). Finally, the isotopic composition of emissions of $CH_4$ from various sources and the observed isotopic composition of $CH_4$ in the atmosphere can be used to constrain the emissions from groups of sources such as from combustion processes (see Stevens and Engelkemeir, 1988; Tyler, 1991).

The increase of $CH_4$ can be caused both by increasing emission rates and decreasing levels of OH. It has been thought for some time that the atmospheric concentrations of OH may be decreasing over the last century and recent decades because of increasing anthropogenic emissions of CO and $CH_4$, which remove OH from the atmosphere. Current information does not favor this mechanism as the major cause of increasing $CH_4$. Increasing emissions from cattle and rice agriculture are the likely cause of the high concentrations of

# TABLE 1
## Estimates of Methane Emissions from Various Anthropogenic and Natural Sources: 1978 to 1988 (in Tg/yr)[a]

| | Ehhalt & Schmitt (1978) | | Donahue (1979) | | Sheppard et al., (1982) | Crutzen (1983) | | Khalil & Rasmussen (1983) | Blake (1984) | | Seiler (1984) | | Crutzen (1985) | | Seiler (1986) | | Bingemer & Crutzen (1987) | | Cicerone & Oremland (1988) | |
|---|---|---|---|---|---|---|---|---|---|---|---|---|---|---|---|---|---|---|---|---|
| Ruminants | 100–220 | 160 | 100–220 | 160 | 90 | 60 | 60 | 120 | 70–160 | 115 | 70–100 | 85 | 60 | 60 | 70–100 | 85 | 70–80 | 75 | 65–100 | 80 |
| Rice paddy fields | 280 | 280 | 140–280 | 210 | 39 | 30–60 | 45 | 95 | 140–190 | 165 | 30–75 | 53 | 120–200 | 160 | 70–170 | 120 | 18–91 | 54 | 60–170 | 110 |
| Biomass burning | | | | | 60 | 30–110 | 70 | 25 | 25–110 | 68 | 50–100 | 75 | 20–70 | 45 | 55–100 | 78 | 30–100 | 65 | 50–100 | 55 |
| Landfills | | | | | | | | | | | | | | | | | 30–70 | 50 | 30–70 | 40 |
| Coal mining | 8–28 | 18 | 16–50 | 33 | | | | | | | 30 | 30 | 34 | 34 | 35 | 35 | 35 | 35 | 25–45 | 35 |
| Natural gas flaring | | | | | 50 | 20 | 20 | 40 | | | 20–30 | 25 | 33 | 33 | 30–40 | 35 | 0–35 | 18 | 25–50 | 45 |
| Automobiles | 1 | 1 | | | | | | | | | | | | | 1–2 | 2 | | | | |
| Other anthropogenic | 7–21 | 14 | 110–210 | 160 | 100 | | | 40 | 62–100 | 81 | | | | | | | | | | |
| dS | 0–50 | 25 | 0–25 | 25 | 18 | 0–25 | 13 | 20 | 0–25 | 13 | 0–30 | 15 | | | 0–10 | 5 | | | | |
| Swamps & marshes | 190–300 | 245 | 200–300 | 250 | 39 | 30–220 | 125 | 150 | 120–190 | 155 | 15–60 | 38 | 70–90 | 80 | 25–70 | 48 | 26–137 | 82 | 100–200 | 115 |
| Lakes | 1–25 | 13 | | | 35 | | | 10 | 13 | 13 | 1–7 | 4 | | | 15–35 | 25 | | | 1–25 | 5 |
| Oceans | 1–17 | 9 | | | 30 | | | 13 | 5–21 | 13 | | | | | | | | | 5–20 | 10 |
| Tropical forests | | | | | 767 | | | | 60–400 | 230 | | | | | | | | | | |
| Tundra | 0–3 | 2 | 3–50 | 27 | | | | 12 | | | | | | | | | | | | |
| Other natural | | | | | | 150 | 150 | 48 | | | 5–15 | 10 | | | | | 0–30 | 15 | 10–100 | 40 |
| Total anthropogenic | 396–550 | 473 | 366–760 | 563 | 339 | 140–250 | 195 | 320 | 297–560 | 429 | 200–335 | 268 | 267–397 | 332 | 261–447 | 354 | 183–411 | 297 | 255–535 | 365 |
| Total natural | 192–345 | 269 | 203–350 | 277 | 871 | 180–370 | 275 | 233 | 198–624 | 411 | 21–82 | 52 | 70–90 | 80 | 40–105 | 73 | 26–167 | 97 | 116–345 | 170 |
| Total | 588–895 | 741 | 569–1110 | 840 | 1210 | 320–620 | 470 | 553 | 495–1184 | 840 | 221–417 | 319 | 337–487 | 412 | 301–552 | 427 | 209–578 | 394 | 371–880 | 535 |
| F | 67–61 | 64 | 64–68 | 67 | 28 | 44–40 | 41 | 58 | 60–47 | 51 | 90–80 | 84 | 79–82 | 81 | 87–81 | 83 | 87–71 | 75 | 69–61 | 68 |

[a] Table from Khalil and Rasmussen (1990c). The budgets of Crutzen (1985) and Seiler (1986) are taken from Bolle et al., (1986).

$CH_4$ observed now, compared to the concentrations of a few centuries ago (see Khalil and Rasmussen, 1985, and Section III.F). Cattle populations have risen by a factor of 2.7 from the turn of the century to 1985, and the hectares of rice harvested have risen by a factor of 2 over the same time. Major increases occurred during the post-war boom era of the 1950s. In recent decades, however, these sources are no longer increasing, probably because of limitations on the land available for rice agriculture and food for cattle. It may be that the increase of $CH_4$ is slowing down because of the lack of increase of these anthropogenic sources, although other explanations are also possible.

In recent years, sources related to energy, particularly leakages from natural gas usage and venting of gases from drilling for oil, may be increasing rapidly and may determine the future concentrations of $CH_4$. For the anthropogenic sources of $CH_4$, the present may be a period of transition between the importance of agricultural sources closely connected to the population and the recent increase of energy sources that are related not only to population but also to increasing per capita demands. The consequence is that the future of $CH_4$ is most unpredictable because the future concentration is uncoupled from the past causes of increase.

## C. NITROUS OXIDE

The environmental role of $N_2O$ emerged during the 1970s. In the stratosphere the reaction production of NO from the reaction of $N_2O$ with $O(^1D)$ can catalytically destroy ozone (McElroy and McConnell, 1971; and Section I). The contribution of $N_2O$ to the greenhouse effect was reported by Wang et al. (1976).

**Concentrations** — The concentrations during recent times are shown in Figures 8 and 9. In Figure 9, data from the various groups are used to construct a composite monthly global average after correcting for calibration differences (for details see Khalil and Rasmussen, 1991a). The concentration of $N_2O$ is increasing in the atmosphere at a rate of about $1.0 \pm 0.1$ ppbv/yr, or 0.26% per year. There are indications that the rate of increase may be speeding up in recent years. The 3-year moving trends of the composite data (Figure 9) are shown in Figure 10. A cycle and increasing trend appear in this figure. The causes of these features are not known (and may be due to averaging and calibration effects).

**Sources and sinks** — In the first global budgets of $N_2O$, the oceans were a large source and the atmospheric lifetime was believed to be ~20 years, although there were no known removal processes that would account for such a short lifetime (Hahn and Junge, 1977). Nitrogen fertilizers were thought to be an important anthropogenic source. Following this period, the ocean source was shown to be much smaller, and burning of fossil fuels, particularly from coal-fired power plants, was identified as the major anthropogenic source (Weiss and Craig, 1976; Pierotti and Rasmussen, 1976). The sinks, which are still accepted, were identified to be photodissociation and chemical reactions in the stratosphere:

$$N_2O + h\nu \rightarrow N_2 + O(^1D)$$
$$N_2O + O(^1D) \rightarrow NO + NO$$
$$\rightarrow N_2 + O_2$$

Photo-dissociation is the principal sink, while smaller amounts are removed by reaction with $O(^1D)$. These sinks lead to an effective atmospheric lifetime of about $150 \pm 30$ years.

During the early 1980s the increasing trends of atmospheric $N_2O$ were established (Weiss, 1981; Khalil and Rasmussen, 1983b), and the preindustrial concentrations were determined to be about 285 ppbv, or about 7% less than the present (Khalil and Rasmussen, 1988b). The patterns of fossil fuel use fitted the observed atmospheric increases in both magnitude and timing (Weiss, 1981; Hao et al., 1987). Thus, for a while it was believed that coal combustion was a primary cause of increasing $N_2O$ concentrations, with small contributions from the use of Nitrogen fertilizers (McElroy and Wofsy, 1987; WMO, 1985; Hao et al., 1987).

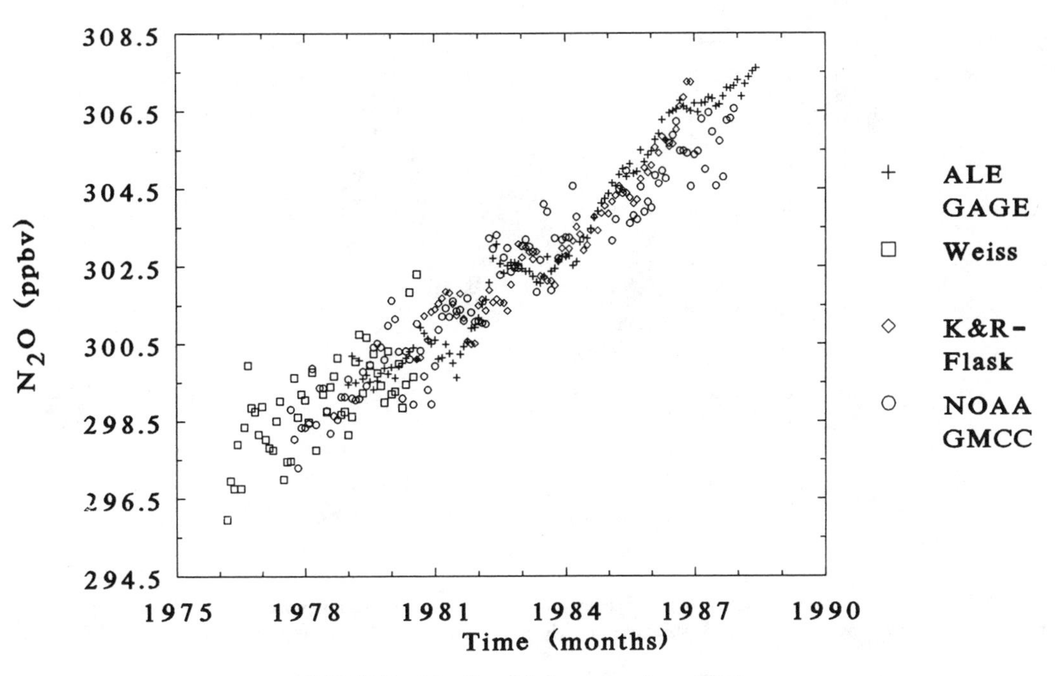


**FIGURE 8.** Monthly global concentrations of $N_2O$.

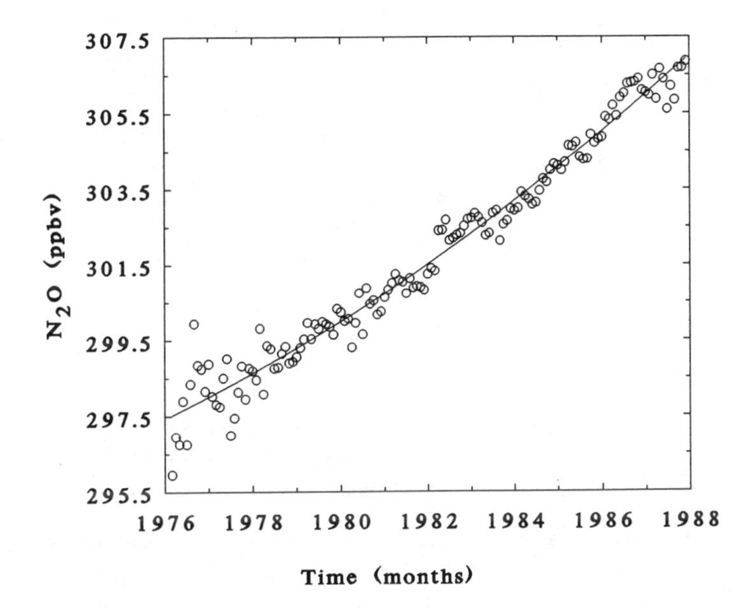


**FIGURE 9.** Monthly global concentrations of $N_2O$ based on the averages of the data shown in Figure 8. The line through the data is a quadratic fit (see Table 4).

Recently, however, it was discovered that the methods used to determine the production of $N_2O$ from coal-fired power plants (and some other combustion sources) may have been affected by a sampling artifact (Muzio and Kramlich, 1988). $N_2O$ can be produced when power plant effluent, including $SO_2$, $H_2O$ and NOx, is stored in containers even for a few h. Subsequent work has shown that when the precursors that make $N_2O$ in containers are eliminated, very little if any $N_2O$ comes from power plants (Linak et al., 1990; Sloan and

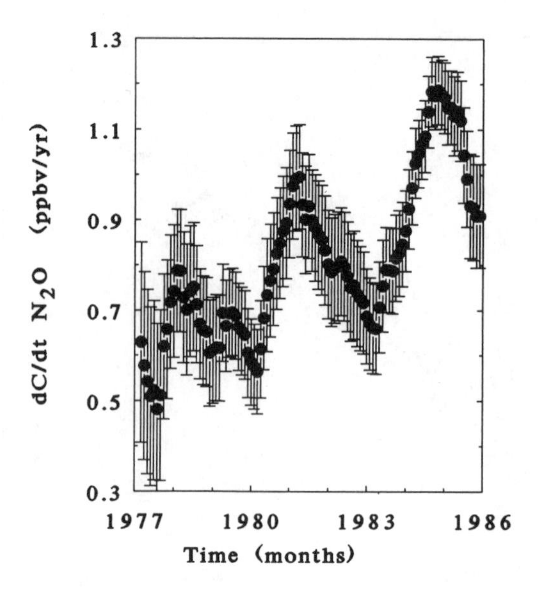


**FIGURE 10.** 3-year moving trends of $N_2O$ concentrations. The increase of $N_2O$ appears to be speeding up in recent years. There is also an apparent 3-year cycle for which there is no explanation at present.

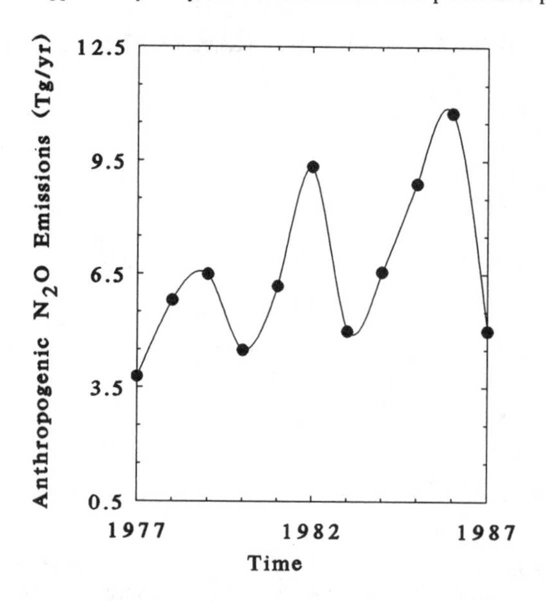


**FIGURE 11.** Estimated emissions of $N_2O$ from all anthropogenic sources.

Laird, 1990; Yokoyama et al., 1991; Khalil and Rasmussen, 1991b; see also Muzio and Kramlich, 1988). The anthropogenic sources and hence the causes of increasing concentrations are therefore in considerable disarray at present. Many small sources have been proposed that may fill the gap left behind by the lack of substantial emissions from high temperature coal combustion.

There are, however, some aspects of the global emission rates that can be determined from the existing data. Khalil and Rasmussen (1991a) used a global two-box model of the troposphere and stratosphere to calculate the emissions necessary to explain the measured concentrations for preindustrial and modern times, based on a range of assumed atmospheric lifetimes consistent with the stratospheric sinks (100 to 200 years). The calculated average global source for preindustrial times is subtracted from the calculated source for recent times.

**TABLE 2**
**Natural and Identified Anthropogenic Sources of $N_2O$ (inTg/yr)[a]**

| | Middle | Range | Uncertainty factor | Ref. |
|---|---|---|---|---|
| | **Anthropogenic sources** | | | |
| Biomass burning | 1.6 | 0.2–3 | 15 | 1 |
| Power plants | 0.0 | 0.0–0.2 | 20 | 2 |
| Nylon manufacture | 0.7 | NK | NK | 3 |
| Nitrogen fertilizer | 1.0 | 0.4–3 | 8 | 4 |
| Sewage | 1.5 | 0.3–3 | 10 | 5 |
| Cattle-agriculture | 0.5 | 0.3–1 | NK | 6 |
| Aquifers-irrigation | 0.8 | 0.8–2 | NK | 7 |
| Automobiles | 0.8 | 0.1–2 | 20 | 6, 8 |
| Global warming | 0.3 | 0.0–1 | NK | 9 |
| Land use change | 0.7 | NK | NK | 10 |
| Atmospheric formation | NK | NK | NK | 11 |
| Total | 8 | 5–10 | | |
| | **Natural sources** | | | |
| Soils | 12[11] | — | — | |
| Oceans | 3[11] | — | — | |
| Total | 15 | | | |

*Note:* NK = not known, uncertainty factor = max/min of the range. Some numbers, especially in the ranges, are rounded. (1) Coffer et al., (1991). Lower limits from Crutzen and Andrea (1990); (2) Linak et al., (1990), Sloan and Laird (1990), Yokoyama et al., (1991), and Khalil and Rasmussen (1991); (3) Thiemens and Trogler (1991); (4) Eichner (1990) and primary references therein; (5) Kaplan et al., (1978); (6) Khalil and Rasmussen (this work); (7) Ronen et al., (1988); (8) EPA (1986); (9) Khalil and Rasmussen (1989); (10) Matson and Vitousek (1990) and Luizao et al., (1989); (11) McElroy and Wofsy (1986) and WMO (1985).

[a] Modified from Khalil and Rasmussen (1991a).

If the preindustrial source is assumed to be natural, the difference gives the anthropogenic source. The difference is shown in Figure 11 based on the composite recent data discussed above (Figure 9). The present (total) anthropogenic source is therefore about 7 ± 1 Tg/yr on average. This calculation is remarkably independent of the assumed lifetime of $N_2O$ between 100 and 200 years. The natural source, deduced from preindustrial concentrations, does depend on the lifetime and turns out to be about 15 Tg/yr if the lifetime is 150 years.

The apportionment of the total natural and anthropogenic emission rates among the various sources is extremely uncertain at present. Table 2 shows a possible budget of $N_2O$ that is consistent with current knowledge of emissions from individual sources.

## D. CHLOROFLUOROCARBONS ($CCl_3F$ AND $CCl_2F_2$)

Although there are many chloro- and CFCs used in industrial processes and consumer goods, trichlorofluoromethane ($CCl_3F_3$; F-11) and dichlorodifluoromethane ($CCl_2F_2$; F-12) are considered to be the most significant in causing global warming and destruction of the $O_3$ layer. The reasons are the large annual production and release of F-11 and F-12 and the long atmospheric lifetimes.

**Concentrations** — The global average concentrations of F-11 and F-12 are shown in Figures 12 and 13 between 1978.58 and 1983.5 (where 1978.0 is January of 1978 and February 1978 is 1978 + $^1/_{12}$ and so on) based on nearly continuous measurements taken at four sites in the middle and tropical latitudes of both hemispheres (Cunnold et al., 1983a, 1983b, 1986). The increasing trends are evident; however, the rates of increase are slowing down. Longer time series of measurements also exist that support the slowdown (Khalil and Rasmussen, 1981). The 3-year moving trends are shown in Figures 14 and 15. The slowdown is thought to be directly related to the sudden freeze in the global production rates around 1973, as shown in Figure 16. Since the fluorocarbons are long-lived, the concentrations are still out of balance with the annual removal rates. Concentrations will therefore continue to rise for many decades, although at diminishing rates, even if the emission rates stay constant.

**Souces and sinks** — The global sources of F-11 and F-12 are believed to be entirely from industrial production and use. Ice core measurements have not shown any consistent concentrations that can be attributed to preindustrial levels indicative of natural production. McCarthy et al. (1977) gave six categories of use for F-11 and F-12. These categories are aerosol propellants, hermetically sealed refrigerators, nonhermetically sealed refrigeration, closed cell foams, open cell foams, and other uses. In addition to these uses, some amount is lost in the manufacturing and storage processes, which are estimated to be 2.5% of the F-12 production and 1.5% of the F-11 production (Gamlen et al., 1986). While the annual industrial production rates are well known, the annual emission rates are complicated by the delays between production and ultimate disposal of refrigerators, air conditioners, or other consumer goods in which the fluorocarbons are used.

The use of F-11 for blowing rigid urethane foam increased greatly during the 1970s and essentially replaced its use as an aerosol propellent. The foam has many uses including insulation for commercial and residential construction, insulation in refrigerators and freezers, packaging, marine flotation, and polystyrene (Gamlen et al., 1986). Unlike the other uses, F-11 is retained in the closed cell foams for 100s of years. The foams thus become a growing reservoir of F-11, from which it will leak very slowly for centuries (see Khalil and Rasmussen, 1986, 1987a).

Both these fluorocarbons are destroyed primarily in the stratosphere where photodissociation processes release the Cl that catalytically destroys $O_3$ (Molina and Rowland, 1974; for detailed analyses see WMO, 1985, 1988, 1989). Some F-11 is taken up by the oceans, but whether this is a reservoir or a net loss is not known (Khalil and Rasmussen, 1984c). In either case the ocean uptake reduces somewhat the effect of F-11 on the ozone layer. In addition to these removal processes, F-11 and F-12, along with other man-made chlorocarbons, are taken up by some soils and probably destroyed by biological or chemical processes (see Khalil and Rasmussen, 1989c). The combined effect of these processes results in atmospheric lifetimes of about 70 years for F-11 and 100 years for F-12 (Cunnold et al., 1986).

## E. CARBON MONOXIDE

CO is a by-product of practically all combustion processes. It is also produced by the atmospheric oxidation of hydrocarbons and particularly $CH_4$. The CO budget is of interest in global tropospheric chemistry principally because of its effect in controlling OH concentrations, as mentioned earlier. The sources and estimates of emission rates are given in Table 3, based on Logan et al., 1981 (see also Khalil and Rasmussen, 1990a). It is removed by reactions with OH radicals and by several less significant processes.

The atmospheric concentrations are probably increasing, as indicated in Figure 17 (Khalil and Rasmussen, 1988a, 1990a). Calculations of CO concentrations from solar spectral observations are also consistent with the trends shown in Figure 17 (Rinsland and Levine,

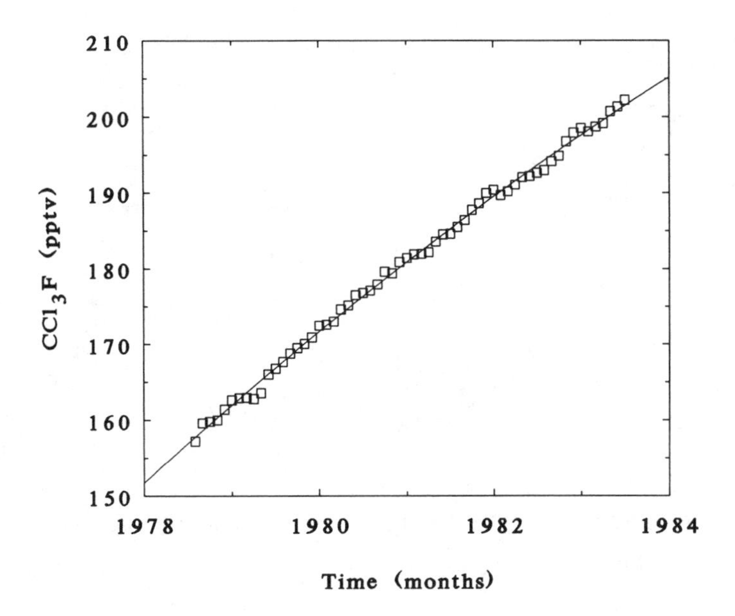


**FIGURE 12.** Monthly global average concentrations of trichlorofluoromethane ($CCl_3F$; F-11).

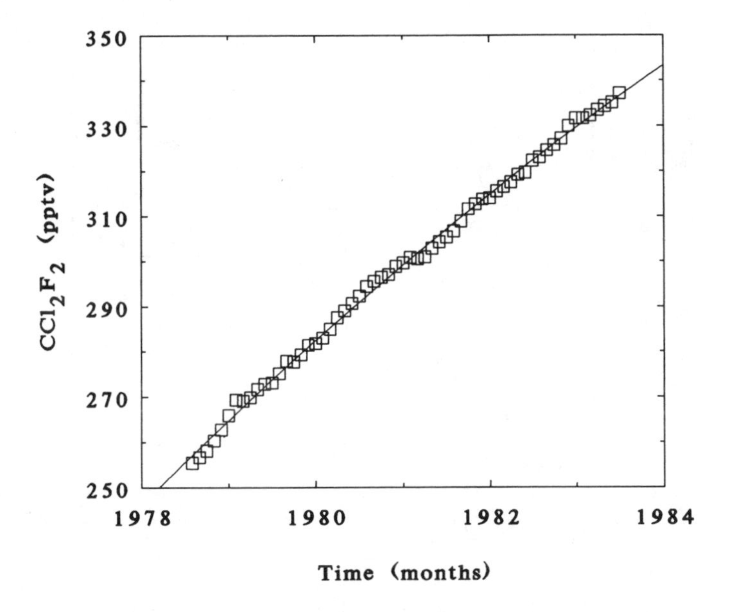


**FIGURE 13.** Monthly global average concentrations of $CCl_2F_2$ (F-12).

1985). It is possible that these trends will not continue in the future and may even be reversed. The reason is that automative and other urban emissions are consistently being reduced in the U.S. and Europe, and there are both natural and environmental constraints being placed on worldwide biomass burning.

## F. METHYLCHLOROFORM AND HYDROXYL RADICALS

$CH_3CCl_3$ is used primarily as a degreasing solvent. It contributes to the greenhouse effect, and because it contains chlorine, it affects the $O_3$ layer. In both these roles, it has a small effect, mostly because it is removed from the atmosphere by reactions with OH. In

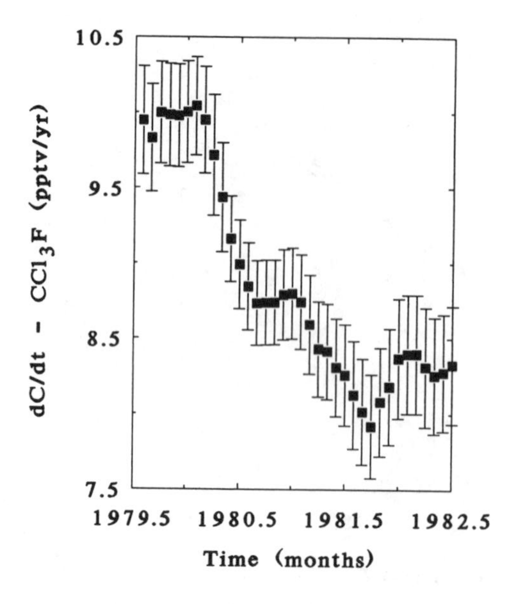


**FIGURE 14.** 3-year moving trends of $CCl_3F$ (F-11). The rate of increase is slowing down.

**FIGURE 15.** 3-year moving trends of $CCl_2F_2$ (F-12). The rate of increase is slowing down.

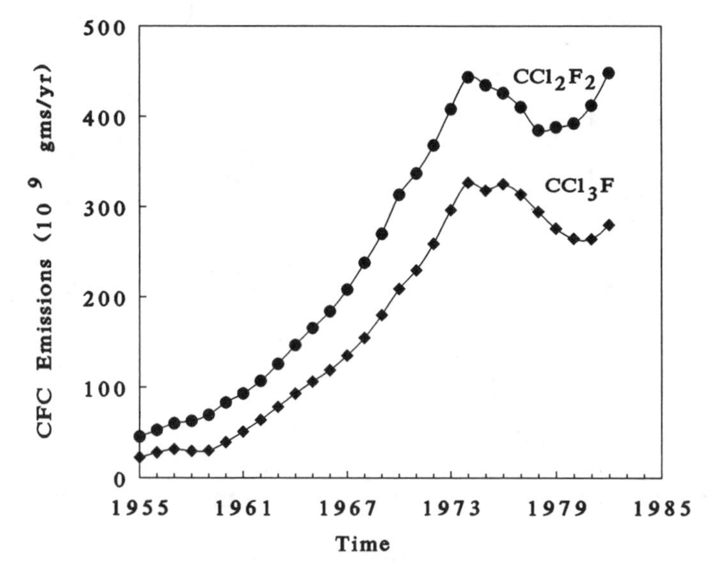


**FIGURE 16.** Estimated emissions of F-11 and F-12.

this sense, it is similar to the new compounds that are being proposed as "environmentally safe" substitutes for F-11 and F-12. The lifetime of $CH_3CCl_3$ is 6 to 7 years (Khalil and Rasmussen, 1984d). The importance of $CH_3CCl_3$ in global tropospheric chemistry is not dependent on its role in causing global warming or depleting stratospheric $O_3$, but rather in the possibility of obtaining accurate estimates of globally averaged OH concentrations.

**Concentrations** — The monthly and globally averaged concentrations of $CH_3CCl_3$ are shown in Figure 18 (Khalil and Rasmussen, 1984d, 1989d; Prinn et al., 1987). The rate of increase is slowing down because of the trends of emissions (Figure 19).

**TABLE 3**
**Sources of Carbon Monoxide (in Tg/yr)**

| | Anthropogenic | Natural | Global | Range |
|---|---|---|---|---|
| Directly from combustion | | | | |
| Fossil fuels | 500 | — | 500 | (400–1000) |
| Forest clearing | 400 | — | 400 | (200–800) |
| Savanna burning | 200 | — | 200 | (100–400) |
| Wood burning | 50 | — | 50 | (25–150) |
| Forest fires | — | 30 | 30 | (10–50) |
| Oxidation of hydrocarbons | | | | |
| Methane | 300 | 300 | 600 | (400–1000) |
| Nonmethane HCs | 90 | 600 | 690 | (300–1400) |
| Other sources | | | | |
| Plants | — | 100 | 100 | (50–200) |
| Oceans | — | 40 | 40 | (20–80) |
| TOTALS (rounded) | 1500 | 1100 | 2600 | (2000–3000) |

*Notes:* (1) table adapted from Logan et al., (1981) and revisions reported by World Meteorological Organization (WMO), (1986). All estimates are in Tg/yr of CO. Tg/yr = Mton/yr = 10 g/yr; (2) all estimates are expressed to one significant figure. The sums are rounded to 2 significant digits; (3) half the production of CO from the oxidation of $CH_4$ is attributed to anthropogenic sources and the other half to natural sources based on the budget of $CH_4$, from Khalil and Rasmussen (1984c); and (4) table taken from Khalil and Rasmussen (1990a).

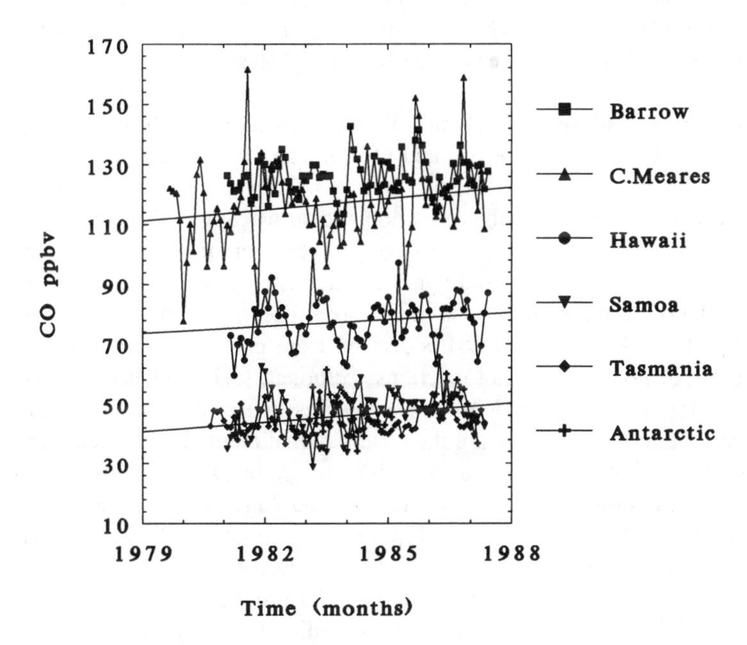


**FIGURE 17.** Trends of CO concentrations at various latitudes.

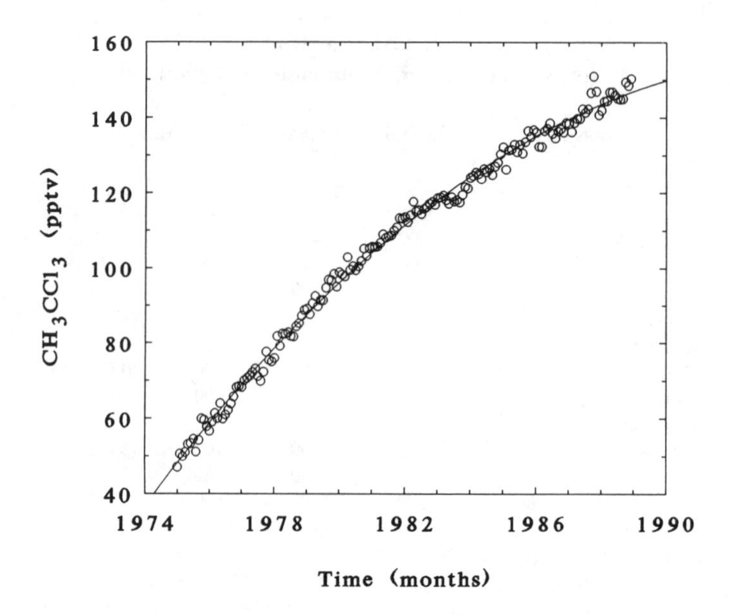


**FIGURE 18.** Monthly global average concentrations of 1,1,1-trichloroethane (methylchloroform; $CH_3CCl_3$).

**Sources and sinks** — Estimates of the industrial production of $CH_3CCl_3$ were reported by Midgley (1989), taking into account previous reported emission rates. Annual emissions data from Neely and Plonka (1978) and Midgley (1989) are shown in Figure 20. For $CH_3CCl_3$, the annual production rate is essentially the same as the release rate since there are no known uses that create sizable reservoirs, as for F-11 and F-12. $CH_3CCl_3$ is removed primarily by reactions with OH radicals. There are, however, other removal processes of lesser magnitude, including the soils and the oceans (see Khalil and Rasmussen, 1984c, 1989c; Butler et al., 1991).

**OH concentrations** — OH radicals are formed in the troposphere primarily by the photolysis of $O_3$ that produces $O_2$ and $O(^1D)$. The $O(^1D)$ reacts with water vapor to make two OH. OH radicals are removed primarily by reacting with CO and $CH_4$. A number of other processes contribute to the formation and destruction of OH (see Lu and Khalil, 1991). OH radicals have atmospheric lifetimes on the order of seconds, on average. The short lifetime, low concentrations, and complex chemistry give OH the appearance of being in a fragile balance. The concentrations of OH are exceedingly small at about $8 \times 10^5$ mol/cm$^3$. Much of what is known about OH is derived from photochemical models based on a large number of potentially important reactions. Although such models are conceptually straightforward, they may not include all the processes that affect OH or may not include these processes correctly. Only now are the observational methods sufficiently developed to hold promise of obtaining the tropospheric climatology of OH concentrations, but this goal is still many years away. Remarkably, the only alternative to photochemical models or direct global measurements is the use of $CH_3CCl_3$ as an indicator of average OH.

**Connection between OH and $CH_3CCl_3$** — The idea is to use the measured concentration and trends of $CH_3CCl_3$ and the known industrial production rate to estimate the net annual loss of $CH_3CCl_3$, which is related to OH according to the following equation (derived from Equation 5).

$$[OH]_{eff} \approx (S_{industrial}/C_* - d\ln C_*/dt - \eta_{o*})/K_{eff*} \qquad (6)$$

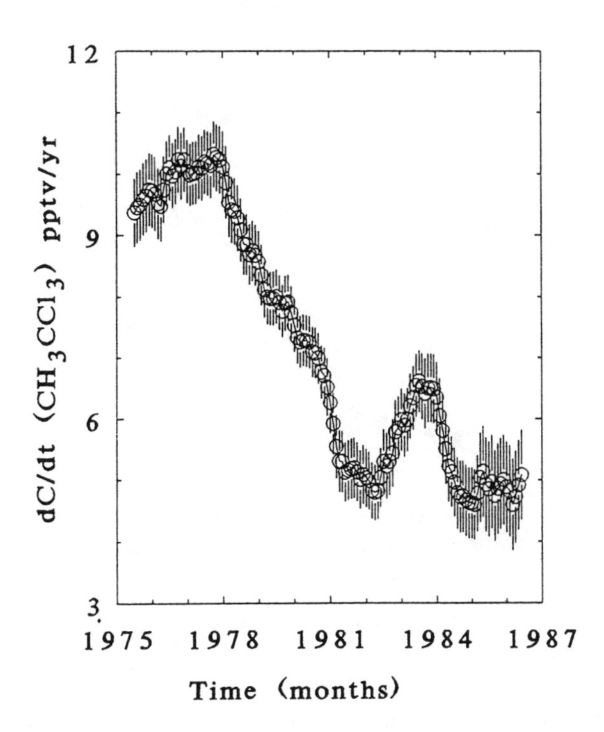


**FIGURE 19.** 3-year moving trends of $CH_3CCl_3$.

Here $C_*$ is the $CH_3CCl_3$ concentration, and $K_{eff}$ is the averaged rate constant for the reaction between OH and $CH_3CCl_3$. Other symbols are the same as in Equation 5. The idea was first suggested by Lovelock (1974) and implemented with the then-available measurements and rate constants by Singh (1977). More refined calculations based on newer measurements of absolute concentrations and reaction rates were reported by Khalil and Rasmussen (1984d). The calculated OH was about $8 \times 10^5$ mol/cm$^3$. Their results were verified recently by Prinn et al. (1987).

**OH trends** — A decrease in the concentration of OH would cause substantial feedbacks on the global climate. Whether OH concentrations are changing is therefore a question of considerable importance in atmospheric chemistry. Because $CH_4$ is known to have increased greatly during the last century and there is indirect evidence that CO has also, it is believed that OH must have decreased (Khalil and Rasmussen, 1985; Levine et al., 1985; Thompson and Cicerone, 1986). If OH decreases, some of the trend of $CH_4$ can be attributed to this process, as mentioned earlier. More importantly, however, the concentrations of many other trace gases would also have increased over the last century. If such trends continued, the indirect effects of decreasing OH on the Earth's climate and atmospheric chemistry could be substantial in the future.

Data from polar ice cores suggested that the concentrations of OH are not as fragile as previously believed (Khalil and Rasmussen, 1989a). During ice ages, the concentrations of $CH_4$ drop to around 350 ppbv, and the concentrations of CO are also likely to be much lower than during warmer times. If the production of OH remained the same as during interglacial periods, the concentration of OH would become much higher during ice ages than during warmer times. Because of the cold temperatures, however, there is less water vapor. There is also less $O_3$ because of reduced emissions of hydrocarbons and other precursors. Therefore, the production of OH would also be reduced. The simultaneous reduction in both the production and destruction of OH would lead to a relatively stable concentration

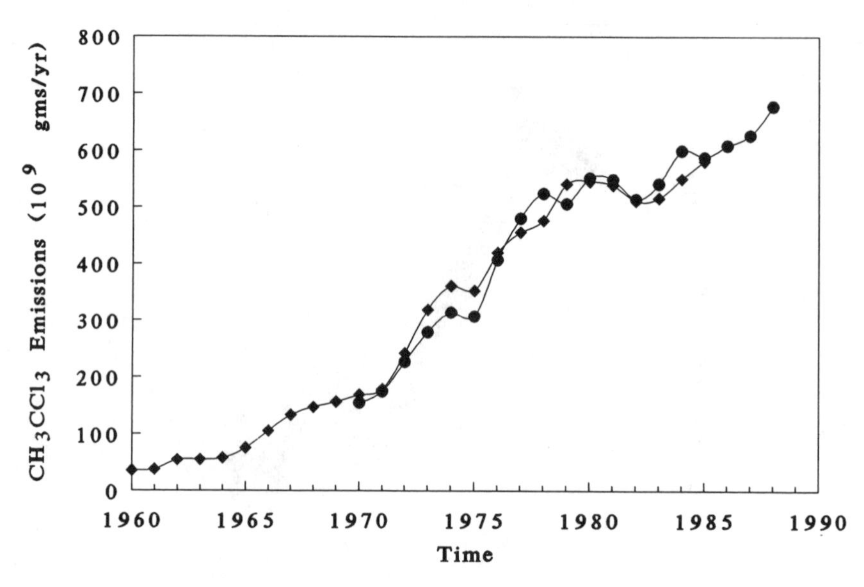


**FIGURE 20.** Estimated industrial emissions of $CH_3CCl_3$. Diamonds are from Neely and Plonka (1978); circles are from Midgley (1989).

of OH. This compensation applies not only to the transition between ice ages and interglacial warm periods, but also to the transition between preindustrial times and the present (Khalil and Rasmussen, 1989a). Model calculations suggest that there is likely to be only ~15% more OH during ice ages than during interglacial times, and that the decrease of OH over the last century is likely to be <5% (Pinto and Khalil, 1991; Lu and Khalil, 1991).

These results suggest that the major cause of the increase of $CH_4$, CO, and possibly other gases over the last century is probably increasing emission rates rather than decreasing removal rates. Nonetheless, the stability of OH can eventually be upset by continued increases of CO, $CH_4$, and other gases. When we may reach this breaking point is currently being investigated.

Since the atmospheric record of $CH_3CCl_3$ spans 13 years (1975 to 1988 in Figure 18), it may be used to obtain information on OH trends during this period. The result of mass balance calculations using Equation 6 and the monthly average data in Figure 18, smoothed with a 12-month filter and the emission rates reported by Midgley (1989), result in an estimate of slightly increasing OH concentrations rather than the expected decrease of OH. The frequency distribution of all possible trends of OH based on the smoothed monthly data of Figure 18 is shown in Figure 21. It shows a small ~0.3%/yr increase of OH during the period of the experiment, but it is not statistically significant. The average concentration of OH deduced from these calculations is about $8.2 \pm 0.8 \times 10^5$ mol/cm$^3$ under the similar assumptions as in Khalil and Rasmussen (1984d); (the ± quoted here does not represent all sources of uncertainty).

The main conclusion that can be drawn from the studies reported by Pinto and Khalil (1991), Lu and Khalil (1991), and the calculations shown above is that there is no observational or theoretical evidence that OH concentrations have changed significantly over the last century or are changing during recent decades. That is not to say that changes of OH levels have not occurred or are not occurring at present, but only that there are no practical means currently available to assess the trends. It is highly unlikely that the question of whether OH is changing will be answered anytime soon. This is because the indirect methods are not precise enough, as already discussed, and direct measurements, when implemented on a global scale, will require decades of reliable measurements to assess any trends of this highly variable atmospheric reactive radical (see Section II.B).

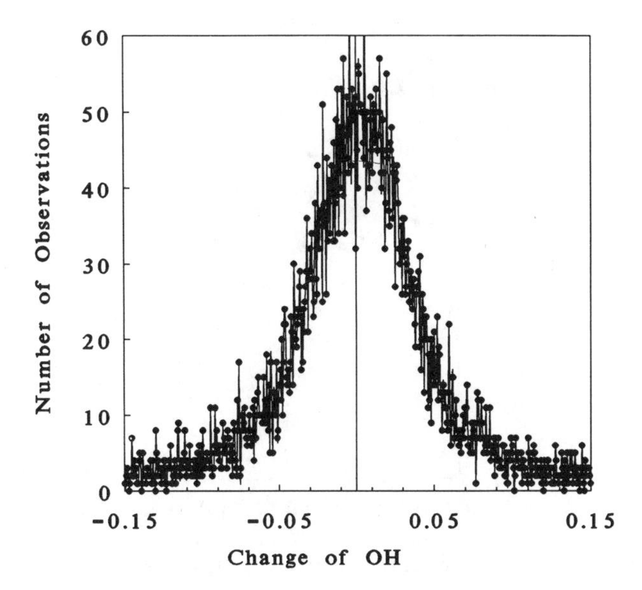


**FIGURE 21.** Trends of OH during the last decade (in $10^5$ mol/month). The results suggest a slight increasing trend but are close to a zero trend.

### G. OTHER GASES

There are many gases that can have the same environmental effects as the main gases discussed here. Long-lived gases with environmental effects similar to F-11 and F-12 include other halocarbons that may become important in the future, such as $C_2F_3Cl_3$ (F-113), $CHClF_2$ (F-22), $CCl_4$, $CF_3Br$, and $CF_2BrCl$.

There are several sulfur-containing compounds that are likely to have significant roles in the global environment. These are dimethylsulfide (DMS), sulfur dioxide ($SO_2$), and carbonyl sulfide (OCS). The sulfur gases generally have the potential of producing fine particles in the atmosphere that may reflect and scatter solar radiation, causing global cooling if concentrations increase (see Charlson et al., 1987).

Finally, there are about 100 $C_2$ to $C_{10}$ hydrocarbons measured in the atmosphere. Only the light ones are globally dispersed. These hydrocarbons can have a significant effect on tropospheric OH and $O_3$ concentrations, because of their direct atmospheric chemical reactions and by the action of their oxidation products.

## IV. COMPARISONS OF TRACE GAS BUDGETS

Further insights in the relative roles of the trace gases considered in this chapter can be obtained by comparing the global budgets. First we consider the interhemispheric gradient. This is the ratio of the Northern Hemisphere concentrations to the Southern Hemisphere concentrations. Monthly values of this ratio are averaged over the length of the records available. The results are shown in Figure 22 in % of excess concentration in the Northern Hemisphere compared to the Southern Hemisphere. This ratio is a function of the atmospheric lifetimes, the ratio of emissions in the Northern and Southern Hemispheres, the rates of increase, and the interhemispheric transport times. Since most anthropogenic emissions are in the Northern Hemisphere, gases with shorter lifetimes are depleted before they can reach the Southern Hemisphere, thus creating larger interhemisphere ratios.

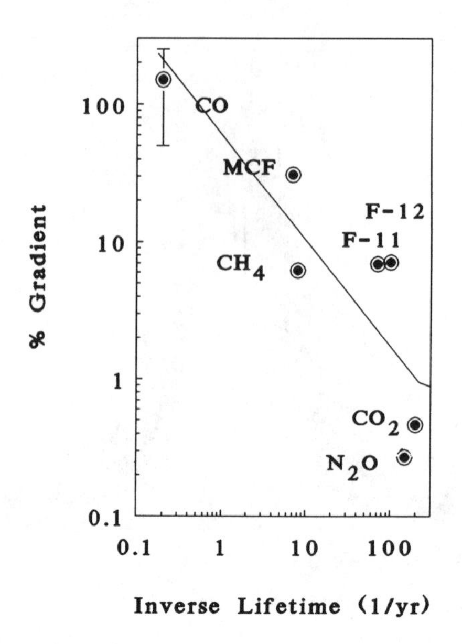


**FIGURE 22.** The % interhemispheric gradient. MCF = methylchloroform. The % gradient is $[(C_n/C_s) - 1] \times 100\%$, where $C_n$ and $C_s$ are the average Northern and Southern Hemispheric concentrations. The average of the ratio $C_n/C_s$ is calculated from monthly data, then the % gradient is calculated. The lifetime is one of several variables that influences this ratio (see text for more details).

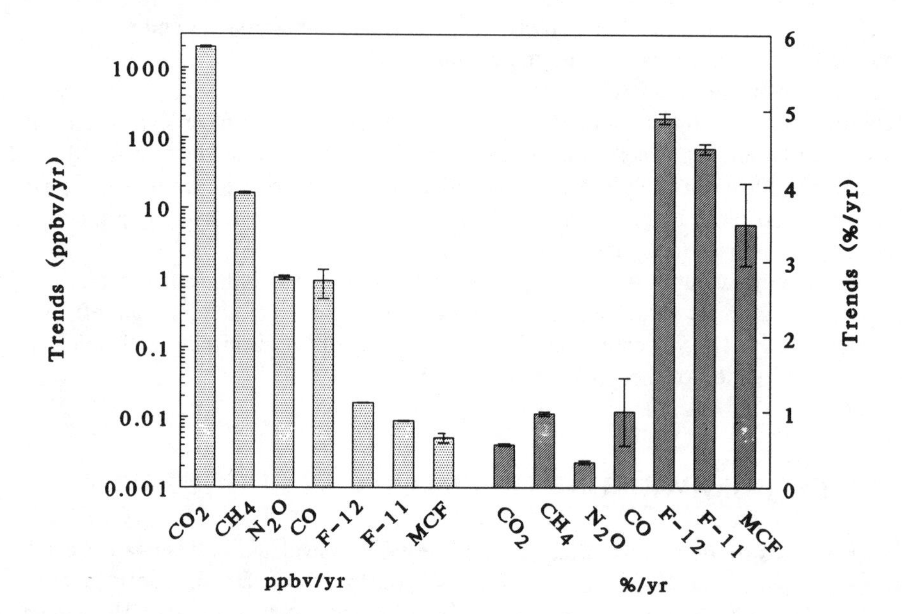


**FIGURE 23.** Comparisons of the trends of trace gases involved in global change. Trends in number of additional molecules added to the atmosphere each year are shown on the left. This trend is most important for causing climatic and other global changes. The change in %/yr is shown on the right hand side. Manmade gases can have very high %/yr increases, but this index is not a robust indicator of global change.

Figure 23 shows a comparison of global rates of increase. On a molecule-to-molecule basis, the $CO_2$ is the most abundant and is increasing at the highest rate. For this reason it has the greatest effect on global warming. $CH_4$ is next, followed by the other trace gases. In terms of relative increases in %/yr, the halocarbons are increasing the fastest, and $CO_2$ is among the slowest increasing gases, as shown in the right half of Figure 23. In spite of the rapid increases of the halocarbons in %/yr, their environmental effects on global warming are not as great as $CO_2$ and $CH_4$ because global warming depends on the absolute atmospheric increases in the number of molecules of a gas and not on the relative increases.

Table 4 is a summary of the mass balances and the physical, chemical, and atmospheric characteristics of the gases that have been discussed in this chapter ($CO_2$, $CH_4$, $N_2O$, $CCl_3F$, $CCl_2F_2$, $CH_3CCl_3$, and CO).

## V. CONCLUSIONS

The foundation for the idea that human activities can cause global environmental changes is inherent in the observed increases of a few key atmospheric trace gases. These gases are $CO_2$, $CH_4$, and $N_2O$, which have existed in the Earth's atmosphere for eons and are now increasing because of increasing population and per capita demand for food and energy. In addition, there are purely man-made gases in the Earth's atmosphere that have reached significant levels and are found everywhere from the North Pole to the South Pole, 1000s of km from the sources of industry and population. The principal gases in this group are $CCl_3F$ and $CCl_2F_2$, although there are dozens more that are at exceedingly low concentrations at present. Increases of other gases, such as CO may indirectly effect the global environment. These gases have been discussed in some detail in this chapter.

Industrially produced gases such as the fluorocarbons are relatively easy to control by legislative actions that can reduce or eliminate their production. Gases such as $CO_2$, $CH_4$, and $N_2O$ are not so easy to control because they are emitted from energy and food production over diffuse areas of the world. It is likely that these gases will continue to rise and cause global change. Whether that global change will be good or bad is not known, but we are likely to find out in a few decades.

## ACKNOWLEDGMENTS

I have benefited from many discussions with Professor R. A. Rasmussen. Ms. M. J. Shearer provided valuable research support. I have also benefited from discussions with my students Y. Lu, R. MacKay, F. Moraes, and W. Zhao. Financial support for this project was provided in part by grants from the Department of Energy (DE-FG06-85ER6031); the National Science Foundation (DPP8717023); and grants to the Andarz Company by the Environmental Protection Agency (Orders No. OD3680NASA and 1D3216NASA). Responsibility for the contents of this chapter rests entirely with the author, and not the sponsors.

**TABLE 4**
**Characteristics of Atmospheric Trace Gases Important in Global Change**

Concentrations[1]

| | | C (present) | C (Preindustrial)[2] | Gradient (%)[3] |
|---|---|---|---|---|
| $CO_2$ | (ppmv) | 350.2 ± 0.3 | 275 | 0.47 ± 0.02 |
| $CH_4$ | (ppbv) | 1686 ± 2 | 700 ± 10 | 6.2 ± 0.2 |
| $N_2O$ | (ppbv) | 302.3 ± 0.2 | 285 ± 1 | 0.27 ± 0.03 |
| $CCl_3F$ | (pptv) | 198 ± 2 | 0 | 6.9 ± 0.3 |
| $CCl_2F_2$ | (pptv) | 329 ± 2 | 0 | 7.1 ± 0.2 |
| $CH_3CCl_3$ | (pptv) | 146 1 | 0 | 30 ± 1 |
| CO | (pptv) | 90 5 | 40 | 250 |

Trends

| | dC/dt | C (Quadratic) = $a + b_0t + b_1t^2$ | | |
|---|---|---|---|---|
| | (ppxv/yr) | a(ppxv)[4] | $b_0$(ppxv/yr) | $b_1$(ppxv/yr$^2$) |
| $CO_2$ | 1.63 ± 0.05 | 315 ± 1 | 0.54 ± 0.01 | 0.00168 ± 0.00004 |
| $CH_4$ | 16.4 ± 0.05 | 1555 ± 10 | 21.2 ± 1.7 | −0.049 ± 0.017 |
| $N_2O$ | 1 ± 0.07 | 298 ± 1 | 0.57 ± 0.08 | 0.020 ± 0.006 |
| $CCl_3F$ | 8.9 ± 0.1 | 156.9 ± 1.1 | 10.3 ± 0.4 | −0.0227 ± 0.0064 |
| $CCl_2F_2$ | 16.1 ± 0.2 | 255.5 ± 1.9 | 18.7 ± 0.7 | −0.0418 ± 0.0112 |
| $CH_3CCl_3$ | 5.1 ± 0.8 | 47 ± 3 | 11.1 ± 0.3 | −0.0236 ± 0.0015 |
| CO | 0.9 ± 0.4 | — | — | — |

Lifetimes and source emissions

| | Lifetime (yr) | Emissions (Tg/yr) Sa | Sn |
|---|---|---|---|
| $CO_2$ | 200 | 30000 | — |
| $CH_4$ | 9 ± 2 | 350 | 200 |
| $N_2O$ | 150 ± 30 | 7 | 15 |
| $CCl_3F$ | 70 ± 20 | 0.29 | 0 |
| $CCl_2F_2$ | 100 ± 50 | 0.41 | 0 |
| $CH_3CCl_3$ | 7 ± | 0.61 | 0 |
| CO | 0.2 | 1500 | 1100 |

Environmental Effects

| | Global warming | | | Ozone change[8] |
|---|---|---|---|---|
| | W/m$^2$/ppxv[5] | Relative molar[6] | W/Trend (%)[7] | < = 0 > |
| $CO_2$ | 0.019 | 1 | 100 | x |
| $CH_4$ | 0.0006 | 20 | 20–30 | x x |
| $N_2O$ | 0.0026 | 200 | 10 | x |
| $CCl_3F$ | 0.000272 | 12,000 | 5 | x |
| $CCl_2F_2$ | 0.000315 | 15,000 | 12 | x |
| $CH_3CCl_3$ | >0 | — | — | x |
| CO | — | — | — | — |

*Notes:*

1. The periods over which data were taken for all calculations in this table are as follows: for quadratic trends the entire periods are used over which data are available. These periods are for F-11 and F-12; 1978.58–1983.5; $CH_3CCl_3$; 1975–1988; $CO_2$; 1958.17–1989; $CH_4$; 1980.67–1988.67; $N_2O$; 1976.17–1987.92. The global average concentrations are reported for as recent a time as possible. $CO_2$ is the average for 1988 and the trend is for the 5 years from 1984.0–1989.0; average $CH_4$ is for 1987.67–1988.67; average F-11 and F-12 are for 1982.5–1983.5 and the trends of $CH_4$, F-11 and F-12 are for the entire period of the data; average $N_2O$ is for 1987 and the trend is for 1983–1987; average $CH_3CCl_3$ is for 1988 and the trend is for 1985–1988; CO trends and concentration are for 1980–1988. The emission rates for $N_2O$ are for 1979–1987; F-11 and F-12 emissions are averages for 1975–1982; $CH_3CCl_3$ emissions are averaged over 1983–1988.
2. Ice core data are taken from Khalil & Rasmussen, 1982, 1987b, 1988b, 1989b; Stauffer et al., 1985; Raynaud et al., 1988; Friedli et al., 1986.
3. The % gradient is defined as $[C_n/C_s)-1] \times 100\%$ where $C_n$ and $C_s$ are the average Northern and Southern Hemispheric concentrations.
4. The units of the trends are for dC/dt, x = m; ppmv/yr for $CO_2$, x = b; ppv/yr for $CH_4$, $N_2O$ and CO and x = t; pptv/yr for $CH_3CCl_3$, $CCl_3F$ and $CCl_2F_2$. The units for the quadratic formulae are x = m for $CO_2$, x = b for $CH_4$, and $N_2O$ and x = t for F-11, F-12, $CH_3CCl_3$.
5. This column shows the increase of radiative forcing per ppxv increase of concentration where x is as before in note #2.
6. This column is the effectiveness of each molecule of a trace gas relative to $CO_2$ for causing global warming at present concentrations. These values are approximate and vary somewhat in different assessments.
7. This column shows the additional radiative forcing from the increase of each trace gas relative to $CO_2$, at present concentrations and rates of increase — the $CO_2$ radiative forcing is normalized to 100% — the annual increase of radiative forcing due to $CO_2$ is 0.03348 W/m$^2$/ppmv. Values for the other gases are divided by this number and multiplied by 100%.
8. Qualitative indication of the effect of trace gases on stratospheric ozone; = 0 means no effect; > means increases $O_3$; and < means the gas depletes $O_3$.

# REFERENCES


Bingemer, H. G. and Crutzen, P. J., The production of methane from solid wastes, *J. Geophys. Res.*, 92, 2181–2187, 1987.

Blake, D. R., Increasing Concentrations of Atmospheric Methane, Ph.D. dissertation, University of California at Irvine, 1984.

Boden, T. A., Kanciruk, P., and Farrell, M. P., *Trends '90,* (Carbon Dioxide Information Analysis Center, ORNL/CDIAC-36 Environmental Sciences Division, Oak Ridge National Laboratory, Oak Ridge, TN, 1990.

Bolle, H.-J., Seiler, W., and Bolin, B., Other greenhouse gases and aerosols, in *The Greenhouse Effect, Climate Change and Ecosystems,* SCOPE 29, J. Wiley & Sons, New York, 1986, chap. 4, p. 157.

Butler, J. H., Elkins, J. W., Thompson, T. M., Hall, B. D., Swanson, T. H., and Koropalov, J. V., Oceanic consumption of $CH_3CCl_3$: Implications for troposphere OH, *J. Geophys. Res.*, 96, 22347–22355, 1991.

Chamberlain, J. W. and Hunten, D. M., *Theory of Planetary Atmospheres: An Introduction to Their Physics and Chemistry,* Academic Press, Orlando, FL, 1987.

Chapman, S., A Theory of Upper Atmospheric Ozone. *Q. J. Rev. Meteorol. Soc.*, 3, 103–105, 1930.

Charlson, R. J., Lovelock, J. E., Andreae, M., and Warren, S. G., Oceanic phytoplankton, atmospheric sulfur, cloud albedo and climate, *Nature (London),* 326, 655–661, 1987.

Cicerone, R. J. and Oremland, R. S., Biogeochemical aspects of atmospheric methane, *Global Biogeochem. Cycles,* 2, 299–327, 1988.

Coffer, W. R., III, Levine, J. S., Winstead, E. L., and Stocks, B. J., New estimates of nitrous oxide emissions from biomass burning, *Nature (London),* 349, 689–691, 1991.

Crutzen, P. J., Atmospheric Interactions — Homogeneous Gas Phase Reactions of C, N and S Compounds, in *The Major Biogeochemical Cycles and Their Interactions,* Bolin, B. and Cooke, R. B., Eds., SCOPE 21, J. Wiley & Sons, New York, 1983.

Crutzen, P. J. and Andreae, M. O., Biomass burning in the tropics: impact on atmospheric chemistry and biogeochemical cycles, *Science,* 250, 1669–1678, 1990.

Cunnold, D. M., Prinn, R. G., Rasmussen, R. A., Simmonds, P. G., Alyea, F. N., Cardelino, C. A., Crawford, A. J., Fraser, P. J. and Rosen, R. D., The atmospheric lifetime experiment. III. Lifetime methodology and application to three years of $CFCl_3$ data, *J. Geophys. Res.*, 88, 8370–8400, 1983a.

Cunnold, D. M., Prinn, R. G., Rasmussen, R. A., Simmonds, P. G., Alyea, F. N., Cardelino, C. A., and Crawford, A. J., The atmospheric lifetime experiment. IV. Results for $CF_2Cl_2$ based on three years data, *J. Geophys. Res.*, 88, 8401–8414, 1983b.

Cunnold, D. M., Prinn, R. G., Rasmussen, R. A., Simmonds, P. G., Alyea, F. N., Cardelino, C. A., Crawford, A. J., Fraser, P. J., and Rosen, R. D., Atmospheric lifetime and annual release estimates for $CFCl_3$ and $CF_2Cl_2$ from 5 years of ALE data, *J. Geophys. Res.*, 91, 10,797–10,817, 1986.

Dickinson, R. E. and Cicerone, R. J., Future global warming from atmospheric trace gases, *Nature (London),* 319, 109–115, 1986.

Donahue, T. M., The atmospheric methane budget, in *Proc. NATO Adv. Study Inst. Atmos. Ozone: Its Variation and Human Influences,* Aikin, A. C., Ed., U.S. Department of Transportation, Washington, D.C., 1979.

Ehhalt, D. and Schmidt, U., Sources and sinks of atmospheric methane, *PAGEOPH,* 116, 452–464, 1978.

Eichner, M. J., Nitrous oxide emissions from fertilized soils: summary of available data, *J. Environ. Qual.*, 19, 272–280, 1990.

Fleagle, R. G. and Businger, J. A., *An Introduction to Atmospheric Physics,* Academic Press, New York, 1980.

Friedli, H., Lotscher, H., Oeschger, H., Siegenthaler, U., and Stauffer, B., Ice core record of $^{13}C/^{12}C$ ratio of atmospheric $CO_2$ in the past two Centuries, *Nature (London),* 324, 237–238, 1986.

Gamlen, P. H., Lane, B. C., Midgley, P. M., and Steed, J. M., The production and release to the atmosphere of $CCl_3F$ and $CCl_2F_2$ (chlorofluorocarbons CFC-11 and CFC-12), *Atmos. Environ.*, 20, 1077–1085, 1986.

Groves, K. S. and Tucks, A. F., Simultaneous effects of carbon dioxide and chlorofluoromethanes on stratospheric ozone, *Nature (London),* 280, 127–129, 1979.

Hahn, J. and Junge, C., Atmospheric nitrous oxide: a critical review, *Z. Naturforsch.*, 32a, 190–214, 1977.

Haigh, J. D. and Pyle, J. A., A two-dimensional calculation including atmospheric carbon dioxide and stratospheric ozone, *Nature (London),* 279, 222–224, 1979.

Hao, W. M., Wofsy, S. C., McElroy, M. B., Beer, J. M., and Toqan, M. A., Sources of atmospheric nitrous oxide from combustion, *J. Geophys. Res.*, 92, 3098–3104, 1987.

Houghton, R. A., Boone, B. D., Fruli, J. R., Hobbie, J. E., Melillo, J. M., Palm, C. A., Peterson, B. J., Shaver, G. R., and Woodwell, G. M., The flux of carbon from terrestrial ecosystems to the atmosphere in 1980 due to changes in land use: geographic distribution of global flux, *Tellus,* 316, 617–620, 1987.

Kaplan, W. A., Elkins, J. W., Klob, C. E., McElroy, M. B., Wofsy, S. C., and Duran, A. P., Nitrous oxide in fresh water systems: an estimate of the yield of atmospheric $N_2O$ associated with disposal of human waste, *Pure Appl. Geophys.*, 116, 424–438, 1978.

Keeling, C. D., Bacastow, R. B., and Whorf, T. P., Measurements of the concentration of carbon dioxide at mauna loa observatory, Hawaii, in *Carbon Dioxide Review 1982,* Clark, W. C., Ed., Oxford University Press, New York, 1982.

Khalil, M. A. K. and Rasmussen, R. A., Decline in the atmospheric accumulation rates of $CCl_3F$ (F-11), $CCl_2F_2$ (F-12), and $CH_3CCl_3$, *J. Air Pollut. Control Assoc.*, 31, 1274–1275, 1981.

Khalil, M. A. K. and Rasmussen, R. A., Sources, sinks, and seasonal cycles of atmospheric methane, *J. Geophys. Res.*, 88, 5131–5144, 1983a.

Khalil, M. A. K. and Rasmussen, R. A., Increase and seasonal cycles of nitrous oxide in the Earth's atmosphere, *Tellus,* 35B, 161–169, 1983b.

Khalil, M. A. K. and Rasmussen, R. A., Modeling chemical transport and mass balances in the atmosphere, in *Environmental Exposure from Chemicals,* Vol. 2, Neely, W. B. and Blau, G. E., Eds., CRC Press, Boca Raton, FL, 1984a, chap. 2, 21–54.

Khalil, M. A. K. and Rasmussen, R. A., Global sources, lifetimes, and mass balances of carbonyl sulfide (OCS) and carbon disulfide ($CS_2$) in the Earth's atmosphere, *Atmos. Environ.*, 18, 1805–1813, 1984b.

Khalil, M. A. K. and Rasmussen, R. A., Latitudinal distributions, sources, and sinks of halocarbons: observations from an oceanic cruise, *Trans. APCA Specialty Conf. Environ. Impact Natural Emissions,* Aneja, V. P., Ed., Air Pollut. Control Assoc., Pittsburgh, 1984c, 219–232.

Khalil, M. A. K. and Rasmussen, R. A., The atmospheric lifetime of methylchloroform ($CH_3CCl_3$), *Tellus,* B36, 317–332, 1984d.

Khalil, M. A. K. and Rasmussen, R. A., Causes of increasing atmospheric methane: depletion of hydroxyl radicals and the rise of emissions, *Atmos. Environ.*, 19, 397–407, 1985.

Khalil, M. A. K. and Rasmussen, R. A., The release of trichlorofluoromethane from rigid polyurethane foams, *J. Air. Pollut. Control Assoc.*, 36, 159–163, 1986.

Khalil, M. A. K. and Rasmussen, R. A., The residence time of trichlorofluoromethane in polyurethane foams: variability, trends, and effects of ambient temperature, *Chemosphere,* 16, 759–775, 1987a.

Khalil, M. A. K. and Rasmussen, R. A., Atmospheric methane: trends over the last 10,000 years, *Atmos. Environ.*, 21, 2445–2452, 1987b.

Khalil, M. A. K. and Rasmussen, R. A., Carbon monoxide in the Earth's atmosphere: indications of a global increase, *Nature (London),* 332, 242–245, 1988a.

Khalil, M. A. K. and Rasmussen, R. A., Nitrous oxide: trends and global mass balance over the last 3000 years, *Ann. Glaciol.*, 10, 73–79, 1988b.

Khalil, M. A. K. and Rasmussen, R. A., Climate induced feedbacks for the global cycles of methane and nitrous oxide, *Tellus,* 41B, 554–559, 1989a.

Khalil, M. A. K. and Rasmussen, R. A., Temporal Variations of Trace Gases in Ice Cores, Dahlem Konf., Berlin, 1988; *The Environmental Record in Glaciers and Ice Sheets,* Oeschger, H. and Langway, C. C., Jr., Eds., J. Wiley & Sons, New York, 1989b, 193–205.

Khalil, M. A. K. and Rasmussen, R. A., The potential of soils as a sink of chlorofluorocarbons and other man-made chlorocarbons, *Geophys. Res. Lett.*, 16, 679–682, 1989c.

Khalil, M. A. K. and Rasmussen, R. A., The Role of Methylchloroform in the Global Chlorine Budget, Presented at the 82nd Annu. Meet., Air & Waste Manage. Assoc., June 1989, 1989d.

Khalil, M. A. K. and Rasmussen, R. A., Global cycle of CO — trends and mass balance, *Chemosphere,* 20, 227–242, 1990a.

Khalil, M. A. K. and Rasmussen, R. A., Atmospheric methane: recent global trends, *Environ. Sci. Technol.*, 24, 549–553, 1990b.

Khalil, M. A. K. and Rasmussen, R. A., Constraints on the global sources of methane and an analysis of recent budgets, *Tellus,* 42B, 229–236, 1990c.

Khalil, M. A. K. and Rasmussen, R. A., The Global Sources of Nitrous Oxide, OGI Rep. 12-91-02, 1991a.

Khalil, M. A. K. and Rasmussen, R. A., Nitrous Oxide Emissions from Coal-fired Power Plants, OGI Rep. 12-91-01, 1991b.

Lacis, A., Hansen, J., Lee, P., Mitchell, T., and Lebedeff, S., Greenhouse effect of trace gases: 1970–1980, *Geophys. Res. Lett.*, 8, 1035–1038, 1981.

Lashof, D. and Ahuja, D., Relative global warming potential of greenhouse gas emissions, *Nature (London),* 344, 529–531, 1990.

Levine, J. S., Rinsland, C. P., and Tenille, G. M., The photochemistry of methane and carbon monoxide in troposphere in 1950 and 1985, *Nature (London),* 318, 254–257, 1985.

Linak, W. P., McSorley, J. A., Hall, R. E., Ryan, J. V., Srivastava, R. K., Wendt, J. O. L., and Mereb, J. B., Nitrous oxide emissions from fossil fuel combustion, *J. Geophys. Res.*, 95, 7533–7541, 1990.

Logan, J. A., Prather, M. J., Wofsy, S. C., and McElroy, M. B., Tropospheric chemistry: a global perspective, *J. Geophys. Res.*, 86, 7210–7254, 1981.

Lovelock, J. E., Methylchloroform in the troposphere as an indicator of OH radical abundance, *Nature (London)*, 267, 32–33, 1974.

Lu, Y. and Khalil, M. A. K., Tropospheric OH: model calculations of spatial, temporal, and secular variations, *Chemosphere*, 23, 397–444, 1991.

Luizao, F., Matson, P., Livingston, G., Luizao, R., and Vitousek, P., Nitrous oxide flux following tropical land clearing, *Global Biogeochem. Cycles*, 3, 281–285, 1989.

MacKay, R. M. and Khalil, M. A. K., Theory and development of a one-dimensional time dependent radiative convective climate model, *Chemosphere*, 22, 383–417, 1991.

Marland, G. and Rotty, R. M., Carbon dioxide emissions from fossil fuels: a procedure for estimation and results from 1950–1982, *Tellus*, 36B, 232–261, 1984.

Matson, P. A. and Vitousek, P. M., Ecosystems approach to a global nitrous oxide budget, *Bioscience*, 40, 667–672, 1990.

McCarthy, R. L., Bower, F. A., and Jesson, J. P., The fluorocarbon-ozone theory. I, *Atmos. Environ.*, 11, 491–497, 1977.

McElroy, M. B. and McConnell, J. C., Nitrous oxide: a natural source of stratospheric NO, *J. Atmos. Sci.*, 28, 1095–1098, 1971.

McElroy, M. B. and Wofsy, S. C., Tropical forests: interactions with the atmosphere, in *Tropical Rain Forests and the World Atmosphere*, Prance, G. T., Ed., Westview Press, Boulder, CO, 1987, 33–60.

Midgley, P. M., The production and release to the atmosphere of 1,1,1-trichloroethane (methylchloroform), *Atmos. Environ.*, 23, 2663–2665, 1989.

Molina, M. J. and Rowland, F. S., Stratospheric sink for chlorofluoromethane: chlorine atom catalyzed destruction of ozone, *Nature (London)*, 249, 810–814, 1974.

Muzio, L. J. and Kramlich, J. C., An artifact in the measurement of $N_2O$ from combustion sources, *Geophys. Res. Lett.*, 15, 1369–1372, 1988.

Neely, W. B. and Plonka, J. H., Estimation of time averaged hydroxyl radical concentrations in the troposphere, *Env. Sci. Technol.*, 12, 810–812, 1978.

NOAA, *Geophysical Monitoring for Climate Change 16*, Bodhaine, B. A. and Rosson, R. M., Eds., U.S. Department of Commerce, Washington, D.C., 1988, 67–76.

Pierotti, D. and Rasmussen, R. A., Combustion as a source of nitrous oxide in the atmosphere, *Geophys. Res. Lett.*, 3, 265–267, 1976.

Pinto, J. P. and Khalil, M. A. K., The stability of tropospheric OH during ice ages, inter-glacial epochs and modern times, *Tellus* 43B, 347–352, 1991.

Prinn, R., Cunnold, D., Rasmussen, R., Simmonds, P., Alyea, F., Crawford, A., Fraser, P., and Rosen, R., Atmospheric trends in methylchloroform and the global average for the hydroxyl radical, *Science*, 238, 945–950, 1987.

Prinn, R., Cunnold, D., Rasmussen, R. A., Simmonds, P., Alyea, R., Crawford, A., Fraser, P., and Rosen, R., Atmospheric emissions and trends of nitrous oxide deduced from 10 years of ALE-GAGE data, *J. Geophys. Res.*, 95, 18,369–18,385, 1990.

Ramanathan, V., Cicerone, R. J., Singh, H. B., and Koehl, J. T., Trace gas trends and their potential role in climate change, *J. Geophys. Res.*, 96, 5547–5566, 1985.

Rambler, M., Margulis, L., Fester, R., Eds., *Global Ecology: Towards a Science of the Biosphere*, Academic Press, New York, 1989.

Rasmussen, R. A. and Khalil, M. A. K., Atmospheric methane in the recent and ancient atmospheres: concentrations, trends, and interhemispheric gradient, *J. Geophys. Res.*, 89, 11,599–11,605, 1984.

Raynaud, D., Chappellaz, J., Barnola, J. M., Korotkevich, Y. S., and Lorius, C., Climatic and $CH_4$ cycle implications of glacial interglacial $CH_4$ change in the Vostok ice core, *Nature (London)*, 333, 655–657, 1988.

Rinsland, C. P. and Levine, J. S., Free tropospheric carbon monoxide concentration in 1956 and 1951 deduced from infrared total column air measurements, *Nature (London)*, 318, 250–254, 1985.

Ronen, D., Magaritz, M., and Almon, E., Contaminated aquifers are a forgotten component of the global $N_2O$ budget, *Nature (London)*, 335, 57–59, 1988.

Seiler, W., Conrad, R., and Schrafee, D., Field studies of methane emissions from termite nests into the atmosphere and measurement of methane uptake from tropical soils, *J. Atmos. Chem.*, 1, 171–186, 1984.

Sheppard, J. C., Westberg, H., Hopper, J. F., Ganesan, K., and Zimmerman, P., Inventory of global methane sources and their production rates, *J. Geophys. Res.*, 87, 1982.

Singh, H. B., Preliminary estimation of average tropospheric HO concentrations in the Northern and Southern Hemispheres, *Geophys. Res. Lett.*, 4, 453–456, 1977.

Sloan, S. A. and Laird, C. K., Measurements of nitrous oxide emissions from P. P. fired power stations, *Atmos. Environ.*, 24A, 1199–1206, 1990.

Stauffer, B., Fischer, G., Neftel, A., and Oeschger, H., Increase of atmospheric methane recorded in antarctic ice core, *Science*, 229, 1386–1388, 1985.

Steudler, P. A., Bowden, R. D., Melillo, J. M., and Aber, J. D., Influence of nitrogen fertilization on methane uptake in temperate forest soils, *Nature (London),* 341, 314–315, 1989.

Stevens, C. M. and Engelkemeir, A., Stable carbon isotopic composition from some natural anthropogenic sources, *J. Geophys. Res.,* 93, 725–733, 1988.

Theimans, M. and Trogler, W. C., Nylon production: an unknown source of atmospheric nitrous oxide, *Science,* 251, 932–934, 1991.

Thompson, A. M. and Cicerone, R. J., Possible perturbations of atmospheric CO, $CH_4$ and OH, *J. Geophys. Res.,* 91, 10,853–10,864, 1986.

Tyler, S., The global methane budget, in *Microbial Production and Consumption of Greenhouse Gases,* Rogers, J. E. and Whitman, W. B., Eds., Am. Soc. for Microbiology, Washington, D.C., 1991.

Wang, W.-C., Hung, Y. L., Lacis, A. A., Mo, T., and Hansen, J. E., Greehouse effects due to man-made perturbations of trace gases, *Science,* 194, 685–690, 1976.

Wang, W.-C. and Molnar, G., A model study of the greenhouse effect due to increasing atmospheric $CH_4$, $N_2O$, $CF_2Cl_2$ and $CF_3Cl_3$, *J. Geophys. Res.,* 90, 12,971–12,980, 1985.

Warneck, P., *Chemistry of the Natural Atmosphere,* Academic Press, Orlando, FL, 1988.

Weiss, R. F., The temporal and spatial distribution of tropospheric nitrous oxide, *J. Geophys. Res.,* 86, 7185–7195, 1981.

Weiss, R. F. and Craig, H., Production of atmospheric nitrous oxide by combustion, *Geophys. Res. Lett.,* 3, 751–753, 1976.

WMO, *Atmospheric Ozone 1985*, WMO, Global Ozone Res. and Monitoring Project — Rep. #16, Geneva, 1985.

WMO, *Report of the International Ozone Trends Panel 1988,* WMO, Global Ozone Res. and Monitoring Project — Rep. #18, Geneva, 1988.

WMO, *Scientific Assessment of Stratospheric Ozone: 1989,* WMO, Global Ozone Res. and Monitoring Project — Rep. #20, Geneva, 1989.

Wuebbles, D. J. and Edmonds, J., *Primer on Greenhouse Gases,* Lewis Publ., Chelsea, MI, 1991.

Yokoyama, T., Nishinomiya, S., and Matsuda, H., $N_2O$ emissions from fossil fuel fired power plants, *Environ. Sci. Technol.,* 25, 347–348, 1991.

# *Section IV: Natural Resource Management Needed to Provide Long-Term Global Energy and Agricultural Uses*

Chapter 18

# SOME CONSIDERATIONS IN APPLYING GIS TECHNOLOGY TO THE GLOBAL WARMING PROBLEM

**Lowell Kent Smith**

## TABLE OF CONTENTS



0-8493-4419-0/93/$0.00 + $.50
© 1993 by CRC Press, Inc.

# I. INTRODUCTION

This chapter deals with the application of GIS technology to a particular problem — the global warming problem. It argues that GIS technology will be an extremely useful means of dealing with the problem and that the technology will be used more extensively in the years just ahead.

This chapter assumes that most readers of this volume will be familiar with global warming issues, but rather unfamiliar with GIS technology, so a brief introduction to the nature of GIS and some closely related technologies is presented. For those wishing a more extensive introduction, the work by Maguire, Goodchild, and Rhind (1991) is highly recommended. The content of this chapter is based chiefly on the author's 20 years of experience in the GIS field.

Some of the perceived advantages of using GIS technology in dealing with environmental problems are sketched out as a way of indicating why this technology should be considered useful in dealing with the global warming problem.

Whereas every application of GIS technology is somewhat different from every other, the 10s of 1000s of projects which have now been carried to completion, and the 1000s of functioning GISs of all sizes throughout the world provide a sound basis for general advice about the creation and use of GIS technology. Put another way, the specific applications of GIS technology may vary, but the underlying principles are, to a great degree, the same.

Since the vast majority of particular applications of GIS technology will thus face many of the same kinds of problems, this chapter indicates the nature of these problems and, where it seems reasonable and possible to do so, suggests solutions which have worked elsewhere and which are likely to work on the global warming problem.

While this chapter is designed to apply specifically to GIS systems, in the context of global warming, it may be of interest to those who must create other kinds of environmentally related information systems.

# II. THE IMPORTANCE OF SYNTHESIS

One of the most common failings in dealing with environmental problems is the lack of a comprehensive, interdisciplinary, synthetic, or holistic view of the problem. While many technologies are directed to the analysis of environmental problems, far fewer are directed to synthesis. Information systems ought to be one of these synthetic technologies. GIS is rapidly becoming just such a synthetic technology, able to bring together the perspectives of many disciplines and information in many forms, fostering a holistic view of environmental and other kinds of problems. It is chiefly because of this characteristic of GIS technology that this chapter belongs in this book.

## III. PROBLEMS ESPECIALLY SUITABLE FOR APPLICATION OF GIS TECHNOLOGY

The kinds of problems to which GIS technology is most usefully applied include the following:

- Problems associated with large geographic areas
- Complex problems
- Problems subject to frequent review
- Multidisciplinary/interdisciplinary problems
- Problems in which much information must be managed
- Controversial or politically complex problems
- Problems which persist over a long time
- Problems requiring difficult, detailed analyses
- Interagency or intergovernmental problems

Based on these criteria, the global warming problem is a prime candidate for the application of GIS technology.

## IV. GIS TECHNOLOGY AND GLOBAL WARMING

This chapter is written based on convictions that GIS technology offers an extremely important means of collecting, storing, analyzing, synthesizing, displaying, and reporting information about the problem of global warming; that it provides a very valuable means for dealing with the problem in an interdisciplinary way; that it provides one effective means for devising broad scale policies with respect to global warming; and that it provides a necessary technology in monitoring and managing the environment so as to deal with the problem. These convictions are based on the extensive use which is now being made of GIS technology in dealing with many other environmental problems.

In many ways the global warming problem is quite similar to other environmental problems. The problem has ecological, economic, political, social, and other aspects; the systems involved are complex and their interactions difficult to understand; and information about critical processes is difficult to acquire and incomplete.

At present, 10s of 1000s of persons and 1000s of organizations are using GIS technology, many of them applying the technology to environmental problems. While many of these problems are uncomplicated and localized in extent, others are complex and global. These GIS applications include, for example, the creation of global GIS databases, application to monitoring, management, decision-making, and policy-making, and the application of large, hierarchical, coupled, interdisciplinary modeling and simulation efforts, including some atmospheric modeling. Based on this extensive body of experience over the last 25 years or so, GIS technology seems suited to be used on the global warming problem.

## V. WHAT IS A GIS?

A Geographic Information System (GIS) is a computer-based information processing system consisting of computer hardware, computer software, and a database. What makes a GIS unique among computer-based information systems is that the information it processes is spatially referenced. GIS information is of two kinds: the geographic, cartographic, or spatial references (e.g., the x,y coordinates of a feature) and the attributes associated with them (e.g., the detailed description of the feature).

While most of the functions of a GIS are supported by general purpose computing hardware and communications equipment (e.g., central processing units, mass storage devices, printers, local area networks, etc.), some specialized peripheral hardware devices are also usually necessary. These include, for example, digitizers and scanners for capturing x,y coordinate-referenced data, and plotters of various types for producing maps.

A full description of GIS software is beyond the scope of this chapter. A modern, full-function GIS probably includes more than 1 million lines of computer code, and is capable of 100s of kinds of cartographic and related manipulations. GIS software is used to capture, store, query, manipulate, analyze, model, and display the spatial and attribute data in the GIS database. GIS software also can provide such functions as network analysis, analysis and display of three-dimensional data, interconversion of data formats, integration of computer-aided drafting (CAD) data, capturing of surveying data, etc. GIS software might include a database management system (DBMS) for maintaining the attribute data and a separate system for dealing with the spatial references. The spatial references are usually in either vector form (points, lines, and polygons) or raster (grid cell) form; these two forms can also be interconverted.

The most familiar products of GIS technology are maps. Many organizations which have previously done their cartographic production work exclusively by manual means are now using GIS technology (e.g., *National Geographic* magazine, the U.S. Geological Survey, and various commercial mapping firms). For cartography, GIS offers rapid map creation, rapid iterative editing, display products which are the equal of most hand drafted maps, ease of updating, ease of rescaling and other fundamental manipulations, etc.

GISs can also be used for statistical analyses, modeling and simulation, network analysis, and a range of other functions which ordinary paper maps do not support.

GISs are often associated with the application of other technologies, i.e., image processing, CAD, photogrammetry, photointerpretation, etc. It is in association with these that GIS evidences its full capabilities as an information integrating technology.

## VI. PERSONNEL, PROCEDURES, AND ORGANIZATIONS

Some people believe that the definition of a GIS should include only computer hardware, software, and data. As a practical matter, such a definition falls short of accurately portraying the reality of GISs. Among the most important elements of successful GISs are the people who support and use them, the procedures they follow in doing so, and the organizations within which such GISs exist.

The creation and long-term maintenance of large GISs are a considerable effort. After the data problem, the human and organizational resources required is probably the biggest problem which must be overcome in creating and maintaining a successful GIS. While many aspects of this problem require fairly routine management efforts, there are some features of a GIS which make it unique; those responsible for a GIS should be aware of these unique features and how to deal with them.

## VII. SOME EXISTING INTERNATIONAL APPLICATIONS OF GIS TECHNOLOGY

While their number is small, global and international GIS databases do exist, and their number is increasing rapidly. GEMS the Global Environmental Monitoring System, and various other global GISs supported by U.N. agencies and other cooperating agencies are perhaps the best known of these.

The Defense Mapping Agency (DMA) in the U.S. is now creating a digital chart of the world based on its series of Operational Navigation Charts (Danko, 1990). The database, of more than 7 gigabytes, will be available on CD-ROM for use on personal computer-based GISs.

At the time of writing this chapter, the U.N. Conference on Environment and Development is considering the creation of its Agenda-21 Information Support System (AGIS), which will probably make use of GIS technology as one of its components. This is to be a global system for use by organizations charged with the responsibility of implementing or monitoring the implementation of Agenda 21, the global environmental charter which is to be one focus of discussion at the Earth Summit in Rio de Janeiro in June of 1992. Among the many problems the AGIS system will deal with is the problem of global warming.

There are other international GISs dealing with such issues as threatened and endangered species, deforestation and the loss of ancient forests, loss of wildlife habitat, etc. These GISs are often maintained by conservation organizations or universities, sometimes in co-operation with government organizations.

## VIII. GIS DATA

The spatially referenced information in a GIS can be of many different types. The cartographic data can be in vector format or in raster or grid cell referenced format, and can include images (e.g., remote sensing), CAD data, video, survey data, scanned documents, textual descriptions, etc. One of the most important characteristics of a GIS is its ability to integrate a wide range of data types into a single system.

### A. THE DATA PROBLEM

As with many environmentally related problems, the chief problem associated with creating and maintaining most GISs is the collection and maintenance of the database. The largest cost of creating and maintaining a GIS is usually the collection of the original source data. This may represent three fourths of the initial cost for a GIS and, if real personnel costs are assigned to this effort, a significant part of the ongoing maintenance of a GIS.

Much of the data which has been collected over the years and is now included in GISs was collected originally for purposes which differ considerably from the purposes for which the data are now used. This means that the data are often being used in ways which were not forseen or even forseeable by those who gathered them. This can cause problems for GIS users. The data may be outdated (obsolete, incorrect, in need of updating or revision); the source maps may have been created on nonstable material (such as paper) which has become distorted; the scale and resolution of the data may be inappropriate for the present uses; the classification systems used originally may be inappropriate or inconsistent with other information. There are other kinds of data problems that must be remedied if GISs are to be useful.

#### 1. Periodic Updating

One approach to keeping a database up-to-date is the periodic update; e.g., as new aerial photography becomes available for a region, it is used to update all the relevant variables in the database. Periodic updating is still perhaps the most common means of updating databases, since so many kinds of spatial information are still collected through special efforts, i.e., overflight missions or field expeditions, rather than routine efforts, i.e., geo-stationary satellites or field monitoring stations.

### 2. Transaction Maintained GISs

Many GISs at the present time are maintained, at least in part, by a process of continuous updating through transactions; a transaction might be the updating of the monitoring information for a particular recording station, or the entry of the most recent text of a revised regulation. Many of the data in a GIS dealing with global warming might be updated by transactions with the database.

## B. DATA ON GLOBAL WARMING

The list of data which are relevant to the global warming problem is extremely long. It would include basic geographical data (elevation, contour, surface hydrology, political and administrative boundaries, etc.), biological data (ecological communities, community respiration, photosynthetic carbon dioxide [$CO_2$] fixation, processes evolving other greenhouse gases, etc.), monitoring data ($CO_2$ levels in the atmosphere, deforestation rates, field measurements of gas evolution, etc.), economic data (prices of fuels, costs of conversion or pollution control, etc.), relevant regulations and laws, political data (pending legislation, organizations active in dealing with the problem, etc.).

It is apparent that data relevant to global warming are relevant to many other environmental, political, and social problems, and that few kinds of data are completely divorced from environmental problems as complex as global warming. This is discouraging, in that one cannot expect to have all the kinds of data wanted; it is encouraging as well, in that many kinds of data have been collected and some of the data needed for the study of global warming are already available.

## C. SURROGATE DATA

In studying complex problems like the environment and those factors (economic, political, social, etc.), which relate to the environment, users will often not have all the data they need and never have all they might like.

This lack of data suggests to some people that many environmental problems are beyond the capabilities of computer analysis; however, another approach is to make use of substitutes of various kinds, especially in performing system modeling and simulation, until the needed data become available. Thus, one might make use of data on waterlogged soils and certain types of wetlands as a way of estimating what areas are likely to harbor those biological processes which produce methane ($CH_4$) gas.

While this is a necessary compromise if some environmental problems are to be addressed at all, the limitations imposed by such an approach must be admitted and kept in mind, otherwise one is tempted to accept the neatly produced final products and forget the limitations of the data on which the products are based.

# IX. GIS STANDARDS

A typical GIS, when hardware, communications, database, software, and other elements are included, may be subject to more than a 1000 individual existing standards produced by one or another national or international standards organization. Despite this, there is, paradoxically, surprisingly little standardization in the GIS field today. To impose standards too early is to straightjacket the development of new technology. Many vendors do not want to adhere to standards which they feel place their products at a competitive disadvantage. Competing standards often force buyers to choose between alternative products, with disastrous results for the losers (witness the Beta vs. VHS standards in videotape). Moreover, producing a data-processing standard is a process that often takes 6 or more years plus the work of numerous committees and, even when the standard is released, this does not mean that it will be adhered to by GIS vendors or users. Perhaps the most important agencies in maintaining standards are large purchasers of data processing goods and services when they require bidders to meet standards in order to compete for contracts.

Nevertheless, in considering the complexity of a major environmental problem like global warming, various kinds of standards would obviously be useful. It would be useful to be able to interconnect any GIS to any other and be able to share data, preferably across the world.

As comprehensive, global analyses of such problems become feasible and customers want to deal with them, the pressure will grow to adopt more GIS standards and find means to enforce them. This is a trend that will grow in the years just ahead.

## X. CREATING A GIS

In order that readers may have a practical sense of what the creation of a GIS application to global warming entails, the process is described below. This account assumes that the GIS in question is a single, coherent system which has been created through a process of design and implementation.

### A. USER NEEDS ANALYSIS

It is always tempting, because of the considerable costs involved in conducting a user needs analysis, to avoid doing so. This is often accomplished by having senior managers make the decisions about what the system ought to be vs. what it should be able to do. Unfortunately, the actual people who will process the information and base their actions and decisions on it often work in ways different from those which were assumed, even by their superiors.

By consulting the actual users and processors of the information, the user needs analysis process attempts to avoid all the pitfalls of the badly designed system. Actual users are interviewed in person, preferably, or by phone or other means where actual visits are impractical. The purpose of these interviews is to discover exactly what each such user or potential user does or would do with the information provided by the system. In particular, the flow of decisions which follow from the use of the system is considered so that the system can be designed to provide what will really be useful. A user needs study might be conducted by the organization acquiring the GIS, but it may be better performed by persons experienced in this kind of work.

### B. DATA SELECTION

It is tempting to assume that which data should go into a GIS is a straightforward question: users, of course, want all the data they can get; or, alternatively, which data to include is obvious, given the problems to be dealt with. In reality, perhaps no question related to the GIS is more difficult to answer and none is more critical to the long-term use and success of the system as a whole. In answering this question, many features of data have to be taken into account: age, availability, costs, comprehensiveness, scale, format, form, ease of maintenance or updating, location, security, etc. While in some cases the data are so sparse that one is forced to accept whatever exists or do without, in most parts of the world today there are choices to be made among alternatives. Making such choices is an extremely sophisticated undertaking, requiring the very best advice and long experience in working with GIS technology. This is an area in which no expense should be spared since the choices, once made, may have an influence on the success of the GIS for years or decades. Skilled and experienced consultants probably provide the best means of accomplishing this task.

In order to make the decisions, existing and available data resources must be inventoried and the alternative sources evaluated and compared. Then the data resources have to be matched against user needs and trial database contents considered. From these analyses flow decisions about the basemaps to use, the preferred scale of mapping, which items to include, how the data need to be classified, where gaps exist, how these gaps are to be filled, etc.

### C. SYSTEM DESIGN

With information available about what users need and what data are available to meet those needs, it becomes possible to size the GIS and determine what its processing, storage, display, and other capabilities will need to be. Hardware and software vendors can often provide useful information on this matter. Experience shows, however, that successful GISs are often very quickly overwhelmed by user requests, so that thought should be given at the outset to how the GIS will be expanded and upgraded in such an eventuality.

### D. BENCHMARK TESTING

One important way to determining whether or not a particular hardware/software system will perform as promised is to conduct an actual test of the system's capabilities. This benchmark test establishes typical processing speeds for the full variety of tasks which the system is designed to perform. While it is possible to design one's own benchmark test, there are people who can provide this service, and using them as consultants is probably worth the cost. Designing tests which are appropriate benchmarks is a high level activity requiring a good deal of experience and judgement, especially since the tests must also be tailored to particular users' applications. While the cost may seem high, in the context of the total cost of the final system the cost of benchmarks is usually quite minimal, and well worth the price.

### E. IMPLEMENTATION PLAN

Given the decisions about the various components of the GIS, an actual phased implementation plan must be designed. This includes actions, a time frame, milestones, total duration, and estimated costs.

### F. DATA STANDARDIZATION

Because source data come in so many forms (differing media, age, format, scale, classification, reliability, completeness, coverage, etc.) GIS data may need to be "standardized" before they are useful and captured in digital form. This process involves correcting as many of the flaws in the data as possible. Because of the complexity of this task, it is, today, largely a manual process of editing and correction in which the most recent and accurate available information is used to correct the flaws in the other data; it relies on the fact that intercomparison of data usually can improve the quality of all the types of data which are compared. This standardization process can save a good deal of time and resources during the subsequent data capture (digitizing, automation, scanning, text entry, etc.) process, but requires skill and experience if it is to be done effectively. There are now many commercial firms throughout the world which offer data-entry services, and some of them perform very well. This is an area in which, as GIS use grows, one can expect to see specialization of function among commercial firms.

### G. APPLICATION PROGRAMING

While a great many GIS software applications already have been programed, and many of these can be gotten for no cost or only nominal costs, it is inevitable that for a complex environmental problem, some new applications programing will be required. This programing might range from the creation of large computer models to the creation of special user interfaces. This kind of programing can often be done by the users themselves, and much of it is getting easier to do as GIS software matures. In implementing a new GIS, time and resources will need to be devoted to this task soon after the system is running and probably periodically thereafter.

### H. TRAINING AND TECHNOLOGY TRANSFER

One of the major problems in the GIS field is the transfer of the technology from organizations and people who have acquired some skill to those who want to learn it. While training and observation provide some background for this transfer, in most cases they are not enough. In particular, in order to be able to do GIS production work efficiently, staff must probably work for some length of time in a functioning production shop in order to attain the necessary skills and experience. Lacking this, new GIS staff should not expect to attain full competence for probably at least 1 year after they begin work with a new system.

The larger problem of efficiently transferring this technology to large numbers of people throughout the world, as would be required for its use on the global warming problem, has not yet been solved. While the use of data gathered and automated by others is probably now within the reach of nearly any intelligent user, the skills necessary to capture such data and prepare it for use still require a good deal of training and experience.

It may be that the quickest way to bring GIS to the large number of people who could profit from its use will be through a class of GIS information-providing organizations which provide GIS information in packaged form. Such a class of "data publishers" is now emerging. Of course the dangers of using data provided in this way are considerable, unless users are also capable of being critical of the published data products they acquire.

### I. PILOT STUDY

It is advisable, when creating a new GIS, to test the complete design of the system by carrying out a pilot project. This pilot should involve the entire operation of the GIS, from data collection to production of final products. One way to accomplish this is to begin the use of the system by performing typical working operations for a discrete area selected from the total area covered by the database. All conditions should be as like the actual working conditions as possible in order to provide a fair test of the system. The results of the pilot should be used to refine the original system design as necessary.

## XI. GIS-RELATED ISSUES

### A. GLOBAL NETWORKS

Among the most important GIS developments for the analysis of global warming will be the increasing use of large or global GIS networks in the years ahead. By making a common information resource available to a large, diverse group of investigators, such networks are likely to stimulate the kind of interdisciplinary, synthetic work which a complex problem like global warming almost certainly demands.

Global networks are still in their infancy. Wideband communications systems, capable of transferring hundreds of megabytes of information in seconds, and doing so at low cost, will be needed before GIS use over global networks is really practical. Fiberoptics or some similar technology may be the key to making such networks feasible.

### B. CD-ROM TECHNOLOGY

Another approach to making GIS (and other) information available to users is through the use of CD-ROM technology or its successors. Instead of the information being made available over the network where it may be constantly updated, the information would be published periodically through the distribution of a new set of CD-ROMs. Given the volume of use which a system dealing with global warming is likely to have, this approach should be economical. The approach has the advantage of avoiding the necessity for networks with wide-bands and high transmission rates. Individual users with little more than a personal computer (PC) and a CD-ROM drive could access enormous databases quite inexpensively. For example, a digital map of the world at 1:1,000,000 will fit on fewer than ten such CDs and the total price for the complete set would probably be under $200 (U.S.).

### C. GIS AND MODELING

One of the most powerful applications of GIS technology is simulation and modeling. Because GISs contain information which is geographically and spatially referenced, they are potentially very useful for dealing with environmental problems which are similarly connected to the real, spatial world. Despite this, relatively little of such modeling using GISs is now done, perhaps because of the lack of needed data, or, alternatively, because of a perceived lack of fundamental understanding of the systems of interest, resulting in a lack of confidence in the ability of GIS models to provide useful and reliable results. That this may be changing is indicated by the occurrence in September 1991 of the First International Conference/Workshop on Integrating Geographic Information Systems and Environmental Modeling. The proceedings of this conference are in preparation.

There are special problems associated with modeling and simulation of global warming using a GIS. Most present modeling is being done at a rather coarse resolution, perhaps chiefly because of the scarcity of reliable data. Models do not take into account the richness of existing data resources, though, admittedly, many kinds of needed information are sparse or nonexistent.

### D. PUBLIC ACCESS

In the years just ahead there is likely to be an explosion on the use of GIS technology, with the number of GIS users quite likely exceeding 5 million within about 1 decade. Most of these users will be nonspecialist members of the public at large. One of the applications they are likely to want to make of GIS technology is query and analysis of environmental problems, including global warming. A major consideration in thinking about future GIS use must therefore be how to accomodate public interest and public access.

### E. FUTURE DEVELOPMENTS IN GIS TECHNOLOGY

Perhaps the strongest force driving the development of GIS technology has been the rapid development of computer hardware; this development has made GIS both possible and increasingly inexpensive. As a rule of thumb, most observers expect that many kinds of hardware capabilities (e.g., processing speed, memory chip size) will double about every 3 years, or that the cost of equivalent computing power will be halved every 3 years. It appears that there will be at least another 5 or 6 such doublings (or halvings) in the next 2 decades, just applying the theoretical understanding that exists today. This means that GIS capabilities, and probably GIS use, will continue to grow at a similar rate, and perhaps faster. As prices for GISs fall, there will inevitably be thresholds crossed which allow whole new classes of users to enter the field. For example, as the price of a complete PC-based GIS hardware and software system falls toward that of today's word processing systems, average citizens will likely become GIS users in great numbers. Such changes are likely to have profound effects. One of these might be greatly increased citizen interest in and monitoring of the global warming problem, with associated political and social effects.

A number of emerging technologies may also have a considerable impact on the use of GIS technology in studying global warming. Parallel processing computers are especially suited, it would appear, to the manipulation of the millions of points, lines, polygons, grid cells, and other elements of a GIS database. The idea of letting a separate processor deal with each grid cell, for example, seems to offer a direct way to speed up GIS processing by many orders of magnitude. At present, however, parallel processing is not widely used in GIS data processing.

Artificial intelligence and expert systems have likewise been minimally applied to GISs, though that is beginning to change. It seems likely that the early applications will be for such mundane problems as designing map legends rather than for the creation of highly intelligent personal assistants for GIS users.

There are innumerable possible applications of nanotechnology to the design and construction of environmental sensing devices. While a discussion of these developments is beyond the scope of this chapter, such applications are likely to revolutionize our study of the environment over the next 20 years or so. Despite this, to date they have had very little effect on the field.

## XII. CONCLUSIONS

The application of GIS technology to the global-warming problem offers great promise, but the implementation of such an application is a very considerable undertaking. Those interested in dealing with global warming would be well advised to consider the use of this technology, and should take into account the large body of GIS experience which has accumulated in the 25 years since GIS use was first introduced.

## REFERENCES

Danko, D. M., 1990. The Digital Chart of the World Project, GIS/LIS 90: Proc. 5th Int. Conf., Anaheim, CA, 1990, 392–401.

Maguire, D. J., Goodchild, M. F., and Rhind, D. W., 1991. *Geographical Information Systems: Overview, Principles, and Applications,* Longman Scientific and Technical Pub., London, U.K.

Chapter 19

# GLOBAL WARMING AND AGRICULTURE: A SURVEY OF ECONOMIC ANALYSES AND PRELIMINARY ASSESSMENTS

**Hayri Önal**

## TABLE OF CONTENTS



0-8493-4419-0/93/$0.00 + $.50
© 1993 by CRC Press, Inc.

# I. INTRODUCTION

Climatic change due to greenhouse warming is an issue of growing importance and has been attracting wider attention among policy makers of developed nations and the scientific community. Atmospheric scientists, environmental scientists, agricultural scientists, and economists have all contributed to this discussion. A large volume of natural sciences research is underway about climatic change; however, few economic analyses have been done so far. The late appearance of economic analyses results from the need for estimates of physical effects of climate change before conducting economic analyses. At present, the issue receives high priority on the research agendas of some leading research institutions.

The impact of global warming may be especially profound on the supply and demand of energy and agricultural products. This chapter focuses on the latter. Economic analysis in this case may include a wide array of issues, including two-way linkages between human activities and climate, worldwide and regional impacts of climate change and interactions between them, evaluation of vulnerable resources, and the role of government interference and control. The existing impact analyses have addressed only the cost-benefit issues at regional and/or national levels. Climate simulation models indicate an average annual warming by about 4°C (9°F) especially in the Northern Hemisphere between 40 to 60° N latitudes, but little effect on the Southern Hemisphere. Therefore, the economic studies conducted so far exclusively covered the agricultural systems of North America and northern regions of Europe. There is a need for doing similar analyses for developing countries as well, especially for those which significantly affect the world supply and demand situation. This is because climate change is a global phenomenon and a strong relationship exists between the economies of developed and developing nations. At present, there is no coordinated research effort to link national research results in a global framework. For that reason, those results are of partial nature in that they do not incorporate the effects of climate change upon the rest of the world. Thus, our knowledge about the worldwide economic effects is still incomplete, and incorporation of the linkages between spatial effects may significantly change the findings of the currently available studies.

The purpose of this chapter is to review the findings of some empirical studies including a regional modeling study in which the author was involved. The review focuses on: (1) effects on agricultural productivity; (2) potential changes in market equilibria; (3) international trade effects; (4) possible environmental and ecological effects; and (5) issues in agricultural resource management and policy implications. Before reviewing those empirical results, however, it is worth mentioning how they were obtained because the findings of individual models are heavily influenced by the underlying assumptions and the methodology used. The next section briefly describes the methodological background of the studies reviewed here.

# II. METHODOLOGY FOR ECONOMIC ANALYSIS OF CLIMATE CHANGE

Economic impact analysis of climate change is a multidisciplinary research which is usually done in three successive stages. In the first stage, a set of plausible climate scenarios, including key climate parameters that affect cropping systems, are generated using simulation results of currently available general circulation models (GCM). The second stage involves the derivation of crop yield response data under the climate conditions specified. Yield responses of crops are obtained either by statistical estimation methods that relate historical yields to observed climatic variation (Thompson, 1969; Santer, 1985; Katz, 1979; Sakamoto, 1981), or by using simulation models that involve mathematical relations between crop yields and climate variable such as altered seasonal temperature, precipitation, and solar radiation. Several crop-weather simulation models have been developed and used for impact analysis (Stapper and Arkin; Maas and Arkin; Allen and Gichuki, Jones and Kiniry; Wilkerson et al.).

Whether a statistical model or a simulation model is used, the crop response model is first validated (tuned) for the particular agricultural system under consideration. Statistical methods are argued to be less appropriate for this purpose due to two reasons. First, the future climatic conditions may fall outside the observed climate variation range and extrapolation may not produce the true effects. Second, an important component of the global warming problem is the increased concentration of $CO_2$ in the atmosphere, which may have substantial direct fertilization effect on crop yields. This factor cannot be included in a statistical estimation model as it has not been observed in the past. Crop-yield responses estimated in this way would, in general, underestimate the true effects. For these reasons, simulation models are usually preferred while generating crop-yield information.

Economists have little to say in either of the first two stages mentioned above. The last stage of the analysis involves modeling the economic components of the system and their interactions, given the climate scenarios and associated crop yield response information. In several case studies the changes in crop yields obtained from simulation models have been directly used to assess the impact of climate change on agricultural crop production (Rosensweig, 1985, 1988; Ritchie et al., 1988; Peart et al., 1988; Newman, 1980; Blasing and Solomon, 1984; Wilks, 1988; Santer, 1985). This approach may be helpful for indicating possible directions of change in agricultural production, but a major shortcoming is the fact that the producer decision-making process is not incorporated. This is because in addition to the physiological effects of climate and other environmental factors, human activity, namely management decisions made by individual producers, affects agricultural production. At firm level, individual producers usually choose the most profitable crop mix alternative(s) given the surrounding physiological and economic conditions. Therefore, in a climatically different environment, the current crop mix patterns may not persist. If individual producers take similar courses of action and switch from one crop set to another, the overall supply of some products will change. In turn, the market equilibrium and prices will be altered. Hence, an appropriate approach has to consider the following four major factors and their relations: (1) potential climate change, especially seasonal temperature and precipitation changes; (2) yield responses of crops under changed climatic conditions; (3) producer behavior; and (4) market conditions. In this regard price endogenous mathematical programming models have been useful tools. In this approach the sum of consumers' and producers' surplus, measured by the areas under the demand functions and above the supply functions, is maximized subject to sectoral and/or regional constraints depicting the production possibility sets. The optimum solution of the model corresponds to the market equilibrium (Takayama and Judge, 1971; McCarl and Spreen, 1980). In this way, market prices are internally determined (endogenized) by simultaneously incorporating the agricultural supply potential and market conditions. Several climate analyses have used this approach (Adams et al., 1988a,b; Dudek, 1988; Önal and Fang, 1991; Binkley, 1988).

As mentioned earlier, the economic analysis of climate change is based on climate scenarios developed by GCMs and crop responses generated by yield response models. Besides the shortcomings that the economic model itself may have, the performance of any economic model largely depends on the performance of the former models, especially the specification of the future climate conditions. Unfortunately, the available GCMs do not produce consistent predictions about the future climate. Although they agree to a large extent on the annual global changes, their spatial and temporal breakdowns vary considerably, which may have very different implications for agricultural productivity. Crop yield response is a complex phenomenon in which several factors are involved, such as the average daily temperature and precipitation during the growing season, solar radiation, and the $CO_2$ concentration. Two climate scenarios with the same annual average temperature and precipitation estimations but with different temporal distributions may have totally different effects on crop yields. Moreover, increased frequency of severe climate events, such as frosts and floods, may also affect crop yields. The predictions of the currently available GCMs regarding the above factors are vague, inconsistent, and not totally reliable. Therefore, while doing economic analysis it is customary to consider several alternative scenarios rather than a single scenario for future climatic conditions. Because of this uncertainty and unreliable data input to the model, the findings of any climate impact analysis should be interpreted carefully. Improved resolution and reliability of the GCMs will enhance the realism of economic models and provide a more accurate understanding of the interaction between increased $CO_2$ concentration and agriculture.

## III. EFFECTS OF CLIMATE CHANGE UPON AGRICULTURAL PRODUCTIVITY AND PRODUCTION PATTERN

Among several aspects of global climatic change, three of them are crucial for agricultural productivity. They include (1) increases in seasonal temperatures; (2) changes in precipitation patterns; and (3) increases in atmospheric $CO_2$ concentration. Several field studies and experimental research conducted under controlled environmental conditions show that both accumulated temperature and precipitation, as well as their seasonal distribution particularly in the growing seasons, significantly affect crop growth and net yields. The effects of these factors were also estimated using statistical methods and were found to be significant (Thompson, 1969). Most of the agricultural impact analyses are based on the predictions of two GCMs, namely the model developed by the Goddard Institute of Space Studies (GISS) (Hansen et al., 1983) and the model developed by the Geophysical Fluid Dynamics Laboratory (GFDL) (Manabe and Wetherald, 1980). In addition, the model developed by Oregon State University (OSU) (Karl et al., 1990) is also used in some case studies. GISS predicts moderate changes in temperature and increased precipitation, while GFDL predicts more severe temperature increase and reduced precipitation, especially during the summer when water demand by crops is greater due to higher temperatures and evapotranspiration. OSU, on the other hand, predicts relatively less temperature and precipitation increases during the crop growing seasons. Tables 1 and 2 present the annual and seasonal climate predictions of these models for the U.S. and some cool regions in the Northern Hemisphere. Consequently, the GISS scenario implies relatively moderate yield changes in U.S. agriculture, while the GFDL scenario shows substantial yield reductions especially in southern regions (Table 3). On the other hand, temperature increases predicted by the GISS model cause the growing season to lengthen. Simulation studies estimate significant yield increases in the cool regions of the Northern Hemisphere (Table 2). The third factor, increased $CO_2$ concentration, is shown to be very important for crop yields since it increases photosynthesis activity by about 30% (Kimball, 1983; Allen et al., 1987) and water use efficiency through reduced evapotranspiration. The combined yield effect of weather and increased $CO_2$ concentration can alter the situation drastically (see Table 3). In most cases the direct fertilization effect of $CO_2$ offsets yield losses due to climate effects, and even increases yields above their present levels. Under the GISS scenario the overall yield effect of climate change upon U.S. agriculture is estimated to be positive (as high as 44% in major agricultural regions). Under the GFDL scenario, however, yield loss would be reduced but not totally offset. The impact of the reduced crop productivity upon U.S. agriculture is estimated as a 10% decline in the overall production under the GISS scenario, and as much as 40% under the drier

**TABLE 1**
**Average Predictions of Three Climate Models for the U.S.**

| | Temperature change (°C) | | Precipitation change (mm/day) | |
|---|---|---|---|---|
| | Annual | Summer | Annual | Summer |
| GISS | +4.32 | +3.50 | +0.20 | +0.24 |
| GFDL | +5.09 | +4.95 | +0.09 | −0.08 |
| OSU | +2.98 | +3.10 | +0.17 | +0.11 |

Modified from Adams, Glyer, and McCarl, (1988).

GFDL scenario. A regional study conducted by Önal and Fang indicates similar results for Illinois agriculture (Table 4). The yield effects reported in Table 2 are estimated without taking the direct $CO_2$ effect into account. Thus, it is likely that the cool regions of northern Europe, Japan, and Canada would enjoy much higher crop yields than shown in Table 2.

The net yield effect of climate change varies between crops and regions depending on two primary factors, water demand and supply, and temperature resistance. In the U.S. corn belt, for example, where soils have greater water-holding capacity, drier climate conditions would affect crops less than in the southern Great Plains and southeastern states where the soil is thinner and/or there is not enough irrigation capacity to compensate for the adverse effects of reduced precipitation (Table 3). Sector models incorporating economic and agronomic factors show that the hypothesized climate conditions may be destructive for the agriculture of the latter regions.

## IV. RESOURCE USE

Among agricultural resources, the two most important are land and water. Simulation studies show that irrigation can mitigate the adverse effects of temperature increases and reductions in rainfall in seasons when soil moisture is limited. The demand for land and irrigation differs among regions depending on the crop yield changes and consequent shifts in the most economical crop mix. The study by Adams et al., 1988b shows that the land use would decline in the southern U.S. because cropping would not be economical due to serious yield losses in some regions, while it would increase in the northern regions. When the direct fertilization effect of $CO_2$ is considered, however, the land use pattern is shown to be virtually the same as the present situation according to the same study. In either case the irrigation demand increases especially in southern regions. The GFDL scenario implies much drier conditions, especially in summer. The irrigation demand is higher under this scenario, ranging between 5 to 160% if the direct $CO_2$ effect is included, and between 40 to 170% increase if the $CO_2$ effect is excluded. Under the GISS scenario, the increase in irrigation demand is moderate but still significant.

Changing climate may also affect the cropping practices and land use. Longer growing seasons and increased temperatures in the fall may allow double cropping and change the crop pattern especially in northern regions. Önal and Fang (1991) find that in northern Illinois, soybeans may be planted after winter wheat as a second crop. This practice is currently used in southern Illinois but not in the northern regions of the state due to unfavorable weather conditions in the fall. Consequently, wheat and double crop soybeans would become a major cropping activity and replace a considerable portion of the corn production in northern Illinois.

**TABLE 2**
**Effects of Climate Change on Some Dryland Crop Yields in Cool Northern Regions**

| Region | (GISS scenario) Growing season: Temperature change | Precipitation change (%) | Crop | Average yield change (%) |
|---|---|---|---|---|
| Canada (Saskatchewan) | 3.5–3.6°C | 9–14 | Spring wheat | −4/−29 |
| Iceland | 4°C | — | Hay, pasture | +66 |
| | | | | +49/+52 |
| Finland | +35% | +59 | Barley, spring wheat | +9/+14 |
| | | | | +10/+20 |
| U.S.S.R. North | 2.2°C | +36 | Winter rye, spring wheat, | +6 |
| Central | 2.7°C | +50 | winter wheat, barley | +3 |
| | 1.5°C | — | | +30 |
| | | | | −4 |
| Japan North | 3.5°C | — | Rice | +5 |
| Central | 3.2°C | — | | +2 |

Modified from, Parry, Carter, and Konijn, (1988).

**TABLE 3**
**Effects of Climate Change on Regional Dryland Corn and Soybean Yields in the U.S.**

| | % change in yield | | | | | | | |
|---|---|---|---|---|---|---|---|---|
| | GISS scenario | | | | GFDL scenario | | | |
| | Corn | | Soybean | | Corn | | Soybean | |
| Region | w/o | w | w/o | w | w/o | w | w/o | w |
| Great Lakes States | −2.2 | 42.0 | −35.7 | 42.0 | −56.8 | −12.7 | −54.1 | 7.8 |
| Corn Belt | −24.1 | 15.3 | −26.8 | 35.2 | −51.4 | −12.3 | −55.8 | −3.7 |
| Northern Plains | −24.5 | 20.0 | −26.0 | 44.2 | −67.9 | −19.2 | −55.8 | 1.4 |
| Southeast | −70.9 | −2.1 | −42.6 | 9.7 | −66.1 | 2.1 | −72.4 | −22.2 |
| Delta | −32.0 | −0.5 | −68.0 | −12.5 | −59.0 | 13.2 | −79.7 | −35.7 |
| Southern Plains | −8.7 | 28.7 | −45.4 | 3.1 | −17.9 | −8.3 | −68.2 | −31.6 |

*Note:* With and without the direct effect of increased $CO_2$ concentration.

Modified from Rosenzweig (1988) and Peart et al. (1988).

# V. ECONOMIC EFFECTS AND NEW EQUILIBRIA IN AGRICULTURAL MARKETS

A direct consequence of altered productivity can be new cropping patterns and spatial shifts in agricultural production. The altered productivity information is used by several simulation studies to predict the crop pattern, aggregate production, and economic effects on various agricultural systems. Most of these studies considered only the climate effect, ignoring the direct fertilization effect of $CO_2$. A common finding of these studies is a northward shift in crop production. Santer (1985), on the basis of the GISS projections, estimates that the biomass potential in the European Economic Community (EEC) would increase by 20% on the average, where the northern EEC countries would experience the most significant increases. Blasing and Solomon (1984) and Newman (1980) estimate that

**TABLE 4**
**Effects of Climate Change on Illinois Agriculture**

| | | | Without $CO_2$ effect | | With $CO_2$ effect | |
|---|---|---|---|---|---|---|
| | **Base[a]** | **OSU** | **GFDL** | **GISS** | **GFDL** | **GISS** |
| **Acreage (1000 acre)** | | | | | | |
| Corn | 9,664 | 9,890 | 7,717 | 7,666 | 10,349 | 7,237 |
| Soybean | 10,015 | 11,994 | 13,615 | 10,931 | 11,311 | 5,491 |
| Wheat | 3,821 | 5,292 | 7,060 | 6,894 | 3,724 | 3,997 |
| **Production (million bushel)** | | | | | | |
| Corn | 1348.0 | 858.9 | 455.6 | 944.7 | 1217.3 | 1517.3 |
| Soybean | 388.1 | 355.0 | 261.1 | 393.4 | 373.7 | 448.3 |
| Wheat | 199.5 | 244.7 | 316.9 | 349.6 | 229.6 | 263.4 |
| **Prices ($/bushel)** | | | | | | |
| Corn | 2.04 | 3.83 | 5.32 | 2.70 | 2.52 | 1.44 |
| Soybean | 3.97 | 5.46 | 9.66 | 4.62 | 4.62 | 1.28 |
| Wheat[a] | 2.77 | 3.98 | 5.07 | 3.98 | 2.77 | 2.00 |
| Production index[b] | 100 | 79 | 59 | 92 | 95 | 116 |
| Income[c] | 1,597 | 3,071 | 3,776 | 2,428 | 2,144 | 387 |

*Note:* With and without the direct effect of increased $CO_2$ concentration under different climate scenarios.

[a] Wheat price was held constant in all runs. The future prices were borrowed from McCarl (personal communication).
[b] Production index is calculated at base prices and endogenous quantities.
[c] Income calculated at endogenous model prices.

Adapted from Fang, (1992).

the U.S. corn belt would shift slightly northward, but the shift would not extend to Canada. Rosenzweig (1985) finds that wheat production in North America would increase, with the largest gains occurring in Canada (contrary to Williams et al., 1988), while Mexico would be vulnerable to high temperature stress. Yet, the major wheat producing regions would remain the same.

If areas with increased production potential offset the production loss in other regions, a significant change in agricultural market conditions may not occur. However, an overall production loss for some crops would lead to new market equilibria where prices of disadvantaged crops would increase, while prices of other crops would decline or remain the same. Several studies incorporated this endogenous price adjustment mechanism while determining the economic impacts of global warming. Using a spatial equilibrium model, Dudek (1988) analyzed the economic effects upon California agriculture. His findings, based on the GISS and GFDL climate scenarios, show drastic yield reductions (between 3 and 40%) especially in the southern and central parts of California, resulting in an overall welfare loss (between 14 to 17%). When the direct fertilization effect of $CO_2$ is incorporated, however, yields of some crops (especially vegetables) are recovered and statewide welfare even slightly increases (1.4%).

Önal and Fang (1991) used the same methodology to investigate the effects of global warming upon Illinois agriculture. Table 4 presents some results of this modeling study.

They find that serious reductions would occur in the area planted to corn, while soybeans and wheat acreage would expand when the direct effect of $CO_2$ is ignored. The GFDL scenario shows dramatic production losses (more than 50%) and consequently all crop prices increase substantially. Incorporation of the direct $CO_2$ effect, however, reverses the situation. Under the moderate GISS climate scenario, the yields and overall production increase. However, under the GFDL scenario, slight production losses still continue for corn and soybeans. These results are consistent with the findings of Adams et al. (1988b) as summarized below. It is quite likely that other regions in the corn belt would experience similar gains and losses.

A more comprehensive assessment of the impacts of climate change upon U.S. agriculture is presented by Adams et al. (1988b). Most of the crop and livestock production, domestic consumption, and international trade of agricultural products have been incorporated in a spatial equilibrium model along with alternative (GISS and GFDL) climate scenarios both with and without the direct effect of increased $CO_2$ concentration. The realism of the detailed findings of the Adams et al. study depends on whether the underlying assumptions reflect the true conditions. However, all alternatives indicate that no matter what climate conditions may be, food supply in the U.S. is not an issue. Some major adjustments, such as regional land use and irrigation expansion, may be needed under the cases showing severe yield reductions (especially the GFDL scenario without the direct $CO_2$ effect). Again, a northward shift in the agricultural supply potential is demonstrated. Depending on the climate scenario assumed and whether the direct effect of $CO_2$ is incorporated, the overall welfare effect may be positive or negative. As summarized in Table 5, if no $CO_2$ effect is incorporated, the net annual welfare loss could be $6.5 billion under the GISS scenario and as much as $35.9 billion under the GFDL scenario. Incorporation of the direct $CO_2$ effect shows moderate gains under the GISS scenario but does not fully offset the losses under the more severe GFDL scenario (about $10.5 billion loss).

In general, demand elasticities of agricultural products are known to be low, which means that small percentage changes in supply may have much larger percentage changes on prices in the opposite direction. This inelastic demand structure causes sharp price variations. While price increases adversely affect consumers' welfare, this situation encourages producers to continue agricultural production and keeps the system viable when it may otherwise collapse. Adams et al. (1988b) find that the decreased production in the U.S. would considerably favor agricultural producers' welfare. Thus, the climate change is not an issue of farm incomes and viability of the agricultural sector. When the direct $CO_2$ effect is considered, the agricultural sector may even benefit from the changed climate.

Increased prices not only help agricultural systems survive but may also affect the production technologies. Namely, some production possibilities which are now economically unattractive may become feasible and attractive alternatives due to increased prices. This may, for instance, increase fertilizer use to offset yield losses or may encourage producers to develop new irrigation systems or improve the existing systems. The study by Adams et al. indicates that irrigated acreage in the entire U.S. would increase by 11% under the GISS scenario and about 40% under the drier GFDL scenario. A dramatic increase in irrigation demand is estimated by Peart et al. (1988) for the southern U.S. (33 and 133% under the GISS and GFDL scenarios, respectively). Dudek (1988) finds opposite results for California agriculture, namely a decrease of 16 to 20% in irrigation demand because the reduced profitability causes some cropping activities to collapse, reducing land and water use. This kind of technical change is heavily influenced by government policies in developing countries. Therefore, climate change may have serious implications on public investment programs in those countries.

New equilibria in agricultural markets, altered income levels, and consumption expenditures for agricultural products may also affect other sectors of an economy. Such second

**TABLE 5**
**Economic Effects of Climate Change on U.S. Agriculture**

| | GISS | GISS + $CO_2$ | GFDL | GFDL + $CO_2$ |
|---|---|---|---|---|
| Crops | | | | |
| Prices | 118 | 82 | 209 | 128 |
| Quantities | 90 | 110 | 61 | 81 |
| Livestock | | | | |
| Prices | 102 | 84 | 135 | 107 |
| Quantities | 99 | 106 | 88 | 98 |
| Change in social surplus ($ billion) | −6.5 | 9.9 | −35.9 | −10.5 |

*Note:* With and without the direct effect of increased $CO_2$ concentration under different climate scenarios. Base index with current climate = 100.

Modified from Adams, Glyer, and McCarl, (1988).

round effects may be captured by linking the agricultural and nonagricultural sector's activities. This may be done in a computable general equilibrium (CGE) framework. However, CGE models are not very suitable for sufficient disaggregation of the agricultural sector. The best approach may be to use a CGE model in conjunction with an agricultural sector model, which may include desired disaggregation regarding the coverage, seasonal and regional breakdown of agricultural activities, and provide aggregate feedback information to the CGE model. So far, a successful application of this methodology has not been reported in the climate change context.

## VI. INTERNATIONAL AGRICULTURAL TRADE

All climate change studies show that a northward shift in agricultural production should be expected due to more favorable climatic conditions in northern regions, while southern regions may suffer from the increased heat and summer dryness. Much of the productive agricultural systems are located in mid-latitudes such as the corn belt in the U.S. and the Ukraine in the U.S.S.R. Therefore, a decreased agricultural potential in these areas may result in a decrease in the marketed surplus of these countries, while the cooler northern regions may increase their share in international trade volume. In a study of the forestry sector of Scandinavian countries, Binkley (1988) shows that Finland may considerably increase its export share in the world market of forestry products, while Sweden would lose shares. Because of substantial productivity increases (between 60 and 135%) in the forestry sectors of northern nations (Canada, Scandinavia, and U.S.S.R.) the prices of forestry products in the world markets decline (about 15 to 20% for coniferous sawlogs and 30 to 50% for pulpwood). Consequently, these countries would gain advantage and increase their sales, while the U.S., EEC, Australia, and Latin America would lose (about 20 to 25%). The U.S. agricultural sector model by Adams et al. (1988b) indicates a decrease in the export volume of the U.S. which may be as much as 70%, most of which is due to declining grain production in the corn belt and Great Plains. The U.S. is a large supplier of these products in the world market. Therefore, serious price increases can be expected from such a supply reduction. Considering the direct $CO_2$ effect, however, alters the situation. Adams et al. (1988b) find that consumers (not only in the U.S. but also abroad) would benefit from the increased supply potential and reduced prices in the world market.

## VII. ENVIRONMENTAL EFFECTS OF CLIMATE CHANGE

Several studies indicate that potential yield losses can be recovered or agricultural production may benefit from the favorable climatic conditions in cooler northern regions by choosing more intensive input applications, especially irrigation and fertilizer use. The central U.S.S.R. study (Pitovranov et al., 1988) shows that to maintain the wheat yield at its present level, 50% more fertilizer must be applied (ignoring the direct effect of $CO_2$). More intensive use of fertilizers and increased precipitation in the spring, a time during which fertilizers are applied, may seriously damage the environment. Soil erosion due to concentrated rainfall or increased irrigation, and groundwater pollution due to leaching are two potential dangers. In many developed countries both of these issues are already recognized among the top environmental issues. Unless the management practices are changed, it is likely that the hypothesized climate change would aggrevate the situation.

Another delicate issue is the possible use of fragile land for crop production, if market prices increase favorably or farm incomes drop to a level that may motivate land owners to cultivate those lands. Economic desirability of using marginal land for crop production and possible measures that can be taken to avoid its uneconomic use are unanswered issues at present. A related problem is the transformation of forests and wetlands to cropland in northern regions. The degree of such transformation, especially deforestation and its possible contribution to climate change, is not fully known yet.

Another environmental issue that needs to be studied is a possible expansion in the insect population that may adversely affect crop yields. Warmer and more humid conditions may cause some insect populations to grow faster, causing more intensive pest management to be required. This, in turn, could worsen the groundwater contamination problem.

## VIII. CONCLUSIONS

The major issue related to the analysis of climatic change is the uncertainty and incompleteness of our knowledge about the future climate. The roles of some important factors, such as clouds and oceans, are not fully understood yet. Even the most sophisticated climate simulation models are not totally convincing because of their simplistic assumptions and failure to simulate the current climate conditions. The latter point is very important and has not been addressed adequately. Economic impact analyses compare the climate-induced agricultural production potential against the current situation. Given that the current climate conditions have not been satisfactorily simulated, it is puzzling to compare the predicted potential in the future against the present situation (Crosson, 1989). It would be more appropriate to make the comparison between the "simulated base case" and the agricultural supply potential under the predicted climate scenarios. Less confidence would be placed on the findings and implications of this approach, however, since the base case would not closely reflect the present situation. To the author's knowledge, no climate analysis has attempted to do this.

Because of the huge uncertainty surrounding the global warming issue, policy-makers of developed nations are reluctant to take immediate actions. On the other hand, some analysts argue that the costs to eliminate or restrict human activities contributing to global warming are so high that the future benefits do not outweigh the present costs at any reasonable discount rate. Some, however, argue that if nothing is done today, climate change may be an irreversible event over centuries ahead. Both of these arguments have economic rationale. Apparently it will take years of more research to resolve these issues. Currently what we know is far less than what we don't know about climate change. Continued scientific research will shed more light on the role of critical factors. Further developments and improvements in the climate simulation models will enable economists to have access to more reliable data. The findings of economic studies would then be more specific and convincing so that policy makers would design appropriate policy changes.

## REFERENCES


Adams, R. M., McCarl, B. A., Dudek, D. J., and Glyer, J. D., Implications of global climate change for western agriculture, *West. J. Agric. Econ.,* 13, 348–356, 1988a.

Adams, R. M., Glyer, J. D., and McCarl, B. A., The Economic Effects of Climate Change on U.S. Agriculture: A Preliminary Assessment, Rep. Congress on the Effects of Global Climate Change, Environmental Protection Agency, Washington, D.C., 1988b.

Allen, L. H., Boote, K. J., Jones, J. W., Jones, P. H., Valle, R. R., Acock, B., Rogers, H. H., and Dahlman, R. C., Response of vegetation to rising carbon dioxide: photosynthesis, biomass and seed yield of soybeans, *Global Biochem. Cycles,* 1, 1–14, 1987.

Allen, R. G. and Gichuki, F. N., Effects of projected $CO_2$-induced changes on irrigation water requirements in the Great Plains states, *Rep. Congress on the Effects of Global Climate Change,* Environmental Protection Agency, Washington, D.C., 1988.

Binkley, C. S., A study of the effects of $CO_2$-induced climatic warming on forest growth and the forest sector. B. Economic effects on the world's forest sector, in *The Impact of Climatic Variations on Agriculture, Vol. 1: Assessments in Cool Temperature and Cold Regions,* Parry, M. L., Carter, T. R., and Konijn, N. T., Eds., IIASA/UNEP, Kluwer Academic, Boston, 1988.

Blasing, T. J. and Solomon, A. M., Responses of the North American corn belt to climatic warming, *Prog. Biometeorol.,* 3, 311–321, 1984.

Crosson, P. R., Climate changes, American agriculture, and natural resources: a discussion, *Am. J. Agric. Econ.,* 71, 1283–1285, 1989.

Dudek, D. J., Climatic Change Impacts Upon Agriculture and Resources: A Case Study of California, Rep. to Congress on the Effects of Global Climate Change, Environmental Protection Agency, Washington, D.C., 1988.

Fang, Yen, Economic Analysis of Greenhouse Effects on Illinois Agricultural Production, unpubl. M.S. thesis, Agricultural Economics Department, University of Illinois at Urbana-Champaign, January 1992.

Hansen, J., Russell, G., Rind, D., Stone, P., Lacis, A., Lebedeff, S., Ruedy, R., and Travis, L., Efficient three-dimensional global models for climatic studies, Models I and II, *Mon. Weather Rev.,* III, 609–662, 1983.

Jones, C. A. and Kiniry, J. R., Eds., *CERES-MAIZE: A Simulation Model of Maize Growth and Development,* Texas A&M University Press, College Station, TX, 1986.

Karl, T. R., Wang, W. C., Schlesinger, M. E., Knight, R. W., and Portmain, D., A method of relating general circulation model simulated climate to the observed local climate. I. Seasonal statistics, *J. Climate,* 3, 1053–1079, 1990.

Katz, R. W., Sensitivity analysis of statistical crop-weather models, *Agric. Meteorol.,* 20, 291–300, 1979.

Kimball, B. A., Carbon dioxide and agricultural yield, an assemblage and analysis of 430 prior observations, *Agron. J.,* 75, 779–788, 1983.

Maas, S. J. and Arkin, G. F., *TAMW: a Wheat Growth and Development Simulation Model,* Program and Model Documentation, No. 80–3, Blackland Research Center, Texas Agricultural Experiment Station, Temple, Texas, 1980.

Manabe, S. and Wetherald, R. D., On the distribution of climatic change resulting from an increase of $CO_2$ concentration in the atmosphere, *J. Atmos. Sci.,* 37, 99–118, 1980.

McCarl, B. A. and Spreen, T. H., Price endogenous sector mathematical programming as a tool for sector analysis, *Am. J. Agric. Econ.,* 62, 87–102, 1980.

Newman, J. E., Climatic change on the growing season of the North American corn belt, *Biometeorology,* 7, 128–142, 1980.

Önal, H. and Fang, Yen, The effects of future climate change upon the agriculture of Illinois, *World Resour. Rev.,* 3, 259–275, 1991.

Parry, M. L., Carter, T. R., and Konijn, N. T., Eds., *The Impact of Climatic Variations on Agriculture, Vol. 1: Assessments in Cool Temperature and Cold Regions,* IIASA/UNEP, Kluwer Academic, Boston, 1988.

Peart, R. M., Jones, J. W., Curry, R. B., Ken, B., and Allen, L. H., Impact of Climate Change on Crop Yield in the Southeastern USA: A Simulation Study, *Rep to Congress on the Effects of Global Climate Change,* Environmental Protection Agency, Washington, D.C., 1988.

Pitovranov, S., Kiselkev, V., Iakimets, V., and Sirotenko, O., The effects of climatic variation on agriculture in the subarctic zone of the USSR, Parry, M. L., Carter, T. R., and Konijn, N. T., Eds., in *The Impact of Climatic Variations on Agriculture, Vol. 1: Assessments in Cool Temperature and Cold Regions,* IIASA/UNEP, Kluwer Academic, Boston, 1988.

Ritchie, J. T., Baer, B. D., and Chou, T. Y., Effects of Global Climatic Change on Agriculture: Great Lake Region, Rep. to Congress on the Effects of Global Climate Change, Environmental Protection Agency, Washington, D.C., 1988.

Rosenzweig, C., Potential $CO_2$-induced climate effects on North American wheat-producing regions, *Clim. Change,* 7, 367–389, 1985.

Rosenzweig, C., Potential Effects of Climatic Change on Agricultural Production in the Great Plains: A Simulation Study, Rep. to Congress on the Effects of Global Climate Change, Environmental Protection Agency, Washington, D.C., 1988.

Sakamoto, C. M., Climate-crop regression yield model: an appraisal, in *Application of Remote Sensing to Agricultural Production Forecasting,* Berg, A., Ed., Balkema, Rotterdam, 1981, 131–138.

Santer, B., The use of general circulation models in climate impact analysis — a preliminary study of the impacts of $CO_2$-induced climatic change on West European agriculture, *Clim. Change,* 7, 71–93, 1985.

Stapper, M. and Arkin, G. F., *CORNF: A Dynamic Growth and Development Model for Maize,* Program and Model Documentation, No. 80–2, Blackland Research Center, Texas Agricultural Experiment Station, Temple, Texas, 1980.

Takayama, T. and Judge, G. G., *Spatial and Temporal Price and Allocation Models,* North-Holland, Amsterdam, 1971.

Thompson, L. M., Weather and technology in the production of corn in the U.S. Corn Belt, *Agron. J.,* 61, 453–456, 1969.

Wilks, D., Estimating the consequences of $CO_2$-induced climatic change on North American grain agriculture using general circulation model information, *Clim. Change,* 13, 19–42, 1988.

Wilkerson, G. G., Jones, J. W., Boote, K. J., and Mishoe, J. W., *SOYGRO V5.0: Soybean Crop Growth and Yield Model,* Technical Documentation, University of Florida, Gainesville, FL, 1985.

Williams, G. D. V., Stewart, R. B., Fautley, R. A., Wheaton, E. E., and Jones, K. H., Estimating the effects of climatic change on agriculture in Saskatchewan, Canada, in *The Impact of Climatic Variations on Agriculture, Vol. 1: Assessments in Cool Temperature and Cold Regions,* Parry, M. L., Carter, T. R., and Konijn, N. T., Eds., Kluwer Academic, Boston, 1988.

Chapter 20

# GREENHOUSE GASES AND AGRICULTURE

**Robert B. Jackson IV**

## TABLE OF CONTENTS



This chapter was completed under the auspices of the U.S. Government and is therefore in the public domain.

# I. INTRODUCTION

Agriculture is crucial to the survival of humanity because of the food and fiber it provides. Favorable climates are crucial to agriculture since they determine where crops can prosper. Scientists now generally agree that the Earth's climate is changing in response to past and present human activities, and that the anticipated changes will have profound effects on global agriculture. Climate change will allow agriculture to prosper in some areas, and cause it to suffer in others. The magnitude and geographic distribution of these climatic changes are of great concern, since most of the world's 5 billion people rely on a relatively few highly productive agricultural regions.

Solar energy drives our Earth's climate. Consequently, activities that can alter the sun-Earth energy balance can potentially change climate. Human activities alter that balance by changing the composition of the Earth's atmosphere through increases or reductions in chemically and radiatively active trace gases (greenhouse gases or GHGs). The human-induced changes to date are the result of fossil fuel use, chlorofluorocarbon (CFC) releases, agricultural practices, land-use modifications, and industrial activities (Figure 1).

Global fossil fuel (coal, petroleum, and natural gas) exploration and use continue to increase, releasing carbon dioxide ($CO_2$) into the atmosphere at a rate of $5.5 \times 10^9$Gt C annually (20.2 Gt $CO_2$). The C polluting fossil fuels (Table 1) and the technology to use them so permeate the world we live in today that it seems unlikely that fossil fuel use will diminish until they become economically impractical, or their supplies become limited.

By contrast, the second major cause of Earth's anthropogenically enhanced greenhouse effect, CFC production and use, has been addressed by the Montreal Protocol (IPCC, 1990). The Montreal Protocol is an international agreement to limit production and use of CFCs. This agreement was reached because CFCs destroy the ozone layer ($O_3$), and are also very strong GHGs.

Agriculture, as indicated in Figure 1, ranks third in its contribution to Earth's anthropogenically enhanced greenhouse effect. Specifically, GHG sources are increased, and sinks are decreased, by conversion of land to agricultural use, using fertilizers, cultivating paddy rice, producing other plant and animal crops, and by creating and managing animal and plant wastes. However, some of these same activities increase GHG sinks and decrease GHG sources so the net effects are not obvious. This chapter will identify the agricultural inputs, outputs, and wastes that alter atmospheric concentrations of carbon dioxide ($CO_2$), methane ($CH_4$), and nitrous oxide ($N_2O$) and, if available data permit, determine agriculture's net impact on GHG fluxes.

# II. INPUTS TO AGRICULTURE

Anthropogenic inputs to agriculture have but one goal: to increase and maintain yields at the highest possible levels. These inputs are generally in the form of energy, chemicals, and a suitable management practice, each of which has an effect on the production or consumption of $CO_2$, $CH_4$, or $N_2O$. Anthropogenic inputs are specified by the available biogenic resources.

Biogenic resources for agriculture are favorable lands, favorable climate, and quality atmosphere. Each of these resources is also affected by anthropogenic inputs. Inadvertent feedbacks occur because the natural flux of C and N among soils and atmosphere, biosphere, and hydrosphere are altered under animal or crop production. When the altered fluxes significantly change climate, land, and atmosphere, the conditions for production are modified, which in turn may require different management strategies and levels of energy and chemical inputs. These different strategies and inputs then further alter the fluxes (see Figure 2).

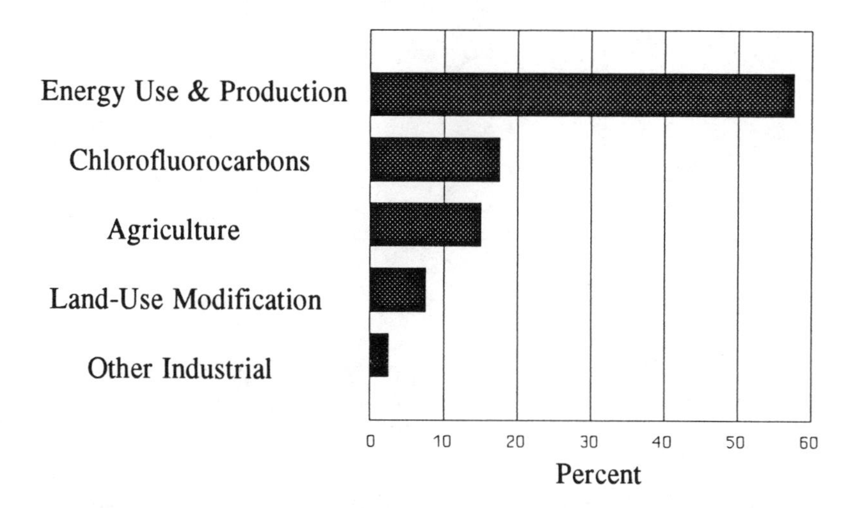


**FIGURE 1.** Anthropogenic contributions to Earth's enhanced greenhouse effect. (From White, R. M., *Sci. Am.*, July 1990. With permission.)

**TABLE 1**
**Average $CO_2$ Emission Factors by Fuel Type[a]**

| Coal | Petroleum | Natural gas | |
|---|---|---|---|
| 23.8 | 19.2 | 13.8 | gC/MJ |

*Note:* See Nomenclature for units.

[a] Carbon Dioxide Information and Analysis Center, (1989).

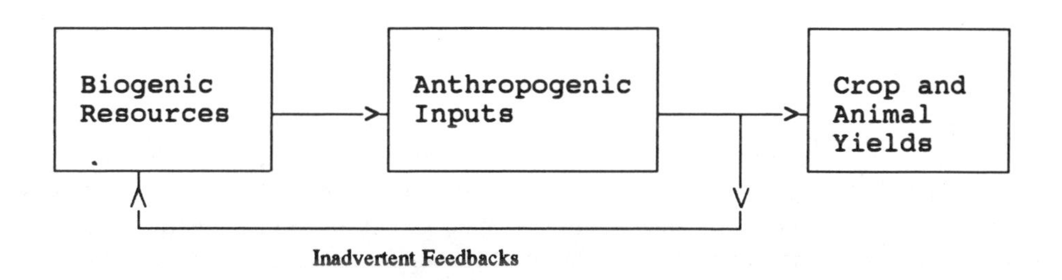


**FIGURE 2.** Agriculture's relationship to its inputs and yields.

## A. ANTHROPOGENIC RESOURCES

### 1. Technology

Changes in agricultural production technology over the last 50 years have enabled more consistent increases in yield than at any other time in history. The most important developments have been mechanization (Figure 3), chemical development and use (fertilizers, fungicides, herbicides, and insecticides), improvements in crop and animal varieties through breeding and genetic engineering, and increased irrigation and drainage. Since C and N (nitrogen) are contained in the produced biomass, technology sequesters atmospheric C and N to the extent that it increases production. However, intensive agriculture is not without

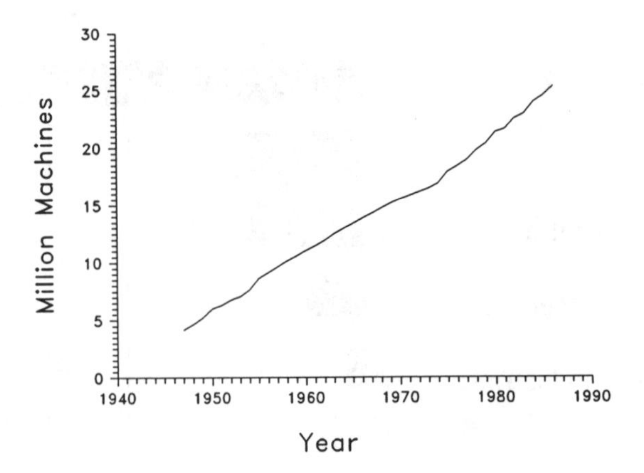


**FIGURE 3.** Number of tractors, harvesters, and threshers in use on the Earth (FAO Production, 1947–1989).

**TABLE 2**
**Energy Required in Crop and Animal Production**

| Agriculture type | Energy demands (GJ/ha) |
|---|---|
| Traditional | 0–5 |
| Transitional | 6–14 |
| Intensive | 15–40 |

Adapted from Slesser, (1975); Stout, (1979); Tivy, (1990).

costs. The critical costs for global change often are loss of soil fertility (which encourages agricultural expansion), increased releases of chemicals into the environment, and large energy requirements.

## 2. Energy

Intensive agriculture requires much more energy than transitional or traditional agriculture (Table 2); however, all three types affect GHGs. In the intensive agriculture of most developed nations, fossil fuels and electricity (often generated from fossil fuels) are the primary providers of energy. In underdeveloped and developing nations, human labor, animal labor, and biomass are standard energy sources, but fossil fuel technologies are eagerly awaited. Once the transition to intensive agriculture is made, a nation's farm output can be spectacular. This greatly encourages change. In America for instance, the labor required to farm an acre has declined 75% since 1940, while farm output per acre has doubled (National Research Council, 1989).

Increases in production require increases in energy and, while C emissions from energy use are accounted for in global fossil fuel emissions estimates, it is important to know how agricultural energy use contributes to those emissions. Unfortunately, national energy statistics classify consumption by sectors, and agricultural energy consumption (like other industries) is not easily extracted.

Increased energy use in agriculture in and of itself is not harmful. However, when the energy supply is fossil fuels, the result for $CO_2$ emissions is clear. When the energy source

**TABLE 3**
**Fertilizer Types Whose Production Processes Affect Chemically or Radiatively Active Trace Gases**

| Fertilizer | Gas affected[a] |
|---|---|
| Nitrogen | Requires $N_2$ |
| | Produces $NH_3$ |
| | Releases $CO_2$ |
| | Produces $CH_4$ |
| | Requires $CO_2$ |
| Phosphorous | Releases $CO_2$ |
| | Releases $N_2$ |
| Potash | None |

Modified from Sittig, (1979).

[a] Process dependent.

is biomass, one must consider that closure of the C and N loops in biomass production and conversion to useable energy has not yet been proven. However, it is clear that, to some extent, atmospheric C and N are recycled. On the other hand, solar and wind energy sources only contribute to GHGs during the manufacture of equipment and not during use. The same is true of nuclear generators. Dams for hydroelectric plants increase the area of submerged land upstream of the dam and decrease the area of submerged land downstream of the dam, which may affect C and N fluxes. However, the energy generation process produces no GHGs.

### 3. Chemicals

Fertilizer production is energy intensive. Many of the different production processes vent $CO_2$ and N gases directly to the atmosphere (Table 3). Crop yields increase substantially as a result of fertilizer applications. For most crops the relationship between yield and fertilizer use is well represented by a mathematical relation that is positive or positive to a point, then negative. Thus, it is no surprise that fertilizer use continues to rise (Figure 4) in response to demands for high agricultural production, sometimes even after the maximum yield point has been exceeded.

Maximum $CO_2$ emissions due to fossil fuel use in fertilizer production can be calculated using literature values for energy requirements (Tables 4 and 5). As before, $CO_2$ emission estimates from fossil fuels already include this energy; however, it is useful to discern their contribution to total emissions.

Water, fertilized soils, and unfertilized soils emit $N_2O$ through nitrification and denitrification processes (IPCC, 1990). The addition of N fertilizers increases the natural $N_2O$ emissions from both soils and water (Li, 1991). The natural production or consumption of $N_2O$ by soils depends on many factors (Table 6) that complicate an accurate flux accounting. The global release of $N_2O$ from fertilized soils, however, has been estimated to be 0.01 to 2.05% of applied N (Bouwman, 1990; Conrad et al., 1983) (Table 7). Thus, fertilizers alter the natural N cycle depending on the amounts that are applied, used by the plant, leached to water, and chemically transformed.

Over long time periods, fertilizer tends to increase soil C and decrease atmospheric $CO_2$ (Li, 1991). Its use increases overall biomass production and growth rates, and some fertilizer production processes actually require $CO_2$. Under certain management conditions (i.e., fallow

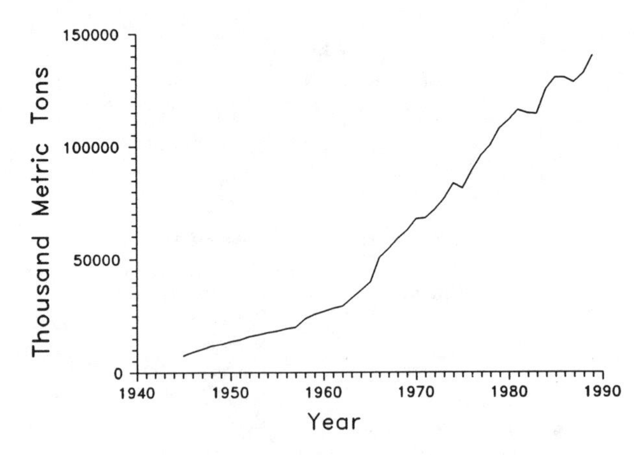


**FIGURE 4.** Global fertilizer use; N + $P_2O_5$ + $K_2O$ (FAO Fertilizer, 1945–1989).

**TABLE 4**
**Maximum Potential $CO_2$ Emissions Due to Fertilizer Production**

| | Global fertilizer consumption[a] (Gt) | | | Emissions[b] (Tg C) | | |
|---|---|---|---|---|---|---|
| **Year** | **N** | **$P_2O_5$** | **$K_2O$** | **N** | **$P_2O_5$** | **$K_2O$** |
| 1985–1986 | 70 | 33 | 26 | 120 | 10 | 6 |
| 1986–1987 | 72 | 35 | 26 | 123 | 11 | 6 |
| 1987–1988 | 76 | 37 | 27 | 130 | 11 | 6 |
| 1988–1989 | 80 | 38 | 27 | 137 | 12 | 7 |

[a] FAO Fertilizer Yearbook, (1989).
[b] Calculated using $CO_2$ emission factor for coal (23.8 g C/MJ).

**TABLE 5**
**Energy Requirements for Fertilizer Production**

| Fertilizer | MJ/t |
|---|---|
| Nitrogen (N) | 72 |
| Phosphorous ($P_2O_5$) | 13 |
| Potash ($K_2O$) | 10 |

Modified from Sittig, (1979).

with cover, conversion to pasture, grassland, or forest), fertilization will result in rapid increases in soil C storage. Whether the stored C becomes labile (subject to change easily) or recalcitrant (stable) is the critical issue. Fertilizer use can also alter the C cycle by influencing net primary production (NPP) of agricultural crops globally.

Fungicide, herbicide, and insecticide (Figure 5) use and fate have yet to be implicated in GHG processes, in spite of the fact that they affect plant growth and the activity and diversity of plant and soil organisms. Either way, pesticide manufacture requires energy, as does the delivery of these chemicals to fields or crops.

**TABLE 6**
**Factors Affecting Production of $N_2O$ in Soils[a]**

| Factor | General tendencies in $N_2O$ production |
|---|---|
| Soil organic matter | Decomposable organic matter is prerequisite for denitrification and nitrification. |
| Soil water | Precipitation stimulates production; alternating wet and dry → increases emissions. |
| Available oxygen | Improved aeration → decreases emissions. |
| Temperature | Increased temperatures → increase $N_2O$; optimum temperature for denitrification is ≈25 to 65°C, and strong diurnal and seasonal variations have been shown. |
| pH | Lower pH → increases ratio of $N_2O/N_2$ |
| N fertilizer | Fertilization → increases emissions; Fertilizer type → unclear effects. |
| Other factors | Flooded soils → decrease $N_2O$; plant canopies alter water & temperature status; plant roots have unclear effects. |

[a] Adapted from Bouwman, (1990); Li, (1991).

**TABLE 7**
**Global Conversion of Fertilizer N to $N_2O$ in Soils**

| Year | Global consumption of N fertilizer[a] ($\times 10^6$ t) | Estimated emissions of $N_2O$ due to N fertilizer (Tg N) |
|---|---|---|
| 1985–1986 | 69.803 | 0.007–1.43 |
| 1986–1987 | 71.555 | 0.007–1.47 |
| 1987–1988 | 75.511 | 0.008–1.55 |
| 1988–1989 | 79.580 | 0.008–1.63 |

[a] FAO Fertilizer Yearbook, (1989).

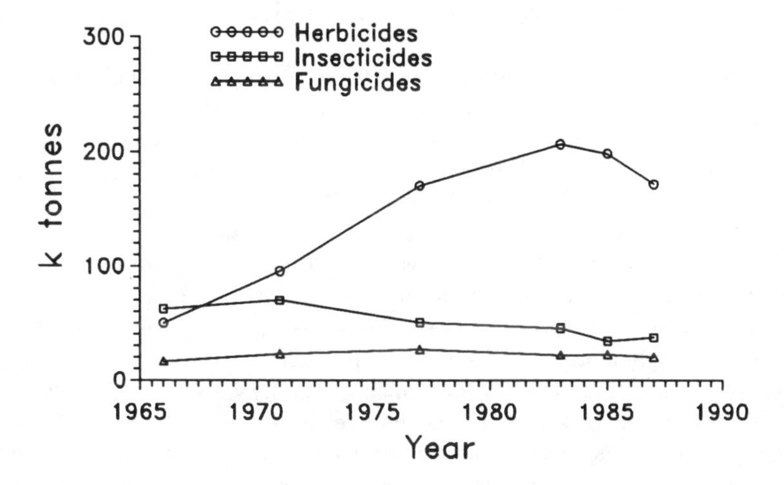


**FIGURE 5.** U.S. fungicide, herbicide, and insecticide consumption. (From National Research Council, *Alternative Agriculture,* National Academy Press, Washington, D.C., 1989. With permission.)

## B. AGRICULTURAL LAND CREATION AND EXTINCTION

Discounting fossil fuel sources and ocean sediments, we recognize that world soils are second only to the oceans in the amount of C they contain. The global range is disputed constantly; however, typically cited values lie between 1000 and 2500 Pg C (Bouwman, 1990).

**TABLE 8**
**Average C Loss from Conversion to Agricultural Land**

| Ecosystem type | No. of studies | Mean loss (%) | Range (%) |
|---|---|---|---|
| Temperate forest | 5 | 34.0 | 3.0–56.5 |
| Temperate grassland | 24 | 28.6 | −2.5–47.5 |
| Tropical forest | 19 | 21.0 | 1.7–69.2 |
| Tropical savanna | 1 | 46.0 | n.a. |

From Bouwman, A. F., *Soils and the Greenhouse Effect,* John Wiley & Sons, Inc., New York, NY, 1990. With permission.

**TABLE 9**
**Site Changes Caused by Agricultural Expansion**

- Removal of original cover
- Incorporation of original cover into soil
- Burning original cover in place
- Burning original cover elsewhere
- Biomass decay on site
- Alterations in site topography & hydrology
- Alterations in site albedo

Agricultural land creation is the conversion of a natural ecosystem to an agricultural ecosystem, which correctly implies that all of the agricultural land on the Earth was, at one time or another, converted from its natural state. Both soil C, and plant C and N in the indigenous vegetation, are usually lost as a result of agricultural land creation (Tables 8 and 9).

In temperate regions today, agricultural expansion is hindered by climate, land availability and compatibility, and competition with commercial, residential, and resource recovery activities. In the past ($\approx$1700 to 1900), when some of these constraints were less limiting, large amounts of C were released because of agricultural expansion, resulting in a maximum of 0.5 Gt C release/yr (as $CO_2$) after about 1800 (IPCC, 1990; Wilson, 1978).

Today, the largest nonfossil fuel source of $CO_2$ to the atmosphere is the conversion of tropical forests to agriculture (Burke and Lashof, 1990); exceeding historical emissions from temperate regions by as much as a factor of four. Houghton et al. (1987, 1988) estimated that tropical $CO_2$ emissions due to land use change are between 0.6 and 2.5 Gt C/yr. Globally, land conversion to agriculture is responsible for a significant portion of the total C emissions from land conversion. Houghton (1986) estimated that between 1860 and 1980, changes in agricultural land area alone accounted for the release of 116 Pg C ($\approx$116 Pg C/120 yr = 0.967 Pg C/yr). One of the inherent difficulties in determining C loss is that the calculations depend on the amount present before conversion. Estimates of the spatial distribution and magnitude of soil C, both before and after agricultural conversion, are constantly disputed.

As for N, Seiler and Conrad (1987) reported that global $N_2O$ loss from soils, due to gain of cultivated land, was 0.4 ± 0.2 Tg N/yr. Vitousek and Matson (as cited in Bouwman, 1990) found that tropical forest soils emit more $N_2O$ after being converted to pasture, and that older pastures emit less $N_2O$ than new pastures. This high initial release may be the result of the changes associated with the missing canopy resulting from exposure to precipitation, wind, and solar radiation.

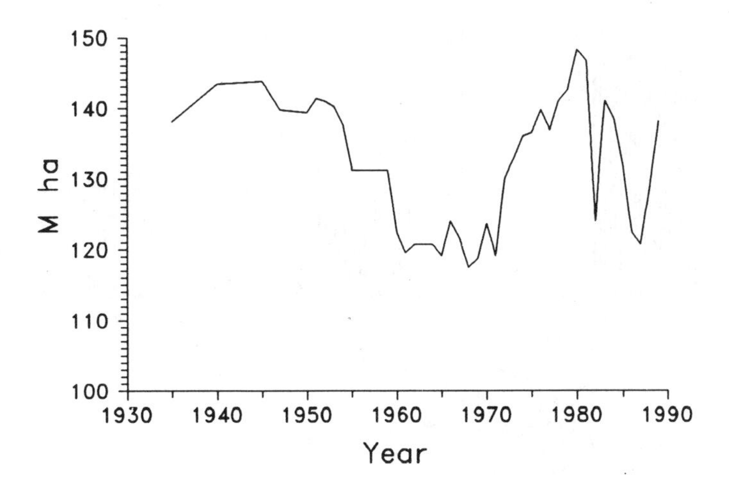


**FIGURE 6.** Area of harvested land in the U.S. (National Research Council, 1989).

C is removed from the atmosphere by plants (as $CO_2$) soil organisms (as CO, $CO_2$, and $CH_4$), and wet and dry deposition and exchange with the oceans and the atmosphere. Abandoned agricultural land tends to maintain levels of C present at abandonment, and even to increase C levels under certain conditions (Schlesinger, 1984).

In the U.S., erosion concerns have led to the conservation reserve program (CRP), a policy that is part of the 1985 Food Security Act that pays farmers not to use highly erodible land for 10 years (U.S. Department of Commerce, 1987). The CRP removes agricultural lands at high risk of erosion from production. In 1987, the program removed 9.87 million acres from production (U.S. Department of Commerce, 1987). Thus, because of the CRP and numerous other reasons, crop and pasture land are always being created and removed from production (Figure 6). In the U.S., up to 3% of agricultural land is routinely kept in a cover crop and neither harvested or used for pasture in any given year. In 1987, this land amounted to almost 20 million acres (U.S. Department of Commerce, 1987).

Agricultural lands also are permanently converted (returned) to grasslands and forests, or sold for residential, industrial, or other development. Although information about the subsequent use of agricultural lands removed from production or the prior use of new agricultural lands globally may not be readily available, the total area gained or lost per year can be calculated from data published by the U.N. (Figure 7).

Information that would be useful in determining the net effect of agricultural land creation and extinction are the gains or losses in kilograms C and N per hectare associated with a gain or loss of crop and pasture. These factors, combined with yearly data of the type given in Table 10, would allow for yearly, country by country assessment of the contribution of agricultural land use change to atmospheric C and N levels.

## C. BIOGENIC RESOURCES

### 1. Soils in Production

Fertile land is the essential ingredient for plant and animal production. Unfortunately, soil changes resulting from mechanization, chemical addition, tillage, drainage, irrigation, and animals are not all desirable. Many of the soil changes are unintended and most are extremely difficult to predict because of the great diversity of management strategies, soil types, crop types, and other heterogeneous factors. Generally, agricultural plant production causes a loss of C (Figure 8), which increases the atmospheric burden (affecting climate) and decreases soil fertility. However, low-C soils can experience slight increases in C after cultivation (Mann, 1986).

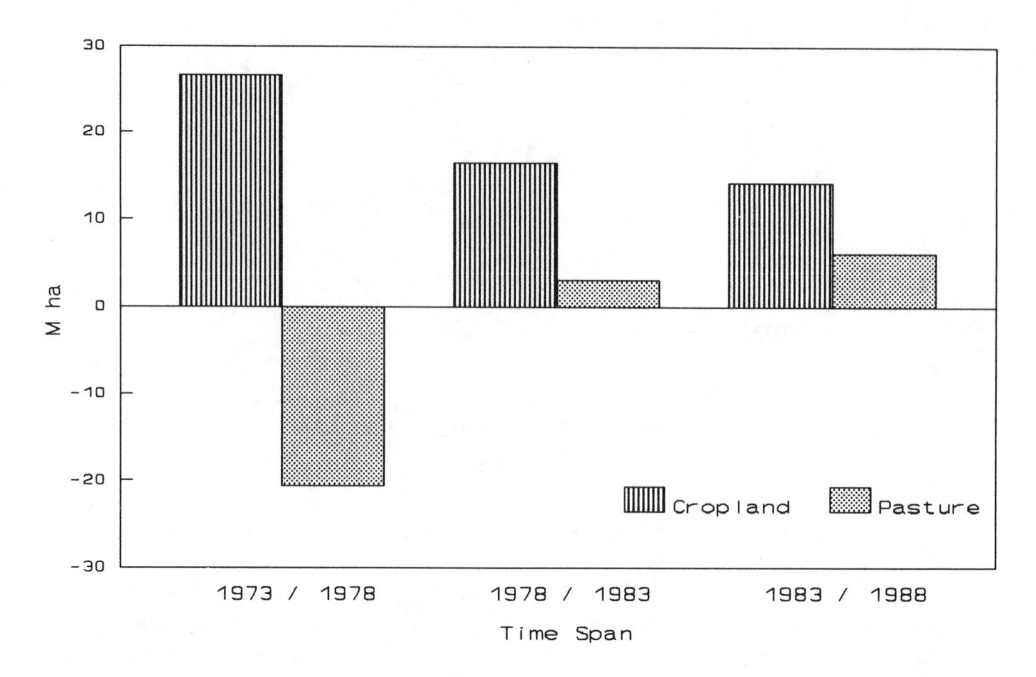


**FIGURE 7.** Global loss or gain of land in crops and pasture (Calculated from FAO Production, 1989).

### TABLE 10
### Global Land Area in Crop and Pasture in 1988[a]

| Total crop & pasture (1000 Ha) | Cropland (%) | Pasture (%) |
|---|---|---|
| 4,687,385 | 31 | 69 |

[a] Calculated from FAO, (1989).

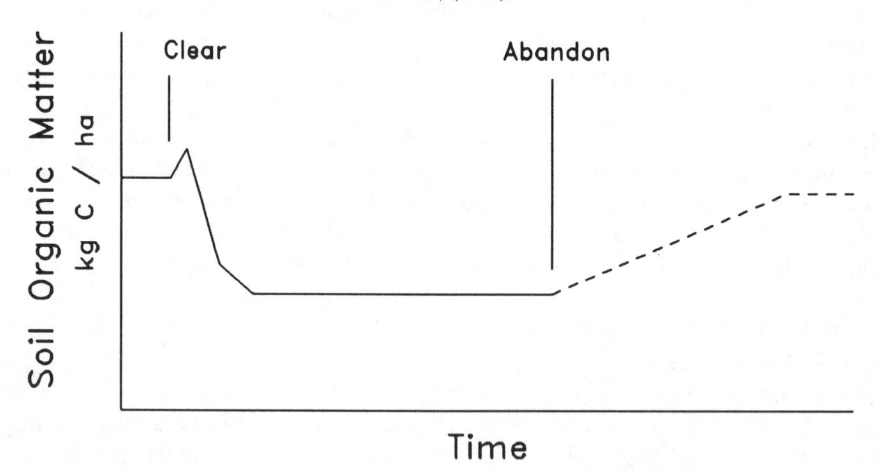


**FIGURE 8.** Simulation of soil C changes due to conversion to agriculture and subsequent abandonment. (From Schlesinger, W. H., Changes in Soil Carbon Storage and Associated Properties with Disturbance and Recovery, in *The Changing Carbon Cycle: a Global Analysis,* Trabalka, J. R. and Reichle, D. E., Eds., Springer-Verlag, New York, 1986. With permission.)

**TABLE 11**
**Soil Changes Induced by Agricultural Plant Production**

| Influencing factor | Changes |
|---|---|
| Continuous disturbance | Initial soil organic matter —<br>high initial SOM → C losses<br>low initial SOM → C accumulation<br>high C:N ratios → more C loss<br>low C:N ratios → more N loss |
| Type of tillage practice | Conventional till → reduced C<br>→ increased erosion losses<br>→ increased hard-pans<br>Minimum or no till → sustained soil C levels<br>→ maintained soil $H_2O$ capacity<br>→ increased leaching potential |
| Presence of plants or crops | Slows erosion losses<br>Increased $CO_2$ respiration<br>Bare soil → increased C loss<br>→ increased transpiration<br>→ increased temperature |
| Type of plant crops | Productivity rate<br>Root mass and depth |
| Machinery use | Increased compaction |
| Irrigation | C gain or loss<br>N loss |
| Drainage of organic soils | C loss<br>Reduced $N_2O$ flux to atmosphere |
| Chemical use | Fertilizers → increased $N_2O$ emissions<br>→ increased NPP<br>Herbicides, pesticides, and fungicides → ? |

Soils are a net global sink for $CH_4$, removing from 15 to 45 Tg of atmospheric C (IPCC, 1990). In general, saturated or flooded lands are sources of $CH_4$, whereas dry or moist soils are sinks (Zepp, 1991). Soils also remove $N_2O$ from the atmosphere (Blackmer and Bremner, 1976); however, the potential global sink has not been quantified accurately (IPCC, 1990). The fractions of $N_2O$ and $CH_4$ soil sinks that are agricultural lands are unknown. Soils may also be a sink for $CO_2$. Schimmel (1987) reported that an agricultural soil was fixing $CO_2$ in the absence of solar radiation (at night). The total annual dark fixation of C measured in that study was 15 $g/m^2$.

Certain types of plant production regimes — such as minimum till, no till, and increased use of crop rotation, strip cropping, and fallow rotation — influence GHG fluxes less than conventional intensive practices. Unfortunately, C and N fluxes vary greatly among management practice, crop type, soil conditions, and climate (Table 11).

Rice paddies are the largest source of agricultural plant production-related $CH_4$ emissions. $CH_4$ is produced by anaerobic decomposition of biomass in flooded rice fields and is transported to the atmosphere through both the rice plants and the water column. According to the U.S. EPA (1990), the U.S., Japan, and Europe are collectively responsible for 3 to 4% of global $CH_4$ rice emissions, and Asia, Africa, and South America are collectively responsible for 96%. Approximately 145 million ha were under rice production globally in 1988, contributing an estimated global $CH_4$ flux of between 25 and 170 Tg $CH_4$/yr (6.3 to 43 Tg C/yr) (IPCC, 1990). The range associated with $CH_4$ flux estimates from paddy rice is large because $CH_4$ production is seasonal and depends on many factors for which information is difficult to obtain regionally (Table 12).

**TABLE 12**
**Variables Affecting $CH_4$ Flux from Rice Paddies**

| Agricultural practices | Soil/paddy characteristics |
|---|---|
| Fertilization | Soil type |
| Water presence & depth | pH |
| Rice plant density | Redox potential |
| Cropping frequency | Temperature |
| Manure or rice straw addition | Nutrients |
| Drainage | Substrate |

Modified from IPCC, (1990).

**TABLE 13**
**Land Changes Induced by Pastured Animals**

| Influencing factor | Changes |
|---|---|
| Movement | Increases soil compaction; distributes seeds. |
| Grazing | May increase or decrease plant density; may increase or decrease plant diversity. |
| Solid waste deposition | Increases soil C and N; alters evolution of C and N gases; increases soil microorganism populations; alters soil moisture status; deposit seeds. |
| Liquid deposition | Increases soil C and N; alters evolution of C and N gases; increases soil microorganism populations; alters soil moisture status and pH. |

Modified from Heady, (1977); Larin, (1962).

Pastured animals affect the land and its vegetation through movement, grazing, and their droppings (Table 13). Animal movements compact and pulverize soil, and animal feces and urine encourage plant growth and distribute seeds. In sufficient quantities, animal feces and urine inhibit plant growth, provide C and N for significant leaching, and can increase evolution of C and N gases from soils. Additionally, the amount and variety of plants may increase or decrease as a result of grazing animals. The net effect of grazing animals on soil C and N is not clear.

As Figure 9 shows, the global land area in crops and pasture is increasing. Currently, given specific crops, cultivated acres, and management practices, the net flux of C and N-containing gases from these soils cannot currently be predicted accurately. In addition, many off-farm factors determine the crops that are planted and the management practices that are used, which then complicates implementation of change. The outside factors include agricultural trade agreements, prevailing prices and costs, taxes, prevailing and anticipated supply and demand, as well as government policies, incentives, subsidies, and laws.

The major impediments to solutions are the lack of credible data sets for calibration and validation of C and N models, the lack of basic knowledge about crop and animal production impacts on soil C and N processes, and the true effects of U.S. agricultural policy in determining which crops are planted and what management practices are used.

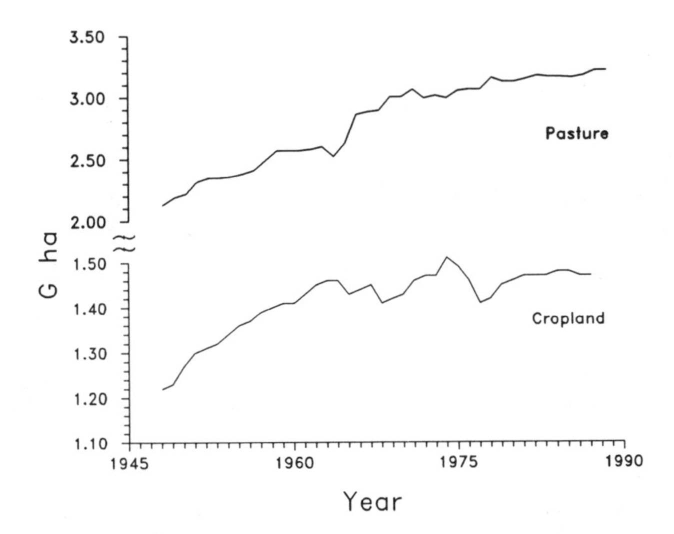


**FIGURE 9.** Global land area in plant crop production and pasture (FAO Production, 1945–1989).

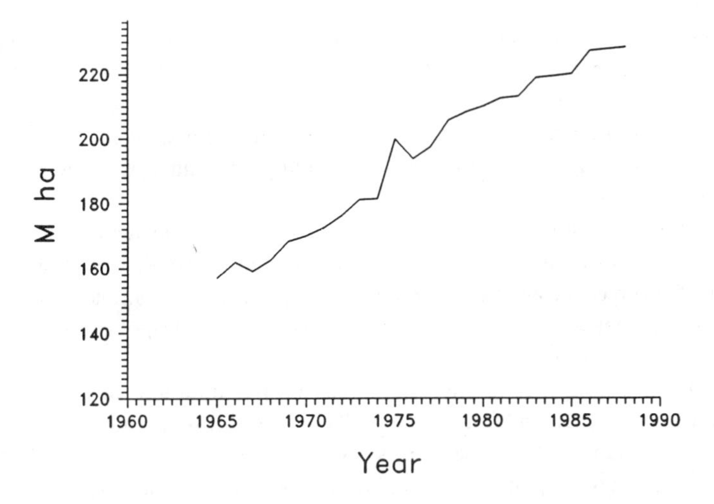


**FIGURE 10.** Global irrigated area (FAO Production, 1965–1989).

## 2. Water

Agriculture accounts for 85% of all consumptive water use, with 94% of that being used for irrigation of crops and pasture, and 4% for animals (National Research Council, 1989). As climate changes, the quality and availability of water will likely be reduced in some regions and increased in others. Approximately half of all agricultural regions rely primarily on irrigation water, whereas other lands require it only as a supplement to precipitation, or not at all. In any case, increasing irrigation usually increases yield if drainage and salination are not significant problems. As a result, irrigation reduces atmospheric C and N to the extent that it increases agricultural NPP. Globally, irrigated land area has increased approximately 25% in the last decade (Figure 10).

$N_2O$ emissions from water are increased after N has leached out of the soil or been carried off in runoff. In water, conversion of N fertilizer to $N_2O$ accounts for up to 1.1 Tg N/yr (IPCC, 1990). Additionally, C from the soil leaches into groundwater, where its fate with respect to GHGs is unknown. In this regard, irrigation affects the GHG flux to the extent that it increases $N_2O$ production and C leaching and erosion.

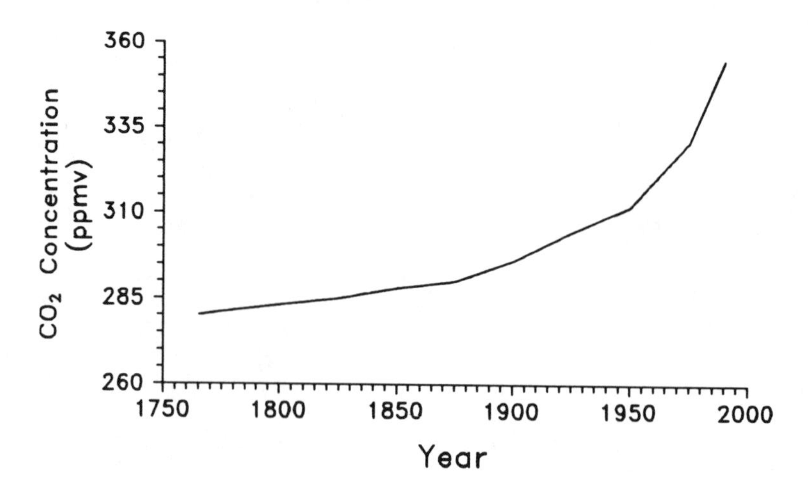


**FIGURE 11.** Atmospheric $CO_2$ concentration. (IPCC — Intergovernmental Panel on Climate Change, *Climate Change: the IPCC Scientific Assessment,* Cambridge University Press, 1990. With permission.)

### 3. Atmosphere

Atmospheric $CO_2$ is essential for plant growth. During photosynthesis, plants remove $CO_2$ from the atmosphere and release $O_2$. During respiration, they release $CO_2$. The gas transfer rates vary according to the plant's photosynthetic pathway ($C_3$, $C_4$, or CAM), the atmospheric concentration of $CO_2$ and $O_2$, and other environmental factors (i.e., sunlight, nutrient availability). Agriculture obviously ties up large amounts of C and N in the plants and animals it produces.

Elevated levels of atmospheric $CO_2$ are beneficial to plant growth (Starr and Taggart, 1989). Some of the observed benefits include a reduced susceptibility to water and light stress, increased biomass production, stimulation of photosynthesis, and lower respiration. However, it is not clear whether or not these effects can be extrapolated to all agricultural ecosystems on a global scale.

At least 40% of global agriculture relies solely on precipitation for water. How precipitation distribution, amounts, and frequencies will change as climate changes remains unclear. Precipitation, like ground and surface waters, increases plant and animal production. Thus, to the extent that it increases biomass, it reduces atmospheric C and N.

Elevated levels of $CO_2$, $CH_4$, and $N_2O$ (Figures 11, 12, and 13) are not favorable for animals or humans, which underscores an important question: how is the atmospheric concentration of oxygen ($O_2$) changing? In addition to the aerobic respiration requirements of humans, animals, and plants, $O_2$ is necessary for fossil fuel and biomass combustion. Fossil fuel use continues to increase, the human population is increasing exponentially, and deforestation and biomass burning continue worldwide. If the level of $O_2$ in the atmosphere were decreasing, it would pose a greater threat to human life than would starvation as a result of adverse climate effects on agriculture.

### 4. Solar

Through plant photosynthesis, our sun provides energy for all life on the Earth. Photosynthesis is affected by a number of factors, with the most critical being the quality and quantity of incident photosynthetically active radiation (PAR). PAR is that portion of the solar spectrum that is most useful for photosynthesis. Its wavelength band lies between approximately 400 and 700 nm. Fortunately, the absorption bands for $CO_2$, $CH_4$, and $N_2O$ are not in the PAR region (Table 14); however, the changes they induce in cloud types,

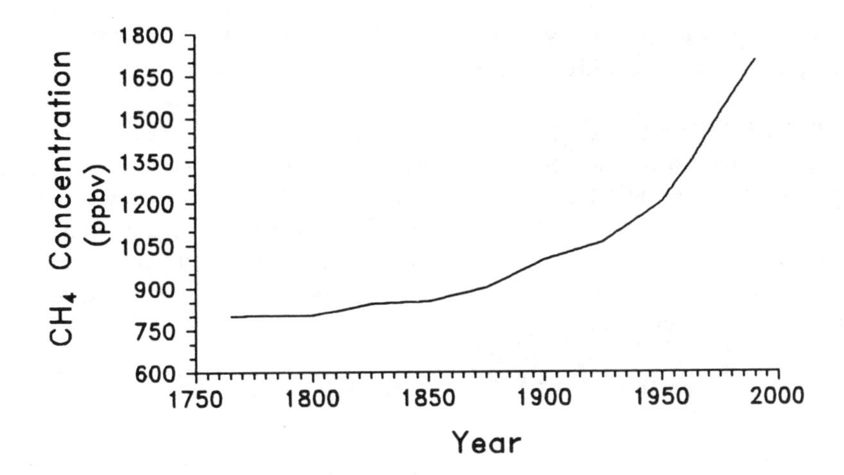


**FIGURE 12.** Atmospheric $CH_4$ concentrations. (IPCC — Intergovernmental Panel on Climate Change, *Climate Change: the IPCC Scientific Assessment,* Cambridge University Press, 1990. With permission.)

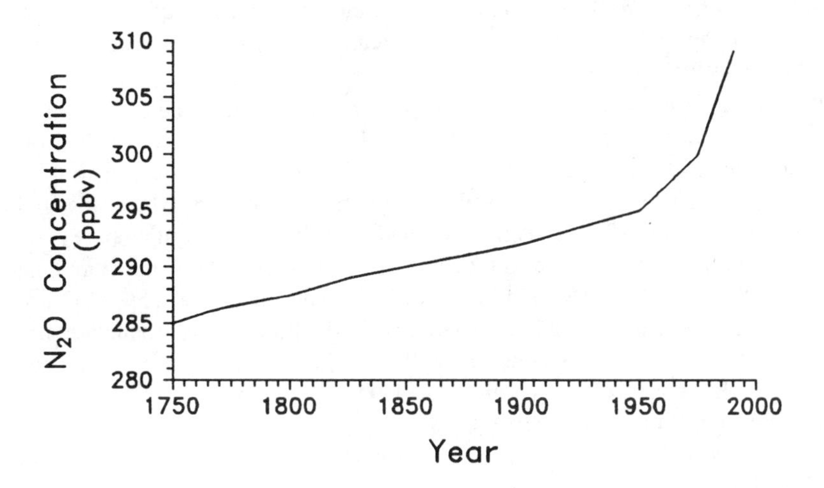


**FIGURE 13.** Atmospheric $N_2O$ concentrations. (IPCC — Intergovernmental Panel on Climate Change, *Climate Change: the IPCC Scientific Assessment,* Cambridge University Press, 1990. With permission.)

**TABLE 14**
**GHGs and Their Approximate IR Absorption Bands**

| GHG | Absorption band(s) ($cm^{-1}$) |
|---|---|
| $CO_2$ | 550–800; 850–1100; 2100–2400 |
| $N_2O$ | 520–660; 1200–1350; 2120–2270 |
| $CH_4$ | 950–1650 |

Adapted from Wuebbles and Edmonds, (1988), for DOE.

cloud frequency of occurrence, and cloud location will affect PAR. If PAR decreases, so will NPP, and likewise, if PAR increases so may NPP.

### D. INPUT EFFECTS SUMMARY

Inputs to plant and animal production are by far the most important source of agricultural contributions to the enhanced greenhouse effect (Table 15). Those inputs whose emissions could be quantified were the inputs that have received maximum attention in the literature. Nevertheless, Table 15 clearly shows that there are many other agricultural inputs that affect the C and N cycle. Values must be placed on these other inputs before the net effect of agriculture can be determined.

## III. OUTPUTS OF AGRICULTURE

### A. PLANTS

Most complex terrestrial plants are 45% C and 1.5% N (by weight) in dry tissue (Starr and Taggart, 1989). Obviously, increasing production efficiencies through anthropogenic or biogenic inputs would result in increased biomass, which means more C and N would be sequestered by agriculture. Responsibility for the fate of C and N sequestered in plant crops does not fall on agriculture, but on food and feed processors, consumers, and the municipalities that treat their waste.

Often in estimating C in vegetation, researchers assign a C value to an area based on a number of parameters. For instance, Melillo et al. (1988) determined a value for crops of $10^6$g C/ha, whereas Olson et al. (1985) determined a range for crops of 1 to 4 $\times$ $10^6$g C/ha. For this analysis, a range of C and N sequestered is obtained by using global plant production for the years 1984 through 1989, and the following assumptions: (1) of the total production weight, 80% is water and 20% is dry matter (i.e., 80% moisture content, wet basis); and (2) of the 20% dry matter, somewhere between 25 and 65% is C and 0.75 to 2.25% is N. The potential C and N removed by agricultural plant production are given in Table 16. The values for C are one order of magnitude smaller than values obtained by multiplying global cropland area (from Table 10) by the factors from Mellillo et al. (1988) or Olson et al. (1985) (1 to 4 $\times$ $10^6$g C/ha).

### B. ANIMALS

Like plant production, animal production also sequesters C and N in the animals themselves. Obviously, increasing production efficiencies through breeding, efficient management, veterinary care, hormones, etc. increases the amounts of C and N sequestered in the animals. Table 17 provides estimates of the C and N content of agricultural animals. It was based on the assumption that 5 to 25% of the total animal production weight is C and that somewhere between 1 and 5% is N. Again, as in the case of plant production, the ultimate responsibility for the fate of C and N sequestered in animal crops does not fall on agriculture, but on the food processing sector, consumers, and the municipalities that process their waste.

Live animals (and humans) produce $CH_4$ as a by-product of digestion (humans produce $\approx$0.05 kg $CH_4$/person/yr) (Crutzen et al., 1986). Estimates of global $CH_4$ production by wild and domestic animals ranges from 60 to 100 Tg $CH_4$/yr ($\approx$44.92 to 74.87 Tg C/yr), which accounts for 15% of global $CH_4$ emissions from all sources (U.S. EPA, 1990). Of the two types of animal digestion, ruminant and nonruminant, the ruminant animals produce much more $CH_4$ per animal than the nonruminant (Table 18, 19, 20).

Determining the actual amounts of $CH_4$ released per animal is difficult because $CH_4$ production depends on animal age, feed quality and amount, genetics, and activity. These data, when available, are country specific and generally unpublished. Additionally, accurate population counts are crucial for estimating overall emissions (Figure 14).

## TABLE 15
## Agricultural Inputs and Atmospheric C and N

| Inputs | Effect on atmospheric C and N: Increase (+) | ± or <> | Decrease (−) | Annual magnitude (Tg C or N) |
|---|---|---|---|---|
| **Anthropogenic** | | | | |
| Technology | | | | |
| Increases outputs | | | −C | <> |
| | | | −N | <> |
| Requires fossil energy | +C | | | <> |
| | | <>N | | <> |
| Increases efficiency | | | −C | <> |
| | | | −N | <> |
| Increases conservation | | | −C | <> |
| | | | −N | <> |
| Energy | | | | |
| Fossil fuel use | +C | | | <> |
| Other energy sources | | ±C | | <> |
| Chemicals | | | | |
| N fertilizer | +N | | | 0.007–1.63 |
| | | ±C | | <> |
| Other fertilizers | | ±C | | <> |
| | | ±N | | <> |
| Other chemicals | | ±C | | <> |
| | | <>N | | <> |
| Agricultural land creation | +C | | | 967 |
| | +N | | | 0.2–0.6 |
| Agricultural land extinction | | ±C | | <> |
| | | ±N | | <> |
| **Biogenic** | | | | |
| Soils under | | | | |
| Rice cultivation | +C | | | 6.3–43 |
| | | ±N | | <> |
| Other plant cultivation | | ±C | | <> |
| | | ±N | <> | |
| Pastured animals | | ±C | | ± |
| | | ±N | | ± |
| Water | | | | |
| Increases production | | | −C | <> |
| | | | −N | <> |
| N leaching | +N | | | up to 1.1 |
| C leaching | | <>C | | <> |
| Atmosphere | | | | |
| Increases production | | | −C | <> |
| | | | −N | <> |
| Solar | | | | |
| Increases production | | | −C | <> |
| | | | −N | <> |
| C Total | +4 | <>1 ±7 | −6 | 973.3–1010 |
| N Total | +3 | <>2 ±5 | −6 | 0.2–3.3 |

*Note:* + Increased atmospheric burden; − decreased atmospheric burden; ± sign undetermined; <> insufficient information.

**TABLE 16**
**Estimated C and N in Harvested Plant Crops**

| Year | Total production[a] (Pg) | Potential C[b] (Tg) | Estimated N[c] (Tg) |
|---|---|---|---|
| 1984 | 3.30 | 165–396 | 5–15 |
| 1985 | 3.34 | 167–401 | 5–15 |
| 1986 | 3.38 | 169–406 | 5–15 |
| 1987 | 3.34 | 167–401 | 5–15 |
| 1988 | 3.28 | 164–394 | 5–15 |
| 1989 | 3.44 | 172–413 | 5–16 |

[a] Calculated from FAO Production, (1989).
[b] Assuming 80% $H_2O$ then 25 to 65% C.
[c] Assuming 80% $H_2O$ then 0.75 to 2.25% N.

**TABLE 17**
**Estimated C and N in Produced Agricultural Animals**

| Year | Total production[a] (Tg) | Potential C[b] (Tg) | Estimated N[c] (Tg) |
|---|---|---|---|
| 1984 | 678.89 | 34–170 | 7–34 |
| 1985 | 695.80 | 35–175 | 7–35 |
| 1986 | 710.20 | 36–178 | 7–36 |
| 1987 | 714.56 | 36–179 | 7–36 |
| 1988 | 729.19 | 37–183 | 7–37 |
| 1989 | 739.41 | 37–185 | 7–37 |

[a] Calculated from FAO Production, (1989).
[b] Assuming 5 to 25% C.
[c] Assuming 1 to 5% N.

**TABLE 18**
**Agricultural Animals**

| Ruminants | Nonruminants |
|---|---|
| Cattle & buffalo | Swine |
| Camels & llamas | Equine |
| Goats & sheep | Poultry |
| Reindeer | Rabbits |

Modified from Tivy, (1990).

The conversion of C to $CH_4$ instead of animal mass is a loss for the rancher, providing motivation for hormones, more efficient breeds and feeds, and other management strategies aimed at increasing the protein produced.

## C. OUTPUT EFFECTS SUMMARY

The two outputs of agriculture are plant crops (for food and fiber) and animals. As Table 21 shows, both outputs appear to be net sinks for C and N. The amounts of C estimated to be in the produced animals were subtracted from the amount of C estimated to have been released as $CH_4$ by the live animals during growth. Were accurate global population data readily available for animals, Tables 19 and 20 would have been used to estimate the

**TABLE 19**
**Estimates of $CH_4$ Production Per Ruminant Animal**

| Animal | Per animal $CH_4$ production (kg/yr) |
|---|---|
| Cattle & buffalo | 35–55 |
| Camels & llamas | ≈58 |
| Goats & sheep | 5–8 |
| Reindeer | ≈15 |

From Crutzen, P. J., Aselmann, J., and Seiler, W., *Tellus* 38B, 275–276, 1986. With permission.

**TABLE 20**
**Estimates of $CH_4$ Production Per Nonruminant Animal**

| Animal | Per animal $CH_4$ production (kg/yr) |
|---|---|
| Swine | 1–1.5 |
| Equine | 10–18 |
| Poultry | ND |

ND — no data given.
From Crutzen, P. J., Aselman, J., and Seiler, W., *Tellus* 38B, 275–276, 1986. With permission.

contribution of agricultural animals to global $CH_4$ emissions. Instead, it was assumed that agriculture was responsible for one half of the total wild and domestic animal C emissions as $CH_4$. The wild and domestic animal source term is from the U.S. EPA (1990), and the agricultural animal sink term was taken from the 1989 data presented in Table 17. A large portion of the harvested plants are used as animal feed, which means that some of the C and N counted as sequestered in the animals had already been counted as sequestered in the plants.

## IV. AGRICULTURAL WASTE

### A. PLANT WASTE

Crop residue yields vary considerably among crop types and harvest methods. Some crops such as hay produce little waste other than the roots that are left in the ground. Before the advent of policy that encourages it, and the availability of chemical fertilizers, farmers faithfully returned residues to the soil, knowing that soil fertility and tilth would be maintained and that wind and water erosion would be reduced (Larson et al., 1978). This is no longer the case in spite of the fact that crop residues, like the harvested crop themselves, contain C and N. According to Larson et al. (1978), during the mid- to late-1970s, more than 363 Tg of crop residue were produced in the U.S. from the nine leading crops alone.

In the previous section, data from global production estimates were used to estimate the amounts of sequestered C and N. Those values, in some instances, included the unusable portion of the crops and, in some instances (perhaps most), did not. In fact, such information is not available except from estimates based on the crop type and harvest method. Estimates

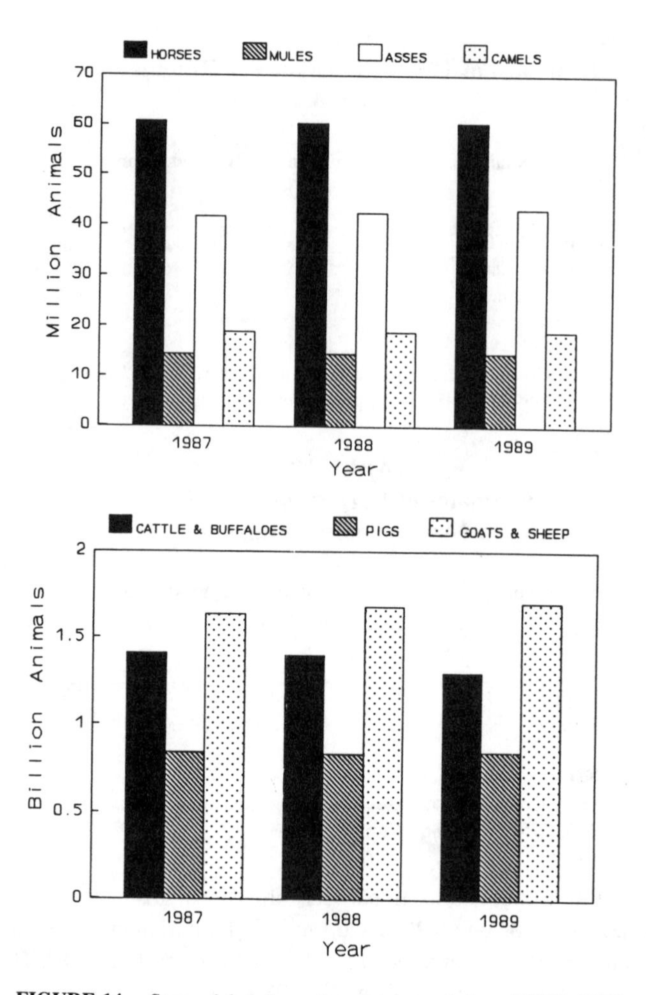


**FIGURE 14.** Some global domestic animal populations (IPCC, 1990).

of sequestered C and N become even more fragmented for horticultural crops and agricultural crops grown on trees, bushes, or vines. For instance, amounts of orchard, bush, or vine crops harvested are generally only the weights of the fruit or usable portion. This means that the C and N retained by the tree, bush, or vine is not accounted for when residue is calculated from yield data. Therefore, agricultural C and N storage calculated in that manner are likely to be lower than is actually the case. Table 22 shows typical ranges of factors used for calculating residue amounts from yield data.

Agricultural plant wastes that remain in the field (residues) are burned, plowed under the soil, left alone, or composted, each of which causes the material to decompose at a different rate. Soil water and N are often the factors that limit the initial decomposition rate (Figure 15). Allison and Cover (1960) reported that under conditions favorable for decomposition, about 50% of the total C of crop residues will be lost as $CO_2$ in 2 to 6 months.

Residue decomposition also increases soil C and N, depending on the amount and composition of material added and the soil environment. One notable study showed that even under conventional tillage and monocropping (continuous corn), soil organic matter (SOM) could increase under proper management (Larson et al., 1972). In the study, the

## TABLE 21
## Agricultural Outputs and Atmospheric C and N

| Outputs | Affect on atmospheric C and N: Increase (+) | ± or <> | Decrease (−) | 1989 Estimated magnitude (Tg C or N) |
|---|---|---|---|---|
| Harvested plants | | | −C | 172 to 413 |
| | | | −N | 5.2 to 15.5 |
| Animals produced | | | −C | 0 to 162[a] |
| | | | −N | 7.4 to 37 |

*Note:* + Increased atmospheric burden; − decreased atmospheric burden; <> insufficient information; ± sign undetermined.

[a] Difference of agricultural contribution of $^1/_2$ of wild and domestic $CH_4$ source: ([44.92 to 74.87 Tg C]0.5) and agricultural animal sink: 37 to 184.9 Tg C; 1989 data from Table 17.

## TABLE 22
## Agricultural Plant Crops

| Crop Type | Harvested Organ | Ratio of Residue[a] to Product ($wt_R/wt_P$) |
|---|---|---|
| Cereal | Seed | 0.5–2.0 |
| Fruit | Seed Coat or Capsule | 0–0.4 |
| Legume | Seed | 0.5–1.5 |
| Root or tuber | Root and Stem | 0–0.2 |
| Vegetable | Leaves and Stem | 0–0.3 |

*Note:* Numbers are highly variable; does not include root remains

[a] Larson et al., (1978); Kossila, (1988).

Adapted from Tivy, J., *Agricultural Ecology,* John Wiley & Sons, Inc., New York, N.Y., 1990. With permission.

corn residues and some added alfalfa residues were plowed into the soil annually (Figure 16). After 11 years, the SOM was a positive linear function of the amount of residue added.

When burned, agricultural wastes immediately release $CO_2$, $CH_4$, and $N_2O$ to the atmosphere, decrease the soil C:N ratio, increase soil pH, and increase soluble soil salts. Burning also induces other changes in soil parameters such as mean temperature, erodability, biological activity, and moisture status (Vallentine, 1989). Using emiprical, double exponential regression equations ($ae^{-kt} + be^{-jt}$), and a two-pool C model (fast and slow pool), Ford (1987) showed that the initial decomposition rate of crop residue is increased when the residues are buried, but that overall decomposition rates are generally unaffected. Other research has shown that decomposition rates are faster on the soil surface than under it (Thomaston, 1984; Parr and Pappendick, 1978). These discrepancies arise because heterogenous conditions — differences in plant chemical and physical structures, soils, climate, microbial processes, nutrients, etc. — always exist among research of this kind, and no one theory has yet been proposed that accounts for all of the variables.

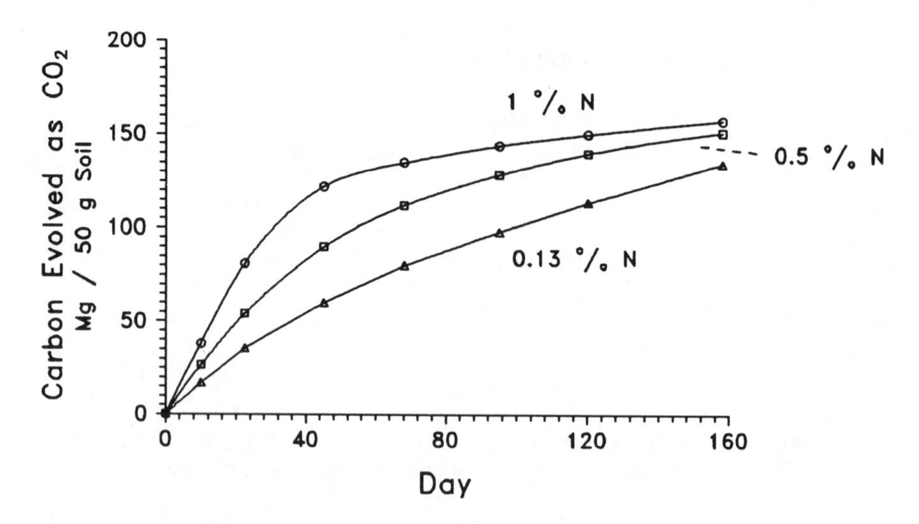


**FIGURE 15.** Decomposition of sawdust in soil at 3 N levels. (From Farr, J. F. and Papendick, R. I., Factors Affecting the Decomposition of Crop Residues by Microorganisms, in *Crop Residue Management Systems,* Oschwald, W. R., Ed., American Society of Agronomy, Madison, WI., 1978. With permission.)

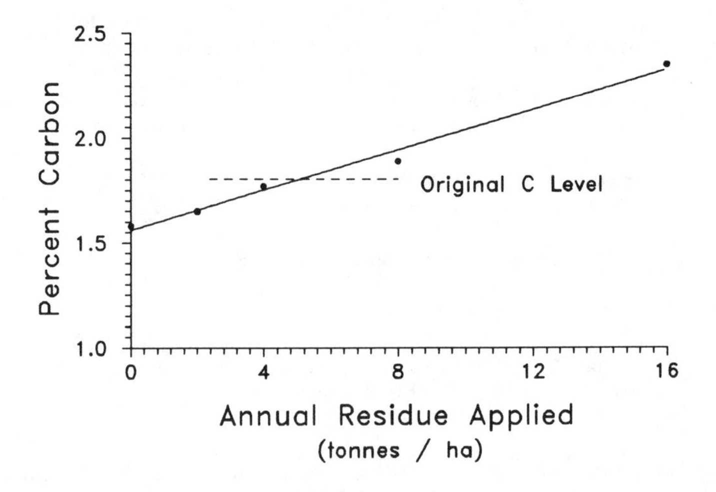


**FIGURE 16.** Soil C as a function of added residue (Larson et al., 1972). (From Larson, W. E., Holt, R. F., and Carlson, C. W., Residues for Soil Conservation, in *Crop Residue Management Systems,* Oschwald, W. R., Ed., American Society of Agronomy, Madison, WI., 1978. With permission.)

Plant waste at processing facilities is converted to animal feed, used as fertilizer, or processed and released to the environment. Plant wastes also can be converted to energy through a number of wet or dry processes that produce heat or fuels (Table 23). However, these wastes are used for fertilizer and feed more than for energy (Stout, 1979). When plant wastes are converted to energy that replaces fossil fuels, they may decrease fossil fuel $CO_2$ emissions.

## B. ANIMAL WASTE

Wherever animals are confined, animal waste must be managed. This occurs in the production of feedlot cattle (a feedlot is a confined space where the manure drops and stays with the animal), dairy operations, confined swine operations, and poultry operations. The amount and composition of waste generated depends on the animal populations and on the

**TABLE 23**
**Processes That Convert Organic Residues to Energy**

| Process | Product |
|---|---|
| Dry | |
| Combustion | Heat & light |
| Pyrolysis | Solid (Charcoal) |
| Gasification | Gas ($H_2S$, CO, $H_2$) |
| Wet | |
| Anaerobic fermentation | Gas ($CH_4$) |
| Alcoholic fermentation | Liquid (alcohol) |

Modified from Stout, (1979).

animal's weight and diet. Waste is collected and either spread on the land or placed into a digester for decomposition and subsequent application to the land.

The IPCC (1990) noted that the $CH_4$ source/sink accounting did not balance — that a source of $CH_4$ had been missed or a sink overestimated. One source of $CH_4$ that was not included in the table was an estimate of $CH_4$ emissions from animal manure. Emissions due to enteric fermentation in animals was included, and ranged from 65 to 100 Tg $CH_4$/yr. The estimates that led to those numbers, and others like them in the literature, determined the percent of gross energy intake converted to $CH_4$ on a per animal basis, and extrapolated to global emissions based on population data. Any emissions from the manure and manure treatment facilities were not considered, and today are estimated to be from 0.8 to 15 Tg $CH_4$/yr globally ($\approx$0.6 to 11.3 Tg C/yr) (ICF Resources, 1990; Verma et al., 1988).

If manure is collected in a pit or a lagoon, $CH_4$ will be produced. In those situations (lagoon, dry pit, feedlot) the instantaneous $CH_4$ flux has been observed to increase when the manure is disturbed (Guenther et al., 1990). Using a sulfur(s) hexaflouride tracer and measuring isotopic ratios of $^{13}C$ and $^{12}C$, Guenther et al. (1990) found that the manure in a feedlot or waste pit (before flushing) produced much more $CH_4$ than the animals themselves. If confirmed, the possibility exists that some of the conversion of feed to $CH_4$ occurs outside of the animal's body (after excretion) and, therefore, can be physically managed.

Manure is frequently flushed into anaerobic or slurry lagoons for decomposition prior to land disposal. Anaerobic lagoons not only break down the waste, they also (given the proper temperatures) produce $CO_2$ and enough $CH_4$ to make recovery economically viable (Safely, 1987, 1988). Historically, biogas from anaerobic lagoons has not been recovered or measured on a large scale.

Land application of animal excrement affects the soil C and N cycle, and its chemical, biological, and physical properties (Barrington et al., 1987). Acceptable land application rates — as determined by waste composition, soil type, topography, climate, application methods, and management objectives — result in rapid decomposition of organic matter releasing $CO_2$. On the other hand, excessive loading rates induce slow decomposition, releasing both $CO_2$ and $CH_4$. Both acceptable and excessive land application rates produce various N and S compounds and increase soil soluble salt levels, which heightens other environmental considerations including N leaching to surface and groundwaters. Table 24 gives the N content of the waste of some domestic animals whose waste often ends up on the land.

**TABLE 24**
**N Content of Some Agricultural Animal Wastes**

| Animal type | N Content (% of dry material) |
|---|---|
| Cattle | 1.2–2.0 |
| Poultry | 3.5–5.0 |
| Swine | 1.9–3.2 |

Modified from Parr and Papendick, (1978).

**TABLE 25**
**Agricultural Wastes and Atmospheric C and N**

| | Affect on atmospheric C and N | | | |
|---|---|---|---|---|
| **Wastes** | **Increase (+)** | **± or <>** | **Decrease (−)** | **Magnitude** |
| Plant waste | | ±C | | ± |
| | | ±N | | ± |
| Animal waste | | ±C | | ± |
| | | ±N | | ± |

*Note:* + Increased atmospheric burden; − decreased atmospheric burden; <> insufficient information; ± sign undetermined.

Land application of animal excrement, as in the case of plant residues, may also increase soil C and N levels. However, the benefits of N increases are restricted by the amount of N that can be utilized by the plants. As for SOM, C added to soils in the form of animal or plant waste will ultimately exist as C in the microbial biomass, the plant biomass, in the soil as slowly degradable humus, and in the atmosphere as $CO_2$ or $CH_4$ (Loehr, 1974). As recently as 1974, none of these C-containing components were believed to have adverse environmental effects and, as a consequence, C was not considered an element of environmental quality concern (Loehr, 1974).

### C. WASTE EFFECTS SUMMARY

Agricultural wastes constitute a large body of C and N whose fate with respect to GHGs is unknown. As with many plant crops, many plant wastes are used as animal feed, and neither the animal or plant wastes are managed to reduce atmospheric fluxes of C and N gases. Table 25 underscores the lack of information about the fate of plant and animal wastes globally, and it shows that the potential exists for agricultural wastes to be managed in ways that increase or reduce atmospheric C and N.

## V. CONCLUSIONS

The first observation of this investigation is the large number of inputs to agriculture as compared to the number of outputs and wastes. Accordingly, four of the six previously identified agricultural practices that contribute to the enhanced greenhouse effect are inputs to crop or animal production (Table 26). Of these inputs, four are C sources and three are N sources (Table 15). Also of the inputs, 12 were categorized as sinks, with 6 C sinks and

**TABLE 26**
**Commonly Reported Agricultural Contributions to the Enhanced Greenhouse Effect**

| Inputs | Outputs | Wastes |
|---|---|---|
| 1. Fossil fuels | Animals | Animal waste |
| 2. N fertilizer | | |
| 3. Ag land creation | | |
| 4. Land under | | |
| Rice cultivation | | |
| Other plant cultivation | | |

**TABLE 27**
**Net Agricultural C and N Affects[a]**

| | Sources (Tg) | Sinks[b] (Tg) | Balance (Tg) |
|---|---|---|---|
| C | 1010 | 172–575 | 435 to 838 |
| N | 3.3 | 12.6–52.5 | −9.3 to −49.2 |

[a] Does not include animal wastes.
[b] Sum of animal and plant crop C or N from Table 21.

6 N sinks (Table 15). Unfortunately, because of a lack of information, none of the suspected input sinks were quantifiable. Additionally, 12 inputs known to alter C and N cycles were identified, but their positive or negative effects were not resolvable. These 12 inputs whose sign with respect to C and N is unknown, and the strength of the identified input sources and sinks, are greatly influenced by the specific agricultural management practices that govern their use. Based on the available data, up to 1010 Tg C and 3.3 Tg N may be released annually because of inputs to agriculture (Table 15). Since C emissions due to agricultural fossil fuel use are not included, the magnitude of these emissions may be underestimates (agricultural fossil fuel emissions are accounted for in energy use and production). However, the 12 unquantified sinks tend to make the numbers appear to be overestimates.

Because of the C and N they contain, agricultural outputs reduce the gross emissions. For instance, the output that contributes to the enhanced greenhouse effect is animal production — $CH_4$ emissions from live animals. However, on balance between the C in the animals and the amount of C they are estimated to have released (not including waste emissions), they are a net C sink (Table 27; see also Table 21). Nevertheless, the relevance for climate change depends on the fate of the C in the animals and its ultimate radiative potential vs. the radiative potential of the $CH_4$ they have generated. Plants, on the other hand, do not generate $CH_4$ or $CO_2$ — they recycle $CO_2$, retaining some of the C. This investigation estimates that 172 to 413 Tg C and 5.2 to 15.5 Tg N are retained annually in crops harvested globally, and 37 to 184.9 Tg C and 7.4 to 37 Tg N in the produced animals.

Evidence also exists to show that animal waste management produces large amounts of $CH_4$. Results have not been extrapolated globally because of a lack of specific information about waste management use and procedures, animal populations, and animal diets. Similarly, animal waste applied to the land can increase soil C and N, increase soil C and N leaching and runoff, change the physical soil properties, and increase evolution of $CO_2$ and $CH_4$. Global extrapolation of local results is then highly uncertain because of lack of knowledge about the use of land application and other heterogeneous factors.

**TABLE 28**
**U.S. 1987 Plant Crop Areas**[a]

| Crop | Million ha |
|---|---|
| Corn | 23.8 |
| All hay | 23.5 |
| Soybeans | 22.4 |
| Wheat | 21.5 |
| Lawn[b] | 10.1 to 12.3 |
| Alfalfa hay | 9.5 |
| Grain sorghum | 3.9 |
| Cotton | 3.9 |

[a] Department of Commerce, 1987.
[b] Modified from Samuelson, (1991).

Agricultural plant wastes increase soil C and N when returned to the field, release $CH_4$ and $CO_2$ as they decompose, produce $CH_4$, $CO_2$, and $N_2O$ when burned, release C and N to the soil when burned, and are used for feed or energy. Global extrapolation of the effects on C and N cycling by plant residues has been hindered by a lack of specific information on the amounts and compositions of crop residues and residue use. Because of the inherent diversity in agricultural production regimes, environments, and other factors, it may be that accurate C and N balances in the agricultural sector are best made at the level of the specific farm in question. Thus, high confidence, national, and global scale assessments will have to be a compilation of C and N balances calculated for every farm.

No aspect of fossil fuel use or CFC production and use — number one and number two cause of the Earth's enhanced greenhouse effect, respectively — is redeeming with respect to global warming. In that regard, agriculture stands alone as the only industry whose net effect may be low or even a zero contribution.

It is interesting to note that the agricultural sector is not alone in its cultivation of plant crops. In the U.S., domestic lawns rank fifth in total cropped area (Table 28). The C- and N-containing lawns are planted, fertilized, watered, and cut for beauty; the harvest is used for nothing.

## VI. ACKNOWLEDGMENTS

I wish to thank the following persons for their critical reviews: Thomas Barnwell, Richard Zepp, Roger Burke, Chuck Steen, Robert Swank, Robert Ryans, Kevin Weinrich, Alan Rowell, George Bailey, Lee Mulkey and David Brown.

# REFERENCES

Allison, F. E. and Cover, R. G., Rates of decomposition of shortleaf pine sawdust in soil at various levels of nitrogen and lime, *Soil Sci.*, 89, 194–201, 1960.

Barrington, S. F., Jutras, P. J., and Broughton, R. S., The sealing of soils by manure. II. Sealing mechanisms, *Can. Agric. Eng.*, 29(2), 105–108, 1987.

Blackmer, A. M. and Bremner, J. M., Potential of soil as a sink for atmospheric nitrous oxide, *Geophys. Res. Lett.*, 3(12), 739–742, 1976.

Bouwman, A. F., *Soils and the Greenhouse Effect,* John Wiley & Sons, Inc., New York, 1990.

Burke, L. M. and Lashof, D. A., Greenhouse gas emissions related to agriculture and land-use practices, in *Impact of Carbon Dioxide, Trace Gases, and Climate Change on Global Agriculture,* Kimball, B. A., Rosenberg, N. J., and Allen, L. H., Eds., Am. Soc. Agron., Special Publ. No. 53, Madison, WI, 1990.

Conrad, R., Seiler, W., and Bunse, G., Factors influencing the loss of fertilizer nitrogen in the atmosphere as $N_2O$, *J. Geophys. Res.*, 88, 6709–6718, 1983.

Crutzen, P. J., Aselmann, I., and Seiler, W., Methane production by domestic animals, wild ruminants, other herbivorous fauna, and humans, *Tellus,* 38B, 271–284, 1986.

FAO Production, *FAO Production Yearbook,* Vol. 43, Food and Agric. Organ. U.N., 1989.

FAO Fertilizer, *FAO Fertilizer Yearbook,* Vol. 39, Food and Agric. Organ. U.N., 1989.

Ford, P. B., Crop Residue Management as an Integral Component of Sustainable Crop Production, M.S. thesis, University of Georgia, Athens, 1987, p. 162–163.

Guenther, A., Zimmerman, P., Hills, A., Westburg, C., and Lamb, B., *Flux Measurements of Methane from Cattle Feedlots,* Natl. Center for Atmos. Res., Atmos. Chem. Div., Boulder, CO, 1990.

Heady, H. F., Management of grazing animals based upon consequences of grazing, in *The Impact of Herbivores on Arid and Semi-Arid Rangelands,* Australian Rangeland Soc., Perth, Western Australia, 1977.

Houghton, R. A., Estimating changes in the carbon content of terrestrial ecosystems from historical data, in *The Changing Carbon Cycle: A Global Analysis,* Trabalka, J. R. and Reichle, D. E., Eds., Springer-Verlag, New York, 1986, 175–193.

Houghton, R. A., Boone, R. D., Fruci, J. R., Hobbie, J. E., Melillo, J. M., Palm, C. A., Peterson, B. J., Shaver, G. R., Woodwell, G. M., Moore, B., Skole, D. L., and Meyers, N., The flux of carbon from terrestrial ecosystems to the atmosphere in 1980 due to changes in land use: geographic distribution of the global flux, *Tellus,* 39B, 122–139, 1987.

Houghton, R. A., Woodwell, G. M., Sedjo, R. A., Detwiler, R. P., Hall, C. A. S., and Brown, S., The global carbon cycle, *Science,* 241, 1736–1739, 1988.

ICF Resources, Inc., *Preliminary Technology Cost Estimates of Measures Available to Reduce U.S. Greenhouse Gas Emissions by 2010,* Submitted to U.S. Environmental Protection Agency, Fairfax, VA, 1990.

IPCC — Intergovernmental Panel on Climate Change, *Climate Change: The IPCC Scientific Assessment,* Cambridge University Press, Cambridge, 1990.

Kossila, V., The availability of crop residues in developing countries in relation to livestock populations, in *Plant Breeding and the Nutritive Value of Crop Residues,* Reed, J. D., Capper, B. S., and Heate, P. J., Eds., Int. Centre for Africa, Addis Ababa, Ethiopia, 1988, 29–39.

Larin, I. V., *Pasture Rotation: System for the Care and Utilization of Pastures,* Israel Program for Sci. Transl. Union Acad. Agric. Sci., 1962, 45–56.

Larson, W. E., Holt, R. F., and Carlson, C. W., Residues for soil conservation, in *Crop Residue Management Systems,* Oschwald, W. R., Ed., Am. Soc. Agron., Special Publ. No. 31, Madison, WI, 1978.

Larson, W. E., Clapp, C. E., Pierre, W. H., and Morachan, Y. B., Effect of increasing amounts of organic residues on continuous corn: organic carbon, nitrogen, phosphorus, and sulfur, *Agron. J.*, 64, 204–208, 1972.

Li, Changsheng, personal communication, The Bruce Co., Washington, D.C., 1991.

Loehr, R. C., *Agricultural Waste Management,* Academic Press, New York, 1974.

Mann, L. K., Changes in soil carbon storage after cultivation, *Soil Sci.*, 142(5), 279–288, 1986.

Melillo, J. M., Fruci, J. R., Houghton, R. A., Moore, B., and Skole, D. L., Land-use change in the soviet union between 1850 and 1980: causes of a net release of $CO_2$ to the atmosphere, *Tellus,* 40B, 118, 1988.

National Research Council, *Alternative Agriculture,* National Academy of Sciences, Washington, D.C., 1989.

Olson, J. S., Watts, J. A., and Allison, L. J., Major World Ecosystem Complexes Ranked by Carbon in Live Vegetation: A Database, NDP-017, Carbon Dioxide Inf. Center, Oak Ridge Natl. Laboratory, Oak Ridge, TN, 1985.

Parr, J. F. and Papendick, R. I., Factors affecting the decomposition of crop residues by microorganisms, in *Crop Residue Management Systems,* Oschwald, W. R., Ed., Am. Soc. Agron., Special Publ., No. 31, Madison, WI, 1978, chap. 6.

Safely, L. M., Jr. and Lusk, P. D., Low Temperature Anaerobic Digester, N.C. Department of Economic and Community Development: Energy Division, Raleigh, NC, and Biological and Engineering Department, North Carolina State University, 1988.

Safely, L. M., Vetter, R. L., and Smith, L. D., Management and operation of a full-scale poultry waste digester, *Poultry Sci. J. Ser. NC Agric. Res. Serv.*, 66, 941–945, 1987.

Samuelson, R. J., The joys of mowing, *Newsweek,* April 29, p. 49, 1991.

Schimmel, S. M., Dark fixation of carbon dioxide in an agricultural soil, *Soil Sci.*, 144(1), July, 1987.

Schlesinger, W. H., Changes in soil carbon storage and associated properties with disturbance and recovery, in *The Changing Carbon Cycle. A Global Analysis,* Trabalka, J. R. and Reichle, D. E., Eds., Springer-Verlag, New York, 1984, 194–220.

Seiler, W. and Conrad, R., Contribution of tropical ecosystems to the global budgets of trace gases, especially $CH_4$, $H_2$, CO, and $N_2O$, in *Geophysiology of Amazonia,* Dickinson, R. E., Ed., John Wiley & Sons, New York, 1987, 133–160.

Sittig, M., *Fertilizer Industry; Processes, Pollution Control, and Energy Conservation,* Noyes Data Corp., Park Ridge, NJ, 1979.

Slesser, M., Energy requirements for agriculture, in *Food, Agriculture, and Environment,* Lenihan, J. M. and Fletcher, W. W., Eds., Academic Press, New York, 1975.

Starr, C. and Taggart, R., *Biology: The Unity and Diversity of Life,* Wadsworth, Belmont, CA, 1989.

Stout, B. A., *Energy for World Agriculture,* Food and Agric. Organ. U.N., Rome, Italy, 1979.

Thomaston, S. W., Crop Residue Decomposition as Affected by Soil Erosion and Tillage, M.S. thesis, University of Georgia, Athens, 1984.

Tivy, J., *Agricultural Ecology,* John Wiley & Sons, New York, 1990.

U.S. Department of Commerce — Bureau of the Census, *1987 Census of Agriculture,* U.S. Summary and Data Rep., Vol. 1–2, Geographic Ser., U.S. Government Printing Office, Washington, D.C., 1987.

U.S. EPA, U.S. Environmental Protection Agency, U.S. Department of Agric. and Intergov. Panel on Climate Change, *Greenhouse Gas Emissions from Agricultural Systems,* Vol. 1–2, U.S. Government Printing Office, Washington, D.C., 1990.

Vallentine, J. R., *Range Development and Improvement,* Academic Press, New York, 1989.

Verma, S. B., Britton, R. A., Janson, M. D., Klopfenstein, T. J., Lauda, S. M., Norman, J. M., Schulte, D. D., Skopp, J. M., and Ullman, F. G., *Bioatmospheric Studies of Trace Gas Dynamics,* Proposal Submitted to the National Science Foundation, 1988.

White, R. M., The great climate debate, *Sci. Am.*, July 1990.

Wilson, A. T., Pioneer agriculture explosion and $CO_2$ levels in the atmosphere, *Nature (London),* Volume 273, 40–41, 1978.

Wuebbles, D. J. and Edmonds, J., *A Primer on Greenhouse Gases,* U.S. Department of Energy, Carbon Dioxide Information and Analysis Center, Oakridge Natl. Laboratory, Oak Ridge, TN, 1988.

Zepp, R., Personal communication, U.S. Environmental Protection Agency, Athens, GA, 1991.

# NOMENCLATURE

| | |
|---|---|
| C | — carbon |
| CFC(s) | — chlorofluorocarbon(s) |
| E | — exa; $10^{18}$ |
| g | — gram |
| G | — giga; $10^9$ |
| GHG(s) | — greenhouse gas or gases |
| ha | — hectare; 10,000 $m^2$ |
| J | — joule |
| k | — kilo; $10^3$ |
| m | — meter |
| M | — mega; $10^6$ |
| n | — nano; $10^{-9}$ |
| N | — nitrogen |
| NPP | — net primary production (dry matter production per unit area and time) |
| P | — peta; $10^{15}$ |
| SOM | — soil organic matter |
| T | — tera; $10^{12}$ |
| t (tonne) | — 1,000,000 g |
| S | — sulfur |

Chapter 21

# GLOBAL WARMING AND THE WATER RESOURCES OF SAHEL (LAKE CHAD) REGION

**Solomon A. Isiorho**

## TABLE OF CONTENTS



0-8493-4419-0/93/$0.00 + $.50
© 1993 by CRC Press, Inc.

# I. INTRODUCTION

There is a global concern for the quantity and quality of water resources as more than half of the world's population has no access to an adequate water supply. This water problem is more acute in developing nations, especially in the Sahel region of Africa. The droughts of the 1960s and 1980s in the Sahel region brought to focus some of their environmental hardships. Within this region is Lake Chad, a "large" shallow (mean depth of ~1.5 m) body of water covering an area of approximately 1800 to 22,000 $km^2$ (Carmouze, 1983). Lake Chad, an old closed-basin lake, owes its existence to the tectonic episodes that occurred during the Cretaceous (Burke, 1976). The lake is located in a semiarid region with a low annual rainfall (~30 cm) and a high evaporation rate (~2 m/year; Roche, 1980; Carmouze, 1983; Eugster and Maglione, 1979). The high evaporative nature, low annual rainfall, and its proximity to the Sahara desert make the surface water unreliable, especially during periods of low annual rainfall in the active watershed region of the lake to the south. About 85% of the input water to Lake Chad comes from the Lagone-Chari river system, 10% from rainfall, and the other rivers supplying the balance (Roche, 1980). The surface water resource situation is becoming more of a problem with the advance of the Sahara desert to the south. Attempts to gather information that would be useful in the management of the water resources of the region include the determination of the relationship between the major surface water of the region (Lake Chad) to the underlying groundwater. Utilizing shuttle radar technology, it was discovered that the Sahel region was more humid in the past than it is at present. This new evidence is based on the existence of fossil river channels buried beneath sand layers. The presence of sand dunes within the lake indicates a dry phase of the lake, which may be due to change in climate.

Closed-basin lakes are of interest because they sometimes possess some peculiarities. These lakes tend to concentrate salts with time (Jones, 1966; Eugster and Jones, 1979; Eugster and Hardie, 1978; Yeretich and Cerling, 1983); however, the open waters of Lake Chad remain relatively fresh (~320 mg/l) Roche (1977). The lake's level has fluctuated annually over 1000s of years, as indicated by measurements of the past few decades (Roche, 1980; Carmouze et al., 1976) or estimated by Klans (1980), Jackel (1984), and others as due mainly to climatic changes. The low salinity of the open waters in Lake Chad has been attributed to the following factors: low volume-to-surface ratio that ensures significant dilution by rainfall, biogeochemical regulations, and seepage through the lake bottom (Carmouze, 1983; Carmouze et al., 1976; Roche, 1977, 1980; Gac et al., 1977; Isiorho and Matisoff, 1990).

Seepage is an important component in the water budget and chemical mass balance of a lake. One major aim of seepage measurements in lakes is to determine the interrelationship between a lake and its surrounding groundwater. It would also enable predictions of the effect and extent of pollution on water quality, on the transfer of solutes between the two bodies of water, and on which areas would be susceptible to pollution. Seepage flux may be estimated by direct methods, using instruments such as seepage meters or minipiezometers which can detect seepage to or from lakes, or it may be estimated using a water balance or a salt budget (Lee, 1977; Lee et al., 1980; Krabbenhoft and Anderson, 1986; Isiorho and Matisoff, 1990). The measurement of seepage rates on a lake water budget should be of great significance in semiarid regions where water supplies pose severe problems. Thus, seepage rates are crucial in the determination of the relationships between surface waters, especially lakes, and groundwaters.

In recent years there has been increased attention to the interaction of lakes and groundwaters. Theoretical studies now provide an understanding of the hydrologic factors that control seepage through lake bottoms (Winter, 1978, 1981, 1986). Seepage has been esti-

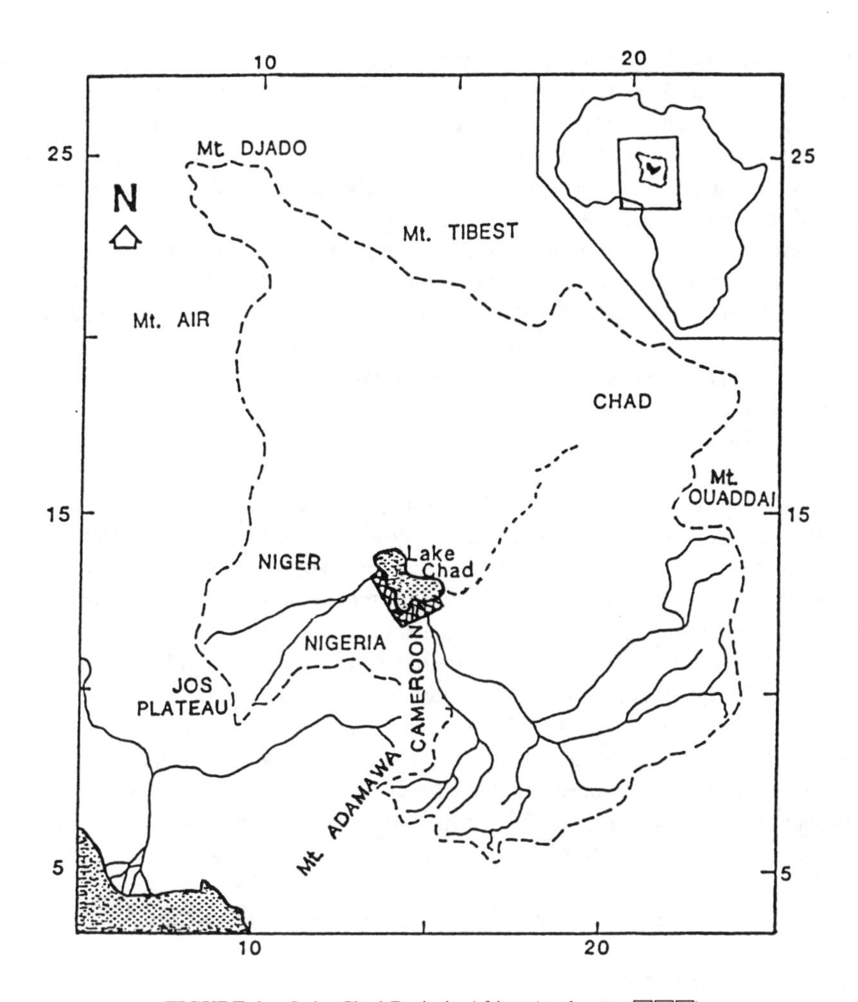


**FIGURE 1.** Lake Chad Basin in Africa; (study area ☒☒☒).

mated for small lakes but none has been made for a large lake. Attempts have been made to examine large lakes, but only small areas of such lakes were investigated. It has been demonstrated by several workers (McBride and Pffannuch, 1975; Wesnner and Sullivan, 1984) that seepage rates tend to decrease with distance from the shoreline into a lake, but has not been found to be so in the southwestern portion of Lake Chad (Isiorho and Matisoff, 1990). The location of the study area is given below.

## II. LOCATION, DESCRIPTION, AND AREAL EXTENT OF STUDY AREA

Lake Chad lies between latitude 12° 20 and 14° 20 N and longitude 13° and 15° E. This lake lies within Cameroon, Chad, Niger, and Nigeria. The watershed bounds an approximate area of 4,000,000 $km^2$, and the Chad Basin height ranges from 300 m at Bongor to about 3 km in the Tibest (Burke, 1976); (see Figure 1). The lake's surface area is large, approximately 30,000 $km^2$, but is usually 20,000 $km^2$; however, areas of $<$10,000 $km^2$ have been reported (Roche, 1980; Carmouze, 1983; Burke, 1976).

The region under study has a savannah climate with a fairly wide seasonal and diurnal range of temperatures. It is semiarid towards the north, with two seasons: a long, dry season, from October to April, when the dry Harmattan wind blows off the Sahara desert with day temperatures of 30 to 36°C (85 to 98°F) and night temperatures of 5 to 11°C (40 to 50°F), and a short, wet season, from about May to August, with daily maximum temperatures of ~34°C (~90°F) and relative humidity of about 40 to 70%. The vegetation in the study area can be described as savannah woodland which is divided into two zones: Sudan savannah to the south and Sahel savannah towards the north.

The data for this paper were gathered west and south of the Chad Basin in Cameroon, Chad, and Nigeria. It is a plain that slopes gently towards the lake and is devoid of rock outcrops and is covered by superficial deposits of sands and clays. The area studied includes the south basin and areas ~80 km from the lake shore. All surface drainage is towards the lake; however, surface drainage never reaches the lake (southwest of the basin). The waters are dissipated in either broad swamps and lost through evaporation and transpiration or by percolation to the underlying sediments.

The lake is generally divided into two basins; a north and a south of roughly equal surface area. They are separated by a line from Baga-Kawa to Baga-Sola as a result of a narrowing coastal periphery at this level. The lowest region of the lake bottom, with an average altitude of 275.5 m, lies in the north basin. The area occupied by this region is about 4000km$^2$ and its position is slightly north of the geographical center of the basin. The slope is more gentle and smoother towards the south than the northeast and west. There are three regions of higher altitudes (278.5 m on the average) in the south basin where the relief is less sharp. Morphometric curves showing the graphical determination of the surface occupied by the lake in the north and south basins, as well as the corresponding volumes, are given in Figure 2 (Carmouze et al., 1976; Lemoalle, 1978; Carmouze and Lemoalle, 1983). From Figure 2, it is observed that the 20,000 km$^2$ area is reached when the lake level is approximately at 282 m, with both the north and south basins having equal volume. The geologic and climatic setting of the region is presented below.

## III. GEOLOGIC AND CLIMATIC HISTORY

### A. GEOLOGIC SETTING

Geophysical evidences provided by Cratchley (1960) suggest that Cretaceous beds filled some broad troughs within the basement. One major trough with a WSW/ENE trend with a maximum depth of over 3500 m was filled with approximately 3000 m Cretaceous sediments (Cratchley et al., 1984). This trough in the Chad Basin has a similar trend to structures in the Benue Trough, and may have been formed either by rift faulting or folding. Furon (1960) and Cratchley et al. (1984) have suggested that the Chad Basin was a tectonic cross-point between a NE-SW trending Tibesti-Cameroon Trough and a NW-SE trending Air-Chad Trough. Immediately to the southwest of Lake Chad, a Maiduguri trough has been confirmed by seismic refraction and gravity methods (Cratchley, 1960). A Maiduguri Trough may represent one continuous fracture extending from the Benue Trough through Lake Chad, to the vicinity of Largean, and has been suggested by Isiorho (1990a). A trans-African lineament from the Atlantic through the Benue trough, going underneath the Quaternary Chad Basin sediments into the Nile delta, has been suggested by Ajakaiye and Burke (1973) and Nagy et al. (1976).

Neotectonic (quaternary) lineaments are located along the directions of positive Bouguer anomalies which indicate the deep structures of the basement (Cratchley et al., 1984). Durand (1982) suggested that the present day rivers (Lagone, Chari, and Kamadogu) and fossil ones are controlled by structural features (lineament). He also noted that the western shorelines of the present day lake, the direction of the Grande Barrier, and the direction of the peri-lacustrine ridge correspond to known structural directions. Numerous other positive bouguer

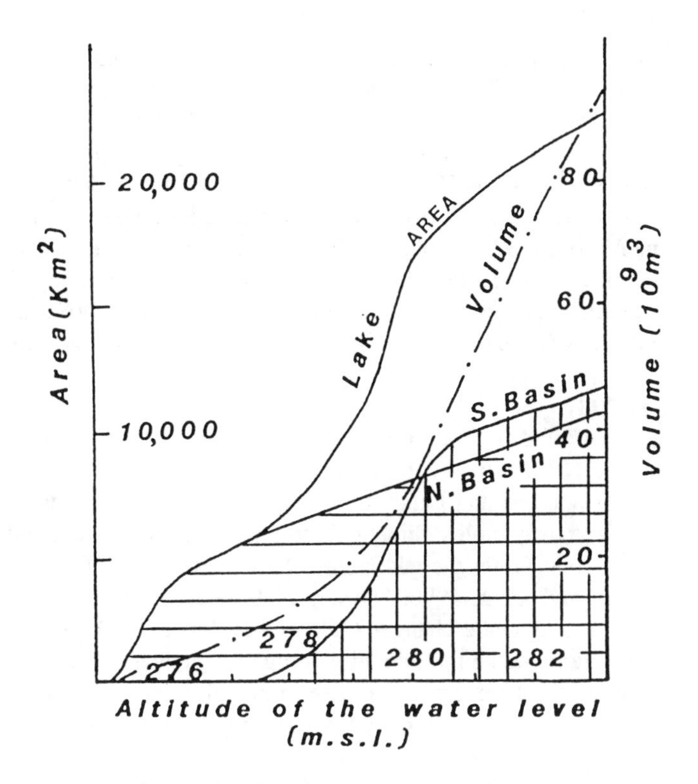


**FIGURE 2.** Morphometric curves for Lake Chad. (Modified after Carmouze and Lemoalle, 1983).

anomalies have been reported near Lake Chad, which tend to suggest that the structural history of this area is extremely complex. Reportedly, there is an overall increase in fault frequency upward from the basement (Avbovbo et al., 1986). It does appear that deep structures are recorded in the present day morphology. Some of these lineaments have been mapped from landsat and radar images and verified in the field by geophysical method (Isiorho, 1990a, Isiorho et al., 1991).

The Chad Formation is the major concern in this study, which is hydraulically connected with Lake Chad. In the southwest part of the Chad Basin, the quaternary age Chad Formation is composed of three aquifers referred to as the Upper, Middle, and Lower Aquifers. The dating of the Chad Formation is based on vertebrate remains of Lower Pleistocene age and diatoms which are older than the Plioceone (Barber and Jones, 1960). The quaternary sediments that are more than 200 m thick are restricted to within the shoreline of a Mega-Chad, indicating a link between subsidence and sedimentation. During the last major wet phase in the Chad Basin (12,000 to 7000 years ago), the lake rose to a height of about 320 m (Carmouze, 1983; Burke, 1976). This 320 m high lake was referred to as Mega-Chad. A Mega-Chad of the magnitude reported by Burke (1976) and others is disputed by Durand (1982), Durand et al. (1984), and Durand and Lang (1986). Durand (1982) criticized the idea of a Mega-Chad of such magnitude in that the ridge is not continuous and the dating of the sediments show that the ridge is younger than the period in which the Mega-Chad existed. The Chad Formation is generally less than 200 m thick outside the shoreline of the Mega-Chad, but thickens rapidly inside the lake to nearly 600 m.

The Chad Formation is not exposed in the study area. It is composed of argillaceous (fine grain) sequence in which arenaceous (coarse grain) horizons occur. The argillaceous sediments represent deposition either in quiet waters or during periods when the discharge of the rivers was low. Beds of pure diatomite, which are a few centimeters to a few meters thick, occur in the study area. They are indicative of conditions of extremely low sedimentation which persisted for long periods of time in the lake (Matheis, 1976); in addition, the silica content of the lake water must have been relatively high (Barber, 1965). The Chad Formation is overlain by post-Chad Formation aeolian sands, fluvial, deltaic, and lacustrine deposits. Most of the fluvial deposits occur along stream valleys which are made up of two units (Hammad and Abdou, 1982): the old alluvium and the young alluvium. The old alluvium consists of deposits of old rivers, while the young alluvium contains recent river beds and flood plains. The detailed geology of the area can be found in reports by Raeburn and Jones (1934), Barber and Jones (1960), Carter et al. (1963) and Avbovbo et al. (1986).

## B. CLIMATIC SETTING

The Chad Basin suffered oscillation between aridity and relative rainfall as indicated by the sand dunes with NE-SW trend. In the study area the dunes are indications of aridity periods (Durand, 1982; Durand and Lang, 1986; Isiorho, 1989). The quaternary history of the basin is characterized by complex climatic fluctuations resulting in wet and dry periods, as indicated by sand dunes within the lake with a NE-SW and NW-SE trend (Servant and Servant, 1983; Durand and Lang, 1986; Isiorho, 1989). Lake Chad has also been said to be a sensitive indicator of climatic changes (Adams and Tetzlaff, 1984) making the Chad Basin a good area for getting data that would help one to simulate what may happen in an inland continental setting in the event of increase in global temperature. The hydrology of any region is tied to the climate, and a brief hydrology of the region is presented below with emphasis on the surface waters.

The main source of surface water is Lake Chad. The volume, area, and depth are controlled by the basin morphology, and the input/output regimes of water. The lake levels in the northbasin changed from about 281.5 to 280.5 m with a maximum of 283.3 in 1976 (Carmouze, 1983). The lake volume varies between $42.5 \times 10^9$ $m^3$ and $91 \times 10^9$ $m^3$, with an average volume of $72 \times 10^9$ $m^3$ (Carmouze, 1983). Figure 3 shows the lake at normal level ($\sim$281 m) in 1969 (Roche, 1977) and at low level ($\sim$279 m) in 1975, 1987, and 1991 (August) from landsat images and field observations. The annual variability of the lake level is indicated by the hydrographs (Figure 4) at Baga. It is observed that at low level, the north basin is virtually dry, which agrees with the morphometric curves of Carmouze and Lemoalle, 1983 (Figure 2). Other surface waters in the Chad Basin are in the form of streams and rivers. Figure 4 shows the hydrographs of a canal at Baga showing the fluctuation in the lake level on an annual basis.

One of the major rivers/streams in the south and northwest of the Chad Basin is the Chari, which flows from the equatorial south of the Chad Basin and Lagone, with a great flood plain in northern Cameroon. During flood periods (in July/September), this flood plain is immersed under 0.7 to 1 m of water. El-Beid River, also known locally as the Ebeji, forms part of the border between Nigeria and the Cameroon Republic. The stream flows most of the year, beginning in June or July and ending the following May. The peak discharge occurs in the November/December period. The El-Beid supplies 0.6 to $2.7 \times 10^9$ $m^3$ of water (average $1.35 \times 10^9$ $m^3$) annually to the lake. Other rivers, such as the Yobe/Kamadogu and the Ngadda, are intermittent rivers. All the rivers contribute about 90% of the total water input to Lake Chad (Roche, 1977; Carmouze et al., 1976; Eugster and Maglione, 1979), leaving 10% to rainfall.

In the study area, the rainfall is approximately 0.31 m/year (Roche, 1977; Eugster and Maglione, 1979; Gac et al., 1977; Adams and Tetzlaff, 1984; Jackel, 1984). The rainfall supplies 2.7 to $8.7 \times 10^9$ $m^3$, with an average of $6.3 \times 10^9$ $m^3$, which is about 10% of the total annual water input to the lake. The rainy season begins in May/June and ends in

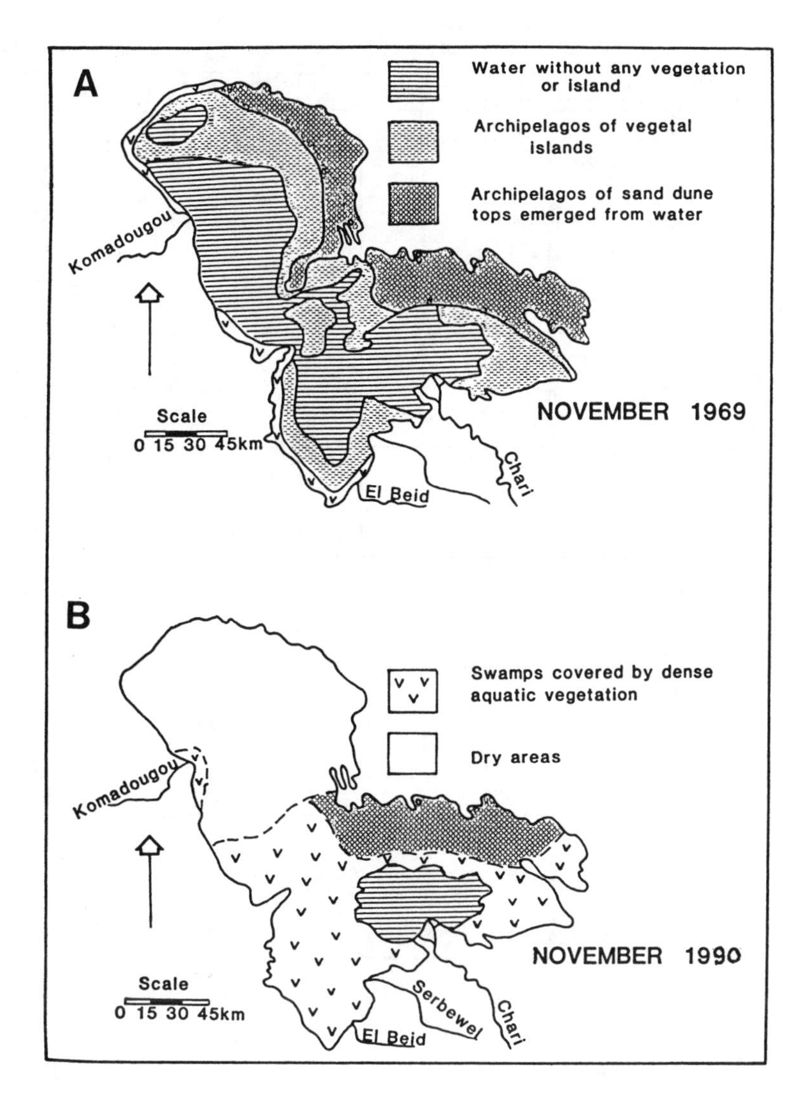


**FIGURE 3.** Lake Chad at normal level (~281 m) November 1969 (after Roche, 1977); and at low level (~278 m) November 1990 (from Landsat) and August 1991 (field observations).

October, with maximum rainfall in August, when the lake receives half of its annual rainfall which decreases considerably from the south to the north. The lake, situated between the isohyets 550 and 200 mm (Figure 5), has an average rainfall of 315 mm (Carmouze and Lemoalle, 1983) at Bol. In the study area, the average rainfall for the past 7 years at New Marte is 430 mm. Because of the high temperature and relatively low humidity of the region, evaporation becomes a significant aspect of the hydrology of the area.

Several different climatic parameters, such as high wind frequency and intensity, high temperatures, low humidity and high isolation, lead to considerable evaporation in the lake. Evaporation, one of the main constituents of the hydrological balance, was evaluated from theoretical formulae with annual estimates ranging from 210 to 220 cm. Calculations from direct measurement made on evaporation tanks, collected by ORSTOM (Office de la Rechercher Scientifique et Technique Outre-Mer) since 1964, near Bol, are considered to give the best value of potential evaporation (Carmouze and Lemoalle, 1983). The evaporation rate for the lake is generally taken to be approximately 2 m (Eugster and Maglione, 1979;

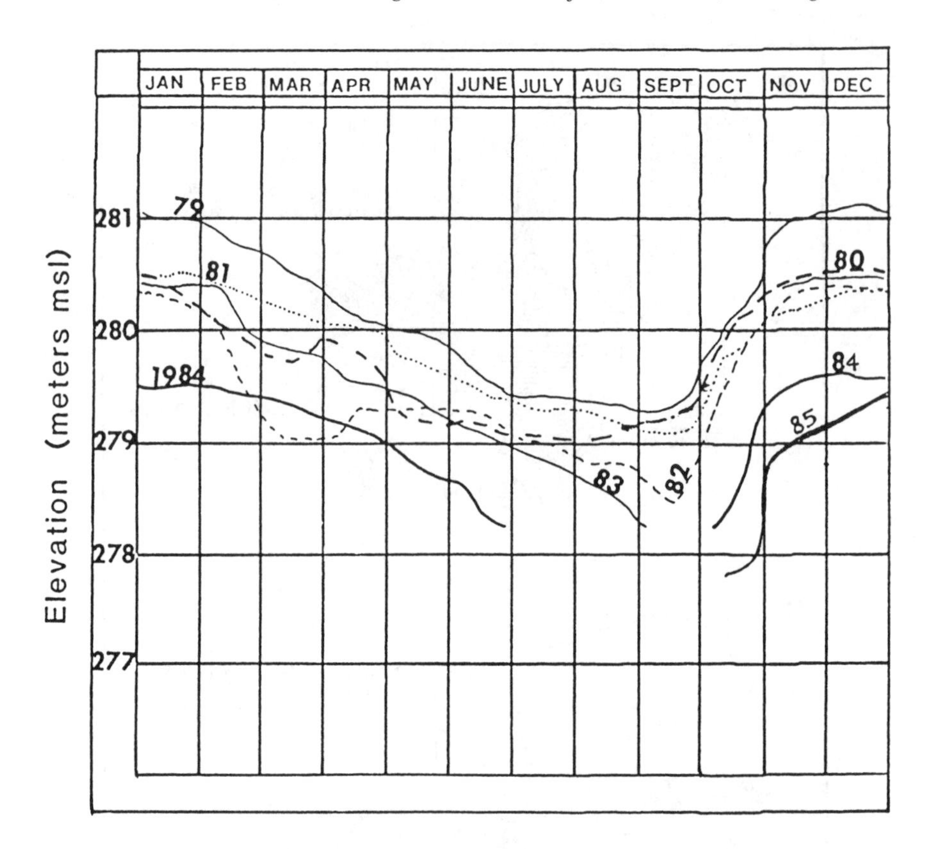


**FIGURE 4.** Hydrographs of a canal at Baga.

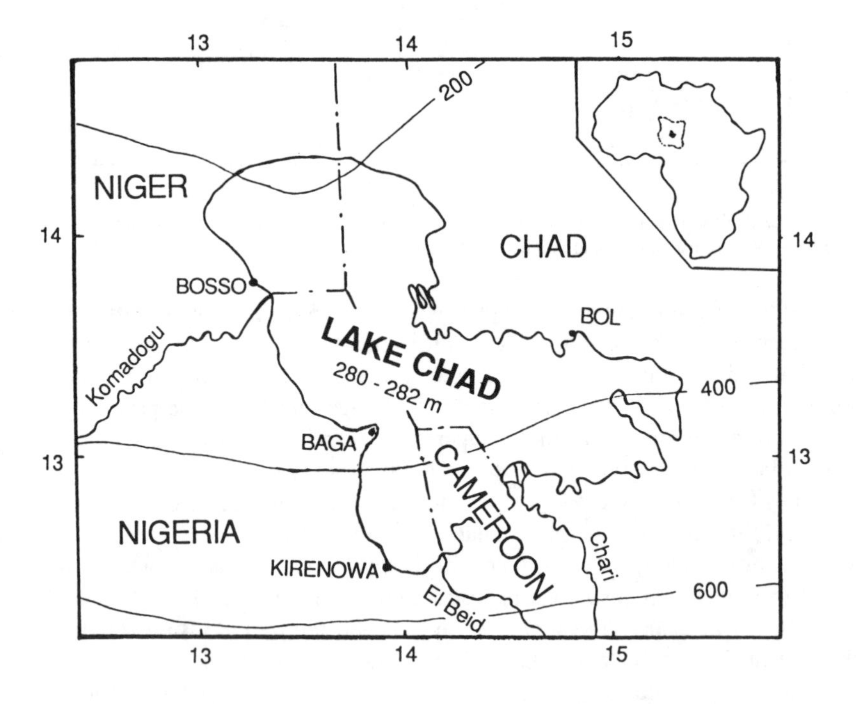


**FIGURE 5.** Map of Lake Chad with isohyets.

Maglione, 1980; Tetzlaff and Adams, 1983). Another source of water in the region is groundwater.

The main source of groundwater in the study area is in the Chad Formation. The Chad Formation aquifer system consists of three types of aquifers: confined, unconfined, or semiconfined; moreover, three distinguished zones of permeable levels are identical to the three aquifer types.

## IV. AIM OF THE STUDY

This chapter looks at some of the information needed to model the quantity and quality of water in the region during "global warming" (an ongoing work in the region). A major objective of this study is to generate information that could be used for water resources management in the Sahel region in the event of global warming. This will depend on the current understanding of the regional hydrology. The role played by seepage in salinity control in Lake Chad, and thus the seepage velocity of the lake's water to the phreatic aquifer, is extremely pertinent. Towards this end, several methods were used to understand the relation between the large surface water (lake) to the underlying groundwater resource of the region. Included were surface geophysical method (electrical resistivity; EM meters), seepage meters, dilution tests, tracing experiments, and chemical and isotopic analyses of waters (lake water, groundwater, rainwater, and river water).

## V. APPROACH

To understand the hydrology of the area, data were collected by several methods that could be helpful in the future management of the water resources of the region. Two factors should make Lake Chad's water saline: its closed basin, and its location in a semiarid region; instead, the water is relatively fresh, a hydrologic anomaly.

The seepage flow of water from the lake was measured using seepage meters. Seepage flux between the groundwater and the overlying water can be measured directly by covering an area of sediment with an open bottom container with time (Lee, 1977). In lake bottoms and streambeds, groundwater flow is upward, downward, or horizontal, but is rarely nonexistent (Lee and Cherry, 1978). Seepage meters have proven useful in quantifying groundwater-surface water interactions and groundwater recharge (Gaudet and Melack, 1981; Lee et al., 1980; Wessner and Sullivan, 1984). Seepage measurements were made at locations as described in Isiorho and Matisoff (1990) and south of the lake.

By knowing the seepage across the sediment-water interface, the hydraulic conductivity (the volume of water that would move through a unit area in a unit time [Freeze and Cherry, 1979]) of the lake sediments was measured by the slug test method. The hydraulic conductivity (or the transmissibility of an aquifer) is determined from the rate of the rise of the water level in a well after a certain volume or "slug" of water is suddenly removed from the well. Solutions to the slug test have been formulated in Hvorsler (1951), Cooper et al. (1967), Papadopulus et al. (1973) and Bouwer and Rice (1976).

The analysis of the slug test data is generally based on the method described in Bouwer and Rice (1976). The hydraulic conductivity, K(m/d) is calculated from the time rate of change in the water level, as in the method described in Boast and Kirkhan (1971), Bouwer (1978), and Todd (1980). Apart from the hydraulic conductivity and the vertical velocity, the horizontal velocity is also determined by using a borehole dilution test.

The average flow velocity of groundwater is the major variable which determines the transport of pollutants and chemicals in an aquifer. The single-well, or borehole dilution test, performed to estimate the horizontal flow veolcity in the phreatic aquifer, is also known as the point-dilution method.

Another estimate of the infiltration (seepage) velocity in the lake was made by inserting core tubes into the lake bed, spiking the overlying water with a chemical tracer (chloride), and analyzing the chloride distribution in the retrieved cores after a known period of time. These are refered to as the tracer experiments, distinct from the borehole dilution tests, which are also a form of tracer experiment.

The working principle of this method is that the horizontal velocity vector is eliminated by the tube walls which only allows the vertical component to be observed. Cores retrieved at the beginning of the experiments served as controls. The advection velocity is calculated from the conservative tracer (NaCl) in the two sets of cores.

Water samples were collected from the lake and the surrounding groundwater for laboratory analysis. They were filtered, collected in 250-ml plastic bottles, acidified, and analyzed for the major ions. Other water samples were collected for isotopic analysis for use in tracing the lake's water within the surrounding groundwater. Certain parameters tested for in the field were pH, acidity, alkalinity, total dissolved solids, dissolved oxygen, carbonate, and sulfate, using portable test kits and field instruments (Hach DR/2000 spectrophotometer and test kits).

The depths to the water table were measured where open wells exist, by using an electric sounder. The electric sounder circuit closes when its tip touches water. The static water well measurements were later combined with resistivity measurements to construct a water table map of the study area. Resistivity is the reciprocal of electrical conductivity and is thus a measure of a material resistance to the flow of electric current. The ability of a rock unit to conduct electricity depends primarily on three factors: the amount of open spaces between particles (porosity), the degree of interconnections between the pore spaces, and the volume and conductivity of the fluid in the pores.

At locations shown in Figure 6, 19 sounding measurements were made. The sounding stations are denoted by "x" in the figure, and well locations are denoted by "w". An additional 100 resistivity stations were established in 1991 in an attempt to verify lineaments and to measure the depth to groundwater in the region. The maximum total electrode distance used at all the stations was ~30 m. The resistivity readings were interpreted using a DC rest computer program. A separate test for the flow direction of the groundwater was also performed using a minipiezometer nest as described in Lee and Cherry (1978) and Isiorho and Matisoff (1990). Lee and Cherry (1978) described the use of a minipiezometer in surface water/groundwater interaction studies.

The minipiezometers were used to calculate the horizontal velocity of water in an area near the lakeshore, and also to determine the flow direction near the lake. A piezometer nest was made of nine tubes, as described in Lee and Cherry (1978) and Isiorho and Matisoff (1990), with the tubes being approximately 5 cm apart. The center minipiezometer was spiked with a conservative tracer, and the surrounding piezometers in the nest were monitored for the tracer plume after the elapsed time. A custom-designed specific chloride ion electrode was used. The flow direction was determined by the direction of the chloride plume. The data obtained for these piezometers were also used to calculate the horizontal velocity near the lake shore, as in the borehole dilution test described earlier. The minipiezometer experiments were performed only at two locations (Duro and Kirenowa). A two-dimensional solute transport model was also used to understand the hydrology of the region.

## VI. RESULTS AND DISCUSSIONS

Seepage meter measurements indicate that groundwater recharge occurs in the lake, at the rate of $1.9 \times 10\ m^{-3}$/day. Figure 7 shows seepage rates at six stations, plus their distance from the shoreline. In the figure, there is no apparent decrease in seepage rate with

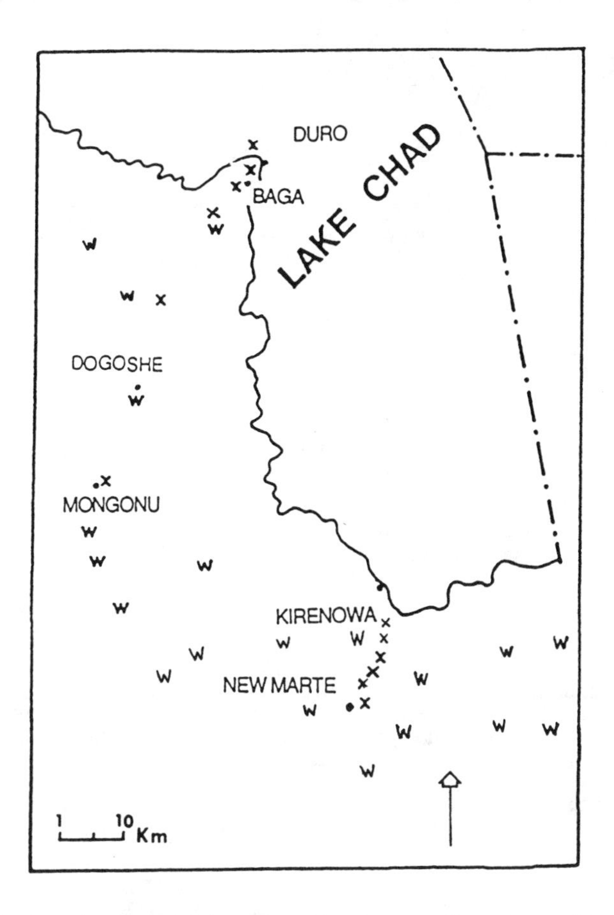


**FIGURE 6.** Map showing resistivity stations (x) and open wells southwest of Chad Basin (w).

distance from the shoreline. The limited data do not support previous reports that seepage rates decrease with distance from the shoreline lakewards. A water table map of the SW portion of the basin based on the data obtained in this study shows the lake water level to be higher than the phreatic aquifer (groundwater). Figure 8 shows a three-dimensional view of the SW lake level; the flat surface to the northwest relative to that of the groundwater, and the slope surface to the southwest. Lake sediments were found to be sandy in most places; the principal component of the samples was quartz grains. The sand grains were either elongated or oblong, with fairly smooth surfaces (Isiorho and Matisoff, 1989). The sediments of the lake bottom to the northeast of the lake consist of mud/clay and mud, where ~75% of water loss from the lake by infiltration into the phreatic aquifer is reported to occur (Carmouze, 1983). It was observed that most of the lake floor is permeable, as indicated by the sand (from grab samples), oolites, and mud, with a small portion of the lake bottom covered only by pure clay (off the River Yobe and northeast of Baga). In contrast, there are deep (~1 m) cracks in the clay sediments that allow water to flow through them. Seepage has been found to play a great role in the water and solute budget of a lake (Wessner and Sullivan, 1984; Brock et al., 1982; Fellows and Brezonik, 1980; Lee et al., 1980). Several workers (Miller et al., 1968; Roche, 1977, 1980; Fontes et al., 1969; Fontes, 1980; Eugster and Maglione, 1979; Schneider, 1967; Maglione, 1980) implied or demonstrated a link between the lake and the surrounding groundwater. Carmouze (1983) and Roche (1980) estimated that about 4 to 9% of the input water to the lake is removed by seepage. Isiorho and Matisoff (1990) estimated 20% of the input water is removed by seepage

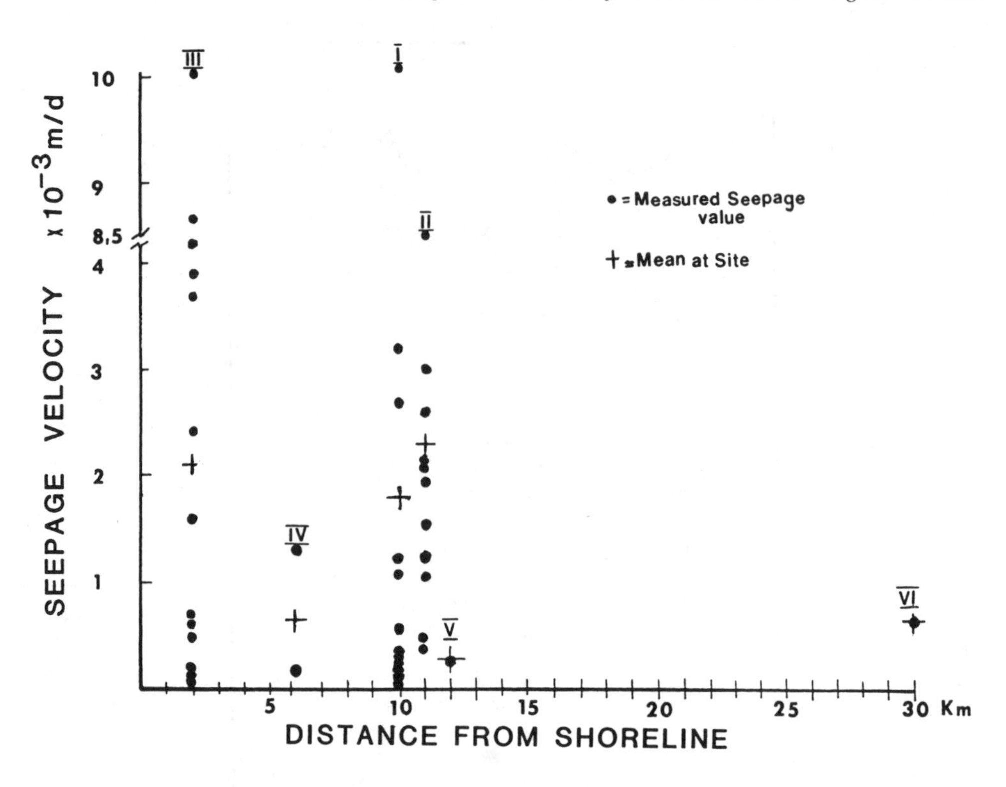


FIGURE 7. Plot of seepage rate with distance from shoreline for six stations.

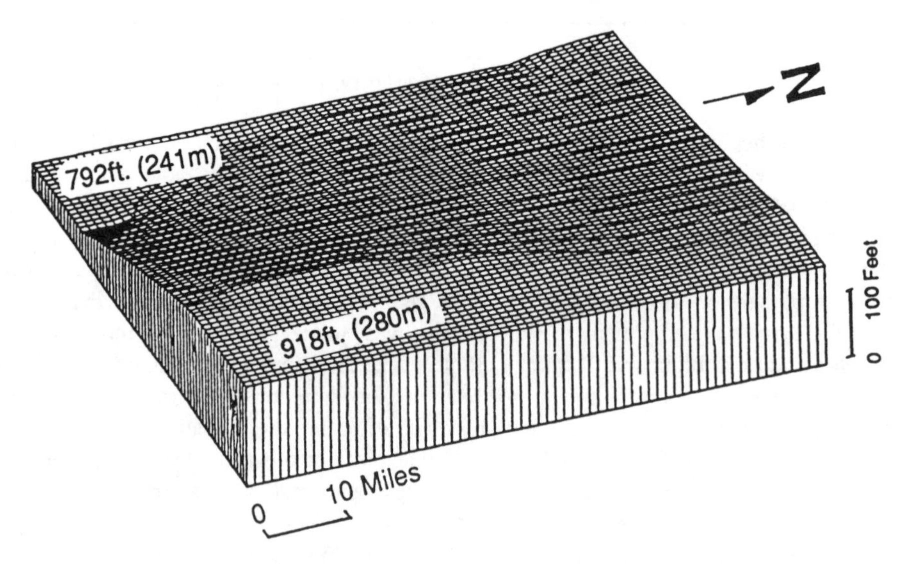


FIGURE 8. Three-dimensional view of the water table southwest of Chad Basin.

from the lake. It is possible that the seepage rate could be higher, as seepage measurements were performed at sites considered not to be high seepage areas (Carmouze, 1983).

Lineament analysis from landsat and shuttle images shows lineaments with a predominant NE-SW trend from the lake to the shoreline (Isiorho, 1990b). Figure 9 shows a lineament map of the SW Chad Basin. Preliminary investigation of lineaments from landsat images

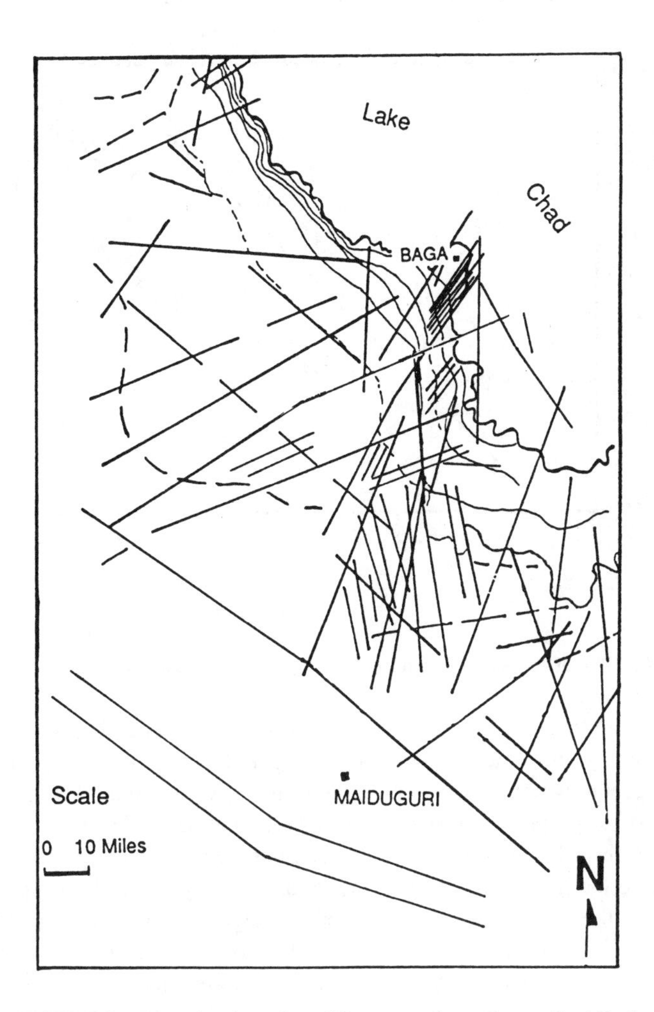


**FIGURE 9.** Map showing selected lineaments in southwest Chad Basin.

indicate a trend from the lake southwestwards. The Chad Basin is said to be a tectonic area with structural features that extend northeast to the Air Plateau and to the southwest towards the Benue Trough, a failed arm of a triple junction. The Chad Basin is structurally linked to the Benue Trough to the south (Ajayi and Ajaikaye, 1981), and it is most likely that this structural linkage plays a role in the basin hydrology. The Lake Chad waters may be linked to the Benue River (Isiorho, 1990b) through these lineaments. It is possible that the lake is linked to the upper aquifer, which in turn is linked to the underlying aquifers. The relationships among the three aquifers and overlying waters are still not well defined, and the hydrology not well understood, even though Lake Chad's waters have been traced 10s, 100s, and 1000s of m away from the shoreline in the Upper Aquifer (Schneider, 1967; Roche, 1980; Isiorho, 1990a,b) using oxygen isotopic compositions.

The hydrogen and oxygen isotopes of water in rain water, open lakewater, and groundwater were determined in order to trace the lake's water in the phreatic aquifer. Results of preliminary isotopic analysis are shown in Figure 10. The $-5\%$ measured for the $^{18}O$ local precipitation is in good agreement with the values ($-4$ to $-5\%$) reported by Roche (1973). The open waters, and the groundwaters within approximately 15 km of the lake, are enriched in $^{18}O$ ($+0.2$ to 3.4/mil), indicating that they are highly evaporated with respect to the local precipitation.

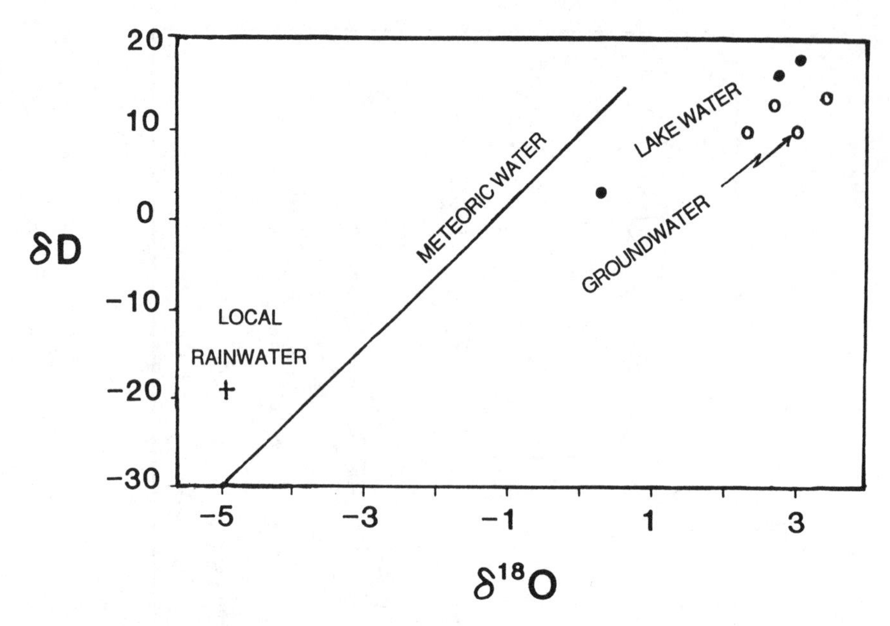


**FIGURE 10.** Plot of oxygen/hydrogen isotopic composition of lake, groundwater, and rainwater.

The linear relationship between ∂D and $\partial^{18}O$ of the lake and groundwater indicates that both the lake and groundwaters have similar origin. There is no change in isotopic composition of the groundwater as it travels through the aquifer (Savin, 1980). Since the Chari/Lagon river system supplies approximately 85% of the annual water (Roche, 1980; Carmouze, 1983) to the lake, and only ~10% is supplied by precipitation, the phreatic aquifer water is therefore thought to come from the lake. This appears to be supported by the low annual rainfall in the region (30 cm) and the very high evaporation rate (200 cm). Isiorho and Matisoff (1989) examined the implication of seepage (i.e., groundwater recharge from the lake) in the basin by calculating the amount of water that may be available from the recharge alone on an annual basis. This amount ($10 \times 10^9$ $m^3$) is considered to be "stored" water which could be made available for domestic, industrial, and agricultural services. Therefore, the lake is the main source of water in the region, either as surface water or groundwater. As more than 90% of the annual water input to the lake comes from the river systems (Chari/Lagone, El Beid, Kamadogu), it is very important to monitor these rivers; any diversion or damming of these rivers could have an adverse effect on the availability of water for the Sahel region of Africa.

During the summer of 1991, water samples were collected from the southern basin. Preliminary results indicate that the total dissolved solids increased with distance from the lake within the groundwater. Attempts were also made to measure the flow direction of the lake water within the lake. This was performed at only two sites within the lake. The shoreline tends to migrate several thousands of meters within a year, especially during periods of low precipitation in the active watershed to the south, making sampling of the lake water difficult; (movement or transportation within the lake, especially when it is dry, is very slow and painful). The decrease in the lake volume has been documented by other workers, especially in the discussion of the geologic history of the region. The chemical composition of both the lake and groundwaters, as well as the hydraulic conductivity values, were used in a computer simulation of the solute transport for the Chad Basin. The simulation was used as another method to assess the significance of seepage in Lake Chad's water budget, as well as in the chemical mass balance calculation. The U.S.G.S. MOC two-dimensional solute transport model gave results that were similar to those obtained from the direct field measurements, as discussed in Isiorho (1989). What are the implications of these data in the water management of the region?

# VII. A REGIONAL APPROACH TO WATER MANAGEMENT

As a result of droughts in the 1960s and 1980s, it was suggested that the lake should be reduced by one half of its present size. Generally the lake is divided into two basins, north and south, of roughly equal surface area and separated by a line from Baga-Kawa to Baga-Sola. The proposal to reduce the surface area of the lake is supposed to save the lake by ''leaving'' water in the south basin for farming. Since the north Chad Basin is supposed to have some silt, it could be a fertile ground for agriculture. The proposal was not based on any detailed hydrological study of the lake. Should the proposal be followed, it would create some seriously adverse ecologic and environmental effects. There would be reduction in groundwater recharge, as there would be less area for groundwater to pass through, and the increase in hydrostatic pressure would not be sufficient enough to offset the reduced recharge surface. The presence of the sandy lake bed, as observed from the grab samples, should allow water to leave the lake, as demonstrated in this study. This recharge water could be considered as ''stored'', and would be available for use. Currently there are plans to construct several dams along the reach of the Chari/Lagone river system. One major problem with this new proposal would be the reduced water input to the lake. Construction of such dams without adequate hydrologic studies could eliminate the lake at a later time in the event of a severe and prolonged drought in the region. The lessons from the high dam across the Nile river should be examined closely. Construction of the Nile dam led to problems of silting, flooding, and starving of the coastal region of sediments, resulting in shore erosion. There were also decreases in sea foods, and a rise in diseases, notably river-blindness. Lifestyles and cultures of the people along the Nile have been changed due to the dam and the above-mentioned flooding that threatened the famous Egyptian monuments. Such unforeseen consequences of the Nile dam project should be kept in mind before damming the Chari/Lagone river system.

There are currently several irrigation farming projects that depend on Lake Chad surface water. The South Chad Irrigation Project (SCIP) is currently cultivating over 100,000 ha of land, and may be expanded to 150,000 ha. To the north of this project is the Baga Polder Project, where much land is likewise being irrigated. These various irrigation projects use some of the least efficient methods. Open drain systems are used to transfer water from the lake to the field. It should be kept in mind that large surface areas are exposed as a result of the irrigation. It could be estimated that about a quarter of the drainage water from the lake is lost to evaporation. This could easily be avoided or reduced by having more efficient irrigation systems, such as drip irrigation.

Boreholes in the area have been sited without adequate hydrologic and hydrogeologic studies. Several of the boreholes tap the confined aquifer, and the water condition is artesian, meaning the water flows out of the borehole. Some of the boreholes are not capped, an as a result water is being wasted through evaporation. There is less infiltration of the wat to the underlying aquifers, as the top soils are very clayey, thereby limiting recharge of aquifer. Attempts should be made to reduce these wastes.

As Lake Chad is shared by four nations, there is the need to have cooperative agree Currently there is a standing committee — the Chad Basin Commission (CBC) — th into agreements. There are no known standing regulations for the water resou various nations within the basin. A first step towards this is to have some form regulations that would be agreeable to the nations. It should be kept in mind the input water to the lakes comes from the south. The Cameroon and Nigerian rivers supply ~90% of the input water, and about 10% of the lake's water is used for irrigation farming to the SW of the basin alone. There are suggestions for an interbasin transfer of water from the Congo to the Chad basin. This suggestion needs to be looked at in terms of the environmental impact it would have in the region. There is a need for the governments of the region to meet in order to effectively manage the water resources of the region based on sound hydrological data. With greater demand for water in the region, more research/study should be encouraged.

## VIII. FUTURE STUDIES

There is an implicit relationship between the frequency of sampling and the groundwater quality information obtained. This author suggests that more water samples should be analyzed to see if a form of pattern can be established in order to properly evaluate the chemistry and water quality of the region. The seepage measurement should be conducted to cover a larger area in order to establish any areas that may be very porous or areas where there may be no seepage. There is no known sedimentation rate for the lake as a whole; for the clay sedimentation rate for the northwest Chad Basin see Mothersill (1975). It is suggested that the sedimentation rate be calculated for the lake using the lead 210 method, as this is a well-established procedure. This may be useful in considering the importance of benthos (bottom-dwelling aquatic life forms) in the transport of solutes between the water and the sediments. Field and laboratory experiments should be designed to evaluate the absorption property of the lake sediments and what role they play in the removal of solutes from the lake. With remote sensing becoming increasingly popular, its use in monitoring the water quality and quantity should be encouraged. The occurrence, movement, and chemistry of the water in the Chad Basin, especially around the lake, have not been properly identified and documented. The author strongly recommends that the detection, location, and management studies of the water resources of the Chad Basin be continued.

A salinity problem reported to the northeast shoreline and in the SCIP area (Bukar and Usman, 1991, personal communication) should be investigated in order to find ways of preventing this occurrence in other areas. In connection with this, a detailed resistivity survey or other geophysical methods should be carried out, especially some 50 km around the lake, together with the installation of monitoring wells for both water samples and water-level measurements. It is also suggested that infiltration rates of rain water to the phreatic aquifer be carried out. It was assumed in this study that groundwater recharge from rainfall is not significant. The relationship of the lineaments to both groundwater exploration and the fate of the recharge water from the lake should be examined.

Preliminary study indicates a relationship between the prominent lineaments in the southwest portion of the basin to the structural features buried deep within the bedrock. Geophysical transect surveys will be conducted primarily normal to the linear trend, and because the interpretation of the electrical geophysical method is not unique, other geophysical methods should be used. Another method that could also be used to verify the linear features is the performance of pumping tests around and near the lineaments. Seepage rates could be used to verify lineaments especially within the lake, as the rates would be expected to be higher along these lineaments, if they are indeed fractures.

## IX. CONCLUSIONS

The results of this study show that seepage is a major controlling factor. It plays the dominant role in controlling the salinity in Lake Chad. The evidences for this comes from the mass balance calculations discussed in Isiorho and Matisoff (1990). Additionally, this is based on other evidences such as the seepage velocity, isotopic tracing, flow direction determination, and the hydraulic relationship between the lake and the phreatic aquifer. Also, seepage accounts for approximately 20% of the input water removed from the lake.

Lake Chad recharges the phreatic aquifer (upper zone), as has been demonstrated through the infiltration (seepage) measurement, using seepage meters and tracer experiments which gave infiltration velocity of $1.9 \times 10^{-3}$ m/d. The water table map derived from this study shows the lake to be at a higher hydraulic level than the phreatic aquifer. The water table suggests groundwater flow to the south and southwest. The flow of water from the lake is radial but the regional groundwater flow near the lake is southwestwards. The sandy shoreline and "leaky" lake bed is continuous with the phreatic aquifer; therefore, it allows the flow of water from the lake to the groundwater. The lake water was traced in the phreatic aquifer about 15 km away from the lake, indicating the recharge of the groundwater is from the lake.

Lake Chad is the main source of water to the phreatic aquifer. From both the seepage measurements and two-dimensional computer model, the amount of water supplied to the phreatic aquifer is approximately $10 \times 10^9$ $m^3$; the lake acts as the baseline for the groundwater.

For the past 20 years, the groundwater level in the phreatic aquifer has remained stable, especially some 80 km away from the lake shoreline, despite the Saharan droughts of the 1960s and 1980s. The present practice of the indiscriminate sitting of boreholes and of allowing groundwater to flow continuously out of uncapped wells should be banned, if proper use will be made of the water resources of this region. A coordinated effort by the areal nations similar to the French (ORSTOM) is recommended for this region. The interbasin water transfer should be examined closely, as this would have serious implications for water resource management.

There is an areal variation in the distribution of the solute in the lake and groundwater. The variation observed in the phreatic aquifer may be reflective of the areal variation observed in the lake water. This means that proper planning for the exploitation of the water resources for this region is greatly needed. The transport of the solutes from the lake to the groundwater occurs by an advection mechanism.

It should be noted that pollution of the lake means pollution of the phreatic aquifer, therefore the lake water quality should be guided from pollution. More field study needs to be considered if the water will be used for industrial and agricultural purposes.

In summary, Lake Chad, as an old, closed-basin lake, serves as the main source of water, (surface water and groundwater) in the Sahel region of Africa. As part of an ongoing study, the lake's water has been identified and traced 20 km away in the surrounding phreatic aquifer. Preliminary data from oxygen/deuterium, chemical, and lineament trace analysis indicate that Lake Chad could account for about 95% of the groundwater recharge of the region. This has great implications for the water resource management of the region. More study is mandatory, and the need for the nations of the region to coordinate their efforts towards effective management of the water resources of the region must be emphasized.

## ACKNOWLEDGMENT

Financial support was received from Earthwatch Research Corps, the National Geographic Society, Indiana University-Purdue University, Fort Wayne, and the National Science Foundation. The following were instrumental in the completion of the present project: Chad Basin Development Authority, CBC, A. Hudson, B. deVeer, SCIP, Prof. G. Matisoff, B. Zanna, Prof. A. Ola, E. Isiorho, and K. Wehn. Others include the Earthwatch "Lake Chad

Project'' volunteers, and R. Geyer for encouragement. Thanks to F. Beeching for editing a version of the manuscript.

## REFERENCES

Adams, L. J. and Tetzlaff, G., Did Lake Chad exist around 18,000 yr BP?, *Arch. Meteorol. Geophys. Bioklimatol. Ser. B,* 34, 299–308, 1984.

Ajaikaye, D. E. and Burke, K., Bouger gravity map of Nigeria, *Tectonophysics,* 16, 103–115, 1973.

Ajayi, C. I. and Ajakaije, D. E., The origin and peculiarities of the Nigerian Benue trough: another look from recent gravity data obtained from middle Benue, *Tectonophysics,* 80, 285–303, 1981.

Avbovbo, A. A., Ayoola, E. O., and Osahon, G. A., Depositional and structural styles in Chad Basin of northeast Nigeria, *AAPG,* 70(12), 1787–1798, 1986.

Barber, W., Pressure water in the Chad Formation of Borno and Dikwa Emirates, Northeastern Nigeria, *Niger. Geol. Surv. Bull.,* #35, 1965, 138p.

Barber, W. and Jones, D. C., The geology and hydrogeology of the Maiduguri, Borno Province, *Rec. Geol. Surv. Niger.,* 1960, 5–20.

Boast, C. W. and Kirkhan, D., Auger hole seepage theory, *Soil Sci. Soc. Am. Proc.,* 1971, 365–373.

Bouwer, H., Surface-subsurface water relations, in *Groundwater Hydrology,* McGraw-Hill, New York, 1978.

Bouwer, H. and Rice, R. C., A slug test for determining hydraulic conductivity of unconfined aquifers with complete or partial penetrating wells, *Water Resour. Res.,* 12(3), 423–428, 1976.

Brock, T. D., Lee, D. R., Janes, D., and Winek, D., Groundwater seepage as a nutrient source to a drainage lake: Lake Mendota, Wisconsin, *Water Resour. Res.,* 16, 1255–1263, 1982.

Burke, K., The Chad Basin: an intra-continental basin, *Tectonophysics,* 36, 192–206, 1976.

Carmouze, J. P., Hydrochemical regulation of the lake, in *Lake Chad: Ecology and Productivity of a Shallow Tropical System,* Carmouze, J. P., Durand, J. R., and Leveque, C., Eds., Dr. W. Junk Publ., The Hague (Netherlands), 1983.

Carmouze, J. P., Golterman, H. L., and Pedro, G., The neoformations of sediments in Lake Chad; their influence on the salinity control, in *Interaction Between Sediments and Fresh Water,* Golterman, H. L., Ed., Dr. W. Junk Publ., The Hague (Netherlands), 1976, 33–39.

Carmouze, J. P. and Lemoalle, J., The lacustrine environment, in *Lake Chad: Ecology and Productivity of a Shallow Tropical System,* Carmouze, J. P. and Leveque, C., Eds., Dr. W. Junk Publ., The Hague (Netherlands), 1983.

Carter, J. D., Barber, W., Tait, E. A., and Jones, D. G., The geology of parts of Adamawa, Bauchi and Bornu Pronvinces in NE Nigeria, *Bull. Geol. Surv. Niger.,* 1963, #30.

Cooper, H. H., Jr., Bredehoeft, J. D., and Papadopulus, I. S., Response of a finite diameter well to an instantaneous charge of water, *Water Resour. Res.,* 3, 263–269, 1967.

Cratchley, C. R., Geophysical Survey of the South-western Part of the Chad Basin, Pap. presented at Conf. on Geol., Kaduna, Northern Nigeria, February, 1960.

Cratchley, C. R., Louis, P., and Ajakaiye, D. E., Geophysical and geological evidence for the Benue Chad Basin Cretaceous rift valley system and its tectonic implications, *J. Afr. Earth Sci.,* 2(2), 141–150, 1984.

Durand, A., Oscillations of Lake Chad over the past 50,000 years: new data and new hypothesis, *Palaeogeogr. Palaeoclimatol. Palaeoecol.,* 39, 37–53, 1982.

Durand, A., Fontes, J. C., Gasse, F., Icole, M., and Lang, J., Le nord-oest du Lac Tchad au Quaternaire: etude de Paleoenvironnements allurianx, eolians, palustres et lacustres, *Palaeoecol. Afr.,* 16, 215–243, 1984.

Durand, A. and Lang, J., Approche critique des methods de reconstitution paleoclimatique: le sahel Nigero-Tchadian depuis 40,000 ans, *Bull. Soc. Geol. France,* (8) t.II, #2, 267–278, 1986.

Eugster, H. P., Chemistry and origin of brines in Lake Magadi, Kenya, *Miner. Soc. Am. Spec. Pap.,* 3, 215–235, 1970.

Eugster, H. P. and Hardie, L. A., Saline lakes, in *Lakes-Chemistry, Geology and Physics,* Lerman, A., Ed., Springer-Verlag, New York, 1978, 237–293.

Eugster, H. P. and Jones, B. F., Behavior of major solutes during closed basin brine evolution, *Am. J. Sci.,* 279, 604–631, 1979.

Eugster, H. P. and Maglione, G., Brines and evaporites of the Lake Chad Basin, Africa, *Geochim. Cosmochim. Acta,* 43, 973–981, 1979.

Fellows, C. R. and Brezonik, P. L., Seepage flow into Florida lakes, *Water Resour. Bull.,* 16(4), 635–641, 1980.

Fontes, J. Ch., Environmental isotopes in groundwater hydrology, in *Handbook of Environmental Isotope Geochemistry,* Fritz, P. and Fontes, J. Ch., Eds., Elsevier, Amsterdam 1980.

Fontes, J. Ch., Maglione, G., and Roche, M. A., Donnees isotopes preliminaires sur les rapports du Lac Tchad avec les nappes de la bordure north-est, *Cah. ORSTOM, Ser. Hydrol.*, VI(1), 17–34, 1969.

Freeze, R. A. and Cherry, J. A., *Groundwater,* Prentice-Hall, Englewood Cliffs, NJ, 1979.

Furon, R., Geologic de l'Afrique, deuxieme edition, *Payot Paris,* 1960.

Gac, J. Y., Droubi, A., Fritz, B., and Tardy, Y., Geochemical behaviour of silica and magnesium during the evaporation of waters in Chad, *Chem. Geol.*, 19, 215–228, 1977.

Gaundet, J. J. and Melack, J. M., Major ions in a tropical African lake basin, *Freshwater Biol.*, 11, 309–333, 1981.

Hammad, F. A. and Abdou, H. F., Hydrogeology, in *Lake Chad Hydrology Literature Survey,* Adeniyi, F. A., Ed., University of Maiduguri Press, 1982.

Hardie, L. A. and Eugster, H. P., The evolution of closed-basin brines, *Miner. Soc. Am. Spec. Publ.*, Washington, D.C., 3, 273–290, 1970.

Hvorsler, J. M., Time Lag and Soil Permeability in Groundwater Observations, Bull. 36, U.S. Corps of Engineers, 1951, 50p.

Isiorho, S. A., Taylor-Wehn, K. S., and Corey, T. W., Abstract: locating groundwater in Chad Basin using remote sensing technique and geophysical method, *Eos, Trans.*, AGU, 72(44), October 29, 1991.

Isiorho, S. A. and Matisoff, G., Groundwater recharge from Lake Chad, *Limnol. Oceanogr.*, 35(4), 931–938, 1990.

Isiorho, S. A. and Matisoff, G., Groundwater seepage and its implication on the water resources planning and management in the Chad Basin, *J. Water Resour.*, 1(1), 210–215, 1989.

Isiorho, S. A., Remote Sensing Applications in Chad Basin, GSA North Central section, Abstr. (with programs), Vol. 21, No. 4, 1989.

Isiorho, S. A., Water Resources Management in Chad Basin, Africa, Abstr. 7th World Congr. on Water Resour., May 13–18, 1991, Rabat, Morocco, 1991a, p. A3, 3–4.

Isiorho, S. A., A Regional Approach to the Water Shortage Problems in the Sahel Region of Africa, Abstr. Global Warming — A Call for Int. Coord. Conf., Chicago, April 8–11, 1991b.

Isiorho, S. A., Fractures in the water budget of Chad Basin, Africa, In proceeding Vol. on Fluid Flow in Fractured Rocks, 1990a, 571–583.

Isiorho, S. A., The hydraulic connection between Lake Chad and the Benue Trough, Abstr., *Eos* 71(17), 507, 1990b.

Jackel, D., Rainfall patterns and lake level variations at Lake Chad, in *Climatic Changes on a Yearly to Millennial Basis: Geological, Historical and Instrumental Records,* Morner, N. A. and Karlen, W., Eds., D. Reidel Publ., Dordrecht, Netherlands, 1984.

Jones, B. F., Geochemical evolution of closed basin water in the western Great Basin, *Proc. 2nd Salt Ohio Geol. Soc.*, 1, 181–200, 1966.

Klans, A., Climatological aspects of the spatial and temporal variations of the southern Sahara margin, in *Palaeoecology of Africa* Vol. 12 Sahara and Surrounding Seas, Sarnthein, M., Seibold, E., and Rognon, P., Eds., Gower Pub., U.K., 1980.

Krabbenhoft, D. P. and Anderson, M. P., Use of a numerical groundwater flow model for hypothesis testing, *Ground Water,* 24(1), 49–55, 1986.

Lee, D. R., A device for measuring seepage flux in lakes and estuaries, *Limnol. Oceanogr.*, 22(1), 140–147, 1977.

Lee, D. R. and Cherry, J. A., A field exercise on groundwater flow using seepage meters and minipiezometers, *J. Geol. Educ.*, 27, 6–10, 1978.

Lee, D. R., Cherry, J. A., and Pickens, J. F., Groundwater transport of a salt tracer through a sandy lakebed, *Limnol. Oceanogr.*, 25(1), 45–61, 1980.

Lemoalle, J., Application des donnees Landsat a l'estimatin de la production du phytoplancton du Lac Tchad, *Cah. ORSTOM Ser. Hydrobiol.*, 7, 95–116, 1978.

Maglione, G., An example of recent continental evaporitic sedimentation, the Chad Basin (Africa), in *Evaporite Deposits. Illustration and Interpretation of Some Environmental Sequences,* ed. Technip. 27, 15, 1980.

Matheis, G. Short review of the geology of the Chad Basin in Nigeria, in *Geology of Nigeria,* Kogbe, C. A., Ed., Elizabethan Publ., Surulere (Lagos), Nigeria, 1976, 289.

McBride, M. S. and Pfannuch, H. O., The distribution of seepage within lakes, *J. Res. U.S. Geol. Surv.*, 3, 505–512, 1975.

Miller, R. E., Johnston, R. H., Olowu, J. A. I., and Uzoma, J. U., Groundwater Hydrology of the Chad Basin in Borno and Dikwa Emirates, Northeastern Nigeria, with Special Emphasis on the Flow Life of the Artesian System, U.S. Geol. Soc., Water-supply Pap. I, 757, 1968.

Mothersill, J. S., Lake Chad: geochemistry and sedimentary aspects of a shallow polymictic lake, *J. Sedimen. Petrol.*, 45(1), 295–309, 1975.

Nagy, R. M., Ghuma, M. A., and Rogers, J. J. W., A crustal suture and lineaments in North Africa, *Tectnophysics*, 31, T67–T72, 1976.

Papadopulus, S. S., Bredehoeft, J. D., and Cooper, H. H., Jr., On the analysis of "slug test" data, *Water Resour. Res.*, 9, 1087–1089, 1973.

Raeburn, C. and Jones, B., The Chad Basin geology and water supply, *Bull. Geol. Surv.*, Nigeria, #15, 1934.

Roche, M. A., Tracage Hydrochimique Naturel du Mouvement des eaux Dans le Lac Tchad, in Hydrology of Lakes Symp. (Hydrologie des lacs), IAHS-AISH Publ. 109, 1973.

Roche, M. A., Lake Chad: a subdesertic terminal basin with fresh waters, in *Desertic Terminal Lakes*, Webster State College, Ogden, UT, 1977, 213–223.

Roche, M. A., Tracage naturel salin et isotopique des eaux du system hydrologique du lac Tchad, *Cah. ORSTOM*, Publ. #117, 1980.

Ronen, D., Magaritz, M., Paldor, N., and Bachmat, Y., The behavior of groundwater in the vicinity of water table evidence by specific discharge profiles, *Water Resour. Res.*, 22(8), 1217–1224, 1986.

Savin, S. M., Oxygen and hydrogen isotope effects in low-temperature mineral interactions, in *Handbook of Environmental Isotope Geochemistry*, Fritze, P. and Fontes, J., Eds., Elsevier, Amsterdam, 1980.

Schneider, J. L., Relation Entre le Lac Tchad et la Nappe Phreatic, A.I.H.S. Symp. Garda, Publ. #79, 122–131, 1967.

Tetzlaff, G. and Adams, L. J., Present-day and early-holocene evaporation on lake Chad, in *Variations in the Global Water Budget*, Street et al., Eds., D. Reidel Publ., Dordrecht, Netherlands, 1983, 347–360.

Todd, D. K., *Groundwater Hydrology*, John Wiley & Sons, New York, 1980.

Wessner, W. W. and Sullivan, K. E., Results of seepage meter and minipiezometer study, Lake Mead, Nevada, *Groundwater*, 22(5), 561–568, 1984.

Winter, T. C., Numerical simulation of steady-state three-dimensional groundwater flow near lakes, *Water Resour. Res.*, 14, 245–254, 1978.

Winter, T. C., Uncertainties in estimating the water balances of lakes, *Water Resour. Bull.*, 17, 82–115, 1981.

Winter, T. C., Effects of ground-water recharge on configuration of the water table beneath sand dunes and on seepage in lakes in the sandhills of Nebraska, *J. Hydrol.*, 86, 221–237, 1986.

Yeretich, R. F. and Cerling, T. E., Hydrochemistry of Lake Turkana, Kenya: mass balance and mineral reactions in an alkaline lake, *Geochim. Cosmochim. Acta*, 47, 1099–1109, 1983.

Chapter 22

# SATELLITE MONITORING OF DESERT EXPANSION

**H. E. Dregne and C. J. Tucker**

## TABLE OF CONTENTS



0-8493-4419-0/93/$0.00 + $.50
© 1993 by CRC Press, Inc.

# I. INTRODUCTION

Vegetation degradation on the border of climatic deserts such as the Sahara, as the result of human abuse of the land, has attracted attention in the scientific and popular press. Common wisdom has held that vegetation degradation was causing expansion of the world deserts. No measurements have been made to justify that conclusion but anectodes abound that support it. Geographers from the U.K. apparently were the first to refer to "desert encroachment" of the Sahara on to adjoining lands on the south side of the Sahara. Repeated assertions that desertification was responsible for desert expansion undoubtedly has convinced many people that it must be true, despite the absence of proof either confirming or denying it.

A study using polar-orbiting meteorological satellites to collect global data on green biomass production was initiated in 1985 by the Goddard Space Flight Center of the National Aeronautics and Space Administration (NASA). The primary objective of the study was to develop a method that would allow measurements to be made of the area of climatic deserts. A secondary objective was to determine whether human-induced desertification was causing a long-term change in vegetation on the borders of deserts. The desert chosen to study first was the Sahara, by far the largest climatic desert and a part of the 14,000 km long chain of arid lands ranging across Africa and Asia, from the Atlantic Ocean to northeastern China. Satellite imagery for the 1980 to 1990 period was analyzed and a normalized difference vegetation index (NDVI) was employed to estimate green biomass production for each of the 11 years.

# II. DESERT ENCROACHMENT

The first known reference to the threat of encroachment of the Sahara onto adjoining pastoral and cultivated lands on its southern boundary was made by Cana (1915). He noted that the Sahara seemed to be encroaching upon the plains of Nigeria and that the cause could be human activity which induced climatic changes. He believed that regenerating the pastoral zone between the desert in the north and the croplands in the south would stop further desert encroachment. The latter view is held by many persons involved in desertification control.

Bovill (1921) concluded that, in certain areas, the Sahara was encroaching on the pastoral and crop lands of West Africa. He contended that man was, to no small extent, responsible for the desiccation of oases and the fertile plains south of the Sahara. To Bovill, the encroachment consisted of sand-dune movement and the loss of vegetation resulting from tree cutting and a resultant climatic change toward increased aridity. He believed that many rivers in Senegal had dried up within the memory of living natives, while others had become more saline. All of this anecdotal evidence pointed to an increasingly bleak landscape with few trees, sparse vegetative cover, more sand dunes, and a marked abandonment of cropland. Those observations convinced Bovill that humans were causing the climate to become drier and the Sahara to expand. On the other hand, Bovill describes the belief of two French and one British geologist that the Sahara was actually retreating, not expanding. He dismissed that contention as being due to the broad-scale perspective of geologists and the fact that they did not have to cope with the practical problems of human-induced aridity.

In a note at the end of his paper, Bovill cites what may be the first estimate of the rate of desert encroachment. He says that an observer named Migeod had calculated that the Sahara was advancing southward at a rate of 200 m/year, or 300 km in 1500 years. There is no indication of how Migeod arrived at his numbers.

## A. STEBBING'S OBSERVATIONS

The article that initiated a lively debate in Great Britain and gave world prominence to the concept of a steadily expanding Sahara was presented by a British forester, E. P. Stebbing, to a meeting of the Royal Geographical Society in 1935 (Stebbing, 1935). Stebbing concluded that the Sahara was steadily encroaching on the pastoral and crop lands on its southern border as the result of tree cutting, overgrazing, and shifting cultivation. He believed that land degradation brought increased desiccation and the formation and expansion of sand dunes marking the leading edge of the Sahara as it moved southward. His conclusions were based on his own observations of conditions in northern Nigeria, southern Niger, and eastern Mali during the winter of 1933 to 1934, as well as on comments made to him.

On his trip, Stebbing met a French political officer who had spent many years in Niger and Mali. The officer confirmed Stebbing's observations on sand-dune movement and offered the opinion that the Sahara had advanced at an average rate of 1 km/year over the past 3 centuries. That rate would be five times the rate Bovill cited, but over a different time period.

Considerable controversy arose over Stebbing's article and a book published by him in 1937 that elaborated upon the Sahara encroachment concept. A reviewer of the book in the December 1937 issue of *The Geographical Journal* categorically denied that there was any evidence of regional desiccation in the Sahel, or of southward shifting of the Sahara border as sand dunes grew and moved. In Stebbing's response to the reviewer's critical remarks, he changed his stand on desiccation (Stebbing, 1938). Instead of representing a climate change, he said, the increased desiccation he had described was actually a reduction in soil moisture due to increased runoff from eroded land. He also deplored his use of the term, "encroaching Sahara", in his famous 1935 paper.

An emphatic denial of Stebbing's desiccation and encroachment thesis came from a joint British-French forestry mission that surveyed the Nigeria-Niger border from December 1936 to February 1937. Their purpose was to determine whether desiccation and desert encroachment were threats to the two colonies, as Stebbing contended (Anonymous, 1973). They concluded that there was " . . . nothing that points to increasing unproductiveness due to an extension of the desert zone in these areas".

## B. ACCEPTANCE OF CONCEPT

In later years, there seemed to be a general acceptance of the proposal that the Sahara was expanding. A U.S. Agency for International Development (USAID) document claimed that the Sahara was moving southward at the amazing rate of 30 mi/year (USAID, 1972). No data were cited to justify a rate that could hardly be credible.

The first and only data that were collected in an attempt to quantify the Sahara expansion so widely believed to have occurred came from a study by Lamprey (1988) in the Sudan. Lamprey participated in a survey that was carried out in 1975 of land degradation in northern Sudan. His report was reprinted in the Desertification Control Bulletin in 1988 in response to requests for copies of the original report.

In his analysis of the land condition in 1975, Lamprey compared conditions then with the desert ecological boundary mapped in 1958 by Harrison and Jackson. Lamprey estimated that the desert vegetation boundary had shifted 90 to 100 km southward between 1958 and 1975. That shift gave an annual rate of 5 to 6 km, a widely quoted number for Saharan desert expansion.

Much evidence of ecological damage done by cultivation and livestock grazing was presented by Lamprey. His report is a comprehensive analysis of land degradation in one section of the Sahara border. Numerous examples were given of sand-dune movement and sand encroachment on better lands in northwestern Sudan.

## C. DOUBTS RAISED

It came as a shock when Hellden (1984, 1988) concluded from his study of the same area in the Sudan that Lamprey's claim of long-term land degradation there could not be substantiated. Hellden used aerial photographs, satellite imagery, and several gound surveys in his study of land conditions from 1962 to 1979. Perhaps most importantly, he was unable to find the sand-dune fields Lamprey had described, and he was not able to find evidence of any major and systematic sand-dune encroachment. As far as Hellden could determine, land conditions in 1979 were essentially unchanged from 1962.

## III. METEOROLOGICAL SATELLITE STUDIES

In 1985, we began a study to determine the boundaries of the major deserts of the world. One of the principal reasons for initiating the research was to find reasonably acceptable methods of measuring a desert's boundaries so that a determination could be made about whether those boundaries were or were not expanding.

### A. DETERMINING DESERT BOUNDARIES

Locating desert boundaries would seem to be simple: select a suitable mean annual precipitation figure, then locate that isohyet on a map; or establish a standard for what constitutes desert vegetation and map the boundaries of that vegetation zone; or even use soils as a criterion since climate is the major soil forming factor.

Unfortunately, none of those criteria is easy to use in the sparsely settled and often little-known borders of the great deserts such as the Sahara or Takla Makan. The first problem is to obtain agreement or, at least, a consensus on what rainfall level, for example, constitutes a satisfactory criterion to delimit even one desert. No such agreement or consensus exists. Probably no scientists would accept a single number, i.e., 100 mm of rainfall, to represent the edge of all deserts. Many scientists would accept the idea that the best criterion would be some ratio of precipitation to potential evapotranspiration. However, just what that ratio should be is controversial. Similarly, although vegetation has much merit for defining ecological zones because it integrates climate and edaphic conditions, mapping vegetation systems on the ground in extensive arid regions is difficult.

### B. DATA AVAILABILITY

The number one problem in using direct measurements of rainfall or vegetation around deserts is the scarcity of data. Rainfall stations are widely scattered and frequently have unreliable records, with long periods of no monthly record at all. Vegetation mapping in the field consists nearly always of mapping vegetation complexes in sample areas, not continuously. Interpolation of vegetative zones is based on topographical differences and rainfall data.

Satellite imagery offers the opportunity to obtain a continuous indirect measurement of vegetative cover, using the NDVI (Tucker et al., 1985). When polar-orbiting meteorological satellites are employed for obtaining the NDVI, as we did, synoptic coverage is possible over areas of several 100–1000 $km^2$. Resolution is sacrificed in order to obtain wide areal coverage with the advanced very-high-resolution radiometer (AVHRR) aboard the meteorological satellites.

### C. INITIAL RESULTS

Our first report on determining the southern vegetation boundary of the Sahara desert used satellite data for the years 1980 to 1985 (Dregne and Tucker, 1988). We noted then that there were unexpectedly large interannual shifts in green biomass production on the south side of the Sahara. Our conclusion was that the massive shifts in vegetation boundaries made it impossible to know, over a short period of observation, whether the shifts were permanent or temporary. We estimated that an observation period of at least 30 to 40 years would be needed before conclusions could be drawn. Certainly, no one could do that in 6 years, no matter how accurate the measurements on the ground.

The interannual shifts in vegetation boundaries that we measured amounted to as much as 250 km. They varied from 50 to 250 km across the southern border of the Sahara. This north-south oscillation in Sahara vegetation boundary did not, for the most part, have any apparent relation to the presence or absence of mountains. The only mountains in the east-west transect in our study were in Ethiopia and along the Sudan-Chad border. In the remainder of the area, only low hills and sand dunes break the monotony of the plain topography of the Sahel. There was no obvious reason why the interannual variation would be 50 km in one place and 250 km in another.

### D. LATER FINDINGS

In 1991, we published the results of the continuation of the study for the period 1980 through 1990 (Tucker et al., 1991). 1980 to 1984 had seen progressively less annual rainfall in the Sahel, with 1984 being the driest year on record. Rainfall had fluctuated from 1985 to 1990, but no year was as dry as 1984 or as wet as 1980. According to Nicholson (1989), Sahel rainfall has been below the long-term average since 1970. Rainfall and NDVI have been shown to be highly correlated (Tucker et al., 1991). The NDVI value we used to represent the boundary of the Sahara is approximately equivalent to the 200 mm isohyet.

The extreme interannual variations in the Sahara vegetation boundary continued to be present for the years 1986 to 1990. Sahara "expansion" occurred from 1980 to 1984, followed by expansion or contraction in subsequent years, depending on the rainfall. Our conclusion about Sahara desert encroachment is that no one knows whether the Sahara is expanding or contracting on a long-term basis. There is simply too much interannual variation to allow a judgment to be made over an 11-year period. We believe that 40 to 50-year period of observation, at least, is necessary for conclusions to be drawn. If there is an expansion or contraction, it will be difficult to assign a reason to the change because weather stations are scarce. The change, if it occurs, could be due to human-induced land degradation or reclamation, or to climate change.

## IV. LOCATION OF SAHEL ISOHYETS

Basic to planning for agricultural development in the drylands especially is knowledge of the local rainfall. Amount, distribution in time, and intensity of rainfall are all important in determining prospects for plant production and the environmental hazard of water erosion. In the Sahel region of Africa, a short wet season is followed by a long dry season. As is true for most summer rainfall areas, rains tend to be of high intensity, short duration, and highly variable in space (one place may be flooded while there is no rain at all a few kilometers away). Isohyets, insofar as their characteristics are known, generally parallel one another, with the lowest rainfall near the Sahara; rainfall increases to the south.

### A. ISOHYET COMPARISON

Figure 1 illustrates how little is known of the location of isohyets in the Sahel region of West Africa and the value of satellite imagery in the construction of isohyets. The two isohyets (100 and 200 mm) drawn in two National Academy of Science (NAS) reports obviously are incorrect. Under no circumstances should the 200 mm isohyet be closer to the Sahara proper than the 100 mm isohyet. That is impossible when the rainfall gradient runs north-south. Yet that is exactly what two rainfall maps produced by the same NAS committee in the same year claims (NAS, 1983a, 1983b). The Tucker et al. (1991) isohyet of 200 mm represents our version of the boundary of the Sahara. It is quite different from the NAS 200-mm isohyet.

The point made here is not whether any one of the isohyets is located properly. Both NAS isohyets cannot be correct. Our 200-mm isohyet is derived from the vegetation-rainfall

**SAHARA BOUNDARY VARIATION, 1980-1990**

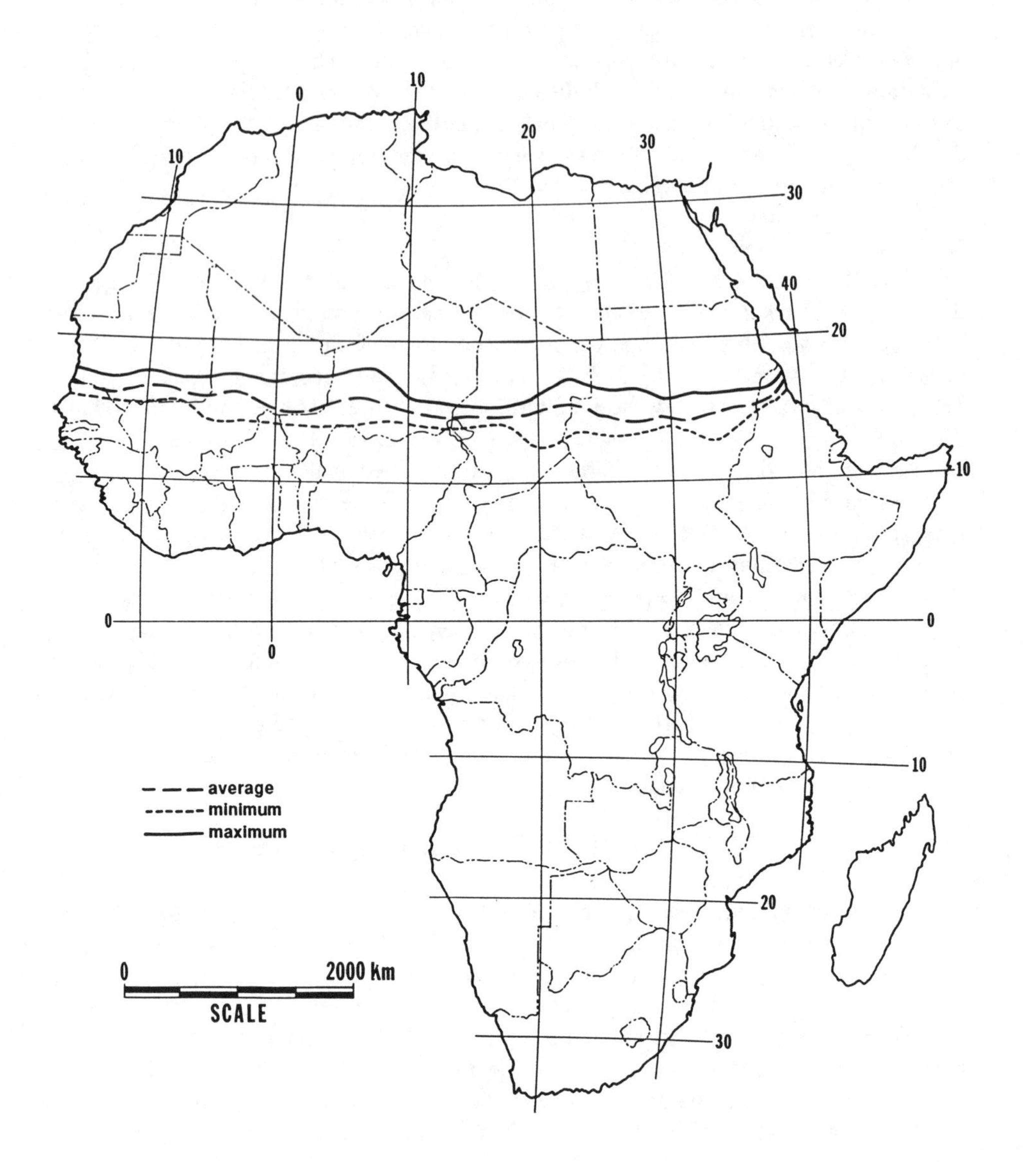


**FIGURE 1.** Mean and extreme values of the normalized difference vegetation index (NDVI) marking the vegetation boundary of the Sahara, 1980 to 1990.

correlation during a long dry period. In all likelihood, the isohyet is too far south, at least relative to the 90-year rainfall record. What the figure implies is that rainfall stations are so widely scattered in sparsely populated areas such as those near deserts that it is hazardous to rely on them when drawing isohyets. The special merit of the satellite-derived 200-mm isohyet is that the data upon which the isohyet is based are continuous across the Sahel. There are no gaps in space, no incomplete records, and no need for interpolations.

## V. RAINFALL VARIATION PATTERNS

Figure 2 is a redrawing of the figure in Tucker et al. (1991) of the most southern (dry year) and northern (wet year) position of the 200-mm isoline during the 1980 to 1990

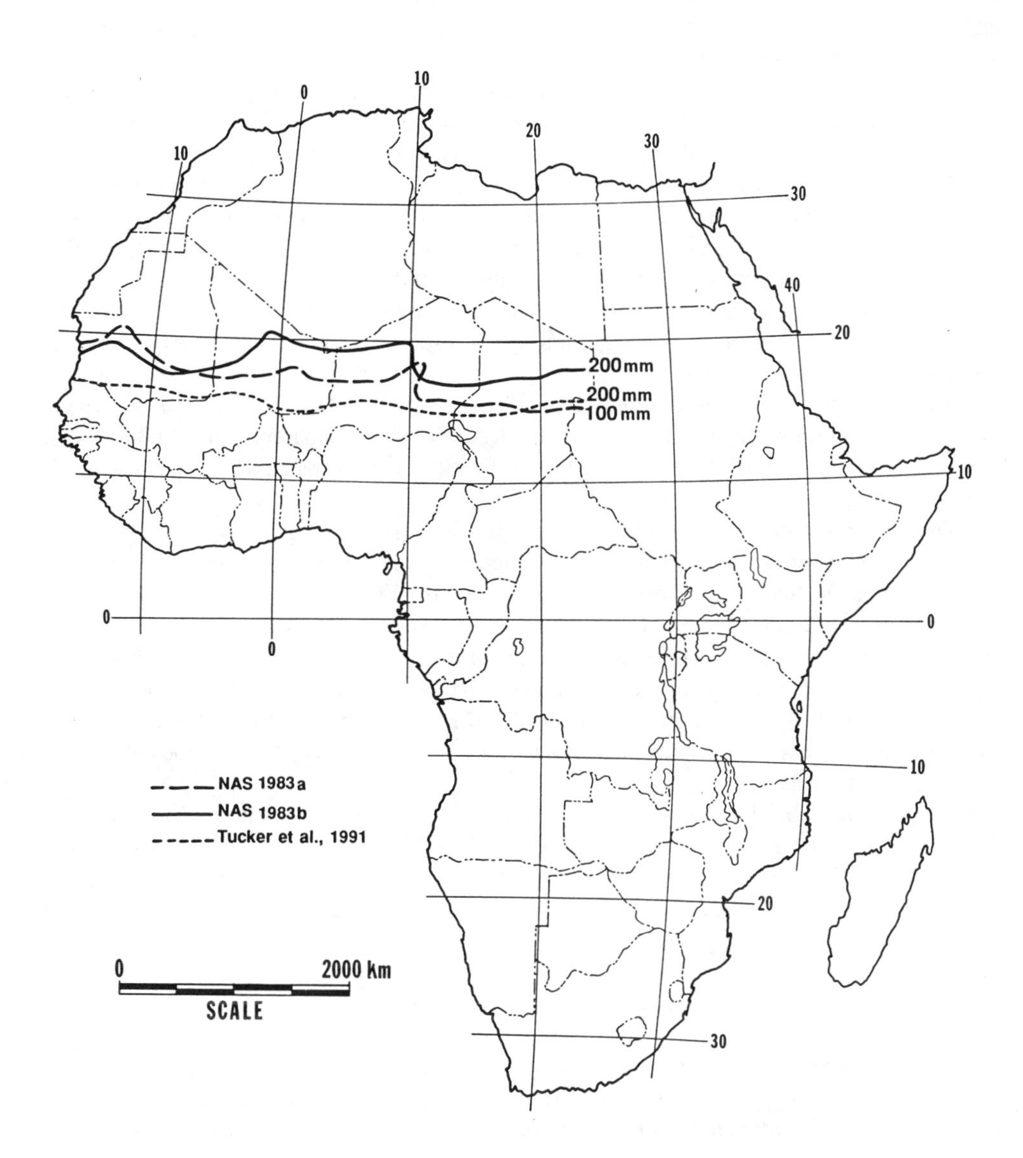


**FIGURE 2.** Comparison of 100- and 200-mm isohyets in West Africa using rainfall (NAS, 1983a, 1983b) and vegetation (Tucker et al., 1991) data.

observation period and the mean for the period. In most years, but not all, the southernmost location of the isoline came in the very dry year of 1984. The northernmost location of the isoline came in one of 4 years (1980, 1981, 1988, or 1989). Those differences in the wettest or driest years accentuate the fact that there can be great differences in rainfall in both space and time in arid regions. On average, 1984 may have been the driest year in 90 years, but that was true for the region as a whole, not at every point.

It is evident from even a casual glance at the figure that the 11-year variation in annual rainfall appears to fit a pattern. Starting on the west in Senegal and Mauritania, the two outside lines (minimum and maximum) are relatively close together for about 900 km.

Moving eastward, there is, next, a long section in Mali and Niger where the lines are farther apart. Continuing eastward, another 900 km section where the extremes are close together can be seen, running from the middle of Niger to the middle of Chad. East of there is a second long section where the minimum and maximum isolines are relatively far apart.

We were intrigued by the regularity of the rainfall difference pattern. There is no obvious (to us) reason for the pattern. While the seeming regularity may be fortuitous, there is always the possibility that it is real. If so, the pattern suggests that a number of useful conclusions may be drawn: (1) that the risk to pastoralists of weather variations is less where the extreme isolines are close together than where they are farther apart — (the reason is that a pastoralist would need to move his herds shorter distances to obtain more favorable forage); (2) that monitoring of land degradation or climate change would achieve results sooner if it were done where the isolines are closer together and interannual variations in rainfall are less; and (3) that something other than topography or distance from oceans is controlling the rainfall pattern at a local scale.

We realize that our study covered only 11 years and that the period may not be typical of a longer time frame. Nevertheless, the occurrence of those apparent anomalies was first noted by us in 1986, and they have remained in the same approximate position in subsequent years.

The likelihood that this rainfall variation pattern is real was given some support in a Sahel climatology study. Monod (1986) has a figure in his paper showing the great variation of the 100-mm isoline that Toupet reported in 1972 for Mauritania. According to Toupet, the 100-mm isoline for 1941 to 1942 was in a line from Fderik to Atar to Boutilimit, about 700 km farther south than the 100-mm isoline for 1951 to 1952; 1941 to 1942 was a wet year; 1951 to 1952 a dry year. Toupet showed that the 100-mm isohyet was located approximately halfway between the 1941 to 1942 and 1951 to 1952 extremes. A short distance (perhaps 200 km) east and west of that peak variation, the isolines of the extremes of rainfall were <250 km apart.

Figuring out why there should be such extreme variation in annual rainfall isolines is difficult in Mauritania, as it is for the rest of the Sahel. There are no mountains anywhere near the region Toupet studied and no large bodies of water. There is no indication that the Atlantic Ocean plays a role, either, since the variation at the coast was much less than in the interior. We do not have answers.

## VI. CONCLUSION

Desert expansion is a concept that has concerned people for at least 75 years. Land conservationists tend to believe that the expansion, if it actually has occurred, is the result of human-induced land degradation on the borders of deserts. Climatologists are inclined to ask whether expansion is simply a response to a short- or long-term climate change.

Our study indicates that no one can say whether the Sahara, the world's largest desert, is or is not expanding. Interannual changes in vegetation boundaries on the south side of the Sahara are so great that only a long-time monitoring project lasting at least 40 or 50 years could determine whether expansion is occurring. Pending such a study, arguments about expansion of the Sahara are fruitless; the data base is inadequate.

We did find intriguing patterns of shifts in vegetation and, indirectly, rainfall across the Sahel. The pattern may be spurious, but we are not yet convinced that it is. If the pattern is real, it has implications that can be important.

The hazard of drawing isohyets on the edges of deserts, where weather stations are few and far between, was illustrated in a comparison of satellite-derived isohyets with those shown in two publications from the NAS. Satellite imagery gives a continuous data set even in the most remote and inhospitable areas, something that weather stations cannot provide.

# REFERENCES


Anonymous, Rapport de la mission forestiere Anglo-Francais Nigeria-Niger (Decembre 1936–Fevrier 1937), *Bois For. Trop.*, 148, 3–26, 1973.

Bovill, E. W., The encroachment of the Sahara on the Sudan, *J. Afr. Soc.*, 20, 174–185, 259–269, 1921.

Cana, F. R., The Sahara in 1915, *Geogr. J.*, 46, 333–357, 1915.

Dregne, H. E. and Tucker, C. J., Desert encroachment, *Desert. Contr. Bull.*, 16, 16–19, 1988.

Hellden, U., Drought Impact Monitoring — A Remote Sensing Study of Desertification in Kordofan, Sudan, Lund Universitets Naturgeografiska Institution, Rapporter och Notiser No. 61, 61 p, 1984.

Hellden, U., Desertification monitoring, *Desert. Contr. Bull.*, 17, 8–12, 1988.

Lamprey, H. F., Report on the desert encroachment reconnaissance in northern Sudan: 21 October to 10 November 1975, *Desert. Contr. Bull.*, 17, 1–7, 1988.

Monod, T., The Sahel zone north of the equator, in *Ecosystems of the World 12B, Hot Deserts and Arid Shrublands, B,* Evenari, M., Noy-Meir, I., and Goodall, D. W., Eds., Elsevier, New York, 1986, 203–243.

NAS, *Environmental Change in the West African Sahel,* National Academy of Sciences, Washington, D.C., 1983a, 96 p.

NAS, *Agroforestry in the West African Sahel,* National Academy of Sciences, Washington, D.C., 1983b, 86 p.

Nicholson, S. E., Long-term changes in African rainfall, *Weather,* 44, 46–56, 1989.

Stebbing, E. P., The encroaching Sahara: the threat to the West African colonies, *Geogr. J.*, 86, 506–524, 1935.

Stebbing, E. P., The advance of the Sahara, *Geogr. J.*, 91, 356–359, 1938.

Tucker, C. J., Vanpraet, C. L., Sharman, M. J., and Ittersum, G., Satellite remote sensing of total herbaceous biomass production in the Senegalese Sahel: 1980–1984, *Remote Sensing Environ.*, 17, 233–249, 1985.

Tucker, C. J., Dregne, H. E., and Newcomb, W. W., Expansion and contraction of the Sahara desert from 1980 to 1990, *Science,* 253, 299–301, 1991.

USAID, Desert Encroachment on Arable Lands, Office of Science and Technology, Agency for International Development, TA/OST 72, Washington, D.C., 1972, 63 p.

Chapter 23

# THE USE OF BIOFUELS TO MITIGATE GLOBAL WARMING

**M. D. Ackerson, E. C. Clausen, and J. L. Gaddy**

## TABLE OF CONTENTS



0-8493-4419-0/93/$0.00 + $.50
© 1993 by CRC Press, Inc.

# I. INTRODUCTION

This planet is habitable because of the warming effect resulting from the presence of trace gases in the atmosphere that absorb and trap longer IR wavelengths reradiated from the Earth's surface. These trace greenhouse gases are composed of about 50% carbon dioxide ($CO_2$), with varying concentrations of water vapor, methane ($CH_4$), halocarbons, nitrogen oxides ($N_2O$), and ozone ($O_3$) (Belton, 1990). It is well established that $CO_2$ concentrations in the atmosphere are rising at the rate of about 0.5%/year (Smith, 1988). $CO_2$ concentrations have risen from 275 ppm in the middle of the last century, to 315 ppm in 1958, and 350 ppm today (Hileman, 1989).

As the levels of greenhouse gases increase, more solar radiation is trapped and the Earth's temperature increases. Measurements show that the average global temperature has risen only about 1°F, but the problem appears to be accelerating. The 5 warmest years have occurred in the last decade, with 1987 the warmest (Wisconsin Energy News, 1988). Although the consequences are not yet clear, many scientists predict radical climatic changes, with melting of the polar ice caps and the creation of vast deserts.

It is recognized that the increase in greenhouse gases is largely due to fossil fuel use, as well as changing land use (Bolin et al., 1986). While deforestation and land exploitation have been responsible for high $CO_2$ emissions in the past, these sources will be comparatively small in the future, since the rate of deforestation will decline. Hence, future trends in the atmospheric $CO_2$ concentration will depend primarily upon fossil energy usage (Smith, 1988). Except for a short period following the 1973 oil embargo, world $CO_2$ emissions from combustion of fossil fuels have increased about 3%/year during the last 40 years, to about 24 billion ton in 1988 (Post et al., 1990). The U.S. consumes one third of the world's energy, and contributes about one fourth of the $CO_2$ emissions, or 6 billion ton/year. Clearly the energy policies of the U.S. will have a significant influence on potential global warming.

# II. THE BIOMASS SOLUTION

The biosphere assimilates $CO_2$ as biomass by photosynthesis. Most of this biomass eventually degrades aerobically to $CO_2$, and the carbon cycle is repeated. The $CO_2$ discharged by combustion of fossil fuels upsets the balance, with the result that when the $CO_2$ assimilation capacity is exceeded (as is currently the case), $CO_2$ accumulates in the atmosphere. If biomass is used as fuel, less fossil fuel is burned and less $CO_2$ is produced. Provided that the biomass used as fuel is replaced each year by photosynthesis, a net reduction in $CO_2$ accumulation results. Therefore, residues (or annual energy crops) that would ordinarily be degraded to $CO_2$, would mitigate global warming when used for energy. This analysis also applies to longer growth cycles, such as trees at steady state, where death and decay equal assimilation and new growth.

Table 1 presents the reported quantities of residues available annually in the U.S. (Sitton et al., 1979). As noted, a total of 1.5 billion ton are available as an energy source each year. If all of these residues were used for energy, they would reduce the $CO_2$ burden by 2 billion ton, while generating about 10 quads of energy (at 50% conversion). In addition, if energy crops were grown on idle arable rangeland and forestland in the U.S. (approximately 300 million acres; Clausen et al., 1977), the remaining 4.0 billion ton of $CO_2$ from fossil combustion in the U.S. could be eliminated and another 20 quads of energy generated. Consequently, the use of biomass as fuel could eliminate $CO_2$ accumulation from the U.S. emissions.

Solid biomass is an inconvenient form of energy. Consequently, biomass must be converted into gaseous or liquid forms of energy to be utilized in conventional energy processes.

**TABLE 1**
**Annual Biomass Residues in the U.S.**

| Residue source | Annual amount (MM ton) |
|---|---|
| Municipal | 200 |
| Agricultural | 300 |
| Forestry | 1000 |
| Total | 1500 |

Over the last 2 decades, considerable research has been devoted to biomass conversion technologies. The purpose of this chapter is to survey these technologies. Systems to produce ethanol, methanol, $CH_4$, and chemical intermediates are examined. Preliminary designs and economic analyses for these processes are developed.

# III. BIOMASS CONVERSION TECHNOLOGIES

Biomass may be used as a solid fuel and burned directly to produce energy. However, efficiencies are low and handling problems are serious. Therefore, only gaseous and liquid fuels will be considered herein, although biomass combustion is certainly encouraged where practical. Since our greatest need is for liquid fuels, attention will be focused on technologies to produce alternative liquid fuels.

# IV. BIOMASS HYDROLYSIS/FERMENTATION

The major components of cellulosic biomass are hemicellulose, cellulose, and lignin. The composition of the biomass resources varies; however, most materials contain 15 to 25% hemicellulose, 30 to 45% cellulose, and 5 to 15% lignin. A recent analysis of corn stover showed hemicellulose — 27%; cellulose — 43%; lignin — 8%; ash — 2% (Industrial Testing Laboratory, 1982). Therefore, 60 to 70% of the biomass is available as carbohydrates, which exist as polymers of hexoses and pentoses. The polymers may be hydrolyzed to monomeric sugars which can be fermented to ethanol or chemicals.

## A. BIOMASS HYDROLYSIS

The carbohydrate hydrolysis can be carried out by contact with cellulase or xylanase enzymes, or by treatment with mineral acids. Enzymatic hydrolysis has the advantage of operating at mild conditions and producing a high-quality sugar product. However, the enzymatic reactions are quite slow, and the biomass must be pretreated with caustic or acid to improve the yields and kinetics. Acid hydrolysis is a much more rapid reaction, but requires higher temperatures or high acid concentrations to achieve good yields. Furthermore, under these conditions, xylose degrades to furfural and glucose degrades to 5-hydroxymethyl furfural (HMF), both of which are toxic to microorganisms. A two-stage acid hydrolysis process enables the hemicellulose and cellulose to be degraded separately under conditions appropriate for each reaction. High yields are obtained and mild conditions are required, with very little resultant degradation. Large quantities of acid are necessary, however, and acid recovery is an important part of this process.

The two major factors which control the hydrolysis reactions are temperature and acid concentration. The sugar concentrations and yields from a typical prehydrolysis and hydrolysis of corn stover from our laboratories are given in Table 2 (Clausen and Gaddy,

**TABLE 2**
**Corn Stover Acid Hydrolysates**

| | |
|---|---|
| **Prehydrolysate** | |
| Xylose, g/l | 27.3 |
| g/100g | 17.5 |
| Glucose, g/l | 10.9 |
| g/100g | 7.0 |
| **Hydrolysate** | |
| Xylose, g/l | 0 |
| g/100g | 0 |
| Glucose, g/l | 62.0 |
| g/100g | 45.5 |
| **Combined** | |
| Xylose, g/100g | 17.5 |
| Glucose, g/100g | 52.2 |

1987). The prehydrolysis step yields 27% of the initial stover as xylose. The combined yield of glucose is 52%. These yields represent nearly complete conversion of hemicellulose and cellulose to sugars.

The process for stagewise acid hydrolysis of corn stover is shown schematically in Figure 1. Corn stover is fed continuously to the prehydrolysis reactor. Residual solids are separated by filtration, washed with fresh acid, and fed to the hydrolysis reactor. Solids from the hydrolysis reactor are filtered, washed, and discarded. Acid and sugars are separated and the acid is returned to the reactors. Two dilute sugar streams result: the prehydrolysate, containing primarily xylose, and the hydrolysate, containing only glucose. These sugars can then be fermented to ethanol or other chemicals.

## B. HYDROLYSATE FERMENTATION

The acid hydrolysis of cellulosic residues, if performed at high temperature, can produce a sugar/degradation product mixture that is difficult, if not impossible, to ferment. However, if the hydrolysis conditions are mild, only small quantities of toxic substances are observed and normal fermentation is expected.

Batch fermentation experiments were carried out with acid hydrolysates at 30°C with *Saccharomyces cerevisiae* ATCC 24860 using various nutrients. As shown in Table 3, the fermentation proceeded well only in the presence of yeast extract. The hydrolysate serves as a carbon source only and does not contain sufficient nutrients for efficient fermentation. However, with a small quantity of yeast extract (0.25%), the hydrolysate can be completely fermented in only 16 h. This successful fermentation is due to the absence of yeast inhibitors, furfural, and HMF, which result from sugar decomposition at high temperatures (Banerjee and Vishwanathan, 1974).

## C. ACID RECOVERY

Acid recovery is an economic necessity in this technology. A complete acid hydrolysis/acid recovery process has been developed for the conversion of agricultural materials to monomeric sugars. The process utilizes concentrated sulfuric or hydrochloric acids at near-ambient conditions for hydrolysis of the hemicellulose and cellulose fractions, and solvent extraction for acid recovery. Following hydrolysis, the solids (lignin and ash) are filtered

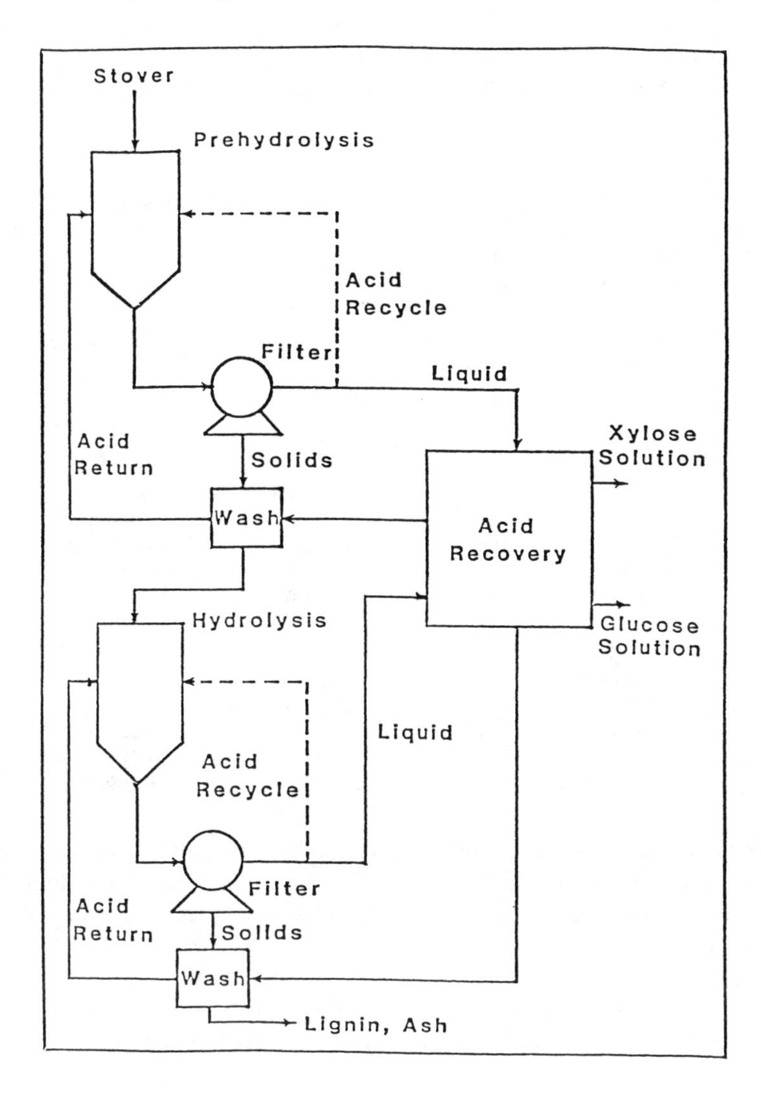


**FIGURE 1.** Schematic of acid hydrolysis process.

**TABLE 3**
**The Effect of Nutrient Additions on Hydrolysate Fermentation to Ethanol**

| | % sugar utilization | | | |
|---|---|---|---|---|
| **Fermentation** | **Hydrolysate plus amino acids & vitamins** | **Hydrolysate plus vitamins & ammon. phos.** | **Hydrolysate plus amino acids, vitamins & ammon. phos.** | **Hydrolysate plus yeast extract** |
| 16 | 15.9 | 21.9 | 27.3 | 97.5 |
| 23 | 19.3 | 24.9 | 35.8 | 97.5 |

**TABLE 4**
**Economics of 20 Million Gal/Year Ethanol Facility**

**Capital cost**

| | **Million $** |
|---|---|
| Feedstock preparation | 3.0 |
| Hydrolysis | 5.0 |
| Acid recovery | 8.5 |
| Fermentation & purification | 8.0 |
| Utilities/offsites | 5.5 |
| | 30.0 |

**Operating cost**

| | **Million $/yr** | **$/gal** |
|---|---|---|
| Corn stover — $20/ton | 5.0 | 0.25 |
| Utilities | 1.5 | 0.08 |
| Chemicals | 1.9 | 0.09 |
| Labor | 2.5 | 0.13 |
| Fixed charges | | |
| Maintenance (4%) | 1.2 | 0.6 |
| Depreciation (10%) | 3.0 | 0.15 |
| Taxes & Insurance (2%) | .6 | 0.03 |
| Pretax profit (48%) | 14.3 | 0.71 |
| | $30.0 | $1.50/gal |

and washed with water. The liquid stream from filtration (containing acid, sugar, and water) is sent to two countercurrent extraction units. The first unit removes acid from the sugars and the second unit removes the first solvent from the acid. Separation of the two solvents occurs by distillation.

The sugar product stream from the first extraction unit contains about 0.2% $H_2SO_4$. This acid is removed by neutralizing with lime. A final produce stream containing 15% glucose is available for fermentation. Higher sugar concentrations (up to 40%) can be achieved with acid recycle.

## D. ECONOMIC PROJECTIONS

To illustrate the economics of this process, a design has been performed for a facility to convert corn stover into 20 Mgal/year of ethanol, utilizing the acid hydrolysis procedures previously described. The capital and operating costs are summarized in Table 4.

Corn stover, in large round bales, would be stored in the field and delivered to the plant site as needed. Feedstock preparation consists of shredding, grinding, and conveying to the reactors. The hydrolysis section consists of continuous reactors. Acid resistant materials of construction are necessary for the hydrolysis and the acid recovery units. Continuous ethanol fermentation is used and the typical ethanol distillation units are included. The total capital cost for this plant is $30 million, including all utilities, storage, and offsites.

The annual operating costs are also shown in Table 4. These costs are also given on the basis of unit production of alcohol. Corn stover is estimated to cost $20/ton. The energy costs are nominal at $0.08/gal. A lignin boiler is used to reduce the energy requirements. Fixed charges are computed as a percentage of the capital investment and total 16% or $0.24/gal. A profit of 48% is possible at the current ethanol price of $1.50/gal.

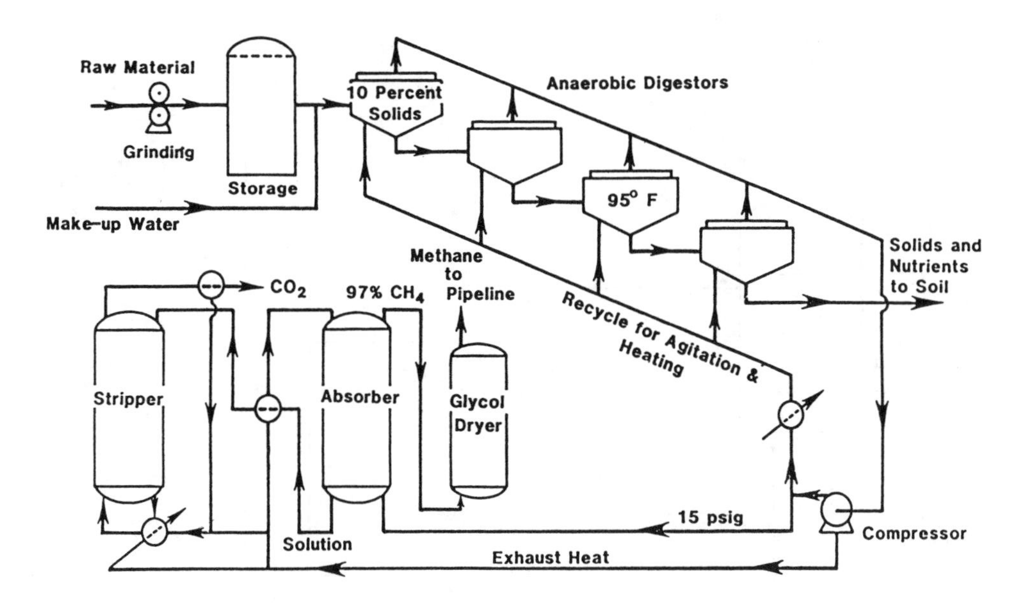


**FIGURE 2.** Process for production of $CH_4$ from biomass.

# V. METHANE/METHANOL FROM BIOMASS

Methane ($CH_4$), or natural gas, may be produced from biomass by anaerobic digestion. A clean-burning medium Btu gas (500 Btu/ft$^3$) is produced, which contains about 50% $CO_2$. Extensive research has been conducted recently on the anaerobic digestion of various residues, such as animal wastes. The University of Arkansas has been investigating the conversion of crop residues and MSW (Municipal Solid Waste). Systems have been developed for high solids conversion with cell recycle, which result in substantially reduced reactor volumes (Clausen and Gaddy, 1985). These data show that yields of about 5 ft$^3$ of $CH_4$/lb of residue result.

MSW consists of paper, yard and food waste, glass, plastic, metals, etc. The carbohydrate fraction, which represents about 70% of the total, can be converted by a mixed culture of bacteria to $CH_4$. (The proposed process is shown in Figure 2.) After separation of the glass, plastic, and metal, the refuse is ground and fed to a series of biological reactors where the microorganisms are maintained at 35°C. The $CH_4/CO_2$ mixture is collected, and $CO_2$ is removed by absorption in diethanolamine. The $CH_4$ is then dried and put into the natural gas pipeline. Alternatively, the lower Btu mixture of $CH_4$ and $CO_2$ may be burned for fuel.

## A. METHANOL PRODUCTION

If a liquid fuel is desired, the $CH_4$ may be converted into methanol. Methanol may be produced by reforming $CH_4$ to $H_2$ and CO, shift conversion of CO to $H_2$, and synthesis of methanol from CO and $H_2$. These catalytic reactions occur at high pressure and temperature with low efficiencies and high capital requirements.

Alternatively, methanol may be produced biologically from $CH_4$. Methanotrophic bacteria oxidize $CH_4$ according to:

$$CH_4 + O_2 + 2H^+ \rightarrow CH_3OH + H_2O \rightarrow HCHO + 2H^+ \rightarrow HCOOH$$
$$+ 2H^+ \rightarrow CO_2 + 2H^+ \qquad (1)$$

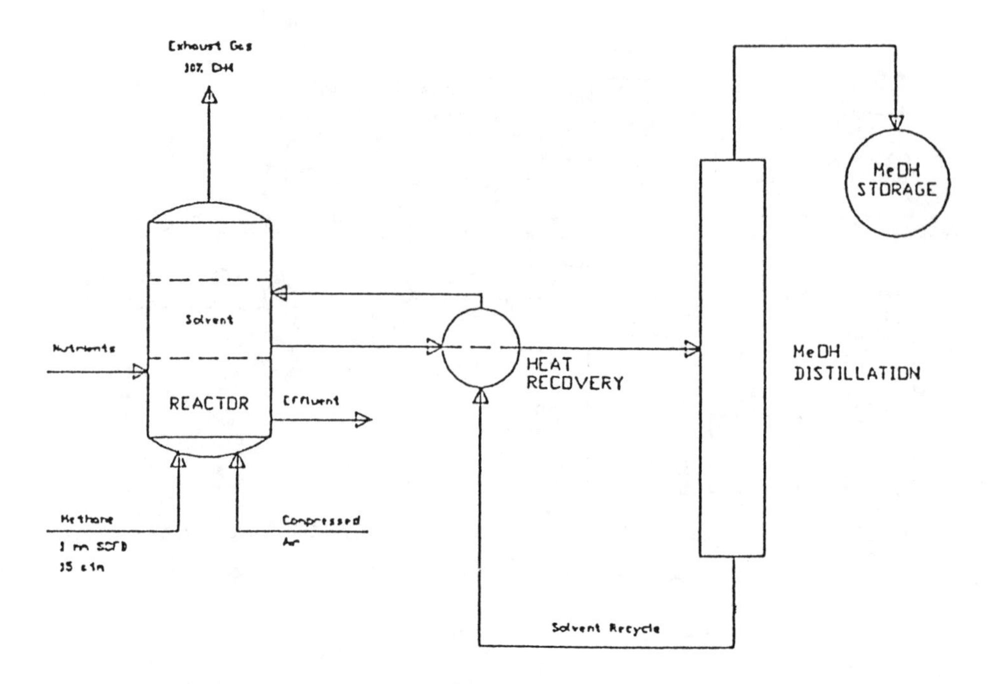


**FIGURE 3.** Schematic of biological process for methanol production.

As noted, subsequent reactions degrade the methanol to formaldehyde, formate, and $CO_2$. Studies in our laboratories with methanotroph isolates have resulted in methanol yields of 60%, very close to theoretical.

Data have been obtained for methanol production in continuous cultures that show high conversions with gas retention times of <2h. These data have been extrapolated into a design for a facility to convert 1 $Mft^3$ of $CH_4$/day. Figure 3 shows the equipment for the biological process to produce methanol. $CH_4$ and air are sparged into the agitated biological reactor containing the methanotrophs. The methanol produced is extracted into a solvent phase. The heavy boiling solvent is heated and distilled to produce MeOH. The methanol-free solvent is cooled and returned to the reactor. Condensed MeOH is stored.

The capital and operating costs for producing MeOH by the process of Figure 3 are shown in Table 5. At 80% conversion and 50% yield, 1.92 MM gal of MeOH are produced. The bioreactor is the predominant capital cost, with an installed cost in carbon steel of about $300,000. The column and exchangers are estimated at $250,000 installed, for a total investment of about $650,000.

Operating costs include natural gas at $1.50/MCF. Nutrient requirements at the long liquid residence time are nominal and solvent make-up is estimated at $7000/h. Utilities, if purchased, include power for air compression and pumping and heat for product recovery. Capital charges include depreciation, interest, and insurance, and are estimated at 30%. The total operating charges are about $785,000/year.

The revenue from methanol production at $0.60/gal is $1.15 M/year. Available profit is $365,000/year, or 56%, before income taxes. This small scale application affords sufficient profit potential to justify commercialization.

## VI. BIOMASS GASIFICATION

Technology for the gasification of biomass has been studied extensively, and small scale commercial systems have been developed (Steinberg, 1988; Bulpitt and Rittenhouse, 1988;

**TABLE 5**
**Economic Projections for Biological Production of Methanol**

| Capital requirements | **M$** |
|---|---|
| Biological reactor | 300 |
| MeOH distillation | 150 |
| Heat exchangers | 100 |
| Contingency | 100 |
| Total | $650 |
| | |
| Operating cost | **M$/yr** |
| Natural gas at $1.50/ MCF | 500 |
| Nutrient chemicals | 10 |
| Utilities | 30 |
| Capital charges (30%) | 195 |
| Total | $785/ yr |
| Revenue | 1150 |
| Potential pretax profit | 365 |
| % | 56 |

*Note:* 1 MM MCF $CH_4$ processed/day, 80% conversion; CSTR reactor = 15 atm; MeOH production — 5260 gal/day, value $0.60/gal.

Stasson and Stiles, 1988). The product is a mixture of CO, $H_2$, and $CO_2$. Typical gas compositions from a steam gasifier show a CO:$H_2$ ratio of 2 with CO compositions of about 40% (Collaninno and Mansour, 1988). The gas has a low heating value and can be burned for fuel. An advantage of this process over hydrolysis or digestion is that the lignin fraction is converted to fuel.

## A. ETHANOL PRODUCTION

The synthesis gas may be converted into liquid fuels by Fischer Tropsch reactions at high temperature and pressure, but with poor selectivity. Recently, an anaerobic mesophilic bacterium was isolated in the University of Arkansas laboratories that produces ethanol from synthesis gas components (Klasson et al., 1990). The bacterium has been identified as a new clostridial strain and named *Clostridium ljungdahlii.* The reactions for production of ethanol and acetic acid are

$$6\,CO + 3\,H_2O \rightarrow CH_3CH_2OH + 4\,CO_2 \qquad \Delta G^\circ = -59.9 \text{ kcal/reaction} \tag{2}$$

$$2\,CO_2 + 6\,H_2 \rightarrow CH_3CH_2OH + 3\,H_2O \qquad \Delta G^\circ = -23.2 \text{ kcal/reaction} \tag{3}$$

$$4\,CO + 2\,H_2O \rightarrow CH_3COOH + 2\,CO_2 \qquad \Delta G^\circ = -37.8 \text{ kcal/reaction} \tag{4}$$

$$2\,CO_2 + 4\,H_2 \rightarrow CH_3COOH + 2\,H_2O \qquad \Delta G^\circ = -18.7 \text{ kcal/reaction} \tag{5}$$

As noted from these equations, all the synthesis gas components can be utilized. Under normal conditions, the bacteria produces more acetate than ethanol. Studies to reduce the

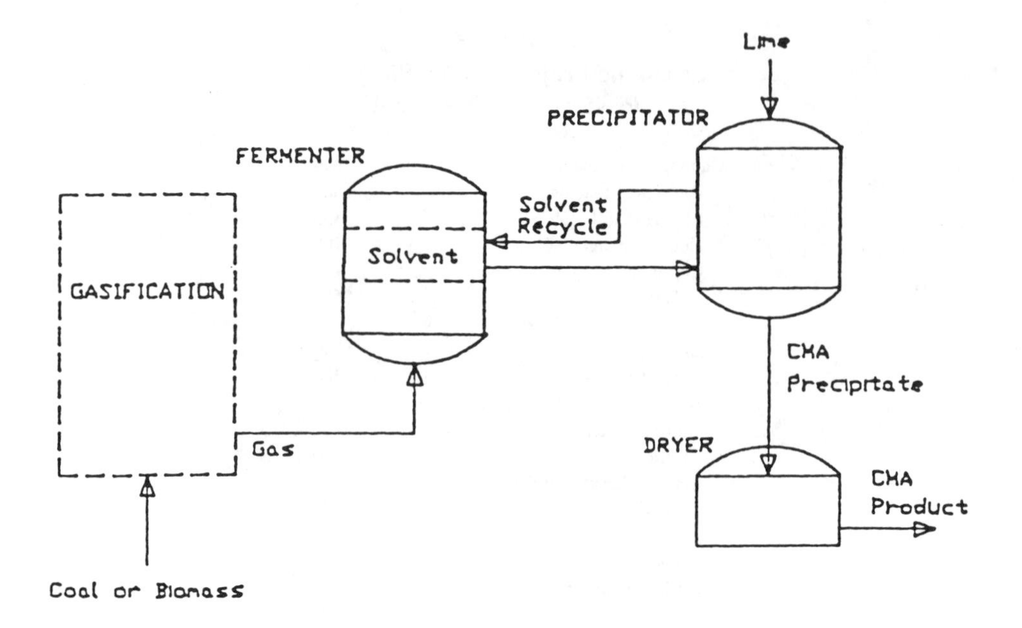


**FIGURE 4.** Schematic of acetic acid/CMA process.

acetate production have shown that increased ethanol ratios result under conditions that limit growth of the organism. Acetic acid production has been eliminated completely in the second reactor of a two-stage system, where conditions unfavorable for growth are maintained.

## B. CHEMICALS PRODUCTION

In some cases, it may be desirable to produce petrochemicals from biomass. Acetic acid is sold as a petrochemical in quantities of 3.4 billion lb annually at a price of \$0.36/lb, or twice the price of ethanol. Several organisms are known to produce only acetic acid from synthesis gas components, according to Equations 4 and 5. The theoretical yield of acetic acid from biomass is 100%.

The use of rock salt and $CaCl_2$ for roadway deicing is the cause of serious environmental and corrosion problems. A mixture of Ca and magnesium acetate (CMA) has been identified as a promising substitute. The replacement of all rock salt with CMA would require 22 billion lb of acetic acid. CMA is made by reacting lime or dolemite with acetic acid.

Figure 4 shows a process description for the production of CMA from biomass or coal. Synthesis gas would be produced and introduced into a fermenter. Heat recovery from the gas stream is necessary to reduce the temperature to fermenter conditions. The recovered heat could be used to dry the CMA.

Acetic acid would be produced in the fermenter from CO and $H_2$. An immobilized cell reactor has been found to maximize cell densities and minimize retention times (Vega et al., 1988, 1987; Sitton and Gaddy, 1980). These systems have been found to be especially suited for fermenting gaseous substrates, such as CO and $H_2$ (Vega et al., 1989). A small nutrient stream would be provided to maintain cell growth and activity. A solvent phase would be utilized in the fermenter to extract acetic acid. The solvent would be removed and mixed with lime and the CMA precipitated. The wet product would be dried and packaged.

The economics of this process have not been determined, since a market price for CMA in large quantities has not been established. Raw material costs are only about \$0.02/lb of acid. It is projected that CMA could be produced for \$0.10 or less with this process.

# VII. SUMMARY

The use of biomass as fuel in this country could eliminate $CO_2$ accumulation from fossil fuel combustion. To achieve maximal use, biomass must be converted into liquid or gaseous fuels. Processes are available to produce ethanol, $CH_4$, methanol, or chemicals. Design and economic projections indicate that significant economic potential exists for commercialization of these technologies at present energy prices.

# VIII. CONCLUSIONS

$CO_2$, produced by aerobic biomass degradation and fossil fuel combustion, accumulates in the atmosphere and will eventually lead to global warming. The substitution of biomass as fuel would alleviate this problem, since all $CO_2$ produced from biofuels must be balanced by photosynthetic consumption. The utilization of biofuels will require economical technology for biomass conversion. This chapter examined technologies for producing liquid and gaseous fuels from biomass based upon hydrolysis, gasification, or anaerobic digestion. Processes for producing ethanol, methanol, and $CH_4$ were described. The design and economics for these technologies are projected based upon current energy prices.

# REFERENCES

Banerjee, N. and Vishwanathan, L., *Proc. Annu. Sugar Conv. Tech. Conf.*, 82, 123277W, 1974.

Belton, K., Global Climate Change, Am. Chem. Soc., Department of Government Relations and Science Policy, Washington, D.C., 1990.

Bolin, B., Doos, B. R., Jager, J., and Warrick, R. A., Eds., *The Greenhouse Effect: Climate Change and Ecosystems,* John Wiley & Sons, New York, 1986.

Bulpitt, W. S. and Rittenhouse, O. C., Proc. Energy Biomass and Wastes, IGT, 1988.

Clausen, E. C., Sitton, O. C., and Gaddy, J. L., Converting crops into methane, *Chem. Eng. Prog.*, 73(1), 71, 1977.

Clausen, E. C. and Gaddy, J. L., Corn Stover as a Chemical Feedstock, Proc. Corn. Util. Conf., St. Louis, 1987.

Clausen, E. C. and Gaddy, J. L., *High Solids Digestion of MSW to Produce Methane,* CRC Press, Boca Raton, FL, 1985.

Collaninno, J. and Mansour, M., Proc. Energy Biomass and Wastes, IGT, 1988.

Industrial Testing Laboratory, Corn Stover Analysis, Rep. by Industrial Testing Laboratory, St. Louis, MO, 1982.

Hileman, B., Global warming, *Chem. Eng. News,* 67(15), 25(March), 1989.

Klasson, K. T., Elmore, B. B., Vega, J. L., Clausen, E. C., and Gaddy, J. L., Biological production of liquid fuels from synthesis gas, *Appl. Biochem. Biotech.*, 24/25, 857, 1990.

Post, W. M., Peng, T. H., Emanuel, W. R., King, A. W., Dale, V. H., and DeAngelis, D. L., The global carbon cycle, *Sci. Am.*, 784, 310(July/August), 1990.

Smith, J. M., $CO_2$ and Climate Change, IEA Coal Research, London, 1988.

Sitton, O. C., Foutch, G. L., Book, N. L., and Gaddy, J. L., Ethanol from agricultural residues, *Chem. Eng. Prog.*, 75(12), 52, 1979.

Sitton, O. C. and Gaddy, J. L., Ethanol production in an immobilized cell column, *Bioeng. Biotechnol.*, XXII, 1735, 1980.

Stasson, H. E. M. and Stiles, H. N., Proc. Energy Biomass and Wastes, IGT, 1988.

Steinberg, M., Proc. Energy Biomass and Wastes, IGT, 1988.

Vega, J. L., Clausen, E. C., and Gaddy, J. L., Performance of immobilized cell reactors, in *Handbook on Anaerobic Fermentation,* Marcel Dekker, New York, 1987, chap. 22.

Vega, J. L., Clausen, E. C., and Gaddy, J. L., Biofilm reactors for ethanol production, *Enzyme Micro. Tech.*, 10(7), 389, 1988.

Vega, J. L., Clausen, E. C., and Gaddy, J. L., Study of gaseous substrate fermentations; continuous culture, *Bioeng. Biotechnol.*, 34, 785, 1989.

Wisconsin Energy News, 1988.

Chapter 24

# PROSPECTS OF AGRICULTURE IN A CARBON DIOXIDE-ENRICHED ENVIRONMENT

**N. C. Bhattacharya**

## TABLE OF CONTENTS



This chapter was completed under the auspices of the U.S. Government and is therefore in the public domain.

# I. INTRODUCTION

When it first became obvious that global carbon dioxide ($CO_2$) was increasing, much attention was focused on possible climatic and societal effects of global warming, melting of polar ice, rising sea levels, and inundation of coastal cities, etc. (Anonymous, 1980). Subsequently, it was realized that $CO_2$ could have numerous direct effects on the growth of plants in agricultural and natural ecosystems (Strain, 1985). While the effects of elevated $CO_2$ on agricultural plants have been studied to a limited extent, the effects on natural ecosystems remain largely unstudied (Strain and Cure, 1985). Having recognized that humans are increasing the carbon (C) content of the global atmosphere, the question of plant responses to $CO_2$ concentration becomes highly important. The summer 1988 drought in the U.S. (Sidey, 1988) and elsewhere in the world (Revkin, 1988) indicated that "global warming" and climatic changes may be on the way, and therefore farmers and agricultural scientists should prepare crops for a future atmosphere of high $CO_2$.

Atmospheric $CO_2$ concentration has increased significantly since the preindustrial revolution, and is rising at an annual rate of 1 $\mu$mol $CO_2$ $mol^{-1}$ (Keeling et al., 1982; Bacastow et al., 1985). These increases in atmospheric $CO_2$ are due primarily to fossil fuel consumption, deforestation in tropical, subtropical, and temperate regions (Strain, 1987), and geochemical changes in carbonic rocks (Berner and Lasaga, 1989). This global increase in $CO_2$ concentration has been well documented (Keeling et al., 1982), and the importance of its effects on growth and yield of plants has been recognized for some time (Wittwer, 1980, 1983, 1986; Kimball, 1983; Strain and Cure, 1985, 1986; Cure and Acock, 1986; Houghton et al., 1990).

The increase in atmospheric $CO_2$ concentration may affect global climates, thus altering climatic zones through changes in the mean annual temperatures and precipitation patterns (Allen, 1989). It is predicted that a high atmospheric $CO_2$, and the corresponding increase in other greenhouse gases, will significantly increase the global temperature (U.S. National Research Council, 1972, 1979, 1982, 1983). Climatologists have estimated this $CO_2$-induced rise in the mean annual global surface temperature with the aid of an "energy-balanced" model (Budyko, 1969; Idso, 1980; Ramanathan, 1981; Lindzen et al., 1982; Chau et al., 1982; MacCracken and Luther, 1985; Mitchell, 1989), and general circulation models (GCMs) (Manabe and Wetherald, 1975; Manabe and Stouffer, 1981). Estimates on the effect of doubling the ambient $CO_2$ concentration differ from model to model. In general, various climate models indicate a 2.8 to 5.2°C increase in the global average surface temperature due to a doubling atmospheric $CO_2$ (Allen, 1989). The magnitude and pattern of climate change associated with a doubling of atmospheric $CO_2$ are predicted to have major effects on the plants and animals of the earth (Baes et al., 1977; Woodwell, 1978). If climate patterns change, as predicted by the global circulation models (Schlesinger and Mitchell, 1985), major adjustments will surely have to be made in management decisions in agriculture, forestry, hydrology, and in habitat and species conservation (Strain, 1987 and references therein). Recently, these concerns over global climate change issues were addressed by scientists, environmental activists, and politicians during a climate change conference, sponsored by the Climate Institute, Washington, D.C. (Allen, 1989).

In this chapter, I have described the direct effects of $CO_2$-enriched environments on selected physiological processes in several plant species including agricultural crops.

# II. $CO_2$ ENRICHMENT AND PHOTOSYNTHETIC PROCESSES

In general, elevated levels of $CO_2$ above the current atmospheric concentration promote photosynthesis, growth, and yield of crops (Strain and Cure, 1986). Also, elevated $CO_2$ causes partial closure of leaf stomata. However, plants also have low $CO_2$ assimilation and growth rates when grown in low temperature regimes even at high $CO_2$ (Markhart et al., 1980). Both short- and long-term effects of $CO_2$ enrichment on $CO_2$ exchange rate, overall growth, and reproductive capability of a variety of plant species have been documented (Jones et al., 1984; Paez et al., 1983; Sionit et al., 1980). Enhancement of net photosynthesis of various species via short-term exposure (a few hours) to high $CO_2$ has been widely reported (Kramer, 1981; see also Strain and Cure, 1986). Generally, photosynthesis is enhanced by elevated $CO_2$ more in $C_3$ plants than in $C_4$ plants, whereas stomatal resistance is increased more in $C_4$ than $C_3$ plants (Kimball, 1986; Kimball et al., 1986, 1987; Hileman et al., 1990).

The response of photosynthesis, transpiration, and stomatal conductance to changes in air temperature, irradiance, ambient $CO_2$ concentration, absolute humidity deficit, and xylem pressure potential was examined in 2-year-old seedlings of *Pinus taeda* reared in a greenhouse (Teskey et al., 1986). The seedlings showed little response to a wide range of absolute humidity deficit (7 to 16 g $m^{-3}$) and temperature (20 to 35°C), but were sensitive to changes in water deficit, irradiance, and $CO_2$ concentration. Net photosynthesis and stomatal conductance were linearly related under all of the environmental conditions measured. However, the gas phase limitation to photosynthesis was generally small (20 to 30%). It was concluded that although the stomatal response was closely coupled to changes in photosynthesis, the internal limitations, rather than the rate of gaseous diffusion of $CO_2$, were primarily responsible for limiting photosynthesis.

The rates of photosynthesis and specific activity of ribulose bisphosphate carboxylase/oxygenase (Rubisco) were investigated by Rowland-Bamford et al. (1991) in rice grown at subambient to superambient (160 to 900 $\mu$mol $mol^{-1}$) $CO_2$ concentrations under computer-controlled sunlit growth chambers. It was shown that photosynthesis increased linearly up to 550 $\mu$mol $mol^{-1}$ $CO_2$, but leveled off at higher $CO_2$ values. The acclimation of photosynthesis beyond 550 $\mu$mol $mol^{-1}$ $CO_2$ was due to significant decreases in Rubisco activity and Rubisco protein in leaf. The inhibition of photosynthesis in response to increased concentrations of $CO_2$ was reported in other crops when grown in a $CO_2$-enriched environment for a long period of time (Delucia et al., 1985; Cure and Acock, 1986; Spencer and Bowes, 1986; von Caemmer and Farquhar, 1984; Peet et al., 1985). The downward regulation of photosynthesis was primarily associated with restricted growth of roots in pots and accumulation of starch in leaves (Delucia et al., 1985). Indeed when Havelka et al. (1984) grew soybean in $CO_2$-enriched open-top field chambers, they reported a significant increase in photosynthesis throughout the growing period of the crop. A similar increase in canopy photosynthesis was reported in soybean grown at high $CO_2$ even when the canopy was fully closed (Campbell et al., 1990). The higher rates of canopy photosynthesis were due to greater substrate-$CO_2$ concentrations. The absence of downward acclimation in photosynthetic capacity was demonstrated by the observation that the soybean canopies grown at 300 and 660 $\mu$mol $mol^{-1}$ $CO_2$ had similar photosynthetic rates when measured at the same $CO_2$ concentration. This was further supported by the similarities in Rubisco activities and activation for plants grown at 330 and 660 $\mu$mol $mol^{-1}$ $CO_2$ (Campbell et al., 1990). Recently, Idso and Kimball (1991) reported a continuous increase in photosynthesis and growth of sour orange (*Citrus aurantium* L.) over a 3-year period in a $CO_2$-enriched environment. According to them, downward regulation of photosynthesis and growth at high $CO_2$ levels may be an experimental artifact of the manner in which most $CO_2$-enrichment studies have been conducted in greenhouse, phytotron, and growth chambers.

## III. $CO_2$ ENRICHMENT AND TEMPERATURE INTERACTIONS

It is evident that agricultural yields in crop plants are dependent on specific temperatures, with each species having its own optimum temperature regime, provided other agroclimatic conditions are not limiting plant growth. In this respect, an increase in any environmental factor such as $CO_2$, irradiance level, or availability of water, affects plant productivity in response to temperature. In general, the interaction of high irradiance and increased $CO_2$ concentration has a positive effect on crop plants, provided the temperature is at optimum levels for plant growth (Cure and Acock, 1986). Limited literature is available on the interactive effects of enriched $CO_2$ with temperature in soybean (Baker et al., 1989; Sionit et al., 1987a, 1987b), okra (Sionit et al., 1981), carrot, cotton, water-hyacinth, water fern, and radish (Idso and Kimball, 1989; Idso et al., 1987), wheat (Wall, 1990; McKinion and Wall, 1990), Pima cotton (Bhattacharya et al., 1991b; McKinion et al., 1991; Reddy et al., 1991; Wall et al., 1991). Pima cotton plants were grown from seedling stage to anthesis in five different temperatures (20/12, 25/17, 30/22, 35/27, and 40/32°C day/night), and two levels of $CO_2$ (350, 700 $\mu$mol mol$^{-1}$) in soil-plant-atmospheric research (SPAR) units under sunlit conditions. One of these investigations demonstrated that the response of Pima cotton to the direct and interactive effects of long-term exposure to different temperature and $CO_2$ treatments is indeed complex and varied. The observed decrease in reproductive mass at 35/27°C, and the failure of reproductive growth at 40/32°C in both the $CO_2$ treatments, indicate a need to select cotton cultivars with higher temperature optima (McKinion et al., 1991). These results also showed that high $CO_2$ and high temperature did not impair photosynthesis; however, formation of squares was inhibited.

Allen (1989, 1990) reviewed various aspects of $CO_2$ interaction with different temperatures on crops and concluded with caution (information was mostly on photosynthesis, biomass production, and limited reproductive growth) that $C_3$ species would benefit with a near doubling of $CO_2$ concentration with increased temperature. Idso et al., (1987) conducted experiments in carrot, radish, water-hyacinth, and *Azolla* throughout the year in open-top chambers enriched with $CO_2$. The average daily temperature ranged from 12 to 34°C. The combined results for the five species clearly indicated a positive growth response as mean air temperature increased. However, based on this study, a general increase in plant growth cannot be deduced because the temperature was not controlled in the open-top chambers.

The effects of temperature and $CO_2$ enrichment on carbon translocation of plants of the $C_4$ grass species *Echinochloa crus-galli* (L.) Beauv. from Quebec and Mississippi were studied under two thermoperiods (28/22 and 21/15°C) at two $CO_2$ concentrations (350, 675 $\mu$mol mol$^{-1}$); (Potvin et al., 1984). The translocation was monitored using radioactive tracing with short-lived $^{11}C$. $CO_2$ enrichment decreased the size of the carbon pool available for export in plants of both populations. Low temperature reduced translocation drastically for plants from Mississippi in normal $CO_2$ concentration, but this reduction was ameliorated at high $CO_2$. Overall, plants from Quebec had a higher $^{11}C$ activity in leaf phloem and a higher percentage of $^{11}C$ exported, whereas these northern adapted plants had lower turnover time and smaller pool size than plants from southern populations (Potvin et al., 1984). In fact, while a fundamental understanding of $CO_2$ interactions with water availability, nutrition level, light, and temperature can most readily be obtained in controlled environments, only field trials of the major crops can validate model predictions with respect to different plant physiological processes.

The effects of high temperature on photosynthesis were observed in cucumber seedlings grown in glasshouses under spring and winter conditions (Challa, 1976). Under spring conditions, the stomata closed after about 8 h of light, causing a decline in $CO_2$ uptake. Under winter conditions, $CO_2$ production rapidly decreased after about 12 h of darkness, as a result of carbohydrate depletion. Reducing the air temperature from 25 to 12°C accelerated plant growth, probably by reducing protein breakdown.

Allen et al. (1985) pointed out that leaf and canopy temperature should rise as $CO_2$-induced stomatal closure express itself under elevated $CO_2$ conditions. Furthermore, leaf or whole canopy energy balance predicts that decrease in leaf or canopy conductance will cause an increase in leaf vapor pressure through the increase in leaf temperature. This effect nullifies part of the direct reduction in transpiration expected, due to a reduction in stomatal conductance. In fact, leaf and boll temperatures were periodically measured in cotton grown under free-air-$CO_2$-enrichment (FACE) system by Kimball and Pinter (1990). In this study, foliage temperature was slightly more in FACE plots than ambient-$CO_2$-grown plants. The bolls tended to be 1.9°C warmer than leaves at the same position and illumination levels in the canopy. No correlation of boll-leaf temperature with boll mass, height of boll above soil, solar radiation, or percent of boll covered by calyx was found. The $CO_2$-induced leaf warming could possibly pose a problem in the future if air temperatures also increase due to the greenhouse effect. However, literature is limited in regard to the effects of rising temperature as a consequence of increased $CO_2$ levels on plants' reproductive growth (Strain and Cure, 1985; Idso et al., 1989; Schneider, 1989). The consequences of high temperature in a future world of high $CO_2$ may contribute to the extinction of some of the plant species in a natural ecosystem, through pollen sterility or restricted growth of pollen tubes for fertilization (Bhattacharya et al., 1992a). In essence, heat-tolerant cultivars need to be developed to overcome adverse effects of a global increase of the Earth's temperature on crop productivity.

Indeed, there is a growing body of literature showing (based on analyses of the widely cited results of GCMs) that there would be benefits from global climate change. For example, on the basis of detailed studies of Japanese agriculture, Yoshino and his colleagues concluded in 1988 that in Japan, scenarios based on a doubling of atmospheric $CO_2$ levels would result in an enormous rice surplus (Ausubel, 1991). The surplus would be even greater if farmers were, as is probable, to adapt their cultivation methods to the new climate by actions such as introducing more heat-tolerant rice varieties with earlier planting dates. Such adjustments might result in increases of rice yields in areas of Japan up to 25%, relative to recent levels.

Increased photosynthesis under high $CO_2$ and high temperatures (above optimum) does not always result in the effective partitioning of photoassimilate for reproductive organ differentiation (Reddy et al., 1991) and in many cases plants remain vegetative. The effects of high temperature on plants' reproductive growth need to be investigated carefully to project beneficial effects of a $CO_2$-enriched environment in a future world. The data on the interactive effects of low temperature and high $CO_2$ are also limited. It is unknown whether high atmospheric $CO_2$ will be able to circumvent adverse effects of a sudden drop in temperature to suboptimal levels in semiarid crops and fruit trees, especially during anthesis and early fruit formation. Idso et al. (1987) calculated a growth modification factor with a 3°C increase in mean air temperature with $+300\ \mu mol\ mol^{-1}$ $CO_2$ above ambient level (increase of growth from 1.30 to 1.56); however, the question of adaptation of plants with time was not taken into consideration. In summarizing research on the interactive effects of high $CO_2$ with different temperatures (suboptimal to high temperature) on food crops and flowering plants, it is evident that in most of the cases, vegetative mass increased proportionately more than reproductive mass (yield). Future research should address some of the consequences of temporal changes in temperature with high $CO_2$ to project crop performance in a future world of high $CO_2$ and increased temperature.

## IV. $CO_2$ ENRICHMENT AND WATER STRESS INTERACTIONS

The stimulation of growth of $C_3$ plants by $CO_2$ enrichment is affected by several environmental restraints. Changes in atmospheric $CO_2$ may be accompanied by rising temperatures and increased water stress in many areas (Tolbert and Zelitch, 1983). These changes will have complex interrelated consequences in photosynthetic carbon metabolism. Plant phenolics appear to exhibit a variety of responses to water stress. The levels of several simple cinnamic and benzoic acid derivatives in wheat (*Triticum aestivum* L.) declined noticeably under drought conditions (Tsai and Todd, 1972). Growth, transpiration, stomatal conductance, xylem potential, osmotic potential, turgor potential, and water use efficiency have been measured against soil water deficit in $CO_2$-enriched environments in wheat (*T. aestivum* L.), sugarbeet (*Beta vulgaris* L.), okra (*Abelmoschus esculentus* [L.] Moench); (Sionit et al., 1982), sweetgum (*Liquidambar styraciflua* L.), Loblolly pine (*Pinus taeda* L.); (Tolley, 1982), pea (*Pisum sativum* L.), and tomato (*Lycopersicon paniculatum* [L.] DC); (Wulff and Strain, 1982). It was concluded from these studies that increasing $CO_2$ concentration resulted in decreased stomatal conductance and transpiration per unit leaf area. Water was conserved and xylem potential remained higher in plants grown at elevated $CO_2$ (Dahlman et al., 1984). Morison and Gifford (1984) studied the response of 16 agricultural and horticultural crop species to limited water supply in pots in separate glasshouses at 340 and 680 $\mu$mol mol$^{-1}$ $CO_2$ at optimum temperature for growth and development. Water use and leaf area development were measured, while soil moisture content declined to about 6% from field capacity. High $CO_2$ increased leaf area from 20 to 75% in all except rice (*Oryza sativa* L.) and cotton (*Gossypium hirsutum* L.).

Rogers et al. (1984) and Sionit et al. (1984) investigated photosynthesis, stomatal conductance, and growth of soybean (*Glycine max* [L.] Merr.) under elevated $CO_2$ concentrations and water deficit conditions in open-top chambers. Soybean growth was greater under $CO_2$ enrichment than in ambient $CO_2$ under the same drought or water deficit conditions. Furthermore, lower rates of water use in high $CO_2$ delayed the effects of severe water stress. Stomatal conductance decreased with high $CO_2$ concentrations in both water-stressed and well-watered plants.

The interactive effects of enriched $CO_2$ and water stress on physiology and biomass production were investigated in sweet potato plants grown in large capacity pots containing sandy loam soil at two concentrations of $CO_2$ and two water regimes (Bhattacharya et al., 1990b). During the first 12 days of water stress, leaf xylem potentials were higher in plants grown in a $CO_2$ concentration of 438 and 666 $\mu$mol mol$^{-1}$ than in plants grown in 364 $\mu$mol mol$^{-1}$. The 364 $\mu$mol mol$^{-1}$ grown plants had to be rewatered 2 days earlier than in the high $CO_2$-grown plants in response to water stress. For plants grown under water stress, the yield of storage roots and root:shoot ratio was greater at high $CO_2$ than at 364 $\mu$mol mol$^{-1}$. These results suggest that even under water stress conditions, the production of photosynthate was higher for plants grown at $CO_2$-enriched environment than ambient $CO_2$-grown plants under the same water stress level. The increase, however, was not linear with increasing $CO_2$ concentrations. In well-watered plants, biomass production and storage root yield increased at elevated $CO_2$. These increases were greater for well-watered plants than for water-stressed plants grown at the same $CO_2$ concentration.

Extensive research has been done on cotton (*Gossypium hirsutum* L.) under enriched $CO_2$ environments in interactive experiments with limited water supply (Kimball et al., 1984, 1985, 1986, 1987, 1991) with regards to growth and development in glasshouses, phytotron, and open-top chambers (Strain and Cure, 1985, 1986). Kimball et al. (1984, 1985, 1986, 1987), with season-long deficit irrigation in cotton, demonstrated that growth response to $CO_2$ was as large or larger under water stress than under well-watered conditions. The growth and development of cotton is significantly affected by water stress (Marani et al., 1985; Krizek, 1986). Severe water stress during early preflowering reduced yield by increasing shedding of squares before they flowered. Water stress late in the flowering stage reduced the rate of flower formation and the retention of bolls (Krizek, 1986). According to Kimball et al. (1986, 1987), both the duration and amplitude of flowering and boll-set cycles were altered by $CO_2$ enrichment, and there were interactions with nitrogen (N) and water stress. Based on 1983 to 1987 studies, including water and nitrogen stress treatments (Kimball et al., 1983, 1986, 1987), a near-doubling of $CO_2$ concentration from 350 to 650 $\mu$mol mol$^{-1}$ increased seed cotton yield by about 60% on the average.

Studies were also conducted on cotton using a FACE system with sufficient or limited water supply to characterize growth and development throughout the growing season (Bhattacharya et al., 1990d, 1991a, 1992b; Mauney, 1990). In one of these studies, xylem potentials, osmotic potentials, and relative water content were measured in ambient and FACE plots during predawn and midafternoon, beginning June 6, 1990 through September 12, 1990. In the second week of August and first week of September, midafternoon xylem potentials decreased to $-2.4$ to $-2.9$ MPa. In September, the predawn and midafternoon xylem potentials were more negative in water stressed than well watered in both ambient and FACE plots, as expected (Bhattacharya et al., 1991a). $CO_2$-enriched plants were drier in the midafternoon compared to ambient-grown plants. The osmotic potentials during predawn and midafternoon followed patterns similar to the xylem potentials. The relative leaf water content (RLWC) results indicate that when plants were growing actively in a vegetative stage, the high $CO_2$ environment caused lower RLWC in water-stressed than in well-watered plots. However, RLWC did not vary much in well-watered and water-stressed in ambient-$CO_2$ control plots during the same growth period. At later stages of growth (square [flower buds] and anthesis), the differences in RLWC were small in water stressed and well watered conditions regardless of $CO_2$ treatment. In summary, several members of plant water status indicated little effect of $CO_2$ concentration on plant water status in a FACE system (Bhattacharya et al., 1991a, 1992b).

In general, a high $CO_2$ environment appears to alleviate water stress effects in several species, e.g., soybean (Rogers et al., 1984), sweet potato (Bhattacharya et al., 1990b), wheat (Sionit et al., 1980), and cotton (Kimball et al., 1991; Bhattacharya et al., 1992b). In a few cases, growth and biomass even increased under water-stressed conditions compared to that seen under well-watered conditions (Kimball, 1985; Gifford, 1979; Paez et al., 1983, 1984; Tolley and Strain, 1984; Conroy et al., 1986, 1988; Wray and Strain, 1986). According to Kimball and Mauney (1992), the probable explanation for growth stimulation under high $CO_2$ and water stress may be through profuse root growth at high $CO_2$ which enables plants to more fully mine a volume of soil and thereby withstand moisture stress.

In conclusion, water use efficiency of plants in a $CO_2$-enriched environment may have beneficial effects in tropical or subtropical regions of the world where water is limited for crop production. It is not certain, however, whether the beneficial effects of a $CO_2$-enriched environment will persist under prolonged periods of drought.

# V. $CO_2$ ENRICHMENT AND MINERAL NUTRIENTS INTERACTIONS

Growth and development of plants in response to elevated $CO_2$ concentrations have been studied in most experiments when mineral nutrients were not limiting. The more rapid growth rate and increased biomass production of plants in response to $CO_2$ enrichment probably results in faster depletion of nutrients in the root media, so one would expect nutrient supply to affect the response of plants to $CO_2$. The effect of nutrient supply and atmospheric $CO_2$ level on growth, total carbon (C) and nitrogen (N) contents, and C/N ratios have been studied in wheat (Sionit et al., 1981). As the supply of soil nutrients decreased and atmospheric $CO_2$ concentration increased, plant tissue became relatively poor in total nitrogen. Based on 1986 to 1987 open-top chamber studies in cotton under low nitrogen in a $CO_2$-enriched environment, Kimball and Mauney (1992) reported increases in biomass by 44 and 51%, with no added nitrogen for the wet and dry condition, respectively. The comparable increases under added nitrogen were 46 and 70%. No significant effects on harvest index, root/shoot ratio, or lint percentage were reported by Kimball and Mauney (1992). Other investigators have found similar or even larger growth responses to $CO_2$ at low nutrients (Kimball and Mitchell, 1979; Peet and Willits, 1984; Zangerl and Bazzaz, 1984; Hocking and Meyer, 1985; Norby et al., 1986). In essence, enhanced growth under low nutrient levels in a high $CO_2$ environment is hard to explain without any data on root growth, root distribution patterns, and soil nitrogen profile. Furthermore, we have limited knowledge on the interactive effects of soil microbes with roots for the effective use of nutrients in a $CO_2$-enriched atmosphere. Sionit (1983) reported that when soybean plants were grown under conditions of low nutrients and elevated $CO_2$, leaf senescence started 3 days earlier, and the number of pods and seeds declined, probably due to an inadequate supply of photosynthate to the seeds. The interaction of $CO_2$ enrichment with nutrient nitrogen and phosphorus deficiency was studied by Goudriaan and de Ruiter (1983) in several crops. Except for faba bean, no $CO_2$ effect existed under shortage.

It is evident from the earlier reports that $CO_2$ enrichment of different crops under limited nitrogen supply resulted in the accumulation of carbohydrates (Alberda, 1965; Dorvat et al., 1972; Williams et al., 1981). Thus, in a high $CO_2$ environment, more mineral nutrients will be essential to utilize the extra carbon available. Some data concerning nitrogen assimilation and $CO_2$ concentration are available in soybean (Hardy and Havelka, 1975a, 1975b; Sheehy et al., 1980; Williams et al., 1981, 1982). It was inferred from one of these studies (Williams et al., 1981) that increasing the $CO_2$ concentration from 320 to 1000 $\mu$mol mol$^{-1}$ in symbiotically grown soybean seedlings under limited nitrogen supply caused no significant increase in the Kjeldahl nitrogen content of the plants. The maximum $CO_2$ treatment effects were observed in plants supplied with a high level of N (8.0 mM $NH_4^+$); dry weight and protein nitrogen content were increased 64 and 20%, respectively. An increase in $CO_2$ to 1000 $\mu$mol mol$^{-1}$ led to significant increases in leaf dry weight and starch content under all nitrogen treatments.

There have been few studies on the interactions of $CO_2$ and nutrients in root crops. Studies with white potatoes on the interactions of enriched $CO_2$ with nitrogen and phosphorus as variables revealed an improved uptake of nitrogen and an increased nitrogen use efficiency with increased $CO_2$. Under phosphorus shortage, no $CO_2$ effect was found (Goudriaan and de Ruiter, 1983). While the interactions of $CO_2$ with nitrogen and phosphorus have received some attention, there have been no published studies on the assimilation of sulfur or other nutrients under $CO_2$ enrichment.

From the preceding reports, it may be inferred that high $CO_2$ environments may be beneficial for growth and development of plants, even under low soil nutrients provided that recycling of nutrients occurs through soil microbial activities, or by profuse growth of roots at the deeper layers of the soil for the removal of nutrients.

## VI. CARBOHYDRATES AND DIETARY FIBER IN ENRICHED $CO_2$

My colleagues and I have investigated the effects of enriched $CO_2$ on the physiological and biochemical changes on sweet potatoes and cowpeas (Bhattacharya et al., 1985a, 1985b; Bhattacharya et al., 1987; Bhattacharya et al., 1990c; Biswas et al., 1985, 1986). Others have studied soybeans (Havelka et al., 1984; Rogers et al., 1984; Vu et al., 1987), cotton (Krizek, 1986 and references therein; Kimball et al., 1986), and rice (Rowland-Bamford et al., 1990). The effect of $CO_2$ enrichment (350, 675, and 1000 $\mu$mol mol$^{-1}$) on carbohydrate concentrations in leaves, stems, roots, and storage roots was investigated in sweet potato (cv. Georgia Jet) at different stages of growth and development. The glucose, sucrose, and starch concentrations in leaves increased during 0 to 35 days after planting, compared to stems and roots receiving increased $CO_2$ concentrations. However, starch and glucose concentrations increased significantly in storage roots during the 50 to 65-day interval, which corresponded with rapid growth of storage roots at high $CO_2$ concentrations (Bhattacharya et al., 1989). In cowpeas, the glucose, sucrose, starch, and protein contents increased slightly in stems during vegetative growth, while these components remained unchanged during reproductive growth in 350, 675, or 1000 $\mu$mol mol$^{-1}$ $CO_2$. $CO_2$ enrichment caused a considerable increase in total carbohydrate in pod walls and seeds during the 71 to 106-day interval, being most pronounced at 675 μmol mol$^{-1}$ $CO_2$-grown plants (Bhattacharya et al., 1990c).

Allen et al. (1982) reported that $CO_2$ enrichment did not produce any significant differences in the total nonstructural carbohydrates (TNC) in leaves or pod walls of soybeans at the final harvest. Seeds from a $CO_2$-enriched atmosphere had more TNC than at 330 μmol mol$^{-1}$ $CO_2$. The TNC in roots at 600 and 800 μmol mol$^{-1}$ $CO_2$-grown plants were higher than those grown at 330 μmol mol$^{-1}$ $CO_2$, whereas in stems, only TNC levels of 600 μmol mol$^{-1}$ $CO_2$ air treatment were significantly higher. Some investigators (Nafziger and Koller, 1976; Finn and Brun, 1982; Mauney et al., 1979; Bhattacharya et al., 1985a) have reported that elevated levels of $CO_2$ increased the starch content of soybean and sweet potato leaves. In rice, the concentration of carbohydrates in the rice was similar in all $CO_2$ treatments at maturity (Rowland-Bamford et al., 1990). However, rice yield was greater at higher $CO_2$ concentrations due to an increase in the number of panicles per plant (Baker et al., 1990). In tomato plants grown at 1000 μmol mol$^{-1}$ $CO_2$, there were increases in leaf carbon fixation rate and starch and sugar accumulation rates compared with 350 μmol mol$^{-1}$ in $CO_2$-grown plants (Ho, 1978). Short-term $CO_2$ enrichment of soybean plants increased the total nonstructural carbohydrates in the leaves and stems plus petioles, but not in the roots and nodules (Finn and Brun, 1982). The TNC and soluble carbohydrates were found to be greater in tomato plants grown in $CO_2$ enrichment (1000 μmol mol$^{-1}$) than plants grown under ambient $CO_2$. Thus, increased carbohydrate concentrations in plant tissues under a high $CO_2$ environment alter N and protein contents in foliage and seeds. The increased C/N ratios have direct effects on the nutritive values of cereals and leguminous crops which are directly related to the supply of proteins for human consumption. It should be emphasized that in a high $CO_2$ environment, more photosynthates are partitioned into carbohydrate compared to amino acids or proteins. One of the explanations for high carbohydrate accumulation may be due to continuous fixation of $CO_2$ during photosynthesis, and limited availability of reduction conditions for protein synthesis and secondary plant products. Future research should address some of the problems related to limited partitioning of extra carbon for the formation of secondary plant products and biomolecules.

## VII. PLANT NUTRIENT STATUS IN ENRICHED $CO_2$

Partial depletion of $CO_2$ from 400 to 200 μmol mol$^{-1}$ in growth chambers was shown to elevate nitrate concentrations in tobacco leaves (Raper et al., 1973), and $CO_2$ enrichment was shown to lower nitrate concentrations in the leaves. Lower nitrogen and other nutrient concentrations have been observed in leaves and litter of plants grown in $CO_2$-enriched atmospheres. Lincoln et al. (1984) observed that plants grown at elevated $CO_2$ concentrations had a high C content, with no increase in total N, resulting in high C/N ratios. Similar results of high C/N ratios are also reported in perennial ryegrass (Overdieck and Reining, 1986) and cowpeas (Overdieck et al., 1988; Mbikayi et al., 1989). At high $CO_2$ levels, N concentrations of plant tissues have been shown to decrease in both nonlimiting nitrate and $N_2$-fixing systems, but to what extent this represents a dilution of nitrogen by carbohydrates is unknown (Strain and Cure, 1985).

The physiological significance of low nutrient concentration of $CO_2$-enriched plants is unclear. Studies conducted in greenhouses with lettuce (Knecht and O'Leary, 1983), and beans (Porter and Grodzinski, 1985) in an enriched $CO_2$ environment resulted in the reduction of nutrient concentrations in different plant parts. Necrosis of leaves was observed in cucumbers resembling potassium (K)-deficiency at high $CO_2$ (Peet et al., 1985). Tissue analysis of cucumber leaves showed consistent deficiency in K and magnesium (Mg) concentrations in the high $CO_2$-grown plants. However, Knecht and O'Leary (1983) reported that lettuce grown under high $CO_2$ levels did not differ nutritionally from plants grown under ambient $CO_2$. A significant decline (about 25%) in the leaf mineral content (N, P, K, Ca, Mg) was observed in 7-day-old bean plants (*Phaseolus vulgaris*) grown under 1200 $\mu$mol mol$^{-1}$ $CO_2$ (Porter and Grodzinski, 1985). "Georgia Jet" sweet potatoes were grown at $CO_2$ concentrations of 354, 431, 506, and 659 $\mu$mol mol$^{-1}$ for 90 days. Elevated $CO_2$ concentrations decreased protein, total carotenoid, and insoluble dietary fiber in sweet potato tubers. An increase in dry matter and a reddish-orange color of sweet potato tubers were observed at 596 and 659 $\mu$mol mol$^{-1}$ in $CO_2$-grown plants (Lu et al., 1986).

Sensory evaluation scores for flavor and moistness indicated that sweet potatoes grown under higher $CO_2$ concentrations were acceptable and not different from the ambient $CO_2$-grown plants (Lu et al., 1986). The negative aspects of greenhouse effects on protein composition in some of the economically important legume, cereal, and root crops have been reported by Bhattacharya et al. (1989), Biswas et al. (1986), Mbikayi et al. (1988), Lincoln et al. (1984), and Overdieck et al. (1988). In general, these reports suggest that with an increase in global $CO_2$, there will be a considerable decrease in proteins in seeds of leguminous, cereals, and root crops. The Third World countries will be affected considerably due to poor nutritive values of the crops such as cowpeas, wheat, and sweet potatoes. Literature is limited on the effects of elevated $CO_2$ on the nutrient quality of edible plants and its significance on human health.

If global climate change occurs as projected by different economic and climatic models, the formation of carcinogenic and/or anticarcinogenic plant metabolites may be affected in response to interaction with high $CO_2$ and other environmental constraints (personal communication — S. Bhattacharya, Mississippi State University and B. R. Strain, Duke University, NC). Recent studies in sweet potatoes indicate that plants grown at 364, 438, and 666 $\mu$mol mol$^{-1}$ $CO_2$ in open-top chambers and imposed water stress 71 days after planting showed differential mineral composition in tubers (Bhattacharya et al., 1990a). The N, K, Ca, and iron (Fe) concentrations in stems and leaves were greater than in storage roots regardless of $CO_2$ concentrations and water status of the plants.

## VIII. NATURAL ECOSYSTEMS AND $CO_2$ ENRICHMENT

The effects of elevated $CO_2$ on unmanaged ecosystems are potentially very important, but remain largely unknown (Strain, 1987; Strain and Cure, 1985). Many of the predictions made on the response to elevated $CO_2$ are based on observations of managed agricultural systems (Kimball, 1983; Rogers et al., 1983, Jones et al., 1985, Sionit et al., 1987a, 1987b). Rangeland ecosystems are among the important natural ecosystems that provide watersheds, recreation areas, and wildlife habitats, and are sources of pastures for grazing animals (Udall, 1966). This type of vegetation covers an area between 790 and 1580 km wide that extends from Mexico to Canada (Loomis et al., 1971). Limited data are available regarding rangeland responses to atmospheric $CO_2$ enrichment (Smith et al., 1987; Riechers and Strain, 1988). According to Sionit (1989), the perennial grasses of the rangeland ecosystems may already be undergoing changes due to the continuous increase in atmospheric $CO_2$ level. Because of the significance of the plant responses to $CO_2$ enrichment, a major research effort is in progress to determine likely responses of the rangeland ecosystems to elevated $CO_2$. Tallgrass Prairie is a major ecosystem type in rangelands of the U.S. that is the subject of research to determine its responses to atmospheric $CO_2$ enrichment (Owensby and colleagues, personal communication).

## IX. IMPLICATIONS OF $CO_2$ ENRICHMENT AND WEEDS

Most studies of plants and plant communities' response to elevated $CO_2$ are being done under weed-free conditions in growth chambers, glasshouses, or open-top field chambers

(Strain and Cure, 1986). Limited studies on weed/crop interactions at different $CO_2$ concentrations were conducted by Patterson and associates at the Duke University Phytotron (Patterson and Flint, 1990). The data from these studies are of limited scope for projecting consequences of weeds in future atmospheres of high $CO_2$ because they are not being validated under most natural conditions. The widely reported differential responses of $C_3$ and $C_4$ plants to $CO_2$ enrichment are particularly important to weed/crop competition in future agroecosystems in a high $CO_2$ environment. According to some estimates, $CO_2$ enrichment may alter the competitive balance between $C_3$ crops and $C_4$ weeds, and between $C_4$ crops and $C_3$ weeds. The effects of $CO_2$-enriched atmosphere on the pattern/shift of agriculture will be equally important in the assessment of weed/crop competition due to indirect effects of increased temperature and aridity. Sasek and Strain (1990) discussed the potential effects of increased $CO_2$ and climate change on the distribution of two exotic weeds of forest ecosystems. The current distribution of both Japanese honeysuckle (*Lonicera japonia* Thumb.) and kudzu (*Pueraria lobata* [Wild.] Ohwi) apparently is limited by low winter temperatures. If these temperatures increase by 3°C as projected, the northern limits of the honeysuckle and kudzu could be extended by several 100 km. Increased $CO_2$ also enhances the growth of these species at suboptimum temperatures, and this effect would contribute to northward expansion. The western limits of kudzu and honeysuckle probably are controlled by moisture availability. These limits might decrease in response to $CO_2$-induced warming, whereas the water use efficiency of both species could be increased in direct response to elevated $CO_2$. Thus, the western limits of distribution of these species are unlikely to change, according to Sasek and Strain (1990).

Apart from potential competitiveness of weeds with agricultural crops, the effective control of weeds may be a problem in an altered environment, where temperature, humidity, and water deficit may change considerably. In such circumstances, methods of application of herbicides, timing of the application, or other procedures may have to be different from the current practices to avoid crop injury and effective control of targeted weeds. The tolerance level of weeds towards herbicides may also change and cause problems in a future world of high $CO_2$ (Decker et al., 1986).

## X. INTERACTIONS OF INSECT HERBIVORES WITH $CO_2$ ENRICHMENT

Temperature frequently plays a prominent role in determining distribution and development of insect pests. Therefore, any increase in temperature will have a significant effect on such pests and their interaction with agricultural crops. A number of important effects of global warming on insect pests have been identified. These include increases in the rate of development and the number of generations produced per year; extension of the geographical range beyond the present margins of distribution; earlier establishment of pest populations in the growing season; and an increase in the risk of invasion by migrant and "exotic" species (Parry et al., 1990). Many of these factors could lead to increases in pest density and damage. Indeed, in the U.S. grain belt, an increase in the overwintering range and population density of the corn earworm (*Heliothis zea*) will increase damage to soybeans, resulting in significant economic loss (Parry et al., 1990; Environmental Protection Agency, 1989). Studies on the interactive effects of herbivores with a $CO_2$-enriched atmosphere have been done for soybean, cotton, and a perennial herb, *Plantago lanceolata* (Lincoln et al., 1984, 1986; Osbrink et al., 1987; Akey and Kimball, 1989; Fajer et al., 1989). Leaves of soybean plants grown under three $CO_2$ regimes (350, 550, and 650 $\mu$mol mol$^{-1}$) were fed to soybean looper larvae (Lincoln et al., 1984). Larvae were fed at 80% greater rates on leaves from the 650 $\mu$mol mol$^{-1}$ treatment than on leaves from the 350 $\mu$mol mol$^{-1}$ $CO_2$ treatment. Variation in larval feeding was related to the leaf content of N and water and specific leaf weight (SLW), each of which was altered by the $CO_2$ growth regime of the soybean plants. This study suggests that the impact of herbivores may increase as the level of atmospheric $CO_2$ rises.

Growth and development were studied for the beet armyworm (BAW), *Spodoptera exigua* (Hübner), a foliage feeder, reared on cotton seedlings grown at 640 and 320 $\mu mol\ mol^{-1}$ and two fertilizer levels (Akey and Kimball, 1989). Under high fertilization, female BAW reared on $CO_2$-enriched seedlings weighed significantly less than those reared on ambient plants, and had a significantly longer development period. Perhaps, more importantly, the survival rate of BAW on high fertilizer-$CO_2$-enriched cotton seedlings was 19.1 compared to 41.6% for controls. Also, females survived males by 2:1 ratio. In contrast, there was not a significant response in growth and development of the cotton pink bollworm (*Pectinophora gossypiella* [Saunders]), a seed feeder, when raised on $CO_2$-enriched mature cotton plants (Akey et al., 1988). In fact, differential responses of some of the insects raised in an enriched $CO_2$ environment are not clearly known. Thus, in a future world of high $CO_2$, leaf- and seed-feeding insects may have significantly changed impact on agricultural production. The slow growth of larvae of a monophagous herbivore, *Junonia coenia* (Lepidoptera: Nymphalidae) in a $CO_2$-enriched environment (700 $\mu mol\ mol^{-1}$) also suggests that interactions between plants and herbivore insects will be modified under the projected $CO_2$ conditions of the 21st century (Fajer et al., 1989). Based on these results, it may be inferred that insects have to eat more foliage to maintain normal growth rates due to low concentrations of nutrients (proteins, carotene, and essential minerals) in plant tissues. In such circumstances, there will be a greater loss of foliage due to insect pests and more so with the increase in temperature expected with the doubling of $CO_2$ concentration in the middle of the 21st century.

## XI. SUMMARY

The $CO_2$ concentration in the atmosphere is steadily increasing. It has been predicted that it will double the preindustrial level (270 $\mu mol\ mol^{-1}$) by the year 2080. Investigations conducted on different food and fiber crops in response to elevated $CO_2$ in phytotrons, glasshouses, open-top chambers, SPAR units, and FACE environments have generally showed increases in growth and yields of most of the crops, although some plants responded negatively to increased concentrations of $CO_2$. The increased growth of plants in a $CO_2$-enriched environment may rapidly deplete nutrients from the soil and consequently, positive effects of $CO_2$ may not persist under low fertility levels. Similarly, interactive effects of high $CO_2$ with high temperature may not be good for all plant species because of specific temperature requirements for each plant. In certain cases, plants may remain vegetative at high temperatures throughout the growth cycle. Therefore, cropping patterns may have to be modified with the increase in atmospheric temperature in the future world of high $CO_2$. Interestingly, water use efficiency of plants in a $CO_2$-enriched environment may have beneficial effects in tropical and subtropical regions of the world where water is limited for crop production. Elevated $CO_2$ in the atmosphere results in increased concentrations of carbohydrates and "dilution" of other metabolites such as chlorophyll, proteins, amino acids, carotene, and reduced nutrients in plant tissues. Increasing atmospheric $CO_2$ may alter plant/herbivore interactions. The impact of leaf-eating herbivores may increase as the level of atmospheric $CO_2$ rises. Furthermore, $C_3$ weeds may grow faster than $C_4$ crops of agricultural importance in a $CO_2$-enriched environment, and vice versa. In unmanaged ecosystems, these effects of elevated $CO_2$ may cause marked changes.

## ACKNOWLEDGMENTS

The author is grateful to Dr. John W. Radin, Dr. Bruce A. Kimball, Dr. Jack R. Mauney, and Dr. David Akey, USDA-ARS, Phoenix, Arizona; Dr. Hugo H. Rogers, USDA-ARS, Auburn, Alabama; Dr. Leon H. Allen, Jr., USDA-ARS, Gainesville, Florida; Dr. Robert L. Musser, Duke University Phytotron, Durham, North Carolina; and Dr. Sheila Bhattacharya, Arizona State University, Tempe, Arizona, for reviewing this chapter and providing suggestions for improvement.

## REFERENCES

Akey, D. H. and Kimball, B. A., Growth and development of the Beet Armyworm on cotton grown in an enriched carbon dioxide atmosphere, *Southeast. Entomol.*, 14, 255–259, 1989.

Akey, D. H., Kimball, B. A., and Mauney, J. R., Growth and development of the pink bollworm *Pectinophora gossypiella* (Lepidoptera: Gelechiidae), on bolls of cotton grown in enriched carbon dioxide atmospheres, *Environ. Entomol.*, 17, 452–455, 1988.

Alberda, T., The influence of temperature, light intensity and nitrate concentration on dry matter production and chemical composition of *Lolium perenne*. L, *Neth. J. Agric. Sci.*, 13, 335–360, 1965.

Allen, L. H., Jr., Global Climate Change and its Impact on Plant Growth and Development, Proc. Plant Growth Regulator Soc. Am., 16th Annu. Meet., August 6–10, 1989, Cook, A., Ed., 1989, 1–13.

Allen, L. H., Jr., Plant responses to rising carbon dioxide and potential interactions with air pollutants, *J. Environ. Qual.*, 19, 15–34, 1990.

Allen, L. H., Jr., Botte, K. J., Jones, J. W., Mishoe, J. W., Jones, P. H., Vu, C. V., and Campbell, W. J., Response of Vegetation to Carbon Dioxide: Effects of Increased Carbon Dioxide on Photosynthesis and Agricultural Productivity of Soybeans, No. 003, U.S. Department of Energy, Carbon Dioxide Research Division, Office of Energy Research, Washington, D.C., 1982.

Allen, L. H., Jr., Jones, P. H., and Jones, J. W., Rising atmospheric $CO_2$ and evapotranspiration, in *Advances in Evapotranspiration,* Proc. of Natl. Conf. Adv. Evapotranspiration, ASAE Pub. 14–85, *Am. Soc. Agric. Eng.*, St. Joseph, MI, 1985, 13–27.

Anonymous, Workshop on environmental and societal consequences of a possible $CO_2$-induced climate change, U.S. Department of Energy Conf., 790413 NTIS, Springfield, VA, 1980.

Ausubel, J. H., A second look at the impacts of climate change, *Am. Sci.*, 79, 210–221, 1991.

Bacastow, R. B., Keeling, C. D., and Whorf, T. P., Seasonal amplitude increase in atmospheric $CO_2$ concentration at Mauna Loa, Hawaii, *J. Geophys. Res.*, 90, 10,529–10,540, 1985.

Baes, C. F., Goeller, H. E., Olson, J. S., and Rotty, R. M., Carbon dioxide and climate: the uncontrolled experiment, *Am. Sci.*, 65, 310–320, 1977.

Baker, J. T., Allen, L. H., Jr., and Boote, K. J., Growth and yield responses of rice to carbon dioxide concentration, *J. Agric. Sci.*, 115, 313–320, 1990.

Berner, R. A. and Lasaga, A. C., Modeling the geochemical carbon cycle, *Am. Sci.*, 20(3), 74–81, 1989.

Bhattacharya, N. C., Biswas, P. K., Bhattacharya, S., Sionit, N., and Strain, B. R., Growth and yield response of sweet potato *Ipomoea batatas* to atmospheric $CO_2$ enrichment, *Crop Sci.*, 25, 975–981, 1985a.

Bhattacharya, S., Bhattacharya, N. C., Biswas, P. K., and Strain, B. R., Response of cowpea (*Vigna unguiculata*) to $CO_2$ enrichment environment on growth, dry matter production and yield components at different stages of vegetative and reproductive growth, *J. Agric. Sci.*, 105(3), 527–534, 1985b.

Bhattacharya, N. C., Bhattacharya, S., Biswas, P. K., and Strain, B. R., Partitioning of photoassimilates into Carbohydrates and Proteins of Cowpea (*Vigna unguiculata*) to $CO_2$ Enrichment Environment During Different Stages of Vegetative and Reproductive Growth, 84th Annu. Meet. Am. Soc. for Hortic. Sci., Hyatt Orlando Hotel, Orlando, FL, November 6 to 12, 1987, 137.

Bhattacharya, N. C., Bhattacharya, S., Strain, B. R., and Biswas, P. K., Biochemical changes in carbohydrates and proteins of sweet potato plants in response to enriched $CO_2$ environment at different stages of growth and development, *J. Plant Physiol.*, 135, 261–266, 1989.

Bhattacharya, N. C., Bhattacharya, S., Hileman, D. R., and Biswas, P. K., Elemental Composition of Sweet Potato in Response to $CO_2$ Enrichment and Water Stress, Annu. Meet. Agron. Soc. Am., San Antonio, TX, 1990a, 118.

Bhattacharya, N. C., Hileman, D. R., Ghosh, P. P., Musser, R. L., Bhattacharya, S., and Biswas, P. K., Interaction of enriched $CO_2$ and water stress on the physiology and production in sweet potato grown in open-top chambers, *Plant Cell Environ.*, 13, 933–940, 1990b.

Bhattacharya, N. C., Bhattacharya, S., and Strain, B. R., Changes in carbohydrates and proteins of cowpeas (*Vigna unguiculata* L.) during different stages of vegetative and reproductive growth in response to enriched $CO_2$ environment, *Am. J. Hortic. Sci.*, 1990c.

Bhattacharya, N. C., Hileman, D. R., Bhattacharya, S., and Biswas, P. K., Xylem Potential Measurements in the FACE Environment, Beltwide Cotton Conf., January 9 to 14, 1990, Las Vegas, NV, 1990d, 34.

Bhattacharya, N. C., McKinion, J. M., Gowda, P., Haluka, G., Sinha, N. K., Hileman, D. R., Kimball, B. A., Lynch, B., Hosny, A. A., Abdel-Al, M. H., and Ismail, M. S., Water Potential of Cotton Grown Under Optimal and Limiting Levels of Water in Free-Air $CO_2$ Enriched Environment, Beltwide Cotton Conf., San Antonio, TX, 1991a, 834–835.

Bhattacharya, S., Luthe, D. S., Bhattacharya, N. C., McKinion, J. M., Reddy, K. R., Wall, G. W., Bridges, S. M., and Hodges, H. F., Heat Shock Induced Protein Synthesis in Pima Cotton Leaves Exposed to Long-Term Temperature Stress and Enriched $CO_2$ Environment, Beltwide Cotton Conf., San Antonio, TX, 1991b, 844.

Bhattacharya, S., Bhattacharya, N. C., and Luthe, D. S., Interactive effects of high temperatures and $CO_2$ concentration on the induction of heat shock proteins in Pima cotton, 1992a.

Bhattacharya, N. C., Radin, J. W., and Kimball, B. A., Xylem Potential Measurements in Cotton Under Optimal and Limiting Levels of Water in Free-Air $CO_2$-Enriched Environment, Plant Physiol. Conf., Beltwide Cotton Conf., January 6 to 10, 1992, Nashville, TN, 1992b.

Biswas, P. K., Hileman, D. R., Allen, J. R., Bhattacharya, N. C., Lu, J. Y., Pace, R. D., Rogers, H. H., Coleman, K. B., Eatman, J. F., Ghosh, P. P., Mbikayi, N. T., Macrimon, J. N., and Menefee, A., Response of Vegetation to Carbon Dioxide: Field Studies of Sweet Potatoes and Cowpeas in Response to Elevated Carbon Dioxide, Rep. 002, U.S. Department of Energy, Carbon Dioxide Research Division, Office of Energy Research Washington, D.C., 1985.

Biswas, P. K., Bhattacharya, N. C., Ghosh, P. P., and Hileman, D. R., Influence of Atmospheric Carbon Dioxide Enrichment on the Biochemical Composition of Roots of Sweet Potato, Caribb. Food Crops Soc., 22nd Annu. Meet., St. Lucia, August 25 to 26, 1986, 30–38.

Budyko, M. I., The effect of solar radiation variations on the climate of the earth, *Tellus,* 21, 611–619, 1969.

Challa, J., An Analysis of the Diurnal Course of Growth, Carbon Dioxide Exchange and Carbohydrate Reserve Content of Cucumber, No. 861, Agric. Res. Serv. Rep., Wageningen, The Netherlands, 1976.

Campbell, W. J., Allen, L. H., Jr., and Bowes, G., Response of soybean canopy photosynthesis to $Co_2$ concentration, light and temperature, *J. Exp. Bot.*, 41, 427–433, 1990.

Clark, W. C., Cook, K. H., Moreland, G., Weinberg, A. K., Rotty, R. M., Bell, P. R., Cooper, L. J., and Cooper, C. L., The carbon dioxide question: a perspective of 1982, in *Carbon Dioxide Review,* Clark, W. C., Ed., Clarendon Press, Oxford, 1982, 3–43.

Conroy, J., Barlow, E. W. R., and Bevege, D. I., Response of *Pinus radiata* seedlings to carbon dioxide enrichment at different levels of water and phosphorus: growth, morphology, and anatomy, *Ann. Bot. (London),* 57, 165–177, 1986.

Conroy, J., Kuppers, M., Kuppers, B., Virgona, J., and Barlow, E. W. R., The influence of $CO_2$ enrichment, phosphorus deficiency and water stress on the growth, conductance and water use of *Pinus radiata* D. Don, *Plant Cell Environ.*, 11, 91–98, 1988.

Cure, J. D. and Acock, B., Crop responses to carbon dioxide doubling: a literature survey, *Agric. For. Meteorol.*, 38, 127–145, 1986.

Dahlman, R. C., Strain, B. R., and Rogers, H. H., Research on the response of vegetation to elevated atmospheric carbon dioxide, *J. Environ. Qual.*, 14(1), 1–8, 1984.

Decker, W. L., Jones, V. K., and Rao, A., The Impact of Climate Change from Atmospheric Carbon Dioxide on American Agriculture, TR031, DOE/NBB-0077, Office of Energy Research, Carbon Dioxide Research Division, Washington, D.C., 1986, 99.

Delucia, E. H., Sasek, T. W., and Strain, B. R., Photosynthesis inhibition after long-term exposure to elevated levels of atmospheric carbon dioxide, *Photosyn. Res.*, 7, 175–184, 1985.

Dorvat, A. B., Deinum, B., and Dirven, J. G. P., The influence of defoliation and nitrogen on the regrowth of rhodes grass (*Chloris gayana* Kunth). II. Etiolated growth and nonstructural carbohydrate, total-N and nitrate-N content, *Neth. J. Agric. Sci.*, 20, 97–103, 1972.

Environmental Protection Agency, The potential effects of global climate change on the United States, *Regional Studies, Draft Report to Congress, U.S. Environmental Protection Agency,* Vol. 1, 1989, 91.

Fajer, E. D., Bowers, M. D., and Bazzaz, F. A., The effects of enriched carbon dioxide atmospheres on plant-insect herbivore interactions, *Science,* 243, 1198–1200, 1989.

Finn, G. A. and Brun, W. L., Effect of atmospheric $CO_2$ enrichment on growth, nonstructural carbohydrate content, and root nodule activity in soybean, *Plant Physiol.*, 69, 327–331, 1982.

Gifford, R. M., Growth and yield of $CO_2$-enriched wheat under water-limited conditions, *Aust. J. Plant Physiol.*, 6, 367–378, 1979.

Goudriaan, J. and de Ruiter, H. E., Plant growth in response to $CO_2$ enrichment at two levels of nitrogen and phosphorus supply. I. Dry matter, leaf area and development, *Neth. J. Agric. Sci.*, 31, 157–169, 1983.

Hardy, R. W. F. and Havelka, U. D., Photosynthesis as a major factor limiting nitrogen fixation by field grown legumes with emphasis on soybeans, in *Symbiotic Nitrogen Fixation in Plants*, Nutman, P. S., Ed., Cambridge University Press, London, 1975a.

Hardy, R. W. F. and Havelka, U. D., Nitrogen fixation research: a key to world food?, *Science*, 188, 633–643, 1975b.

Havelka, U. D., Ackerson, R. C., Boyle, M. G., and Wittenbach, V. A., $CO_2$-Enrichment effects on soybean physiology. I. Effects of long-term $CO_2$ exposure, *Crop Sci.*, 25, 1146–1150, 1984.

Hileman, D. R., Gowda, P., Bhattacharya, N. C., Biswas, P. K., Kimball, B. A., Nakayama, F., and Peresta, G., Single Leaf and Canopy Photosynthesis in Ambient and FACE Conditions, Beltwide Cotton Conf., January 9 to 14, 1990, Las Vegas, NV, 1990, 34–35.

Ho, L. C., The regulation of carbon transport and the carbon balance of mature tomato leaves, *Ann. Bot. (London)*, 42, 155–164, 1978.

Hocking, P. J. and Meyer, C. P., Responses of Noogoora Burr (*Xanthium ocidentale* Bertol.) to nitrogen supply and carbon dioxide enrichment, *Ann. Bot. (London)*, 55, 835–844, 1985.

Houghton, J. T., Jenkins, G. J., and Ephramus, J. J., Eds., *Climate change: The IPCC Scientific Assessment*, Intergov. Panel on Climate Change, World Meteorological Organization, United Nation Environmental Programe, Cambridge University Press, London, 1990.

Idso, B. I. and Kimball, B. A., Downward regulation of photosynthesis and growth at high $CO_2$ levels — no evidence for either phenomen in three-year study of sour orange trees, *Plant Physiol.*, 96, 990–992, 1991.

Idso, S. B., *Carbon Dioxide: Friend or Foe? An Inquiry into the Climatic and Agricultural Consequences of the Rapidly Rising $CO_2$ Content of the Earth's Atmosphere*, IBR Press, Tempe, AZ, 1980, 1–92.

Idso, S. B. and Kimball, B. A., Growth response of carrot and radish to atmospheric $CO_2$ enrichment, *Environ. Exp. Bot.*, 29(2), 135–139, 1989.

Idso, S. B., Kimball, B. A., and Anderson, M. G., Greenhouse Warming Could Magnify Positive Effects of $CO_2$ Enrichment on Plant Growth, CDIAC Communications, Carbon Dioxide Inf. Analysis Center, Oak Ridge Natl. Laboratory, Oak Ridge, TN, 1989.

Idso, S. B., Kimball, B. A., Anderson, M. G., and Mauney, J. R., Effects of atmospheric $CO_2$ enrichment on plant growth: the interactive role of air temperature, *Agric. Ecosystem Environ.*, 20, 1–10, 1987.

Jones, D. H., Allen, L. H., Jr., Jones, J. W., Boote, K. J., and Campbell, W. J., Soybean canopy growth, photosynthesis, and transpiration responses to whole season carbon dioxide enrichment, *Agron. J.*, 76, 633–637, 1984.

Jones, P., Allen, L. H., Jr., Jones, J. W., and Valle, R., Photosynthesis and transpiration responses of soybean canopies to short- and long-term $CO_2$ treatments, *Agron. J.*, 77, 119–124, 1985.

Keeling, C. D., Bacastow, R. B., and Whorf, T. T., Measurement of the concentration of carbon dioxide at Mauna Loa Observatory, Hawaii, in *$CO_2$ Review*, Clark, W. C., Ed., Oxford Press, London, 1982, 377–385.

Kimball, B. A., Carbon dioxide and agricultural yield: an assemblage and analysis of 430 prior observations, *Agron. J.*, 75, 779–788, 1983.

Kimball, B. A., Adaptation of vegetation and management practices to a higher carbon dioxide world, in *Direct Effects of Increasing Carbon Dioxide on Vegetation*, Strain, B. R. and Cure, J. D., Eds., DOE/ER-0238, Carbon Dioxide Research Division, U.S. Department of Energy, Washington, D.C., 1985, 185–204.

Kimball, B. A., $CO_2$ Stimulation of growth and yield under environmental restraints, in *Carbon Dioxide Enrichment of Greenhouses*, Vol. 11, *Physiology, Yield, and Economics*, Enoch, H. Z. and Kimball, B. A., Eds., CRC Press, Boca Raton, FL, 1986, 53–67.

Kimball, B. A. and Mauney, J. R., Response of cotton to varying $CO_2$, irrigation, and nitrogen: yield and growth, *Agronomy*, 1992.

Kimball, B. A. and Mitchell, S. T., Tomato yields from $CO_2$-enrichment in unventilated and conventionally ventilated greenhouses, *J. Am. Soc. Hortic. Sci.*, 104, 515–520, 1979.

Kimball, B. A. and Pinter, P. J., Jr., Leaf and Boll Temperatures on the FACE Experiment. Special Session — Free Air Carbon Dioxide Enrichment (FACE), Beltwide Cotton Conf., Las Vegas, NV, 1990, 720.

Clawson, K. L., Reginato, R. J., and Idso, S. B., Effects of Increasing Atmospheric $CO_2$ on the Yield and Water Use of Crops, No. 021, U.S. Department of Energy Series, Response of Vegetation to Carbon Dioxide, Agricultural Res. Serv., U.S. Department of Agriculture, Washington, D.C., 1983.

Kimball, B. A., Mauney, J. R., Guinn, G., Nakayma, F. S., Pinter, P. J., Jr., Clawson, K. L., Idso, S. B., Butler, G. D., Jr., and Radin, J. W., Effects of Increasing Atmospheric $CO_2$ on the Yield and Water Use of Crops, No. 023, Response of Vegetation to Carbon Dioxide, U.S. Department of Energy, Carbon Dioxide Res. Div., U.S. Department of Agriculture, Agricultural Res. Serv., Washington, D.C., 1984, 60 pp.

Kimball, B. A., Mauney, J. R., Guinn, G., Nakayama, F. S., Idso, S. I., Radin, J. W., Hendrix, D. L., Butler, G. D., Jr., Zarembinski, T. I., and Nixon, P. E., III, Effects of Increasing Atmospheric $CO_2$ on the Yield and Water Use of Crops, No. 027, Response of Vegetation to Carbon Dioxide, U.S. Department of Energy, Carbon Dioxide Res. Div., U.S. Department of Agriculture, Agricultural Res. Serv., Washington, D.C., 1985.

Kimball, B. A., Mauney, J. R., Radin, J. W., Nakayama, F. S., Idso, S. B., Hendrix, D. L., Akey, D. H., Allen, S. G., Anderson, M. G., and Hartung, W., Effects of Increasing Atmospheric $CO_2$ on the Growth Water Relations, and Physiology of Plants Grown Under Optimal and Limiting Levels of Water and Nitrogen, Rep. #039, U.S. Department of Energy, Carbon Dioxide Res. Div., Office of Energy, Washington, D.C., 1986.

Kimball, B. A., Mauney, J. R., Akey, D. H., Hendrix, D. L., Allen, S. G., Idso, S. B., Radin, J. W., and Lakatos, E. A., Effects of Increasing Atmospheric $CO_2$ on the Growth, Water Relations, and Physiology of Plants Grown Under Optimum and Limiting Levels of Water, Rep. #049. U.S. Department of Energy, Carbon Dioxide Res. Div., Office of Energy, Washington, D.C., 1987.

Kimball, B. A., Akey, D. H., Mauney, J. R., Idso, S. B., Allen, S. G., Hendrix, D. L., and Radin, J. W., Elevated $CO_2$: Modeling Crop Response, Interaction with Temperature, Effects on Trees and Insects, No. 052, Response of Vegetation to Carbon Dioxide, U.S. Department of Energy, Carbon Dioxide Res. Div., U.S. Department of Agriculture, Agricultural Res. Serv., Washington, D.C., 1988.

Kimball, B. A., Mauney, J. R., La Morte, R. L., Guinn, G., Nakayama, F. S., Radin, J. W., Lakatos, E. A., Mitchell, S. T., Parker, L. L., Peresta, G. J., Nixon, P. E., Savoy, B., Harris, S. M., MacDonald, R., Pros, H., and Martinez, J., Response of Cotton to Varying $CO_2$, Irrigation, and Nitrogen: Data for Growth Model Validation, Carbon Dioxide Information Analysis Center report, Oak Ridge Natl. Laboratory, Oak Ridge, TN, 1991.

Knecht, G. N. and O'Leary, J. W., The influence of carbon dioxide on the growth pigment, protein, carbohydrate and mineral status of lettuce, *J. Plant Nutr.*, 6, 301–312, 1983.

Kramer, P. J., Carbon dioxide concentration, photosynthesis, and dry matter production, *Bioscience,* 31, 29–33, 1981.

Krizek, D. T., Photosynthesis, dry matter production and growth in $CO_2$-enriched atmospheres, in *Cotton Physiology,* Mauney, J. R. and Stewart, J. McD., Eds., Natl. Cotton Council of America, Memphis, TN, 1986, chap. 17.

Lincoln, D. E., Sionit, N., and Strain, B. R., Growth and feeding response of *Pseudoplusia includens* to host plants grown in controlled carbon dioxide atmospheres, *Environ. Entomol.*, 13, 1527–1530, 1984.

Lincoln, D. E., Couvert, D., and Sionit, N., Response of an insect herbivore to host plants grown in carbon dioxide enriched atmospheres, *Oecologia,* 69, 556–560, 1986.

Lindzen, R. S., Hon, A. Y., and Farel, B. F., The role of connective model choice in calculating the climate impact of doubling $CO_2$, *J. Atmos. Sci.*, 39, 1189–1205, 1982.

Looms, C. P., Gates, C. W., and Rose, J. K., Agriculture, United States, *Encyclopedia Britanica,* 1, 387, 1971.

Lu, J. Y., Biswas, P. K., and Pace, R. D., Effect of elevated $CO_2$ growth conditions on the nutritive composition and acceptability of baked sweet potatoes, *J. Food Sci.*, 51, 358–359, 1986.

MacCracken, M. C. and Luther, F. M., Eds., Detecting the Climatic Effects of Increasing Carbon Dioxide, DOE/ER-0235, U.S. Department of Energy, Carbon Dioxide Res. Div., Washington, D.C., 1985.

Manabee, S. and Stouffer, R. J., Summer dryness due to an increase of atmospheric $CO_2$ concentration, *Clim. Change,* 3, 347–386, 1981.

Manabee, S. and Wetherald, R. T., The effects of doubling $CO_2$ concentration on the climate of a general circulation model, *J. Atomos. Sci.*, 32, 3–15, 1975.

Marani, A., Baker, D. N., Reddy, V. R., and McKinion, J. M., Effect of water stress on canopy senescence and carbon exchange rates in cotton, *Crop Sci.*, 25, 798–802, 1985.

Markhart, A. H., III, Peet, M. M., Sionit, N., and Kramer, P. J., Low temperature acclimation of root fatty acid composition, leaf water potential, gas exchange and growth of soybean seedlings, *Plant Cell Environ.*, 3, 435–441, 1980.

Mauney, J. R., Growth and Yield of Cotton in the FACE Experiment. Special Session: Free Air Carbon Dioxide Enrichment (FACE) Applied to Cotton, Beltwide Cotton Conf., Las Vegas, NV, 1990, 716.

Mauney, J. R., Guinn, G., Fry, K. E., and Hesketh, J. D., Correlation of photosynthetic carbon dioxide uptake and carbohydrate accumulation in cotton, soybean, sunflower and sorghum, *Photosynthetica,* 13, 260–266, 1979.

Mbikayi, N. T., Hileman, D. R., Bhattacharya, N. C., Ghosh, P. P., and Biswas, P. K., Effects of $CO_2$ Enrichment on the Physiology and Biomass Production in Cowpeas (*Vigna unguiculata* L.) Grown in Open-Top Chambers, Proc. Int. Congr. Plant Physiol., New Delhi, India, Vol. 1, 640–645, 1988.

McKinion, J. M. and Wall, G. W., $CO_2$ Enrichment and Temperature Effects on Spring Wheat. I. Leaf and Spikelet Primordia Initiation, Annu. Meet. Agron. Soc. Am., San Antonio, TX, 1990, 19.

McKinion, J. M., Reddy, K. R., Wall, G. W., and Bhattacharya, N. C., Growth Response of Pima Cotton to $CO_2$ Enrichment During Vegetative Period, Beltwide Cotton Conf., San Antonio, TX, 1991, 841.

Ming-Dah, Chou, Li, Pen, and Arking, A., Climate studies with a multilayer energy balance model. Part II. The role of feedback mechanisms in the $CO_2$ problems, *J. Atmos. Sci.*, 39, 2657–2666, 1982.

Mitchell, J. F. B., The "greenhouse" effect and climate change, *Rev. Geophysics*, 27, 115–139, 1989.

Morison, J. I. L. and Gifford, R. M., Plant growth and water use with limited water supply in high $CO_2$ concentrations. II. Plant dry weight partitioning and water use efficiency, *Aust. J. Plant Physiol.*, 11, 375–384, 1984.

Nafziger, E. D. and Kroller, H. R., Influence of leaf starch concentration on carbon dioxide assimilation in soybean, *Plant Physiol.*, 57, 560–563, 1976.

Norby, R. J., O'Neill, E. G., and Luxmoore, R. J., Effects of atmospheric $CO_2$ enrichment on the growth and mineral nutrition of *Quercus alba* seedlings in nutrient-poor soil, *Plant Physiol.*, 82, 83–89, 1986.

Osbrink, W. L. A., Trumble, J. T., and Wagner, R. E., Host suitability of *Phaseolus lunata* for *Trichoplusia ni* (Lepidoptera: Noctuidae) in controlled carbon dioxide atmospheres, *Environ. Entomol.*, 16, 639–644, 1987.

Overdieck, D. and Reining, E., Effect of atmospheric $CO_2$ enrichment on pernnial ryegrass (*Lolium perenne*) and white clover (*Trifolium repens* L.) competing in a managed model-ecosystems. I. Phytomass production, *Acta Oecologica*, 7, 26–35, 1986.

Overdieck, D., Reid, C., and Strain, B. R., The effects of preindustrial and future $CO_2$ concentrations on growth, dry matter production and the C/N relationship in plants at low nutrient supply: *Vigna unguiculata* (Cowpea), *Abelmoschus esculentus* (Okra) and *Raphanus sativus* (Radish), *Angew. Bot.*, 62, 119–134, 1988.

Paez, A., Hellmers, J., and Strain, B. R., $CO_2$ enrichment, drought stress and growth of Alaska pea plants (*Pisum sativum*), *Plant Physiol.*, 58, 161–165, 1983.

Paez, A., Hellmers, H., and Strain, B. R., Carbon dioxide enrichment and water stress interaction on growth of two tomato cultivars, *J. Agric. Sci.*, 102, 687–693, 1984.

Parry, M. L., Porter, J. H., and Carter, T. R., Climatic change and its implication for agriculture, *Outlook Agric.*, 19, 9–15, 1990.

Patterson, D. T. and Flint, E. P., Implications of increasing carbon dioxide and climate change for plant communities and competition in natural and managed ecosystems, in *Impact of Carbon Dioxide, Trace Gases, and Climate Change on Global Agriculture*, Kimball, B. A., Rosenbery, N. J., and Allen, L. H., Jr., Eds., ASA Special Publ. No. 53, Am. Soc. Agron., Madison, WI, 1990.

Peet, M. M., Huber, S. C., and Patterson, D. T., Acclimation to high $CO_2$ in monoecious cucumbers. II. Carbon exchange rates, enzyme activities and starch and nutrient concentrations, *Plant Physiol.*, 80, 63–67, 1985.

Peet, M. M. and Willits, D. H., $CO_2$ enrichment of greenhouse tomatoes using a closed-loop heat storage: effects of cultivar and nitrogen, *Sci. Hortic.*, 24(2), 1–32, 1984.

Porter, M. A. and Grodzinski, B., $CO_2$ enrichment of protected crops, *Hortic. Rev.*, 7, 345–399, 1985.

Potvin, C., Goeschl, J. D., and Strain, B. R., Effects of temperature and $CO_2$ enrichment on carbon translocation of plants of the $C_4$ grass species *Echinochloa crus-galli* (L.) Beauv. from cool and warm environments, *Plant Physiol.*, 75, 1054–1057, 1984.

Ramanathan, V., The role of ocean-atmosphere interactions in the $CO_2$ climate problem, *J. Atmos. Sci.*, 38, 918–930, 1981.

Raper, C. D., Jr., Weeks, W. W., and Downs, R. J., Chemical properties of tobacco leaves as affected by carbon dioxide depletion and light entensity, *Agron. J.*, 65, 988–992, 1973.

Reddy, K. R., McKinion, J. M., Wall, G. W., Bhattacharya, N. C., Hodges, H. F., and Bhattacharya, S., Effect of Temperature on Pima Cotton Growth and Development, Beltwide Cotton Conf., San Antonio, TX, 1991, 841.

Revkin, A. C., Endless summer: living with the greenhouse effect, *Discover*, 9, 50–61, 1988.

Riechers, G. H. and Strain, B. R., Growth of blue grama (*Bouteloua gracilis*) in response to atmospheric $CO_2$ enrichment, *Can. J. Bot.*, 66, 1570–1573, 1988.

Rogers, H. H., Bingham, G. E., Cure, J. D., Smith, J., and Surano, K. A., Responses of selected plant species to elevated carbon dioxide in the field, *J. Environ. Qual.*, 12, 569–574, 1983.

Rogers, H. H., Sionit, N., Cure, J. D., Smith, J. M., and Bingham, G. E., Influence of elevated carbon dioxide on water relations of soybeans, *Plant Physiol.*, 74, 233–238, 1984.

Rowland-Bamford, A. J., Allen, L. H., Jr., Baker, J. T., and Boote, K. J., Carbon dioxide effects on carbohydrate status and partitioning in rice, *J. Exp. Bot.*, 41, 1601–1608, 1990.

Rowland-Bamford, A. J., Baker, J. T., Allen, L. H., Jr., and Bowes, G., Acclimation of rice to changing atmospheric carbon dioxide concentration, *Plant Cell Environ.*, 14, 577–583, 1991.

Sasek, T. W. and Strain, B. R., Implications of atmospheric $CO_2$ enrichment and climatic change for the geographical distribution of two introduced vines in the U.S.A., *Clim. Change*, 16, 31–51, 1990.

Schneider, S. H., The greenhouse effect: science and policy, *Science*, 243, 771, 1989.

Schlesinger, M. E. and Mitchell, J. F. B., In projecting climatic effects of increasing carbon dioxide, MacCracken, M. C. and Luther, F. M., Eds., U.S.-DOE/ER-0237, *Natl. Tech. Int. Serv.*, 1985, 259–272.

Sheehy, J. E., Fishbeck, K. A., Delong, T. M., Williams, L. E., and Phillips, D. A., Carbon exchange rates of shoots required to utilize available acetylene reduction capacity in soybean and alfalfa root nodules, *Plant Physiol.*, 66, 101–104, 1980.

Sidey, H., The big dry, *Time,* 132, 12–18, 1988.

Sionit, N., Response of soybean to two levels of mineral nutrition in $CO_2$ enriched atmosphere, *Crop Sci.*, 23, 329–333, 1983.

Sionit, N., Influence of Atmospheric $CO_2$ Enrichment on Rangeland Forage Quality and Animal Grazing, U.S. DOE Project, Carbon Dioxide Res. Div., Washington, D.C., 1989.

Sionit, N., Hellmers, H., and Strain, B. R., Growth and yield of wheat under carbon dioxide enrichment and water stress conditions, *Crop Sci.*, 20, 687–690, 1980.

Sionit, N., Strain, B. R., Beckford, H. A., Environmental controls on the growth and yield of okra. I. Effects of temperature and of $CO_2$ enrichment at cool temperature, *Crop Sci.*, 21, 885–888, 1981.

Sionit, N., Hellmers, H., and Strain, B. R., Interaction of atmospheric carbon dioxide enrichment and irradiance on plant growth, *Agron. J.*, 74, 721–726, 1982.

Sionit, N., Rogers, H. H., and Bingham, G. E., Photosynthesis and stomatal conductance with $CO_2$ enrichment of container and field grown soybeans, *Agron. J.*, 76, 447–451, 1984.

Sionit, N., Strain, B. R., and Flint, E. P., Interaction of temperature and $CO_2$ enrichment on soybean: growth and dry matter partitioning, *Can. J. Plant Sci.*, 67, 59–67, 1987a.

Sionit, N., Strain, B. R., and Flint, E. P., Interaction of temperature and $CO_2$ enrichment on soybean: photosynthesis and seed yield, *Can. J. Plant Sci.*, 67, 629–636, 1987b.

Smith, S. D., Strain, B. R., and Sharkey, T. D., Effects of $CO_2$ enrichment on four Great Basin grasses, *Func. Ecol.*, 1, 139–143, 1987.

Spencer, W. and Bowes, G., Photosynthesis and growth of water hyacenth under $CO_2$ environment, *Plant Physiol.*, 82, 528–533, 1986.

Strain, B. R., Physiological and ecological controls on carbon sequestering in ecosystems, *Biogeochemistry,* 1, 219–232, 1985.

Strain, B. R., Direct effects of increasing atmospheric $CO_2$ on plants and ecosystems, *Trends in Ecol. Evol.*, 2, 18–21, 1987.

Strain, B. R. and Cure, J. D., Direct effects of increasing carbon dioxide on vegetation, DOE/ER-0238, U.S.-DOE, Carbon Dioxide Res. Div., Office of Energy Research, Washington, D.C. and Duke University, Durham, NC, 1985.

Strain, B. R. and Cure, J. D., Direct effects of atmospheric $CO_2$ enrichment on plants and ecosystems: a bibliography with abstracts, ORNL/CDIC-13, U.S. Department of Energy, Washington, D.C., 1986, 197 pp.

Teskey, R. D., Fites, J. A., Samuelson, L. J., and Bongarten, B. C., Stomatal and nonstomatal limitations to net photosynthesis in *Pinus taeda* L. under different environmental conditions, *Tree Physiol.*, 2, 131–142, 1986.

Tolbert, N. E. and Zelitch, I., in *$CO_2$ and Plants: The Response of Plants to Rising Levels of Atmospheric Carbon Dioxide,* Lemmon, E. R., Ed., Waterview Press, Boulder, CO, 1983, 21–64.

Tolley, L. C., The Effects of Atmospheric Carbon Dioxide Enrichment, Irradiance and Water Stress on Seedling Growth and Physiology of *Liquidambar styraciflua* and *Pinus taeda,* Ph.D. dissertation, Duke University, Durham, NC, 1982.

Tolley, L. C. and Strain, B. R., Effects of $CO_2$ enrichment and water stress on growth of *Liquidambar styraciflua* and *Pinus taeda* seedlings, *Can. J. Bot.*, 62, 2135–2139, 1984.

Tsai, S. D. and Todd, G. W., Phenolic compounds of wheat leaves under drought stress, *Phyton,* 30, 67–75, 1972.

U.S. National Research Council, *Genetic Vulnerability of Major Crops,* National Academy of Sciences, Washington, D.C., 1972.

U.S. National Research Council, *Carbon Dioxide and Climate: A Scientific Assessment,* National Academy of Sciences, Washington, D.C., 1979.

U.S. National Research Council, *Carbon Dioxide and Climate: A Second Assessment,* National Academy of Sciences, Washington, D.C., 1982.

U.S. National Research Council, *Changing Climate,* National Academy of Sciences, Washington, D.C., 1983, 216–234.

Udall, S. L., Hour of decision, *J. Soil Water Conserv.*, 21, 3–4, 1966.

von Caemmer, S. and Farquhar, G. D., Effects of partial defoliation, changes of irradiance during growth, short-term water sress and growth, at enhanced p ($CO_2$) on the photosynthetic capacity of *Phaseolus vulgarius* L., *Planta,* 160, 320–329, 1984.

Vu, J. C. V., Allen, L. H., Jr., and Bowes, G., Drought stress and elevated $CO_2$ effects on soybean ribulose bisphosphate carboxylase activity and canopy photosynthetic rates, *Plant Physiol.*, 83, 573–578, 1987.

Wall, G. W., $CO_2$ Enrichment and Temperature Effects on Spring Wheat. II. Culm Development, p. 24, Annu. Meet., Agron. Soc. Am., San Antonio, TX, 1990, p. 24.

Wall, G. W., Reddy, K. R., McKinion, J. M., Bhattacharya, N. C., Bhattacharya, S., and Hodges, H. F., Photosynthesis Response of Pima Cotton to Temperature and Carbon Dioxide Enriched Environment, Beltwide Cotton Conf., San Antonio, TX, 1991, 831–833.

Williams, L. E., Delong, T. M., and Phillips, D. A., Carbon and nitrogen limitations on soybean seedling development, *Plant Physiol.*, 68, 1206–1209, 1981.

Williams, L. E., DeLong, T. M., and Phillips, D. A., Effect of changes in shoot carbon exchange rate on soybean root nodule activity, *Plant Physiol.*, 69, 432–436, 1982.

Wittwer, S. H., Carbon dioxide and climate change: an agricultural perspective, *J. Soil Water Conserv.*, 35(3), 116–120, 1980.

Wittwer, S. H., Rising atmospheric $CO_2$ and crop productivity, *Hortic. Sci.*, 189, 666–673, 1983.

Wittwer, S. H., Worldwide status and history of $CO_2$ enrichment — an overview, in *Carbon Dioxide Enrichment of Greenhouse Crops*, Enoch, H. Z. and Kimball, B. A., Eds., CRC Press, Boca Raton, FL, 1986, 3–15.

Woodwell, G. M., The carbon dioxide question: will enough carbon be stored in forests and the ocean to avert a major change in climate?, *Sci. Am.*, 238, 34–43, 1978.

Wray, S. M. and Strain, B. R., Response of two old field perennials to interactions of $CO_2$ enrichment and drought stress, *Am. J. Bot.*, 73, 1486–1491, 1986.

Wulff, R. and Strain, B. R., Effects of carbon dioxide enrichment on growth and photosynthesis in *Desmodium paniculatum*, *Can. J. Bot.*, 60, 1084–1091, 1982.

Zangerl, A. R. and Bazzaz, F. A., The response of plants to elevated $CO_2$. II. Competitive interactions among annual plants under varying light and nutrients, *Oecologia*, 62, 412–417, 1984.

# *Section V: Legal Policy and Educational Considerations Required to Properly Evaluate Global Warming Proposals*

Chapter 25

# BEYOND VIENNA AND MONTREAL: A GLOBAL FRAMEWORK CONVENTION ON GREENHOUSE GASES*

**David A. Wirth and Daniel A. Lashof**

## TABLE OF CONTENTS



* Copyright © 1992 by David A. Wirth and Daniel A. Lashof. With permission.

[W]e are handicapped . . . by policies based on old myths, rather than current realities.

Senator J. William Fulbright[1]

# I. INTRODUCTION

More than 20 years ago, the "greenhouse" effect was identified to policymakers at the highest level of the U.S. government as an environmental threat of potentially mammoth proportions.[2] Since then, greenhouse gases (GHGs) that cause global warming have accumulated in the atmosphere at an unprecedented rate.[3] Only now, more than two decades later, have policy responses been contemplated and negotiations on a multilateral convention on climate change initiated.

Recent progress in negotiating and implementing an international regime for regulating chlorofluorocarbons (CFCs), halons, and other chemicals that deplete the stratospheric ozone layer has fueled optimism about the potential for success of international undertakings to protect the climate from global warming. Emissions of CFCs and halons are governed by the Vienna Convention for the Protection of the Ozone Layer (Vienna Convention)[4] and the Montreal Protocol on Substances That Deplete the Ozone Layer (Montreal Protocol).[5] The Vienna Convention is often described as a "framework" agreement whose purpose is to establish an institutional basis for cooperation in research, exchange of information, and discussion of substantive policy measures like those contained in the Montreal Protocol. Following the entry into force of amendments adopted in 1990,[6] the Montreal Protocol will require a total phase-out by industrialized countries in consumption of 19 of these chemicals by the end of the century.

The connection between stratospheric ozone depletion and "greenhouse" warming is particularly significant. CFCs themselves are potent contributors to global warming. Despite their relatively low concentrations compared to other GHGs, CFCs and halons are responsible for 12% of the global warming potential of current GHG emissions.[7] Per molecule in the atmosphere, these chemicals are up to 20,000 times more potent in absorbing infrared radiation (heat energy) than carbon dioxide ($CO_2$), the most important GHG originating from human activities.[8] Although the Vienna Convention and Montreal Protocol help to address the global warming problem in an indirect, incremental manner by controlling emissions of CFCs and halons, those instruments are far from a comprehensive GHG regime. In particular, the Montreal Protocol does not specify that alternatives to the CFCs and halons controlled by the agreement must be or even should be greenhouse-friendly.[9]

Major reductions in emissions of GHGs other than CFCs, including $CO_2$, methane ($CH_4$), and nitrous oxide ($N_2O$), are necessary to assure the stability of the biosphere. National commitments by individual countries and concerted action by groups of large emitting nations, such as the Group of Seven (G-7) major industrialized nations, are crucial for achieving progress toward meaningful reductions in GHG emissions. Binding multilateral instruments are also needed to attack global warming on a global scale. New international institutions and decision-making processes may be desirable or even essential.

The desirability of a "framework" or "umbrella" treaty analogous to the Vienna Convention, with associated ancillary agreements analogous to the Montreal Protocol, has dominated the discussion of multilateral change instruments for some time. Unfortunately much of that interchange, especially in the U.S., has crystallized, if not ossified, around a rigid interpretation of the precedential importance of the stratospheric ozone negotiations.[10] The purpose of this article is to examine the implications of the Vienna-Montreal model and to stimulate debate on the form and substance of a future global GHG regime that would include all GHGs, not just CFCs.

# II. THE EARLY NEED FOR MULTILATERAL GREENHOUSE GAS CONTROLS

"The Earth is one but the world is not."[11] With these words, the World Commission on Environment and Development underscored the principal difficulty in formulating a concerted attack on international environmental threats in a world where the primary actors are independent, sovereign, coequal States. Among these environmental threats, few, if any, rival the greenhouse effect. Without attempting a comprehensive review of the scientific, policy, and legal issues associated with global warming,[12] it is important to highlight the overarching requirements that an international strategy on this compelling issue must address.

## A. AN INTERNATIONAL ISSUE

Greenhouse warming, like stratospheric ozone depletion, is a global problem. The most important greenhouse gases—$CO_2$, CFCs, $CH_4$ (methane), and $N_2O$ (nitrous oxide)—remain in the atmosphere for many years after being emitted.[13] As a result, their atmospheric concentrations are essentially the same in all regions of the world. Emissions anywhere on the planet have the same impact on climate, regardless of their geographic origin.[14].

Global warming and ozone depletion share a number of other characteristics with significant policy consequences. In contrast to some other international issues like acid rain, regional solutions, while incrementally helpful, cannot resolve these problems entirely. Current patterns of industrialization result in an enhanced greenhouse effect and stratospheric ozone depletion. Both threaten long-term, potentially catastrophic harm, whose precise delineation is complicated by a range of uncertainty.

Multilateral cooperation is even more important for global warming than for ozone depletion. GHG emissions are more varied and more widely distributed around the globe than the CFC emissions that cause stratospheric ozone depletion. Although $CO_2$ and methane emissions of fossil fuel origin, such as gasoline consumption and natural gas leaks, are highly concentrated in the industrialized countries that are the dominant CFC consumers, $CO_2$ releases from deforestation and methane emissions from rice paddies and domestic animals emanate largely from developing countries. For example, the U.S., Japan, and the European Community accounted for 70% of global CFC production in 1985. This same configuration of countries accounts for only about 40% of total GHG emissions.[15]

No comprehensive solution is possible without the active participation of developing countries and a GHG agreement must address their special needs. On the one hand, developing countries have caused little of the problem and industrialized countries must bear the bulk of the responsibility. On the other hand, as their economic development accelerates, Third World countries may account for the bulk of GHG emissions by the middle of the next century. Moreover, developing countries, with fewer resources to adapt to environmental disturbances, stand to suffer disproportionately from rapid climate change. An international solution that provides incentives for the participation of developing countries while equitably distributing the responsibility for implementing solutions is essential to combating greenhouse warming.

## B. WINNERS AND LOSERS

Although the buildup of GHG concentrations is uniform around the globe, the impacts of the resulting climate change will vary from region to region. This has led to the erroneous suggestion that there will be "winners" and "losers" from global warming.[16] So long as this notion persists, there is a serious risk that broad international agreement on environmentally meaningful reductions in GHG emissions will be stymied. The assumption that there will be winners from global warming is often based on a comparison of current GHG levels with a future, hypothetical climate regime in equilibrium, with $CO_2$ concentrations at double their preindustrial levels. This arbitrary and totally unrealistic scenario was developed solely for the convenience of climate modelers, who needed simple assumptions for their calculations.

The very concept of "winning" implies the existence of a stable warmer climate, which will not occur unless the current warming trend is halted. There is no natural endpoint to climate disruption from the greenhouse effect. Moreover, no single country will be able to guarantee that the phenomenon is arrested at an optimal point for that country. The only way to ensure that there will be *any* winners is to guarantee that *all* countries are winners by reversing the global buildup of GHGs in the atmosphere.

The long atmospheric lifetimes of GHGs necessitate major reductions in emissions from current levels. Even after these reductions, atmospheric concentrations of GHGs will fall only very slowly. The heat capacity of the oceans will further delay the climatic response by decades more.[17] Indeed, temperatures might continue to rise for many years even after the elimination of all anthropogenic GHG emissions.[18] Moreover, the required changes in utility and transportation infrastructure will take years or even decades to accomplish following decisions to embark on a GHG emissions reduction program.[19] The likelihood of positive feedbacks, through which the warming itself further accelerates GHG emissions rates, raises the further frightening possibility that human efforts to reduce emissions could be overwhelmed by natural processes such as the release of methane from Arctic regions. Once such a crisis has been reached, it will be too late to act.

## C. POLICY IMPLICATIONS OF SCIENTIFIC UNCERTAINTIES

Considering both what is known and what is not known about global warming, prudent public policy demands the implementation of measures to avoid the most serious risks. The question of how to react to scientific uncertainties was an explicit component of the Montreal ozone negotiations. Policy discussions concerning global warming, however, have generally not been guided by this crucial principle.

For instance, a widely quoted statement avers that the radiative equivalent of doubling the concentration of $CO_2$ would "most likely" result in a global warming of 1.5 to 4.5°C. This statement is based on a series of assessments by the National Academy of Sciences and the U.S. Department of Energy.[20] However, when biogeochemical feedback processes[21] are incorporated into climate models and standard deviations are accounted for, the temperature rise resulting from an initial doubling of $CO_2$ concentrations might increase to more than 6°C. A warming of as much as 8 to 10°C cannot be ruled out.[22]

A useful way to think about the policy implications of this scientific uncertainty is to examine the policies needed to limit climatic change to a given level. For example, consider the policy aim of confining warming commitments to a target of 2.5°C above preindustrial levels by the year 2030. Achieving this goal would result in an average global temperature at or below the maximum global temperature experienced during the last several million years. This limit is also consistent with the goal of preventing the maximum rate of warming from exceeding 0.1°C per decade, proposed by seminal international policy workshops held in Villach and Bellagio.[23]

Table 1 shows the current warming commitment and the $CO_2$ concentration limit that would be required to prevent a global warming of more than 2.5°C for various climate sensitivities to doubling $CO_2$ between 1.5° and 6°C.[24] The current warming commitment is the global temperature increase above preindustrial levels that would occur in equilibrium if GHG concentrations were frozen at today's levels. The $CO_2$ concentration limit is based on the assumption that other GHG concentrations can be stabilized at today's levels. The last column shows the total amount of $CO_2$ that could be emitted between now and when $CO_2$ concentrations are stabilized — for example the year 2030 — assuming that 55% of the emitted $CO_2$ remains in the atmosphere over this period. This analysis shows that if the

**TABLE 1**
**Carbon Budget to Limit Warming to 2.5°C Above Preindustrial Levels**

| Climate sensitivity ($2 \times CO_2$) | Current warming commitment | $CO_2$ Concentration limit | Carbon budget (Billion tons) |
|---|---|---|---|
| 6°C | 3.1°C | 330 ppm | — |
| 4.5 | 2.3 | 360 | 30 |
| 3.0 | 1.5 | 440 | 340 |
| 1.5 | 0.8 | 760 | 1600 |

*Note:* Data based on research and calculations by Dr. Daniel Lashof.

climate system turns out to be quite sensitive to increases in GHG concentrations, it may already be impossible to prevent unprecedented climatic change. With even a modest climate sensitivity, aggressive policies to eliminate fossil fuel dependence still would be required.

Given this situation, the only prudent policy is to minimize the risk of catastrophic climatic change by reducing $CO_2$ emissions to allow atmospheric concentrations of this gas to begin declining at the earliest possible date. It is both necessary and feasible to set a policy course consistent with preventing $CO_2$ concentrations from exceeding 400 parts per million (ppm). This could prevent a long-term warming commitment of more than 2.5°C so long as climate sensitivity turns out to be below 3.6°C. If subsequent scientific studies show that the climate system is definitely much less sensitive than this value, then this constraint might be relaxed. A mid-course correction of this sort would have few if any adverse economic consequences, as most if not all of the policies needed to achieve this target will prove beneficial in their own right. On the other hand, any delay in establishing policies consistent with the above goal will be extremely costly, both economically and environmentally, if it is subsequently shown that the climate system is at least this sensitive. For example, the continent-size ozone "hole" over Antarctica, which appeared suddenly and was not predicted by scientific models,[25] compellingly demonstrates that uncertainties cut both ways. Similarly, an unexpectedly high potential for alarming losses of stratospheric ozone over North America recently has been reported.[26] Scientific projections may not only overstate risks, but also understate them, and the potential for unpleasant surprises always exists.

The integration of scientific uncertainty into policy decisions has now been sufficiently widely accepted that the doctrine has acquired a name: the "precautionary principle."[27] However, the U.S. has relied on scientific and economic uncertainties as a justification for *rejecting* actions needed to protect the climate unless such actions are independently warranted.[28] This misleadingly-dubbed "no regrets" policy turns the salutary precautionary principle on its head.

## III. RECENT STEPS TOWARD A GREENHOUSE GAS CONVENTION

In the past several years, there has been a great deal of international activity on scientific, technical, and policy aspects of the greenhouse issue. Multilateral attention to the causes, consequences, and control of global warming has accelerated. As a result, a number of significant international undertakings relevant to the form, content, and timing of multilateral GHG instruments have been initiated.

## A. THE INTERGOVERNMENTAL NEGOTIATING COMMITTEE (INC)

The Intergovernmental Negotiating Committee for a Framework Convention on Climate Change (INC), established by U.N. General Assembly resolution[29] and reporting directly to that body, is the forum in which negotiations on the anticipated framework convention on climate change are taking place. According to the General Assembly's mandate, negotiations on the agreement are to be complete and the convention ready for signature by the U.N. Conference on Environment and Development, to be held in Brazil in June 1992. The work of the INC has been distributed between two working groups, one on commitments and one on mechanisms. The fourth session of negotiations, held in Geneva in December 1991, produced a unified negotiating text, but considerable disagreement on fundamental issues remains. In particular, the inclusion of targets and timetables for emissions of $CO_2$ by industrialized countries was strenuously opposed by the U.S., which is now effectively isolated on this question.[30] The 12 European Community States have called for a specific agreement by June 1992 for "the stabilization of $CO_2$ emissions by the year 2000 in general at 1990 levels by industrialized countries individually or jointly."[31] All other members of the Organization for Economic Cooperation and Development, except the U.S. and Turkey, have taken similar positions.[32] The stances of these countries differ somewhat in specificity with regard to dates, base years, and coverage, which varies from $CO_2$ alone to all GHGs not controlled by the Montreal Protocol. For example, Japan has adopted an ambiguous formulation, calling for the convention to require that: "Industrialized countries, in particular, shall make the best effort aimed at stabilizing emissions of $CO_2$ or $CO_2$ and other greenhouse gases not controlled by the Montreal Protocol as soon as possible, for example by the year 2000 in general at 1990 levels, recognizing the differences in approach and in starting-point in the formulation of objectives."[33] This roadblock in the negotiations has, moreover, effectively stymied any discussion of obligations for developing countries, which, as described above, are essential to the success of a multilateral strategy. At the fourth round of negotiations, in December 1991, it became apparent that the interests of developing countries are highly divergent and that the Third World could not effectively negotiate as a bloc. While petroleum-exporting States like Saudi Arabia and large developing countries like China, India, and Brazil have demonstrated varying levels of skepticism about the proposed convention, small island States like Vanuatu have pressed for a strong instrument. Even if the U.S. were to remove its objection to the stabilization goal for $CO_2$ by industrialized countries, either by modifying its policy to accommodate that goal or acknowledging that it would not sign and ratify a convention with that requirement,[34] it is probably already too late to make significant progress on the considerably more difficult but equally compelling question of the role of developing countries in a "global bargain."

## B. THE INTERGOVERNMENTAL PANEL ON CLIMATE CHANGE (IPCC)

The IPCC is now the principal ongoing multilateral vehicle for scientific and technical assessment of the greenhouse issue. The IPCC, which met for the first time in November 1988, was created under the auspices of the U.N. Environment Programme (UNEP) and the World Meteorological Organization (WMO) with an initial mandate to study the climate change issue and to report to the Second World Climate Conference (SWCC) in fall 1990. More than 35 countries participate in IPCC activities, which are distributed among three working groups: a science working group; a working group studying social and environmental impacts of climate change; and a response strategies working group charged with identifying policies to adapt to and mitigate the effects of global warming.

The first phase of the IPCC's work, which predated the current negotiations on a framework climate convention, concluded in 1990 with the adoption of the Panel's First Assessment Report, consisting of reports from the three working groups.[35] At the Second

World Climate Conference[36] representatives of 137 countries considered further steps on the global warming issue in light of the IPCC's report. Even now that formal convention negotiations have begun, as described above, under the auspices of the INC, the IPCC continues to provide technical background information for the climate treaty discussions. In particular, the science working group (Working Group I) is scheduled to supplement its 1990 Scientific Assessment Report in early 1992. This revision will update information on selected topics including global warming potentials, national emissions inventories, model validation, and climate observations.

## C. NONBINDING DECLARATIONS AND STATEMENTS

Over the past several years, a number of multilateral conferences have proven to be important occasions for stimulating international debate on global warming. The resulting instruments include:

- The final statement of an international meeting hosted by the government of Canada in mid-1988;[37]
- Three major U.N. General Assembly resolutions;[38]
- The conclusions of a group of international lawyers convened by the government of Canada in February 1989;[39]
- The declaration of an international meeting attended by seventeen heads of State in the Hague in March 1989;[40]
- The communiqués of the G-7 industrialized countries from their three most recent annual gatherings;[41]
- Declarations of ministerial conferences hosted by the Dutch government in November 1989[42] and the Norwegian government in May 1990,[43] respectively; and
- The SWCC Ministerial Declaration.[44]

There are numerous examples in international environmental law of multilateral meetings where political authorities establish an agreed action agenda, which can then serve as a basis for drafting binding legal obligations in a subsequent treaty negotiation. For example, the need for the 1990 amendments to the Montreal Protocol was established by a meeting in London attended by representatives of 124 nations in March 1989[45] and a statement of 80 countries at the first meetings of the parties to the Protocol in May 1989.[46]

Remarkably early in the succession of nonbinding instruments on climate change, consensus crystallized around the need for a multilateral climate convention. As discussed above, however, similar unanimity on the content of that agreement has yet to emerge. The SWCC Ministerial Declaration,[47] among the most recent and most influential of these precatory statements of purpose, reiterates the necessity for a global framework convention on climate change, noting that "the ultimate global objective should be to stabilise greenhouse gas concentrations at a level that would prevent dangerous anthropogenic interference with climate."[48] Significantly, the Declaration also stresses the need to stabilize emissions of non-CFC greenhouse gases and expressly identifies a number of countries — including the 12 European Community States, Austria, Canada, Finland, Iceland, Japan, New Zealand, Norway, Sweden, and Switzerland — that have committed to achieving that goal.[49] The SWCC final statement also emphasizes the need for funding to assist developing countries in carrying out their obligations under the proposed convention.[50]

## IV. AN ENVIRONMENTALLY MEANINGFUL GREENHOUSE GAS CONVENTION

The rapidity and magnitude of environmentally meaningful actions is the ultimate test of any combination of national and international policy responses to the threat of global warming. In strong contrast to the high-level political consensus on the need for multilateral instruments, there has been relatively little attention to the environmental goals a GHG convention should seek to accomplish. The following sections discuss benchmark tests relevant to the ongoing negotiations.

### A. THE OZONE PRECEDENT

High expectations about the prospects for a convention to limit emissions of GHGs other than CFCs and halons regulated by the Montreal Protocol arise in large measure from recent progress in negotiating and implementing the earlier instrument. It is therefore natural that discussions concerning multilateral instruments on climate explicitly have relied on the structures established in the ozone negotiations that led to the Vienna Convention and the Montreal Protocol. However, there is a serious risk that this precedent may be misinterpreted to impede rather than advance environmentally meaningful actions.

States negotiating under UNEP auspices to reduce threats to the stratospheric ozone layer made an explicit decision to undertake a two-component process. One product was to be a so-called ''framework'' multilateral convention of an essentially procedural character that would establish an institutional basis for cooperation in research, exchange of information, and discussion of substantive policy measures. Ancillary agreements known as ''protocols'' containing substantive regulatory measures would be appended to this convention.

The process-oriented ozone umbrella treaty evolved into the Vienna Convention concluded in March 1985. The allusions to a ''framework'' convention in the SWCC Ministerial Declaration and earlier conference statements are conscious references to this instrument. The Vienna Convention itself contains no substantive requirements for specific regulatory actions to protect stratospheric ozone. Instead, it embodies only a vague, unenforceable exhortation to protect the stratospheric ozone layer through the implementation of ''appropriate measures.''[51]

The history of the ozone negotiations demonstrates that process and substance on global atmospheric issues are not mutually exclusive, but intimately interrelated. Negotiations on a CFC protocol, which eventually became the Montreal Protocol, proceeded simultaneously with deliberations on the convention up to the adoption of the convention in early 1985. For a considerable portion of the negotiations, when consideration of the two instruments proceeded in tandem, a number of countries, including the U.S., called for a mandatory CFC protocol to which all parties to the convention would have to adhere.[52] When negotiations on the CFC protocol broke down, the Convention alone was adopted.[53] Renegotiation of the protocol after a scheduled one-year ''cooling off'' period coincided with an upsurge in public concern about the Antarctic ozone hole. This heightened attention to the stratospheric ozone problem broke the deadlock and facilitated adoption of the Montreal Protocol in September 1987.[54]

The procedural ''framework'' mechanisms for exchange of information and cooperation in research analogous to those institutionalized by the Vienna Convention are already in place internationally for GHGs in the form of the IPCC. The IPCC has also performed another function often ascribed to the Vienna Convention: laying the groundwork for substantive action through preliminary discussions. In addition, by facilitating an extended discussion of scientific and policy questions in advance of the adoption of legally binding commitments, the IPCC process serves very much the same function as the one-year "cooling off" period that preceded renegotiation of the CFC protocol. Accordingly, because of the IPCC's important work, the need for a strictly procedural "framework" is considerably lessened, if not eliminated altogether. Moreover, identifying a "framework" convention as an interim goal that must precede consideration of environmentally efficacious targets could seriously undercut the considerable momentum already generated on this issue. All these considerations strongly suggest that a GHG convention could and should be more aggressive than the Vienna Convention. In particular, there is an urgent necessity for early consideration of substantive, environmentally meaningful goals.

## B. $CO_2$ TARGETS DETERMINED BY ENVIRONMENTAL NECESSITY

As discussed above,[55] multilateral GHG instruments should establish a global goal of reducing as rapidly as possible emissions of GHGs sufficient to reverse their current buildup in the atmosphere. Given that the model of a convention with ancillary protocols has already been adopted, the ongoing convention negotiations would presumably be the earliest opportunity to set global emissions levels, at least for $CO_2$, consistent with this goal. Unfortunately, even the most aggressive proposals currently on the table, which call only for stabilization of $CO_2$ emissions from industrialized countries, still will allow significant increases in atmospheric concentrations of $CO_2$.

Of all GHGs, $CO_2$ is responsible for the largest portion of the global warming potential accumulating in the atmosphere. Apart from CFCs and halons, $CO_2$ is also the GHG for which emissions reduction options are most fully developed. Many of these options are cost-effective or involve net benefits.[56] To that extent, those options are analogous to the ban on aerosol uses of CFCs adopted in the U.S.[57] and other countries[58] in the 1970s, which produced savings to industry and the public as a result of the substitution of less expensive propellants.[59]

Consequently, halting the buildup of $CO_2$ in the atmosphere must be the first priority for multilateral climate instruments. For instance, a recent IPCC science assessment noted that a cut in $CO_2$ emissions of at least 60% would be required just to stabilize atmospheric concentrations of this gas.[60] As discussed above,[61] a stringent but achievable target that lowers the risk of catastrophic climate change is the following: limitation of global emissions to assure that atmospheric concentrations of $CO_2$ never exceed 400 ppm, with concentrations of $CO_2$ firmly established on a declining trajectory by the year 2030.

To achieve this goal, global emissions of carbon (as $CO_2$) from all sources, which now total 6.4 to 8.3 gigatons (Gt) per year, would have to be limited to a total budget of approximately 200 Gt for the 40 years — an average 5 Gt per year — between 1991 and 2030. To ensure declining $CO_2$ concentrations after 2030, the total global emissions rate for $CO_2$ would have to be no more than 1 to 3 Gt of carbon per year by that time. Interim goals, analogous to the 20% reduction target identified by the 1988 Toronto conference,[62] could and should also be established in a GHG convention to facilitate smooth, measured, and steady progress toward the ultimate aim. Because emissions from fossil fuel combustion can be most easily controlled and verified, $CO_2$ releases from the burning of oil, natural gas, and coal for industrial purposes or to generate electricity should be the subject of near-term, specific reduction targets for industrialized countries. However, the long-term global goal should include both industrial and biotic $CO_2$ emissions of anthropogenic origin. The climate system does not distinguish between industrial releases and those of biotic origin, such as forest clearing and burning, which currently account for 10 to 30% of $CO_2$ emissions.[63] Furthermore, including both industrial and biotic emissions of anthropogenic origin in a $CO_2$ agreement allows for a balancing of obligations, benefits, and other considerations of equity in a broader context. If handled properly, inclusion of biotic sources and sinks[64] within the climate convention can create incentives for reforestation and can contribute to protection of biological diversity. For this to occur, however, the agreement must place the highest priority on preserving primary forests and must contain safeguards against environmentally destructive forestry practices like large-scale monoculture tree farms designed to create "sink plantations."

An attractive conceptual framework is the apportionment of responsibilities based on national carbon "budgets" calculated according to a specified formula.[65] Two fundamental criteria appear to be relevant to the calculation of carbon budgets. The first is national population, probably the variable most closely connected with a "need" to emit $CO_2$, particularly as a result of energy consumption. India, for example, has insisted that the climate treaty articulate the principle that emissions from all States converge at a common per capita level. To guarantee the intended environmental results from the treaty and to ensure that budgets remain a stationary figure against which to measure future emissions, population should be calculated as of a fixed base year, such as 1988.

Apportioning $CO_2$ emissions budgets strictly on the basis of national population, however, is unlikely to be practicable. For instance, reductions of total $CO_2$ emissions from the U.S. of approximately 75% would be required merely to bring per capita releases of this gas down to the current global average of approximately 1.3 metric tons per person.[66] Although the U.S. and other disproportionately large emitters must use their best efforts to reduce, even strenuous measures would probably not produce annual cuts in emissions in excess of 5% from a fixed baseline. Consequently, a second criterion — current emissions, or a measure correlated with current emissions such as gross national product (GNP) — should also be a component of the budget calculation. A formula that accounts for both these factors is probably the most equitable and practical. Although the mix could be a subject of further discussion, an apportionment of 50% of the global budget between 1990 and 2030 based on population and the remaining 50% based on current emissions or GNP appears simultaneously to achieve three goals: (1) environmentally meaningful, mandatory $CO_2$ emissions reductions in industrialized countries; (2) an equitable distribution of obligations based on population; and (3) recognition of the lead time necessary to reverse implicit commitments to continued $CO_2$ emissions resulting from prior infrastructure investment decisions.[67] For example, the U.S. might receive a budget of 25 billion tons of carbon to emit cumulatively between 1990 and 2030. The budget concept will also assure that countries will receive "credit" for any emissions reductions they make even before the entry into force of the agreement.

Articulating a global goal in terms of a worldwide carbon budget implies, but does not require, subsequent apportionment of national obligations by means of national carbon budgets. An alternative approach is to frame the overall global endpoint in terms of percentage reductions from a base year. This might but does not necessarily imply national obligations framed in terms of percentage reductions from the base year, similar to the strategy adopted in the Montreal Protocol. An appropriate goal would then be an overall reduction in global $CO_2$ emissions levels of 20% from 1988 to 2005 and an 80% reduction from 1988 to 2030. Although phrased in different terms, this is equivalent to the carbon budget concept as a response to the magnitude of the global warming problem.

### C. RESOURCE TRANSFERS

The Montreal Protocol creates a special exemption for developing countries. Provided that the annual CFC consumption of these countries does not exceed 0.3 kilograms per capita, the Montreal Protocol entitles them to a ten-year exemption from the agreement's control measures. After this grace period, the Montreal Protocol requires of developing countries the same uniform percentage reductions in total national consumption that are required of all parties. Under the rubric of ''common but differentiated responsibilities,'' an analogous approach to the obligations of developing countries appears to be well accepted in the climate convention negotiations. In particular, a number of developing countries have emphasized the necessity for additional assistance, supplemental to existing foreign aid, to offset the incremental costs those countries will incur in fulfilling the obligations in the proposed GHG convention.

The Montreal Protocol contains explicit provisions for aid to developing countries to underwrite the dissemination of alternative technologies such as substitute chemicals and process changes for the production of alternatives to ozone-depleting CFCs.[68] The 1990 amendments to the Montreal Protocol create a special Multilateral Fund specifically for this purpose.[69] Even before the entry into force of those amendments, the Multilateral Fund is already in operation on an interim basis with resources of approximately $200 million of voluntary contributions. The Multilateral Fund may reach as much as $240 million when India becomes a party to the Montreal Protocol. The Multilateral Fund is administered by an executive committee of 14 parties consisting of 7 developed and 7 developing countries and is implemented by UNEP, the U.N. Development Programme (UNDP), and the World Bank.[70]

A related undertaking, known as the Global Environment Facility (GEF),[71] is a voluntary $1.3 billion, three-year joint pilot undertaking of UNEP, UNDP, and the World Bank. The GEF, administered by the World Bank, is designed to provide concessional financing for environmentally beneficial undertakings. It addresses four areas that might not be eligible or attractive for multilateral bank financing or that might not otherwise be a high priority for governments of developing countries: (1) limiting GHG emissions through, for example, projects designed to improve energy efficiency; (2) preserving biodiversity (variety of species); (3) combating the degradation of international water resources; and (4) mitigating ozone depletion from CFCs and other man-made chemcials. The bulk of assistance on the stratospheric ozone issue is administered through the Multilateral Fund of the Montreal Protocol. The major exception is those countries, primarily in eastern Europe, that are eligible for GEF financing but that do not qualify as a low-consuming developing country under the Montreal Protocol. Virtually all potential donor countries insist on using the GEF as the principal conduit for assistance to developing countries under the anticipated climate convention, thereby further magnifying the precedential significance of the Facility.

Resource transfers to assist poorer countries are likely to be at least as important for GHG agreements as they are for ozone depletion. Technical assistance grants and concessional loans may be necessary for up-front, start-up costs associated with forms of assistance largely unfamiliar to development aid agencies. The dissemination of alternative, environmentally benign options relying on wind, solar, biomass (plant material and organic waste), tidal, and geothermal energy sources may require infusions of new capital. Reforesting and conserving existing forest resources in tropical countries will necessitate additional foreign exchange. One way to finance these resource transfers would be a requirement for countries to contribute to a fund in proportion to their $CO_2$ emissions. In countries with market economies, these contributions could be financed by a tax on fossil fuel use.[72]

The Multilateral Fund under the Montreal Protocol and the GEF, while providing instructive examples in related areas, must be examined carefully as precedents for resource transfers pursuant to a GHG convention. First, in contrast to both the Multilateral Fund and the GEF, mandatory contributions of money may be necessary to meet the expenses incurred by poorer countries in shouldering the obligations of a climate convention and related protocols or other instruments. Second, developing countries have stressed the need for additional resources supplemental to development assistance already provided by donor countries to help them combat global climate change. Without contesting this assertion, the goals and quality of existing foreign aid must also be reexamined to assure consistency with the purposes of these additional flows of resources while simultaneously allowing Third World countries to meet their development goals. Because donors had little if any involvement in the financing of industries that produce or use CFCs and other ozone-destroying chemicals, this was hardly an issue with respect to stratospheric ozone depletion. By contrast, existing development assistance from multilateral and bilateral donors in such environmentally sensitive sectors as energy and forestry are substantial.[73] All resource transfers should contain conditions to ensure environmental quality and cost-effectiveness as measured by environmental impact assessment and least cost energy planning methodologies. To improve the prospects for long-term environmental sustainability, adequate input from recipient country governments and, especially, the public in receiving countries — the intended beneficiaries of this increased assistance — should be assured.[74]

## D. TRADING EMISSIONS ALLOCATIONS

Calculating emissions reductions in terms of a comprehensive ''bundle'' of GHGs and allowing international trading of emissions allowances recently have been advocated as mechanisms for maximizing economic efficiency within the environmental constraints established by multilateral GHG instruments. The U.S. in particular has with great vigor insisted that a GHG convention must satisfy these requirements.[75] While intellectually attractive on the grounds of economic efficiency, these proposals for trading among gases and between countries involve serious practical concerns as to their implementation. Insistence on trading of either sort, particularly in advance of agreement on global $CO_2$ targets, could become a serious barrier to achieving genuine $CO_2$ emissions reductions.

In principle, it should be possible to agree on the contributions of various GHGs to climate warming, through an analysis of chemical and physical properties such as absorption strength and atmospheric life-times.[76] Permitting trading among gases, however, as described below, would ignore the very real differences among those chemicals from a policy point of view. One unfortunate consequence could be disruption and unnecessary delay in the process of reaching agreement on global goals for those chemicals, such as $CO_2$, for which control options are readily available.

The Montreal Protocol as amended provides an example of trading in the ozone depletion context. It limits consumption and production of each of three ''baskets'' of chemicals consisting of five CFCs identified in the original 1987 agreement, ten other fully halogenated CFCs added by the 1990 amendments, and three halons, respectively.[77] The Montreal Protocol specifies controls on production and consumption not of each chemical within the basket, but of the basket as a whole, with the contribution of each chemical to calculated levels of production and consumption weighted according to its ozone-depleting potential. This formula permits each country to determine for itself the reductions required in consumption and production of each controlled substance, so long as the weighted levels of consumption and production of each basket conform to the Montreal Protocol's requirements.

Although the Montreal Protocol allows trading among CFCs and halons that deplete ozone, similar trading among gases that contribute to global warming is not necessarily good policy. CFCs and halons are strictly man-made and emanate from readily identifiable and controllable sources. By contrast, control options for the various GHGs are at substantially different levels of development. Of non-CFC greenhouse gases, $CO_2$ is the chemical for which the policy options are clearest. By contrast, baseline emissions of methane, the next most important gas from the point of view of contributions to the global warming problem, are highly uncertain. Although emissions reduction techniques are being developed for specific sources, comprehensive targets, covering all methane sources, individually or as a component of a multigas ''bundle,'' would be very difficult to establish now or in the near future. Depending on the distribution of initial allocations, trading among gases could also create disincentives for early participation by low-$CO_2$, high-methane-emitting developing countries, which may have little leverage in the negotiations. Given the disparate state of development in emissions monitoring and control options for various GHGs, delays in international progress on those portions of the global warming problem that are easier to solve — particularly $CO_2$ and selected industrial sources of methane, such as coal mining and land-fills — are a likely consequence of premature implementation of this bundling approach.

Trading in emissions allowances among countries, as opposed to among gases, presents different problems. The 1990 amendments to the Montreal Protocol permit wholesale trading in production of the chemicals regulated by that agreement, "provided that the total combined calculated levels of production of the Parties concerned for any group of controlled substances do not exceed the production limits set out . . . for that group."[78] Parties to the Protocol must document any international trading arrangements by notifying the Secretariat to the Protocol.

International trading of emissions allocations for GHGs could provide a mechanism for resource transfers to developing countries, provided that agreement could be reached on an equitable allocation of initial allowances. A variant of this concept is now reflected in the draft negotiating text, which contemplates bilateral or multilateral arrangements through which groups of countries can meet their treaty obligations cooperatively. Criteria for these arrangements are to be established by the conference of the parties to the convention. Presumably, this strategy has its greatest application to situations in which industrialized countries find it more economically efficient to pay for emissions reductions in developing countries than in their own.

While these tradeable offset proposals may have theoretical appeal, the practical obstacles to the successful implementation of such a system are formidable.[79] For example, an institutional structure to administer and oversee the trading system very likely would be necessary. A streamlined, informal procedure for adjudicating allegations of noncompliance included in the 1990 amendments to the Montreal Protocol[80] may represent some progress in this direction. Nonetheless, a supervisory mechanism to assure that countries use the proceeds of emissions trades for investments consistent with future global emissions budgets, not currently included in the Montreal Protocol, probably also would be required.

## E. INSTITUTIONAL ISSUES

The international legal system, as currently structured, assumes interaction among co-equal, sovereign States. States can create legally binding obligations through treaties, which are analogous to contracts, that require the express consent of the States concerned. Sovereign States can be bound absent express assent through long-standing custom and practice. The creation and identification of these customary international legal obligations, however, can be very slow and subject to considerable dispute. Moreover, customary norms, even if they existed, are unlikely to be sufficiently specific to protect the biosphere adequately from the worst effects of global climate change.[81]

Accordingly, international solutions to the greenhouse problem are most likely to come, if at all, from a multilateral treaty-making process. Any country may decline to be bound by a multilateral agreement merely by withholding its consent. Moreover, decisions at most international conferences are taken by "consensus," which in practice implies unanimity. Any single reluctant country can eviscerate or thwart an effective agreement. Consequently, effective international solutions to global environmental problems can be held hostage to the national imperatives of virtually every country. The necessity for consensus in multilateral processes can create a built-in inertia, which may produce disappointing "least common denominator" results that are not responsive to a particular problem.[82]

Of course, progress can be made within the confines of exisiting international structures. Nonetheless, the magnitude and urgency of the greenhouse warming threat may overwhelm the capacity of existing international mechanisms to respond effectively. For this reason, there recently have been calls for nonconsensus decision-making procedures and new institutions that would exercise some of the sovereign prerogatives of States. For instance, the Declaration of the Hague asserts the need for a new international body that would operate pursuant to "such decision-making procedures as may be effective even if, on occasion, unanimous agreement has not been achieved."[83]

The history of the Montreal Protocol demonstrates both the limitations and possibilities of international procedures. On the one hand, the Montreal process shows how particular countries can impede the purposes of an international agreement. Although India now has indicated its intention to accept the obligations of the revised Montreal Protocol when the 1990 amendments enter into force, for some time there was considerable concern about that country's commitment to the agreement. Without the participation of large, populous developing countries, atmospheric chlorine levels would continue to increase, and the likelihood of a return to pre-Antarctic hole atmospheric concentrations in the foreseeable future would be virtually nil.

On the other hand, the process for reassessing the Montreal Protocol's efficacy is a modest step toward international approaches that transcend the confines of the consensus model. For environmental issues like stratospheric ozone depletion, in which the scientific knowledge underlying treaty provisions is in a constant state of evolution, the reassessment of international obligations is often desirable if not necessary. Under customary international law, however, an amendment to a multilateral treaty is binding only on those States that indicate their affirmative intent to accept those new obligations, ordinarily through ratification of the amendment.[84] Consequently, there is a serious risk that repeated amendment of an agreement in light of new scientific developments will result in the creation of classes of parties, each with its own configuration of obligations depending upon the amendments to which it has acceded. This danger is particularly grave in the case of complex, delicately-balanced regulatory regimes like the Montreal Protocol.

While few would argue that the international community will or even should adopt majority or supermajority voting procedures in the near future, certain discrete areas may be especially fertile ground for departures from the consensus model through the adoption of more streamlined, quasi-legislative processes that nonetheless afford individual States guarantees that their needs will be met. One area that is ripe for a deviation from the consensus prnciple is that of amendments to existing multilateral agreements. The Montreal Protocol expressly departs from the customary rule for amending multilateral agreements by specifying expressly that "adjustments" to the agreement's reduction schedule, which are binding on all States party to the instrument, may be adopted by a two-thirds majority instead of by consensus.[85] At the time of the first review and assessment of the Montreal Protocol's efficacy in 1990, the precise meaning of "adjustment" was not clear. The parties to the Montreal Protocol adopted an interpretation worthy of Solomon, in which revisions to the reduction schedules for the eight chemicals originally covered by the agreement are subject to the nonconsensus adjustment process, but the addition of new chemicals requires a full-blown amendment. Nonetheless, the adoption of binding rules by qualified majority has been firmly established in the environmental sphere and may be a useful precedent in the greenhouse context.

## V. CONCLUSION

Preserving the integrity of the climate system requires early, environmentally meaningful reductions in emissions of GHGs on a multilateral basis. There is now an international consensus that a principal component in the mechanism for accomplishing this task will be a ''framework'' convention, or a multilateral treaty. Even as a first step, a framework convention should articulate a multilateral GHG control strategy incorporating specific national commitments by industrialized countries, while simultaneously encouraging additional unilateral action.[86] A framework convention should not be viewed as a significant objective in and of itself, but only as an interim step in the implementation of concrete emissions reductions. A convention should include specific targets, at a minimum for $CO_2$, sufficient to preserve the integrity of the climate system with an adequate margin of error. It expressly should provide for resource transfers to developing countries and address the need for new international institutions and decision-making procedures.

Unfortunately, the U.S., a leader on the stratospheric ozone issue, has dragged its feet on the global warming agreement. Indeed, the moniker ''no regrets'' adopted by the Bush Administration to describe its current policy demonstrates the inherently flawed bias of that position. The only ''regrets'' contemplated by the Administration's policy are those that result from environmental protection measures that subsequently prove unnecessary. The costs of uncertainty and delay in the event of unanticipated catastrophes like the ozone hole or major losses of stratospheric ozone over North America — or even of widely-accepted, if uncertain, predictions of climate disruption — are dismissed altogether.

''Delays have dangerous ends,''[87] wrote Shakespeare. This prudent advice is nowhere more relevant than for global warming. Procrastination today will cost dearly — perhaps not tomorrow, but certainly for the tomorrows of our children.

## NOTE ADDED IN PROOF

This article was drafted and went to press in early 1992, before the final rounds of negotiations on, and adoption of, the United Nations Framework Convention on Climate Change.* The Convention was opened for signature at the United Nations Conference on Environment and Development in Rio de Janeiro and, as of June 29, 1992, had been signed by 155 states and the European Economic Community.

The objective of the Convention is set out in article 2, which states that the agreement's principal purpose is to accomplish

> stabilization of greenhouse gas concentrations in the atmosphere at a level that would prevent dangerous anthropogenic interference with the climate system.

The main commitments in the agreement, which are clearly insufficient to accomplish this goal, are contained in article 4 of the Convention. The provisions addressing substantive limitations on emissions of greenhouse gases, whose negotiation is discussed in this article, apply to industrialized countries and are found in paragraph 2, subparagraph (a), which provides in part as follows:

> [National] policies and measures will demonstrate that developed countries are taking the lead in modifying longer-term trends in anthropogenic emissions [of greenhouse gases] consistent with the objective of the Convention, recognizing that the return by the end of the present decade to earlier levels of anthropogenic emissions of carbon dioxide and other greenhouse gases not controlled by the Montreal Protocol [on Substances That Deplete the Ozone Layer] would contribute to such modification . . .

Subparagraph (b) specifies that

* 31 I.L.M. 851 (1992).

> each of [the industrialized country] Parties shall communicate, within six months of the entry into force of the Convention for it and periodically thereafter . . . detailed information on its policies and measures referred to in subparagraph (a) above . . . with the aim of returning individually or jointly to their 1990 levels . . . anthropogenic emissions of carbon dioxide and other greenhouse gases not controlled by the Montreal Protocol . . .

Despite its curious bifurcation between these two provisions, the goal of emissions stabilization at 1990 levels by the year 2000 — among the most hotly debated aspects of the negotiations, as described in the preceding article — is nonetheless identified in the Convention. Subparagraph (a) designates the target date of "the end of the present decade" in the context of specific control measures, while the base year of 1990 is set out only in subparagraph (b), which is confined to reporting. This convoluted phraseology, and in particular the language that "recogniz[es]" instead of requires emissions stabilization as a goal, is largely attributable to the continued resistance of the U.S. to precise numerical requirements. For instance, after the conclusion of the U.N. conference, the Administrator of the Environmental Protection Agency (EPA) acknowledged that "the U.S. stood alone in resisting a commitment to targets and timetables for reducing $CO_2$ emissions." Memorandum from William K. Reilly, Administrator, Environmental Protection Agency, to all EPA Employees (July 15, 1992).

Like the Montreal Protocol, article 4(2)(d) calls for a review of the substantive adequacy of these requirements at the first session of the conference of the parties, whose precise timing depends on the date of the Convention's entry into force but which is likely to occur in 1994. A second review is to be concluded by the end of 1998, with periodic reevaluations thereafter. The Convention does not specify the availability of binding, non-consensus decision-making procedures analogous to the "adjustment" mechanism in the Montreal Protocol. Although there was some discussion of the desirability of majority voting for certain items of business before the conference of the parties — identified by article 7, paragraph 2 as the "supreme body of [the] Convention" — that issue was not explicitly addressed in the text of the Convention and will have to wait for the adoption of the rules of procedure of the conference of the parties anticipated by article 7(2)(k).

The negotiators left unresolved the similarly controversial issue of "bundling" greenhouse gases, also described by the U.S. as a "comprehensive" approach. The conference of the parties is directed by article 4(2)(c) to address "the respective contributions of [greenhouse] gases to climate change" — an activity that might be interpreted as a precursor to a more detailed regime of trading among gases. Article 4(2)(a) anticipates at least some international trading of emissions by permitting parties to "implement . . . policies and measures jointly with other Parties." The Convention does not set out a mechanism for monitoring these international exchanges. Instead, article 4(2)(d) states that the conference of the parties at its first session "shall . . . take decisions regarding criteria for joint implementation."

Article 3 sets out basic principles to be applied in the implementation of the Convention, presumably including any ancillary instruments or protocols that might be adopted in the future. Paragraph 3 articulates a precautionary approach, to be applied under conditions of scientific uncertainty "[w]here there are threats of serious or irreversible damage." Based on principles of equity and "common but differentiated responsibilities," industrialized countries in paragraph 1 commit to "take the lead in combating climate change and the adverse effects thereof."

Article 11 establishes a financial mechanism to implement a commitment in article 4, paragraph 3 according to which developed countries agree to "provide such financial resources, including for the transfer of technology, needed by the developing country Parties to meet the agreed full incremental costs of" performing certain of the obligations of the

Convention intended to help mitigate climate change. These resources will be administered, at least on an interim basis, by the Global Environmental Facility described in the article proper.

## ACKNOWLEDGMENT

This work was supported by grants from the Creswell Foundation and the Rockefeller Foundation's Study Center in Bellagio, Italy, which sponsored research on this article through Professor Wirth's appointment as a scholar-in-residence. The authors gratefully acknowledge the advice and assistance of Dr. John E. Bardach, Mary M. Brandt, James A. Losey, Alan S. Miller, Glenn T. Prickett, and Kevin C. Wells. This article appears simultaneously in a symposium entitled "Confronting Global Warming" in 2 *Transnational Law & Contemporary Problems* 79 (1992). Previous versions of this article appeared in David A. Wirth and Daniel A. Lashof, Beyond Vienna and Montreal — Multilateral Agreements on Greenhouse Gases, 19 *Ambio* 305 (1990); and David A. Wirth and Daniel A. Lashof, Beyond Vienna and Montreal — Multilateral Agreements on Greenhouse Gases, in *Greenhouse Warming: Negotiating a Global Regime* 13–24 (Jessica T. Mathews ed., 1991).

## ENDNOTES

1. 110 CONG. REC. S6227 (daily ed. Mar. 25, 1964) (statement of Sen. Fulbright).
2. COUNCIL ON ENVTL. QUALITY, FIRST ANNUAL REPORT 93–105 (1970).
3. INTERGOVERNMENTAL PANEL ON CLIMATE CHANGE, CLIMATE CHANGE: THE IPCC SCIENTIFIC ASSESSMENT xv-xvii (J. T. Houghton et al. eds., 1990) [hereinafter IPCC SCIENTIFIC ASSESSMENT].
4. Vienna Convention for the Protection of the Ozone Layer, Mar. 22, 1985, T.I.A.S. No. 11097, reprinted in [Reference File] Int'l Env't Rep. (BNA) 21:3101 (Jan. 1989), and in 26 I.L.M. 1529 (1987) [hereinafter Vienna Convention].
5. Montreal Protocol on Substances That Deplete the Ozone Layer, Sept. 16, 1987, S. TREATY DOC. NO. 10, 100th Cong., 1st Sess (1987), *reprinted in* 26 I.L.M. 1550 (1987) [hereinafter Montreal Protocol], *amended and adjusted,* S. TREATY DOC. NO. 4, 102d Cong., 1st Sess. (1991), *reprinted in* [Reference File] Int'l Env't Rep. (BNA) 21:3151 (Mar. 1991), *and in* 30 I.L.M. 539 (1991), *and in* 1 Y.B. INT'L ENVTL. L. 591 (1990) [hereinafter Montreal Protocol *as amended*]. For a history of the negotiations leading to the Montreal Protocol and its 1990 revisions, see generally RICHARD E. BENEDICK, OZONE DIPLOMACY: NEW DIRECTIONS IN SAFEGUARDING THE PLANET (1990); SHARON ROAN, OZONE CRISIS: THE 15-YEAR EVOLUTION OF A SUDDEN GLOBAL EMERGENCY (1989); Elizabeth P. Barratt-Brown, *Building a Monitoring and Compliance Regime Under the Montreal Protocol,* 16 YALE J. INT'L L. 519 (1991); Dale S. Byrk, *The Montreal Protocol and Recent Developments to Protect the Ozone Layer,* 15 HARV. ENVTL. L. REV. 275 (1991); Jamison Koehler & Scott A. Hajost, *The Montreal Protocol: A Dynamic Agreement for Protecting the Ozone Layer,* 19 AMBIO 82 (1990); Peter M. Morrisette, *The Evolution of Policy Responses to Stratospheric Ozone Depletion,* 29 NAT. RESOURCES J. 793 (1989); Recent Developments, 29 HARV. INT'L L.J. 185 (1988). *See also infra* text accompanying notes 52–54.
6. The Montreal Protocol as adopted in 1987 requires a 50% reduction in consumption of five CFCs, as measured against the base year of 1986, by the end of the century. The original Montreal Protocol also freezes consumption of three halons at 1986 levels. In June 1990, the parties to the Montreal Protocol adopted amendments and adjustments. *See infra* text accompanying note 85 (discussing difference between amendments and adjustments). The adjustments, which became effective on March 7, 1991, accelerate the reduction schedule for the eight CFCs and three halons covered by the original agreement, requiring a total phase-out in consumption of those chemicals by the end of the century. The accompanying amendments, which were scheduled to enter into force on January 1, 1992 provided they had been ratified by 20 parties to the original Montreal Protocol, would establish similar requirements for 10 additional fully halogenated CFCs, carbon tetrachloride ($CCl_4$), and methyl chloroform ($C_2H_3Cl_3$). Nonetheless, as of early 1992, the 1990 Amendments had not entered into force because they had received fewer than the necessary 20 ratifications. Recent scientific reports have documented the potential for unexpectedly high losses of stratospheric ozone over North America. *See, e.g., Ozone-Hole Conditions Spreading: High Concentrations of Key Pollutants Discovered over U.S.*, WASH. POST, Feb. 4, 1992, at A1. In response, the President recently announced that the United States will accelerate the

ban on ozone-depleting chemicals. *See* White House Press Release, 28 WEEEKLY COMP. PRES. DOC. 249 (Feb. 17, 1992); *see also U.S. to End CFC Production 4 Years Earlier Than Planned: Schedule for Other Ozone-Protecting Action Reexamined,* WASH. POST, Feb. 12, 1992, at A2.

7. IPCC SCIENTIFIC ASSESSMENT, supra note 3, at 51–54. The value of 12% is based on the Global Warming Potential (GWP) of 1990 emissions, calculated over a 100-year time horizon. CFCs' share of the radiative forcing increase between 1980 and 1990 is 24%. *Id.* at 55 tbl.2.2. This differs from the GWP-based value of 12% because changes in radiative forcing over a given period do not fully reflect the differences in atmospheric lifetimes among the GHGs. *See, e.g., id.* at 54–57; Daniel A. Lashof & Dilip R. Ahuja, *Relative Contributions of Greenhouse Gas Emissions to Global Warming,* 344 NATURE 529 (1990). The above values do not account for the radiative effect of ozone depletion caused by CFCs. A very recent U.N. Environment Programme (UNEP) report has suggested that the cooling effect of ozone depletion in the lower stratosphere observed during the 1980s may have offset the radiative forcing increase due to the direct effect of the CFC concentration increase observed during that same period. *Executive Summary: WMO/UNEP Scientific Assessment of Stratospheric Ozone,* World Meteorological Organization, WMO No. 473 (Oct. 22, 1991) [hereinafter WMO Ozone Assessment]. No estimate of the net GWP of CFCs, including the ozone depletion effect, is available at this time.
8. IPCC SCIENTIFIC ASSESSMENT, *supra* note 3, at 62–63.
9. *But see* Clean Air Act Amendments of 1990, Pub. L. No. 101–549, § 612, 104 Stat. 2399, 2648-70 (1990), 42 U.S.C.S. § 7671k (Law. Co-op. Supp. June 1991) (safe alternatives policy containing directive to Environmental Protection Agency to promulgate regulations making it unlawful to replace named ozone-depleting chemicals with substitutes that may have adverse effects on human health or environment if environmentally benign alternatives available); 57 Fed. Reg. 1984 (1992) (request for data and advance notice of proposed rulemaking pursuant to § 612).
10. *See, e.g.,* James K. Sebenius, *Crafting a Winning Coalition: Negotiating a Regime to Control Global Warming, in* GREENHOUSE WARMING: NEGOTIATING A GLOBAL REGIME 69, 70 (Jessica Mathews ed., 1991) ("the current 'general-framework-convention-followed-by-specific-protocols' approach to addressing climate change has practically assumed the status of conventional wisdom").
11. WORLD COMM'N ON ENV'T AND DEV., OUR COMMON FUTURE 27 (1987) (report of independent World Commission on Environment and Development, chaired by Dr. Gro Harlem Bruntland, Prime Minister of Norway, prepared at request of U.N. General Assembly).
12. *See generally* IPCC SCIENTIFIC ASSESSMENT, *supra, note* 3; INTERGOVERNMENTAL PANEL ON CLIMATE CHANGE, CLIMATE CHANGE: THE IPCC IMPACTS ASSESSMENT (W. J. McG.Tegart et al., eds., 1990) [hereinafter IPCC IMPACTS ASSESSMENT]; INTERGOVERNMENTAL PANEL ON CLIMATE CHANGE, CLIMATE CHANGE: THE IPCC RESPONSE STRATEGIES (1991) [hereinafter IPCC RESPONSE STRATEGIES]; ROBERT H. BOYLE & MICHAEL OPPENHEIMER, DEAD HEAT: THE RACE AGAINST THE GREENHOUSE EFFECT (1990); THE CHALLENGE OF GLOBAL WARMING (Dean E. Abrahamson ed., 1989); GREENHOUSE WARMING: NEGOTIATING A GLOBAL REGIME (Jessica T. Matthews ed., 1991); INT'L COUNCIL OF SCI. UNIONS, UNITED NATIONS ENV'T PROGRAMME & WORLD METEOROLOGICAL ORGANIZATION, REPORT OF THE INTERNATIONAL CONFERENCE ON THE ASSESSMENT OF THE ROLE OF CARBON DIOXIDE AND OF OTHER GREENHOUSE GASES IN CLIMATE VARIATIONS AND ASSOCIATED IMPACTS (1986) (report of conference held under auspices of World Climate Programme at Villach, Austria, Oct. 9–15, 1985); FRANCESCA LYMAN ET AL., THE GREENHOUSE TRAP (1990); IRVING MINTZER, A MATTER OF DEGREES (1987); NAT'L ACAD. OF SCI., POLICY IMPLICATIONS OF GREENHOUSE WARMING (1991); NAT'L RES. COUNCIL, OZONE DEPLETION, GREENHOUSE GASES, AND CLIMATE CHANGE (1989); NAT. RES. DEF. COUNCIL, COOLING THE GREENHOUSE: VITAL FIRST STEPS TO COMBAT GLOBAL WARMING (1989); POLICY OPTIONS FOR STABILIZING GLOBAL CLIMATE (Daniel A. Lashof & Dennis A. Tirpak eds., 1990) (report to Congress of U.S. Environmental Protection Agency) [hereinafter POLICY OPTIONS]; THE POTENTIAL EFFECTS OF GLOBAL CLIMATE CHANGE ON THE UNITED STATES (Joel B. Smith & Dennis A. Tirpak eds., 1989) (report to Congress of U.S. Environmental Protection Agency); UNITED NATIONS ENV'T PROGRAMME & WORLD METEOROLOGICAL ORGANIZATION, DEVELOPING POLICIES FOR RESPONDING TO CLIMATIC CHANGE (1988) (report of converence held at Villach, Austria, Sept. 28-Oct. 2, 1987 and Bellagio, Italy, Nov. 9–13, 1987) [hereinafter Bellagio Report]; U.S. CONGRESS, OFF. OF TECH. ASSESSMENT, CHANGES IN STRATOSPHERIC OZONE AND GLOBAL CLIMATE (1986) (4-volume proceedings of conference held at Leesburg, Virginia, June 16–20, 1986); David A. Wirth, *Climate Chaos,* 74 FOREIGN POL'Y 3 (1989).
13. IPCC SCIENTIFIC ASSESSMENT, *supra* note 3, at 7, 48, 59.
14. *See id.* ch. 1.
15. POLICY OPTIONS, *supra* note 12, at 148.
16. *E.g.,* Thomas C. Schelling, *Climatic Change: Implications for Welfare and Policy,* in CHANGING CLIMATE: REPORT OF THE CARBON DIOXIDE ASSESSMENT COMMITTEE 454 (1983).
17. POLICY OPTIONS, *supra* note 12, at 114.
18. *Id.*

19. Replacing coal-fueled power plants with renewable energy systems and building high-speed rail to reduce the need for air travel are examples of necessary changes. *See infra* note 56 (discussing the National Academy of Sciences' proposal of such measures).
20. IPCC SCIENTIFIC ASSESSMENT, *supra* note 3, at 145; NATIONAL ACADEMY OF SCIENCES, CHANGING CLIMATE 28 (1983); Frederick M. Luther, *Projecting the Climatic Effects of Increasing Carbon Dioxide: Volume Summary*, in PROJECTING THE CLIMATIC EFFECTS OF INCREASING CARBON DIOXIDE 261, 226 (Michael C. MacCracken & Frederick M. Luther eds., 1985) (Department of Energy report).
21. Biogeochemical feedback processes are changes in emissions or removal of GHGs induced by global warming.
22. Daniel A. Lashof, *The Dynamic Greenhouse: Feedback Processes That May Influence Future Concentrations of Atmospheric Trace Gases and Climatic Change,* 14 CLIMATIC CHANGE 213, 238–39 (1989).
23. Bellagio Report, *supra* note 12, at 25.
24. Climate sensitivity is defined here as the eventual increase in global average temperature that would result from doubling the atmospheric concentration of $CO_2$ from preindustrial levels. The range of climate sensitivities evaluated here includes the most likely range according to the IPCC, as well as a higher value based on the risk of positive biogeochemical feedbacks.
25. *See* J. C. Farman et al., *Large Losses of Total Ozone in Antarctica Reveal Seasonal* $C10_x/NO_x$ Interaction, 315 NATURE 207 (1985).
26. *See supra* note 6.
27. *See, e.g.,* Daniel Bodansky, *Scientific Uncertainty and the Precautionary Principle,* ENV'T, Sept. 1991, at 4; James Cameron & Juli Abouchar, *The Precautionary Principle: A Fundamental Principle of Law and Policy for the Protection of the Global Environment,* 14 B. C. INT'L & COMP. L. REV. 1 (1991). The precautionary approach has firm roots in domestic environmental law. *See, e.g.,* Ethyl Corp. v. EPA, 541 F.2d 1, 28 (D.C. Cir. 1976) (en banc) (''Where a statute is precautionary in nature, the evidence difficult to come by, uncertain, or conflicting because it is on the frontiers of scientific knowledge, the regulations designed to protect the public health, and the decision that of an expert administrator, we will not demand rigorous step-by-step proof of cause and effect. Such proof may be impossible to obtain if the precautionary purpose of the statute is to be served.''); Reserve Mining Co. v. EPA, 514 F.2d 492, 528 (8th Cir. 1975) (en banc) (''In the context of [the Federal Water Pollution Control Act], we believe that Congress used the term 'endangering' in a precautionary or preventive sense, and, therefore, evidence of potential harm as well as actual harm comes within the purview of that term.'').
28. *See e.g.,* U.S. GEN. ACCT. OFF., GLOBAL WARMING: ADMINISTRATION APPROACH CAUTIOUS PENDING VALIDATION OF THREAT (1990); D. Allan Bromley, *The Making of a Greenhouse Policy,* ISSUES IN SCI. AND TECH., Fall 1990, at 55, 57–59; C. Boyden Gray & David B. Rivkin, Jr., *A ''No Regrets'' Environmental Policy,* 83 FOREIGN POL'Y 47 (1991) (discussing the Bush Administration's ''no regrets'' policy).
29. G. A. Res. 212, U.N. GAOR, 45th Sess., Supp. No. 49A, at 148, U.N. Doc. A/45/49 (1990), *reprinted in* 21 ENVTL. POL'Y & L. 76 (1991) (establishing the Intergovernmental Negotiating Committee).
30. *See, e.g., Global Warming Rift Threatens Treaty: U.N. Talks Close with Industrialized Nations, Third World at Odds,* WASH. POST, Feb. 28, 1992, at A3. There recently has been a marked trend toward very specific regulatory regimes that have measurable procedural and substantive standards for implementation by individual States. In addition to the Montreal Protocol, *supra* note 5, the Protocol to the 1979 Convention on Long-Range Transboundary Air Pollution on the Reduction of Sulfur Emissions or Their Transboundary Fluxes by at Least Thirty Per Cent, July 8, 1985, reprinted in [Reference File] Int'l Env't Rep. (BNA) 21:3021 (Mar. 1989), *and in* 27 I.L.M. 707 (1988) [hereinafter Sulfur Protocol], as its name suggests, requires each State party to accomplish a uniform percentage cutback in pollution, measured from an agreed base year, by a firm deadline. The Protocol to the 1979 Convention on Long-Range Transboundary Air Pollution Concerning the Control of Emissions of Nitrogen Oxides or Their Transboundary Fluxes, Oct. 31, 1988, *reprinted in* 18 ENVTL. POL'Y & L. 228 (1988), *and in* [Reference File] Int'l Env't Rep. (BNA) 21:3041 (Jan. 1989), *and in* 28 I.L.M. 214 (1989), sets out highly specific technology-based standards for pollution control within the context of an overall emissions limitation. The Basel Convention on the control of Transboundary Movements of Hazardous Wastes and Their Disposal, Mar. 22, 1989, S. TREATY DOC. NO. 5, 102d Cong., 1st Sess. (1991), *reprinted in* 19 ENVTL. POL'Y & L. 68 (1989), *and in* [Reference File] Int'l Env't Rep. (BNA) 21:3701 (May 1989), *and in* 28 I.L.M. 657 (1989), which enters into force on May 5, 1992, mandates detailed procedures governing the export of municipal trash and toxic waste. Of course, targets and timetables are well accepted as regulatory strategies within the domestic context. For instance, the Clean Air Act Amendments of 1990, Pub. L. No. 101–549, 104 Stat. 2399 (1990), 42 U.S.C.S. §§ 7401–7671q (Law. Co-op. Supp. June 1991), incorporate numerous deadlines and numerical goals. In particular, to combat acid rain those Amendments require a 10 million ton reduction in sulfur dioxide emissions by the year 2000. Clean Air Act Amendments of 1990, Pub. L. No. 101–549, §401, 104 Stat. 2585 (1990), 42 U.S.C.S. § 7651 (Law. Co-op Supp. June 1991).
31. Intervention by the Presidency on Behalf of the European Community and Member States, Nairobi, Kenya (Third Session of the INC) (transcript available at Natural Resources Defense Council, Washington, D.C.).

32. *See* Ministerial Declaration, Second World Climate Conference (statement from international meeting in Geneva, Oct. 29-Nov. 7, 1990), *reprinted in* 20 ENVTL. POL'Y & L. 220 (1990), *and in* 1 Y.B. INT'L ENVTL. L. 473 (1990) [hereinafter SWCC Ministerial Declaration].
33. Statement of Japan at the Third Session of the INC, Nairobi, Kenya, *reprinted in Japan Switches to 'Best Effort'*, ECO, Sept. 9, 1991, at 1 (cooperative newsletter published by nongovernmental environmental groups at major international conferences since the Stockholm Environment Conference in 1972).
34. For instance, U.S. acid rain policy at the time of the conclusion of the Sulphur Protocol, *supra* note 30, was inconsistent with the requirements of that instrument. Instead of impeding the negotiations by insisting on an agreement consistent with domestic policy, the U.S. declined to accept the obligations in a more ambitious instrument that was responsive to the needs and desires of the other parties to the negotiation.
35. *See* IPCC SCIENTIFIC ASSESSMENT, *supra* note 3; IPCC IMPACTS ASSESSMENT, *supra* note 12; IPCC RESPONSE STRATEGIES, *supra* note 12.
36. *See* SWCC Ministerial Declaration, *supra* note 32.
37. The Changing Atmosphere: Implications for Global Security (statement from international meeting sponsored by Government of Canada in Toronto June 27–30, 1988), *reprinted* in 5 AM. U.J. INT'L L. & POL'Y 515 (1990) [hereinafter Toronto Conference Statement]. More than 300 individuals from 48 countries, including government officials, scientists, and representatives of industry and environmental organizations attended this conference. The following recommendations from the conference were prominent: (1) the necessity for a "comprehensive global convention as a framework for protocols on the protection of the atmosphere"; (2) the establishment of a "World Atmosphere Fund financed in part by a levy on the fossil fuel consumption of industrialized countries" to facilitate technology transfer to Third World countries; and (3) a reduction in "$CO_2$ emissions by approximately 20% of 1988 levels by the year 2005" as an initial goal. *Id.* paras. 30, 5, 22.
38. G.A. Res. 212, *supra* note 29; G.A. Res. 207, U.N. GAOR, 44th Sess., Supp. No. 49, at 130, U.N. Doc.A/44/49 (1990), *reprinted in* 20 ENVTL. POL'Y & L. 43 (1990); G. A. Res. 43/53, U.N. GAOR, 43rd Sess., Supp. No. 49, at 133, U.N. Doc. A/43/49 (1989), *reprinted in* 5 AM. U.J. INT'L L. & POL'Y 525 (1990), *and in* 19 ENVTL. POL'Y & L. 27 (1989), *and in* 28 I.L.M. 1326 (1989). *See also* G.A. Res. 169, U.N. GAOR, 46th Sess., U.N. Doc. A/46/729 (1991). *See generally* Frederic L. Kirgis, Jr., *Standing to Challenge Human Endeavors That Could Change the Climate,* 84 AM. J. INT'L L. 525 (1990) (discussing G.A. Resolution 43/53).
39. Protection of the Atmosphere: Statement of the Meeting of Legal and Policy Experts (statement from international meeting sponsored by government of Canada in Ottawa, Feb. 20–22, 1989), *reprinted in* 5 AM. U.J. INT'L L. & POL'Y 529 (1990), *and in* 19 ENVTL. POL'Y & L. 78 (1989).
40. This conference was attended by representatives of 24 countries, including 17 heads of state. The resulting declaration emphasizes the desirability of the "negotiation of the necessary legal instruments to provide an effective and coherent foundation, institutionally and financially," for a new institutional authority charged with "combating any further global warming of the atmosphere." The conference recognized the need for this authority to apportion "fair and equitable assistance" to those countries that are asked to bear an "abnormal or special burden, in view of the level of their development and actual responsibility for the deterioration of the atmosphere . . . " Declaration of the Hague (statement from international meeting sponsored by government of the Netherlands in the Hague, Mar. 11, 1989), *reprinted in* 5 AM. U.J. INT'L L. & POL'Y 567 (1990), *and in* 19 ENVTL. POL'Y & L. 78 (1989), *and in* 30 HARV. INT'L L.J. 417 (1989), *and in* 12 Int'l Env't Rep. (BNA) 215 (1989), *and in* 28 I.L.M. 1308 (1989).
41. The governments of many of the world's largest emitters of $CO_2$ and other GHGs — Canada, France, Germany, Italy, Japan, the U.K., and the U.S. — are represented at the annual meeting of heads of State of the seven major industrialized nations. The communiqué from the gathering in Paris in July 1989 declared the following:

    We believe that the conclusion of a framework or umbrella convention on climate change to set out general principles or guidelines is urgently required to mobilize and rationalize the efforts made by the international community . . . . Specific protocols containing concrete commitments could be fitted into the framework as scientific evidence requires and permits.

    Economic Declaration (statement of Group of Seven major industrialized nations in Paris, July 16, 1989), *reprinted in* 5 AM. U.J. INT'L. L. & POL'Y 571 (1990), *and in* 19 ENVTL. POL'Y & L. 183 (1989), *and in* 28 I.L.M. 1293 (1989), *and in* 25 WEEKLY COMP. PRES. DOC. 1101 (July 17, 1989). This commitment was reaffirmed at the next two G-7 summits, held in Houston and London in July 1990 and 1991, respectively. *See* Houston Economic Declaration (statement of Group of Seven major industrialized nations in Houston, July 11, 1990), *reprinted in* 26 WEEKLY COMP. PRES. DOC. 1073 (July 16, 1990), *and in* N.Y. TIMES, July 12, 1990, at A15; Economic Declaration: Building World Partnership (statement of Group of Seven major industrialized nations in London, July 17, 1991), *reprinted in* 27 WEEKLY COMP. PRES. DOC. 968 (July 22, 1991), *and in* N.Y. TIMES, July 18, 1991, at A12.
42. This conference of 68 environment ministers stressed the necessity for the adoption of a framework convention "as early as 1991 if possible and no later than at the Conference of the United Nations on Environment and

Development in 1992.'' In addition, the conference endorsed the ambitious goal of reversing deforestation to make forests a net sink for carbon by early in the next century, to be accomplished by ''[a] world net forest growth of 12 million hectares a year . . . .'' Noordwijk Declaration on Atmospheric Pollution and Climatic Change (statement of ministerial conference sponsored by government of the Netherlands in Noordwijk, Nov. 7, 1989), *reprinted in* 5 AM. U.J. INT'L L. & POL'Y 592 (1990), *and in* 19 ENVTL. POL'Y & L. 229 (1989), *and in* 12 Int'l Env't Rep. (BNA) 624 (1989).

43. Bergen Ministerial Declaration on Sustainable Development in the ECE Region (statement of ministerial conference sponsored by Government of Norway in Bergen, May 16, 1990), *reprinted in* 20 ENVTL. POL'Y & L. 100 (1990). This regional meeting, held under the auspices of the U.N. Economic Commission for Europe, was attended by representatives of 34 governments from Europe and North America and was an intermediate juncture between the release of the report of the World Commission on Environment and Development and the 1992 U.N. Conference on Environment and Development. The conference reaffirmed its

> full support for the early completion of the work on a framework convention on climate change and the development of protocols dealing with, *inter alia,* greenhouse gases and forestation, with a view to signing not later than at the 1992 Conference on Environment and Development.

*Id.*

44. SWCC Ministerial Declaration, *supra* note 32.
45. *See* 19 ENVTL. POL'Y & L. 45 (1989).
46. Helsinki Declaration on the Protection of the Ozone Layer (statement from first meeting of parties to Vienna Convention on the Protection of the Ozone Layer and Montreal Protocol on Substances that Deplete the Ozone Layer, May 2, 1989), *reprinted in* 5 AM. U.J. INT'L. L. & POL'Y 570 (1990), *and in* 12 Int'l Env't Rep. (BNA) 268 (1989), *and in* 28 I.L.M. 1335 (1989).
47. SWCC Ministerial Declaration, *supra* note 32.
48. *Id.* para. 10.
49. *Id.* para. 12.
50. *Id.* paras. 15, 17, 19.
51. Vienna Convention, *supra* note 4, art. 2.
52. See BENEDICK, *supra* note 5, ch. 5.
53. *Id.* ch. 6.
54. *Id.* at 74.
55. *See supra* part II.B-C.
56. A recent report issued by the National Academy of Sciences proposed the following policies for reducing $CO_2$ emissions from the U.S.: (1) encouraging improvements in automobile fuel efficiency through a combination of regulation and tax incentives; (2) nationwide building codes to improve energy efficiency; (3) strengthened mandatory appliance efficiency standards; (4) restructuring public utility regulation to encourage energy utilities to promote efficiency and conservation; (5) greater federal and state support for mass transit; (6) public education and information programs targeted at conservation and recycling; (7) increased federally-sponsored research and development of energy-efficient and energy-conserving technologies; and (8) utilization of federal procurement programs as demonstration projects for best-practice technologies and energy conservation programs. *See* NAT'L ACAD. OF SCI., *supra* note 12.
57. 40 C.F.R. pt. 762; 43 Fed. Reg. 11,318 (1978); 43 Fed. Reg. 11,301 (1978) (banning nonessential aerosol uses of CFCs).
58. A number of countries besides the U.S., including Canada and the Nordic nations, enacted similar controls on nonessential aerosol uses of CFCs. By contrast, the European Community established a limit, considerably above then-existing levels, on total production of CFCs. *See* THOMAS B. STOEL ET AL., FLUOROCARBON REGULATION ch. 4 (1980).
59. *See, e.g., id.* at 47–48 (1980) (discussing economic assessment of banning aerosol uses of CFCs); Daniel J. Dudek et al., *Cutting the Cost of Environmental Policy: Lessons from Business Response to CFC Regulation,* 19 AMBIO 324, 326 (1990) (aerosol ban in United States involved net saving to industry and public because prices of alternatives less than those of CFCs); John S. Hoffman, *Replacing CFCs: The Search for Alternatives,* 19 AMBIO 329, 331 (1990) (aerosol ban in U.S. involved net savings).
60. IPCC SCIENTIFIC ASSESSMENT, *supra* note 3, at 5, 18.
61. *See supra* p. 88.
62. *See* Toronto Conference Statement, *supra* note 37.
63. IPCC SCIENTIFIC ASSESSMENT, *supra* note 3, at 13, 17.
64. ''Sink'' refers to a mechanism resulting in the removal of $CO_2$ from the atmosphere.
65. *See, e.g.,* JOSHUA M. EPSTEIN & RAJ GUPTA, CONTROLLING THE GREENHOUSE EFFECT: FIVE GLOBAL REGIMES COMPARED (1990) (Brookings Occasional Papers Series); FLORENTIN KRAUSE ET AL., ENERGY POLICY IN THE GREENHOUSE (1989).

66. EPSTEIN & GUPTA, *supra* note 65, at 18–25.
67. This formula can be made consistent with the call for convergence at a common per capita emissions level by incorporating a transition to national allocations by a strictly per capita formula beginning in some future year, such as 2030.
68. Montreal Protocol *as amended, supra* note 5, art. 5.
69. *Id.* art. 10.
70. *Id.* art. 10.
71. *See generally Documents Concerning the Establishment of the Global Environment Facility,* 30 I.L.M. 1735 (1991).
72. The European Community, for example, is now considering a proposal for a tax on energy, a portion of which might be levied based upon carbon content. *See* E.C. Doc. SEC(91)1744 (Oct. 14, 1991) (Commission communication to Council concerning Community strategy to limit carbon dioxide emissions and to improve energy efficiency).
73. For example, the four multilateral development banks of which the U.S. is a member — the World Bank (including the International Bank for Reconstruction and Development, the International Development Association, and the International Finance Corporation), the African Development Bank, the Asian Development Bank, and the Inter-American Development Bank — approved about $5 to $6 billion per year in energy lending in the 1980s. Energy is the second largest lending sector at the World Bank. However, "[e]nd-use energy-efficiency has accounted for [only] about 1% of the World Bank's total energy lending since 1980 . . . ." MICHAEL PHILIPS, THE LEAST COST ENERGY PATH FOR DEVELOPING COUNTRIES: ENERGY EFFICIENT INVESTMENTS FOR THE MULTILATERAL DEVELOPMENT BANKS 52, 59 (1991).
74. *See, e.g.,* Pat Aufderheide & Bruce M. Rich, *Environmental Reform and the Multilateral Banks,* 5 WORLD POL'Y J. 301 (1988); Bruce M. Rich, *The Emperor's New Clothes: The World Bank and Environmental Reform,* 7 WORLD POL'Y J. 305, 323–29 (1990); Bruce M. Rich, *The Multilateral Development Banks, Environmental Policy, and the United States,* 12 ECOLOGY L.Q. 681, 741–45 (1985); David A. Wirth, *Legitimacy, Accountability, and Partnership: A Model for Advocacy on Third World Environmental Issues,* 100 YALE L.J. 2645 (1991).
75. *See generally* AMERICA'S CLIMATE CHANGE STRATEGY: AN ACTION AGENDA (1991) (U.S. comprehensive approach); A "COMPREHENSIVE" APPROACH TO ADDRESSING POTENTIAL GLOBAL CLIMATE CHANGE: REPORT OF THE TASK FORCE ON THE COMPREHENSIVE APPROACH TO CLIMATE CHANGE (1991) (report of interagency task force chaired by Department of Justice); A "COMPREHENSIVE" APPROACH TO ADDRESSING POTENTIAL GLOBAL CLIMATE CHANGE (1990) (discussion paper prepared by U.S. Government for IPCC plenary meeting); Gray & Rivkin, *supra* note 28.
76. *See, e.g.,* IPCC SCIENTIFIC ASSESSMENT, *supra* note 3, at 47–61; Lashof & Ahuja, *supra* note 7. Accounting for the indirect effect of each gas, however, is proving to be quite difficult. For example, it is now believed that the GWP of $NO_x$ given in IPCC Working Group I is wrong, and that the net GWP of CFCs should be lowered due to the cooling effect of the ozone depletion that they cause. WMO Ozone Assessment, *supra* note 7. The current uncertainty regarding GWP values is an additional reason to avoid intergas trading, at least for now.
77. Montreal Protocol *as amended, supra* note 5, annexes A-B. Carbon tetrachloride and methyl chloroform are each regulated separately, and no trading between them and other chemicals is permitted. *See supra* note 6.
78. *Id.* art. 2, para. 5. *Cf. id.* art. 1, para. 8 (defining "industrial rationalization" as "the transfer of all or a portion of the calculated level of production of one Party to another, for the purpose of achieving economic efficiencies or responding to anticipated shortfalls in supply as a result of plant closures.") *See* 40 C.F.R. § 82.9(b)-(c) (1991) (allowing increases in production allowances upon documentation of corresponding offset by another Protocol party); 56 Fed. Reg. 49,548 (1991) (notice of proposed rulemaking to conform EPA regulations to 1990 Protocol amendments and stratospheric ozone provisions of Clean Air Act Amendments of 1990, Pub. L. No. 101–549, §§ 601 & 602, 104 Stat. 2399, 2648–70 (1990), 42 U.S.C.S. §§ 7671–7671q (Law. Co-op. Supp. June 1991). *Cf.* Clean Air Act Amendments of 1990, Publ. L. No. 101–549, § 401, 104 Stat. 2399, 2584–2631 (1990), 42 U.S.C.S. §§ 7651–7651o (Law. Co-op Supp. June 1991) (adding acid rain control program allowing trading of emissions allocations among sources to Clean Air Act). With certain exceptions for extremely low-producing countries, the original 1987 agreement limited the level of transferred production to 15% of a country's weighted 1986 production.
79. *See, e.g.,* DAVID G. VICTOR, TRADEABLE PERMITS AND GREENHOUSE GAS REDUCTIONS: SOME ISSUES FOR U.S. NEGOTIATORS (1990) (John F. Kennedy School of Government, Harvard University, discussion paper No. G-90-06).
80. *See* Report of the Second Meeting of the Parties to the Montreal Protocol on Substances that Deplete the Ozone Layer, UNEP/OzL.Pro.2/3, Annex III (June 1990) (establishing five-member Implementation Committee to rule on cases of asserted failure to implement Montreal Protocol), *reprinted in* 1 Y.B. INT'L ENVTL. L. 591 (1990).

81. *See, e.g., Developments in the Law — International Environmental Law,* 104 HARV. L. REV. 1484, 1521–50 (1991).
82. *See* PETER H. SAND, LESSONS LEARNED IN GLOBAL ENVIRONMENTAL GOVERNANCE 14–15 (1990).
83. Declaration of the Hague, *supra* note 40.
84. *See* Vienna Convention on the Law of Treaties, May 22, 1969, art. 40, para. 4, 1155 U.N.T.S. 331, *reprinted in* 63 AM. J. INT'L L. 875 (1969), *and in* 8 I.L.M. 679 (1969). This instrument, although not in force for U.S., has been accepted by the Executive Branch as a codification of customary international law regarding international agreements. *See* 1 RESTATEMENT (THIRD) OF THE FOREIGN RELATIONS LAW OF THE UNITED STATES pt. III, introductory note (1987).
85. Montreal Protocol, *supra* note 5, art. 2, para.9.
86. U.S. experience with the Montreal Protocol compellingly demonstrates the need for the creation of incentives for additional national measures beyond the international minimum. In response to the argument of some commenters that the Clean Air Act contained more demanding requirements for the regulation of ozone-depleting chemicals than the Montreal Protocol, the Environmental Protection Agency, in its final regulation implementing the Montreal Protocol, interpreted the international instrument as establishing not only an international minimum level for national regulations, but also, based on prudential and strategic considerations, an affirmative limitation on the stringency of domestic measures. 53 Fed. Reg. 30,566, 30,569, 30,573–74 (1988) (final regulation implementing Montreal Protocol). This position of the Executive Branch has now been reversed by Congress, which has enacted legislation regulating ozone-depleting chemicals more strictly than the Montreal Protocol by requiring the following: (1) a larger number of intermediate reduction steps; (2) a phase-out of some alternatives to substances controlled in the Montreal Protocol; (3) the introduction of a recycling program; and (4) an additional requirement, not found in the Montreal, Protocol, specifying that substitutes for substances controlled by the Montreal Protocol must be environmentally benign. Clean Air Act Amendments of 1990, Pub. L. No. 101–549, §§ 601–602, 104 Stat.2399, 2648–70 (1990), 42 U.S.C.S. §§ 7671–7671q (Law. Co-op Supp. June 1991) (adding new §§ 601–618).
87. WILLIAM SHAKESPEARE, HENRY VI, Part I, act III, sc. 2.

Chapter 26

# THE DEVELOPING LAW OF THE ATMOSPHERE AND THE 1992 RIO DE JANEIRO CONVENTION

**Justin Lancaster**

## TABLE OF CONTENTS



0-8493-4419-0/93/$0.00 + $.50
© 1993 by CRC Press, Inc.

# I. INTRODUCTION

The potential impact of global warming on society began to be addressed earnestly by researchers in the 1970s.[1] This attention expanded dramatically in the 1980s[2] with social impacts of climate change receiving intensified scrutiny by the National Academy of Sciences[3] and within governmental agencies.[4]

During the past three years, strong international pressure has developed to agree upon steps to reduce the future threat of potential climate change. A Framework Convention on Climate Change was presented for signature at the United Nations Conference on Environment and Development, convened in Rio de Janeiro, Brazil, in June of 1992. The U.S., while an active leader in global warming research and international dialogue on the subject, did not lead the charge for agreement to strongly limit emissions of greenhouse gases — in fact, the U.S. position bordered on being recalcitrance.

In this article I argue that an early international agreement to reduce greenhouse gas emissions is desirable because it may forestall catastrophic effects, and it may encourage greater energy efficiency in human technology and social practice, which is beneficial in its own right. I stress that economic cost-benefit analysis cannot value the impact of environmental surprise and that the risk of surprise has yet to be assessed adequately. I suggest that the scope of the agreement should be broadened to explicitly treat biogeochemical cycles, forests, and climate as three equal parts of the global management problem. I examine the history of international cooperation leading up to the Convention with brief discussion of some alternate paths rejected. I discuss some of the specific protocols being considered for inclusion in the Convention, including the controversial carbon tax, as well as provisions for institutional setting, enforcement, and ongoing review. I suggest that the convention might have effectively employed UNEP as the Secretariat rather than initiate a new international institution. Finally, I briefly describe the history of the U.S. role in this arena and some of our own legislative developments, followed by some comments on the strongly differing situation of developing countries.

# II. IMMEDIATE NEED FOR A CONVENTION

Early agreement to reduce greenhouse gas emissions is urged by at least three considerations. First, more rapid and complex change could occur than has been previously anticipated by the "consensus" view. Second, current scientific uncertainty prevents states from yet knowing who will be "winners"; thus, consensus may be more easily achieved today than decades hence. Third, a fundamental need exists to regulate energy use.

On the other hand, many factors inhibit early agreement on global warming response. Just as $CO_2$ itself is invisible and odorless, so the risk of permanently altering world climate eludes normal human sensitivity. The greenhouse problem is global and it spans generations. It relates directly to a natural, historical pattern of energy consumption. Its environmental costs and benefits are difficult to analyze in light of its scientific complexity. The potential economic costs of some responses seem enormous.[5]

## A. THE PROBLEM: ENERGY, BIOGEOCHEMISTRY, AND GLOBAL ENVIRONMENTAL CHANGE

Global environmental change is human society altering the Earth's biogeochemical cycles.[6] Deforestation is a key aspect of this change as it alters carbon and water cycles. Another geochemical perturbation, infusion of halocarbons into the stratosphere,[7] threatens the ozone layer, an impact unforeseen before 1973.[8] Spewing carbon dioxide into the atmosphere threatens global warming, while adding nitrogen and sulfur oxides into the troposphere threatens lakes and forests. Global environmental change is caused by a growing population rapidly consuming energy to drive an accelerating technology. Cutting forests, changing agricultural techniques, and manufacturing modern chemicals are merely a few of the multiple paths by which energy expenditure is stressing the global environment.

Global warming may provide severe ecological and societal stresses, particularly in the coastal zone.[9] Combining sea level rise with other direct consequences of warming could increase the impact of coastal storms.[10] Other impacts could present concerns for public health and resources, particularly when considered in combination with ozone depletion.[11] In terms of direct health effects, for instance, summer deaths owing to heat waves, even with full acclimatization, are expected to exceed the estimated decline in winter deaths.[12] Available fresh water will be at greater risk where supplies are already rationed.[13]

How bad is the worst-case scenario? We do not know; this is a key point. We cannot, therefore, assess its probability. Nor can we begin to describe economic costs of potential damage under the worst-case scenario. Also, thresholds may exist beyond which corrective actions become ineffectual. We do know that feedbacks between the atmosphere and biosphere are nonlinear, sensitive to initial conditions, and capable of enormous amplifications. The Earth's climate system is a fluid, chemical system, ''poised'' by an energetic throughflow. The biosphere, also, is a *dynamic* chemical structure balanced within the climate system. The human component of the biosphere, more than any other, has perched itself atop a specific, supporting pathway of energy flow, i.e., consumption of fossil fuels.[17]

We know that complex feedbacks in the Earth's system can produce unexpected and potent responses.[18] Possible biogeochemical feedbacks could exacerbate the effects of forest loss and ozone depletion,[19] and this could result in a more rapid global warming than is predicted by purely physical models.[20] When these feedbacks are incorporated into climate models, the most likely global mean temperature increase (resulting from an initial forcing equivalent to a doubling of $CO_2$) becomes greater than 6°C.[21] It already appears that global warming could outpace the ability for some tree species to survive in some regions,[22] and more complex interactions with the problem of ozone depletion may emerge. A potential contribution to the stratospheric ozone problem alone could warrant controlling atmospheric $CO_2$ emissions.[23] CFCs directly affect global warming as absorbers of infrared energy, but it is possible for systemic feedbacks to provide indirect effects, as well.[24] It is too soon to tell what feedbacks will occur, but governments might sensibly choose to avoid the risk that feedbacks could be more detrimental than we now estimate. If we need not move blindly into a doubled $CO_2$ future, as those who urge energy-efficient technologies claim, then we should not want to, even with uncertainty about global warming impacts, because significant reasons other than just the threat of climate change urge our restraint. Such burning, for instance, has created air pollution and acid deposition, problems that need solutions before being made worse. Two other aspects of our fossil fuel use, however, not frequently discussed, may be even more critical.

First, as a global society, eventually we must grip the fact that our luxury fuels, i.e., oil and gas, which represent a billion-year heritage of ancestral solar energy storage, are finite resources. A shift away from fossil fuels may become more difficult each decade the shift is postponed, because we will be forced to burn increasing amounts of coal until alternative sources fill the gap.[25] We will need a good portion of these fossil energies just to build the alternative energy technologies and infrastructure, i.e., nuclear, solar, wind, tidal, and biomass. The amount of our shrinking, finite resource that is needed for a transition to a long-term energy supply has not been calculated. Future generations will shake their heads in disbelief at the ease and rapidity with which we will have spent our fossil treasure. We should take every step we can to conserve it.

Second, our momentous growth affects social stability. The rapid intensification of energy use in human society has changed social structure, and it might well create further change faster than our ability to cope.[26] A structure dependent on steadily increasing energy flow becomes vulnerable to major disruptions in energy supply. We must establish throttles on the increasing flows of energy through society that are designed not for profit, which favors acceleration, but for long-term stability and public benefit.

Society's rampant energy use is caused by a fundamentally natural growth principle — an energy-technology feedback. Our environmental problems are intimately intertwined with this mutually reinforcing dynamic between energy acquisition, industrial metabolism and technological power. This feedback, which bounds social evolution and governmental control, impacts political systems and stability more directly than through its environmental manifestations alone.[27] Wars over energy, for instance, are a more immediate threat to human health and welfare than global warming. It is not just the uncertain extent of climate change, then, that presents risks that are difficult to assess. It is important to realize that changing energy use and perturbing global biogeochemical cycles pose their own serious concerns for humanity, separate from their relevance to global warming.[16] Even change within predicted trends could trigger shifts in biogeochemical and energy processes that could, particularly if compounded by internal societal stresses, present us with severe challenges in the long term.

On balance, it is time to work for international agreement on response strategies. The question then becomes, "how great is the unknown risk, to warrant how great a response cost"? The dilemma seems to be that at least the first part of the question cannot yet be answered. Having raised the seriousness of the problems involving energy, biogeochemistry, and global environmental change, I will now turn to the effect that uncertainties have on policy formulation.

## B. SCIENTIFIC UNCERTAINTY AFFECTING POLICY

Countries will be reluctant to limit economic expansion because of uncertain warnings about climate change. Recently, as they struggle to evaluate the appropriateness of various response strategies, economists have grappled with the problem that impacts from global warming are uncertain. There is no solid basis for believing, some argue, that a modest degree of warming would be disadvantageous for the world as a whole.[28] Uncertainty does not mean that there is not a stable scientific consensus. The estimates of the extent and rate of global warming have not fluctuated significantly over the past decade, remaining in the range of 1.5 to 4.5°C by roughly the year 2035, with a "best guess" increase of about 2.5 degrees.[29] Most scientists agree that over the past 100 years global climate has warmed by about 0.5°C.[30] However, a lively controversy has emerged about whether this warming can be connected to the theoretical greenhouse forcing. Also, computer modeling of future climate cannot provide reliable predictions about regional impacts. These uncertainties are being urged as support for a policy of deferring responses to limit $CO_2$ emissions.[31]

Strategic considerations can weigh heavily in the response decision. It can be argued that waiting until the data show which countries will be more severely impacted would allow more appropriate controls and adaptive measures to be tailored most effectively. Some warn of the potential danger that nations could set a limit on $CO_2$ emissions only to have subsequent research show that higher $CO_2$ could be beneficial.[32] Others argue that if the change in climate will occur so slowly that we can easily adapt, then there is no pressing need for a growth-restricting response.

In 1987, Task Teams on Implications of Climate Change were established for six regions covered by the UNEP Regional Seas Programme and a regional impact assessment has been completed for the Mediterranean region.[14] Impacts of sea level rise for the Pacific Islands are potentially quite profound, particularly for the Marshall Islands, Tuvalu, and Kiribati, with severe impacts also forecast for Tonga, Cook Islands, and French Polynesia.[15]

In one sense, present scientific uncertainty may encourage early intergovernmental agreement, because it is likely to make negotiations less biased now, with individual nations unable to exactly calculate their positions.[33] This could make it more likely for countries to consent to common goals. The more that $CO_2$ levels increase, and the closer we come to seeing which countries lose and which benefit, the more difficult it will be to get those who benefit to enter negotiations; the political trade-offs will become increasingly costly. A sensible approach would be to use the uncertainty that exists now to push for a limiting scheme that will spread the cost and the impact over many nations who later might be less willing to join an agreement. The choice and timing of these responses must, to a large degree, be based on social value judgment, rather than any objective measure of scientific certainty.[34] The debate is really over how, given all the uncertainties, we should respond to the possibility of climate change. Beyond establishing the facts and assessing the uncertainties, science can contribute relatively little.[35]

The economist tends to address environmental problems such as these with a cost/benefit analysis.[36] Her typical approach is to minimize the sum of the costs of abatement and the value of the harm,[37] choosing optimal pollution levels where further abatement would cost the polluter more than the victim would otherwise pay for a purer resource.[38] This approach favors the industrial value of a resource over its ecological or aesthetic value, because the latter are so often undeterminable. This problem is compounded by human tolerance for a degraded environment; we tend to accept tons of industrial wastes from factories that create intentionally short-lived machines, or containers to be sold half-empty, or other products to satisfy wants and whims, rather than needs and health. The aesthetic and ecological values of the public natural environment are simply not easily valued in the marketplace. William Nordhaus has put it succinctly:

> "The threat of an unforeseen calamity argues for a more aggressive action than a plain-vanilla cost-benefit analysis would suggest . . . The cost-benefit analysis ignores uncertainty, instead following a 'best guess' analysis, appropriate so long as risks are symmetrical and when uncertainties are unlikely to be resolved in the forseeable future. Neither fits the greenhouse problem. Risks are asymmetrical — the larger the change the larger the aversion . . . Adding stresses to an already stressed, and unknown, system poses tremendous risk."[39]

It is critical to distinguish surprises from foreseeable, low-frequency catastrophes. Both are within the realm of uncertainty. Surprises, such as CFCs attacking ozone and DDT thinning pelican egg shells, are unforeseeable. Foreseeable catastrophes include Bhopal, Chernobyl, and the *Exxon Valdez,* and although the moment and location are not exactly predictable in cases such as these (including aircraft disasters), rough probabilities can be determined for their occurrence and worst-case scenarios formulated. Cost-benefit analysis can sometimes provide some guidance in the foreseeable cases. Surprises, however, cannot be evaluated, so that an economic analysis cannot provide a rational decision about increasing or reducing exposure to their risk. A better decision may then be based on some instinct or emotion, e.g., caution or fear.

Given our ignorance of the biogeochemical feedbacks in the Earth system, and given the recent surprise in the Antarctic ozone hole, the non-market cost of impoverishing the biosphere by increasing multiple stress factors cannot be reasonably estimated. Nonetheless, economists have estimated damage costs for some climate change impact scenarios and then balanced these against projected costs of response strategies. Forgiving for a moment the potential invalidity of assessing the cost of harm from global warming, we can briefly discuss uncertainty in terms of the economics associated with potential response options.

## C. UNCERTAIN ECONOMICS OF RESPONSE OPTIONS

Potential policy responses to global warming include further research, adaptation, intervention (engineering countermeasures),[40] and preventive limitation. The costs of research and some programs of intervention are fairly predictable, but a good deal of uncertainty surrounds the options for adaptation and preventive limitation. Adaptation costs are uncertain for the most part because the extent of the environmental change is so unknown. Adaptation to a 3°C global warming will be dramatically greater than to a 1° warming. Still, estimates have been made based on the "best guess" impact scenario. Nordhaus, for one, believes that very few economic activities are dependent on the weather or climate, these being primarily agriculture, fisheries and forestry. A 20% decline in agriculture, he argues, would be less than 1% of GNP, and it is unlikely that impacts on health would equal impacts of job exchange, eating, smoking, and drinking.[41] He points out that adaptation as a strategy can be automatic, market effected, and regulated. It has the advantage of being able to wait for many years. The time scale of adaptations is much less than the time scale of climate change,[42] thus we need to begin adaptation now only if a long lead time is needed, i.e., if a penalty exists for delay and if it would be an economical step anyway.

Much of the focus of a global convention to respond to climate change is on the costs of limiting emissions of greenhouse gases.[43] Balancing abatement against adaptation, Thomas Schelling argues that controlling the emissions of fossil fuels to protect the atmosphere would be a cost for any nation over the short term, so that "no nation will find it in the purely national interest to engage on its own in substantial abatement."[44] The assumption made is that abatement will have to be induced by taxing fuels in order to inhibit use, with the consequently higher efficiency costs for transportation and industry putting a damper on GNP. Nordhaus concludes that changing climate makes no difference for the developing countries, and thus the international effort will not lead to serious greenhouse abatement commitments but will become absorbed in institutional arrangements.[45] Given this, the only plausible steps are pursuing national undertakings to reduce emissions, and financing efforts to reduce emissions later in developing countries.[46]

In formulating a response strategy, Nordhaus, balancing preventive cost vs. damage caused without prevention, determines that costs at \$10 to \$50 per ton $CO_2$ equivalent are manageable (i.e., 0.5 to 2.5% of global income). A 10% reduction in emission would cost about \$10/ton, or \$6 billion annually for a 0.6 billion ton reduction, whereas a 50% reduction would cost about \$130/ton, or \$390 billion annually (in taxes, but about \$180 billion annually in resource cost, or about 1% of global output).[47] He suggests that reducing CFC emissions at less than \$5/ton $CO_2$ equivalent is more cost-effective, as would be reforestation with a total cost for a 15% reduction costing \$6 billion/yr, or \$4/ton C equivalent.[48] He stresses that a gradual phase-in and efficient design of response policy is worth 0.3 to 0.5% in economic growth per year.[49] Nordhaus would promote a healthy economy, internalize external costs, broaden market scope to communicate signals of scarcity, and ensure high savings rate to provide investment in infrastructural change. He argues that an efficient policy response framework would: (1) improve knowledge; (2) develop new technologies for low GHG/energy output (safe nuclear, solar and conservation), carbon sequestration, and climate engineering; (3) initiate "all-weather" policies, such as restricting CFCs, curbing deforestation, and increasing fuel efficiencies; and (4) impose a penalty on greenhouse gas emissions of around \$5 per ton carbon equivalent, to produce about 12% reduction. This results in a \$3.50 per ton tax on coal, \$0.58 per barrel on oil, and 1.4 cents per gallon of gasoline.[50]

The problem with Nordhaus's approach is that this carbon tax is not sufficient to produce a meaningful reduction. Perhaps there is some middle ground. Three countries have already imposed carbon taxes: Finland (\$6.50/ton C), Sweden (\$62/ton) and the Netherlands (\$1.50/ton).[51] Recommending a tax of \$50 per ton carbon equivalent, producing almost a 25%

reduction from 1989 levels by year 2000, might be politically acceptable, being about a $35 per ton coal, $5.80 per barrel oil, and 14 cent rise in gas price (in 1990 we saw the public easily accept a 10–20 cent increase in gas prices owing to the Kuwait crisis, with minor protests at the 40 to 50 cent increases). Although Schelling calls this level of cost politically unmanageable, with warnings about equitable distribution of costs,[52] annual U.S. revenues of $100 billion could allow reduction in capital gains taxes ( − 20 billion) and income taxes ( − 20 billion), while augmenting, by $5 billion each support for infectious disease research, general health care, agricultural development, education, public transportation, justice system, urban renewal, basic science, space program, the arts, and environmental protection (including $1 billion each to scientific research, regulatory monitoring, enforcement, cleanup, and support for clean, efficient technologies). This still leaves $5 billion annually to administer the tax, reduce the deficit or augment foreign aid. The pressure for higher efficiency would drive technology and consumer patterns toward conservation, preserving energy resources for future wiser generations.

The carbon tax presents many complexities. Thus far, it has been envisioned as a *specific* tax, i.e., fixed per weight of carbon in fuel produced, rather than an *ad valorem* tax, i.e., based on value or price at consumer end.[53] The precise form of effects depends on disbursement design, whether consumption-based or production-based, and upon elasticity parameters used in models. A global carbon tax base for 1990 to 2030 could be as high as $43 trillion, or about 10% of the projected gross world product (GWP) for that 40 year period. If a portion of this tax base flows to developing countries, then their participation in the scheme is likely.[54] Other problems involve whether revenues should be nationally or internationally collected, and whether per capita emission targets or proportional emissions cuts should be employed. Per capita targets, for instance, may cause relocations of industry to the developing countries. Also, incidence effects of trade in oil products dominate those of downstream products.[55]

A carbon tax may or may not be regressive. Because heating fuel, electricity and gasoline are necessities, the distributional burden falls on low-income groups. The extent is diminished by using, for the basis of the computation, total household expenditure (lowest spending households pay for energy roughly 2 times greater percentage of total expenditures than do highest spending households) versus income ranking (lowest incomes pay for energy 4 to 6 times greater percentage of income than do highest income households).[56] It has been argued that a carbon tax should not exceed the equivalent tax on CFCs.[57] Note, however, that phase-out of CFC's already has been agreed to by a rather small group of industries, with the tax not being necessary to force the phase-out. Comparability with a carbon tax here is not necessary, because these are distinct products, no shifting of use would occur from one to the other because of tax rates differing.

We can describe multiple benefits deriving from a tax on carbon. The tax would:

1. Reduce $CO_2$ emissions and thereby slow global warming
2. Reduce $SO_x$ and $NO_x$, by reducing fuel emissions and shifting use away from coal, thereby reducing acid deposition
3. Encourage use of renewable energy sources and efficient technologies, thereby conserving hydrocarbon stock for future generations.
4. Provide control over the energy-technology feedback, thereby moving governments closer to long-term stability.

Replacement technologies and reduction technologies are becoming available to promise make major inroads against carbon emissions,[58] but many of these steps, although more efficient in producing electricity from fuel, require enormous capital expenditures to put in place, which initially must be considered a cost that damps economic growth.

The cost to the U.S. economy, or to the global economy, of reducing carbon emissions is quite uncertain. The above estimates by Nordhaus and Schelling are strongly contested by Amory Lovins, who argues that lowering of $CO_2$ emissions can be achieved at savings to the national economy, merely by making available to consumers technologies that are state-of-the-art in energy efficiency. Lovins argues that it is cheaper today to save fuel than to burn it.[59] The pollution avoided by not burning the fuel can, therefore, be achieved not at a cost but a profit. He says that if the U.S. economy were to strive to meet the Toronto $CO_2$ goals through promoting energy efficiency gains, $200 billion would be saved.[60] By year 2005, world primary energy requirements are expected to increase by about 55% over 1987 levels.[61] By choosing, or encouraging, most-efficient technologies, an emerging consensus of economists at the recent 1990 Dahlem Workshop in Berlin have said that 50% reductions in world energy use is attainable. Consumer behavior, however, may still tend toward increased energy use as leisure time increases.[62]

Clearly, further policy research on the costs and benefits of meeting an emissions limitation target is needed. Still, it appears that most economists would be comfortable with a $5 to 10/ton C tax, if for no other reason than to send a signal to encourage conservation.

Now, I will leave the turmoil of uncertainty in science and economics to look at a very certain record of international cooperation in environmental matters, beginning many decades ago amongst scientists, and accelerating recently as international lawyers have molded common concerns into protective legal instruments.

# III. INTERNATIONAL COOPERATION IN ENVIRONMENTAL PROTECTION

## A. HISTORICAL SCIENTIFIC COOPERATION

In the 1700s, inspired by Francis Bacon teaching that cooperative science was a path to understanding nature, international programs began to emerge, a landmark being the 1751 venture to track the transit of the planet Venus. Within 30 years, a meteorological observing network was established from Greenland to Russia, published in the *Ephemerides Societatis Meteorologico Palitinae*. In 1875, an Australian geologist named Edward Suess coined the term "biosphere". Seven years later came the First International Polar Year, with 24 countries participating to make basic meteorological and astronomical measurements in the arctic region.

In the 20th century, the Second Polar Year involved 40 countries and made much more extensive investigations of solar activity and magnetics in Antarctica. During the late 1950s, more than 70 countries participated in the International Geophysical Year, including 20,000 scientists in the work. In the 1980s, there was a tremendous explosion of international science to study the global environment. The International Biosphere Program (IBP), the Man and the Biosphere (MAB), the International Decade of Ocean Exploration (IDOE), and the International Geosphere-Biosphere Program (IGBP) are the most recent expressions of a long track record of international cooperation in science, and toward developing a focus on the entire global system.

The international organizations involved with scientific aspects of $CO_2$ and global warming are numerous. The United Nations Environment Program (UNEP), the World Meteorological Organization (WMO) and the International Council of Scientific Unions (ICSU) have emerged as the three lead agencies. The Scientific Committee on Problems of the Environment (SCOPE), under ICSU, has published invaluable references on the biosphere,[63] owing much to Bert Bolin from Sweden, who is now chair of the Intergovernmental Panel on Climate Change (IPCC). The Committee for Climate Change in the Oceans (CCCO) is an outgrowth of the Scientific Committee for Ocean Research (SCOR), and both are involved in international collaborative science concerning global warming.

The modern focus on "global change" burst onto the scene as a result of an ICSU conference in 1983, where plans were being laid for the IGBP.[64] Through the 1980s global warming and global change crept from the realm of science into the social limelight. To a large extent the history of international recognition of the global warming problem and the beginnings of consensus-building culminating in the 1990 IPCC Report, reach back to the formation of the early international scientific associations, such as the Intergovernmental Oceanographic Commission and the International Council of Scientific Unions. This development was quite intended.[65]

## B. PREVIOUS INTERNATIONAL ENVIRONMENTAL LEGAL REGIMES

One of the earliest international environmental agreements was between Canada and the U.S. concerning migratory birds.[66] International environmental legal regimes[67] have been adopted to control law of the oceans (UNCLOS), long-range transboundary air pollution (CL-RTAP), monitoring of the stratosphere (TAMS), and protection of the ozone layer (CPOL). These precedents provide differing models for a developing Law of the Atmosphere. The U.N. Convention on the Law of the Sea (UNCLOS) was negotiated over a period of more than twenty years and involved nearly all countries. That Convention establishes an ongoing international regime with jurisdiction over issues concerning the high seas, the sea bed, straits, and aspects of environmental protection. Why not follow a parallel course to create a Law of the Atmosphere?

A number of distinctions can be drawn between ocean issues and global warming that make UNCLOS a less than ideal model for reaching that agreement on trace gases. While preventing pollution of the atmosphere obviously concerns anybody who breathes, the complexity of uses for the atmosphere does not approach the multiple jurisdictional and use problems comprising ocean policy. And, while it could be argued that more countries will emit trace gases than will exploit the oceans, so that a large conference is even more necessary for trace gases, the counterargument is that emissions from many developing countries will remain insignificant for many decades. The technical aspects of the climate warming problem should not be more difficult to resolve than the question of ocean exploitation, particularly because a large part of the $CO_2$ problem has already received careful review by an international body, namely the IPCC.

There is one reason why a comprehensive Law of the Atmosphere regime should be easier than UNCLOS to complete quickly. The comprehensive UNCLOS agreement had none of the sense of urgency driving it forward that surrounds the global warming issue. No threats to public welfare existed if the UNCLOS agreement were to fail. A Law of the Atmosphere, on the other hand, is compelled by a deep sense of concern on the part of some countries that significant environmental threats are looming — inundation of the Maldive Islands, for instance. On the other hand, there are also confounding impediments in the atmospheric negotiation that are much more serious than those faced by UNCLOS. Countries were attracted to the evolving ocean regime because it endorsed significant expansion of state territorial claims, without much sacrifice, if any, for a state choosing to sign. A climate protocol that attempts to limit carbon emissions will ask potential sacrifices of all signatories (unless one is convinced that energy efficient technologies will keep everyone a "winner"). While developing countries may stand to gain from The Global Environmental Facility Fund (GEF), the industrialized nations gain nothing economically by signing.[68]

A more important consideration, perhaps, than the character of the negotiating process, is the character of the new emerging structure. The new ocean regime may become necessary to license and tax exploiters of the sea's resources beyond national jurisdiction. Such a structure should not be necessary in the case of trace gas emissions, however, because existing national agencies already have in place legal structures to control industrial pollution.

The Convention on Long-Range Transboundary Air Pollution (CL-RTAP) is another potential model for trace gas agreement. Initiated by the Scandinavian countries in 1972, and signed into effect in 1979, the convention defines air pollution as . . . "the introduction by man, directly or indirectly, of substances or energy into the air resulting in deleterious effects of such a nature as to endanger human health, harm living resources, eco-systems and material property, and impair or interfere with amenities and other legitimate uses of the environment" . . . and it defines transboundary air pollution to be . . . "air pollution of one state which has adverse effects on another state at such a distance that it is generally possible to distinguish the contribution of individual emission sources or groups of sources."

CL-RTAP was designed to cover $S0_2$, one of the chemical culprits leading to acid deposition. Historically, industrially-produced $CO_2$ has not been classed as a pollutant; but, if research can demonstrate that it and other trace gases result in deleterious effects on human health, then one can imagine fitting these gases under the definition of long-range transboundary air pollutants. CL-RTAP provides for air quality management, exchange of information and implementation of a monitoring and evaluation program,[69] but the convention does not provide a timetable, nor any mechanism for enforcement, and therefore it is really more of a pledge and a political commitment than a controlling agreement. CL-RTAP was formulated under the Economic Commission for Europe (ECE). Their pollution monitoring program, Earthwatch, was formed in January, 1978, in collaboration with UNEP and the WMO.[70]

The Tripartie Agreement on Monitoring the Stratosphere (TAMS) was signed in 1976 by the U.S., U.K. and France.[71] In its analysis and reporting, TAMS involves the World Meteorological Organization (WMO) and the United Nations Environment Program (UNEP). TAMS provides a good model for integrating a trilateral agreement with U.N. agencies, but, as will be discussed below, the climate convention is not limited to so few parties.

More recently, the 1985 Vienna Convention for the Protection of the Ozone Layer is perhaps now the best model of an atmospheric protection regime upon which to base an upcoming climate protocol.[72] Four years after Rowland and Molina discovered the threat that CFCs pose to stratospheric ozone, UNEP convened a meeting to adopt a "World Plan of Action on the Ozone Layer," and created an international Coordinating Committee on the Ozone Layer (CCOL). When the Vienna Convention was signed in 1985 the parties had essentially failed to agree on the amount and type of future controls that should be imposed on CFC production and use. Thus, the Conference committed to developing such controls in a future Protocol.[73] The surprise discovery of the Antarctic ozone hole induced agreement on the Montreal Protocol on Substances that Deplete the Ozone Layer only two years later, in September 1987.[74]

The Vienna Convention/Montreal Protocol model, that of umbrella framework followed by specific enforcing provisions later is the path toward which the Climate Convention is moving. The attractiveness of this model is that it made it much more likely that an agreement would be signed in 1992, albeit a rather vacuous one, giving strong direction to an ongoing negotiation for workable protocols with teeth. Trying to incorporate stringent provisions in the 1992 Convention would have doomed it to failure, as time was too short for the negotiations to close the considerable gap in philosophy exhibited.

It should be mentioned that another international body, the Organization for Economic Cooperation and Development (OECD), has been actively involved with the $CO_2$ issue. OECD has addressed the area of transboundary pollution, conducting specific studies of $S0_2$ and mercury in 1973 and 1974, respectively. A decade ago, at its 1978 environment committee meeting, the OECD recognized the potential concern of the $CO_2$ build-up from coal production.[75] While neither the ECE nor OECD would provide an appropriate forum for advancing the trace gas agreement, it is conceivable that these organizations could play some formal role in an ongoing management regime.

## C. STEPPING TOWARD THE CLIMATE CONVENTION

Shortly after the 1977 publication of *Energy and Climate,* by the U.S. National Research Council, and the 1978 OECD environment committee meeting, The United Nations Environment Programme (UNEP), the World Meteorological Organization (WMO) and the International Council of Scientific Unions (ICSU) hosted the First World Climate Conference in 1979, followed by a first joint conference on $CO_2$ in Austria in November 1980.[76] These organizations traditionally have shared the study of the $CO_2$ problem, with WMO and ICSU emphasizing the scientific aspects and UNEP evaluating the potential impacts on society. ICSU formed the Scientific Committee on Problems of the Environment (SCOPE), which has focused intensively on carbon-cycle research. More recently, this triad carried its joint forum further toward an early agreement on $CO_2$.

International consensus on global warming emerged at a joint conference of WMO, UNEP, and ICSU held in Villach, Austria in October, 1985, following the 1983 publication of *Changing Climate* by the U.S. National Academy of Sciences and the 1985 EPA report *Can We Stop a Greenhouse Warming*? In response to the recommendations of this joint conference, two international workshops for Developing Policies for Responding to Future Climatic Change were held in 1987. The first 1987 workshop, again in Villach, examined how climatic change could affect various regions of the Earth and explored the technical, financial, and institutional options available to cope with global warming. The second 1987 workshop, in Bellagio, Italy, further explored policy steps and institutional arrangements that should be considered in the near term. The policy recommendations included working toward a Law of the Atmosphere as a global commons, and developing strategies to achieve limitation of emissions, adaptation to warming, and institutional changes necessary for an effective response.[77]

In late October, 1987, between the two international workshops mentioned above, the First North American Conference on Preparing for Climate Change was held in Washington, D.C., sponsored by the Climate Institute, to address scientific and political implications of global warming and ozone depletion. Here, Howard Ferguson, head of Canada's Atmospheric Environment Service, restated the need for a ''Global Law of the Atmosphere''.[78] In the spring of 1988, James Hansen of NASA presented his now famous temperature curve, showing that global mean temperatures had increased about 0.6°C over the past 100 years, with four of the warmest years on record being in the 1980s. He stated that this was not within the realm of interannual or decadal noise within 99% confidence limits, which created quite a stir, particularly a few months later as a major U.S. drought hit that summer. At the June 1988 Toronto Conference on global warming, co-sponsored by UNEP and Canada, Ferguson pressed again for a global warming convention, and officials resolved to work for a 20% reduction in fossil fuel use by year 2005. This Conference called upon governments to work toward an *Action Plan for the Protection of the Atmosphere* and to establish a *World Atmosphere Fund,* essentially a tax on fossil fuel consumption to pay for measures to protect the atmosphere. The Conference urged pursuing an atmospheric convention at upcoming global meetings, including the Second World Climate Conference in Geneva, June 1990, with a view to having principles ready for consideration at the U.N.-sponsored Intergovernmental Conference on Sustainable Development slated for 1992.[79] A media splash hit *Newsweek* and *Time* and brought the greenhouse effect into the public mind where it had not been previously. Instead of person of the year in January 1989, *Time* carried a front page titled ''Planet of the Year'', showing the Earth wrapped in a plastic cover. This public awareness quickly transformed into a national pressure.

Following the Toronto conference, another international workshop was convened in September, 1988, by the Woods Hole Research Center, to consider and clarify appropriate steps toward an International Convention Stabilizing the Composition of the Atmosphere. Proposed measures included shifting from fossil fuels to solar energy, preserving the Earth's remaining primary forests, reforestation and implementation of the Montreal CFC Protocol. The workshop recommended relying on the U.N. system, and proposed a Declaration of Principles be adopted by the U.N. as a first step. Appointing a U.N. Commission and a Special Task Force on Global Climate, with UNEP as the lead agency, was also recommended. An attempt to codify an international Law of the Atmosphere, along the lines of the U.N. Law of the Sea, was deemed inadvisable over the short term, although the idea of a World Atmosphere Fund was endorsed.[80]

In November, 1988 the Intergovernmental Panel on Climate Change (IPCC) was formed in Geneva, involving 35 nations.[81] Three working groups were established, the first to review greenhouse effect science, the second to analyze impacts on environment and society, and the third to develop governmental response strategies for delaying or mitigating the adverse impacts of climate change. In December, U.N. General Assembly Resolution resolution 43/53 requested the IPCC to anticipate a climate convention. A second Canadian meeting of experts was convened in Ottawa in February of 1989, which concluded that: " . . . (international) conventions with appropriate protocols are needed as a means to ensure rapid international action to protect the atmosphere and limit the rate of climate change."[82] The March 1989 Declaration of the Hague, attended by 17 heads of state called for a new institutional authority to combat global warming, which would develop instruments and define standards, monitor compliance, and apportion assistance to developing countries most severely impacted by climate change.[83]

The call for a framework convention was reiterated in May in Helsinki and at a UNEP General Council meeting in Nairobi, where 44 nations initiated plans for negotiations to begin soon after the 1990 Second World Climate Conference, and again two months later at the Paris G-7 Summit.[84] In November, 1989, a Conference of 68 environment ministers was convened in Noordwijk, the Netherlands, urging the adoption of a framework climate convention "no later than at the Conference of the United Nations on Environment and Development in 1992." An attempt to include a call for specific emissions reductions by year 2000 was blocked by the U.S., U.S.S.R., and Japan. At the U.S./Soviet Summit in Malta in December 1989, President Bush offered to host the first session of negotiations on a framework convention on global climate change,[85] and four months later, in April 1990, Bush hosted the White House Conference on Economic and Scientific Issues Relating to Global Warming. The final IPCC meeting was held in Sundsval in August of 1990, and the resulting report was presented at the Second World Climate Conference held in Geneva in October/November 1990.

The Second World Climate Conference provided a formal marker between a two-year study by the IPCC and the negotiations on a global convention that began in early 1991. While mostly a formality in terms of science, the Conference allowed serious discussion about the problems of developing countries. The main Conference Statement, said that scientists agree that global warming is real and the regional impacts are unknown, but that ignorance of consequences must not be an excuse for inaction. Second, the Conference called for a global climate observing system to be established. Third, the Conference urged the creation of several regional research centers focused on global climate research, with particular support for scientists and data from developing nations. The Task Group 12 (Synthesis) recommended that research support should approach 0.2 to 0.3% of world economic output by the year 2000, or roughly $30 to 45 billion annually, which is well over 20 times what is now being spent.[86] Finally, the Conference called on the IPCC Task Force to remain active to support the negotiations and other preparations for the 1992 Brazil Convention. Formal negotiations on the Climate Convention began in Washington, D.C., on February 4, 1991.

# IV. SHAPING A NEW CONVENTION

## A. PARTIES AND ALTERNATIVE LEGAL MECHANISMS

Although the Framework Convention on Climate Change was open to all the nations of the world at Brazil in 1992, it is worthwhile to discuss whether or not there would have been potential benefits in considering a smaller subset of states to advance negotiations on certain protocols, or even to enter into separate multi-lateral treaties for appropriate purposes. For instance, should the U.S. pursue a less formal commitment between the U.S., Russia, and China to restrict emissions from coal burning over the very long term? With these three countries holding the vast majority of the world's coal reserves the issue of long-term climate control might be benefitted by this small group forming a separate understanding specific to coal production. Various concerns can be raised about choosing the U.N. forum over more direct arrangements between fewer nations.[87] The U.N. Law of the Sea Conference has shown the complexity of a negotiation encompassing 150 nations. A similar political bargaining concerning energy development could become obstructive. This agreement could easily take five to ten years to hammer out, perhaps longer. To avoid repeating the protracted UNCLOS debate, perhaps a suitable participation would be the first three countries by $CO_2$ emissions and a fourth delegation to represent other industrialized nations, together with the group of coal producers and OPEC delegations, and then a delegation to represent those countries who will suffer the impacts. In this way the ongoing work to formulate protocols can be done by perhaps a dozen delegations, rather than ten times that many.

That the majority of countries should have limited representation for more efficient negotiations would not prevent a world-wide signatory, however. The Antarctic Treaty, The Limited Test-Ban Treaty and the Non-Proliferation Treaty, for instance, were negotiated by a few parties and signed by more. A trilateral agreement could later become a world-wide protocol.

Clearly, the global warming issue is not merely a problem for the industrialized nations to solve on their own. U.S., Japan and the European Economic Community accounted for 70% of global CFC production in 1985, but only 40% of greenhouse gases.[88] China will become an increasingly stronger source of emissions in the decades ahead, as will other newly industrializing nations. The impact of $CO_2$ will be global, which argues for a worldwide negotiation. Perhaps the UNEP/WMO/ICSU team is most suitable for that role, having proved themselves capable with the Vienna Convention/Montreal Protocol model. In addition to thinking carefully about the most appropriate parties and forum to advance a global climate protocol, it will also be useful to think creatively about mechanisms for achieving international cooperation.

Alternative legal mechanisms are suggested by international lawyer Peter Sand, who notes that the existing pattern for creating international environmental regimes via an *ad hoc* agreement followed by treaty and ratification faces two drawbacks. One drawback is that consensus reflects minimum standards. Second, the time lag for ratification delays implementation. Ways to improve the situation, he suggests, include selective incentives,[89] differential obligations,[90] regionalization,[91] and promoting over-achievement.[92] The problem of ratification speed can be improved by provisional treaty application, soft-law options,[93] or delegated lawmaking.[94] Some of these techniques may find application in the new Framework Convention on Climate Change.

Two selective incentives, useful in the Climate Convention, would be differing access to the GEF and arrangements for technology transfer to developing countries. Differential obligations are also likely, with industrial nations paying more dollars up front and being obligated to cut their emissions more. Regionalization might not be applicable here, unless it manifests as a formalized 'bubble' provision. For instance, all ECE countries might be considered as one entity for setting emissions quotas. In a sense, the over-achievement technique has already been initiated by Sweden, Finland, and the Netherlands leading the international community with their early carbon taxes, although this "club" might become more influential if it included more states and a uniform taxing scheme amongst the entire group.

## B. OBLIGATIONS TO REDUCE EMISSIONS

Should the Convention attempt to set a ceiling on atmospheric greenhouse gas concentrations? Keeping $CO_2$ under 400 parts per million (in air, by volume) has been suggested,[95] meaning that emissions would have to cease climbing within about 40 years. Should emissions quotas be set for each country? Based on what criteria? Previous emissions baseline? GNP? Or should a per capita standard be set? A formulation based on 50% population and 50% GNP has been proposed.[96] But it is easy to imagine keen debate over this and other suggestions. Other variables, such as historical baseline, growth trend, climate, available technology, industrial mix, and available and non-fossil fuel sources could be factored in. Tradeable emission allowances have been suggested also, perhaps allowing groups of countries to negotiate exchanges under a "bubble" or a country to "bundle" various gas emissions in terms of meeting a single carbon emission equivalence.[97] The latter case might be attractive to the U.S., as it could allow enormous credit to be given for CFC reductions, an idea that might seem less than fair to some other countries.

A carbon excise tax is likely to be proposed, because it would provide conservation incentive and support for non-fossil fuel sources, but it is likely also to be unpopular because the revenues would be so large.[98] Many countries will resist an international body collecting revenues in excess of 1% GWP.[99] The definition of "carbon use" is likely to become controversial — should a tax apply as carbon crosses national borders? Should there be distinction between various levels of refinement? A production-based formula, or an "extraction" tax, with OPEC exporters keeping the revenues, would draw different support than a consumption-based tax, where oil importers would gain more revenues.[100] If the industrial countries who now underprice fossil fuels, such as the U.S., will have trouble accepting a tax approaching $0.50/gallon on motor fuel,[101] and if a tax ten times less would be trivial in effect, then it would make sense to propose an intermediate value initially, perhaps $0.25/gallon, and monitor the economic consequences of this response for a five-year period. Even better may be a marginal tax on emissions in excess of a quota, generating lesser revenue and stronger incentives.[102]

Efficiency standards should be proposed in conjunction with other limitations strategies, but may be difficult to apply to industrialized and developing countries equally, unless phased-in as part of a program to transfer efficient technologies from industrial nations to those countries that have none. Fleet mileage standards and household electricity use efficiencies would be worth considering.[103]

## C. DEFORESTATION AND REFORESTATION

It is likely that a resolution to create a protocol on forests will be introduced, aiming to both reduce deforestation and increase reforestation. A Tropical Forestry Action Plan has been developed by FAO, UNDP, World Bank, WRI, and country representatives, and it is likely that an effort to support part of this plan with proceeds from the GEF will be on the table. It could be proposed that the debt-for-nature swaps employed so effectively by the World Wildlife Fund,[104] be expanded formally within UNEP or a new institution. An industry group with headquarters in Japan, the International Tropical Timber Organization (ITTO), which grew out of UNCTAD 1985, will be paying close attention.

Reforestation programs are not trivial to establish. Although trees are rather cheap to plant, problems may arise with government subsidies for tree planting (i.e., it's easy to cheat, and the government expense is high contrasted against the low cost for the private investor).[105] To minimize stress to the forest generally, a ''maximum ecological tolerance'', essentially an upper bound to warming rates, could be proposed (e.g., 0.1°C per decade for forests).[106]

## D. OTHER PROVISIONS

**Energy, technology and efficiency** — A provision is necessary to establish some guidance for a massive transfer of technology to the developing world to hasten their development toward greater energy efficiency. Important items such as electricity meters, thermostats, insulating technologies, and timed switches could be very beneficial and not difficult to provide. Aid to build better roads, so that more fuel-efficient cars can be used, might be a more difficult program.

**Improved research programs and information exchange** — The Second World Climate Conference spoke strongly for improved global climate monitoring and building a small number of comprehensive regional research centers to enhance this research and to convey information to developing countries. Particularly important is research into feedbacks among biogeochemical cycles and feedbacks linking these cycles to human actions and/or climate, with a view to discovering, as early as possible, all potential surprises. Also included will be social science research, especially economic and policy analysis, and risk assessment.

## E. ENFORCEMENT, INSTITUTIONAL SETTING, AND THE ATMOSPHERE FUND

What level of commitment should be demanded in an early protocol? It will do no good to just agree that $CO_2$ should be limited, because a treaty only binds sovereign states so long as it is in their interest to stay bound.[107] Mutual adherence is unlikely to overcome the powerful, fundamental tendency for society to use more energy, rather than less.

What enforcement is desirable and what flexibility should be maintained? Enforcement, given the absence of any international regulatory power, must be left to legislative provisions within each signatory state. Thus, the protocol must offer sufficient value to induce passage of effective internal legislation, and then to maintain states' commitment to internal enforcement. The regime will persist if the parties continue to believe that the benefits of membership outweigh the costs.[108] The challenge, then, will be to translate the shared concern for a problem into a set of exchanges that will ensure that each party will stay bound. If we consider the U.S., U.S.S.R. and China to be the most important parties to bind over the long term, following depletion of oil and gas, then a shared interest in developing alternatives to coal production, sufficient to support a steady-state energy supply in these countries could be valuable enough to bind all signatories, particularly if the funding mechanism is attractive, i.e., fair and workable. A moderate tax on $CO_2$ emissions could be applied to the GEF and the penalty for non-compliance could be exclusion from WAFT benefits.

Should an existing international organization take primary responsibility for administering this new environmental management regime, or should a new international institution be established to set global air quality standards for trace gases, push for their implementation at the national/regional level, and enforce compliance?[109] It has been suggested that development of an institutional framework for global environmental management remain at a rudimentary stage.[110] The 1989 Declaration of the Hague called for a "new institutional authority" to set and implement environmental standards, but licensing by national institutions would still be required, and vetoes to the court's jurisdiction could still block claims.[111]

Given compatible procedures between nations, regulatory function need not require new international institutions. Rather, nations can use a variety of regulatory mechanisms, such as mutual recognition,[112] model diffusion,[113] alert diffusion,[114] and epistemic networks.[115] In the context of this Convention, standards pertaining to energy-efficient technologies could be mutually recognized. Models for exacting some sort of a carbon use tax might easily diffuse through the community of nations. UNEP could easily alert states regarding ecosystem surprise without resorting to a new institution. And the IPCC already provides an epistemic network for building consensus on procedures and standards. If enforcement were to appear inadequate, then it is conceivable that an aggrieved state might take direct action to protect the health of its citizenry. A country that could show it is adversely affected by trace gas emissions might bring an action against pollutors before the International Court of Justice based on the principle of international common law, *sic utere tuo ut alienum non laedes.*[116] Of course, problems of proof would make such an action difficult. Further, the alleged pollutor must submit to the jurisdiction of the Court. However, most transnational environmental regimes have learned to avoid the adversarial state-liability approach, and instead have used or developed different methods to ensure compliance with treaty obligations.[117]

Imaginative approaches to compliance control can be drawn from the rich institutional and procedural experience of existing transnational regimes, including local remedies,[118] complaints and custodial actions,[119] and environmental audits.[120] Again, UNEP would be a logical host for these functions. An interesting possibility would be to enlist a coalition of non-governmental organizations (NGOs) to conduct and publicize audits. Some will favor publicity and persuasion over financial penalities within the first decade,[121] but stronger compliance incentives will follow if payments to and from the GEF were connected to adherence under audit.

### F. ONGOING REVIEW

Ongoing review will be necessary because the international response will need to be tailored to an updated knowledge of biogeochemistry and climate change. Quotas might have to be renegotiated, changes in technology transfer may become desirable, or compliance mechanisms might be found to be inadequate. The UNEP Secretariat can be expected to monitor progress, critique and report, and could be granted some rule-making authority. Major changes, however, will require reconvening the signatories.

Built-in review schedules have appeared in a number of recent environmental agreements: the 1987 Montreal Protocol established a four-year cycle, initiated in 1990 by four assessment panels and an intergovernmental working group; the 1988 Sofia Protocol to the CL-RTAP provides for regular review under a similar group; and the 1989 Basel Convention sets up a six-year evaluation cycle.[122] A reasonable review period for the Climate Change Convention might be every five years for the first two decades, and every ten years thereafter.

## V. THE U.S. PERSPECTIVE

### A. HISTORY OF INVOLVEMENT

The scientific research that has led to the current global warming concern has been primarily supported by the U.S. Government. The National Oceanic and Atmospheric

Administration has been a key contributor since the 1960s to the university-based research that is responsible for our current knowledge about greenhouse gases. Other agencies stepped up their global warming research role considerably in the 1980s, particularly the Environmental Protection Agency (EPA) and the Department of Energy (DOE), which launched a ten-year carbon dioxide research program in 1980 following passage of the 1980 Energy Security Act. The U.S. collaborated internationally, forming a bilateral Agreement of Co-operation in the Field of Environmental Protection with the Soviet Union in the late 1970s, specifically calling for the $CO_2$-related research.[123]

In 1983 the National Academy of Sciences published a comprehensive assessment of global warming in the volume titled *Climate Change* and two years later the EPA completed its assessment titled *Can We Stop a Greenhouse Warming?* The Global Climate Protection Act of 1987 required the EPA to develop a coordinated national policy on global climate change, resulting in two major reports, the *1989 Draft Report to Congress on Policy Options for Stabilizing Global Climate* and the *1989 Potential Effects of Global Climate Change on the United States.*[124]

The first nationwide assessment of the primary impacts of a 1 meter sea level rise on the U.S. recently gave estimated costs of $270 to 475 billion, ignoring future development.[125] Nordhaus has estimated the annual impact of global warming on the U.S. economy, ranging from a benefit of $4 billion to costs of $20 billion, with $7 billion per year as a central estimate, or approximately 0.3% GNP. Thus, his worst case is about 1% GNP.[126] The steady expansion of governmental agencies to support science relating to global change[127] has progressed to the point where the President's Committee on Earth Sciences of the Federal Coordinating Council for Science, Engineering, and Technology recently recommended more than $1 billion dollars to study Global Change in the 1992-93 budget.

## B. LEGISLATIVE ACTIVITY

A smaller number of Congressional hearings in the 1970s related to global environmental change paved the way for dozens of hearings on this topic in the House and Senate during the 1980s. As the decade closed, more than half a dozen bills had been introduced proposing mitigative response to anticipated global warming. The American Bar Association has paid heed to the topic, as well.[128] Two broad categories can be described for U.S. legislation introduced in Congress relating to global warming. The bills introduced to the Senate and the House contained either research/assessment provisions or substantive provisions.[129]

One of the research bills, S.169 the Global Change Research Act of 1990, introduced by Fritz Hollings (D-SC) became law in November, 1990.[130] The Act requires the establishment of a U.S. Global Change Research Program aimed at understanding and responding to global change, creates a 10-year research council under the Federal Coordinating Council for Science, Energy, and Technology (FCCSET), to (1) study effects and remedies, including the cumulative effects of human activities and natural processes on the environment, (2) promote discussions toward international protocols in global change research, (3) coordinate research amongst federal agencies, and (4) develop an agenda for global warming response.

One of the substantive bills, S.333, the Global Environmental Protection Act, introduced by Senator Leahy (D-VT), would have mandated specific reductions in $CO_2$ emissions.[131] Another, H.R. 4805, would call for a carbon tax of $15/ton coal, $3.25/barrel oil and $0.40/MCF on natural gas, well below what is needed to stabilize $CO_2$ emissions at 1988 levels by year 2000, but perhaps effective in encouraging consumer awareness about efficiency.[132]

The governments of three states, Oregon, Vermont, and New York, moved to implement global warming responses.[133] Oregon passed a law to require a state strategy to reduce greenhouse emissions by 20% from 1988 levels by 2005.[134] Vermont issued an executive order calling for reducing greenhouse gas emissions and acid rain precursors by 15% below current levels by the year 2000.[135] New York's executive order sets a goal of 20% reduction in $CO_2$ emissions by 2008.[136]

## VI. THE DEVELOPING WORLD PERSPECTIVE

It is generally recognized that the stresses of global environmental change will place greater burden on the developing countries than on the industrial countries.[137] Because these countries have a lesser ability to apply capital and technology for adaptation, they are more vulnerable to climate impacts. Drought, sea level rise, and coastal storms are felt more harshly by peoples in Nigeria, the Maldives, and Bangladesh, respectively, than in the U.S., France, and Britain. Capital and technology allow for water storage and distribution, for concrete seawalls and for solid houses to withstand storms.

In addition to suffering more profoundly from the potential impacts of global warming, developing nations, including China and the states Eastern Europe, India, and the former Soviet Union, are not likely to be able to afford the costs to abate greenhouse gas emissions, particularly as they must cope with other environmental problems that are greater.[138] Despite the fact that historically developing nations have contributed little to the greenhouse gas burden, these countries cannot be left out of the developing regime to manage emissions. Developing nations, most of whom did not ratify the Montreal Protocol, account for about 12% of the world's CFC consumption, with expected increase of 6% annually to year 2000 and perhaps 13% annual increases for India and China.[139] India (56%) and China (80%) rely heavily on coal, China subsidizing production from about $3 to $7/ton. In India, production losses are about $8/ton.[140] A global solution that includes the developing nations will have to include transfers of capital and energy-efficient technology from the industrial nations to the Third World. If China and India are to be asked to restrain from burning coal, then the extra cost of a more expensive energy source, at least in the short term, has to be provided them.

An *ad hoc* Sub-Group of the IPCC Task Force formed in 1989 to address ways to increase the participation of the developing countries in IPCC activities. One testimonial to their success is the strong and clear message of the Second World Climate Conference that their scientists be made a strong component of a network of regional research centers to gather and interpret data from the expanding Global Climate Observing Program. The next challenge will be to ensure that (protocols following) the Convention on Climate Change signed in Brazil in 1992 contain strong provisions to transfer the technologies that are so critically needed if the developing nations are to meet successfully their own role in mitigating global warming.

## VII. CONCLUDING REMARKS

While my own "best guess" is that we will not see a global mean warming greater than 3.0°C for an equivalent $CO_2$ doubling, I am too concerned about the possibility for surprise to comfortably recommend a path of adaptation. The inherent weakness of these cost-benefit analyses is more profound than just saying that "non-market costs" require more analysis. The important point is that threats to ecosystem resilience in the face of unknown stress cannot be valued economically.

The upcoming international agreement in 1992 should address continued research, a goal for atmospheric composition, emissions limitations, alternative energy sources and adaptive strategies. Even if not a formal "law of the atmosphere," the agreement must still specifically require enforcement of guidelines through national Clean Air Act mechanisms, or "harmonization" of legislation. It should link compliance with a share of the GEF to provide incentive for maintaining commitment. The time to begin applying the brakes on our energy use has clearly arrived. The momentum of our energy-consuming society threatens to make some measure of climate change unavoidable. The energetic cost of society reacting to the greenhouse effect has to be added to the energy we will need to react to other problems. We will likely have to set priorities, as our budget for response may not allow optimal responses to every looming ecological problem.

Human society as it faces global warming might be analogous to the prehistoric organism developing an eye. Most likely, survival initially became favored for the organism who followed its very first visual warnings only with a humble and hasty retreat. It would be later organisms, after investing much more time and energy into their ability to see, who would be able to derive advantage through responses specifically tailored to the changing scene. As we pride ourselves with our growing perception of a chemical system that was beyond our senses until only recently, we must not neglect the possibility that the greenhouse effect could be more complex and more rapidly destabilizing ecologically than we have estimated. Given that our biogeochemical understanding still may be in its infancy, immediately reducing our energy use is a humble and wise precaution.

Global warming is really a symptom of a more serious challenge facing human society — the feedback between energy and technology. In attempting to understand the momentum of social development and the patterns of evolving international relations, we must concentrate not only on the trend in energy use, which is a focal point in the global warming debate, but on the purposeful force underlying these energy flows. It is this force that governments must manipulate artfully; but, first, it must be recognized and understood.

## ENDNOTES

1. (Kellogg and Schneider, 1974; Bach et al., 1979).
2. (Kellogg and Schware, 1981; MacDonald, 1982; Clark, 1982; Chen et al., 1983; Kates et al., 1985; Edmonds et al., 1986; and Bolin et al., 1986).
3. (NRC, 1983, 1989).
4. (EPA, 1983, 1989; DOE, 1985).
5. (Revelle, 1980).
6. Hydrogen (H), carbon (C), nitrogen (N), oxygen (O) and sulfur (S) are the elements of prime concern.
7. Halocarbons include those carbon species bearing chlorine, fluorine, bromine, and iodine, which include the chlorofluorocarbons (CFCs; Freon-11,12, etc.), the bromofluorocarbons (halons), methyl iodide, and carbon tetrachloride, among others.
8. *See* Lovelock, J. E., Maggs, R. J., and Wade, R. J., *Halogenated Hydrocarbons in and over the Atlantic at* p. 194, SCIENCE, Vol. 241, pp. 194–196, 1973. "The presence of these compounds constitutes no conceivable hazard; indeed the interest lies in their potential usefulness as inert tracers for the study of mass transfer processes in the atmosphere and the oceans."
9. *See* Wigley and Warrick, 1990: A rise in global mean sea level is one of the most certain consequences of global warming. *See* Warrick and Oerlems, 1990: Predicted is an 18-cm rise by year 2030 and a 44-cm rise by year 2070, and likely to exceed 1 meter during the 100 years following. *See* Leatherman, 1989. Sea level rise will inundate and displace wetlands and lowlands. *See* IPCC, 1990. Other potential impacts of sea level rise are increased coastal erosion, drowning of some barrier reefs, intrusion of salt water into coastal groundwater supplies, altered tidal ranges in rivers and bays, and altered sediment deposition patterns. *See* Titus et al., 1984. Development immediately landward of estuaries can affect the ability of the marsh to adjust to a rising sea. Bulkheads set too close can prevent the inland edge of the marsh from advancing. At least two jurisdictions (New Jersey and Massachusetts) already prohibit construction of such bulkheads.
10. *See* Mehta and Cushman, 1989. Effects on wave heights from rising sea level are expected to be small for deep-sloping shelves with shoreline recession, but in areas of shallow shelf and fixed shoreline wave heights may increase by 5–10% for a 0.5 meter rise in sea level. The reach of storm surge will be exacerbated by the rising mean sea level. *See* Emanuel, 1987. Predicted increases in storm force from warming would add to storm surge, while greater erosion would remove wave protection. Higher sea level would slow drainage at the outlet of flooded rivers. Increasing storm frequency could multiply the above effects. *See* Flynn et

al., 1984. Sea level rise holds implications for hazardous waste sites located in the coastal floodplains, where the expansion of the 100-year storm surge area may require costly floodproofing for sites now located just outside this zone.

11. *See* Damkaer, 1987. Increasing penetration of solar UV-B may hurt phytoplankton productivity. *See* Roberts et al., 1986, and *see* Giannini, 1986, 1990, Immunosuppression in humans from increasing UV-B levels could lead to greater susceptibility to infection. *See* (Curson, 1989; Liehne, 1989; Shope, 1990). Flies, mosquitos, and bats carrying infectious Leishmaniasis, encephalitis, and rabies, respectively, could move with warming toward higher latitudes. *See* (Longstreth, 1989; Gillett, 1981). In the U.S., three of the numerous mosquito-borne diseases have been considered to be significant risks following climate change (malaria, dengue fever, and arbovirus-induced encephalitis), while two others (yellow fever and Rift Valley fever) present possible risks. *See* (White and Hertz-Picciotto, 1985). An airborne fungus, *Histoplasma capsulatum,* reaches about 500,000 people annually, causing illnesses in roughly one third of them, ranging from respiratory disorder to death. This and another soil fungus, *Coccidioides immitis,* are easily carried in dust storms, and climate change could affect their range.
12. *See* (Haines, 1990). *See* (Oeschli and Buechley, 1970). In Los Angeles, 4 straight days of temperatures exceeding 100°F will cause 50 extra deaths per day.
13. Three straight years of drought have prompted California cities to impose water-rationing schemes by 1990–1991. *See* (Gleick, 1989). Warmer winters in the California Sierra will bring higher rain to snow ratio, decreased snowpack, and faster spring snowmelt, which will reduce water available for agriculture. *See* (Frederick and Gleick, 1989). California has a small storage volume relative to its consumer demand, making the water supply very vulnerable to climate change.
14. (Sestini et al., 1989).
15. *See* (Gable and Aubrey, 1989). Loss of land area (28%) and fresh water (50%) owing to inundation and salt water intrusion, respectively, have been modeled for the Laura section of the Majuro Atoll in the Marshall Islands, based on a 1.0 meter rise in sea level. *See* (Miller and Mackenzie, 1988). Accompanied by likely increases in storm damage, these impacts may make living on this and other atolls difficult or even impossible. *See* Ramakrishna, Third World Countries in the Policy Response to Global Climate Change, in J. Legget, Ed., *GLOBAL WARMING: THE GREENPEACE REPORT,* 1990, quoting Ernest Beni, Vanuatu's principal delegate, before the IPCC Response Strategies Working Group, Geneva, Oct. 1989," . . . We, the peoples of the South Pacific Region, appeal to you in a common voice, the voice of those who may become the first victims of global warming . . . to ensure the survival of our cultures and our very existence and to prevent us from becoming 'endangered species' or the dinosaurs of the next century . . . ". The Japanese will spend $240 million to bolster against the rising sea their southernmost reef, Okinotorishima.
16. Lashof at 239. (Wirth and Lashof, 1990) *and see* Lashof, 1989, — adding biogeochemical feedbacks with a gain of 0.16 to the gain estimated on the basis of geophysical feedbacks gives an overall gain of 0.80, which would increase the climate sensitivity to an initial doubling of $CO_2$ from 3.5 to 6.3°C. In a high-gain system, the total amplification is very sensitive to small additional gains. *And at* 238 — Thus, while there are many uncertainties, there is the potential for the biogeochemical feedbacks discussed here to substantially increase the overall sensitivity of the climate system, so that a global warming of even 8 to 10°C owing to an equivalent $CO_2$ doubling cannot be ruled out.
17. *See* Lancaster, 1989, *at* 64. Ilya Prigogine was awarded the Nobel prize in the mid-1970s for describing the sensitivity of such energetic chemical systems to minor perturbations in chemical or energy balance. We do not yet know the resiliency of our climate and biogeochemical system to recent and present anthropogenic perturbations. Short-term climate variations, such as El Nino/Southern Oscillation events give us a minor measure of system response within narrow parameters of perturbation. *See* Keeling, Bacastow, Lancaster, Whorf, and Mook, *Evidence for Accelerated Releases of Carbon Dioxide to the Atmosphere, Inferred from Direct Measurements of Concentration and* $^{13}C/^{12}C$ *Ratio,* Testimony before Senate Committee on Energy and Natural Resources Hearing on Trends in Carbon Dioxide Emissions, June 22, 1989: "This recent evidence means we cannot rule out the possibility that a natural, positive feedback could be accelerating the Greenhouse Effect over the short term." We do not yet know the mechanism underlying this biogeochemical-climatic linkage. Until all of the important sources and sinks, and all of the important potential feedback mechanisms, for all of the greenhouse gases have been identified, it would be irresponsible to claim that any reliable assessment of long-term global warming impacts on the global biosphere, including its human component, could be made.
18. Over ten years after the initial surprise that CFCs posed a threat to the ozone layer, another complete surprise arose — the ozone hole in Antarctica — resulting from an unforeseen amplification of chemical reactions by the presence of ice particles and certain meteorological patterns.
19. (Kellogg, 1983; Lashof, 1989).
20. *See* Schimel, Biogeochemical Feedbacks in the Earth System, in *Global Warming: the Greenpeace Report* 68–82, J. Leggett, Ed., 1990. Nitrogen fertilization appears to inhibit methane uptake by soils (J. Melillo,

personal communication, 1991). Reductions in sulfur oxide emissions mandated by legislative response to atmospheric deposition (1990 Clean Air Act Amendments) could reduce a cooling effect from sulfate-induced cloud formations. Nitrous oxide ($N_2O$) has a greenhouse potential between 150–300 times that of $CO_2$, and its sources are poorly understood. Unexplained surges in emissions have been reported in the tundra region, where warming potential may be greatest in decades ahead. *See* Lancaster, Ph.D. thesis, 1990, Scripps Inst. of Oceanography. The prospect of a drastic runaway $CO_2$ emission can be considered very improbable because we can estimate that the readily available carbon reservoir in the biosphere, which is about 1200 Gigatons carbon, is only about enough to increase atmospheric $CO_2$ levels another 1.5 times. Yet, it is still possible that sufficient greenhouse gases ($CO_2$, $N_2O$ and $CH_4$) are available to lead us into an equivalent tripling or quadrupling of the $CO_2$ baseline, if multiple, presently unforeseen, feedback mechanisms are triggered.

21. (Wirth and Lashof, 1990) at p5.
22. (Roberts, 1988).
23. If moisture in the high atmosphere aids the reactions that destroy ozone, then global warming, by evaporating more moisture from the ocean, could worsen ozone destruction. Also, global warming at high latitudes may release more methane from tundra soils, with water formed from methane oxidation enhancing high-altitude ice cloud formation. Further, greenhouse theory predicts the stratosphere will cool as the troposphere warms. If cooling occurs at an altitude low enough to freeze the increased moisture in the presence of ozone, then even more ice clouds might be produced.
24. For instance, damage to the ozone layer over Antarctica could hurt phytoplankton productivity in the Southern Ocean, thereby reducing the ocean's uptake of $CO_2$. See Oppenheimer and Boyle at p. 69, 1990.
25. In our Buberesque locomotive we will have to keep our foot partially on the accelerator even as we apply the brakes.
26. Buber, 1970: " . . . an instant ago you saw no less than I that the state is no longer led; the stokers pile up coal, but the leaders merely seem to rule the racing engines. . . . They will tell you they have adjusted the apparatus to modern conditions, but you will notice that henceforth they can only adjust themselves to the apparatus, as long as that permits it." This metaphore, drawn from the coal-fired technologies of the early 20th century, will become more apt as we return to coal in the 21st century.
27. *See* (Lancaster, 1989), developing idea that an energy-technology feedback is a basic principle underlying the steady incorporation of energy into structure, which describes evolution.
28. *See* Rathjens, *Energy and Climate,* at 172 in J. T. Mathews, Ed., 1991, *below*, note 110.
29. The greenhouse theory itself, i.e., that $CO_2$ and other trace gases will impede infrared radiation escaping from the Earth surface and thus warm the atmosphere, is not uncertain. In fact, we employ this same principle to measure atmospheric $CO_2$ concentrations in the laboratory. The main force of the expected global warming, however, stems from a positive feedback involving water: an initial warming of about 1°C evaporates more water from the ocean and that water vapor itself absorbs more infrared energy, heating the atmosphere more, and so on. Uncertainty stems from not knowing if, when and where, in the ever-moving atmosphere, the increased water vapor will condense to form clouds. Different cloud heights, densities and extent provide potential cooling by increasing albedo (reflecting visible sunlight to space before it can hit the ground).
30. IPCC, 1990a, Working Group I. It is very probable that there has been a real, although irregular, increase of global surface temperature since the late 19th century, at about 0.45 +/−0.15°C, with an estimated small (less than 0.05°C) exaggeration owing to urbanization. *See* Solow, 1990 at 23. "Despite problems in data collection and interpretation, there appears to have been a mean global warming of 0.3–0.5°C over the past century."
31. *See* Nierenberg, Jastrow, Seitz, *Scientific Perspectives on the Greenhouse* 1989 (Marshall Institute, Wash., D.C.). The controversy involves unknown impacts of the long-term solar cycle, statistical criticism of recent conclusions regarding long-term warming and weaknesses of computer simulations of climate. *See also* letters of Lindzen, Nierenberg, Jastrow, Baliunas, Stuiver, and Roberts at 14–16 *Science,* Vol. 247, 1990. (**see also recent article in SCIENCE**)
32. Solow at 29, 1990. "The global warming problem is the potential massive economic dislocation that could occur from an extreme response to stabilize $CO_2$ concentrations."
33. *See* Glantz, Price, and Krenz, (Eds.) 1990, *On Assessing Winners and Losers in the Context of Global Warming,* Report of a Workshop held at St. Julians, Malta, 18–21, June, UNEP and National Center for Atmospheric Research, Boulder, Colorado, 44 pp.
34. Schneider (1989).
35. *See* Solow at p30, 1990.
36. When one's only tool is a hammer, every problem becomes a nail.
37. In water pollution cases, value of the harm is the amount the downstream victim would pay to remove contaminants from the resource.
38. (Smetz, 1972).

39. *See* Nordhaus at 23–24 and 32–34.
40. *See* Nordhaus, at 23–24. Climatic engineering includes, among other strategies, modifying albedo, increasing the ocean sink for carbon (putting iron in the ocean has a crude cost estimate of $10 per ton carbon, but a large uncertainty as to effectiveness), changing the hydrological cycle, or loading particulates into the stratosphere. *See* (Barbier et al., 1990) Overall global warming options to reduce greenhouse gases include: (1). Reduce emissions from fossil fuels by (a) slowing economic activity, (b) slowing population growth, and (c) substituting technologies/increasing efficiency; (2). Slow deforestation by direct regulation and deb-for-nature swaps; (3). Invest in downstream sinks to sequester carbon by reforestation and stimulating ocean productivity.
41. Nordhaus at 8–9.
42. Nordhaus at 25. The past 80 years have seen the map of Europe redrawn thrice and U.S. power density increase from 1.5 HP/capita to 130 HP/capita.
43. *See* IPCC, 1990a. To stabilize emissions in 2005 at 1980 levels, $CO_2$ emissions would have to be reduced by 50–80%, methane by 10–20%, nitrous oxide 80–85%, and CFCs by 75–100%. *See* Barbier et al., 1990, at 6. The IPCC is now recommending $CO_2$ emissions be held to 1990 levels, which requires a 20% reduction in their projected levels for 2005.
44. Schelling, at 2. It is important to avoid substantial efficiency costs or output losses today in pursuit of uncertain future benefits (Lave, 1990).
45. Nordhaus at 16.
46. Nordhaus at 17.
47. Nordhaus at 17–18; Whaley and Wigle at 5; To meet 50% reduction would require $460 per ton carbon (compare $180/ton for Nordhaus, above) at 5.
48. Nordhaus at 21.
49. Nordhaus at 22–23.
50. Nordhaus at 35–40; Based on carbon content, tax coal 100%, oil 80%, gas 58%, shale oil 115% (coal 24 kg/gigjoule; oil 20 kg; gas 14 kg) (Flavin, 1990; Edmonds and Reilly, 1983).
51. Poterba at 5.
52. *See* Schelling at 3.
53. Poterba at 3. (cite for coal distributions, or World Resources Inst, 1990–91.p3).
54. Whaley and Wigle at 5.
55. Whaley and Wigle at 5–6.
56. *See* Poterba at 12.
57. *See* Poterba at 35.
58. *See* Barbier et al., 1990 at 11–13. Replacement technologies (100% reduction): Electricity generation (hydropower, hydrothermal, tidal power, biomass, passive solar and small remote photovoltaic systems are in place, with wind, solar thermal, and ethanol in transition); Building/industrial and other stationary uses; transportation; Switching to hydrogen/methane economy; CFC substitution; methane substitution. Reduction technologies: fuel mix, fluidized bed coal combustion, combined cycle gas turbines, ocean thermal energy conversion.
59. *See* A. Lovins, The Role of Energy Efficiency, in *Global Warming: the Greenpeace Report* 193–223, J. Leggett ed. 1990.
60. Rather than $200 billion in cost, as is suggested by Schelling, to achieve stabilizing 50% emissions reductions through carbon taxes, without acccounting for more efficient technologies.
61. Barbier et al., 1990 at 9, citing the International Energy Agency, *Greenhouse Gas Emissions: The Energy Dimension* (working paper for the White House Conference on Science and Economics Research Related to Global Change, 17–18 April, 1990).
62. *See* Cherfas, Skeptics and Visionaries Examine Energy Savings, January 1991, *Science,* Vol 251, at 154–56. Reporting on Dahlem Conference, Berlin, December 10–14, 1990 (Proceedings to be published by John Wiley & Sons, Chicester, England).
63. *See* SCOPE 13, 16, 21, 23 and 29 on the carbon cycle and global warming.
64. The phrase 'global change' can be attributed to Herb Freidman at an annual NRC meeting in 1983 of the National Research Council, calling for ''a bold holistic venture in global research, a study of whole systems of interdisciplinary science in an effort to understand global changes.'' This language was then incorporated into the title of a 1983 ICSU report titled, *Toward the International Biosphere-Geosphere Program: A study in Global Change*. In 1984 was the ICSU symposium on global change: ''Interaction between the physical and living world as a focus of a global change program, otherwise named the IGBP''.

65. *See* excerpt of letter from R. Revelle to M. Visscher, Feb 4, 1950, Archives of the Scripps Institution of Oceanography:
"1. The international scientific unions are in a sense laboratory models for the development of international democratic procedures. Their scientific activities, administrative decisions and interchange of ideas result from discussions, conflicts, compromise and agreements between active scientists at all levels of age and accomplishments, acting as individuals and not as representatives of a national interest. In his famous Federalist Paper No. 15, Alexander Hamilton pointed out that individuals are the only proper objects of government, and it is only by development of the concept of individual citizenship in the world community that progress will be made towards effective world government.
2. The international scientific societies have long been and continue to be leaders in the development of a world point of view as opposed to national points of view. Through their experiences at meetings of such bodies, younger scientists will gain the understanding and impetus to become forceful advocates of international cooperation in their own communities. Thus international thought and action will evolve from below rather than be imposed from above.
3. The maintenance of peace and security must be dynamic rather than static in our rapidly changing world. Scientific discovery and the resulting technological advances are perhaps the most important cause of change in the world today. In order to play a role in the forefront of changing ideas, UNESCO must keep abreast of advancing science. The international scientific unions are by far the best medium to serve UNESCO in this regard through advisory councils, committees of experts, symposia and documents. But they can only do so if they are strong and healthy.
4. Peace is indivisible. While areas of poverty and misery exist there can be no lasting security. Poverty and misery can be temporarily alleviated through material aid, but in the long run only scientific discovery and application can produce a remedy. In general this remedy can be applied only by the peoples themselves, but the peoples must have the necessary tools if they are to work out their own salvation. A broadening and deepening of international scientific cooperation is necessary for the distribution of some of these essential tools.
5. The work of the international scientific unions has demonstrably led in the past to steady programs and accomplishment. Such progress even though slow is more desirable than spectacular short term projects, which are liable to leave little lasting effect."
66. Sand, 1991, at p239.
67. Sand, at 239. Regime defined as "norms, rules and procedures agreed to regulate an issue area."
68. Unless one believes that treaty compliance is a more politically saleable way to push dollar-saving energy-efficient technology onto a reluctant public back home.
69. (ECE, 1979; Heywood, 1979).
70. The WMO grew out of the International Meteorological Society — est. 1873 — and was integrated with the U.N. in 1951.
71. (Weiss, 1980).
72. *See* Vienna Convention for the Protection of the Ozone Layer, opened for signature Mar. 22, 1985, UNEP/IG.53/5/Rev.1 (1985), reprinted in 26 I.L.M. 1516 (1987) [hereinafter Vienna Convention] (ratified by the U.S., Aug. 27, 1987). For comprehensive background on the ozone problem and related legal developments *see* J. Brunnee, *Acid Rain and Ozone Level Depletion: International Law and Regulation* 263 (1988). *See generally* Rummel-Bulska, The Protection of the Ozone Layer Under the Global Framework Convention, in *Transboundary Air Pollution* 281 (C. Flinterman ed. 1986). *Also see generally* Doolittle, Underestimating Ozone Depletion: the Meandering Road to the Montreal Protocol and Beyond, in 16 *Ecology Law Quarterly* 407 (1989); and *see* Noble-Allgire, The Ozone Agreements: a Modern Approach to Building Cooperation and Resolving International Environmental Issues, in *14 So. Illinois Univ. Law J.* 265 (1990).
73. *See* Doolittle, *supra* at 421. *See* Sand, at 41, Protecting the Ozone Layer: The Vienna Convention is Adopted, 27 *Envt* 19, 40 (1985).
74. Montreal Protocol on Substances that Deplete the Ozone Layer, opened for signature, Sept. 16, 1987, reprinted in 26 I.L.M. 1550(1987). *See generally* Sorenson, International Agreements — Montreal Protocol . . . , in 29(1) *Harv. Intl. L. J. 185 (Winter, 1988). See generally* Ogden, The Montreal Protocol: Confronting the threat to Earth's Ozone Layer, in *63 Wash. L. R.* 997(1988). 24 nations signed the Montreal Protocol calling for a 50% reduction by 1999 in CFCs, with 20% by 1993. The U.S. ratified on April 21, 1988 and EPA issued regulations on August 1, 1988 (Federal Register, Aug. 12, 1988). Article 5 of the Protocol allows countries with consumption less than 0.3 kg/capita to delay 10 yrs, so long as they do not exceed that rate of consumption. (U.S. consumption exceeded 1.0 kg/cap in 1985). Industrialized countries also agreed to provide technical assistance and financial aid in support of developing countries' efforts to adopt alternatives. *See* Poterba, 1991, at 35, for CFC tax table, sliding rate scale 1990–2000, giving implied

equivalent $CO_2$ tax rate. *See* Barbier et al., 1990, at 6. Major reductions in CFCs are still considered to be the most cost-effective control option for controlling greenhouse emissions. The Montreal Protocol has targets for reducing CFC consumption to 15% of 1986 levels by 1998 for developed countries and by 2008 for developing countries. CFCs are now still increasing −5% annually instead of 20% annually previously (R. Weiss, personal communication, 1991, Scripps Inst. Oceanography).

75. (Weiss, 1980).
76. (WMO, 1980).
77. (Jaeger, 1988).
78. (Tangley, 1988).
79. (Toronto, 1988). The Toronto conference statement: ''Humanity is conducting an unintended, uncontrolled, globally pervasive experiment (borrowing from the statement by Revelle and Suess in the 1950s) . . . ultimate consequences second only to a nuclear war . . . ''.
80. (Woods Hole, 1988). Perhaps this fund could be called the World Atmosphere Fund-in-Trust, or WAFT — to bear or carry through the air or over water.
81. *See generally,* Fitzgerald, The Intergovernmental Panel on Climate Change: Taking the First Steps Towards a Global Response, in *14 So. Illinois Univ. L. J.* 231(1990). *See* Wirth and Lashoff, Beyond Vienna and Montreal: Multilateral Agreements on Greenhouse Gases, (1990) submitted to *Ambio.*
82. Wirth and Lashof, 1990, at 8.
83. *See* Declaration of the Hague at 12 Int'l Env't Rep. (BNA) 215, 1989.
84. *See* G-7 Economic Declaration, 19 Envtl Policy and Law 183 (1989).
85. *See* Wirth and Lashof at 2.
86. Conclusion no. 3 concerns uncertainty and inaction, Recommendation no. 4 concerns funding for research and policy development, Recommendations of Task Group 12 (Synthesis), Second World Climate Conference, S&T/SWCC/no.12, (submitted by Mr. T. Malone, Chairman of the Task Group), 3 November, 1990.
87. For instance, Alexandre Timoshenko, of the USSR's Institute of State and Law, has expressed distress that only 58 of the 170 UN delegates belong to UNEP (Henson and Hively, 1989), raising concern about the representativeness of UNEP and commitment of the global community to this agency.
88. Wirth and Lashof at 3.
89. *See* Sand, International Cooperation, in *Preserving the Global Environment: the Challenge of Shared Leadership* 242–244 (J. T. Mathews, Ed., 1991). These incentives create post-treaty asymmetry. See Montreal Protocol art. 2(5), 2(6), and 2(8), and art. 5. Examples are access to funding (World Heritage Fund), resources (Fisheries catch quotas and 1988 Wellington Convention), markets (1973 Washington Convention controlling crocodile hides) and technology (1985 Vienna Convention, CFCs; 1988 Sofia Protocol, nitrogen oxides; and 1989 Basel Convention, hazardous wastes).
90. *Id.* at 244–46. Examples of these structured asymmetries are the 1976 Bonn Convention to protect the Rhine River (partitioning costs to the Netherlands, Germany, France, and Switzerland), and the 1977 Mediterranean Convention Trust Fund (partitioning monies based on economic, geographic and demographic criteria). Problems can arise with this approach; for instance, in the 1988 Sofia Convention the definition of ''critical load'' left open variable translation into national abatement targets.
91. *Id.* at 246–48. Custom-built asymmetrical regimes where trade-offs exist, with integration and solidarity.'' . . . restricting membership should raise the standard, particularly where such restriction adds an element of geographic or other affinity between members.'' Examples are the 1974 Helsinki and Paris Conventions on the Baltic and North Seas, respectively, the 1976 Mediterranean Convention and the 1989 Kuwait Protocol on pollution from offshore mining.
92. *Id.* at 248–50. Here optional additional protocols to apply higher standards or more stringent controls by a smaller ''club'' can cause a ''bandwagon effect''. Examples are the upwardly mobile target in the 1985 Helsinki Protocol on $SO_x$ and the 1989 Helsinki Declaration on CFCs, formalized in the 1990 London Protocol.
93. *Id.* at 252–54. This procedure, used for 1988 Sofia Declaration on Reduction of $NO_x$, was initiated by Governing Council of UNEP inviting a Council of Experts to adopt a statement saying what countries ''should'' do, which statement is then approved by UNEP. At this stage, or after going another step to become a Resolution in the U.N. General Assembly, it can be referenced by governments as a ''jointly honored principle of cooperation.''
94. *Id.* at 254–56. Examples are the World Health Organization, the World Meteorological Organization, the Intergovernmental Maritime Organization and the independently elected European Parliament (which has an Environment Committee).
95. Wirth and Lashof, at 13.
96. Wirth and Lashof, at 13.
97. Wirth and Lashof at 14. Schelling at 19–21.
98. Schelling at 18.
99. Poterba at 2–3.

100. (Whaley and Wigle, 1990) at 13–15.
101. *See* Schelling at 18. With the U.S. consuming 250 billion gallons per year, the tax would be $125 billion, not including coal and gas.
102. *Id.* at 18.
103. *Id.* at 19.
104. *See generally,* Hamlin, Debt-for-Nature Swaps: a New Strategy for Protecting Environmental Interests in Developing Nations in 16 *Ecology L. Quart.* 1065 (1989).
105. Poterba at 33.
106. Barbier et al., at 7.
107. *see* note 108.
108. *Id.* Chayes and Chayes, *below* at 289.
109. Weiss, 1980; Tickell, 1977.
110. *See* Chayes and Chayes, Adjustment and Compliance Processes in International Regulatory Regimes, in *Preserving the Global Environment: the Challenge of Shared Leadership* 280–244 (J. T. Mathews ed. 1991).
111. *Id.* at 259. At the 1983 Cartagena Convention (Caribbean Regional Seas Conv.), the U.S. introduced the concept of "veto available, but waiver possible upon later signing", which was also applied to the 1985 Vienna and 1989 Basel Conventions.
112. *Id.* at 259–261. For instance, uniform vaccination certificates under the WHO regulations, oil transport certificates under IMO's MARPOL, and environmental licensing of imported cars under the 1958 Geneva Agreement concerning the Adoption of Uniform Conditions of Approval and Reciprocal Recognition of Approval for Motor Vehicle Equipment and Parts.
113. *Id.* at 261–265. For instance, environmental impact procedures, modeled after the 1969 U.S. National Environmental Policy Act, were mimicked in Colombia (1974) and France (1976), eventually diffused to Algeria (1983), East Germany (1988) and West Germany (1990). Similar diffusion has occurred for the concept of effluent fees (initiated in the Ruhr River Basin in 1904), gas emissions fees (from East Germany in 1969), and environmental labels (West Germany, 1978).
114. *Id.* at 265–267. Dissemination of published alerts regarding hazardous compounds, exemplified by UNEP adopting a provisional notification scheme in 1984, now implemented by 75 countries. Similarly, Appendix III optional, unilateral listing of endangered species under art. 5 of the CITES treaty, now numbering 240 taxa.
115. *Id.* at 267–268. Ongoing communities of individuals entrusted with implementation in their national settings build consensus on procedures and standards. The network of meteorological stations involved in the Convention on Long-Range Transboundary Air Pollution constitutes such a group.
116. (Bramsen, 1972). "One country's use of a resource shall not be such as to injure a neighbor's use."
117. *See* Sands, *supra* at 269. Such litigation is uncommon although state responsibility for environmental harm dates back at least to the 1941 U.S.-Canadian Trail Smelter arbitration. Those claims, initiated in 1926 for immediate harm to identified U.S. victims closeby a Canadian single point source, took 15 years to settle. Modern claims, concerning long-term and long-range effects of multiple pollutants from multiple sources, are probably too complex for resolving by this process. *See also* (NAPAP reference → Canada-U.S. acid deposition claim) (Sand, 1991).
118. *Id.* at 270–71. The Trail Smelter arbitration was forced because local remedy by private action in Canadian courts was blocked in the 1930s by a House of Lords Rule to refuse jurisdiction over suits based on damage to foreign land. Since then, however, numerous local remedies have become available to resolve transboundary problems, aided by the OECD adoption of the 1976–77 Recommendations on Equal Right of Access and Nondiscrimination in Relation to Transfrontier Pollution.
119. *Id.* at 271–73. The 1987 Montreal Protocol foresees nonjudicial complaints filed with the UNEP secretariat to notify of non-compliance. European Economic Community environmental standards can be enforced by a "custodial" procedure initiated within the EEC Commission, under art. 169 of the 1957 Rome Treaty. Half of the 460 proceedings in 1989 were started by complaints of private individuals and non-governmental organizations (often Greenpeace or Friends of the Earth) sent to the EEC "complaint registry" in Brussels.
120. *Id.* at 273–76. Occupational environmental standards have been maintained by the International Labour Organization (ILO) through a system of periodic reporting combined with regular auditing by an independent technical committee and a public debate of the audit results, rather than through the formal dispute mechanisms established by the relevant multilateral conventions. Similar approaches are used by the U.N. Commission on Human Rights with their country reports and public hearings, and by the Executive Body of the Convention on Long-range Transboundary Air Pollution with its published reviews. Nongovernmental Organizations (NGOs) are playing an increasingly important role by publishing their own audits of implementation of international agreements. See also Chayes and Chayes, *infra* note 110 at 295–307.
121. Schelling at 25.
122. *See* Sands, *supra* at 277.

123. *See* Dougher, 1983. One of three cooperative projects implemented in 1983 was entitled "Effects of Atmospheric Pollution on Climate." *See* Lancaster, 1985. This project included a joint expedition in the Pacific on the Soviet research vessel *Akademic Korolyev* during October-December 1983, in which the Carbon Dioxide Research Group at the Scripps Institution and other U.S. researchers participated with the Soviets, with the aid of the U.S. National Oceanic and Atmospheric Administration (NOAA).
124. These reports are difficult to obtain. The Draft Report was regularly cited, despite its bold markings to the contrary. The Final Report is now available. The Report on Effects is rather expensive: its dozen volumes are available at about $30–40 apiece from the National Technical Information Service.
125. Or, roughly, $10 billion per year spread over 25 to 50 years, or $5 billion per year over 50 to 100 years. *See* Titus et al., 1991. Although constructing levees and bulkheads may be cost-effective mitigating steps, the environmental consequence of such action could be loss of most of the nation's wetland shorelines, which may be unacceptable. *See* (Smith and Tirpak, 1989). In 1986, Congress asked EPA to prepare two reports on global warming, one to be an examination of policy options to stabilize greenhouse gases and the other to examine health and environmental effects of climate change in the United States. This second report, *The Potential Effects of Global Change on the United States* (1989), examined potential changes in hydrology, sea level rise, agriculture, forests, aquatics, air quality, human health, and infrastructure in the Southeast, Great Lakes, California and Great Plains regions of the United States.
126. *See* (Nordhaus, 1990) at 10–14. 1981 dollars and GNP at $2.4 trillion. *See generally* Miller, Policy Responses to Global Warming in *14 So. Illinois L. J.* 187(1980), for an excellent analysis of the United States policy and recent legislation.
127. particularly the Department of Energy's 10-year research program on carbon dioxide and the greenhouse effect, the National Oceanic and Atmospheric Administration's Office of Global Programs, and NASA's expanding program in earth sciences, including the recent "Mission to Planet Earth" project.
128. The ABA Section of Natural Resources, Energy, and Environmental Law, for example, now has a Special Committee on Global Climate and a Special Committee on International Environmental Law; while the International Environmental Law Committee of the ABA's Section of International Law and Practice recently held a conference entitled Global Change and International Law (February, 1989) and has formed a Subcommittee on Climate Change. *See International Law News,* Vol. 19, No. 3, Summer, 1990, at 3.
129. *See generally, ABA,* 1990, *Natural Resources < Energy and Environmental Law, 1989: The Year in Review* (Section of Natural Resources, Energy, and Environmental Law, American Bar Association, and Nat'l Energy Law and Policy Institute, University of Tulsa College of Law, Report by Fleming, Hollis, and Gallant) at 175–180, giving excellent summary. *See also* Miller, *supra,* note 1.1, at 227.
130. Global Change Research Act of 1990 (Pub. L. 101–606, approved 11/16/90) introduced Wed, January 25, 1989; 135 Cong. Rec. 523(1989); Reported to the Senate in May, 1989 by the Committee on Commerce, Science and Transportation (S. Rep. No. 40, 101st Cong. 1st Sess. (1989)); Committee on Environment and Public Works requested sequential referral in June, 1989, where it lay dormant for the remainder of 1989. Signed in the HOUSE 11/07/90.
131. Global Environmental Protection Act, S.133, 101st Cong., 1st. Sess.; 135 Cong. Rec. 1069; introduced 2/2/89. Leahy's bill was placed within S. 2386 by Daschle (D-SD), int. 3/30/90; referred to Senate Committee on Agr, Nutr, and Forestry 3/30/90.
132. Poterba at p7.
133. *See* Miller at 228, *supra,* note 126. It is interesting to consider whether these state provisions might be vulnerable to a federal pre-emption challenge, on the basis that the restrictions might not be able to be tied to the health of the citizens of the state, and that the federal government has acted internationally to not support emissions limitations. Probably, the federal government would be deemed to have not spoken on this issue, as no statues or regulations have issued. But if Congress were to adopt a greenhouse strategy that explicitly rejected emissions limitations, could states maintain their positions? Would they have to prove a potential harm? Could they, given present scientific uncertainties?
134. Ore. ALS 466; 1989 Ore. Laws 466, 1989 Ore. S.B. 576 (amending O.R.S. 469.060).
135. Vermont Executive Order No. 79 (Oct. 23, 1989).
136. New York Exec. Order No. 118 (Dec. 28, 1988).
137. *See generally,* Usher, Climate Change and the Developing World, in *So. Illinois L. J.* 257(1990). "[Developing countries] may well be the principal recipients of the worst that climate change might offer." at 259. *Also see generally,* Ramakrishna, North-South Issues, Common Heritage of Mankind and Global Climate Change, in 19(3) *Millenium: J. Intl Studies* 429. " . . . when issues of global environmental protection are competing for scarce resources needed to address the very means of improving living conditions, . . . , the developing countries are hard pressed to pay much attention to environmental issues." *And see generally,* Ramakrishna, at 431, and Woodwell & Ramakrishna, The Warming of the Earth: Perspectives and Solutions in the Third World, in 16(4) *Env'l Conservation. See* Schelling at 11.
138. Schelling at 10.

139. Barbier et al. at 4.
140. *See* EPA, 1989. Policy Options for Stabilizing Global Climate.

# REFERENCES


Bach, W., Pankrath, J., and Kellogg, W., eds., 1979, *Man's Impact on Climate.* Proceedings of an International Conference held in Berlin June 14–16, 1978, Elsevier Pub., New York.

Barbier, E. B., Burgess, J. C., and Pearce, D. W., 1990, Slowing Global Warming: options for Greenhouse Gas Substitution, Paper presented at the Conference on Economic Policy Responses to Global Warming, sponsored by the Istituto SanPaolo di Torino, October 5–6, 1990, Rome.

Bolin, B., Doos, B. R., Warrick, R. A., and Jaeger, J., eds., 1986, *The Greenhouse Effect, Climatic Change, and Ecosystems,* Scope 29, Scientific Committee on Problems of the Environment, International Council of Scientific Unions (ICSU), John Wiley & Sons, New York.

Bramsen, C. B., 1972, Transnational Pollution and International Law, in *Problems in Transboundary Pollution,* OECD, August, 260.

Buber, M., 1970, *I and Thou,* Scribner, New York, 97.

Clark, W. C., ed., 1982, *Carbon Dioxide Review:* 1982 Oxford University Press, New York.

Chayes, A. and Chayes, A. H., 1991, Adjustment and Compliance Processes in International Regulatory Regimes, in: (Mathews, ed. 1991), 280–308.

Chen, R. S., Boulding, E., and Schneider, S. H., eds., 1983, *Social Science Research and Climate Change: an Interdisciplinary Appraisal,* D. Reidel, Boston.

Curson, P., 1989, Human Health in the Greenhouse Era, in *Planning for the Greenhouse,* School of Earth Sciences, MacQuarie University, Australia (cited in Haines, 1990).

Damkaer, D. M., 1987, Possible influences of solar UV radiation in the evolution of marine zooplankton, in *The Role of Solar Ultraviolet Radiation in Marine Ecosystems,* J. Calkins, ed., Plenum Press, New York.

DOE, 1985, *The Potential Climatic Effects of Increasing Carbon Dioxide,* MacCracken, M. C. and Luther, F. M., eds., United States Dept. of Energy, Office of Energy Research, DOE/ER-0237, December, Washington, D.C.

Dougher, 1983, International Carbon Dioxide-Related Activities: the International Organizations Involved and U.S. Bilateral Arrangements. Prepared for the U.S. Dept. of Energy, Office of Energy Research, DOE/NBB-0039, Washington, D.C.

ECE, 1979, Long Range Transboundary Air Pollution, (Draft Convention, ECE), *Environmental Policy and Law,* 5, No. 2.

Edmonds, J. A., Reilly, J., Trabalka, J. R., Reichle, D. E., Rind, D., Lebedeff, S., Palutikof, J. P., Wigley, T. M. L., Lough, J. M., Blasing, T. J., Solomon, A. M., Seidel, S., Keyes, D., and Steinberg, M., 1986, *Future Atmospheric Carbon Dioxide Scenarios and Limitation Strategies,* Noyes Pub., Park Ridge, New Jersey.

Edmonds, J. and Reilly, J., 1983, Global Energy and #$CO_2$# to the Year 2050, *The Energy Journal,* Vol. 4, No. 3, 21–47.

Emanuel, K. A., 1987, The dependence of hurricane intensity on climate, *Nature,* Vol. 326, 483–485.

EPA, 1983, *Can We Delay a Greenhouse Warming?,* Prepared by Seidel, S. and Keyes, D., U.S. Environmental Protection Agency. Wash., D.C., September.

EPA, 1989, *Policy Options for Stabilizing Global Climate,* Lashof, D. A. and Tirpak, D. A., eds., Draft Report to Congress, U.S. Environmental Protection Agency, February.

Flavin, C., 1990, Slowing Global Warming, in: *State of the World,* Brown, L. R., ed., W. W. Norton, New York.

Flynn, T. J., Walesh, S. G., Titus, J. G., and Barth, M. C., 1984, Implications of Sea Level Rise for Hazardous Waste Sites in Coastal Floodplains, in (see Barth and Titus, 1984), 271–294.

Gable, F. J. and Aubrey, D. G., 1989, Changing Climate and the Pacific, *Oceanus,* Vol. 32, No. 4, pp. 71–73.

Giannini, S. H., 1990, Effects of UVB on Infectious Diseases, in: *Global Atmospheric Change and Public Health,* White, J. C., ed., Elsevier, New York, 33–45.

Gillett, J. D., 1981, Increased Atmospheric Carbon Dioxide and Spread of Parasitic Disease, in *Parasitological Topics: a Presentation Volume to P. C. Garnham FRS,* Society of Protozoologists, special publication, Vol. 1, pp. 106–111.

Glantz, M. H., Price, M. F., and Krenz, M. E., eds., 1990, *On Assessing Winners and Losers in the Context of Global Warming,* Report of a Workshop held at St. Julians, Malta, 18–21 June, 1990, UNEP and National Center for Atmospheric Research, Boulder, Colorado.

Glantz, M. H. and Feingold, L. E., eds., 1990, *Summary Report of the Climate Variability, Climate Change and Fisheries Project,* ESIG/NMFS/EPA Study, Environmental and Societal Impacts Group, National Center for Atmospheric Research, Boulder, Colorado.

Gleick, P. H., 1989, Climate Change, Hydrology, and Water Resources, *Reviews of Geophysics*, Vol. 27, No. 3, 329–344.

Gleick, P. H. and Maurer, E. P., 1990, *Assessing the Costs of Adapting to Sea Level Rise. A Case Study of San Francisco Bay,* Stockholm Environment Institute, Stockholm, Sweden.

Haines, A., 1990, Implications for Health, in *Global Warming: the Greenpeace Report,* Leggett, J., ed., Oxford Univ. Press, New York, chap 7, 149–162.

Henson, R. and Hively, W., 1989, Toward an International Law of the Atmosphere, *Amer. Scientist,* 77, 324–326.

Heywood, T. A., 1979, Environmental Modification: Convention on Long-Range Transboundary Air Pollution, *21 Harvard Int. Law Jour.* 536–8, citing U.N. Doc. ECE/HLM.1/R.1 (1979), reprinted in *18 Int. Law Mats.* 1442 (1979).

IPCC, 1990, *Potential Impacts of Climate Change,* Report prepared for Intergovernmental Panel on Climate Change by Working Group II, June 1990, WMO and UNEP, Geneva, Switzerland.

IPCC, 1990a, Working Group I, etc.

Jaeger, J., 1988, Developing Policies for Responding to Climatic Change. A Summary of Discussions and recommendations of the workshops held in Villach (28 September - 2 October 1987) and Belagio (9–13 November), under the auspices of the Beijer Institute, Stockholm, World Meteorological Organization/Technical Document No. 225, April 1988.

Kalkstein, L. S., Davis, R. E., Skindlov, J. A., and Valimont, K. M., 1986, The impact of human-induced climate warming upon human mortality: a New York case study, in *Proc. Int. Conference on Health and Environmental Effects of Ozone Modification in Climate Change,* Washington, D.C. (cited in Haines, 1990).

Kates, R. W., Ausubel, J. H., and Berberian, M., eds., 1985, *Climate Impact Assessment* SCOPE 27, Scientific Committee on Problems of the Environment, International Council of Scientific Unions (ICSU), John Wiley & Sons, New York.

Kellogg, W. W., 1983, Feedback mechanisms in the climate system affecting future levels of carbon dioxide, *JGR,* 88, C2, 1263–1269.

Kellogg, W. W. and Schneider, S. H., 1974, Climate Stabilization: For better or for worse? *Science,* 186, 1163–1172.

Kellogg, W. W. and Schware, R., 1981, *Climate Change and Society: consequences of Increasing Atmospheric Carbon Dioxide.* Westview Press, Boulder, Colorado.

Lancaster, J., 1985, Carbon Dioxide Variations, Ocean Air Sampling near Hawaii and Southern California from July 1983 to September 1984. Master's Thesis, University of California, San Diego.

Lancaster, J., 1989, The Theory of Radially Evolving Energy, *Int. J. General Systems,* 16, 43–73.

Lashof, D. A., 1989, The Dynamic Greenhouse: Feedback Processes that may Influence Future Concentrations of Atmospheric Gases and Climatic Change, *Climatic Change,* 14, 213–242.

Lave, L. B., 1990, The Greenhouse Effect: What Government Actions are Needed?, *Journal of Policy Analysis and Management,* Vol. 7, pp. 460–470.

Leatherman, S. P., 1989, Impact of Accelerated Sea Level Rise on Beaches and Coastal Wetlands, in *Global Climate Change Linkages,* White, J. C., ed., Elsevier, New York, 43–57.

Liehne, P. F. S., 1989, Climatic Influences on Mosquito-Borne Diseases in Australia, in *Greenhouse Planning for Climate Change,* Pearman, G. R., ed., Division of Atmospheric Research, CSIRO, Australia.

Longstreth, J. A., 1989, Human Health, in *The Potential Effects of Global Climate Change on the United States,* Smith, J. B. and Tirpak, D., eds., U.S. Environmental Protection Agency, Washington, D.C.

MacDonald, G. J., 1982, *The Long-term Impacts of Increasing Atmospheric Carbon Dioxide Levels,* Ballinger, Cambridge, Mass.

Matthews, J. T., ed., 1991, *Preserving the Global Environment,* W. W. Norton, New York, 236–279.

Mehta, A. J. and Cushman, R. M., eds., 1989, *Workshop on Sea Level Rise and Coastal Processes,* (Palm Coast, Florida, March 9–11, 1988) DOE/NBB-0086, U.S. Dept. of Energy, Washington, D.C.

Miller, D. L. R. and Mackenzie, F. T., 1988, Implications of Climate Change and Associated Sea-Level Rise for Atolls, in *Proc. 6th Intl. Coral Reef Symposium,* Australia, Vol. 3, 519–522 (Hawaii Institute of Geophysics Contribution No. 2063).

Nordhaus, W. D., 1990, A Survey of the Costs of Reduction of Greenhouse Gases, *The Energy Journal,* Fall, 1990.

Nordhaus, W. D., 1991, Economic Approaches to Greenhouse Warming, in *Economic Policy Responses to Global Warming,* Dornbusch, R. and Poterba, J., eds., (available through W. Nordhaus, Dept. of Economics, Yale University).

NRC, 1983, *Changing Climate.* Report of the Carbon Dioxide Assessment Committee, National Research Council, National Academy Press, Washington, D.C.

NRC, 1989, *Ozone Depletion, Greenhouse Gases and Climate Change,* Proceedings of a Joint Symposium by the Board on Atmospheric Sciences and Climate and the Committee on Global Change, National Research Council, National Academy Press, Washington, D.C.

Oeschli, F. W. and Buechley, R. W., 1970, Excess Mortality Associated with Three Los Angeles September Hot Spells, *Environmental Research,* Vol. 3, 277–284 (cited in White and Hertz-Picciotto, 1985).

Poterba, J. M., 1991, Designing a Carbon Tax, in: *Economic Policy Responses to Global Warming,* Dornbusch, R. and Poterba, J., Eds., (see Nordhaus, 1991).

Rathjens, G. W., 1991, Energy and Climate Change, in Mathews, ed., 1991, 154–186.

Revelle, R., 1980, $CO_2$-Induced Climate Change — Asking the Right Questions, paper presented at the Amer. Geophys. Union, San Francisco.

Roberts, L., 1988, Is There Life After Climate Change?, *Science,* 242, 1010–1012.

Roberts, L. K., Samlowski, W. E., and Daynes, R. A., 1986, The Immunological Consequences of Ultraviolet Radiation Exposure, *Photodermatology,* Vol. 3, 284–298.

Sand, P. H., 1991, International Cooperation, in: Matthews, J. T. ed., *supra.*

Schneider, S. H., 1989, The Greenhouse Effect: Science and Policy. *Science,* 243, 771–781.

Schelling, T. C., 1991, International Burden Sharing and Coordination: prospects for Cooperative Approaches to Global Warming, in *Economic Policy Responses to Global Warming,* Dornbusch, R. and Poterba, J., eds., (see Nordhaus, 1991).

Schimel, D., 1990, Biogeochemical Feedbacks in the Earth System, in *Global Warming: The Greenpeace Report,* Leggett, J., ed., Oxford University Press, New York, chap. 3, 68–82.

Seitz, F., Bendelsen, K., Jastrow, R., and Neirenberg, W. A., 1989, *Scientific Perspectives on the Greenhouse Problem: executive Summary,* George C. Marshall Institute, Washington, D.C.

Sestini, G., Jeftic, L., and Milliman, J. D., 1989, *Implications of Expected Climate Changes in the Mediterranean Region: an Overview,* UNEP Regional Seas Reports and Studies No. 103, United Nations Environment Programme, Nairobi.

Shope, R. E., 1990, Infectious Diseases and Atmospheric Change, in *Global Atmospheric Change and Public Health,* White, J. C., ed., Elsevier, New York, 47–54.

Smetz, H., 1972, Alternative Economic Policies of Unidirectional Transfrontier Pollution, in *Problems in Transboundary Pollution,* OECD, Aug. 1972.

Smith, J. B. and Tirpak, D. A., eds., 1989, *Potential Effects of Global Climate Change on the United States: Appendix H, Infrastructure,* U.S. Environmental Protection Agency, Washington, D.C.

Solow, A. R., 1990, Is There a Global Warming Problem? Paper presented at the Conference on Economic Policy Responses to Global Warming, sponsored by the Istituto SanPaolo di Torino, October 5-6, 1990, Rome.

Tangley, L., 1988, Preparing for climate change. *BioScience,* 38, no. 1, 14–18, (January).

Tickell, C., 1977, *Climate Change and World Affairs,* see 1986 Rev. Ed., Harvard University (Cambridge) and University of America Press (Lanham, MA).

Titus, J. G., Henderson, T. R., and Teal, J. M., 1984, Sea Level Rise and Wetlands Loss in the United States, *Nat. Wetlands Newsletter,* Vol. 6, No. 5, Sept-Oct.

Titus, J., 1991, Greenhouse Effect and Coastal Wetland Policy: how Americans Could Abandon an Area the Size of Massachusetts, *Environmental Management Journal,* (in press, Jan/Feb).

Toronto, 1988, *The Changing Atmosphere: implications for Global Security.* Conference Statement, Toronto, Ontario, Canada, June 27–30.

Warrick, R. A. and Oerlemans, J., 1990, Sea Level Rise, in *Scientific Assessment of Climate Change,* Report prepared for Intergovernmental Panel on Climate Change by Working Group I, June 1990, WMO and UNEP, Geneva, Switzerland, 261–285.

Weiss, E. B., 1980, *International Legal and Institutional Implications of an Increase in Carbon Dioxide,* prepared for the American Association for the Advancement of Science, Washington, D.C.

Whaley, J. and Wigle, R., 1991, The International Incidence of Carbon Taxes, in: *Economic Policy Responses to Global Warming,* Dornbusch, R. and Poterba, J., eds.

White, M. R. and Hertz-Picciotto, I., 1985, Human Health: Analysis of Climate Related to Health, in *Characterization of Information Requirements for Studies of Carbon Dioxide Effects: Water Resources, Agriculture, Fisheries, Forests and Human Health,* White, M. R., ed., U.S. Department of Energy, DOE/ER-0236, December, Washington, D.C., 171–221.

Wigley, T. M. L. and Warrick, R. A., eds., 1990, *United Nations Environmental Programme/CEP/USEPA International Workshop on Climatic Change, Sea Level, Severe Tropical Storms and Associated Impacts,* UNEP, Nairobi.

Wirth, D. A. and Lashof, D. A., 1990, Beyond Vienna and Montreal: Multilateral Agreements on Greenhouse Gases, *Ambio,* (in press), (republished in this volume).

WMO, 1980, Assessment of the Role of $CO_2$ on Climate Variations and Their Impact, meeting report, WMO/UNEP/ICSU, Austria.

Woods Hole, 1988, Steps Toward an International Convention Stabilizing the Composition of the Atmosphere. Draft Summary of Workshop on Global Climatic Change held September 13–14, 1988, at the Woods Hole Research Center in Woods Hole, Massachusetts.

Yohe, G. W., 1990, Uncertainty, Climate Change and the Economic Value of Information: an Economic Methodology for Evaluating the Timing and Relative Efficacy of Alternative Responses to Climate Change, in *Useable Knowledge for Managing Global Climatic Change,* (Clark, W. C., Ed.), The Stockholm Environment Institute, Stockholm, 101–134.

# APPENDIX I
# SIGNIFICANT EVENTS IN THE EVOLUTION OF THE FRAMEWORK CONVENTION ON GLOBAL CLIMATE CHANGE

**1957** Revelle and Suess ''grand geophysical experiment'' *Tellus* 9, 18–27
**1960** Keeling first notes $CO_2$ increase. *Tellus*, 12, 2, 200–203
**1968** Paris, Biosphere Conference
**1972** Stockholm Conference and formation of UNEP
**1979** First World Climate Conference
**1983** NAS/NRC publishes ''Changing Climate'', Neirenberg ed.
**1985** EPA publishes ''Can We Stop a Greenhouse Warming?''
(Mar) Vienna Convention on Protection of Ozone
(Oct) Villach, Austria, 29 countries reach consensus on global warming
**1987** Montreal Protocol on CFCs
**1988** (Jun) Toronto Conf., 46 nations propose 20% reduction by 2005, WAF and Convention
(Nov) Intergovernmental Panel on Climate Change (IPCC) forms in Geneva, 35 nations attend
(Dec) U.N. Gen'l Assembly Res. 43/169 requests IPCC to anticipate Convention
**1989** (Feb) Ottawa Meeting of Legal and Policy Experts, Convention needed
(Mar) Hague Declaration, Convention needed
(May) Helsinki Declaration. In Nairobi, UNEP plans negotiations, 44 nations attend
(Jul) Paris, G-7 meeting for ''green summit''. Convention needed.
(Nov) Noordwijk, Call for 20% reduction, blocked by U.S., U.S.S.R., and Japan
**1990** (Apr) White House Conference on Economics and Science Relating to Greenhouse Warming
(Aug) Sundsval, IPCC final meeting
(Nov) Second World Climate Conference
**1991** (Feb) Washington, Negotiations start for Convention at UNCED, Brazil, 1992
**1992** (Jun) Rio de Janeiro, Convention opened for signature at UNCED

Chapter 27

# GLOBAL WARMING: THE ENERGY POLICY CHALLENGE

**Christopher Flavin and Nicholas Lenssen**

## TABLE OF CONTENTS



0-8493-4419-0/93/$0.00+$.50
© 1993 by CRC Press, Inc.

## TABLE OF CONTENTS



0-8493-4419-0/93/$0.00 + $.50
© 1993 by CRC Press, Inc.

# I. INTRODUCTION

Scientific discoveries over the past decade suggest that the world's fossil fuel-based economy now threatens catastrophic changes in the global climate. Such changes could derail the world economy and disrupt its environmental support systems. If we are to ensure a healthy and prosperous world for future generations, only a few decades remain to redirect the energy economy.

Carbon is a basic component of all fossil fuels that is released whenever they are burned. This carbon is converted to carbon dioxide, the leading greenhouse gas, which is building steadily in the atmosphere — now at 354 ppm. Computer models of the world's climate that have been endorsed by the Intergovernmental Panel on Climate Change (IPCC) project an unprecedented warming of between 1.5 and 4.5°C by the middle of the next century.

A scientific study released in 1990 by the United Nations-commissioned IPCC confirmed that a rapid and highly disruptive increase in global temperatures would likely occur unless emissions are cut. Upon releasing the report, Dr. John Houghton, head of the British Meteorological Service, noted that it represented "remarkable consensus," with fewer than 10 of 200 scientists dissenting. Although greenhouse gas concentrations rise slowly, future climate disruptions are likely to be abrupt and catastrophic. Already, the past decade has seen the warmest global average temperatures recorded during the past century, a mild foretaste of what may be ahead. Combustion of all the world's remaining fossil fuels would raise the concentration of carbon dioxide as much as tenfold, compared with the mere doubling that now concerns scientists. Slowing global warming inevitably means placing limits on the use of fossil fuels.[1]

Ignored by policy makers as recently as 1988, the risk of climate change, and the enormous damage to agriculture and the rest of the environment it could bring, is now a force that drives energy policy in many nations. Some 23 countries and the European Community as a whole have announced intentions to limit their carbon emissions. Although some are planning just to slow the rate of growth, others, such as Germany, plan reductions of 25 to 30% and, by implication, their use of fossil fuels. At the Second World Climate Conference, held in Geneva in November 1990, 137 nations agreed to draft a treaty by 1992 to slow global warming. Treaty negotiations began in early 1991, with plans to have it ready for signing by heads of state at the U.N. Conference on Environment and Development in Brazil in June 1992.[2]

Scientists have concluded that stabilizing the climate will ultimately require reducing global carbon dioxide emissions by 60 to 80%. The wealthy nations, which currently produce most of the carbon dioxide, would have to make even more dramatic cuts to allow for population and economic growth in the Third World. These changes imply the development of a far different energy system. But few political leaders have any notion of an economy not based on fossil fuels. Indeed, the inability of societies to redirect their energy course is as much a failure of vision as of policy.[3]

Government efforts to map out such a course have been based on the mistaken assumption that future energy systems must follow a centralized, fossil-fuel-dependent path. The most recent meeting of the World Energy Conference, a gathering of government officials and experts, concluded that energy needs three decades from now will be 75% higher than current levels, and will be met mainly by coal, oil, and nuclear power. The World Energy Conference scenario would lead to soaring carbon dioxide emissions and accelerated global warming. While traditional energy planners may view such a future as logical, close examination raises doubts about its desirability or even feasibility.[4]

To stabilize the climate, the world soon will have to reduce its consumption of fossil fuels drastically from the current level of 6 billion ton annually; this entails not only improving energy efficiency but also developing major new energy sources. The technologies are at hand to greatly expand the use of renewable energy in the next few decades. We have constructed a practical energy scenario for the year 2030 that involves a 55% cut in carbon dioxide emissions, greatly improved energy efficiency, and an energy production system that relies heavily on solar energy, geothermal energy, wind power, and the energy of living plants. The year 2030 can be viewed as a mid-point in a long-term energy transition — enough time to develop major new energy systems, but not to eliminate fossil fuels entirely.

## II. GLOBAL $CO_2$ EMISSIONS

In order to achieve atmospheric stability in the long run, $CO_2$ emissions will eventually need to be trimmed back to the level at which the oceans can absorb it. (They can only take up one to two billion tons/year, approximately the level of emissions in 1950.) To achieve the needed two-thirds reduction means restructuring the global energy economy over the coming decades — greatly improving energy efficiency and shifting from heavy reliance on oil and coal to dependence on renewable energy sources and natural gas.

Global $CO_2$ emissions have tripled since 1950. However, emission growth rates have slowed from 4.6% annually in the 1950s and 1960s to 2.5% in the 1970s and 1.2% in the 1980s. The main reasons: higher oil prices, which encouraged greater energy efficiency, saturation of some energy markets, and slowed economic expansion in developing countries.[5] Like the 7% drop in $CO_2$ emissions from 1979 to 1983, the recent decline represents both efficiency gains and declining economic activity. Energy efficiency continues to improve by about 1% annually in industrial countries, responsible for more than half of $CO_2$ emissions. The 1991 recession had an additional impact.

In Eastern Europe, the decline is more dramatic. Between 1987 and 1990, emissions fell 28%. In the Soviet republics, the decline did not begin until 1989, and is still picking up speed — down 4% by 1990, and likely to be down another 5% or more by the end of 1991. These figures reflect the fact that the current industrial base of the Soviet republics and Eastern Europe has no hope of being competitive, and is rapidly being shut down. As they proceed with economic reforms, further declines in $CO_2$ emissions are inevitable, and may plummet if the region's intertwined energy economy breaks down. Taken together, these trends could cut Soviet and East European emissions 30% by the mid-1990s. While emissions will eventually rebound, they may never regain the 1987 peak, let alone exceed it by 20%, as the OECD suggested in a 1991 report. The reason: in order to be competitive, these nations will have to adopt energy technologies that are far more efficient than those in use today.

The industrial market countries of North America, Europe, and Japan, remain the key to atmospheric stability since they dominate global $CO_2$ emissions.

Cutting their emissions in the near future is critical; however, even developing countries that have so far contributed only minimally to global warming, cannot be ignored. Although the Third World contains more than 75% of the world's population, $CO_2$ emissions per capita are between one fiftieth and one fifth that of industrial nations. However, developing countries are bound to produce a rising share of carbon dioxide emissions during the next several decades. Unless they gain access to new efficiency and renewable energy technologies, their $CO_2$ emissions will surge, offsetting declines in industrial nations. As in industrial nations, such technologies are also the key to developing countries' economic competitiveness as well as their domestic environmental quality.

## III. ENERGY EFFICIENCY

Energy efficiency is without doubt the cornerstone of any global energy system that does not threaten the world's climate. Overall, the world will have to produce goods and services with one third to one half as much energy as now. The 21 industrial market nations that belong to the International Energy Agency have lowered their energy use per unit of gross national product 24% since 1973, and opportunities for further improvement abound. The Soviet Union, Eastern Europe, and developing countries have an even larger untapped potential for efficiency gains.[6]

Technologies are already available that could quadruple the efficiency of most lighting systems and double the fuel economy of new cars. Efficiency improvements in areas such as lighting, electric motors, and appliances could reduce the need for power by 40 to 75% — at less than half the cost of power from new plants. Already, two electric utilities — the Southern California Edison Company and the Los Angeles Department of Water and Power — plan to rely on investments in improved efficiency in their customers' buildings and factories to reduce $CO_2$ emissions from power plants by 19% over the next 20 years.[7] Heating and cooling needs of buildings could be cut even further through improved furnaces and air conditioners, as well as better insulation and windows. Over the next 30 years, industrial countries could reduce their energy use per capita by at least half. In developing countries, improved efficiency could allow energy consumption per capita to remain constant or increase modestly while their economies grow.[8]

For societies eventually to achieve the 60 to 80% reduction in global carbon emissions needed to stabilize climate, however, the mix of fuels will also need to be shifted. Even among the fossil fuels there are wide variations in environmental impact. One key distinction is that oil contains 17% less $CO_2$ per unit of energy than coal, while natural gas has 43% less. Coal is far and away the most environmentally damaging of the fossil fuels, both from an air pollution and a climate staindpoint.[9] The use of coal will almost certainly need to be greatly reduced in the future. Expanded use of oil shale and synthetic fuels derived from coal can be ruled out entirely due to their high carbon content. For most nations, natural gas will likely be the predominant fossil fuel, as it produces much more energy per kilogram of carbon released than do the other fossil fuels.[10]

## IV. RENEWABLE ENERGY RESOURCES

The world will also soon need to begin turning to non-carbon producing energy sources in order to meet the needs of growing populations. Since nuclear power has failed to meet basic economic and technical requirements in most nations, the best available alternative is the various renewable sources of energy. Such a course would entail expanding the use of energy from biomass (biological sources such as wood or agricultural wastes) and hydropower, but more importantly would require large contributions from solar, wind, and geothermal sources. The technologies to harness these resources are now ready for widespread use — if they receive effective support in the years ahead.[11]

Renewable energy resources are actually far more abundant than fossil fuels. The U.S. Department of Energy estimates that the annual influx of accessible renewable resources in the U.S., for example, is more than 200 times its use of energy, and more than 10 times its recoverable reserves of fossil and nuclear fuels. Harnessing these resources will inevitably take time, but according to a new study by U.S. government scientific laboratories, renewables could supply the equivalent of 50 to 70% of current U.S. energy use by the year 2030.[12]

### A. BIOMASS AND HYDROPOWER

Contrary to popular belief, renewables — primarily biomass and hydropower — already supply about 20% of the world's energy. Biomass alone meets 35% of developing countries' total energy needs, though often not in a manner that is renewable or sustainable in the long term. And in certain industrial countries, renewables play a central role: Norway, for example, relies on hydropower and wood for more than 50% of its energy.[13]

Steady advances have been made since the mid-1970s in a broad array of new energy technologies that will be needed if the world is to increase its reliance on renewable resources. Many of the machines and processes that could provide energy in a solar economy are now almost economically competitive with fossil fuels. Further cost reductions are expected in the next decade as these technologies improve. As leading solar scientists, Carl Weinberg and Robert Williams, wrote in *Scientific American:* "Electricity from wind, solar-thermal, and biomass technologies is likely to be cost-competitive in the 1990s; electricity from photovoltaics and liquid fuels from biomass should be so by the turn of the century." The pace of deployment, however, will be determined by energy prices and government policies. After a period of neglect in the eighties, many governments are now supporting new energy technologies more effectively, which may signal the beginning of a renewable energy boom in the years ahead.[14]

## B. SOLAR ENERGY

Direct conversion of solar energy will likely be the cornerstone of a sustainable world energy system. Not only is the sunshine available in great quantity, it is more widely distributed than any other energy source. Solar energy is especially well suited to supplying heat at or below the boiling point of water (used largely for cooking and heating), which accounts for 30 to 50% of energy use in industrial countries and even more in the developing world. A few decades from now, societies may use the sun to heat most of their water, and new buildings may take advantage of natural heating and cooling to cut energy use by more than 80%.[15]

Solar rays are free and can be harnessed with minor modifications in building construction, design, or orientation. In Cyprus, Israel, and Jordan, solar panels already heat between 25 and 65% of the water in homes. More than 1 million active solar heating systems, and 250,000 passive solar homes, which rely on natural flows of warm and cool air, have been built in the U.S. Advanced solar collectors can produce water so hot — 200°C — that it can meet the steam needs of many industries. Indeed, using electricity or directly burning fossil fuels to heat water and buildings may become rare during the next few decades.[16]

Solar collectors, along with other renewable technologies, can also turn the sun's rays into electricity. In one design, large mirrored troughs are used to reflect the sun's rays onto an oil-filled tube that produces steam for an electricity-generating turbine. A southern Californian company, Luz International, generates 354 MW of power with these collectors and has contracts to install an additional 320 MW. The newest version of this "solar thermal" system turns 22% of the incoming sunlight into electricity. Spread over 750 ha, the collectors produce enough power for about 170,000 homes for as little as $0.08/kWh, already competitive with generating costs in some regions.[17]

Future solar thermal technologies are expected to produce electricity even more cheaply. Parabolic dishes follow the sun and focus sunlight onto a single point where a small engine that converts heat to electricity can be mounted, or the energy transferred to a central turbine. Since parabolic dishes are built in moderately sized, standardized units, they allow for generating capacity to be added incrementally as needed. By the middle of the next century, vast areas of arid and semiarid countryside could be used to produce electricity for export to power-short regions.[18]

Photovoltaic or solar cells, which convert sunlight into electricity directly, almost certainly will be ubiquitous by 2030. These small, modular units are already used to power pocket calculators and to provide electricity in remote areas. Within a generation, solar cells could be installed widely on building rooftops, along transportation rights-of-way, and at central generating facilities. A Japanese company, Sanyo Electric, has incorporated them into roofing shingles.[19]

Over the past two decades, the cost of photovoltaic electricity has fallen from $30 a kWh to just $0.30. The forces behind the decline are steady improvement in cell efficiency and manufacturing, as well as rising demand. These cost reductions mean that in rural areas, pumping water with photovoltaics is already often cheaper than using diesel generators. Solar cells are also the least expensive source of electricity for much of the rural Third World; more than 6,000 villages in India now rely on them, and Indonesia and Sri Lanka also have initiated ambitious programs.[20] By the late 1990s, when solar cell electricity is expected to cost $0.10/kWh, some countries may be turning to photovoltaics to provide power for well-established grids. By 2030, they could provide a large share of the world's electricity — for as little as $0.04/kWh.[21]

## C. WIND POWER

Another form of solar energy, wind power, captures the energy that results from the sun's unequal heating of the earth's atmosphere. Electricity is generated by propeller-driven mechanical turbines perched on towers located in windy regions. The cost of this source of electricity has fallen from more than $0.30/kWh in the early 1980s to a current average of just $0.08. By the end of the 1990s, the cost is expected to be around $0.05. Most of the price reductions have come from experience gained in California, which accounts for nearly 80% of the world's wind-produced electricity. Denmark, the world's second-largest wind energy producer, received about 2% of its power from wind turbines in 1990.[22]

Wind power has a huge potential. It could provide many countries with one fifth or more of their electricity. Some of the most promising areas are in northern Europe, northern Africa, southern South America, the U.S. western plains, and the trade-wind belt around the tropics. A single windy ridge in Minnesota, 160 km long and 1.6 km wide, could be used to generate three times as much wind power as California gets today. Even more productive sites have been mapped out in Montana and Idaho.[23]

## D. BIOENERGY

Living green plants provide another means of capturing solar energy. Through photosynthesis, they convert sunlight into biomass that, burned in the form of wood, charcoal, agricultural wastes, or animal dung, is the primary source of energy for nearly half the world — about 2.5 billion people in developing countries. Sub-Saharan Africa derives some 75% of its energy from biomass, most of it using primitive technologies and at considerable cost to the environment.[24]

Many uses of bioenergy will undoubtedly increase in the decades ahead, though not as much as some enthusiasts assume. Developing nations will need to find more sophisticated and efficient means of using biomass to meet their rapidly increasing fuel needs. With many forests and croplands already overstressed, and with food needs competing for agricultural resources, it is unrealistic to think that ethanol distilled from corn can supply more than a tiny fraction of the world's liquid fuels. And shortages of irrigating water may complicate matters, especially in a rapidly warming world. In the future, ethanol probably will be produced from agricultural and wood wastes rather than precious grain. By employing an enzymatic process, rather than inefficient fermentation, scientists have reduced the cost of wood ethanol from $4 a gallon to $1.35 over the past 10 years, and expect it to reach about $0.60 gallon by the end of the 1990s.[25,26]

More efficient conversion of agricultural and forestry wastes to energy could boost biomass energy's role in the future. Wood stoves that double or treble today's efficiency levels already exist, and better designs are under development. For modular electricity generation, highly efficient gas turbines fueled by biomass can be built even at a very small scale. Some 50,000 MW of generating capacity, 75% of Africa's current total, could come from burning sugarcane residues alone. In the future, integrated farming systems, known as agroforestry, could produce fuel, food, and building materials.[27]

### E. HYDROPOWER

Hydropower now supplies nearly one fifth of the world's electricity. Although there is still ample growth potential, particularly in developing countries, environmental constraints will greatly limit such development. Small-scale projects are generally more promising than the massive ones favored by governments and international lending agencies. Smaller dams and reservoirs cause less social and ecological disruption. In deciding which hydropower resources to develop, issues such as land flooding, siltation, and human displacement will play an important role. These considerations will likely keep most nations from exploiting all of their potential.[28]

### F. GEOTHERMAL ENERGY

Another important element of a renewable-based system is geothermal energy — the heat of the earth's core. This is not strictly a renewable resource, however, and it needs to be carefully tapped so as not to deplete the local heat source. Since geothermal plants can produce power more than 90% of the time, they can provide electricity when there is no sun or wind. Geothermal resources are localized, through found in many regions. Worldwide, more than 5,600 MW worth of geothermal power plants have been built. El Salvador gets 40% of its electricity from the earth's natural heat, Nicaragua 28%, and Kenya 11%. Most Pacific Rim countries, as well as those along East Africa's Great Rift Valley and around the Mediterranean, could tap geothermal energy. Virtually the entire country of Japan, for example, lies over an enormous heat source that one day could meet much of the country's energy needs.[29]

While fossil fuels have been in storage for millions of years, renewable energy is in constant flux — replenished as the sun shines. While not a constraint in the near future, the intermittent nature of sunshine and wind means that the large-scale use of renewables will need to be backed by some form of energy storage. Developing new and improved storage systems is therefore one of the key challenges in building a sustainable energy economy. The best option identified so far is to convert renewable power to a gaseous form that is easy to transport and store — such as hydrogen. Hydrogen is a simple gas composed of a single atom that can be substituted for most other energy carriers. It can be used to generate electricity, power industry, run home appliances, fuel automobiles, and even fly commercial aircraft. Burning it produces virtually no air pollution or greenhouse gases. Furthermore, hydrogen can be produced easily by running an electric current through water, a process known as electrolysis. Solar, wind, or geothermal plants in remote regions can produce hydrogen that is easily stored and then transported to cities with little energy loss.

Building the infrastructure of a solar-hydrogen economy will take several decades at least. One of the first cornerstones that must be laid is to increase the overall energy efficiency of the economy. By lowering the amount of power needed to heat and light buildings, run a computer, or drive a car by 50 to 90%, the feasibility and economics of using solar hydrogen are greatly enhanced. In a truly sustainable energy economy, people would live in efficient solar homes, the need for traveling by automobile would be reduced through improved urban design, and factories would run on a fraction of today's voracious energy requirements.

## V. POTENTIAL ROLE OF NATURAL GAS

Making the shift to renewable energy will also require a transitional fuel. Petroleum products such as gasoline are unlikely ever to be made from renewable energy, and trying to electrify the entire economy would be expensive and impractical. The better course is to reinforce a recent trend — the substitution of natural gas for oil, coal, and even electricity. New natural gas technologies make it possible to boost conversion efficiencies considerably, cutting air pollutants 90 to 99% and carbon dioxide ($CO_2$) emissions 30 to 65% from those typical of today's fossil-fuel-based engines and power plants.[30]

Whereas a decade ago, many countries seemed to be running out of natural gas, geologists now believe that it is available in sufficient quantity to be the bridge fuel to a renewable energy economy. Natural gas is not only more abundant than oil, it is dispersed more widely around the world. In many countries, reserves have barely begun to be tapped. And as concern about the environment mounts and new policy measures such as carbon taxes are enacted, the use of gas is likely to accelerate. If a more heavily gas-based economy does take hold, solar hydrogen can be added gradually to the mix of fuels, making a smooth shift to a sustainable energy economy.

Natural gas is the simplest of the hydrocarbons. Methane, its main ingredient, is one of the most basic organic molecules, made up of a carbon atom surrounded by four hydrogen atoms. The methane generally is accompanied by varying amounts of $CO_2$ and other flammable gases such as ethane and propane. The simplicity of natural gas, particularly the fact that it is usually low in sulfur, is one reason it causes less air pollution than other hydrocarbons.

The main uses of natural gas today are as an industrial energy source and as a heating fuel in residential and commercial buildings. These applications can both be expanded and made more efficient thanks to a new generation of gas furnaces, water heaters, and air-conditioning systems. Heating with natural gas is in most circumstances less polluting — and contributes less to global warming — than heating with electricity produced from coal or oil. In addition, two major elements of the global energy economy stand out for their potential to use large additional amounts of natural gas: the electric power industry and transportation. Together, they are responsible for more than half of the carbon emissions from fossil fuels. By substituting natural gas for oil in the transportation section and for coal in power generation, major reductions in air pollution and greenhouse gas emissions are possible.[31]

Estimating the ultimate scale of the worldwide natural gas resource base is still difficult, but available data suggest that resources are sufficient to at least double world use of gas during the next 20 to 30 years and then to sustain that level for at least a couple of decades. Since world oil production is likely to grow little from the current level, and could well begin to decline soon after the turn of the century, natural gas will almost certainly soon become the most important fossil fuel — available in sufficient quantity to replace many existing uses of oil and coal.[32]

Relying on natural gas as a substitute for oil and coal will allow a rapid reduction in air pollution and greenhouse gas emissions during the next two decades. But it cannot, by itself, remove the specter of global warming. For that to happen, the long-term transition to an energy-efficient economy based on renewable resources must be achieved. The year 2010 can be viewed as a midpoint in the energy transition. It is far enough off to allow extensive change, but near enough to be anticipated with some reliability, since the world is unlikely to be relying on any completely new energy source or technology within a mere two decades. Although renewable energy can play a much larger role by then, it cannot by itself displace sufficient coal or oil to achieve the desired cuts in carbon emissions.

With modest encouragement, it now appears that the use of natural gas may double by the year 2010, and not begin to decline until at least 2030 as resources are gradually depleted. If combined with a 25% cut in oil use in 2010 (driven largely by rising oil prices) a 75% cut in coal use (dictated by ecological constraints, carbon taxes, and an effort to replace coal-fired power plants), and a 75% increase in renewables, the result would be a 20% decline in global carbon emissions. World primary energy use would fall under this scenario, but delivered energy services could be increased more than 50% as a result of improved energy efficiency. Meanwhile, renewable energy sources can begin to make a contribution in the nineties, with growth accelerating in the decades ahead.[33]

# VI. POTENTIAL ROLE OF HYDROGEN

The transition to hydrogen may be made easier by the fact that it can be manufactured directly from natural gas as well as derived from a broad range of renewable energy sources. In addition, hydrogen can be mixed with methane in up to a 1:10 ratio without altering today's gas pipelines, furnaces, and burners. As natural gas reserves are gradually depleted and prices rise, hydrogen can be eased into the mix bit by bit. Just as renewable electricity is quietly fed into today's electric grids, so could solar hydrogen gradually supplant natural gas. A 1:7 hydrogen-natural gas mixture, dubbed ''hythane'' by Denver engineer Frank Lynch, cut auto emissions of hydrocarbons at least in half and of nitrogen oxides by 75% compared with natural gas. Although carbon monoxide levels increased, they still complied with California's ultra-low 1997 emission standards, the toughest in the world.[34]

Solar hydrogen could eventually become the foundation of a new global energy economy. All the world's major population centers are near sunny and wind-rich areas. Moreover, land constraints do not appear likely to affect the potential. Most hydropower projects require at least 20 times as much land per kilowatt-hour as solar power does. Calculations for the U.S. show that one quarter of today's electricity could be supplied by deploying solar generators on an area of 15,000 square kilometers — equivalent to 8% of the land use by the U.S. military.[35]

The U.S. southwest, for example, could supply much of the country either with electricity or hydrogen. The pipeline routes that now link the gas fields in Oklahoma or Wyoming with the industrial midwest and northeast could one day carry hydrogen. (The pipelines themselves will need to be modified or rebuilt, since hydrogen is more chemically active than natural gas and leaks much more easily.) Although renewable energy sources are somewhat regionally concentrated, they are far less so than oil. From the windy high plains of North America, hydrogen could flow to the eastern seaboard; from the deserts of western China it could move to the populous coastal plane; and from the Australian outback, hydrogen could power that nation's southern cities.

For Europe, solar-hydrogen plants could be built in southern Spain or North Africa. From the latter, hydrogen would be transported along existing pipeline routes into Spain via Gibraltar or into Italy via Sicily. Within Europe, today's pipelines and electrical networks would make it relatively easy to distribute solar energy. To the east, Kazakhstan and the other semiarid Asian republics might supply much of the Soviet Union's energy. In India, the sun-drenched northwestern Thar Desert is within 1,000 miles of more than a half-billion people. Electricity for China's expanding economy could be generated in the country's vast central and northwestern desert regions.

# VII. GOVERNMENTAL POLICY CONSIDERATIONS

The path to a sustainable energy economy now appears clearer than ever before. Sufficient renewable energy resources have been identified, and the needed technologies are within reach. But the route is still a steep one, made more difficult by a lack of vision and by the powerful lobbying of today's energy industries. The key challenge, then, is political: how to reshape energy policy to encourage a sweeping energy transition in the decades ahead.

Governments have a long history of clumsy or counterproductive energy reform efforts. Certainly, the aggressive and monolithic attempt to jumpstart the nuclear industry in the 1960s is not a model. Government policymakers would be better off setting the overall goals and policy context rather than trying to micro-manage the details. For example, national targets for greenhouse gas emissions or oil consumption can help focus governmental and private planning efforts. Japan's goals for lowering petroleum dependence led to far-reaching improvements in energy efficiency in the 1980s, and Europe's recently enacted carbon targets may have a similar effect.[36]

Beyond that, specific policies regulating emissions of some pollutants and taxing those of others will tend to shift the world toward improved efficiency and more environmentally sustainable energy sources. The 1990 Clean Air Act Amendments in the U.S., for example, may tilt the electric power industry away from coal by placing constraints on sulfur dioxide emissions. Some state air pollution laws may do the same. Legislatures and parliaments around the world are now considering a bold new approach — taxing a range of pollutants, including carbon, so as to discourage their production. Several European countries have already enacted taxes on carbon, and the European Commission has proposed a Community-wide combination energy-carbon tax equivalent to $10 per barrel of oil.[37]

Research and development is another field in which governments have a large role to play, both directly and in concert with industry. Some of the bloated research budget that currently goes to military hardware could be redirected to wind turbines, solar photovoltaics, improved auto engines, hydrogen storage systems, fuel cells, etc. However, only if the funds are channeled on a competitive basis to various companies, including innovative smaller ones, will real progress be made. If budgets are tight, massive engineering projects such as supercolliders and fusion could be trimmed in favor of more decentralized technologies. Germany has already organized a joint government-industry hydrogen program, with a federal budget of $58 million in 1991. Among its results are solar electrolysis projects in Germany and Saudi Arabia, and prototype hydrogen-powered cars built by BMW and Mercedes-Benz.[38]

Electric and gas utilities may also help bridge the way to a sustainable energy economy. Although they have lost their monopoly in power plant construction — having been joined by independent power companies — they can be reinvigorated as suppliers of efficient energy services, investing in improved light bulbs, electric motors, home insulation, efficient gas furnaces, and so on. In the U.S., an estimated $2 billion worth of low-interest capital was invested by utility companies in end-use efficiency in 1991. In some California utilities, such investments now outstrip those in new power plants. And as environmental costs are incorporated in utility decision making, these institutions could even help customers invest in more-efficient homes or less-polluting cars.[39]

## VIII. CONCLUSIONS

The challenge ahead is unprecedented. Just as we face, for the first time, planetary environmental changes that could take decades to reverse, so we face the extraordinary need to consciously reshape our energy systems in the years ahead. The magnitude of what is required is comparable to what was accomplished during the past several decades in adopting today's oil and electricity based economies. It remains to be seen whether societies will rise to this new global economic challenge.

## ENDNOTES

1. Intergovernmental Panel on Climate Change (IPCC), ''Policymakers' Summary of the Scientific Assessment of Climate Change,'' Report to IPCC from Working Group I, Geneva, June 1990; Roger Milne, ''Pressure Grows for U.S. to Act on Global Warming,'' *New Scientist,* June 2, 1990; IPCC, ''Policymakers' Summary of the Potential Impacts of Climate Change,'' Report of the Working Group II to IPCC, Geneva, undated; carbon emissions estimate from Worldwatch Institute, based on Gregg Marland et al., *Estimates of* $CO_2$ *Emissions from Fossil Fuel Burning and Cement Manufacturing, Based on the United Nations Energy*

*Statistics and the U.S. Bureau of Mines Cement Manufacturing Data* (Oak Ridge, Tenn.: Oak Ridge National Laboratory, 1989), and on BP, *BP Statistical Review;* "Final Conference Statement: Scientific/Technical Sessions," Second World Climate Conference, Geneva, November 7, 1990; fossil fuel resources from Eric Sundquist, "Geological Perspectives on Carbon Dioxide and the Carbon Cycle," in E. T. Sundquist and W. S. Broecker, eds., *The Carbon Cycle and Atmospheric $CO_2$: Natural Variations Archean to Present* (Washington, D.C.: American Geophysical Union, 1985); conversion to carbon dioxide from Bert Bolin et al., eds., *The Greenhouse Effect, Climate Change and Ecosystems* (SCOPE 29) (Chichester, England: John Wiley & Sons, 1986), and Peter Brewer, Woods Hole Oceanographic Institution, "Global Change: The Relevance of Oceanic Chemical, Geological and Biological Processes," Testimony before the Committee on Commerce, Science, and Transportation, U.S. Senate, April 11, 1989; current $CO_2$ concentrations from Charles D. Keeling, "Measurements of the Concentration of Atmospheric Carbon Dioxide at Mauna Loa Observatory, Hawaii, 1958–1986," Final Report for the Carbon Dioxide Information and Analysis Center, Martin-Marietta Energy Systems Inc., Oak Ridge, Tenn., April 1987; C. Keeling, Scripps Institution of Oceanography, private communication, July 31, 1989.

2. Christopher Flavin, *Slowing Global Warming,* Worldwatch Paper 91 (Washington, D.C.: Worldwatch Institute, October 1989); Tim Worf, Scripps Institution of Oceanography, La Jolla, Calif., private communication, September 26, 1991; Schmidt, "Industrial Countries' Responses"; "The Carbon Club," *Atmosphere,* June 1991.
3. Roger S. Carlsmith et al., "Energy Efficiency: How Far Can We Go?" Oak Ridge National Laboratory, Oak Ridge, Tenn., prepared for the Office of Policy, Planning and Analysis, U.S. Department of Energy (DOE), in support of the National Energy Strategy, January 1990; Birgit Bodlund et al., "The Challenge of Choices: Technology Options for the Swedish Electricity Sector," in Thomas Johansson et al., *Electricity: Efficient End-Use and New Generation Technologies, and Their Planning Implications* (Lund, Sweden: Lund University Press, 1989); William U. Chandler, ed., *Carbon Emissions Control Strategies: Case Studies in International Cooperation* (Washington, D.C.: World Wildlife Fund and The Conservation Foundation, 1990); "Canadian Energy Ministers Fail to Agree on Reduction in Carbon Dioxide," *International Environment Reporter,* September 1989; Commission of the European Communities, Directorate-General for Energy, "Energy for a New Century: The European Perspective," *Energy in Europe,* special issue, July 1990.
4. "Cost of Change Could Be Great," *Financial Times,* October 13, 1989; Worldwatch Institute estimates, based on DOE, Energy Information Administration (EIA), *International Energy Outlook 1990* (Washington, D.C.: 1990), and on Frank Barnaby, "World Energy Prospects," *Ambio,* Vol. 18, No. 8, 1989.
5. Worldwatch Institute estimates, based on Marland et al., *Estimates of $CO_2$ Emissions,* and on BP, *BP Statistical Review.*
6. IEA, *Energy Policies and Programmes, 1989;* William U. Chandler et al., "Energy for the Soviet Union, Eastern Europe and China," *Scientific American,* September 1990; José Goldemberg and Amulya Reddy, "Energy for Development," *Scientific American,* September 1990; José Goldemberg et al., *Energy for Development* (Washington, D.C.: World Resources Institute, 1987).
7. Schmidt, "Industrial Countries' Responses"; Matthew L. Wald, "Two Big California Utilities Plan to Cut $CO_2$ Emissions," *New York Times,* May 21, 1991.
8. Christopher Flavin and Alan Durning, *Building on Success: The Age of Energy Efficiency,* Worldwatch Paper 82 (Washington, D.C.: Worldwatch Institute, March 1988); Arnold P. Fickett et al., "Efficient Use of Electricity," *Scientific American,* September 1990; CEC, *Conservation Report, 1990,* staff draft (Sacramento, Calif.: August 1990); Rick Bevington and Arthur H. Rosenfeld, "Energy for Buildings and Homes," *Scientific American,* September 1990; José Goldemberg et al., *Energy for a Sustainable World* (Washington, D.C.: World Resources Institute, 1987); Amulya K. N. Reddy et al., "Comparative Costs of Electricity Conservation: Centralised and Decentralised Electricity Generation," *Economic and Political Weekly,* June 2, 1990.
9. Daniel A. Lashoff and Dennis A. Tirpak, eds., *Policy Options for Stabilizing Global Climate* (Washington, D.C.: EPA, 1990); Gregg Marland, "Carbon Dioxide Emission Rates for Conventional and Synthetic Fuels," *Energy,* Vol. 8, No. 12, 1983.
10. Christopher Flavin, *Slowing Global Warming: A Worldwide Strategy,* Worldwatch Paper 91 (Washington, D.C.: Worldwatch Institute, October 1989); Worldwatch Institute estimates, based on J. M. O. Scurlock and D. O. Hall, "The Contribution of Biomass to Global Energy Use," *Biomass,* No. 21, 1990, on BP, *BP Statistical Review,* on Marland et al., *Estimates of $CO_2$ Emissions,* and on Nigel Mortimer, "Proposed Nuclear Power Station Hinckley Point C," Proof of Evidence, Friends of the Earth U.K., London, undated; Gregg Marland, "Carbon Dioxide Emission Rates for Conventional and Synthetic Fuels," *Energy,* Vol. 8, No. 12, 1983.
11. For renewable energy potential see Idaho National Engineering Laboratory (INEL) et al., *The Potential of Renewable Energy: An Interlaboratory White Paper,* prepared for the Office of Policy, Planning and Analysis, DOE, in support of the National Energy Strategy (Golden, Colo.: Solar Energy Research Institute

(SERI), 1990); the three U.S. scenarios presented in this report envision renewable energy supplies growing by approximately 300%, 450%, 600% over 1988 levels by 2030.

12. Accessible resources are that portion of the "total resource base that can be exploited with currently available technology or technology that will soon be available," from Meridian Corporation, "Characterization of U.S. Energy Resources and Reserves," prepared for Deputy Assistant Secretary for Renewable Energy, DOE, Alexandria, Va., June 1989; INEL et al., *The Potential of Renewable Energy;* DOE, EIA, *Annual Energy Review 1989* (Washington, D.C.: 1990).
13. Scurlock and Hall, "The Contribution of Biomass to Global Energy Use"; Norwegian figure is based on Norwegian Central Bureau of Statistics, *Natural Resources and the Environment, 1989* (Oslo: 1990), wherein more than 45% of total supply is from hydroelectric power and 5% is from biomass.
14. Carl J. Weinberg and Robert H. Williams, "Energy from the Sun," *Scientific American,* September 1990; INEL et al., *The Potential of Renewable Energy;* Christopher Flavin and Rick Piltz, *Sustainable Energy* (Washington, D.C.: Renew America, 1989); DOE, *Energy Technologies & the Environment* (Washington, D.C.: 1988); Peggy Sheldon, Luz International Limited, Los Angeles, Calif., private communication and printout, August 28, 1990; Susan Williams and Kevin Porter, *Power Plays* (Washington, D.C.: Investor Responsibility Research Center, 1989); Nancy Rader et al., *Power Surge* (Washington, D.C.: Public Citizen, 1989); "Country Profiles: Denmark"; "Spain Resurrects Funding Programme," *European Energy Report,* July 13, 1990; "West Germany Announces $3bn Plan for Research and Technology," *European Energy Report,* March 9, 1990.
15. Low-temperature heat is Worldwatch Institute estimate, based on Amory B. Lovins, *Soft Energy Paths: Toward a Durable Peace* (Cambridge, Mass.: Ballinger Publishing Company, 1977), on John Hebo Nielsen, "Denmark's Energy Future," *Energy Policy,* January/February 1990, and on DOE, EIA, *Annual Energy Review 1989*; Bevington and Rosenfeld, "Energy for Buildings and Homes"; Solar Technical Information Program, *Energy for Today: Renewable Energy* (Golden, Colo.: SERI, 1990).
16. Joyce Whitman, *The Environment in Israel* (Jerusalem: Environmental Protection Service, Ministry of the Interior, 1988); Mark Newham, "Jordan's Solution Circles the Sky," *Energy Economist,* June 1989; Cynthia Pollock Shea, *Renewable Energy: Today's Contribution, Tomorrow's Promise,* Worldwatch Paper 81 (Washington, D.C.: Worldwatch Institute, January 1988); Eric Young, "Aussies to Test Novel Solar Energy Collector," *Energy Daily,* May 3, 1990; Solar Technical Information Program, *Energy for Today: Renewable Energy.*
17. Sheldon, private communication and printout; Don Logan, Luz International Limited, Los Angeles, Calif., private communication, September 26, 1990; Bureau of the Census, U.S. Department of Commerce, *Statistical Abstract of the United States 1990* (Washington, D.C.: U.S. Government Printing Office, 1990).
18. INEL et al., *The Potential of Renewable Energy.*
19. Steven Dickman, "The Sunny Side of the Street . . . ," *Nature,* May 3, 1990; "Sanyo Develops Solar Cell Shingles," *Independent Energy,* April 1989.
20. DOE, *Energy Technologies and the Environment;* DOE, *Photovoltaic Energy Program Summary* (Washington, D.C.: 1990); Ken Zweibel, *Harnessing Solar Power: The Photovoltaics Challenge* (New York: Plenum Publishing, 1990); Meridian Corporation and IT Power Limited, "Learning from Success: Photovoltaic-Power Water Pumping in Mali," prepared for U.S. Committee on Renewable Energy Commerce and Trade (CORECT), Alexandria, Va., February 20, 1990; Maheshwar Dayal, Secretary, Department of Non-Conventional Energy Sources, New Delhi, India, private communication, July 13, 1989; "Indonesia Installs First Solar Village, Schedules Total of 2,000," *International Solar Energy Intelligence Report,* February 9, 1990; "A New Group of Sun Worshippers," *Asiaweek,* October 12, 1990.
21. Zweibel, *Harnessing Solar Power;* INEL et al., *The Potential of Renewable Energy.*
22. Flavin and Piltz, *Sustainable Energy;* INEL et al., *The Potential of Renewable Energy;* Danish experience from Paul Gipe, "Wind Energy Comes of Age," Gipe & Assoc., Tehachapi, Calif., May 13, 1990.
23. U.S. Windpower, Inc., "The Design Specifications for a Wind Power Plant in Patagonia Using U.S. Wind Turbines," Livermore, Calif., January 1989; Christopher Flavin, *Wind Power: A Turning Point,* Worldwatch Institute (Washington, D.C.: Worldwatch Institute, July 1981); "Minnesota Resource Greater than Previously Reported," *Wind Energy Weekly,* American Wind Energy Association, Washington, D.C., July 5, 1990.
24. P. J. de Groot and D. O. Hall, "Biomass Energy: A New Perspective," prepared for the African Energy Policy Research Network, University of Botswana, Gaborone, January 8, 1990.
25. Lester R. Brown, *The Changing World Food Prospect: The Nineties and Beyond,* Worldwatch Paper 85 (Washington, D.C.: Worldwatch Institute, October 1988); Sandra Postel, *Water for Agriculture: Facing the Limits,* Worldwatch Paper 93 (Washington, D.C.: Worldwatch Institute, December 1989).
26. Biofuels and Municipal Waste Technology Program, Office of Renewable Energy Technologies, DOE, *Five Year Research Plan: 1988–1992, Biofuels: Renewable Fuels for the Future* (Springfield, Va.: National Technical Information Service, 1988); Norman Hinman, SERI, Golden Colo., private communication, August 25, 1989.

27. P. P. S. Gusain, *Cooking Energy in India* (Delhi: Vikas Publishing House, 1990); Eric D. Larson et al., ''Biomass Gasification for Gas Turbine Power Generation,'' in Johansson et al., *Electricity: Efficient End-Use and New Generation Technologies;* Eric D. Larson et al., ''Biomass-Gasifier Steam-Injected Gas Turbine Cogeneration for the Cane Sugar Industry,'' presented at the Energy from Biomass and Wastes XIV conference, Lake Buena Vista, Fla., January 29–February 2, 1990; United Nations, *1988 Energy Statistics Yearbook.*
28. United Nations, *1988 Energy Statistics Yearbook;* Satyajit K. Singh, ''Evaluating Large Dams in India,'' *Economic and Political Weekly,* March 17, 1990.
29. Donald Finn, Geothermal Energy Institute, New York, private communication and printout, March 16, 1990; United Nations, *1988 Energy Statistics Yearbook;* Phillip Michael Wright, ''Developments in Geothermal Resources, 1983–1988,'' *The American Association of Petroleum Geologist Bulletin,* October 1989; New Energy Development Organization, ''The Map of Prospective Geothermal Fields in Japan,'' Tokyo, 1984, as cited in Michael J. Grubb, ''The Cinderella Options: A Study of Modernized Renewable Energy Technologies,'' Part 2: Political and Policy Analysis, *Energy Policy,* October 1990.
30. See Table 3-2 for references for these figures.
31. OECD, *Greenhouse Gas Emissions The Energy Dimension* (Paris: OECD/IEA, 1991).
32. Worldwatch Institute estimates based on gas resource estimates cited in this section and on the expectation that the recent growth rate in world gas production of 3.5% per year will continue.
33. This scenario is a Worldwatch Institute projection, using Oak Ridge and British Petroleum figures; it is not a prediction but simply an indication of what the effect of a major shift in the fossil fuel mix might be.
34. Lynch and Egan, ''An Introduction Strategy for Hythane.''
35. The photovoltaic panels would actually cover just 15% of the land area; John Schaefer and Edgar DeMeo, Electric Power Research Institute, ''An Update on U.S. Experiences with Photovoltaic Power Generation,'' Proceedings of the American Power Conference, April 23, 1990; U.S. land area used by the military from Michael Renner, ''Assessing the Military's War on the Environment,'' in *State of the World 1991* (New York: W. W. Norton Co., 1991); Dennis Anderson, World Bank, private communication, October 16, 1991.
36. Organisation for Economic Co-operation and Development, *Energy Policies and Programmes of IEA Countries* 1989 Review (Paris: 1990).
37. Richard H. Hilt and Marie L. Lihn, ''The Clean Air Act's Impact on Natural Gas Markets,'' *Public Utilities Fortnightly,* October 15, 1991; Kevin Commins, ''Clean Air Act Could Devastate Ill. Coal, Producer Says,'' *Journal of Commerce,* April 10, 1991; Meridian Corporation, ''Energy System Emissions and Material Requirements''; Pace University Center for Environmental Legal Studies, *Environmental Costs of Electricity* (New York: Oceana Publications, 1990); Tom McNiff Jr., ''Coal-Fired Plant Plans Stock Conflict in Mass.,'' *Journal of Commerce,* August 8, 1991; ''Carbon Taxes,'' *Global Environmental Change Report,* June 24, 1991; Andrew Hill, ''EC Energy Tax Would Put $10 on Barrel of Oil,'' *Financial Times,* August 23, 1991.
38. Grasse and Olster, *Hysolar Results and Achievements 1985–1989;* Carl-Jochen Winter, ''Canned Heat Solar Energy Could be Stored and Distributed as Hydrogen,'' *The Sciences,* March/April 1991; ''Germany Spends $58 Million on H, Plus PV Outlays,'' *The Hydrogen Letter,* October 1991.
39. ''UDI: Gas is Favorite Fuel of Utilities, IPPs;'' Peter Keat, ''Selling Efficient Kilowatt-hours,'' *Public Power,* September/October 1991; Ralph Cavanagh, Senior Attorney, Natural Resources Defense Council, San Francisco, Calif., private communication, October 21, 1991.
40. ''Ministerial Declaration of the Second World Climate Conference''; Bill Hare, Australian Conservation Foundation, Melbourne, private communication and printout, October 17, 1990; ''Austria to Reduce $CO_2$ Emissions 20% by 2005,'' *Global Environmental Change Report,* September 14, 1990; ''Canada to Stabilize $CO_2$ Emissions at 1990 Levels by 2000,'' *Global Environmental Change Report,* June 22, 1990; Denmark Ministry of Energy, ''Danish Efforts to Reduce Energy Consumption and the Emission of Greenhouse Gases,'' Copenhagen, June 7, 1990; ''Country Profiles: Denmark,'' *European Energy Report,* May 1990; Emmanuele D'Achon, First Secretary, Embassy of France, Washington, D.C., private communication and printout, October 10, 1990; ''Germany and the Greenhouse: A Closer Look,'' *Global Environmental Change Report,* August 17, 1990; ''West German Environment Minister Proposes Tax on Carbon Dioxide Emissions,'' *International Environment Reporter,* September 26, 1990; ''East Germany: Country will Comply with CFC Ordinance of West Germany, Seeks Smaller $CO_2$ Cut,'' *International Environment Reporter,* July 1990; Japan Council of Ministers for Global Environment Conservation, ''Action Program to Arrest Global Warming,'' *Global Environmental Change Report,* December 22, 1989; Bert Metz, Counselor for Health and Environment, Royal Netherlands Embassy in the U.S., ''The Dutch Policy on Global Warming,'' Testimony presented to the Canadian House of Parliament, Standing Committee on Environment, Ottawa, January 23, 1990; ''New Zealand Announces $CO_2$ Reduction Target,'' *Global Environmental Change Report,* August 17, 1990; Gunner Mathisen, Secretariat for Climate Affairs, Ministry of the Environment, Norway, private communication, January 30, 1990; The Ministry of Environment and Energy, *Action for a Common Future: Swedish National Report for Bergen Conference, May 1990* (Stockholm: 1989); ''Swiss Act on Global Warming with Major New $CO_2$ Tax,'' *European Energy Report,* November 2, 1990; United Kingdom

Department of the Environment, *This Common Inheritance: Britain's Environmental Strategy* (London, September 1990); "European Councils Set $CO_2$ Targets and Agree Gas Transit Directive," *European Energy Report,* November 2, 1990; IPCC, "Policymakers' Summary of the Scientific Assessment of Climate Change"; Finland and Iceland also have set goals for greenhouse gas emissions, but details were not available at press time.

41. IPCC, "Policymakers' Summary of the Scientific Assessment of Climate Change"; U.S. Environmental Protection Agency (EPA), *Policy Options for Stabilizing Global Climate,* draft (Washington, D.C.: EPA, 1989).
42. Mark A. DeLuchi, "Hydrogen Vehicles," in Daniel Sperling, ed., *Alternative Transportation Fuels: An Environmental and Energy Solution* (Westport, Conn.: 1990).

Chapter 28

# JAPAN AND THE GLOBAL ENVIRONMENT

**Alan S. Miller and Curtis Moore**

## TABLE OF CONTENTS



0-8493-4419-0/93/$0.00+$.50
© 1993 by CRC Press, Inc.

Chapter 28

# JAPAN AND THE GLOBAL ENVIRONMENT

**Alan S. Miller and Curtis Moore**

## TABLE OF CONTENTS



0-8493-4419-0/93/$0.00 + $.50
© 1993 by CRC Press, Inc.

# I. INTRODUCTION

In many areas, the word most often used to describe Japanese policy is "enigma". In some ways, Japan's record on environmental policy also has elements of mystery and contradiction. On the one hand, Japan's history and culture often are associated with a reverence for nature. Indeed, Japan does lead the world in certain environmental areas, such as reduction of conventional air pollutants and compensation of air pollution victims. On the other hand, Japan has been widely criticized for its poor record in preserving its domestic environment, contribution to tropical deforestation, and unwillingness to protect endangered species.

Today, the international community clamors for Japan to take its share of responsibility, as an economic superpower, for the global environment. To secure its place in world affairs, Japan slowly has begun to respond to this pressure on issues ranging from ivory importation to reduction of CFC emissions. There is some hope that the government's particular willingness to address global warming may be a sign of significant changes to come in Japanese environmental policy. However, international pressure remains on the many environmental issues Japan has yet to resolve, including tropical deforestation and financing of Third World development projects that harm the environment.

Japan's environmental policy is most effective when government and industry cooperate to find technical solutions to environmental problems. Although in recent years Japan's energy consumption has risen sharply, the Japanese have developed numerous technologies to reduce pollution and increase economic growth by improving energy efficiency. It is in these technologies that Japan has made the greatest strides and has the most to offer the global environment.

# II. AN OVERVIEW OF ENVIRONMENTAL POLICY IN JAPAN

## A. ENVIRONMENTAL VALUES IN JAPAN

It is generally thought that the Japanese, because of their religious beliefs, value nature much more than Westerners, a cultural stereotype that has some historical basis. Joseph Kitagawa notes in his book *On Understanding Japanese Religion:* "Buddhism affirms the sacrality [*sic*] of the world of nature. The feature is probably the most basic to the Japanese Buddhist understanding."[1] Edwin Reischauer adds in *The Japanese,* "Early Shinto centered around the animistic worship of natural phenomenon — the sun, mountains, trees, water, rocks, and the whole process of fertility".[2] Although Shintoism and Buddhism thus appear to be far more sympathetic to nature than the Judeo-Christian tradition of the West, these traditions have very little relevance to current policy.

Japan and the U.S. begin with different views of nature, which has had profound effects on their environmental policies. Destruction of the environment began to attract widespread attention in Japan in the 1950s, when industrial pollution started to affect human health.[3] In the U.S., however, environmentalists have focused on conservation of the wilderness since the 19th century. Americans assign strong symbolic and cultural significance to their national parks, for example. The Japanese hold more utilitarian views, considering nature a resource for man to enjoy.[4] The management of Japanese national parks generally is oriented toward facilitating maximum visitation and recreational activity, possibly including resort development. The Shiga Heights area, home for tribes of mountain monkeys, had one hotel before designation as a national park. It now has 22 ski resorts and 101 hotels.[5] In the U.S., there is a great deal of public support for protecting even the remote parks of Alaska, although few Americans expect to go there and the land may contain substantial oil reserves. Threats to the forests of Hokkaido and the coral reefs in Okinawa, both of which the average Japanese is likely to visit, evoke much less concern in Japan. Perhaps the most glaring example is Mount Fuji, revered for centuries in Japanese painting and poetry. The mountain today reveals a different attitude among the Japanese toward nature: "There is a huge parking area with shops, restaurants, horses to ride and photographers. Coachloads arrive, crowds emerge and immediately line up to have their photograph taken . . . when the sun lights Fujisan, it reveals the tin cans and other rubbish left behind by the visitors who had come to see Japan's symbol of ultimate purity."[6]

In a 1989 14-nation survey commissioned by the United Nations Environment Program (UNEP), only 44% of the Japanese public polled expressed a willingness to contribute money or labor to improving their environment. Other nations in the survey expressed a 60% to 100% willingness to contribute to betterment of the environment. Although they expressed great apprehension about global climate change, Japanese respondents cited polluted drinking water as their foremost environmental concern, followed by the pollution of lakes and rivers.[7] However, interestingly in a country where pollution has had devastating effects on human health, the Japanese believe far less often than other respondents that environmental deterioration threatens public health.

## B. LACK OF A STRONG CITIZEN'S LOBBY

The modern Japanese environmental movement began in the early 1960s with anti-pollution citizen movements that opposed crippling and even lethal industrial pollution. These anti-pollution movements had a significant impact on the government's overall environmental policies. The government tightened environmental regulations significantly and in some areas, particularly control of traditional air pollutants, Japan became a world leader. However, in contrast to experiences in the U.S. and Europe, Japanese environmentalists have failed so far to build a strong national movement able to influence the national political agenda. The UNEP poll showed that most Japanese do not believe voluntary agencies or community organizations in Japan play an important part in resolving environmental problems, which they view as the responsibility of the government. Only 26% of Japanese respondents, compared to 80% to 90% of other respondents, expressed support for leaders of such organizations.[8] Eugene Linden of *Time Magazine* notes, "If anything will hold back progress (on environmental issues), it will be Japan's lack of environmental activists and experts. Only about 15,000 Japanese — most of them bird watchers — belong to conservation groups, and the country does not have an extensive network of environmentalists, like those who monitor policies in the U.S. and Western Europe."[9]

A brief examination of the Japanese environmental movement's evolution will elucidate its current status and prospects for change. The first environmental activists were pollution victims and their families, and others who lived in contaminated areas.[10] A significant proportion of these activists were women. (See Appendix A.) They organized in response to local problems, often health or nuisance related (e.g., noise levels). While these efforts often provided relief, they did not lead to larger coalitions of similarly affected parties or provide the basis for a national environmental movement.[11] Indeed, as Frank Upham argues, the creation of dispute resolution procedures and an administrative system for compensating victims effectively preserved bureaucratic control.[12]

Environmental groups became notably less active and influential in the early 1970s as public attention shifted toward the oil shocks and national political scandals in the Tanaka administration.[13] The environmental movement was unable to establish an effective bureaucratic foothold despite the creation in 1971 of an environmental agency with its own minister.[14] (See Section C.) For example, neither the citizen's movements nor JEA could persuade the Diet to pass a general environmental impact assessment bill. The courts, which had provided environmentalists with a series of important victories in the 1960s, also retreated in several major cases that could have provided an avenue for environmental litigation on a wide range of issues.[15] Although the outcome of the pollution trials of the 1970s generally favored plaintiffs, court decisions of the 1980s have tended to favor industry.[16] The movement continues to face opposition from industry, government, and the ruling Liberal Democratic Party (LDP).

The feeling that the government is too big to fight contributes to the prevailing public apathy about the environment[17] and the inaccessibility of important government documents reinforce this perception. The national government has no freedom of information (FOI) act. Some towns and prefectures do have FOI ordinances, but these often unduly limit access to information. According to the *Japan Times,* ''Politicians have begun to define as public documents only those papers that have been officially stamped by a set number of officials. Bureaucrats have learned to avoid stamping sensitive documents to keep them out of public hands.'' An environmental protection committee in the town of Oiso opposing the construction of a chemical research laboratory by the Showa Denko company was refused access to municipal and prefectural data on the project. The committee also was denied the town mayor's letter to the company, urging Showa Denko to approve the project quickly, presumably to head off local opposition. In another case, the Japanese Citizen's Movement for a Freedom of Information Law finally obtained data on Japanese nuclear reactors through the U.S. Freedom of Information Act, having been denied the information from the Japanese government. The government provides such information to the U.S. Nuclear Regulatory Commission but not to Japanese citizens. Fortunately, some LDP members are pushing for an FOI law. Several factors, including U.S. pressure and support from the Japan Committee for Economic Development work in favor of the legislation's passage.[18]

As discussed in Section D, there are some signs of change in the public's attitude toward the environment.

## C. THE ROLE OF THE ENVIRONMENTAL AGENCY

The Japanese Diet established the Japanese Environmental Agency (JEA) in 1971, largely in response to pollution victims' demands that the government take a more responsible approach toward the environment. The agency's mandate is to coordinate and administer programs to prevent environmental pollution and to protect nature.[19] JEA has limited authority over a wide range of environmental law, including pollution control for individual factories, toxic wastes, and the regulation of sewage, waste disposal, marine pollution, and agricultural chemicals.[20] JEA also manages Japan's national parks.

Although JEA provides an important focal point for environmental advocates and analysis, the agency has much less power than the Ministry of International Trade and Industry (MITI) and other established agencies with missions to bolster economic growth.[21] Some scholars include JEA among the ''relatively ignored structures of Japanese politics''[22] and many believe that JEA cannot afford to offend industry.[23]

By the early 1980s, paralleling developments in U.S. environmental politics, the environmental agenda lost some of the political support it had attained in the previous decade. The government successfully campaigned to reduce public concern over pollution, noting reduction in the levels of sulfur oxide, nitrogen dioxides, and photochemical smog, but downplayed problems with toxic wastes and chemicals, including dioxin, water pollution, and contamination of drinking water.[24] Japanese participation in international environmental negotiations was, and continues to be, dominated by economic ministries.

The environmental budget grew steadily during the 1970s, but MITI and the Ministry of Construction, rather than JEA, received much of this funding. Industrial pressure forced JEA to reduce some emission control standards[25] and cut certain programs instituted in the 1970s to strengthen environmental administrations in cities and prefectures.[26] Due to opposition from industry and other ministries, JEA most notably failed to persuade the Diet to pass environmental impact legislation, a priority for the agency since the early 1970s. JEA proposed but failed to secure five separate impact bills by 1980.[27] A Cabinet Resolution finally implemented a weaker impact requirement in 1984.

Despite its weaknesses, JEA gradually is becoming a more established and accepted part of the government. Until the 1980s, former members of MITI and the Ministry of Finance composed most of the agency's senior directors. Many JEA officials retained their old allegiances, anticipating they would return to their previous agencies.[28] Today, however, many JEA officials expect to finish their careers there. Moreover, in the annual contest among government agencies to recruit top graduates from Tokyo University, a growing number have made JEA their first choice.

Apart from a shift in national values toward environmental protection, JEA's biggest problem may be the lack of an organized environmental movement that can forcefully counter the tightly organized industrial lobby. In the mid-1980s, industry argued that Japan had reduced air pollution so effectively that emissions no longer significantly contributed to health problems. Industry successfully lobbied the government to amend its pollution compensation system to halt all new designations of air pollution ''victims''. This system, which compensates people suffering health problems from air pollution, is unique to Japan.[29] Environmental opposition was too weak and the amendments were enacted in September 1987.

## D. RECENT DEVELOPMENTS IN ENVIRONMENTAL ACTIVISM IN JAPAN

An important indicator of new interest in conservation and wildlife protection may be recent environmental protests opposing several dams, an airport project in Okinawa that threatens a coral reef, and housing for U.S. military personnel in the Ikego Forest.

During 1989, campaigns were mounted to halt construction of dams on the Kamp River in Kyoto, the Shimanto River in Shikoku, and the Nagara River, running from the Japan Alps to Ise Bay.[30,31] The battle for protection of the Nagara in particular has become a major organizing issue.[32] (See Appendix B.) The proposed dam will kill most of the 60 species of fish in the river, including a species of trout called *ayu,* a delicacy for which the river is known. The Nagara is the only river where enough *ayu* exist for cormorant fishing, a unique sport under the patronage of the Imperial Household Agency since the Meiji Restoration.[33,34] ''Tradition still holds that part of the first catch each season lands on the Emperor's table. Because of this legacy, the Nagara River is one of the cleanest rivers running through a major urban area in Japan today, and one of the last major rivers not to be damned.''[35]

Despite vehement public opposition, dam construction began in July 1989. Several critics submit that the government has an unstated reason for continuing with the project. The dam, they charge, would help alleviate international (primarily U.S.) pressure on Japan to increase its spending on domestic projects and employ more foreign goods and services.[36]

When local issues did not attract much national attention, environmental groups found it easier and more effective to solicit help from abroad than rely on Japanese support to protest government policies. In these cases, foreign pressure has benefited the environmental movement. For example, Japanese conservationists solicited help from foreign scientists and international environmental organizations to protect a rare coral reef off the Island of Ishigaki in Okinawa where an airport was planned.[37] The government later moved the construction site to a nearby location.

In another instance, environmentalists entreated American officials and environmental groups to oppose a Japanese government plan allowing the U.S. military to build housing for its personnel in the Ikego Forest near the town of Zushi. The Natural Resources Defense

Council made efforts on their behalf. The issue even generated a letter from the head of the Smithsonian to the Secretary of Defense asking why the U.S. government was building housing in an environmentally sensitive area. Ikego shelters a number of endangered species and is one of the few forests still standing in the Tokyo area. Support for the forest's preservation was so strong that the head of the opposition movement defeated Zushi's pro-housing mayor in his 1986 bid for reelection.[38]

There are indications of growing public and corporate interest in global environmental issues. New environmental fads have fluorished favoring sprays that do not contain chlorofluorocarbons, biodegradable plastic shopping bags, a new brand of beer called "Earth Beer", and clothing with wildlife pictures on them.[39] In Osaka and Tokyo in 1989, citizen conferences on the international environment attracted considerable public interest and media coverage. A government-sponsored conference on the environment that was closed to the public provided the impetus for these citizen meetings. Over a thousand people, many students and housewives, attended the Osaka meeting. Similar meetings have been held in the past, including several on tropical rainforest issues, but none have been nearly so large.

On the corporate level, a 1990 *Japan Economic Journal* survey found that 60 of 113 major companies had established or planned to establish environmental committees or departments. In response to increasing pressure from American and European environmentalists, most of Japan's largest trading companies have created environmental sections. They appraise the firm of tighter domestic and international environmental standards, help improve its image, and find business opportunities for new environmental technologies and products.[40]

In the remainder of this chapter, we focus in more detail on the strengths and weaknesses of Japan's environmental record and suggest indications of change that give hope of a broader and more forward-looking environmental policy in the future.

## III. THE EVOLUTION OF JAPANESE POLICY TOWARD INTERNATIONAL ENVIRONMENTAL AGREEMENTS

With few exceptions, Japan has been slow to recognize and respond to international environmental problems and agreements.[41] Local concerns historically have taken precedence over international environmental issues. This is due in part to the history of the anti-pollution movement. Pollution victims had a war to wage at home and had neither energy nor resources to address international issues as well. However, a July 1990 poll conducted by the Prime Minister's office found that almost 60% of respondents believe that the government's first priority should be preservation of the global environment, particularly protection of the ozone layer and tropical forests, and reduction of $CO_2$ emissions.[42] These results contrast with a previous poll conducted by the same office in February 1989, in which only one fifth of the respondents indicated a strong personal interest in global environmental issues.[43] Growing domestic concern for the global environment, combined with pressure from abroad, has forced the Japanese government to reconsider its position on many environmental issues, including drift nets, endangered species, and global atmospheric pollution.[44]

### A. DRIFT NETS

Of all nations, Japan uses the largest drift nets for fishing.[45] Squid and tuna fishermen cast these huge nylon nets that cover an area of ocean up to 40 miles wide and 40 miles deep. Environmentalists have called drift net fishing "aquatic strip mining"[46] because the nets, set out overnight, randomly kill dolphins, seals, sea turtles, sea birds, and other marine animals. U.S., Soviet, and Canadian fishing industries also complain of the massive loss of sea trout and salmon.[47] In 1989, environmental activists won a U.S. Supreme Court case to ban Japanese fishing boats in U.S. waters. Japan subsequently agreed to have 32 American observers aboard 460 Japanese squid boats to determine the number of sea animals being killed.[48]

In 1989, fifteen countries in the South Pacific called for a total ban on drift net use. Japan subsequently implied that the move might jeopardize Japanese aid and trade to these countries, a form of "checkbook diplomacy" that provoked an immediate negative response.[49] On September 29, a subcommittee of the U.S. House of Representatives passed legislation to ban drift nets worldwide. Japan then announced its intention to reduce the number of ships permitted to use drift nets by two thirds.[50] In July 1990, the Foreign Ministry announced that the government will suspend drift-net fishing in the South Pacific for the 1990–91 season. The suspension, beginning in November, will last until "regulatory measures for drift-net fishing are established".[51]

## B. ENDANGERED SPECIES

Japan has been the largest ivory importer, accounting for 38% of the world's total. In June 1989, MITI announced an ivory import ban from September 20 through the end of 1989. The move followed European and U.S. bans to save the African elephant from extinction and preceded the probable signing of a worldwide agreement.[52] On October 30, the Japanese government announced a total ban on ivory trade.[53]

The ivory ban is suggestive of the gradual change in Japan's attitude toward endangered species. In 1980, Japan signed the Washington Convention on International Trade in Endangered Species of Wild Fauna and Flora, but had more exemptions to the agreement than any other signatory.[54] Some of those exemptions, particularly on whales, remain the subject of severe international criticism. However, in 1987, Japan enacted a law on the domestic trade of endangered species to enforce the Convention within the country.[55] Also since 1987, Japan has agreed to prohibit imports of the green sea turtle, the musk deer, and the desert monitor lizard. Japan has responded so slowly to halting this trade in part because MITI, the ministry most interested in economic development, has authority over endangered species policy.[56]

## C. GLOBAL ATMOSPHERIC POLLUTION

In September 1987, more than 20 countries signed the Montreal Protocol on Substances that Deplete the Ozone Layer, agreeing to reduce CFC emissions 50% over the next decade. Japan, which consumes more than 10% of the world's CFCs, opposed the treaty until only a few months before its conclusion. Japanese industry was skeptical, environmentalists were largely uninvolved, and government research was limited.

Ultimately, Japan took action largely because the U.S. and Europe had agreed to do so and non-signatories risked adverse trade consequences designated in the Protocol and in several bills pending in the U.S. Congress.[57] Once Japan did agree, however, it moved quickly to implement the agreement. MITI requested several million dollars for fiscal year (FY) 1990 to develop CFC substitutes.*[58] Several Japanese automakers promise to phase out CFCs in automobile air conditioners by 1993.[59] Japan supported an accelerated and complete international phase out of CFCs adopted in June 1990.

Unlike its response to most other environmental issues, Japan has shown relatively quick progress in supporting measures to prevent global warming. Japan announced in June 1990 that it will stabilize its $CO_2$ emissions by the year 2000 "at the lowest possible level". (The government also calls for increased research and monitoring of climate change, and the

* For next year, MITI has requested $48 million for research on the global environment, an increase of 17% from the present budget. Much of this money will be allocated to development of "environment friendly" technology, including CFC substitutes, technology to absorb and use $CO_2$, and biodegradable plastics. ("A Darker Shade of Green," *Nature,* Aug. 30, 1990, p. 783).

development of technologies transferrable to other countries that reduce greenhouse gas emissions.)[60]

In October 1990, the Japanese government announced that it would stabilize emissions at a level equivalent to 1990 emissions, adjusted for population growth. The decision, while not formally binding, was followed by a 15-page statement of policies to promote efficiency and reduced fossil fuel consumption. The policy is more aggressive than was expected by most observers as it means Japan will have to sharply reduce recent trends toward growth in fossil fuel consumption. Politically, the announcement also means that Japan has joined the European nations, which have already adopted targets for stabilizing $CO_2$ emissions. This leaves the U.S. as the only major industrialized country without such a policy. One reason for this change may be the recognition by some Japanese companies that they stand to benefit from the sale of technologies to improve energy efficiency.

Several American environmental groups have expressed concern that Japan's greenhouse gas emissions will increase in response to the U.S.-Japanese Structural Impediments Initiative (SII). The initiative, intended to reduce trade friction between the two countries, calls for infrastructural improvements and promotion of leisure activities in Japan that will lead to greater general consumption and therefore may increase greenhouse gas emissions.[61]

Japan accounts for about 5% of the world's $CO_2$ emissions, following the U.S., the Soviet Union and China.[62] Japan has been well represented and active at the principle governmental meetings on climate change. With the support of the Prime Minister's office, the Japanese government hosted a major international meeting on the global environment in September 1989 in Tokyo. MITI, the Ministry of Transportation, the Science and Technology Agency, and others have made large funding requests for global warming. Manufacturers see potential market opportunities in environmental regulation, such as increased sales of fuel cells.

A 1989 JEA report described climate change as a serious threat to Japan's import supply and suggests some understandable self-interest motivating Japan's growing attention to global warming. Japan is highly dependent on foreign oil, timber, and some food staples such as soybeans. Thus the impact of climate change on foreign markets greatly concerns Japan. For example, the 1973 drought in the Soviet Union led that country to increase its imports of U.S. wheat, which resulted in higher prices in the world market. Also in 1973, a poor harvest of sunflowers in the Soviet Union and of peanuts in India led to panic in the Japanese soybean market. Since economic activity in coastal areas results in approximately 90% of Japan's GNP, a rise in the sea level due to global warming could have serious consequences for Japan.[63]

## D. FOREIGN AID

Growing international pressure has forced Japan to assume more responsibility for the global environment. With 10 of the world's largest 11 banks, and 53 of the 100 largest companies,[64] Japan is now the leading financial contributor to developing countries.[65] However, much of Japan's aid is tied — going toward infrastructure projects, such as mining operations, intended to benefit Japan.

The Ministry of Foreign Affairs states that the Japanese government has made "contributions to the globe a national goal." Further, Japan "should take initiatives in the development of international strategies and systems to support remedial efforts in protection of the ozone layer, global warming, protection of tropical forests, prevention of desertification, elimination of acid rain, and protection of endangered wildlife species."[66] To combat its image as being internationally irresponsible, the Japanese government aimed to double its giving in U.S. dollars between 1986 and 1990.[67] It met this goal by 1989, and also reached its 5-year target of providing $40 billion in development assistance.[68]

The Official Development assistance (ODA) budget recently has grown considerably — 6.5% between 1987 and 1988 and an additional 7.8% between 1988 and 1989. The 1989 budget totalled $11 billion, $2 billion more than the U.S. budget.[69] In September 1989, Prime Minister Kaifu also pledged approximately $2.25 billion in environmental aid alone over the next three years.[70] Mexico will receive the first $1 billion in credits over three years to fight air pollution.[71]

Despite these increases, a 1989 World Wildlife Fund report (*Timber from the South Seas*) charges than Japan's percentage of aid still lags behind that provided by most members of the Development Assistance Committee (DAC) of the Organization for Economic Co-operation (OECD). Japan ranks 15th among the eighteen member nations. DAC countries aim to contribute 0.7% of their gross domestic product (GDP), while Japan donated only 0.31% of its GDP in 1988.[72] However, for 1988–1998, ODA plans to increase the proportion of aid as a percentage of GDP to 0.35% and increase giving to $50 billion.

According to *Timber from the South Seas,* a central problem with Japanese development aid is that no single administrative body has the experience or the authority to supervise all the development programs. Moreover, there is no general law on foreign aid that would help the many aid institutions coordinate their efforts.[73] While the Ministry of Foreign Affairs is the official coordinating body, numerous ministries and agencies are involved in the decision making process. Japan's International Cooperation Agency, the Overseas Economic Cooperation Fund, The Ministry of Agriculture, Forestry, and Fisheries, and the Ministry of Finance each participate in some aspect of project selection, definition, appraisal, monitoring or funding. As a result, implementation, appraisal and monitoring of on-going projects and evaluation of finished projects is poor.[74]

Because many development offices are new or growing, they often are inexperienced and have yet to gain the stature of other long-time administrations.[75] Many aid agencies are severely understaffed. Richard Forrest of the National Wildlife Federation notes that although the Japanese government prides itself on the small size of its bureaucracy, this " . . . would not seem to be an asset for the world's largest and fastest growing development financing program, which needs careful control and monitoring. There are reports that Japan cannot even disburse all of its promised ODA budget every year, much less study in detail the effects of the funding, due to lack of staff."[76] Moreover, Japanese officials and ministers are not always apprised of how ODA funds will be used — further evidence that Japan's aid programs are inadequately supervised. The government also does not require environmental impact assessments of its own programs or those of the private sector.[77] ODA only recently initiated environmental policy guidelines for its assistance. Environmentalists fear that inadequate supervision and evaluation will lead the government to finance Japanese corporations intending to gain a foothold in markets abroad.[78] For example, JATAN, the Development and Cooperation Program, has subsidized the clearing of roads for logging by Japanese companies in developing countries. On the island of Sarawak, those companies used the roads for timber exploitation, which destroyed the centuries-old habitat of the Penan people.[79]

In another case, the Overseas Economic Cooperation Fund is financing the construction of a large portion of a hydroelectric dam on the Narmada River in central India. Environmentalists charge that the project would force 250,000 people to relocate and cause undue harm to forests and wildlife. Under domestic and international pressure, the Foreign Ministry announced in May that funding will stop until the Indian government completes an environmental assessment of the project and makes adequate plans for relocating displaced persons.[80]

Few of the problems the Japanese development program faces are unique. No one body coordinates environmental policy or provides overarching environmental guidelines for the

U.S. government either. The U.S. Agency for International Development (AID) also has contributed billions of dollars to environmentally destructive programs, like the early 1980s AID program to construct dams in Sri Lanka, which later inundated forests and tea farms. AID began to assess the environmental impacts of its projects after several such environmental catastrophes. However, other U.S. executive branch agencies involved in development assistance do not yet require environmental impact assessments. Agencies like the U.S. Treasury Department that administer U.S. lending to the World Bank and other international lending institutions have only begun this process.[81]

As discussed above, Japan's aid policies will evolve as it establishes a position in international affairs commensurate with it's new economic power. By some accounts, "Tokyo's hesitation to act has less to do with an insider's lack of concern for the outsider's problem and more to do with never before being faced with a leadership role in these areas."[82]

## E. TROPICAL DEFORESTATION: A CASE STUDY OF EVOLVING JAPANESE POLICY

Tropical deforestation is one of the international environmental issues of greatest relevance to Japan. Forests cover 25 million hectares, or 66%, of its land surface, which means that Japan is among those countries with considerable forestry resources.[83] However, the Japanese log relatively little domestic timber. Officials assert that most of Japan's forests are on mountainous terrain (and therefore expensive to harvest) and do not provide the hardwoods most wanted to meet the growing demand for building materials and furniture. However, some foreign authorities challenge these arguments, noting that some of the tropical forests Japan logs are equally mountainous.

Unlike most other international environmental issues, tropical deforestation has attracted modest public interest in Japan, and an active citizen organization addresses the issue. The Japanese government also has taken a more active role than it has toward other international environmental issues, providing a secretariat for the International Tropical Timber Organization (ITTO) in Yokohama.

Japan has been the world's leading tropical timber importer since the 1960s. (See Appendix C.) Economic growth has fueled much of this enormous timber and paper consumption. The high price of indigenous timber, easy access to inexpensive Southeast Asian wood, and the high quality of tropical hardwoods makes them especially attractive. The World Wildlife Fund argues, "Alternative materials exist, are already in use, and are assuming a growing importance."[84] However, according to Japanese wood processing industry, a switch from tropical timbers to alternative temperature woods would be enormously expensive.[85]

A September 1989 report by the Ministry of Foreign Affairs downplays Japan's role in tropical deforestation, asserting that Japan imported only 1% to 2% of tropical hardwoods in 1986 from Central and South America and Africa, and only 2% of its tropical timbers from other Asian countries. The report argues that exports contribute very little to the total deforestation problem; "The main cause of deforestation is the destructive slash-and-burn method of agriculture the poor farmers in developing nations use to maintain their lives."[86]

While the government may be correct when viewing the problem globally, it does not effectively respond to concerns about the impacts of extensive logging in particular areas. Commercial stands are being depleted worldwide; the 33 developing nations that are currently net exporters of forest products may be reduced to fewer than 10 by the year 2000.[87] The more difficult issue may be what role Japanese firms, and indirectly the Japanese government, should take in demanding better management by exporting nations, particularly since they are increasingly taking control of harvests and processing in order to export more high-value wood products and fewer logs.

There are some indications that the Japanese government is becoming more sensitive to tropical deforestation. In response to White House inquiries, Japan aggressively denied official support for construction of a highway in the western Amazon but the possibility of unofficial private support remains. Deforestation issues also were highlighted in the Environment Agency's 1988 White Paper on the international environment and at the September 1989 Tokyo Conference on Environment and Sustainable Development.

The Japanese Environment Agency claims that its international efforts will include ''protection of the environment in developing countries, and in particular to the protection of tropical forests and the ozone layer.''[88] Prime Minister Kaifu has pledged part of Japan's 3-year $2.1 billion environmental and package to tropical rainforest preservation.[89] Responding to accusations that grants sometimes have supported environmentally destructive activities (e.g., logging roads have reportedly been built with foreign-aid funds), Japan recently took steps to apply environmental requirements to foreign-aid programs.[90] The Foreign Ministry also has requested about $30 million in aid funds for ITTO programs on forest preservation and technical assistance.[91].

These financial contributions may be part of a policy designed to promote tropical forest preservation while allowing growing hardwood imports. While seemingly inconsistent, this approach is similar to that of many developing countries which assert some commercial logging is essential to economic growth and consistent with ''reasonable'' conservation goals.[92] As developing countries increasingly seek to control the use of their resources, technical assistance and multilateral pressure for structural reforms may be the most effective levers industrialized countries can use to promote tropical forest protection.[93]

Some Japanese industrialists and traders realize that they cannot count on tropical timber imports indefinitely, and accordingly, have begun to prepare for the future. The government and forestry industry have sponsored research on technologies and products based on greater use of softwoods for some time. The logging traders also have given token amounts ($71,000) to help protect the rain forests.

## IV. ENERGY AND ENVIRONMENTAL POLICY

Although weak concerning conservation of nature, Japanese environmental policy excels in an area the U.S. does not — pollution reduction and energy efficiency technologies. Government cooperation with industry to promote these technologies has reduced air pollution and increased energy efficiency tremendously. Japan, one of the worst polluters in the 1960s, became the world leader in clean air and now exports large quantities of pollution control technology. From 1973 to 1986, Japan also cut energy consumption 50% per unit of GNP. However, Japanese energy consumption has increased significantly since 1986 due to declining energy prices, strong economic growth, new government policies and changing attitudes toward conservation.

### A. ENERGY EFFICIENCY

Japan's economy is one of the most efficient in the world, clearly demonstrating that efficiency goes hand in hand with economic growth (see Figure 1). However, although Japan still uses half as much energy as the U.S. to produce one unit of gross domestic product,[94] different lifestyles, climate, population patterns, and other factors make comparisons between the absolute energy consumption of Japan and other nations risky. The Japanese use less central heating, live in a much smaller area, and rely significantly more on rail and subway sysems for commuting than Americans do. Thus, differences in energy consumption are not as great as they first appear. Even so, Japan's continued progress in efficiency during a period of rapid economic growth is impressive and a model for all

| | Jap. | Can. | U.S. | Fr. | U.K. | Nl. | Swe. | FRG |
|---|---|---|---|---|---|---|---|---|
| Population '87 $10^6$ | 122.1 | 25.9 | 243.8 | 55.6 | 56.9 | 14.7 | 8.4 | 61.2 |
| Area $10^3$ $Km^2$ | 378 | 9,976 | 9,373 | 547 | 245 | 37 | 450 | 249 |
| GDP '87 US$ (in thousands) | 2,376 | 373 | 4,497 | 873 | 575 | 214 | 137 | 1,117 |
| GNP/Capita '87 | 15,760 | 15,160 | 18,530 | 12,790 | 10,420 | 11,860 | 15,550 | 14,400 |
| Ave. Growth Rate/yr. GDP - '65 - '87 | 4.2% | 2.7% | 1.5% | 2.7% | 1.7% | 2.1% | 1.8% | 2.5% |
| Commercial Energy Consumption/Capita '87 Eq. Kg. Petroleum | 3,232 | 9,156 | 7,265 | 3,729 | 3,805 | 5,198 | 6,453 | 4,531 |
| Annual Growth Rate of Energy Consumption '80 - '87 | 1.7% | 0.9% | 0.1% | 0.6% | 1.1% | 1.3% | 2.3% | 0.2% |
| Energy Saving Index 1975=100 '87 | 71 | 85 | 77 | 94 | 86 | 88 | 99 | 86 |
| Total Primary Energy Requirement (TPER) '87 Mtoe | 371 | 240 | 1865 | 206 | 484 | 65 | 56 | 271 |
| TPER/GDP | 0.2624 | 0.6435 | 0.4409 | 0.3788 | 0.4311 | 0.5058 | 0.5448 | 0.4189 |
| TPER/Population | 3.0442 | 9.3636 | 7.6490 | 3.7113 | 3.6654 | 4.4932 | 6.7210 | 4.4399 |
| $CO_2$/TPER | 0.734 | 0.507 | 0.741 | 0.544 | 0.809 | 0.738 | 0.311 | 0.761 |
| $CO_2$/Population '85 | 1.91 | 4.14 | 4.98 | 1.78 | 2.63 | 3.32 | 2.09 | 3.00 |
| Total $CO_2$ Emission $10^6$ t | 230 | 106 | 1,183 | 97 | 146 | 49 | 18 | 183 |
| $SO_x$ Emission $10^3$ t | 1,079 | 3,938 | 20,700 | 1,845 | 3,580 | 230 | 264 | 2,640 |
| $SO_x$/GDP '85 | 0.81 | 11.38 | 5.24 | 3.62 | 7.88 | 1.84 | 2.63 | 4.22(x $10^{-3}$) |
| $SO_x$/Population '85 | 8.93 | 15.29 | 85.5 | 33.42 | 63.63 | 15.86 | 31.34 | 43.27(x $10^{-3}$) |
| $SO_x$/TPER '86 | 0.35 | 16.89 | 11.49 | 9.19 | 17.38 | 3.57 | 4.77 | 9.75(x $10^{-3}$) |
| $NO_x$ Emission $10^3$ t | 1339 | 1940 | 19300 | 1652 | 1965 | 537 | 300 | 2967 |
| $NO_x$/GDP | 1.04 | 5.61 | 4.89 | 3.24 | 4.33 | 4.30 | 3.00 | 4.75(x $10^{-3}$) |
| $NO_x$/Population '85 | 11.07 | 75.48 | 80.75 | 29.93 | 34.78 | 37.03 | 35.71 | 48.67(x $10^{-3}$) |
| $NO_x$/TPER '85 | 3.57 | 8.22 | 10.71 | 8.23 | 9.54 | 8.44 | 5.42 | 10.97(x $10^{-3}$) |

**FIGURE 1.** International comparison of $CO_2$, $SO_x$, $NO_x$, energy and background data. (From *Japan and the Global Environment*, The Center for Global Change. With permission.)

Japan achieved its greatest energy savings between 1979 and 1986, when it cut energy consumption 20%. The government proudly notes that from FY 1973 to FY 1986, the GNP grew 63% while energy demand grew only 6.2%.[95] In addition to high tax on gasoline, incentives for specified conservation investments included accelerated depreciation or tax credits, reduced property taxes, and loans. Small businesses can obtain energy audits at no charge, and all factories above a minimum size must have a licensed energy engineer on site to promote energy efficiency. Minimum efficiency standards also apply to some industrial processes, building, automobiles, and appliances, supported by consumer labeling.[96]

How and why did the Japanese achieve this progress? In a recent study,[97] the Association for the Conservation of Energy suggests several reasons for Japan's success in attaining energy efficiency and economic growth, which provides lessons for other OECD nations. First, the Japanese government provides financial incentives, regulations, standards, information and education to help companies reach the highest possible energy efficiency standards. Second, a single government agency, MITI, administers a comprehensive national program for energy efficiency and has the power to ensure that all sectors adopt its conservation measures. Third, all companies larger than a minimum size must have licensed energy managers, who supervise the firm to ensure maintenance of the highest energy efficiency standards. Fourth, to preserve efficiency gains, the Japanese upgrade energy efficiency standards for appliances and buildings periodically. Finally, the government has a long-term plan to obtain significant increases in national energy efficiency.

Japan's experience shows that economic growth does not depend on growth in energy supply. Indeed, improved energy efficiency improves a nation's competitiveness in the international marketplace. Another important lesson is that Japan did not rely solely on higher prices to spur efficiency improvements.

However, lower oil prices, the strength of the economy since 1986, and the stability of the yen have led to a declining interest in energy efficiency in Japan in recent years.[98] OECD data indicates that the national energy efficiency index for Japan has not risen since 1986 and that demand for energy has been increasing since 1987.[99] Japan's current increased energy use typifies the response of most industrialized nations experiencing declining energy prices. These policies may now change in response to the Gulf crisis and renewed concern about oil security.

Although official government policy extols energy conservation as a primary goal, the Japanese government has cut several initiatives and relaxed some conservation laws. Moreover, growth in personal income and the declining real price of oil has led to growing demand for automobiles, particularly larger automobiles, while the average fuel efficiency of new cars has declined steadily since 1982. Not only has the government not attempted to combat this trend with higher gasoline taxes, it removed a commodity tax on larger cars in 1988 as part of recent tax reform legislation. Electricity prices, among the highest of any industrialized country, also have been reduced to reflect the declining cost of fuel.[100] Japan could make significant energy savings in appliances, automobiles, lighting, and buildings, where efficiency either has declined or increased little. However, until fuel prices rise, or until the government upgrades its energy policies, it is likely that demand will continue to increase while efficiency lessens.[101]

The government and private companies have taken some limited measures to stem energy consumption due to the Iraqi invasion of Kuwait. They are encouraging citizens and employees to turn air conditioners up to 80°, turn off one third of the lights in offices, and adhere to an 80 km/hr speed limit on highways. Even some corporations are "reexamining investment blueprints with an eye toward energy efficiency." However, these measures are not expected to yield large energy savings. The government seems to be relying on years of energy planning to help the economy through the crisis.[102]

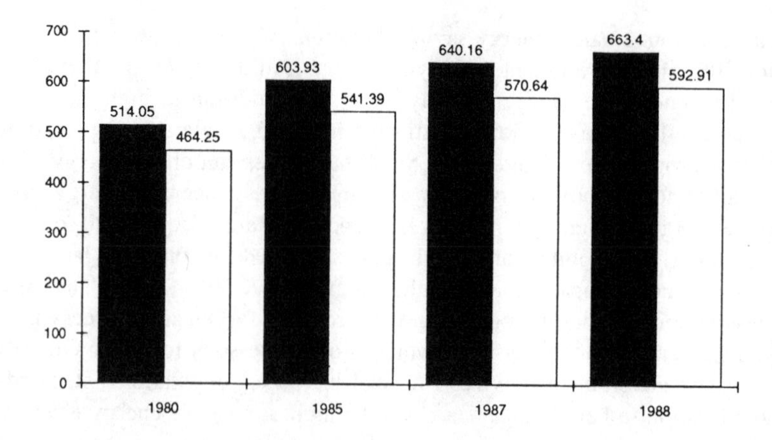


**FIGURE 2.** Japan's total electricity output and consumption. (■) Output (thermal, nuclear, and hydroelectric) and (□) consumption measured in billions of kilowatt hours. Japanese industry and households contributed equally to the rise in total consumption in 1988. (Adapted from *JEI Report,* No. 40A, Oct. 20, 1989. With permission.)

The government's campaign to reduce the personal savings rate and increase consumption — a policy designed party to appease the U.S. government — also has contributed to growing energy use. With the recent conclusion of the U.S.-Japan SII talks promoting domestic measures to increase consumption, this trend may only worsen.[103] The campaign comes at a time when the Japanese have decided to enjoy the benefits of economic growth. Energy-intensive luxury items and home appliances, such as electric bread makers, full-size refrigerators and microwave ovens, have become fashionable. The Japanese use more disposable goods today such as wooden chopsticks and paper towels. More choose to drive rather than ride the subway and buy bigger houses that cost more to heat in winter and cool in summer. Increasingly, Japanese youngsters are now taking *''asa-shan''* — morning showers to shampoo their hair — in addition to the traditional evening bath.

According to Osamu Maeda, a Rikkyo University sociologist, ''In the process of economic growth, the people have gotten used to their modern amenities and have grown more distant from nature . . . They have lost some of their sensitivity to the environment.''[104] A Japan Economic Institute report adds, ''There appears to be no place for a public campaign to promote energy conservation in this new consumer environment.''[105] If anything, new government policies that relax restrictions on building large retail stores in urban areas will accelerate this trend.

The growth in energy use may be due to government miscalculation rather than intention. MITI assumed that with a stronger yen/dollar ratio, Japan's economy would shift from energy intensive industries and to increased consumption of imports and less energy-intensive consumer goods. This has not occurred, and energy growth accelerated from 0.4% in FY 1986 to almost 5% in FY 1987[106] and rose another 5.4% in FY 1988.[107] Industrial growth and higher consumer spending has resulted in greater electricity production and consumption.[108] (see Figure 2) The sharp increases in energy consumption in 1988 and the beginning of 1989 have forced MITI to address energy issues more seriously.

MITI's latest energy plan (1990) focuses heavily on energy conservation. MITI intends to use various energy conservation measures to improve the GNP base unit (primary energy supply quantity/GNP) by 2.0% annually, with a total improvement of 36% by the year 2010, and to cut the energy GNP elasticity ratio (growth rate of primary energy supply/growth rate of the economy) for energy demand by 0.58%. To attain these two goals, MITI aims to reduce energy use by government and industry 11.2% from what it would be in the year

2010 given current trends.* This target is consistent with a reduction in the energy used per unit of GNP of 2% per year. The plan requires a substantial improvement in cooling and heating efficiency in residential use, vehicle fuel efficiency, and improved efficiency in power generation facilities.

MITI projected a growth rate in energy consumption between 1988 and 2000 of 1% for the industrial sector, 0.9% for the manufacturing sector, 2.7% for the residential and commercial sector, and 1.9% for the transportation sector. Between 2000 and 2010, however, the annual growth rate in energy consumption for each of these sectors is predicted to drop 0.6% for the industrial sector, 0.6% for the manufactoring sector, 2.0% for the residential and commercial sector, and 1.3% for the transportation sector.

To combat global warming and to reduce its dependence on foreign energy imports, thereby improving Japan's energy security, MITI plans to promote increased use of new energies, and hydro-electric, geothermal, and nuclear power to 27% of the energy supply by 2010. (see Figure 2) In 1988, these non-fossil energies provided 15% of Japan's energy supply. MITI will encourage research and development of alternative energy from methanol and solar power, with the aim of dramatically cutting solar facility costs, and install new facilities for generating power from liquified natural gas (LNG). MITI plans to increase the share of LNG in gasoline used in motor vehicles in urban areas from 71% to 85% by encouraging small- and medium-sized gas companies to convert to natural gas.[109]

The government views coal as a stable resource and a major oil alternative. In the wake of the 1973 and 1978 oil crises, the Japanese government adopted an aggressive policy to develop alternatives to oil. The attraction of coal is clear: there are abundant world reserves located in countries such as Australia, China, and the U.S. which can supply Japan conveniently via sea routes. Many of Japan's suppliers also have stable governments, and some allow foreign ownership of coal fields or the companies that control them. Thus, coal ensures Japan a relatively secure fuel supply. However, MITI believes it will be difficult to increase Japan's coal use in the mid- to long-term, given the $CO_2$ emissions that arise from coal burning.[110] To combat these emissions, the ministry proposes for FY 1991 to cut the budget for coal liquefaction and gasification substantially, from 24,901 million yen to 16,296 million yen.[111]

MITI also hopes to increase the share of power generated from nuclear power plants from 27% to 43% by the year 2010.[112] Japan has maintained an enviable nuclear safety and reliability record, ironically using U.S. reactor technology. However, Japan still is searching for a final solution to disposal of its waste. Growing public opposition may make it difficult to obtain new sites for reactors, which will force utilities to move further away from demand centers and thereby increase costs. Such factors as Three Mile Island, Chernobyl, the increasing influence of women, who tend to oppose nuclear power, and an awareness of public opposition to nuclear power in the U.S. and Europe have helped fuel anti-nuclear sentiments in Japan.[113] (See Appendix D).

## B. JAPANESE ENERGY TECHNOLOGY AND THE ENVIRONMENT

Sustained by low oil prices, Japanese industry grew rapidly during the 1960s. Then, shocked by the oil crisis of 1973 when prices rocketed upwards, government and industry undertook massive conservation efforts that reduced energy use sharply and quickly.

These programs addressed virtually every aspect of Japanese activity, ranging from home refrigerators to steel mills. Conservation began with simple acts, such as greater use of insulation, and it progressed to complex and expensive undertakings including the development of alternative energy technologies. Japan now consumes less energy per unit than

* MITI notes, however, that if global temperatures were to rise significantly in the near future, requiring Japan to stabilize its $CO_2$ emissions quickly, MITI might have to restructure its energy policies.

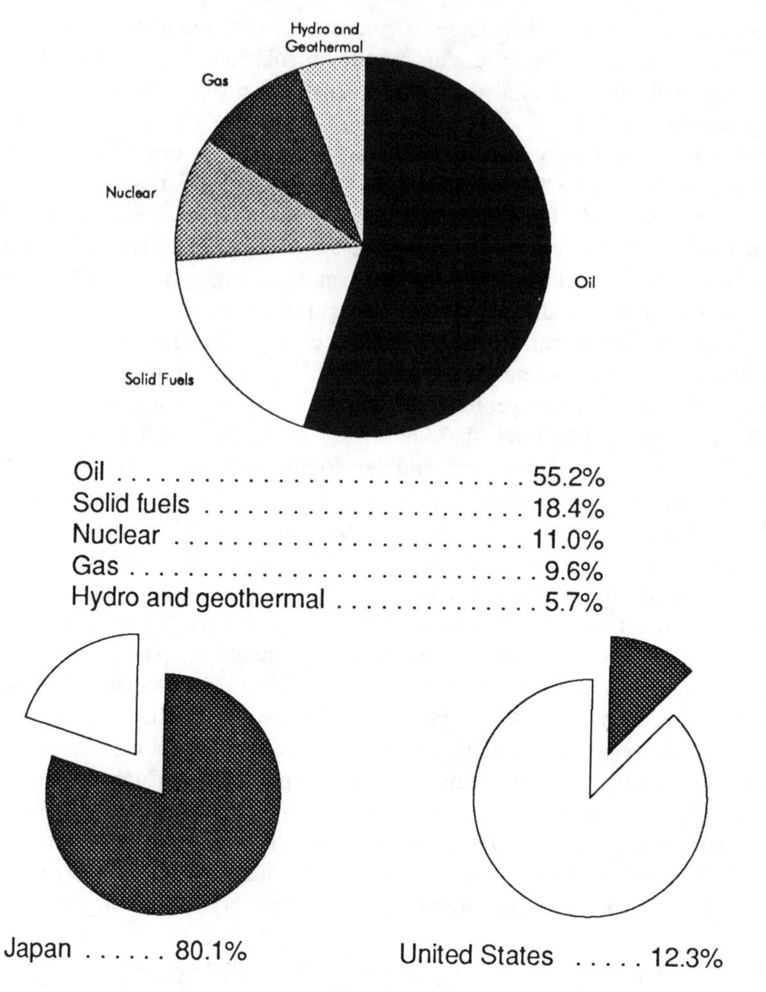


**FIGURE 3.** (A) Japan's energy supply in 1986; (B) comparison of Japanese and U.S. dependence on energy imports. (Adapted from *Science and Technology in Japan,* March, 1990. With permission.)

any other industrialized nation. This results from a concerted effort by Japanese industry, spurred by government demands and cooperation.*

Japanese goverment and industry have developed a menu of technologies and practices that demonstrate that pollution — even carbon dioxide ($CO_2$) — can be cut substantially in ways that increase efficiency and lower costs. This experience challenges the conventional wisdom among scientists, engineers, and politicians that pollution is the inevitable consequence of industrial productivity. The reverse may be true: the path to true productivity may be one that leads to zero pollution and 100% efficiency.

Japanese industry has succeeded in reducing energy consumption in three fundamental ways. First, improved measurement and control devices, such as exhaust gas analyzers with information feedback mechanisms that automatically adjust boiler air-fuel ratios, have been installed to minimize energy consumption. Second, waste heat collection systems are being used to capture and reuse heat that otherwise would be vented into the atmosphere. The Japanese use a variety of devices ranging from heat exchangers to automatic frequency controls for electric pumps and blowers.

* Information in this section is based on extensive site visits and interviews conducted by the authors in January 1989.

Finally, the Japanese have reduced energy consumption dramatically by changing the production process itself. For example, steel can be rolled into a product as it comes from the blast furnace without being cooled, or it can be allowed to cool, inspected for defects, then reheated for rolling. The former process, now used at nearly all Japanese steel mills, reduces energy consumption enormously.

Following is a discussion of certain Japanese innovations in energy efficiency and pollution reduction. From the demand side, we review advances made at Nippon Steel and Toyota Motor companies, whose new technologies reduce pollution and energy consumption significantly. From the supply side, we discuss innovations in fuel cells, nuclear reactors, combined-cycle power plants and coal combustion technologies that reduce pollution and provide energy more efficiently, thus reducing the country's dependence on imports and improving its energy security.

## C. THE DEMAND SIDE

### 1. Nippon Steel, Keihen Works

When it began operations in 1976, Nippon Steel's Keihen Works was arguably the most efficient steel mill in the world. Yet during the next decade, the plant reduced energy consumption by another 30%, enabling the complex to produce more steel plate and tubing with less energy and pollution than any mill in history. The highly computerized and automated plant, located on an island in Tokyo Bay, turns out 6 million tons of steel per year.

Each ton of steel from the Keihen Works requires 5.7 million kcal of energy, substantially less than the average at mills in the U.S. and Western Europe. Stringent pollution controls cut $SO_2$ emissions from the plant by more than 90% and $NO_X$ by 80%, with comparable controls on the superfine and frequently toxic dust and soot that characterize steel production. Nippon Steel officials estimate that the Keihen Works is the third or fourth most efficient mill in Japan. Without the pollution control devices, it likely would be Japan's most efficient.

### 2. Automotive Efficiency

Automotive fuel efficiency has increased about 30% since the 1973 oil crisis. The bulk of this improvement has been attributed to decreases in bodyweight, adoption of aerodynamically superior designs, and improvements in engine technology.

Toyota, the world's largest auto manufacturer, is working on several pollution reduction and efficiency enhancement technologies. One is improvement of the lean-burn engine, which simultaneously reduces $NO_X$ emissions while increasing fuel economy. Toyota has the reputation among automotive experts of offering the best of the lean-burn engines. With current technology, the company can meet Japanese and European standards, but its engines fail under the more stringent U.S. procedure. Toyota is presently developing ways for lean burn engines to meet some governments' stricter $NO_X$ requirements while preserving fuel economy gains.

As discussed above, fleet efficiency has been declining since 1982 due to growing demand for larger cars, declining oil prices, and the elimination of some incentives for small cars. This led Toyota to drop production of its smallest car due to lack of demand.

## D. THE SUPPLY SIDE

### 1. Fuel Cells

Fuel reacts with itself in a fuel cell to generate electricity. Producing virtually no sulphur dioxide ($SO_2$) or oxides of nitrogen ($NO_X$), this technology holds extraordinary potential for reducing air pollution. Fuel cells can run on currently available fossil fuels and their derivatives such as natural gas and methanol, or on hydrogen, which might be produced through the use of solar or nuclear-generated electricity. Because fuel cells are also more efficient than conventional energy technologies, they have the potential to significantly reduce $CO_2$ emissions by minimizing fossil fuel consumption. Unlike most other energy technologies, fuel cells may be versatile enough for small-, medium-, or large-scale applications, ranging from automobiles to central power plants. They make almost no noise, as there are no pistons or controlled explosions of the sort that make gas and diesel engines run. It may be possible to locate smaller scale units, ranging from 10 to 50 megawatts (MW), in or near city centers, eliminating the need for transmission lines and minimizing the cost of land.

Fuji Electric is the largest fuel cell producer in Japan. As of January 1989, Fuji had 11,000 kilowats (kW) of fuel cell projects underway in Japan and 13 projects in the U.S. and Europe. These projects include commercial electricity generation, production of a fuel cell-powered for lift and development of a fuel cell-powered bus for Georgetown University in Washington, D.C.

Fuji has completed construction and testing of a 1-MW power plant and plans to undertake a 5-MW demonstration plant. Fuji believes it can cut production costs enough to compete with coal-fired plants (whose conventional cost is roughly $2,000 per kilowatt) by mass production of standardized components. Fuji has made enormous progress toward commercializing fuel cell technology for generating electricity from natural gas since winning a four-company competition supported by the Japanese government. It now has a semi-automated production facility for the manufacture of 50-MW fuel cells for use in medium-size commercial and residential buildings. These will be turn-key units with error detection systems built in to minimize maintenance and service requirements. Tokyo Gas Company has made a substantial contribution to the development cost in return for a guarantee that they will receive 50-kW fuel cells with a minimum 5-year operating life at $1,800/kW by 1995. The company benefits by selling natural gas to Fuji.

**2. Nuclear Power**

An experimental 2.8-MW demonstration breeder reactor, the ''Monju'', is under construction and projected to achieve criticality in 1992. Officials expect three progressively more powerful demonstration reactors to follow Monju, culminating in 1500-MW scale commercial plants sometime between 2010 and 2030. Nuclear power is already Japan's leading source of electric power, accounting for more than one fourth the nation's electricity output.

As of June 1989, 37 nuclear reactors were running in Japan, producing 26.6% of the country's electric energy.[114] As discussed previously, growing questions about the political future of nuclear power in Japan may jeopardize proposals for additional plants. Japan now imports the fuel to supply these reactors. However, if plans for the development of fast breeder reactors proceed on schedule, the nation will become an exporter of fuel within a generation — possibly becoming the first nation in history to bootstrap itself from energy buyer to energy seller.

**3. Combined Cycle Power Plants**

TEPCO began construction of the world's first large scale combined cycle power plant in April, 1982. Perched on the edge of Tokyo Bay, Futsu is today one of the world's largest power plants and almost certainly the cleanest. It produces 2000 MW of gas-fired electricity but emits almost no $SO_2$ and less than one sixth of the nitrogen based pollution allowed from new plants in the U.S.

Three factors account for the extraordinary performance of the Futsu plant. First, it burns liquefied natural gas, one of the cleanest fuels available. Second, it uses a combined cycle system, burning the gas in one turbine, then using the exhaust gases to power a second turbine run by steam. Third, selective catalytic reduction, an add-on device for pollution control, cleanses the exhaust gases of $NO_X$. Although other power plants have employed one or two of these approaches, Futsu is the first to use all three. The combination makes the plant a model of simultaneous pollution reduction and increased efficiency.

**4. Technologies to Reduce Pollution from Coal Combustion**

To increase efficiency and reduce air pollution, Japan's Electric Power Development Company (EPDC)* is investing in fluidized bed combustion (FBC), an inherently cleaner combustion process for coal. A finely powdered mixture of coal and limestone is suspended in mid-air by blowing air through it at tremendous velocities. The cooler and more complete combustion which results not only lowers levels of both $NO_X$ and $SO_2$ but allows the use of a wide range of different fuels. FBC can be coupled with highly efficient turbines to reduce air pollution still further.

Since 1980, EPDC has been testing a process to powder coal, then mix it with water to form a combustible slurry. As a liquid, coal could be more easily transported, loaded, and stored. The company is also attempting to develop methods of de-watering low-quality, subbituminous and brown coals, which contain too much water to be transported or burned efficiently. The reserves of this fuel are believed to be practically inexhaustible. Its use would increase $CO_2$ emissions, but potentially much less than existing coal combustion technologies because of its greater efficiency. The reduction in $CO_2$ is proportional to the improvement in efficiency; improvements of 10 to 20% or more are technically feasible.

**E. DISCUSSION**

Japan has much to offer the world in the development of innovative technologies that clean the environment and bolster economic growth. The prospect of future international environmental accords could have economic benefits for Japan because of its position in technological leadership. A global warming agreement, for example, might help promote markets for high-efficiency appliances and industrial systems. Similarly, pressuring developing countries to control their pollution could increase orders for Japanese pollution control systems. At a March 1990 presentation in Washington, a MITI official described plans to develop somewhat less effective but much less expensive pollution control technology for sale to developing countries. Already, discussion of a "green industry" has begun to appear in the Japanese press, but so far these considerations seem to be less politically salient than the possible costs of environmental controls on the Japanese economy.[115] Indeed, Japan could do much more to publicize its activities to provide a model for newly industrializing countries.

## V. CONCLUSION

Japan's environmental policy is weakest in conservation of nature and protection of the global environment; strongest in pollution reduction and energy efficiency technologies. It is with these technologies that Japan can make important contributions to resolving global environmental problems. However, lack of domestic support for conservation and NGO activities remains a serious limitation on Japan's involvement in domestic and international environmental issues. Japanese technology can benefit the world significantly, but it cannot address all environmental issues. Ironically, Japan is among the most polluted industrialized nations and has aggressively developed and marketed some of the world's most advanced pollution control technologies, but fails to systematically adopt them at home. Since World War II, Japanese politics have been oriented towards increasing the wealth of the country. Today, how Japan addresses environmental issues may relate to its need to find a moral

* EPDC is a government-funded corporation created in 1952 whose objective is to develop power resources that privately-owned companies might find technically or financially daunting.

basis for involvement in international affairs that transcends the single-minded pursuit of economic wealth.

## ACKNOWLEDGMENTS

The authors gratefully acknowledge research and writing assistance by Susan Conbere. We would like to thank Harumi Befu, Miwako Kurosaka, Pat Murdo, Richard Forrest, Martha Harris, Margaret McKean, Yasuhiro Shimizu, and Jennifer Whitaker for their comments on the draft of this paper.

This work was supported by the U.S. Environmental Protection Agency: Identification Number CX-815954-01-0.

## ENDNOTES

1. Joseph Kitagawa, *On Understanding Japanese Religion,* Princeton, N.J.: Princeton University Press, 1987, p. 268.
2. Edwin Reischauer, *The Japanese,* Tokyo: Charles E. Tuttle Company, 1986, p. 217.
3. Margaret McKean, *Environmental Protest and Citizen Politics in Japan,* Berkeley, CA: University of California Press, 1981, p. 137.
4. Alan Miller, "Three Reports on Japan and the Global Environment", *Environment,* Vol. 31: No. 6, July/August 1989, p. 25.
5. Jo Stewart-Smith, *In the Shadow of Fujisan: Japan and its Wildlife,* Middlesex, England: Viking Penguin, 1987.
6. *In the Shadow of Fujisan: Japan and its Wildlife.*
7. United Nations Environment Programme Survey, May 1989.
8. United Nations Environment Programme Survey, May 1989.
9. "Putting the Heat on Japan," *Time,* July 10, 1989, p. 52.
10. Jon Woronoff, *Politics the Japanese Way,* New York: St. Martin's Press, 1986, pp. 264, 265.
11. "Politics of the Environment," *American Behavior Scientist,* May-June 1974, p. 764. *See also* McKean, *Environmental Protest and Citizen Politics in Japan* and J. Gresser, K. Fujikura, and A. Morishima, *Environmental Law in Japan,* Cambridge, MA: MIT Press, 1981.
12. F. Upham, *Law and Social Change in Japan,* Harvard University Press, 1987.
13. Scott Flanagan, Ellis Krauss, and Kurt Steiner, editors, *Political Opposition and Local Politics in Japan,* Princeton, NJ: Princeton University Press, 1980, pp. 225, 226.
14. Ibid, p. 14.
15. *Law and Social Change* in Japan.
16. Norrie Huddle and Michael Reich, *Island of Dreams: Environmental Crisis in Japan,* Cambridge, MA: Schenkman Books, Inc., revised edition, 1987, pp. 22, 23.
17. Telephone conversation with Margaret McKean, Sept. 14, 1989.
18. "Right-to-Know Laws Fail to Provide Info Access," *The Japan Times Weekly International Edition,* July 30–August 5, 1990, pp. 1, 8.
19. Environment Agency, Government of Japan, *Introduction to the Environment Agency of Japan* (no date).
20. Kelley, Stunkel, and Wescott, "Politics of the Environment," *American Behavioral Scientist,* May-June 1974, p. 766.
21. "Politics of the Environment," pp. 765, 766.
22. T. J. Pempel, *Policy and Politics in Japan: Creative Conservatism,* Philadelphia: Temple University Press, 1982, p. 237.
23. "Politics of the Environment," p. 766.
24. *Politics the Japanese Way,* p. 269.
25. *Policy and Politics in Japan,* p. 234.
26. *Island of Dreams: Environmental Crisis in Japan,* 9, 10.
27. *Policy and Politics in Japan,* p. 234.
28. *Politics the Japanese Way,* p. 266, 267.
29. *See generally* T. Nakemata and C. du Florey, *Health Effects of Air Pollution and the Japanese Compensation Law,* Batelle Press, 1987.

30. "Environment Blossoms as Japan Issue," *Los Angeles Times,* August 16, 1989, pp. 1, 12.
31. "Grassroots Groups Fight to Save Japan's Last Wild River," *World Rivers Review,* March/April 1989, p. 7.
32. "Nagara River Dam Project Hits Logjam of Criticism," *The Japan Times,* April 25, 1989, p. 4.
33. Manju Tanaka, "Save the Nagara River," *The Japan Times,* May 17, 1989.
34. "Grassroots Groups Fight to Save Japan's Last Wild River," p. 7.
35. "Nagara River Dam Project Hits Logjam of Criticism," p. 4.
36. "Grassroots Groups Fight to Save Japan's Last Wild River," p. 7.
37. "Environment Blossoms as Japan Issue," p. 1.
38. "Japan's Environmentalists," *Environmental Action,* July/August 1986, p. 21.
39. "The Greening of Japan," *Newsweek,* August 6, 1990, p. 6.
40. "Environmental Action on Corporate Agenda," *Japan Economic Journal,* July 21, 1990, pp. 1, 4.
41. *Environmental Protest and Citizen Politics in Japan,* pp. 136, 137.
42. "Majority Urge Government to Give Priority to Environment," Kyodo News Service, July 29, 1990.
43. *Japan Quarterly,* April-June 1989, p. 237.
44. P. Murdo, "Japan's Environmental Policies: The International Dimension," *JEI Report,* No. 10A, March 9, 1990.
45. "Putting the Heat on Japan," *Time,* July 10, 1989, p. 52.
46. *Japan Economic Institute (JEI) Report,* No. 37B, Sept. 29, 1989, p. 11.
47. "Fish Mining on the Open Seas," *Time,* June 5, 1989.
48. *JEI Report,* No. 22B, June 9, 1989, p. 14.
49. P. Murdo, "Japan's Environmental Policies."
50. "Strip Mining the Seas," *Washington Post,* Sept. 23, 1989, p. A22.
51. "Drift Net Fishing to be Halted," *Japan Times Weekly International Edition,* July 30-August 5, 1990, p. 3.
52. *JEI Report,* No. 37B, p. 12.
53. "Japanese to Stop Ivory Trade," *Washington Post,* Oct. 31, 1989, p. A14.
54. "Putting the Heat on Japan," p. 51.
55. *Japan Quarterly,* January-March 1988, p. 112.
56. "Putting the Heat on Japan," pp. 51, 52.
57. This conclusion is based on interviews and related research conducted in Japan by one of the authors, Alan Miller, in 1987.
58. Jon Choy, "Initial FY 1990 Budget Requests Up," *JEI Report,* No. 35B, Sept. 15, 1989, p. 9.
59. "Nissan's Efforts to Reduce the Use of Chlorofluorocarbons," *Nissan News,* Aug. 7, 1989.
60. P. Murdo, "Japan Ranked Low on Environmental Scorecard," *JEI Report,* No. 27B, July 13, 1990, pp. 8, 9.
61. Letter to Prime Minister Toshiki Kaifu from Lynn Greenwalt, National Wildlife Federation, on behalf of the Natural Resources Defense Council, et al., Aug. 7, 1990.
62. Shuzo Nishioka, "The Japanese Response to Global Warming: Background, Policy & Research Work," National Institute for Environmental Studies, Environment Agency of Japan, June 1989, p. 2.
63. "The Japanese Response to Global Warming," p. 4.
64. "Global Finance," *Wall Street Journal,* Sept. 23, 1988, Section 3.
65. Richard Forrest, "Japanese Economic Assistance and the Environment: The Need for Reform," National Wildlife Federation, Nov. 1989, p. 7.
66. "Japan's Approach to Environmental Issues of the Globe," Overseas Public Relations Division, Japanese Ministry of Foreign Affairs, Sept. 1989, p. 1.
67. "Japan's Foreign Aid Policies," p. 4.
68. Francois Nectoux and Yoichi Kuroda, World Wildlife Fund, *Timber from the South Seas: An Analysis of Japan's Tropical Timber Trade and its Environmental Impact,* April 1989, p. 88.
69. *Timber from the South Seas,* p. 88.
70. "Japan's Approach to Environmental Issues of the Globe," p. 3.
71. "Japan Plans $1 Billion in Aid for Mexico to Combat Severe Air Pollution," *Washington Post,* August 30, 1989, p. A37.
72. *Timber from the South Seas,* p. 88.
73. Ibid., pp. 87, 88.
74. Ibid., p. 91.
75. Ibid., p. 90.
76. Richard Forrest, "Japanese Economic Assistance and the Environment: The Need for Reform," revised edition, National Wildlife Federation, November 1989, p. 28.
77. Ibid., p. 92.
78. Ibid., p. 87.
79. Ibid., p. 94.

80. ''Japs Give in to Pressure on Narmada,'' *The Daily,* May 31, 1990, p. 18.
81. ''How the US Can Take the Lead in the Third World,'' *Time,* Oct. 23, 1989, p. 63.
82. *JEI Report* No. 10A.
83. Ichiro Yano, ed., *Nippon: A Charted Survey of Japan, 1987–88,* Tokyo: The Kokusei-Sha Corporation, 1987, p. 143.
84. Ibid., p. 6.
85. *Timber from the South Seas,* p. 5.
86. ''Japan's Approach to Environmental Issues of the Globe,'' p. 14.
87. R. Repetto, *The Forest for the Trees?,* World Resources Institute, 1988, pp. 5–8.
88. Global Environmental Conservation Office, Japanese Environment Agency, ''Global Environmental Conservation Policy in Japan,'' Sept. 1989, pp. 2–5.
89. *Japan Economic Institute Report,* No. 36B, Sept. 22, 1989, p. 12.
90. P. Murdo, ''Japan's Environmental Policies.''
91. Ibid.
92. For representative statements from Indonesia, see *News & Views Indonesia,* Oct.-Nov. 1989 L (available from the Embassy of Indonesia).
93. *The Forest for the Trees.*
94. *World Resources 1990–91,* New York: Oxford University Press, 1990, p. 25.
95. ''Japan's Energy Conservation Policy,'' Ministry of International Trade and Industry, April 1988.
96. Jon Choy, ''MITI To Revise Energy Demand Outlook,'' JEI Report, No. 19B, May 12, 1989, pp. 5–6.
97. Linda Taylor, et al., ''Lessons from Japan: Separating Economic Growth from Energy Demand,'' Association for the Conservation of Energy, London, England, Spring 1990, pp. 15–17.
98. Jon Choy, ''Japan's Energy Policy: 1988 Update,'' *Japan Economic Institute (JEI) Report,* NO. 40A, Oct. 20, 1989, p. 1.
99. Association for the Conservation of Energy, *Lessons from Japan: Separating Economic Growth from Energy Demand,* London, Spring 1990, p. 11.
100. ''Public Utilities Apply to MITI for Approval to Cut Rates and to Introduce New Concept to Rate System,'' *Japan Petroleum and Energy Weekly,* Nov. 2, 1989, pp. 2–4.
101. *Lessons from Japan,* p. 13.
102. Jon Choy, ''Mixed Japanese Response to Energy Conservation,'' *JEI Report* No. 33B, August 24, 1990.
103. Letter to Prime Minister Toshiki Kaifu from Lynn Greenwalt, National Wildlife Federation, Aug. 7, 1990.
104. ''Japanese Lifestyle Discourages Environmental Concern,'' *The Japan Times,* July 28, 1989, p. 13.
105. *JEI Report* No. 40A, p. 7.
106. *JEI Report* No. 19B, pp. 5–6.
107. ''Nuclear Power Development and Use in Japan,'' *Science and Technology in Japan,* March 1990, p. 9.
108. *JEI Report* No. 19B, pp. 5–6.
109. ''Long-Term Forecast of Energy Supply and Demand,'' Draft, Ministry of International Trade and Industry, June 1990.
110. ''Long-Term Forecast of Energy Supply and Demand,'' 1990.
111. ''A Darker Shade of Green,'' Nature, Aug. 30, 1990, p. 783.
112. ''Long-Term Forecast of Energy Supply and Demand,'' 1990.
113. Takao Tomitate, ''Political Evolution of International Arguments on Global Warming,'' *Energy in Japan,* Feb. 1990; Toyoaki Ikuta, ''Energy: Recent Trends and Future Prospects,'' *Energy in Japan,* Oct. 1989.
114. ''Nuclear Power Development and Use in Japan,'' p. 10.
115. M. McQuillan and R. Uland, ''The Coming of the Greens,'' *Japan Economic Journal,* Dec. 23, 1989, p. 26.
116. Ibid., p. 28.
117. *Environmental Protest and Citizen Politics in Japan,* p. 109.
118. Ibid.
119. Karl van Wolferen, *The Enigma of Japanese Power: People and Politics in a Stateless Nation,* New York: Alfred A. Knopf, 1989, p. 52.
120. *The Enigma of Japanese Power,* p. 52.
121. Telephone conversation with Margaret McKean, Sept. 14, 1989.
122. ''Kaifu Drops Women From Cabinet, Defies Factions' Leaders,'' *Washington Post,* Feb. 2, 1990, p. A16.
123. Michael K. Blaker, ''Japan at the Polls:'' *The House of Councilors Election of 1974,* Washington, D.C.: American Enterprise Institute for Public Policy Research, 1976, p. 35.
124. Gerald Curtis (author of *The Japanese Way of Politics,* 1988), lecture, School of Public Affairs, University of Maryland at College Park, Sept. 21, 1989.
125. Telephone conversation with Margaret McKean, Sept. 14, 1989.
126. ''Grassroots Groups Fight to Save Japan's Last Wild River,'' p. 7.
127. ''Save the Nagara River.''
128. ''Grassroots Groups Fight to Save Japan's Last Wild River,'' p. 7.

129. "Save the Nagara River."
130. *Timber from the South Seas,* p. 5.
131. Michael Browning, "Japanese Wood Appetite Threatens World Forests," *Journal of Commerce,* April 26, 1989.
132. "Japan to Push Ahead with Nuclear Power," *International Herald Tribune,* Dec. 3, 1988.
133. *Japan Quarterly,* January-March 1989, p. 110.
134. "Hokkaido Rejects Nuclear Waste Site," *Japan Times Weekly International Edition,* July 30-August 5, 1990, 2.
135. "Two Fukushima Towns Voice Concern On N-Plant Restart," *Mainichi Daily News,* April 19, 1990, p. 14.

# APPENDIX I
# THE INFLUENCE OF JAPANESE WOMEN ON ENVIRONMENTAL POLICY

Women composed the largest numbers in the anti-pollution movements of the 1960s and 1970s and have provided much of the leadership for the anti-nuclear movement of the 1970s and 1980s.[116] It may be that women are accepted in Japanese society as environmental activists because they generally are viewed as caretakers of family and community. Housewives belong in large numbers to the neighborhood women's associations, called Fujinkai, to the larger Shufuren, the Housewives' Association, and to the Japanese league of Women Voters. Chifuren, the largest women's organization is an umbrella organization that channels volunteer work to women through the Fujinkai. Chifuren members frequently have addressed issues on the local level.[117] Today's women's groups focus less on pollution and more on dangerous and overpriced consumer goods. Chifuren successfully fought against dangerous food additives, for consumer labeling on fruit juices and for lower-priced televisions and cosmetics in the 1960s and 1970s.[118] Having met some of its initial goals, the organization has been noticeably less active in the 1980s.[119]

Today, because of advances in time-saving goods for the home, housewives have even more leisure time to devote to community service and political activity. Karl van Wolferen, author of The Enigma of Japanese Power calls Japanese housewives "a potentially important political presence."[120] Their recent political history indicates that if women's roles change in Japan, they are likely to have greater influence over environmental policy.

Today, however, Japanese women "have a lower political profile than in almost any other democratic country."[121] It is unclear whether women are gaining greater status and their environmental impact may be minimal for some time to come. In 1989, Prime Minister Kaifu appointed two women — an unprecedented number — to his Cabinet at a time when the LDP was least popular with female voters. One appointment was Mayumi Moriyama, a leading Diet member elevated to chief cabinet secretary. Following his successful reelection in 1990, however, Kaifu replaced both women with men. Mrs. Moriyama protested that she had been used to attract the female vote and dropped when the election was over.[122]

Surprisingly, there are fewer women in the Japanese Diet today than in the 1950s. (It should be noted, however, that there are far more women per capita in the Japanese Diet than in the U.S. Congress). Although more women than men vote in Japan, " . . . few women run for election and still fewer win, in spite of the fact that women constitute a majority of the electorate."[123] It may be that widespread opposition to the LDP in 1989 spurred the election of several women to the Diet. However, it is likely that Japanese women will continue to wield their greatest political influence in the voting booth. Exit polls from 1989 show that women overwhelmingly joined men to express their dissatisfaction with Prime Minister Uno, linked to a sex scandal, and the LDP's 1989 consumer tax.[124] Some observers conclude that the influence of Japanese women on politics is on the rise, but as one Japanologist said, "When you start low, there's only one way to go."[125]

## APPENDIX II
## THREE GROUPS PROTESTING THE NAGARA RIVER DAM

Indicative of the new trend in wildlife protection, three environmental groups have formed to protect the Nagara. One, the "Let Us Love the Nagara River Association", is a grassroots environmental organization. Another group's members use the river for recreation. The last is composed of professors from Kyoto University who seek to preserve the river's marine life.[126] These groups charge that the dam project is antiquated and unnecessary. The government devised the plan in the 1960s primarily to provide water to factories anticipated to be built in the cities of Nagoya and Yokkaichi.[127] But in the 1970s, water conservation efforts and the decision of many Japanese industries to locate abroad reduced the projected demand for water.[128] The government maintains that the dam still would be useful to avert inundation of valuable farmland, although other measures already have been taken to prevent flooding.[129]

## APPENDIX III
## JAPAN'S TIMBER CONSUMPTION

In 1986, Japan consumed 20% or 15.7 million cubic meters, of the world's total tropical woods, increasing its consumption in 1987 to 20.6 million cubic meters. Japan imports most of its tropical timber from southeast Asian countries still sensitive to Japan's wartime occupation and exploitation of resources. In 1987, Papua New Guinea and two Malaysian states, Saraw and Sabah, provided Japan with 96% of its tropical timber imports.[130] Because it is depleting its Asian resources, Japan is seeking timber from other countries, especially Brazil.[131]

## APPENDIX IV
## DOMESTIC OPPOSITION TO NUCLEAR POWER

Japan's nuclear program is now the world's fourth largest behind the U.S., France, and the U.S.S.R. Japan is one of the few countries with a continued commitment to nuclear power and some officials, including the former Environment Minister, have touted it as a way to reduce greenhouse gas emissions.[132] Although Japan's government is among the least open to popular influence of the world's representative governments, the level of opposition to nuclear power has caused considerable concern within the government and even some industry officials have begun to question the likelihood of new plant construction. Support for continued construction of nuclear power plants has fallen from 62% in 1979 to 29% in 1988. Sixty-two percent of respondents to a September, 1988 *Asahi Shimbun* poll expressed concern that a serious nuclear accident might occur in Japan.[133]

The Hokkaido assembly's vote against a government plan to build a nuclear waste disposal laboratory in northeast Hokkaido demonstrates the level of opposition. The laboratory would develop technology for treating nuclear waste, but residents fear that the government will choose the area as a disposal site.[134] In Fukushima, almost 2,400 residents attended town meetings to oppose the restart of a Tokyo Electric Power nuclear reactor that shut down in January 1989 when a recycling pump broke. Plant officials cannot determine if small metal fragments remain in the reactor.[135]

Chapter 29

# REGULATORY ASPECTS OF ACID RAIN

**Partha R. Dey, Eugene E. Berkau, and Karl B. Schnelle**

## TABLE OF CONTENTS



0-8493-4419-0/93/$0.00 + $.50
© 1993 by CRC Press, Inc.

Chapter 29

# REGULATORY ASPECTS OF ACID RAIN

**Partha R. Dey, Eugene E. Berkau, and Karl B. Schnelle**

## TABLE OF CONTENTS



0-8493-4419-0/93/$0.00+$.50
© 1993 by CRC Press, Inc.

# I. INTRODUCTION

On November 15, 1990, President Bush signed the 1990 Clean Air Act (CAA) amendments into law. This was a historical document which marked the beginning of a concerted effort to address a most pressing environmental problem of this century, namely acid rain.

Acid rain is the generic term used to describe the phenomenon by which sulfur dioxides ($SO_2$) and nitrogen oxides ($NO_x$) react in the atmosphere in the presence of sunlight to form acids which are scrubbed out of the atmosphere during a precipitation event. When this happens the pH of the precipitation falls considerably below 7.0. Years of research have shown that acid rain has a very detrimental effect on soils, vegetation, and marine life.[1-5] The large amounts of $SO_2$ and $NO_x$ being released by coal-fired utility boilers have largely incriminated utility companies as being the culprits.[6] Most of the research work has been in Canada because the direction of the jet stream across the U.S. is such that the emissions from the midwestern and northeastern U.S. are carried into southeastern Canada.[7,8] An interim report from the National Acid Precipitation Assessment Program (NAPAP) has assessed that power plants contribute up to 65% of the national annual emissions of $SO_2$, and up to 29% of the $NO_x$ emissions.[9] It is for these reasons that acid rain control has been given such a priority by legislators.

# II. ACID RAIN AND THE 1990 CLEAN AIR ACT AMENDMENTS

The new Clean Air Act (CAA) amendments have laid down a precise timetable according to which the $SO_2$ and $NO_x$ emissions from coal-fired utility boilers will have to be reduced. The 1990 CAA amendments require a 10 million ton reduction in $SO_2$ emissions, and a 2 million ton reduction in $NO_x$ emission, using 1985 through 1987 as the baseline years for comparison.

Phase I of the CAA amendments requires electric power generation boilers above 100 MW to restrict $SO_2$ emissions to 2.5 lb/mmBtu by 1995. Phase II of the CAA amendments takes effect from 2000, and requires all utility units >25 MW to restrict $SO_2$ emissions to 1.2 lb/mmBtu. Within 18 months of the enactment of the CAA amendments, $NO_x$ emissions from tangentially fired boilers will be restricted to 0.45 lb/mmBtu, and dry bottom wall-fired boilers to 0.50 lb/mmBtu. EPA will establish similar $NO_x$ emission limits for wet bottom wall-fired boilers, cyclones, and all other types of boilers, not later than January 1, 1997.

An important feature of the CAA amendments is that it allows emissions trading. It provides for "allowances", an allowance being worth 1 ton of $SO_2$, which are fully marketable. In other words, a utility which controls its $SO_2$ emissions beyond the mandatory limit may sell the additional tonnage to another utility, which will then be treated as having achieved the corresponding amount of $SO_2$ reduction.

The following section addresses the manner in which the practical aspects associated with designing and implementing an acid rain control strategy may be handled. It also provides a glimpse of the cost-benefits possible by using an innovative approach to acid rain control — adopting a systems approach making use of natural gas technologies. Natural gas is a clean burning fuel with negligible $SO_2$ emissions, comparatively lower $NO_x$ emissions, and lower $CO_2$ emissions than an equivalent amount of coal. Complete combustion of natural gas can be achieved easily and without the production of ash. The main drawback of natural gas is its higher price per unit heat content as compared to coal. This drawback can be circumvented by using Natural Gas Reburn (NGR), which is a technology that substitutes 15% of the coal in a coal-fired boiler with natural gas, resulting in 15% $SO_2$ and particulates emissions reduction, a 50% $NO_x$ emissions reduction, and a 43% $CO_2$ emissions reduction.

## III. ACID RAIN CONTROL STRATEGY EVALUATION PROCEDURE

The first step in the selection of an acid rain control strategy is to determine the degree of control required for the system or population of boilers under consideration. One can get an idea of the extent of the problem from the fact that the coal-fired utility boilers in Ohio presently use coal emitting 4.5 lb/mmBtu of $SO_2$ on average. The 1995 $SO_2$ emission limit mandated by the 1990 CAA amendments is only 2.5 lb/mmBtu of $SO_2$, which requires at least 44% $SO_2$ emissions reduction by that time. By the year 2000, the $SO_2$ emission limit will be only 1.2 lb/mmBtu, which requires 73% $SO_2$ emissions reduction by that time.

The degree of control is defined as the annual $SO_2$ and $NO_x$ emissions reduction required, and has units of ton/year. This quantity can be calculated on a boiler-by-boiler basis by applying $SO_2$ and $NO_x$ emission limits on the individual boilers making up a population of boilers.

### A. EMISSION LIMITS AND REQUIRED REDUCTIONS

We carried out a study of a number of $SO_2$ and $NO_x$ emissions reduction scenarios.[10] This study was completed before the enactment of the 1990 CAA amendments, and in the absence of any other guidelines, we chose the New Source Performance Standards (NSPS) as the emission limits in order to calculate the emission targets to be achieved by the control strategy.[11] The NSPS standards are 1.2 lb/mmBtu for $SO_2$ emissions, and 0.6 lb/mmBtu for $NO_x$ emissions (0.8 lb/mmBtu for $NO_x$ emissions from cyclones and wet-bottom boilers).

The $SO_2$ and $NO_x$ emissions from any given coal-fired boiler can be calculated by multiplying the total heat input to the boiler by the $SO_2$ emission factor (e.g., 5.9 lb/mmBtu for a boiler using 3.4% sulfur (S) coal with a heating value of 11,255 Btu/lb) for the boiler. The $SO_2$ emission factor is calculated from the coal S content and heating value, and the total heat input is calculated from the total coal use, knowing its heating value. The data required for this calculation are available from the Energy Information Administration (EIA) in Form-767, which every utility has to file with this agency.[12]

Next, by substituting the coal $SO_2$ and $NO_x$ emission factors with the emission limits under the NSPS regulations, the controlled $SO_2$ and $NO_x$ emissions in ton/year from the boiler can be determined.

By repeating the above procedure for every boiler in the utility boiler population (e.g., 101 boilers in Ohio), the statewide uncontrolled and controlled $SO_2$ and $NO_x$ emissions can be determined. Therefore, the required reductions in emissions can be determined by taking the difference between uncontrolled and controlled S emissions.

### B. STRATEGY COST ESTIMATION PROCEDURE

A control strategy is basically a mix of technologies applied in turn to each of the boilers in the population under consideration, until the emissions reduction target is met. Diverse technologies, such as wet flue gas desulfurization (FGD), dry FGD, limestone injection multistage burners (LIMB), selective catalytic reduction (SCR) for $NO_x$, low-$NO_x$ combustion (LNC — includes low-$NO_x$ burners [LNB] and overfire air [OFA]) NGR, spray dryer-absorber (SDA), coal switching, and physical coal cleaning (PCC) may be used to achieve the $SO_2$ and $NO_x$ emissions target.

The choice of a particular technology or mix of technologies is influenced by a number of parameters which include the total emissions reduction desired, process economics, physical constraints of space available at the plant site, and political constraints such as jobs affected by using a particular technology. Various technologies have different $SO_2$ and $NO_x$ reduction capabilities. The annual operating costs associated with any given technology can vary widely; certain technologies such as wet FGD require a lot of space to be retrofitted to a given boiler, and switching to a coal with a lower sulfur content will affect the jobs of coal-miners employed in mining a high-sulfur coal. These are among a few of the more important considerations in choosing a given technology.

The technology mix need not be the same for every boiler. However, the technology mix should be applied in decreasing order of $SO_2$ + $NO_x$ uncontrolled emissions because the cost effectiveness or $/ton of $SO_2$ + $NO_x$ removed generally increases with decreasing $SO_2$ + $NO_x$ removal, all other factors being constant (i.e., if the same technology mix is successively applied to boilers emitting smaller and smaller quantities of $SO_2$ + $NO_x$). Also, the technologies having the lowest costs associated with it would be applied first.

The study under consideration combined various available technologies such as coal switching, LIMB, LNC, NGR, and FGD in order to achieve the necessary $SO_2$ and $NO_x$ emissions reduction. The $SO_2$ and $NO_x$ emissions reductions, and the assumptions associated with each of the above-mentioned technologies are listed in Table 1. The following text illustrates how a given strategy is applied. All the necessary calculations were performed by using the Integrated Air Pollution Control Systems (IAPCS-3) computer model.[13] IAPCS-3 is a computer model developed for the EPA, which is designed to make cost estimations for implementing various pre-, *in situ*, and post-combustion technologies in coal-fired utility boilers. In the following section, general principles have been defined along with comments specific to the sample strategy. Let us assume that the sample strategy is coal switching (OH2.5) + NGR + FGD.

The procedure for applying a strategy to a mix of boilers is to continue applying various technologies to each of the boilers until the emissions reduction target is achieved. In this case, coal switching would be the first technology to be evaluated since it has the lowest capital and annual costs associated with it, compared to the other technologies in the strategy. Coal switching would be applied to every boiler where it resulted in $SO_2$ emissions reduction, and the $SO_2$ emissions reduction would be calculated. $NO_x$ emissions would not be affected because the total heat input to the boiler would remain constant.

Next, $NO_x$ control technologies would be applied to the highest $NO_x$ emitting boilers, until the $NO_x$ emissions target was achieved. This is done by recalculating the $NO_x$ emissions from the population of boilers after every $NO_x$ control technology application and stopping if the emissions target has been achieved. A general principle adopted by us was that $SO_2$ emissions would not be examined until after $NO_x$ emissions reduction targets had been achieved. This is because if NGR is used to achieve $NO_x$ emissions targets after $SO_2$ emissions targets have already been achieved, then the accompanying 15% $SO_2$ reduction resulting from 15% gas use would cause the $SO_2$ reductions target to be exceeded beyond that required. Next, various $SO_2$ control technologies would be applied to the highest $SO_2$ emitting boilers until the $SO_2$ emissions target is also achieved.

Finally, the IAPCS-3 model would be used to estimate costs for each boiler where at least one control technology, other than coal switching alone, was used. Coal switching costs would be determined from the Rubin coal washing model which is a collection of statistical multivariate regression models which estimate the properties and cost of washed coal in the U.S. for different levels of washing.[14] These models allowed washed coal characteristics to be predicted from data on raw coal origin, heating value, ash, and S content. The costs for the overall control strategy for the entire system of boilers would then be calculated by summing the costs for each boiler, using a spreadsheet designed for this purpose.

From the description of the application of the procedure, it should be apparent that each boiler controlled would not necessarily have the same control configuration. However, the technology applied to the population of boilers would always be selected from only those technologies contained in the strategy under consideration. This means that for the sample strategy, OH2.5, NGR, and FGD would not necessarily be applied to every boiler in the

**TABLE 1**
**Technologies Used To Achieve $SO_2$ and $NO_x$ Emissions Reduction**

| Technology | $SO_2$ | $NO_x$[a] | Assumptions |
|---|---|---|---|
| NGR | 15 | 50 | • Gas used is 15% of total heat input.<br>• Gas-to-coal fuel price differential is $1.00/10$^6$ Btu.<br>• NGR is applicable to all types of boilers. |
| LNC | — | 50 | • LNC is not applicable to cyclones and tangentially fired boilers. |
| LIMB | 50 | — | • LIMB is applicable to all types of boilers.<br>• ESP upgrade with flue gas conditioning and humidification is required.<br>• Ca/S ratio of 2.0 is used. |
| FGD | 90 | — | • Adipic acid-enhanced, LIMB, FGD technology is used.<br>• Ca/S stoichiometric ratio of 1.07 is used.<br>• FGD is applicable to all boilers.<br>• Retrofit difficulty factors based on size and age of boiler is calculated by the IAPCS-3 model. |
| OH2.5 | variable | — | • An adequate supply of Ohio-2.5 coal can be obtained by washing raw Ohio coal using state-of-the-art PCC technology.<br>• A premium of $5.37/ton for the washed Ohio coal (Rubin model calculation).<br>• Washed Ohio coal is not used in wet bottom boilers due to ash slagging problems. |

[a] % reduction.

population under consideration. These technologies would not even necessarily be applied concurrently to any single boiler. However, the technology mix for any given boiler would never use anything other than OH2.5, NGR, and FGD.

## IV. ACID RAIN AND THE STATE OF OHIO

The State of Ohio was a logical choice for a study of an acid rain precursor ($SO_2$ and $NO_x$) emissions reduction scenario. Ohio is the highest $SO_2$ emitting region in the U.S., and it has a large utility boiler population.[15] Tables 2 and 3 list the total and power plant $SO_2$ and $NO_x$ emissions from the 12 highest polluting states in the U.S. The data show that Ohio was the highest $SO_2$ emitter in 1980 and 1985. Ohio was also the second highest $NO_x$ emitter in 1980 and 1985. Of the total $SO_2$ emissions in Ohio, 83% originates from power plants. Similarly, of the total $NO_x$ emissions in Ohio, 44% originates from power plants. Of the total $SO_2$ emissions originating from power plants in the U.S., 92% can be attributed to coal-fired power plants. This indicates the magnitude of the problem in Ohio.

Ohio is currently the focus of a number of studies by the government and private sector, and has recently completed an agreement with New York on a national acid rain program.[16]

Ohio has also enacted clean coal technology (CCT) programs similar to the Department of Energy's CCT program, and is also providing $100 million in matching funds to support CCTs that promote the use of indigenous Ohio high-sulfur coal.

The acid rain control strategy evaluation procedure explained in the earlier section was implemented in Ohio. The statewide emissions limit was calculated on the basis of the current NSPS. These standards require an $SO_2$ emissions limit of 1.2 lb/mmBtu, and an $NO_x$ emissions limit of 0.6 lb/mmBtu on all types of coal-fired utility boilers except cyclones which have a limit of 0.8 lb/mmBtu. The NSPS emissions limits were applied to all the 101 Ohio utility boilers, and the resulting statewide $SO_2$ and $NO_x$ emissions reductions were calculated. Using these emissions limits led to 70% reduction in $SO_2$ emissions, and 32%

**TABLE 2**
**Power Plant and Total $SO_2$ Emissions by States**

| State | Million ton/yr | | | |
|---|---|---|---|---|
| | 1980 power plants | Total | 1985 power plants | Total |
| Ohio | 2.02 | 2.43 | 2.01 | 2.43 |
| Indiana | 1.43 | 1.75 | 1.35 | 1.67 |
| Pennsylvania | 1.36 | 1.74 | 1.14 | 1.50 |
| Missouri | 1.07 | 1.21 | 0.85 | 1.00 |
| Illinois | 1.04 | 1.35 | 0.98 | 1.27 |
| Kentucky | 0.95 | 1.08 | 0.78 | 0.90 |
| West Virginia | 0.88 | 1.02 | 0.86 | 0.99 |
| Tennessee | 0.82 | 0.99 | 0.73 | 0.89 |
| Georgia | 0.67 | 0.77 | 0.89 | 0.96 |
| Florida | 0.65 | 0.89 | 0.41 | 0.61 |
| Michigan | 0.52 | 0.75 | 0.37 | 0.60 |
| Alabama | 0.50 | 0.73 | 0.52 | 0.74 |
| Total | 11.9 | 14.7 | 10.9 | 13.6 |
| Total (all states) | 15.9 | 23.6 | 14.7 | 21.5 |

*Note:* Coal-fired power plants accounted for 14.6 of the 15.9 million tons in 1980.

Modified from NAPAP Report, Emissions and Controls, Vol. 2, September 17, (9), 1987. With permission.

**TABLE 3**
**Power Plant and Total $NO_x$ Emissions by States**

| State | Million ton/yr | | | |
|---|---|---|---|---|
| | 1980 power plants | Total | 1985 power plants | Total |
| Texas | 0.49 | 2.51 | 0.61 | 2.47 |
| Ohio | 0.47 | 1.06 | 0.46 | 1.02 |
| Pennsylvania | 0.36 | 0.90 | 0.34 | 0.86 |
| Illinois | 0.33 | 0.88 | 0.29 | 0.81 |
| Indiana | 0.32 | 0.68 | 0.36 | 0.71 |
| West Virginia | 0.27 | 0.40 | 0.29 | 0.42 |
| Kentucky | 0.25 | 0.56 | 0.27 | 0.58 |
| Michigan | 0.23 | 0.63 | 0.24 | 0.62 |
| Missouri | 0.22 | 0.49 | 0.20 | 0.45 |
| Florida | 0.22 | 0.65 | 0.25 | 0.66 |
| Georgia | 0.21 | 0.53 | 0.28 | 0.58 |
| North Carolina | 0.20 | 0.47 | 0.17 | 0.42 |
| Total | 3.6 | 9.8 | 3.8 | 9.6 |
| Total (all states) | 5.8 | 19.4 | 6.3 | 19.1 |

*Note:* Highway vehicles accounted for 7.2 and 6.8 million tons in 1980 and 1985, respectively.

Modified from NAPAP Report, Emissions and Controls, Vol. 2, September 17, (9), 1987.

**TABLE 4**
**$SO_2$ and $NO_x$ Control Costs for 70% $SO_2$ Reduction and 32% $NO_x$ Reduction**

| | | $SO_2$ control costs | | | $NO_x$ control costs | | |
|---|---|---|---|---|---|---|---|
| No. | Strategy | Capital M$ | Annual M$ | C.E. $/TON | Capital M$ | Annual M$ | C. E. $/TON |
| 7 | NGR + FGD | 3024 | 832 | 539 | 107 | 88 | 567 |
| 8 | NGR + LIMB | 2451 | 819 | 532 | 107 | 88 | 567 |
| 6 | LNC + LIMB | 2667 | 862 | 562 | 191 | 28 | 181 |
| 5 | LNC + FGD | 3106 | 845 | 552 | 191 | 28 | 181 |
| 3 | LNC + OH2.5 + FGD | 2103 | 753 | 491 | 191 | 28 | 181 |
| 1 | NGR + OH2.5 + FGD | 1864 | 691 | 451 | 107 | 88 | 567 |
| 4 | LNC + OH2.5 + LIMB | 1061 | 706 | 460 | 191 | 28 | 181 |
| 2 | NGR + OH2.5 + LIMB | 904 | 645 | 419 | 107 | 88 | 567 |

*Note:* C.E. = Cost effectiveness in dollar/ton.

**TABLE 5**
**Total Control Costs for 70% $SO_2$ and 32% $NO_x$ Reduction**

| No. | Strategy | Capital M$ | Annual M$ | C.E. $/ton $SO_2$ + $NO_x$ |
|---|---|---|---|---|
| 7 | NGR + FGD | 3131 | 920 | 542 |
| 8 | NGR + LMB | 2558 | 907 | 534 |
| 6 | LNC + LMB | 2858 | 890 | 527 |
| 5 | LNC + FGD | 3297 | 873 | 518 |
| 3 | LNC + OH2.5 + FGD | 2294 | 781 | 462 |
| 1 | NGR + OH2.5 + FGD | 1971 | 779 | 461 |
| 4 | LNC + OH2.5 + LMB | 1252 | 734 | 434 |
| 2 | NGR + OH2.5 + LMB | 1011 | 733 | 433 |

*Note:* C.E. = Cost effectiveness in dollars/ton.

reduction in $NO_x$ emissions for the entire state. Therefore, these were chosen as the emissions reduction targets for each control strategy that was studied.

The eight strategies used to achieve 70% $SO_2$ emissions reduction and 32% $NO_x$ emissions reduction, along with the corresponding $SO_2$ and $NO_x$ control costs, are shown in Tables 4 and 5. From these tables it could be concluded that using natural gas technologies in combination with other low-cost technologies (such as coal-switching and LIMB) were a lot more cost-effective than using big budget control options like wet FGD. All costs in Tables 4 and 5 are in constant 1988 dollars, assuming a 30-year plant life for amortization purposes, an annual inflation rate of 6%, and a discount rate of 11%.

## V. CONCLUDING REMARKS

The preceding sections have introduced the reader to the phenomenon of acid rain, and its ramifications on the utility industry. It has been shown how the utility companies would have to reduce $SO_2$ and $NO_x$ emissions in a phased manner in order to comply with the 1990 CAA amendments. A procedure to form and evaluate an acid rain control strategy has been demonstrated. The various considerations in implementing a cost-effective emissions reduction scenario have been reviewed. Preliminary cost estimates have shown that natural gas can form an integral part of a cost-effective $SO_2$ and $NO_x$ emissions reduction strategy.

As newer technologies evolve, the actual control costs may be significantly affected but the method of evaluation remains the same.

## REFERENCES

1. Franklin, C. A., Burnett, R. T., Paolini, R. J., and Raizenne, M. E., Health risks from acid rain, *Environ. Health Perspect.*, 63, 155–68, 1985.
2. Rennie, P. J., Evidence for Effects on Canadian Forests, Symp. Air Pollut. Effects for Ecosystems, St. Paul, MN, 1985, pp. 111–122.
3. Foster, N. W., Acid precipitation and soil solution chemistry within a maple-birch forest in Canada, *For. Ecol. Manage.*, 12, 215–31, 1985.
4. Hern, J. A., Chemical effects of simulated acid precipitation on two Canadian shield forest soils, *Diss. Abstr. Int. B*, 47, 5, 1985.
5. Bertram, H. L., Das, N. C., and Lau, Y. K., Precipitation chemistry measurement in Alberta, *Water, Air, Soil Pollut.*, 30, 1–2, 1986.
6. Hidy, G. M., Hansen, D. A., Henry, R. C., Ganesan, K., and Collins, J., Trends in historical acid precursor emissions and their airborne and precipitation products, *JAPCA*, 34(4), 333–354, 1984.
7. Whelpdale, D. M. and Bottenheim, J. W., Recent Canada-USA transboundary air pollution studies, *Stud. Environ. Sci.*, 21, 357–363, 372–373, 1982.
8. Whelpdale, D. M. and Barrie, L. A., Atmospheric monitoring network operations and results in Canada, *Water, Air, Soil Pollut.*, 18, 1–3, 1982.
9. NAPAP Interim Assessment: The Causes and Effects of Acidic Deposition, Vol. 2: Emissions and Control, National Acid Precipitation Assessment Program, 1986.
10. Dey, P. R., ''Environmental and Economic Benefits of Natural Gas Use for Pollution Control'', Ph.D. dissertation, Vanderbilt University, 1990.
11. 40 CFR 60. Code of Federal Regulations, Title 40, Part 60, Standards of Performance for New Stationary Sources, Office of Federal Register, Washington, D.C., July 1, 1988.
12. Steam-Electric Plant Operation and Design Report, Form EIA-767, Energy Information Administration, 1986.
13. PEI Associates, Inc., User's Manual for the Integrated Air Pollution Control System Performance and Costing Model: Version III, EPA Contract No. 68-02-4284, September, 1988.
14. Rubin, E. S. and Skea, J. F., A model of coal cleaning for sulfur emissions reduction, *JAPCA*, 37(2), 149–157, 1987.
15. Acid Rain and Transported Air Pollutants: Implications for Public Policy. Office of Technology Assessment Report, June 1984.
16. PEI Associates, Inc., Evaluation of Acid Rain Controls for Ohio Electric Utility Coal-Fired Power Plants, prepared for Ohio Department of Development, Coal Development Office, OCDO Grant #10-86-038, PH 3708, September 1988.

Chapter 30

# A STRATEGY FOR GLOBAL ENVIRONMENTAL EDUCATION AT THE UNIVERSITY

**S. Taseer Hussain and Raymond L. Hayes**

## TABLE OF CONTENTS



-8493-4419-0/93/$0.00 + $.50
© 1993 by CRC Press, Inc.

Chapter 30

# A STRATEGY FOR GLOBAL ENVIRONMENTAL EDUCATION AT THE UNIVERSITY

**S. Taseer Hussain and Raymond L. Hayes**

## TABLE OF CONTENTS



0-8493-4419-0/93/$0.00 + $.50
© 1993 by CRC Press, Inc.

## SYNOPSIS

The Earth's environment is a dynamic system that is affected both by natural phenomena and by human activity. The changes occurring in the global environment are bound to have serious consequences for all its inhabitants. Therefore, the world is rapidly becoming interdependent. Multidisciplinary scientific efforts must be directed toward understanding these global environmental changes. These efforts will require sufficient funds to attract scientists into global environmental research and to disseminate new knowledge to future scholars and to the general public alike. The federal government has a definite role to play in this effort and should allocate sufficient funds to initiate and sustain these programs. Unfortunately, such funds are not currently budgeted.

The academic department, as the basic structural and functional unit of the American university system, is most appropriate to ensure environmental educational goals. We propose the establishment of a novel Department of Global Environment at every university. That department must be multidisciplinary in nature and must accumulate a critical mass of scholars from all relevant traditional disciplines in the arts and sciences to generate knowledge, to educate students, and to provide advisory services to policy makers.

The student product of this department should receive a broad-based education and should emerge as an informed individual who possesses sufficient skills to achieve sustainable communities. That student should also be equipped to assume leadership and to formulate policy about global environmental issues. Our investment in education may well be the only way to secure a future for humanity and for the natural world as we now know it.

## I. INTRODUCTION

The time scale of global climate changes is too protracted for us to perceive a direct and personal threat. These changes are expected to operate over a time scale exceeding the balance of the lifetime of most living humans. Although we are able to express concern for our living relatives, it is difficult for the human mind to project beyond one or two generations into the future. The biological time scale which we are accustomed to is very short relative to the rate of change in the global climate. The adult lifetime of an American may be roughly 50 to 60 years, but the changes of climate sufficient to be problematic will probably occur over intervals that might exceed double that period. Therefore, we are not mentally prepared for such extended planning. It is difficult to appeal to the individual, even on a reasonable level, concerning effects that will materialize so slowly.

> Our civilization is threatened today by changes taking place over periods of years and decades, but exchanges over a few years or decades are too slow for us to perceive readily. That is a time scale too leisurely for a nervous system attuned to bears, branches, burglars, and downpours. At the same time, these changes are much too rapid to alow processes of biological or cultural evolution to adapt people to them. We are out of joint with the times. (Ornstein, 1990.)

There may be a psychological basis for our difficulty as humans to respond promptly and responsibly to global climate change data. It is proposed that a time-dependent mismatch might prevent our development of an appropriate perspective.

The one advantage that our species does have, however, is a capacity for long-term memory. Through the written record and through worldwide communication, we are able to make comparisons over time and distance. We are able to generate a sense of threat from events far removed from us and from a knowledge of conditions which may have been described several years previous to our lifetimes. It is on that level that we must be content to begin the process of sensitization and instruction about global climate change.

The primary objective of any educational program about global environmental change should be a heightened appreciation for the fragile nature of our surroundings. The human species must accept responsibility for the fate of the Earth. We must hold ourselves accountable as the only species having a long-term memory about the status of the planet. No other living being is endowed with the capacity to generate environmental wisdom, recall, or consciousness. In fact, few other species have a life span of sufficient length to permit the accumulation of experience necessary to provide such awareness.

## II. THE REALITY OF GLOBAL CLIMATE CHANGE

Throughout geologic time the Earth's climate has fluctuated with irregularity. By examining the geologic record, we can begin to see the world's climate from a longer time perspective. Evidence of past climatic conditions is preserved in the geologic record in marine and continental sediments, glacier ice, peat and coal deposits, cave deposits, coral reefs, and other stratified deposits (Ager, 1988). The two most common approaches to obtaining paleoclimatic information from sediments are those based on paleontology and oxygen isotopic analyses. Large scale migrations of fauna and flora (both marine and terrestrial) covering a broad geographic area can be correlated with variations in environmental conditions. Oxygen isotopes have permitted detailed interpretation of surface and bottom water temperature history throughout the Cenozoic.

The geologic record indicates that there is no single cause of climate change. The climate varies on all time scales due to natural causes, and all of them are not fully understood. These climatic changes and cycles are driven by a variety of phenomena, both extrinsic and intrinsic to the Earth. The major shifts during the Cenozoic era, as far as we can tell, are driven more by intrinsic factors such as the opening or closing of oceans and sea ways, continental positions, and mountain systems. For instance, the collision of Afro-Arabia and India with Europe and Asia, beginning in the Eocene and almost completed by the end of the Pliocene (60 to 2 million years ago), radically changed the shape of the Earth, built major new mountain systems from the Himalayas to the Alps, and altered atmospheric and oceanic circulation patterns. Ultimately, a cooling trend was initiated which resulted in an Earth with significantly more ice and a new climatic pattern driven by fluctuations in ice volume.

About 2.5 million years ago, the climate system of the Earth cycled between long intervals of coolness and shorter segments of warmer climate. The reasons for this shift are not completely known. The closing of the Isthmus of Panama and the opening of the Bering Strait about 3 million years ago altered oceanic circulation patterns. These events probably contributed to the change in global climate patterns (Ager, 1988). These changes may have triggered the onset of a major ice age as a combined effect of marine and continental ice generation systems.

During the past 1 million years, conditions were often 4 to 5°C cooler than now. There were brief warm intervals around 325,000 and 110,000 years ago. Then between 4000 to 8000 years ago, a global warm-up occurred that was above the present mean temperature by as much as 1.5°C. This was followed by a cold period until about 2500 years ago when another warm interval preceded the Little Ice Age, beginning about 650 years ago (Gates, 1989).

## III. HUMAN IMPACT ON GLOBAL CLIMATE CHANGE

If global climate changes are natural phenomena, then why should we worry about such fluctuations? The answer to this question requires an understanding of human impact upon the climate systems of the world. Although humans are considered part of the total planetary ecosystem, we differ from other living organisms in our ability to manipulate and even change the global environment. For instance, human activity has already created new genotypes of both plants and animals by the process of human selection, as distinguished from natural selection. We have created our own evolutionary driving forces and in doing so have created new ecosystems. At the present time, our species *Homo sapiens,* has the highest biomass of any animal, and our population is rapidly expanding (Simmons, 1989). Several of the Earth's ecosystems now are dependent upon human society for their continued functioning and integrity. Human-derived impacts upon nature have been of two types: intentional and accidental. We sometimes refer to the intentional category as our ''environmental management''.

The history of human technology seems to start around 2 million years ago at Olduvai in East Africa. Human civilization at that time utilized stone chopping tools. This primitive technology was primarily used for hunting wild animals and for butchering these prey. Exactly how much early hominids affected their environment is not fully known, however, we do know that their populations were small. It follows that the environmental influence of such limited populations could not have been very great. There is no strong evidence of open fires in human history at that time. The fire is well documented from the Upper Paleolithic period, and even then much of the evidence comes from caves, not open fires (Perles, 1975). The concept of the preservation of natural surroundings existed in the human mind at that time. The primary source of energy required by human populations of this period was from the sun. Agricultural activity suddenly appears in the prehistoric record, perhaps as a result of the domestication of plants and animals.

Up to the 19th century, most industrial activity of humans was confined to small processing plants around which small towns grew. In the early 1800s, city dwellers constituted 2 to 3% of the total human population; now this figure has risen to 42% (Simmons, 1989). Like the control of fire and the invention of agriculture, industrialization represents a turning point in the history of human relationships with nature. An essential factor from this industrial period has been the manufacturing of goods by machines for sale to other communities. Consequently, the structure of environmental relationships over most of the world has been changed drastically by the availability of coal, oil, and natural gas.

More recently, the atomic age emerged when the first controlled fission of atomic nuclei in a chain reaction was achieved by Enrico Fermi (Chicago, December 2, 1942). This milestone event led to both the use and spread of atomic fission as a source of electrical energy, as well as to the propagation of atomic fusion through nuclear weapons (Simmons, 1989).

Industrial development and utilization of fossil fuels have led to the deterioration of our global environment. According to *Time* magazine (January 2, 1989), the U.S. consumes one fourth of the world's energy per year. Yet for a given amount of energy, the U.S. produces less than half as much economic output as Japan and West Germany. Although the U.S. accounts for less than 5% of the global population, it generates 15% of the world's sulfur dioxide ($SO_2$) emissions and 25% of the nitrogen oxide ($N_2O$) and carbon dioxide ($CO_2$) releases. Each American produces an average of 3.5 lb of garbage/day.

Global circulation models predict an average increase in the Earth's temperature between 1.5 and 4.5°C if effective values for $CO_2$ in the atmosphere double the preindustrial level of 280 ppm. The current content of $CO_2$ in the atmosphere is 353 ppm. At the present emission rate for greenhouse gases, we will reach that doubling level by the year 2030. The current average global temperature is 0.3 to 0.6°C higher than the temperature in preindustrial times (IPCC, 1990). The geological record also informs us that the average atmospheric temperature has varied primarily due to changes in the $CO_2$ level (Budyko et al., 1988).

Therefore, because of the emission of harmful gases, deforestation, water shortage, and many other influences, the temperature of the Earth is gradually and progressively rising. This feature alone is expected to accelerate melting of the polar ice caps and ultimately lead to sea level rises. Excessive heat and humidity in many parts of the world will disrupt ecosystems and contribute to the evolution and proliferation of microorganisms with the potential to generate human diseases.

It is clear that the global climate change we are experiencing is real. We need to understand this fact. Furthermore, we need to mount a reasoned response to this fact without delay. The security of our planet is at stake if the global temperature rises by several degrees during the next century. If we are able to improve energy efficiency by a mere rate of 2%/ year, the global average temperature would be maintained within 1 degree of present levels (Ruckelshaus, 1989). The U.S. and the rest of the developed world should invest funds toward educating people and solving problems related to the control of global climate change, and should offer leadership to developing nations with the goal of realizing a stabilized and sustainable world for all nations and peoples.

## IV. THE NEED FOR EDUCATION ABOUT GLOBAL CLIMATE CHANGE

''Efforts to prepare for climate change can be only as enlightened as the people who must carry them out . . . '' (Titus, 1989). The facts concerning issues of global climate change must be made common knowledge to the general public. The science of global climate change and its control is quite complex and definitely multidisciplinary. This knowledge must be disseminated effectively to all segments of the population, beginning at the earliest levels of education and continuing through the full range of university levels. Informal education to heighten environmental awareness at the grass roots level will require the concerted activity of an enlightened local leadership. At the present time, isolated educational programs exist, but they are limited in substance, quality, and approach. Our ultimate goal in education should be to develop a public which understands the implications of global climate change and has modified its collective behavior so as to minimize or obviate contributions to all processes that accelerate global climate change and its associated problems. Through a comprehensive program of education and research, we as humans should be able to develop public administrators, teachers, scholars, and scientists who will provide responsible leadership for the public. Our scientists, for example, will have to be capable of participating in multidisciplinary research in this area. From the advancement of our knowledge of global climate change, we will be better able to understand the workings of nature and to assess the impacts of human habitation upon natural processes.

Global climate change will require major public policy initiatives in both the U.S. and other countries. Governmental as well as nongovernmental organizations from the local through international level must explain issues to the public in order that their policies might be accepted and followed. The vehicles for transmission of this set of explanations should be newspapers and magazines, radio and television, motion pictures, small group public forums, and community group functions. In a more organized manner, these issues can be explained and discussed at the institutional level. State and municipal governments can sponsor meetings and issue press releases concerning global climate change.

School systems can disseminate knowledge in lectures and course formats through community colleges and adult education programs. Museums and science centers can play a significant role in this campaign of public education by translating the results of technical studies into dynamic exhibits and brochures. Natural history museums around the world have already begun to prepare extensive exhibits in the areas of global climate change and its awareness. Churches, mosques, synagogues, and various other religious centers can discuss these issues through sermons or group discussions.

Many broad-based strategies for responding to the urgencies of global climate change and its effects have been proposed. All of these emphasize the need to stabilize or reduce greenhouse gas emissions. Also, they recognize the need to increase efficiency of energy utilization. The allocation of additional funding to alleviate problems created by global climate changes appears inevitable. All of these measures should certainly be taken to engage the necessary response to a critical situation. However, demands for a major series of changes in societal behavior through education and the deliberate production of an enlightened leadership does not appear as a prominent component of short- or long-range planning. Several federal agencies will develop environmentally oriented scientists to assist in graduate student training. Training for elementary and high school science teachers and undergraduate teaching faculty will also be provided. However, federal funds budgeted for these activities in fiscal year 1992 are insufficient to assure success in such a monumental undertaking. For example, the U.S. Department of Energy budgeted only 2.5 million dollars in fiscal year 1991 and only 4.0 million dollars in fiscal year 1992. In our opinion, sufficiency of any effort to intervene in the elimination of unnecessary anthropogenic effects on climate changes demands both firm financial and educational commitments.

## V. THE ROLE OF THE AMERICAN UNIVERSITY SYSTEM

In order to be assured of an environmentally literate population, we will have to initiate extensive educational and alternative behavioral programs. Public and private institutions of higher learning, namely our colleges and universities, must dedicate their resources to this need as well. The university, in our judgment, represents the best institutional site to generate manpower needs of an environmentally conscious public. This pool of leaders collected within and emanating from the university would include persons from the fields of education, business, science, law, sociology, the health professions, public administration, and the entertainment arts. These professionals would function as role models to implement new tactics for disseminating information and to assure that we all act in accord with a desire to minimize all negative human influence upon the world's environment.

The cultivation of the human mind is a difficult, and often risky, endeavor. This awesome responsibility is, at least in part, entrusted to the modern university. As an institution with the potential to encompass the collective wisdom of humankind, the university represents the most appropriate location in today's society for the formulation and promulgation of environmental literacy. To be literate with reference to the global environment begins with the possession of essential information. Equipped with a basic understanding, the environmentally literate individual is better able to reason through scientific complexities to reach intelligent judgements or decisions. Proper choices of lifestyle and natural resource utilization then become decisions accessible to the individual. These decisions, however, depend upon sensitivities derived from firm knowledge and a clear understanding of the consequences of our actions.

The process of becoming educated in human society should be continuous throughout the human's extended lifetime. The formal educative period for the socialization of most Americans represents a gradual experience which spans the interval between late in the first decade through the end of the second decade of life. Those of us who continue into the university system usually enter at the age of 17 to 20 and spend approximately 4 years in pursuit of the baccalaureate degree. It is during the university years that we have matured sufficiently to be receptive to intellectual stimulation and concerns beyond our own egocentric interests. The university phase of our formal educative exposure, then, offers the most opportune time for us to develop the global perspective of ecological awareness. It is during that period that global climate change and its broader implications for life on Earth may best be communicated as a component of the academic experience.

The methods to be used for the education of an environmentally conscious person begin with didactic and literary presentations of information, coupled with practical experiences. The scope of information necessary to appreciate the workings of the planet and the influence of human activities upon the global ecosystem is diverse and broad. The integration of that information should not be expected before the full picture is provided. However, beyond the factual aspects of this training, direct engagement in a series of experiences is necessary to illustrate the reality of the association between human dependencies and their environmental impacts. The laboratory experience should be continual throughout the educational process, since it emphasizes the active interaction which we conduct with our surroundings.

There is an urgent need to instill in the younger generation of American citizens an appreciation for man's influence upon the world's environment. No other institution, in our opinion, is as capable of transmitting this information responsibly as is the American university. Furthermore, we believe that this subject should be required for all students who enter an undergraduate degree-granting program of study. Global ecology is as basic a skill as any other element of the academic program which might be considered essential for the acquisition of a university degree. It is generally accepted that all recipients of higher educational credentials should master certain fundamental skills such as language, communication, and mathematics. Furthermore, subjects like history, philosophy, and science are considered hallmarks of an educated person. While each essential subject in its own right should be considered basic knowledge, responsible citizens of today's world require, in addition, more specific and practical knowledge, especially in the scientific domain. To become environmentally literate should be one of the goals of a proper university education.

The American university system recognizes the baccalaureate degree as the first and most basic certification in higher education. The student traditionally spends a minimum of 4 years in that program. The first year may be highly structured as a set of courses intended to bridge the high school experience and to begin exposure into collegiate level work. Subsequent courses are selected in progressive order to extend and expand the student's knowledge base. The student usually identifies and declares a major field of concentration which imposes additional minimum expectations. A specified amount of credit hours or individual courses in one's major field will be required for graduation. Minor fields of concentration may also be chosen, with lesser commitment of time and course load. For the baccalaureate degree with honors in the progressive university, a student may be required to do a project involving independent laboratory or library research. This experience would follow successful completion of formal courses and would require the approval of the faculty of the student's chosen field or at least the mentor with whom the student will work. The award of the degree represents the fulfillment of all prerequisites necessary for graduation from the institution, supported by a departmental faculty.

The traditional configuration of the university is organized around the artificial subdivision of the faculty into academic units called departments. Each department is entrusted with the administration of a set portion of the university curriculum. The faculty of the department represent the individuals who determine the content, instructional mode, and evaluative standard for courses offered through the department.

In all American colleges and universities, common departments are recognized. Liberal arts departments include such traditional titles as the Departments of Mathematics, Physics, Biology, Chemistry, Philosophy, and History. Some novel departments have amalgamated titles, such as Departments of Life Sciences (Zoology and Botany), Human Ecology (Sociology, Home Economics), or International Studies (Asian Studies, African Studies,

American Studies, etc.), probably as cost-efficient entities. Although the administrative organization of academic disciplines has seen some revision in different colleges through time, the academic program content has remained quite constant throughout the American university system as a whole.

Dating from the 19th century in America, the purpose of a university education was to produce a ''man of learning'' who would have an ''uplifting and unifying influence on society'' (Haskell, 1977). In the 20th century, this learned man has become the scientist who has exchanged his general citizenship for membership in ''the community of the competent'' (Haskell, 1977). This concept entails the evaluation of the scientist through peer review. The scientist has not been evaluated through an openly competitive forum by all who would wish to challenge him. Such a cloistered existence has had both fortunate and unfortunate consequences. In one respect, the progression of scientific accomplishment has been rapid and significant. On the other hand, there has developed a closed society of scientists who speak a different language and operate independently from the general public. It is no wonder, therefore, that at this time when the scientist is attempting to communicate a concern about global crises, the public is not responsive or reactive. There is a built-in inertia resulting from years of neglect and noncommunication. This chasm must be crossed before any meaningful dialogue is to occur naturally between scientist and policy-maker.

## VI. THE DEPARTMENT OF GLOBAL ENVIRONMENT

In recognition of a need to enable students to achieve environmental literacy, a new order of departmental organization is needed which would represent an effort to advance our collective wisdom about the global environment as a fundamental academic discipline. We envision a novel Department of Global Environment that would incorporate scholars of diverse orientations. Those faculty would share a common focus and contribution, namely to convey an appreciation for and to extend research in global climate change to all degree candidates.

The unique Department of Global Environment, as the functional unit for planning and implementation of a new university curriculum, would be staffed by scholars whose interactive assemblage would assure more than the mere summation of contributions from the component personalities. Environmental literacy is dependent upon a wide range of fundamental knowledge. The scientific basis of literacy in this area is complicated because it includes a firm grasp of geography and meteorology, along with an ability to understand natural science, chemical, and physical processes. However, the full appreciation of this area also requires the development of an ability to apply reason and logical analysis that emerges from a comprehensive and fundamental liberal arts education. Natural and physical scientists, philosophers, historians, lawyers, public health managers, economists, public policy planners, and administrators, among others, would be expected to combine their talents as faculty for this departmental entity (Table 1). Cross-fertilization of ideas within that group should be enlightening and progressive, since many of these scholars have been denied intimate sharing of knowledge through the old university organization.

As a new component of the university system, the Department of Global Environment would be integrated into the full range of university functions. Undergraduate students, graduate students, and postdoctoral fellows would be processed through experiences in this composite research, teaching, and service unit. The academic program of this department would provide visible and active engagement in studies relating to the world's environment. Concepts taught through the curriculum would range from the Gaia hypothesis through the precautionary principle. The collection of faculty of that unique department would assume all of the rights and privileges enjoyed by any other traditional department. These faculty

**TABLE 1**
**The University Department of Global Environment**

| Natural sciences | Health sciences | Physical sciences | Social sciences |
|---|---|---|---|
| Life sciences | Environmental diseases | Chemistry | Economics |
| Earth sciences | Public health | Computer sciences | Law |
| Atmospheric sciences | Medical sciences | Physics | Administration |
| Oceanography | | Mathematics | Public policy |
| | | Engineering | Psychology |
| | | | Sociology |
| | | | Philosophy |
| | | | History |
| | | | Politics |

should become multidisciplinary in order to meet their commitment to the departmental mission and identity, rather than to persist as disciplinarians in accord with their primary educational experiences.

In the traditional academic department, independent faculty focus on small aspects of knowledge within a very narrow field of research. This tradition is established during graduate education and continues throughout one's academic career. As a matter of fact, some faculty members never grow out of their dissertation subject. In their ivory towers they remain convinced that theirs is the most significant research effort. With the passage of time, however, they miss the reality of new questions to answer and modern directions to follow for the generation of significant innovations in knowledge. Those outdated faculty with their narrow and over-specialized foci soon find themselves standing apart from the mainstream of their own discipline. Academic specialization provides the impetus for establishing an early scientific reputation, but also limits intellectual growth. It must be remembered that knowledge is dynamic, and an individual academician's inquires must fit into the broader status of the discipline at all times. The faculty must be able to expand their horizons in response to new knowledge by adjusting behavior and skills. We propose that the faculty of this new Department of Global Environment would constantly interact with each other and with faculty from other departments. The new knowledge generated by each faculty member would be most current, practical, and novel. These contributions would be expected to integrate into the holistic perspective which is demanded for resolving global environmental concerns.

Justification for the establishment of a new departmental unit in the area of environmental change and its concerns lies in the recognition that this is a vital component of basic education for all university students. Not only is this knowledge of significance to any citizen of the world, but it will not subside or diminish in significance as time passes. If anything, the need to know this information will surely mount for the next century or more.

The discharge of educational responsibilities for global climate change cannot be done through a center or institute structure. First, most of these are research entities without sufficient academic function to be of value. For instance, the Center of Global Change located at the University of Maryland states its mission to be that it ''disseminates its research to the policy, science, and business communities through publications, conferences, and other educational activities'' (Mission Statement). In the absence of student contact, information transfer would not occur to the next generation. The typical center, even though university-based, is a soft-money unit. Usually, that facility is supported by federal or private contract or grant funds. Centers are not permanent entities and are subject to extrinsic fluctuations. Preferably, the responsibility to educate students should emanate from a hard-budgeted component of the university, assuring continuity of funding and administrative support from the institution, such as a department.

The changes which we recommend in this proposal to develop a new departmental organization in accord with practical and pragmatic objectives will not be easy to accomplish. Faculty will probably experience discomfort with these reorganizations. However, the faculty are likely to recognize the merits of this approach with consideration of the concept. They will, however, expect assurance that other aspects of their contracts with the university are not to be disturbed by any major restructuring. Teaching load, research time, and other details of professional commitment should not be seriously altered in any new allocations.

## VII. THE STUDENT PRODUCT

The student enrolled in this multidisciplinary academic department should be capable of informed decision-making with respect to issues about global warming and other consequences of global climate change. That individual should also be able to appreciate human impacts upon the global climate system. Above all, that student should be able to justify and explain human behavior required to best sustain global environmental conditions. In order to cultivate a student who is literate about the environment and man's influences upon it in an interdependent planet, our educational system needs radical revision. We propose that undergraduate curricula be decompressed and made broad based. Students should be exposed as undergraduates to a wide spectrum of courses both in the arts and sciences, and there should be one baccalaureate degree for all undergraduates. Individuals who successfully complete this level of education would be able to find employment as environmental paraprofessionals. An analogous situation exists in the health professions, where physician's assistants, physical and occupational therapists, and many other allied health career employees assume independent responsibilities for the performance of many patient-related duties. They free the physician to conduct specialized procedures, and they provide valuable data in support of the patient's treatment.

The independent health paraprofessional has been given an increasingly significant role, especially during the last decade. Similar trends are now appearing in the legal profession, where legal paraprofessionals are assuming responsibilities previously carried by attorneys. At this point in time, we can only extrapolate into the future about the environmental paraprofessional, but those individuals would be expected to contribute progressively more responsibilities over the next century to essential environmental services. An environmental paraprofessional with a baccalaureate degree in global environment would be an enlightened individual who would assist more advanced environmental specialists in various global environmental concerns.

Masters degree candidates in science should also be expected to be less focused so that the student might be exposed to all disciplines needed to generate a global perspective and to be scientifically well rounded. If there was one common masters degree for all science majors, then the specialization phase of scientific education would begin with pursuit of the doctoral degree. Following this change in emphasis, we would educate an academic leadership which is broad based, multidisciplinary, and yet capable of advancing knowledge in a specialized field. These individuals will then be employed as advanced environmental specialists.

There is a pressing need to bring the scientist, the policy manager/administrator, and the policy maker together to assure continuity of responsibile action and relevant pragmatism concerning current and future needs for environmental resource management and its conservation.

Efforts are under way now to bridge the gap and assure the necessary linkages to permit scientific information flow into the realms of federal and state legislation, as well as into societal norms of proper attitudes about the environmental impact of fossil fuel utilization and energy consumption. To ask the public to respond to global events which operate on scales too long or large to be immediately appreciated is made worse because of the lack of an atmosphere of trust and understanding.

This educational component, when discharged through our university systems, will exert a global influence because of the tradition of non-U.S. citizens passing through the American university for education and training. Those foreign students in our system will hopefully return to their native lands with skills acquired in the U.S. and as leaders in their own communities they will be able to support the global agenda.

Modified behavior of not just one individual, but of populations of individuals, is necessary to effect any sufficient reduction of the threats of negative human influence upon the world's environment. However, the placement of well-educated role models in visible positions of leadership within our business, political, scientific, and educational communities will yield responsible public policies and effective environmental management programs. Career opportunities both in government and the private sector should emerge from an awareness of the need for personnel to direct this effort.

## VIII. CONCLUSION

We have presented both a mechanism and a justification for the creation of a new multidisciplinary academic department within our system of higher education to focus upon teaching, research, and community service about global environmental issues. The significance of this novel Department of Global Environment lies in its access to a student population, and in its ability to mold the thinking and decision-making of our future business, scientific, and political leadership. The development of an informed public will require time. However, investment in the creation of a knowledgeable public is an investment which is certain to equip future generations of humans to coexist in harmony with the natural world. The educational commitment represents a necessary complement to other efforts toward reaching a responsible and effective solution to the current trends and threats about global warming. By blending this educational objective with practical actions, including a reduction in atmospheric emissions and more efficient energy utilization, we will have produced a complete strategy for international coordination on global environmental policy.

## REFERENCES

Ager, T. A., Climate change in the geologic past. Information on selected climate and climate-change issues, *U.S. Geol. Surv.* Open-File Rep., 88-718, 22–26, 1988.

Budyko, M. I., Golitsyn, G. S., and Izrael, Y. A., *Global Climate Catastrophes,* Springer-Verlag, London (transl. by Yanuta, V. G.), 1988.

CEES Report, Our Changing Planet, The FY 1992 U.S. Global Change Research Program, 1991, 86.

Gates, D. M., Climate change . . . a slow burn, *Anthroquest,* 39, 3–6 1989.

Haskell, T. L., In *The Emergence of Professional Social Science: the American Social Science Association and the Nineteenth Century Crisis of Authority,* University of Illinois Press, Urbana, IL, 1977.

IPCC — Intergovernmental Panel on Climate Change, *Scientific Assessment of Climate Change, Summary and Report,* World Meteorological Organization, U.N. Environmental Program, Cambridge University Press, 1990.

Mission Statement, Center for Global Change, University of Maryland, College Park, MD.

Ornstein, R., Why we don't listen . . . , in *Greenhouse Glasnost,* Minger, T. J., Ed., Ecco Press, New York, 1990.

Perlès, C., L'homme Prehistorique et le Feu, *La Recherche,* 60, 829–839, 1975.

Planet of the Year, *Time,* 133(1), 24–73, 1989.

Ruckelshaus, W. D., Toward a sustainable world, *Sci. Am.,* 261, 166–174, 1989.

Simmons, I. G., *Changing the Face of the Earth,* Basil Blackwell Publ., New York, 1989.

Titus, J., Preparing for climate change, in *The Potential Effects of Global Climate Change in the United States,* Smith, J. B. and Tirpak, D., Eds., U.S.E.P.A., Office of Policy, Planning, and Evaluation (PM-221), 1989, chap. 19.

# *Index*

# INDEX

## A

## D



## E

## F

## H

## I



## J



## K



## L

## P

## R



## S

## T

## U



## V



## W

## X



## Y